Harminder Pal Singh, PhD
Daizy Rani Batish, PhD
Ravinder Kumar Kohli, PhD
Editors

Handbook of Sustainable Weed Management

Pre-publication
REVIEWS,
COMMENTARIES,
EVALUATIONS . . .

"Although weeds constitute only a very small part of earth's land plants, they are one of the problematic groups of existing plants competing mainly with crops in agroecosystems and causing economic losses. Herbicides help greatly in weed management, but they are accused of causing environmental and human health problems; we should also not forget their cost in money and energy. *Handbook of Sustainable Weed Management* deals with how mankind can manage weeds in a sustainable way and suggests possible solutions to the problem of weeds while minimizing the risks to the environment and human health. The book covers many topics related to sustainable weed management and it will be useful to scientists working within this area. This book shows that weed management is an important science combining several strategies and taking into consideration weed population dynamics. It thoroughly covers allelopathy and cover crops as tools for sustainable weed management."

Kiriaki L. Kalburtji, PhD, BSc
Professor of Agricultural Ecology,
Faculty of Agriculture, Aristotle University,
Thessaloniki, Greece

More pre-publication
REVIEWS, COMMENTARIES, EVALUATIONS . . .

"*Handbook of Sustainable Weed Management* comprises twenty-five well-written chapters and the collective contribution of forty-nine leading international experts in the field of integrated weed management. There are no single publications to date that cover the range of important topics contained in this book, and it represents a valuable new information resource for students, teachers, researchers, and practitioners of sustainable weed management. Beginning with conceptual chapters on weeds and basic approaches to IWM and sustainability issues, the book details approaches on weed suppression with cover crops in various cropping systems, crop rotation and tillage effects on weed suppression, and biological control and/or biologically derived herbicides and allelopathy for weed management. Other chapters cover topics that have until now been largely confined to scientific journals. Such timely topics are made accessible in the *Handbook of Sustainable Weed Management* and include the ecology and potential exploitation of weed seed predation for weed management, development of weed-suppressive crop cultivars, mowing for weed management, herbicide fate in high-residue, sustainable agroecosystems, impacts of herbicide-resistant crops and strategies for managing herbicide-resistant weeds, and recent advances in parasitic weed research. There are also chapters on integrated weed management in pasture systems and turfgrass, and finally, two thought-provoking chapters on approaches to developing IWM and implementation of IWM through learning groups. The latter presents guidelines, principles, and practices developed for group participation by professionals and enhance IWM implementation by enlarging the knowledge base and social infrastructure that currently impede its widespread adoption. A comprehensive list of references is included at the end of each chapter, and numerous examples, data tables, figures, and diagrams are presented throughout the book to reinforce key concepts. Overall, this book presents the most comprehensive and cutting-edge information available on the scientific principles, biological processes, and how-to aspects of sustainable weed management."

S. K. Harrison, PhD
Professor of Weed Science,
Ohio State University, Columbus

Food Products Press®
An imprint of The Haworth Press, Inc.
New York • London • Oxford

NOTES FOR PROFESSIONAL LIBRARIANS AND LIBRARY USERS

This is an original book title published by Food Products Press®, an imprint of The Haworth Press, Inc. Unless otherwise noted in specific chapters with attribution, materials in this book have not been previously published elsewhere in any format or language.

CONSERVATION AND PRESERVATION NOTES

All books published by The Haworth Press, Inc., and its imprints are printed on certified pH neutral, acid-free book grade paper. This paper meets the minimum requirements of American National Standard for Information Sciences-Permanence of Paper for Printed Material, ANSI Z39.48-1984.

Handbook of Sustainable Weed Management

FOOD PRODUCTS PRESS®

Crop Science
Amarjit S. Basra, PhD
Editor in Chief

Plant-Derived Antimycotics: Current Trends and Future Prospects edited by Mahendra Rai and Donatella Mares

Concise Encyclopedia of Temperate Tree Fruit edited by Tara Auxt Baugher and Suman Singha

Landscape Agroecology by Paul A. Wojtkowski

Concise Encyclopedia of Plant Pathology by P. Vidhyasekaran

Molecular Genetics and Breeding of Forest Trees edited by Sandeep Kumar and Matthias Fladung

Testing of Genetically Modified Organisms in Foods edited by Farid E. Ahmed

Fungal Disease Resistance in Plants: Biochemistry, Molecular Biology, and Genetic Engineering edited by Zamir K. Punja

Plant Functional Genomics edited by Dario Leister

Immunology in Plant Health and Its Impact on Food Safety by P. Narayanasamy

Abiotic Stresses: Plant Resistance Through Breeding and Molecular Approaches edited by M. Ashraf and P. J. C. Harris

Teaching in the Sciences: Learner-Centered Approaches edited by Catherine McLoughlin and Acram Taji

Handbook of Industrial Crops edited by V. L. Chopra and K. V. Peter

Durum Wheat Breeding: Current Approaches and Future Strategies edited by Conxita Royo, Miloudi M. Nachit, Natale Di Fonzo, José Luis Araus, Wolfgang H. Pfeiffer, and Gustavo A. Slafer

Handbook of Statistics for Teaching and Research in Plant and Crop Science by Usha Rani Palaniswamy and Kodiveri Muniyappa Palaniswamy

Handbook of Microbial Fertilizers edited by M. K. Rai

Eating and Healing: Traditional Food As Medicine edited by Andrea Pieroni and Lisa Leimar Price

Handbook of Plant Virology edited by Jawaid A. Khan and Jeanne Dijkstra

Physiology of Crop Production by N. K. Fageria, V. C. Baligar, and R. B. Clark

Plant Conservation Genetics edited by Robert J. Henry

Introduction to Fruit Crops by Mark Rieger

Sourcebook for Intergenerational Therapeutic Horticulture: Bringing Elders and Children Together by Jean M. Larson and Mary Hockenberry Meyer

Agriculture Sustainability: Principles, Processes, and Prospects by Saroja Raman

Introduction to Agroecology: Principles and Practice by Paul A. Wojtkowski

Handbook of Molecular Technologies in Crop Disease Management by P. Vidhyasekaran

Handbook of Precision Agriculture: Principles and Applications edited by Ancha Srinivasan

Dictionary of Plant Tissue Culture by Alan C. Cassells and Peter B. Gahan

Handbook of Potato Production, Improvement, and Postharvest Management edited by Jai Gopal and S. M. Paul Khurana

Handbook of Sustainable Weed Management

Harminder Pal Singh, PhD
Daizy Rani Batish, PhD
Ravinder Kumar Kohli, PhD
Editors

Food Products Press®
An Imprint of The Haworth Press, Inc.
New York • London • Oxford

For more information on this book or to order, visit
http://www.haworthpress.com/store/product.asp?sku=5412

or call 1-800-HAWORTH (800-429-6784) in the United States and Canada
or (607) 722-5857 outside the United States and Canada

or contact orders@HaworthPress.com

Published by

Food Products Press®, an imprint of The Haworth Press, Inc., 10 Alice Street, Binghamton, NY 13904-1580.

Cover design by Marylouise E. Doyle.

Library of Congress Cataloging-in-Publication Data

Handbook of sustainable weed management / Harminder Pal Singh, Daizy Rani Batish, Ravinder Kumar Kohli, editors.
p. cm.
Includes bibliographical references and index.
ISBN 13: 978-1-56022-956-8 (hc. : alk. paper)
ISBN 10: 1-56022-956-X (hc : alk. paper)
ISBN 13: 978-1-56022-957-5 (pbk. : alk. paper
ISBN 10: 1-56022-957-8 (pbk. : alk. paper)
1. Weeds—Control. 2. Organic farming. I. Singh, Harminder Pal. II. Batish, D. (Daizy). III. Kohli, R. K.

SB611.H36 2005
632'5—dc22

2004028883

CONTENTS

ABOUT THE EDITORS

Harminder Pal Singh, PhD, is a weed ecologist at Panjab University in Chandigarh, India. Specializing in experimental ecology and the management of noxious weeds, Dr. Singh's interests include weed-weed, weed-crop, and tree-crop interactions in agroecosystems. He has been awarded the Young and Deserving Scientist Award by the Asian Pacific Weed Science Society. He is also the recipient of a Young Scientist Research Project awarded by the Department of Science and Technology of the Government of India, the Young Scientist Award in Life Sciences (Panjab Academy of Sciences), and the Young Scientist Award (Indian Science Congress Association). He has published forty research papers, co-authored one book, and co-edited one book.

Daizy Rani Batish, PhD, a weed ecologist, is a Reader (associate professor) in the Department of Botany, Panjab University, India. A dedicated and popular teacher with a background in plant biochemistry, she teaches ecology and environmental botany. She has published thirty-five research papers and seven articles in addition to co-editing four books. Dr. Batish specializes in the ecology, biology, and management of noxious invasive weed species. Her research interests include chemical ecology and allelopathic interactions of the exotic invasive weeds and their management with natural products. She has been awarded the Rajib Goyal Young Scientist Award in the field of Environment Sciences.

Ravinder Kumar Kohli, PhD, is an ecophysiologist of distinction and serves as Professor Botany and Coordinator of the Centre for Environment Studies at Panjab University, India. Dr. Kohli guides research in ecology and biology of invasive plants and their biochemical interactions. His research interests also include plantation forestry and the impact of exotic species on native vegetations. For his research contributions, he was awarded the Nanda Memorial National Young Scientist Award and honored by the government of Chandigarh for disseminating environmental awareness among the masses. Recently the Ministry of Environment and Forests for the Government of India bestowed upon Dr. Kohli the coveted B. P. Pal National Environment Fellowship Award to study the impact of exotic invasives on the native vegetation of Himachal Pradesh in India. He has published more than 100 research papers and has edited or authored a number of books on environmental themes.

CONTRIBUTORS

Mohamed A. Abouzeid, PhD, Lecturer of Microbiological Sciences, Department of Microbiology, University of Ain Shams, Cairo, Egypt.

Kassim Al-Khatib, PhD, Professor, Department of Agronomy, Kansas State University, Manhattan, Kansas.

Randy L. Anderson, PhD, Research Agronomist, U.S. Department of Agriculture, Agricultural Research Service, Mandan, North Dakota.

Karen L. Bailey, PhD, Research Scientist, Agriculture and Agri-Food Canada, Saskatoon, Saskatchewan, Canada.

R. Becker, PhD, Professor, Department of Agronomy and Plant Genetics, University of Minnesota, St. Paul, Minnesota.

Hugh J. Beckie, PhD, PAg, Research Scientist, Agriculture and Agri-Food Canada, Saskatoon, Saskatchewan, Canada.

R. E. Blackshaw, PhD, Weed Scientist, Agriculture and Agri-Food Canada, Lethbridge, Canada.

Hana Blissett, MS, Graduate Student, Department of Agronomy and Plant Genetics, University of Minnesota, St. Paul, Minnesota.

Rick A. Boydston, PhD, Plant Physiologist, U.S. Department of Agriculture, Agricultural Research Service, Prosser, Washington.

S. M. Boyetchko, PhD, MSc, Research Scientist, Agriculture and Agri-Food Canada, Saskatoon, Saskatchewan, Canada.

Douglas D. Buhler, PhD, Chair and Professor, Department of Crop and Soil Science, Michigan State University, East Lansing, Michigan.

Nilda R. Burgos, PhD, Associate Professor, Department of Crop, Soil, and Environmental Sciences, University of Arkansas, Fayetteville, Arkansas.

Rakesh S. Chandran, PhD, Extension Weed Specialist and Assistant Professor, IPM Coordinator, West Virginia University, Morgantown, West Virginia.

G. P. Dawra, PhD, Professor, Department of Botany, Meerut University, Meerut, India.

D. A. Derksen, PhD, Weed Scientist, Agriculture and Agri-Food Canada, Brandon, Manitoba, Canada.

William W. Donald, PhD, Research Agonomist, U.S. Department of Agriculture, Agricultural Research Service, Columbia, Missouri.

Antonio Evidente, PhD, Full Professor of Organic Chemistry, Dipartimento di Scienze del Suolo, della Pianta e dell'Ambiente, Università di Napoli Federico II, Portici, Italy.

David R. Gealy, PhD, Plant Physiologist, U.S. Department of Agriculture, Agricultural Research Service, Dale Bumpers National Rice Research Center, Stuttgart, Arkansas.

Gurjeet S. Gill, PhD, Senior Lecturer, Agronomy and Farming Systems, Adelaide University–Roseworthy Campus, Australia.

J. Gunsolus, PhD, Professor, Department of Agronomy and Plant Genetics, University of Minnesota, St. Paul, Minnesota.

M. L. Hoffman, PhD, Research Associate, Department of Crop and Soil Sciences, Michigan State University, East Lansing, Michigan.

N. Jordan, PhD, Professor, Department of Agronomy and Plant Genetics, University of Minnesota, St. Paul, Minnesota.

Clifford H. Koger, PhD, Weed Biologist, U.S. Department of Agriculture, Agricultural Research Service, Southern Weed Science Research Unit, Stoneville, Mississippi.

Robert J. Kremer, PhD, Microbiologist, U.S. Department of Agriculture, Agricultural Research Service, Cropping Systems and Water Quality Unit; Adjunct Professor of Soil Science, Environmental Soil Science Program, School of Natural Resources, University of Missouri, Columbia, Missouri.

Yong In Kuk, PhD, Research Professor, Biotechnology Research Institute, Chonnam National University, Kwangju, Korea.

Ricardo Labrada, PhD, Weed Officer, FAO Plant Protection Service, Viale delle Teme di Caracalla, Rome, Italy.

A. Légère, PhD, Weed Scientist, Agriculture and Agri-Food Canada, Sainte-Foy, Canada.

Matt Liebman, PhD, Professor, Department of Agronomy, Iowa State University, Ames, Iowa.

Martin A. Locke, PhD, Research Leader and Soil Scientist, U.S. Department of Agriculture, Agricultural Research Service, National Sedimentation Laboratory, Water Quality and Ecological Processes Research Unit, Oxford, Mississippi.

John B. Masiumas, PhD, Associate Professor, Department of Natural Resources and Environmental Sciences, University of Illinois, Urbana, Illinois.

Fabián D. Menalled, PhD, Assistant Professor, Department of Agronomy, Iowa State University, Ames, Iowa.

Karen A. Moldenhauer, PhD, Professor, Crop Soils and Environmental Sciences, University of Arkansas Rice Research and Extension Center, Stuttgart, Arkansas.

J. R. Moyer, PhD, Weed Scientist, Agriculture and Agri-Food Canada, Lethbridge, Canada.

Edmund K. Mupondwa, PhD, Research Economist, Agriculture and Agri-Food Canada, Saskatoon, Saskatchewan, Canada.

Jamal R. Qasem, PhD, Professor of Weed Science, Plant Protection Department, Faculty of Agriculture, University of Jordan, Amman, Jordan.

Krishna N. Reddy, PhD, Plant Physiologist, U.S. Department of Agriculture, Agricultural Research Service, Southern Weed Science Research Unit, Stoneville, Mississippi.

Emilie E. Regnier, PhD, Associate Professor, Department of Horticulture and Crop Science, The Ohio State University, Columbus, Ohio.

Karen A. Renner, PhD, Professor and Extension Specialist, Department of Crop and Soil Sciences, Michigan State University, East Lansing, Michigan.

E. N. Rosskopf, PhD, Research Scientist, U.S. Department of Agriculture, Agricultural Research Service, Fort Pierce, Florida.

Simon F. Shamoun, PhD, Research Scientist and Adjunct Professor, Natural Resources Canada, Canadian Forest Service, Victoria, British Colombia, Canada.

S. Simmons, PhD, Professor, Department of Agronomy and Plant Genetics, University of Minnesota, St. Paul, Minnesota.

Brian M. Sindel, PhD, Associate Professor of Weed Science, University of New England, Australia; Member, Cooperative Research Centre for Australian Weed Management.

Ronald E. Talbert, PhD, University Professor, Department of Crop, Soil, and Environmental Sciences, University of Arkansas, Fayetteville, Arkansas.

A. G. Thomas, PhD, Weed Scientist, Agriculture and Agri-Food Canada, Saskatoon, Saskatchewan, Canada.

G. C. Turnbull, MSc, Research Officer, Dow AgroSciences Canada, Winnipeg, Canada.

P. R. Watson, MSc, Research Associate, Agriculture and Agri-Food Canada, Brandon, Canada.

Mark A. Weaver, PhD, Research Microbiologist, U.S. Department of Agriculture, Agricultural Research Service, Southern Weed Science Research Unit, Stoneville, Mississippi.

S. White, MS, Research Fellow, Department of Agronomy and Plant Genetics, University of Minnesota, St. Paul, Minnesota.

Robert M. Zablotowicz, PhD, Research Microbiologist, U.S. Department of Agriculture, Agricultural Research Service, Southern Weed Science Research Unit, Stoneville, Mississippi.

Preface and Acknowledgments

Weeds constitute one of the small but problematic groups of plants exhibiting a great taxonomic diversity. They are the least cared for, yet they grow luxuriantly and abundantly in different ecosystems owing to their unique set of traits that make them stronger than the nurtured crops. Weeds, especially the invasive ones, due to their greater adaptability are very successful and grow at the cost of other vegetation. They are a nuisance, particularly in agroecosystems, where they compete with crop plants for resources and cause huge economic losses. Not long ago, weeds used to be controlled by simple methods that did not affect the sustainability of croplands. However, with passage of time, weed management, like other sectors, progressed from simple to more advanced techniques and from nonchemical to chemical means. Thus, especially during the later half of twentieth century, weed management relied greatly on herbicides to provide an easy, efficient, and reliable method of weed control. The success story of the herbicides was, however, short-lived when it was realized that indiscriminate use of herbicides is counterproductive to environmental and human health. Nevertheless, it would not be wise to kill or eradicate weeds, as that would mean deliberate genetic erosion in the modern era of rapid biodiversity loss causing imbalance in the natural ecosystems. Thus an urgent need exists to get rid of adverse effects of weeds without affecting the natural balance. Management of weeds should, therefore, be achieved through strategies that do not affect the sustainability of agroecosystems and the life support system.

The *Handbook of Sustainable Weed Management* has been prepared to include different aspects of weed management relevant to the scope of modern weed science such as cultural practices, cover crops, crop rotation designs, potential of herbicide-resistant crops, bioherbicides, allelopathy, microorganisms, and integrated weed management. It includes a total of 25 chapters on such aspects. The first two chapters are mainly attributed to the weeds and the various possible management practices being undertaken. Chapters 3 through 5 deal with the use of cover crops in general, and those of *Brassica* and grass-legume mixture, in particular, for the management of weeds. The role of rye (*Secale cereale* L.) in sustainable weed management under vegetable cropping system is the focus of Chapter 6. Crop rotation is another important aspect that can help not only in managing weeds but also maintaining the sustainability of agroecosystems. Chapters 7 and 8 highlight the importance of crop rotation designs and its combination with tillage in reducing the weed populations. Further, in Chapter 9, the role of allelopathy and allelochemicals for achieving the weed management is discussed. The chapter also gives details on the crop genotypes with greater allelopathic potential, various potential allelochemicals, microorganisms, and their products that hold a great promise for the future. Rice cultivars from around the world have been screened at the U.S. Department of Agriculture–Agricultural Research Service (USDA-ARS) for their allelopathic potential vis-à-vis weed-suppressing ability. Chapter 10 deals with different strategies for improving weed-suppressing ability of various rice cultivars. Chapter 11 provides detailed insight into the ecological management of weed seeds through predation. Mowing as a strategy of mechanical weed control has been addressed in detail in Chapter 12. Because herbicides have a major influence on modern agriculture, Chapter 13 provides insight into the fate of herbicides in soil under conservation tillage and cover cropping systems. Due to the ill effects associated with the use of synthetic herbicides, efforts are being made worldwide to search for newer and safer methods of weed management. Mycoherbicides from fungal pathogens and their phytotoxins offer a great choice in this respect. Chapter 14 reviews various strategies for developing bioherbicides for weed management, whereas the next chapter discusses the various commercial, technical, and biological considerations for developing microbial weed control products. The prospects of the biological control in managing weeds in forest ecosystems have been discussed in the Chapter 16. The next chapter deals with the potential of phytopathogenic fungi and their phytotoxins for weed management. Chapter 18 highlights the potential of soil or rhizosphere microorganisms in weed management. Further, in order to enhance efficiency of herbicides, herbicide-resistant crops are being developed. Chapter 19 highlights the potential of herbicide-resistant crops in this direction. Because excessive use of herbicides has led to the development of herbicidal-resistant weeds, a separate chapter has been devoted to discuss the various strategies for managing such weeds. Chapters 21, 22, and 23 discuss the control of parasitic weeds, weeds in pasture, and turfs, respectively. Chapters 24 and 25 deal with the concepts and challenges of integrated weed management (IWM) and the knowledge networks and guidelines for implementing IWM.

The organization of the book has been carefully conceived to include different aspects of weed management. All the chapters have been reviewed thoroughly by at least two experts in the respective fields and were accepted only after ensuring the incorporation of suggestions of experts. We are hopeful that this book will be of immense help to weed scientists, agronomists, agriculturists, horticulturists, farmers, extentionists, teachers, and students in the field of weed science.

This handbook has been possible only because of the help and active support of many people. We are extremely indebted to Dr. Amarjit Singh Basra, editor in chief of the *Journal of Crop Production,* for his idea of coming out with such a compendium and for his constant help and support during this project. Thanks are also due to all the reviewers for their thorough, critical, and constructive reviews in spite of their busy schedules. These include Stephen O. Duke, Paul E. Hatcher, D. B. Saxena, Brien Henry, David Kemp, Kerry Harrington, Peter Dowling, Wondie Mersie, Kevin Gibson, J. Scott Ebson, Roch Gaussoin, Charles Staver, David Mortensen, Larry Zibilske, A. C. Kennedy, Allen Barker, Nancy Creamer, Bruce A. Auld, Louise Morin, Gavin Ash, Anthony Caesar, Paul Marino, Antonio DiTommaso, Stephen D. Miller, John J. Obrycki, Larry K. Binning, Pamela Hutchinson, John Byrd, Milton McGriffen, Stephen Weller, Stephen Moss, James E. Retzinger, Frank Young, Steve Sutherland, Gary Peterson, Charles L. Mohler, John R. Teasdale, Jeff Gunsolus, John Cardina, Reid Smeda, and Bernal Valverde. Finally, we thank all the students in our group and our family members for their help and cooperation during the completion of this project.

Chapter 1

Weeds and Their Management: Rationale and Approaches

R. K. Kohli
Daizy R. Batish
H. P. Singh

WHAT ARE WEEDS?

Defining *weeds* is perhaps one of the most complex and intriguing problems in plant science. A number of definitions and views have been put forward, but none have met with universal satisfaction. Etymologically, a weed means a plant growing at a place where not required and/or interfering with growth of cultivated plants. Scientists have tried to define weeds in anthropomorphic, biological, and ecological terms. Such eminent botanists as Harper (1960) and Salisbury (1961) defined weeds as nuisances or plants growing where not required. Harlan and deWet (1965), King (1966), and Zimdahl (1999) have attempted to compile various definitions of the weed. Most of the definitions, however, suggest that these are objectionable unwanted plants interfering with the activity of humans and cultivated plants. Many ecologists have also tried to define weeds from different perspectives. Bunting (1960) defined weeds as pioneers of secondary succession, and weeds in an arable field is a special case. Pritchard (1960) referred to weeds as opportunistic species that follow human disturbance from the habitat. Rejmánek (1995) opined that the terms *weeds, invaders,* and *colonizers* are often overlapping and may be considered as one. Begon, Harper, and Townsend (1996) described weeds as successful predators arising out of a mismatch between habitat and types of growing plant. According to Crawley (1997) a weed is a problem plant whose abundance is more than the specific level. However, the specific level remains to be determined and varies from one place to another. Mohler (2001) defined weeds ecologically as the plants that are most successful in colonizing disturbed but potentially productive sites and are able to maintain their abundance even under repeated disturbed conditions.

Clearly, all the definitions and descriptions of weeds point to the negative aspects, particularly with reference to their interference with cultivated plants. On the other hand, Hammer, Gladis, and Diederichsen (1997) considered weeds to be important genetic resources and wild relatives of crop plants. Likewise, weeds have also been viewed as valuable agroecosystem components that provide useful services (Alstrom, 1990; Mapes, Basurto, and Bye, 1997). Presence of weeds in less density may even exert positive influence on crop yield through competition and chemical stimulation (Müller-Schärer, Scheepens, and Greaves, 2000). Thus weeds can be viewed from different perspectives depending upon the geographical area, environmental conditions, and cultural practices. Many of the plants unwanted today under one situation because of the lack of adequate information about their importance may turn out to be worth culturing tomorrow. Further, weed species such as pigweed (*Amaranthus* spp.), lambsquarters/goosefoot (*Chenopodium* spp.), and some minor millets recognized as weeds in some societies are cultivated by the others.

Even the origin of word *weed* in the English language is unknown, as it lacks self-description or intrinsic meaning, although its roots in Germanic and Romance languages have been traced (King, 1966). According to Zimdahl (1999) weeds are the plants with unknown etymology and multiple origins. They may evolve in natural environments possibly because of excessive human interference. Weeds may be native or nonnative invasive plants introduced by humans (for example, *Lantana* as an ornamental plant in Australia and India, etc.) or entered accidentally (for example, ragweed parthenium, *Parthenium hysterophorus* L., in India). Baker (1974) pointed out that some exotic plants are introduced deliberately for the purpose of fiber, ornamental, medicine, and erosion control or timber, while others may enter accidentally and later escape and become weeds. Sources of accidental introduction include impure crop seeds, adhesion to domesticated animals, and soil surrounding roots of nursery stocks (Baker, 1986). Agriculture, logging, and grazing further enhance the establishment of exotics by creating disturbed sites for their colonization (Sakai et al., 2001).

In agricultural fields, weeds seem to have coevolved with crop plants from prehistoric times as evidenced by pollen analysis techniques indicating that both share common evolutionary lines (Hammer, 1988). Distribution of weeds is determined by various environmental and biological characteristics. Human activities are mainly responsible for regional patterns and have certainly played an important role in their spread. Plant species are also affected when their habitats are disturbed (Harlan and deWet, 1965). In fact, in a balanced and healthy ecosystem weeds do not exist. They originated only after humans disturbed the balance of natural ecosystems, and weeds have now become their integral components. In agroecosystems, they have evolved due to continuous selection pressure imposed by humans, technological advancement, and/or through agricultural practices. Various production practices such as use of crop rotation, tillage, intercropping, and herbicide applications have further contributed to the weed infestation in the arable land. The role of humans in selecting crop plants vis-à-vis evolution of weeds is clear from the fact that over 40 percent of the world's total weed species belongs to Asteraceae (sunflower family) and Poaceae (grass family), which happen to provide over half of world's food and food products. Nearly 44 percent of the world's worst weeds belong to family Poaceae, which happens to provide eight major crops, namely, wheat, maize, rice, sorghum, barley, millet, oat, and sugarcane.

WHY ARE WEEDS SUCCESSFUL?

Weeds assume large proportions of the area of their invasion. They possess certain traits or characteristics that make them ideal for proliferation. Traits such as the ability to reproduce at a faster rate, rapid growth from seedling to sexual phase, phenotypic plasticity, and high tolerance to environmental heterogeneity are associated with weedy plant species (Baker, 1974). Weed populations are highly adaptable to production system through herbicidal resistance and shifts in their populations. Pattersen (1985) highlighted numerous agronomical, reproductive, and physiological characteristics that make weeds successful species. Weeds possess adaptive strategies that determine their survival, production, and success in a particular environment (Holt, 1988). On the basis of intensity of stress and disturbance for successful establishment of a given area, plants can be stress tolerators, competitors, or ruderals (Grime, 1979). Weeds, however, fall into two combined categories; they can be competitive ruderals or stress-tolerant competitors (Grime, 1979). They take advantage of human-made habitats and are highly responsive to changes in environmental conditions in such a beneficial manner, enabling them to survive and grow in nature (Grime, 1979). Many annual, biennial, and even perennial weed species found on arable fertile land are known as competitive ruderals that grow rapidly and competition in them occurs before flowering. Stress-tolerant competitors, primarily trees or shrubs and even some perennial herbs, are characterized by rapid dry matter production, large stem extension, and high leaf area production. In addition to growth strategies, weeds possess several other characteristics that make them successful colonizers of a given area (see Exhibit 1.1).

EXHIBIT 1.1. Characteristics that favor the successful establishment of weeds

Regenerative Phase	Established Phase
• Well-adapted seed dispersal, aided by wind, water, animals, and humans • Prolonged viability of propagules with dormancy (sometimes running up to thousands of years) so as to survive under unfavorable conditions • Seed resistance to animal digestion • Seed polymorphism • Establishment of seedlings in a wide range of environment conditions	• Efficient resource capturing • Efficient photosynthetic pathways and water uptake • Abundant seed production • Inbreeding through autogamy or agamospermy • Cross-pollination (if at all occurs) through wind or generalist pollinators • Higher rate of vegetative reproduction, if perennial • Strong interspecific competition

Source: Adapted from Holt, 1988.

IMPACT OF WEEDS

The impact of weeds in agro- and other ecosystems has been highlighted by several researchers (King, 1966; Crafts and Robbins, 1973; Shaw, 1988; Zimdahl, 1999; Liebman, Mohler, and Staver, 2001). Most of the impacts of weeds are negative, though some reports from India, Mexico, and other parts of world point out that they can be useful as food, fodder, fiber, and medicines (Altieri, 1988; Liebman, 2001). Weeds constitute only 0.1 percent of earth's land plants, yet they cause enormous economic loss. Zimdahl (1999) divided harms of weeds into nine categories: competition with crop plants, increased production costs, reduced quality of crops and animals, increased processing costs, problems of water management, human health problems, reduced land value, reduced crop choice, and lower aesthetic value.

Weeds have a significant economic impact on agricultural production (Buhler, 1999) as evidenced by the efforts spent on their management. In 2002 global herbicide sales were nearly US$28 billion, constituting 47 percent of the total agrochemicals used worldwide (Agrow, 2003). The loss of crop yield due to weeds is enormous. Lacey (1985) reported that sometimes weeds cause even 100 percent crop loss, leading to an adverse effect on national economy. In the tropics and subtropics, in addition, parasitic weeds, especially *Orobanche* and *Striga* spp., pose a great threat to agriculture and have been referred to as the third world's most devastating weeds (Gressel, 1992). In sub-Saharan Africa, *Striga* infests nearly 20 to 40 million hectares (ha) of farmland, causing an annual economic loss of nearly US$1 billion and affecting the livelihood of nearly 100 million people (Kanampiu, Friesen, and Gressel, 2003). The reduction in crop yield may also be attributed to the allelopathic property exhibited by a number of weeds, especially the aggressive ones. Such species of weeds as *Polygonum* and *Chenopodium,* because of their growth habit, make agricultural operations difficult. Further, the crop contaminated with the weed seeds is considered to be of poor quality, e.g., presence of seeds of wild oat (*Avena fatua* L.) in wheat and barley seeds. Poisonous weeds such as poison ryegrass (*Lolium temulentum* L.) or corn cockle (*Agrostemma githago* L.) further aggravate this problem and lead to additional costs. They also affect animal productivity by causing fodder scarcity and poisoning effects. In the scarcity of fodder, if cattle ingest them, these plants may cause unwanted flavors in the milk and meat. They may also serve as reservoirs or carriers of harmful pests and diseases. Weeds reduce efficiency of land and human beings. They are known to cause a number of health problems to livestock and human beings; for example, weeds such as ragweed parthenium and billy goat weed (*Ageratum conyzoides* L.) cause asthma, skin allergy (dermatitis), hay fever, and bronchitis. Further, increasing resistance and cross-resistance among weed species toward herbicides has resulted in a number of resistant biotypes and thus shifts in weed populations. As per Heap (2005), 296 weed biotypes belonging to 178 weed species have been identified worldwide, posing a continuous threat to agricultural production.

Although a number of weeds are known to cause economic loss, it is difficult to determine their costs in terms of money and energy on a global scale. Thus variable statistics are available regarding the costs of their management and no accurate estimates of weed costs are available (Zimdahl, 1999). However, Appleby, Müller, and Carpy (2000) claimed that global economic loss due to weeds is more than US$100 billion. The losses are even greater in the developing and underdeveloped countries. Akobundu (1991) reported that for hand weeding 1 ha of cropland, 25 to 120 days of human work are required. In the United States alone weeds cost nearly 12 percent loss of crop yield, costing about US$33 billion (Piementel et al., 2001).

WEED MANAGEMENT

The effort to control weeds is as old as agriculture itself. Humans, however, were familiar with weeds even before the dawn of agriculture, as several aboriginal nomadic tribes suffered from allergies, hay fever, and other health problems caused by poisonous plants. Weed management has progressed from bare hands to tools to animal power to machine power and finally to chemicals and integration of such powers. A brief history of weed control is given in Table 1.1. The future, however, lies in ecological integrated weed management employing various cultural, chemical, biological, and mechanical methods. Weed management strategies have evolved with the advancement of agricultural technology, shifts in weed flora, and the

TABLE 1.1. Progress of weed control methods since prehistoric times.

Time period	Method of weed management	Energy input
15,000-10,000 B.C.	Manual	Neolithic human
6000 B.C.	Crude wooden hand tool	Human
3000 B.C.	Hand sickles	Human
1200 B.C.	Salt and ash	Chemical (first indirect use?)
1000 B.C.	Animal force	Human + animals
950 B.C.	Burning	Human
A.D. 1000-1300	Crop rotation	Cultural
1730s	Cropping in rows, horse hoeing	Human + animals
1896-1920	Use of inorganic salts such as copper sulfate, iron sulfate, dilute sulfuric acid, arsenicals, and petroleum	Largely human + animals with limited chemical energy input
1890	Copper sulfate	Chemical
1896	Bordeaux mixture	Chemical
1896	Iron sulfate	Chemical
1899	Crusher boards (to remove water hyacinth)	Mechanical
1902	Use of insects (for *Lantana*)	Biological
1920s	Use of machines	Mechanical
Mid-1940s	Use of synthetic chemicals, especially the hormonal and organic	Chemical (largely replaced the other types)
1980s	Integrated weed management	Chemical + mechanical + biological + cultural
Since 1990s	Advanced agricultural technology, biotechnology, and alternate options, such as reduced herbicide inputs, sophisticated machine-held tools/cultivators, genetically improved crop cultivars (HRCs, allelopathy, etc.)	Innovations in chemical + mechanical + biological + cultural

formation of weed biotypes with herbicidal resistance. This emphasizes reducing weed density and thus minimizing the weed crop competition—unlike weed control, which tends to eliminate the existing weed population (Zimdahl, 1995). Thus weed scientists prefer to use the term *weed management* as an alternative to *weed control,* as the latter results in the loss of genetic diversity thereby threatening ecosystem integrity.

Today, chemical methods have largely replaced the other methods of weed management. Although fraught with accompanying problems of pollution of soil and groundwater, and toxicity of food products, herbicides are valuable and important tools that have provided major benefit to the production systems (Buhler, 1999). However, new and alternate options would provide farmers with more flexibility for enhancing the effectiveness of the herbicides. These include use of herbicidal mixtures, varietal mixtures, synergists, herbicide antidotes, breeding of herbicide-tolerant and more competitive crops, allelopathy, and genetic engineering (Gressel, 1992, 2002). Liebman, Mohler, and Staver (2001) have stressed the need for ecological approaches in future weed management, such as reducing weed density, reducing interference, and replacing aggressive weeds with less harmful ones (called *weed chasing*).

Various Weed Management Strategies

A number of weed management strategies have been followed, but none alone probably provides a satisfactory solution to the weed problems. Weed management should therefore be viewed as an integrated science combining several strategies together. Efforts should be made to use preventive tools to minimize weed density to a level that does not cause apparent harm to the crop. Thus achieving complete weed control becomes an unrealistic objective. Broadly, four methods are employed for the management of weeds: *mechanical, cultural, biological,* and *chemical*. Each of these methods has certain advantages and disadvantages (see Figure 1.1). Among these, mechanical methods are one of the oldest, involving physical removal of weeds by soil disturbance prior to planting or by hand weeding or hoeing during crop growth. On the other hand, cultural methods are applied largely during the active growth period of the crop. These include crop rotation; use of cover; smother and green manure crops; crop residues; crop genotypes with better competitive and allelopathic ability; and manipulations of sowing or planting date, crop density, and crop pattern. These are effective when they are able to enhance the differential development between crop and weeds to the advantage of the former (Mohler, 1996). Intercropping with crops of different forms, phenology, and physiology creates

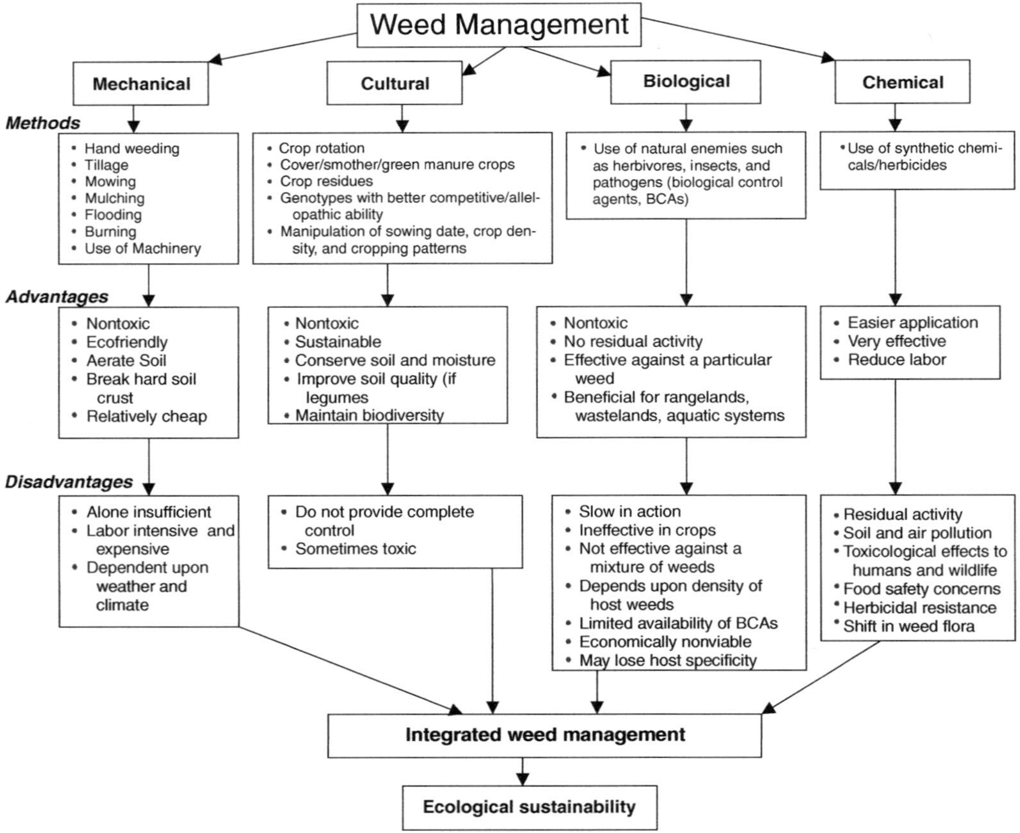

FIGURE 1.1. Weed management strategies—their advantages and disadvantages.

different patterns of resource availability, especially light to weeds, and thus can help in managing them. These methods were in use in traditional agroecosystems, but with the modernization of agriculture and herbicide application these declined. Nevertheless, to restore sustainability of agroecosystems, these practices have resurfaced, though with several innovations. Biological weed management suppresses weed recruitment, growth, and reproduction by promoting herbivory and diseases using natural enemies such as herbivores, insects, and pathogens and damaging weeds below economic injury level (Boyetchko, 1997). The impact of a biological control agent may be direct by harming vital tissues and functions of the host or indirect by causing stress on the weeds and thus reducing their competitive ability. For effective biological control, an organism must be host specific, able to reproduce quickly enough to suppress the weed, and be adaptable to new environments. In addition, these should be capable of escaping from the effect of predation, diseases, competition, and chemical interference (Newman, Thompson, and Richman, 1998). Biological control has been successfully used against invading plant species threatening endangered ecosystems, habitats, and species (Crawley, 1989). However, its application in a managed ecosystem is difficult because of greater disturbances and rapid control pressures required in a relatively shorter cropping season.

Use of chemical methods is probably a twentieth-century technology when the herbicides, especially the hormonal ones discovered in 1940s, revolutionized agriculture. Their improved efficacy and production of herbicide-resistant crops (HRCs) have further expanded this revolution and become an important tool of modern weed management. They are widely used not only in the developed nations but also in developing nations such as India. A 350 percent increase in the use of herbicides occurred from 1971 to 1987 for control of weeds in rice and wheat fields in India (Alstrom, 1990). Numerous herbicide families are known that differ widely in respect to spectrum, unit activity, crop safety, toxicology, and environmental effects. However, increasing herbicidal resistance, environmental, and toxicological concerns have put a question mark on their large-scale use (Burnside, 1993; Heap, 2003). Thus, even after over 50 years of herbicidal use, success has not been achieved in weed management (Holt, 1994). Therefore, efforts should be made either to search for more benign chemicals or to reduce the rate of herbicidal application.

INTEGRATED WEED MANAGEMENT

For an effective and sustainable weed management, strategies cannot be designed in isolation; rather an integrated approach is required. Integrated weed management (IWM) is defined as an integration of effective, environmentally safe, and socially acceptable control tactics that reduce weed interference below the economic injury level (Thill et al., 1991). IWM approaches involve plant breeding, fertilization, cropping patterns, chemical, mechanical, and biological weed control, and soil management so as to reduce weed interference and enhance yield (Swanton and Weise, 1991). IWM is a part of integrated pest management (IPM), which first appeared in literature in 1967 and emphasizes scientific research and development related to pest control. It is based on two key elements, namely, multiple control tactics and integration of pest biology knowledge for the management of pests (Buhler, 2002). Even the United Nations Conference on Environment and Development (UNCED) held in Rio de Janeiro, Brazil, in 1992 recognized IPM as a strategy for sustainable agricultural production (UNCED, 1992). However, unlike IPM, IWM lacks target-oriented strategies specific to weeds, and most of its concepts and approaches have been borrowed from the other pest disciplines. Therefore, the principles unique to weed science have yet to be developed (Cardina et al., 1999).

HERBICIDE-RESISTANT CROPS: BENEFITS AND HARMS

It is now being realized that use of herbicides is inevitable even following integrated approaches. However, their judicious use can enhance crop productivity. A number of herbicides are available that can control weeds effectively, but lack of selectivity limits their use to only the preemergence stage. It is therefore necessary to produce herbicide-tolerant or -resistant crops so as to use some nonselective herbicides with greater flexibility and broad spectrum. This concept has gained greater attention in the recent years, particularly due to progress in molecular genetics. Crops can be made tolerant to herbicides through breeding methods, cell culture selection, protoplast fusion, and genetic engineering. Employing these techniques, a number of crops such as oil seed rape, potato, cotton, canola, maize, broccoli, and tobacco have been made resistant (both transgenically as well as nontransgenically) to herbicides such as glufosinate, glyphosate, imidazolines, sulphonyl urea, triaznes, and bromoxynil (Duke, 1999; Scheffler et al., 2001; Duke et al., 2002; Gressel, 2002). Worldwide, in 2004 herbicide-resistant transgenic crops were planted on 81 million ha (James, 2004). HRCs have both benefits and harms (Wyse, 1992a). When combined with environmentally and toxicologically sound herbicides under an IWM program HRCs can effectively increase crop production (Giaquinta, 1992). These can reduce risks of crop injury and allow more effective use of postemergent-type herbicides. However, their release requires careful scrutiny and adequate field-testing since there are risks of introgression of resistant crop genes and transgenes from HRCs to related weed species (Darmency, 1996). It is feared that herbicide-resistant crops would lead to rapid evolution of herbicide-resistant weeds or "super weeds" that will be difficult to manage. In view of the uncertainty linked with harms and benefits of HRCs, a prior assessment of their economical, ecological, and social consequences should be conducted (Radosevich, Gheresa, and Comstock, 1992).

HERBICIDE ANTIDOTES AND SYNERGISTS

It has long been known that addition of chemicals known as *herbicide antidotes* or *safeners* can protect crops from injury. Antidotes are chemically diverse groups of compounds that increase herbicide tolerance in cereals and dicotyledonous plants (Hatzios, 1991). These have no herbicidal activity of their own but appear to cause rapid degradation of other herbicides (Köcher et al., 1989). The reasons for such action are unknown. However, they enhance the expression of herbicide detoxifying enzymes such as glutathione-*S*-transferase, thereby increasing the synthesis of glutathione (Hatzios, 1991; DeRidder et al., 2002; Hirase and Molin, 2003). Use of naphthalic anhydride, fenclorim, oxabetrinil, and dichlormid to protect the crop against soil-applied chloroacetamides and thiocarbamates in cereal crops is well-known (Walton and Casida, 1995; Scott-Craig et al., 1998). However, they have been successful only in a few crops and with a few herbicides.

In contrast to antidotes, synergists such as azoles, tetcyclasis, and chelators of copper enhance the activity of herbicides and thus reduce their rate of application by a factor of ten or even more (Gressel and Shaaltiel, 1988; Gressel, 1990, 1992; Bewick et al., 1991). However, tridiphane—the only commercial synergist that enhanced the activity of atrazine—has been withdrawn (Gressel, 1992).

WEED MANAGEMENT—SOME ALTERNATE APPROACHES

The best-adapted weed species continue to proliferate and perpetuate because of the simplified cropping systems and selection pressure favoring them and their biotypes (Holt, 1994). Thus, keeping in mind the ill effects of herbicides, future weed management strategies should be cautiously designed. Buhler (1999) stressed the need for developing alternatives and options for maintaining the effectiveness of herbicides and to reduce their usage. Wyse (1992b) emphasized the use of principle-based research that provides basic knowledge about weed ecology and biology to develop new weed control technology. It helps in understanding the very basics of weed problems—time of maximum weed interference, weed population dynamics, and regulation of weed seed germination that can help in their control. Weed science research, therefore, can be divided into *weed control science and technology* (including weed control through chemical, biological, and tillage methods) and *weed science principles* (including research on weed biology and ecology). The discipline of weed science is now evolving more toward ecological and biological research on weeds, economical and alternate weed management methods, scouting, modeling, and integrated systems (Burnside, 1993). To take timely action for weed control, efforts should therefore be made to understand why weed problems arise and how can we avoid them. Unfortunately, little has been done in this regard.

Weed science must also put emphasis on weed population dynamics that provide thresholds for weed management (Norris, 1999). Greater emphasis should therefore be given to the reduction in weed emergence, production of propagules, and minimizing weed competition with crops. Manipulations of soil environment and cropping systems can also help in suppressing weeds by reducing seed bank persistence, seedling establishment, and interference (Gallandt, Liebman, and Huggins, 1999). Buhler (1999) suggested various long-term (weed biology and ecology, site-specific management, redesigning the cropping system, new control methods such as allelopathic and competitive cultivars, managed competition, and biological control) and short-term (scouting, tillage, integrated methods, reduced rate of herbicides) strategies for the management of weeds through various alternative methods. For the success of alternative strategies, all phases of science and technology need to be taken into account by weed scientists (Buhler, 1999). Thus greater focus is now placed on developing alternative strategies with low input after thorough assessment.

For designing ecological-based weed management strategies, there is a need for protection of ecological niches from invasion, growing crops with divergent life cycles, maintaining the crop-weed balance in agroecosystems, and enhancing crop competitiveness against weeds (Liebman, Mohler, and Staver, 2001). Liebman (2001) suggested that weed density should be reduced to tolerable levels to reduce amount of damage by encouraging shift in weed flora, which are less aggressive and easier to manage. There is also a need to generate knowledge on how many weeds can be tolerated at the economic threshold and their various desirable and undesirable effects so as to develop ecological threshold (Hurle, 1997).

Use of crop genotypes with greater competitive and allelopathic ability also forms a promising strategy for the future so that they fit well in the cover cropping or crop rotation models. Such varieties due to their fast growth can overshadow weeds and thus be helpful in reducing weed interference. These could serve as the most economical, effective, and feasible component of IWM for all crops (Pester, Burnside, and Orf, 1999).

However, in some situations herbicides are needed to kill cover crops before planting the next season's crops. Stanilaus and Cheng (2002) have reported that cover crops can be transgenically designed that can undergo self-destruction and thus reduce the need for use of herbicides. Further, a number of other allelopathic crops and their allelochemicals could also help in reducing weed interference at desirable levels (Batish et al., 2001). Use of allelochemicals as alternatives to synthetic herbicides is also a very useful strategy that can solve some of the problems associated with the use of synthetic herbicides or at least can reduce their usage. Molecular genetic approaches for the improvement of allelopathic crops can further expand their use and enhance efficiency (Foley, 1999; Scheffler et al., 2001; Duke et al., 2002; Singh, Batish, and Kohli, 2003). Gressel (1992) emphasized that genetic engineering coupled with chemistry is required for (1) control of parasitic higher weeds, (2) weeds in the wheat fields that have developed multiple and cross resistance to all wheat selective herbicides and to replace herbicides, and (3) weeds in corn fields because of the imposition of restrictions and cancellation of herbicide use. Of late, due to the fears of gene flow from more competitive transgenic crops, a failsafe mechanism for gene containment in transgenic crops is being developed (Daniell, 2002).

The virulence of biological control agents such as mycoherbicide for weed management can also be increased trangenically by introducing genes for new phytotoxin production. For example, transfer of protein phytotoxin Nep1 encoding the *NEP1* gene from *Fusarium* spp. into *Colletotrichum coccodes* (Wallr.) S. Hughes increases its virulence to velvetleaf (*Abutilon theophrasti* Medik.) by ninefold (Cohen et al., 2002). Likewise, new genes can be introduced for the production of higher level of plant hormones such as auxins. For example, the introduction of two genes of indole-3-acetamide pathway into *F. oxysporum* Schltdl.: Fr. and *F. arthrosporioides* Sherb. resulted in the production of more auxins, and these transformants were more effective in controlling the serious parasitic weed Egyptian broomrape (*Orobanche aegyptiaca* Pers.) (Amsellem, Cohen, and Gressel, 2002). However, these two strategies carry the risk of changing the host range of microbes in unexpected and undesirable ways.

FUTURE DIRECTIONS

Areas that need to be investigated on a priority basis for designing future weed management strategies are outlined here:

1. Understanding the biology, ecology, and genetics of weeds in terms of their germination potential, life cycle pattern, seed production, ecological amplitude, habitat range, population dynamics, factors affecting growth, and survival strategies
2. Revival of traditional practices with innovations suitable to present needs such as inter cropping, polyculturing, crop rotation, use of cover, green manure, and smother crops in such combinations that can reduce weed density to a manageable level
3. Promoting ecological principles of weed management such as replacement of weed flora with less aggressive ones and proper land management
4. Convergence of traditional agricultural practices with ecological, economical, and sociological perspective
5. Strengthening IWM through various approaches depending upon the geographical area, climatic conditions, and type of weed and crop combinations
6. Modification of soil environment that make conditions unsuitable for weeds and weed seeds in the soil
7. Use of biotechnological approaches for crop improvement transgenically and nontransgenically so as to generate crops with greater allelopathic/competitive ability and to produce herbicide-tolerant and -resistant crops to expand the use of some broad-spectrum postemergent herbicides with less residual activity
8. Transgenic improvement of biological control agents (mycoherbicides) to enhance their virulence
9. Development of multiple pathogen strategy for controlling a wide variety of weeds

REFERENCES

Agrow (2003). Agrochemical sales flat in 2002. *Agrow: World Crop Protection News,* February 26-March 28. Available at <http://ipm.osu.edu/trans/043_141.htm>.

Akobundu, I.O. (1991). Weeds in human affairs in sub-Saharan Africa: Implications for sustainable food production. *Weed Technology* 5: 680-690.

Alstrom, S. (1990). *Fundamentals of weed management in hot climate peasant agriculture.* Uppasla, Sweden: Swedish University of Agricultural Sciences.

Altieri, M.A. (1988). The impact, uses, and ecological role of weeds in agro-ecosystems. In M.A. Altieri and M.Z. Liebman (Eds.), *Weed management in agroecosystems: Ecological approaches* (pp. 1-6). Boca Raton, FL: CRC Press, Inc.

Amsellem, Z., B.A. Cohen, and J. Gressel (2002). Engineering hypervirulence in a mycoherbicidal fungus for efficient weed control. *Nature Biotechnology* 20: 1035-1039.

Appleby, A.P., F. Müller, and S. Carpy (2000). Weed control. In F. Müller (Ed.), *Agrochemicals* (pp. 687-709). New York: Willey VCH.

Baker, H.G. (1974). The evolution of weeds. *Annual Reviews of Ecology and Systematics* 5: 1-24.

Baker, H.G. (1986). Patterns of plant invasion in North America. In H.A. Mooney and J.A. Drake (Eds.), *Ecology of biological invasions of North America and Hawaii* (pp. 44-57). New York: Springer-Verlag.

Batish, D.R., H.P. Singh, R.K. Kohli, and S. Kaur (2001). Crop allelopathy and its role in ecological agriculture. *Journal of Crop Production* 4: 121-161.

Begon, M., J.L. Harper, and C.R. Townsend (1996). *Ecology: Individuals, populations, and communities,* Second edition. Boston, MA: Blackwell Scientific Publishers.

Bewick, T.A., W.M. Stall, S.R. Kostewicz, and K. Smith (1991). Alternatives for the control of paraquat tolerant American black nightshade *(Solanum americana). Weed Technology* 5: 61-65.

Boyetchko, S.M. (1997). Principles of biological weed control with microorganisms. *HortScience* 32: 201-205.

Buhler, D.D. (1999). Expanding the context of weed management. *Journal of Crop Production* 2(3): 1-8.

Buhler, D.D. (2002). Challenges and opportunities for integrated weed management. *Weed Science* 50: 273-280.

Bunting, A.H. (1960). Some reflections on the ecology of weeds. In J.L. Harper (Ed.), *The biology of weeds* (pp. 11-26). Oxford: Blackwell Scientific Publishers.

Burnside, O.C. (1993). Weed science—The stepchild. *Weed Technology* 7: 515-518.

Cardina, J., T.M. Webster, C.P. Herms, and E.E. Regnier (1999). Development of weed IPM: Levels of integration for weed management. *Journal of Crop Production* 2(3): 239-267.

Cohen, B.A., Z. Amsellem, R. Maor, A. Sharon, and J. Gressel (2002). Transgenically enhanced expression of indole-3-acetic acid confers hypervirulence to plant pathogens. *Phytopathology* 92: 833-841.

Crafts, A.S. and W.W. Robbins (1973). *Weed control.* New Delhi, India: Tata McGraw-Hill Publishing Co., Ltd.

Crawley, M.J. (1989). The success and failures of weed biocontrol using insects. *Biocontrol News and Information* 10: 213-223.

Crawley, M.J. (Ed.) (1997). *Plant ecology.* Oxford, UK: Blackwell Scientific Publishers.

Daniell, H. (2002). Molecular strategies for gene containment in transgenic crops. *Nature Biotechnology* 20: 581-586.

Darmency, H. (1996). Potential disadvantages of herbicide-resistant crops in weed resistance management. In H. Brown, G.W. Cussans, M.D. Devine, S.O. Duke, C. Fernandez-Quintanilla, A. Helweg, R.E. Labrada, M. Landes, P. Kudsk, and J.C. Streibig (Eds.), *Proceedings of the 2nd International Weed Control Congress* (pp. 1175-1181). Denmark: DJF, Flakkebjerg.

DeRidder, B.P., D.P. Dixon, D.J. Beussman, R. Edwards, and P.B. Goldsbrough (2002). Induction of glutathione S-transferases in *Arabidopsis* by herbicide safeners. *Plant Physiology* 130: 1497-1505.

Duke, S.O. (1999). Weed management: Implications of herbicide resistant crops. In P.L. Traynor and J.H. Westwood (Eds.), *Ecological effects of pest resistance genes in managed ecosystems* (pp. 21-25). Blacksburg, VA: Information System for Biotechnology, Virginia Polytechnic Institute and State University.

Duke, S.O., F.E. Dayan, A.M. Rimando, K.K. Schrader, G. Aliotta, A. Oliva, and J.G. Romagni (2002). Genetic engineering crops for improved weed management traits. In K. Rajasekaran, T.J. Jacks, and J.W. Finley (Eds.), *Crop biotechnology* (pp. 52-66). American Chemical Society Symposium Series 829. Washington, DC: American Chemical Society.

Foley, M.E. (1999). Genetic approaches to the development of cover crops for weed management. *Journal of Crop Production* 2(3): 77-94.

Gallandt, E.R., M. Liebman, and D.R. Huggins (1999). Improving the soil quality: Implications for weed management. *Journal of Crop Production* 2(3): 95-122.

Giaquinta, R.T. (1992). An industry perspective on herbicide-tolerant crops. *Weed Technology* 6: 653-656.

Gressel, J. (1990). Synergizing herbicides. *Reviews of Weed Science* 5: 49-82.

Gressel, J. (1992). Addressing real weed science with innovations. *Weed Technology* 6: 509-525.

Gressel, J. (2002). *Molecular biology of weed control*. London: Taylor and Francis.

Gressel, J. and Y. Shaaltiel (1988). Biorational herbicides synergists. In P.A. Hedin, J.J. Menn, and R.M. Hollingworth (Eds.), *Biotechnological approaches to plant protection* (pp. 4-24). Washington, DC: American Chemical Society.

Grime, J.P. (1979). *Plant strategies and vegetation processes*. London: John Wiley.

Hammer, K. (1988). Präadaptationen und die Domestikation von Kulturpflanzen und Unkräutern. *Biologisches Zentralblatt* 107: 631-636.

Hammer, K., Th. Gladis, and A. Diederichsen (1997). Weeds as genetic resources. *Plant Genetic Resources Newsletter* 111: 33-39.

Harlan, J.R. and J.M.J. deWet (1965). Some thoughts about weeds. *Economic Botany* 19: 16-24.

Harper, J.L. (Ed.) (1960). *The biology of weeds*. Oxford: Blackwell Scientific Publishers.

Hatzios, K.K. (1991). An overview of mechanism of action of herbicide safeners. *Zeitschrift für Naturforschung* 46c: 819-827.

Heap, I. (2005). International Survey of Herbicide Resistant Weeds. Available at <www.weedscience.org/in.asp>.

Hirase, K. and W.T. Molin (2003). Sulfur assimilation in plants and weed control: Potential targets for novel herbicides and action sites of certain safeners. *Weed Biology and Management* 3: 147-157.
Holt, J.S. (1988). Ecological and physiological characteristics of weeds. In M.A. Altieri and M.Z. Liebman (Eds.), *Weed management in agroecosystems: Ecological approaches* (pp. 7-24). Boca Raton, FL: CRC Press, Inc.
Holt, J.S. (1994). Impact of weed control on weeds: New problems and research needs. *Weed Technology* 8: 400-402.
Hurle, K. (1997). Concepts in weed control—How does biocontrol fit in it? *Integrated Pest Management Reviews* 2: 87-89.
James, C. (2004). Global status of commercialized biotech/GM crops: 2004. ISAAA Briefs No. 32. Ithaca, NY: ISAAA.
Kanampiu, F., D. Friesen, and J. Gressel (2003). A new approach to *Striga* control. *Pesticide Outlook* 14(2): 51-53.
King, L.J. (1966). *Weeds of the world: Biology and control.* New Delhi, India: Wiley Eastern Private Limited.
Köcher, H., B. Büttner, E. Schmidt, K. Lötzsch, and A. Schule (1989). Influence of HOE 70542 on the behavior of fenoxaprop-ethyl in wheat. In *Proceedings 1989 Brighton Crop Protection Conference—Weeds,* Volume 2 (pp. 495-500). Farnham, Surrey, UK: British Crop Protection Council.
Lacey, A.J. (1985). Weed control. In P.T. Haskell (Ed.), *Pesticide applications: Principles and practices* (pp. 456-485). Oxford: Oxford University Press.
Liebman, M. (2001). Weed management: A need for ecological approaches. In M. Liebman, C.L. Mohler, and C.P. Staver (Eds.), *Ecological management of agricultural weeds* (pp. 1-39). Cambridge, UK: Cambridge University Press.
Liebman, M., C.L. Mohler, and C.P. Staver (Eds.) (2001). *Ecological management of agricultural weeds.* Cambridge, UK: Cambridge University Press.
Mapes, C., F. Basurto, and R. Bye (1997). Ethnobotany of quintonil: Knowledge use and management of edible greens *Amaranthus* spp. (Amaranthaceae) in the Sierra Norte del Puebla, Mexico. *Economic Botany* 51: 293-306.
Mohler, C.L. (1996). Ecological basis for the cultural control of annual weeds. *Journal of Production Agriculture* 9: 468-474.
Mohler, C.L. (2001). Weed life histories: Identifying vulnerabilities. In M. Liebman, C.L. Mohler, and C.P. Staver (Eds.), *Ecological management of agricultural weeds* (pp. 40-98). Cambridge, UK: Cambridge University Press.
Müller-Schärer, H., P.C. Scheepens, and M.P. Greaves (2000). Biological control of weeds in European crops: Recent achievements and future work. *Weed Research* (Oxford) 40: 83-98.
Newman, R.M., D.C. Thompson, and D.B. Richman (1998). Conservation strategies for the biological control of weeds. In P. Barbosa (Ed.), *Conservation biological control* (pp. 371-396). San Diego: Academic Press, Inc.
Norris, R.F. (1999). Ecological implications of using thresholds for weed management. *Journal of Crop Production* 2(3): 31-58.
Pattersen, D.T. (1985). Comparative ecophysiology of weeds and crops. In S.O. Duke (Ed.), *Weed physiology,* Volume 1, *Reproduction and ecophysiology* (pp. 101-130). Boca Raton. FL: CRC Press, Inc.

Pester, T.A., O.C. Burnside, and J.H. Orf (1999). Increasing crop competitiveness to weeds through crop breeding. *Journal of Crop Production* 2(3): 59-76.

Piementel, D., S. McNair, J. Janecka, J. Wightman, C. Simmonds, C. O'Connell, E. Wong, L. Russel, J. Zern, T. Aquino, and T. Tsomondo (2001). Economic and environmental threats of alien plant, animal, and microbe invasions. *Agriculture, Ecosystem and Environment* 84: 1-20.

Pritchard, T. (1960). Race formation in weedy species with special reference to *Euphorbia cyparissias* L. and *Hypericum perforatum* L. In J.L. Harper (Ed.), *The biology of weeds* (pp. 61-66). Oxford, UK: Blackwell Scientific Publishers.

Radosevich, S.R., C.M. Gheresa, and G. Comstock (1992). Concerns a weed scientist might have about herbicide-tolerant crops. *Weed Technology* 6: 635-639.

Rejmánek, M. (1995). What makes a species invasive? In P. Pyšek, K. Prach, M. Rejmánek, and M. Wade (Eds.), *Plant invasions—General aspects and specific problems* (pp. 3-13). Amsterdam: SPB Academic Publishing.

Sakai, A.M., F.W. Allendorf, J.S. Holt, D.M. Lodge, J. Molofsky, K.A. With, S. Baughman, R.J. Cabin, J.E. Cohen, N.C. Ellstrand, et al. (2001). The population biology of invasive species. *Annual Reviews of Ecology and Systematics* 32: 305-332.

Salisbury, E.J. (1961). *Weeds and aliens.* London: Collins.

Scheffler, B.E., S.O. Duke, F.E. Dayan, and E. Ota (2001): Crop allelopathy: Enhancement through biotechnology. *Recent Advances in Phytochemistry* 35: 257-274.

Scott-Craig, J.S., L. Poduje, J.E. Casida, and J.D. Walton (1988). Herbicide safener binding protein of maize: Purification, and cloning and expression of an encoding cDNA. *Plant Physiology* 116: 1083-1089.

Shaw, W.C. (1988). Integrated weed management systems technology for agroecosystem management. In N.B. Mandava (Ed.), *Handbook of natural pesticides: Methods,* Volume 1, *Theory, protection and detection* (pp. 123-129). Boca Raton, FL: CRC Press, Inc.

Singh, H.P., D.R. Batish, and R.K. Kohli (2003). Allelopathic interactions and allelochemicals: New possibilities for sustainable weed management. *Critical Reviews in Plant Sciences* 22: 239-311.

Stanilaus, M.A. and C.L. Cheng (2002). Genetically engineered self-destruction: An alternative to herbicides for cover crop systems. *Weed Science* 50: 794-801.

Swanton, C.J. and S.F. Weise (1991). Integrated weed management: The rationale and approach. *Weed Technology* 5: 657-663.

Thill, D.C., J.M. Lish, R.H. Callihan, and E.J. Bechinski (1991). Integrated weed management—A component of integrated pest management: A critical review. *Weed Technology* 5: 648-656.

United Nations Conference on Environment and Development (UNCED) (1992). Promoting sustainable agriculture and rural development: Section I. Integrated pest management and control in agriculture. In Agenda 21 *Proceedings of United Nations Conference on Environment and Development.* Rio de Janeiro, Brazil, June 3-14, Chapter 14, paragraphs 14.74-14.82. New York: Division for Sustainable Development, Department of Economic and Social Affairs, United

Nations. Available at <http://www.un.org/esa/sustdev/documents/agenda21/english/agenda21chapter14.htm>.

Walton, J.D. and J.E. Casida (1995). Specific binding of a dichloroacetamide herbicide safener in maize at a site that also binds thiocarbamate and chloroacetanilide herbicides. *Plant Physiology* 109: 213-219.

Wyse, D.L. (1992a). Future impact of crops with modified herbicide resistance. *Weed Technology* 6: 665-668.

Wyse, D.L. (1992b). Future of weed science research. *Weed Technology* 8: 162-165.

Zimdahl, R.L. (1995). Weed science in sustainable agriculture. *American Journal of Alternative Agriculture* 10: 138-142.

Zimdahl, R.L. (1999). *Fundamentals of weed science*. San Diego, CA: Academic Press, Inc.

Chapter 2

Weed Management: A Basic Component of Modern Crop Production

Ricardo Labrada

INTRODUCTION

Increased world population will demand more food production, which can be achieved only by increasing crop yields. There are few possibilities to incorporate new areas into crop production. As an example of the problem in the case of cereals, Pinstrup-Andersen (2001) stated that demand for cereals to feed livestock will increase considerably in coming decades, especially in developing countries, in response to the high demand for livestock products. Between 1993 and 2020, it is estimated that developing countries' demand for cereals for animal feed will double, while demand for cereals for food for direct human consumption will increase by 47 percent. These expected increases in cereal demand will be met primarily by productivity increases. Because most of the growth in cereal areas will be concentrated in the relatively low-productivity cereals of sub-Saharan Africa, increases in cultivated area will contribute less than 20 percent of the increase in global cereal production in the period from 1993 to 2020.

From this estimate it is clear that increased crop yields can be achieved only by a sustainable approach, i.e., more production with rational use of available resources, which also implies responsible use of land and water and enhanced food diversity. Efforts are needed to reduce crop losses due to pests through the use of resistant crop varieties, pesticides, and better cultural practices; herein integrated pest management (IPM) is expected to have a major role in reducing losses caused by pests, without harmful side effects (Pinstrup-Andersen and Cohen, 1998).

In this context, weeds are considered an important biotic constraint to food production (Parker and Fryer, 1975; Akobundu, 1987). In developed

This chapter does not reflect fully the FAO's (Food and Agriculture Organization of the United Nations) views on this matter.

economies the importance of weed control is evident due to the fact that present herbicide use exceeds the amount of insecticides and fungicides used. In several developing countries, weed management is not yet a priority due to the high incidence of insects and pathogens, which normally cause more spectacular losses within a few days, and lack of understanding of problems caused persistently by weeds.

This chapter explores the importance of weed management and how crop production can be improved with the introduction and use of technologies for better weed control.

INTEGRATED PEST MANAGEMENT

Since the mid-1960s, integrated pest management has been promoted by the FAO worldwide as a preferred strategy for pest control. IPM means the careful consideration of all available pest control techniques and subsequent integration of appropriate measures that discourage the development of pest populations and keep pesticides and other interventions to levels that are economically justified and reduce or minimize risks to human health and the environment. IPM emphasizes the growth of a healthy crop with the least possible disruption to agroecosystems and encourages natural pest control mechanisms (FAO, 2003).

There is general agreement that IPM must be science based and economically viable for farmers. Prevention is an important requirement as well as pest control during the crop cycle. The preventive approach requires regular evaluation of pest density and risk of entry of new pests into the crop. Common control strategies follow:

- *Cultural methods,* which include crop rotation, resistant cultivars, use of cover crops, intercropping, mechanical control, and efficient water management and fertilizer application
- *Biological control,* either the classic method of introducing new exotic organisms evaluated prior to their introduction or the augmentative method through rearing and release of existing beneficial natural enemies
- *Chemical control,* or the rational and judicious use of pesticides
- *Other methods,* including the use of soil solarization and other physical methods, which remain under investigation

The IPM approach has had a major impact on insect and mite control with some other positive results in disease control, but with little effect on weeds. The reasons for this lack of success with weeds are as follows:

1. Little understanding of the problems caused by weeds exists, particularly in developing countries.
2. Few or no ecological weed elements are included in IPM programs.
3. There is a pronounced tendency in industrial countries to use herbicides as the sole control option on large crop areas or to rely on hand weeding in poor or subsistence sites.

Devising integrated weed management (IWM) strategies that address a diversity of weed species with a diversity of life history traits is a difficult task (Mortensen, Bastiaans, and Sattin, 2000). In the context of IPM, particular importance must be given to the improvement of weed management, which is a minor component of the majority of available programs on IPM. Auld (1994) classifies the impact of weeds on any agricultural system as follows:

1. Reduced agricultural output (quantity and/or quality)
2. The costs involved in a given level of control of weeds within the existing farming system
3. A change in the farming system to a new system (e.g., new crop) brought about mainly because of particular weeds
4. External costs caused by weeds spreading across farm boundaries

While most other pests are under control in various crops, weeds continue to be a limiting factor to crop yields. Weeds directly affect crop growth and development by competing effectively for light, nutrients, and water; by releasing substances from their roots and leaves that are toxic to various crop plants; and by serving as hosts of many pest organisms that pose serious risks to crop production. Weeds also seriously affect inland water bodies and related activities (irrigation, navigation, fisheries, and the energy potential of hydroelectric stations).

Weeds cause important social problems as well. Small-scale farmers in developing countries spend more than 40 percent of their labor time on weeding, which is primarily the labor of women and children in the farmer's family. Children involved in this task often cannot attend school regularly (Labrada and Parker, 1994).

COMPONENTS OF INTEGRATED WEED MANAGEMENT

The inclusion of a weed component in IPM curricula requires the study of various elements of weed ecology and biology and interference with crops to actually achieve long-term weed control effects. Sustainable weed management should foresee the control of weeds during the so-called periods of weed competition but should also guarantee the reduction of viable weed seed bank in the soil.

Basic components of knowledge for sound weed management include

1. regular monitoring of weed infestation, which may include the detection of herbicide-resistant weeds in areas regularly treated with herbicides;
2. biology and ecology of the prevalent weed species;
3. identification of critical periods of competition and periods of massive weed rain seed; and
4. technically effective, economically viable, and environmentally safe control methods.

Monitoring of Weed Infestation

A fundamental basis for sound weed management is to know the species present and their level of infestation. The composition of weed communities varies greatly and is closely linked to cropping history. Regular evaluation of weed infestation in agricultural fields is advisable, particularly in areas currently treated with herbicides. In these surveys weed identification plays an important role. It is essential to be able to identify perennial or parasitic weeds, which will not respond to the standard traditional weeding practices, while more precise determination (even of annual weed species) can be vital for the optimum selection of any control method, including herbicides. Obviously, agricultural extension workers and other agents working in the field should conduct weed surveys using less tedious methods and based on visual assessment to enable them to evaluate large crop areas.

In this context, it is no less important to know about the weed seed bank as a primary source of weeds in cultivated soils. Because the persistence of most weed species is contingent upon the presence of a buried seed bank, knowledge of seed bank size and the flux of seeds within it, in relation to land cultivation practices, is essential (Forcella and Burnside, 1993). If the seed bank is identified well in advance, it is possible to predict weed infestation at least in the medium term. However, existing methods to estimate weed seed bank have always been based on large numbers of samples (e.g., soil cores) and counting visually seeds of weed species, which is a tedious and slow process. However, a strong correlation has been found between seedlings emerged in the greenhouse and seeds extracted in the laboratory for the most abundant weed species (Forcella, 1998; Rahman, James, and Grbavac, 2001).

Forcella (1993) indicated that the "seedling emergence" method of seed bank analysis avoids the problems of seed-soil separation and seed identification inherent in the "seed extraction" method. In this method, multiple soil cores from a field may be aggregated, thoroughly mixed, spread as a 2 to 5 cm deep layer of soil onto a single tray, and stored in conditions that promote germination. Emerged seedlings are counted by species after two or three weeks. After counting seedlings, soil in the trays is dried, thoroughly remixed, and the cycle repeated. For weeds of arable soils, two or three cycles are often sufficient to estimate seed banks. The seedling emergence method makes the collection of seed bank data less time-consuming and cheaper to carry out.

Another method for predicting weed "pressure" was proposed by Gordon Harvey (1993). Harvey considered the best predictor of future weed problems to be a knowledge of previous weed problems. By keeping long-term records of weed problems in replicated untreated check areas within each field, growers should be able to not only anticipate the severity of weed infestations, but also recognize early changes in species composition of weed populations. He finally proposed monitoring weed infestation by making weed pressure (WP) ratings. WP is defined as the visual estimate of the percentage with which weeds contribute to the total volume of both crop and weeds in a plot.

Other Elements of Weed Ecobiology

Weed management means control of weed population and, to this end it is important to estimate some elements of population equilibriums and life-stages for regulating population size. Understanding the ecobiological behavior of weeds is a prerequisite for developing guidelines for environmentally sound weed control (Zimdahl, 1995).

Mortimer (1998) suggests that ecological studies of weeds and weed management may be conducted with a diversity of objectives in mind, for example,

1. autoecological comparison of weed species;
2. measuring the impact of new technologies on the dynamics (change in abundance) of "hard-to-kill" species; and
3. quantifying weed community structure and analyzing its causes.

In practical terms, it is necessary to have information on the behavior of prevailing weeds. Data are needed on germination of diaspores, phases when the plant starts fruit setting (and/or emissions of new vegetative organs in the case of perennial species), factors affecting its growth and development, and production of reproductive organs. In conclusion a picture of the cycle (see Figures 2.1 and 2.2) should be completed to the further extent possible.

Weed germination patterns generally result in cohorts of seedlings emerging over an extended period of time and are heavily influenced by weather conditions, soil type, and cropping system (Vleeshouwers, 1997). Here it is important to know the number and periods of weed flushes. In hot climatic countries, several weed flushes may occur during the crop cycle, and knowledge of this is essential when designing better control measures to prevent weed seed bank buildup.

Critical Periods of Interference

Weeds interfere severely with the growth and development of crops. Interference comprises *competition* with crops for light, nutrients, and water, and *allelopathy,* which refers to the adverse effect of substances produced by one plant on the growth and development of other plant(s). When referring to interference some authors simply use the term *competition.*

For understanding competition, it is common to study the so-called critical period of weed competition, which has been defined as the period during which weeds must be controlled to prevent yield losses (Sattin, Berti, and Zanin, 1998). Knowledge of critical periods of weed competition may help to determine when to implement control measures in order to minimize yield losses for many crops (Zimdahl, 2004).

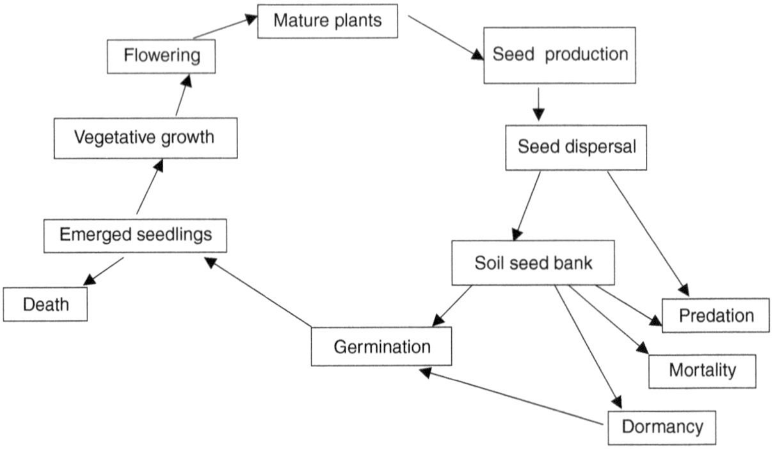

FIGURE 2.1. Cycle of an annual weed.

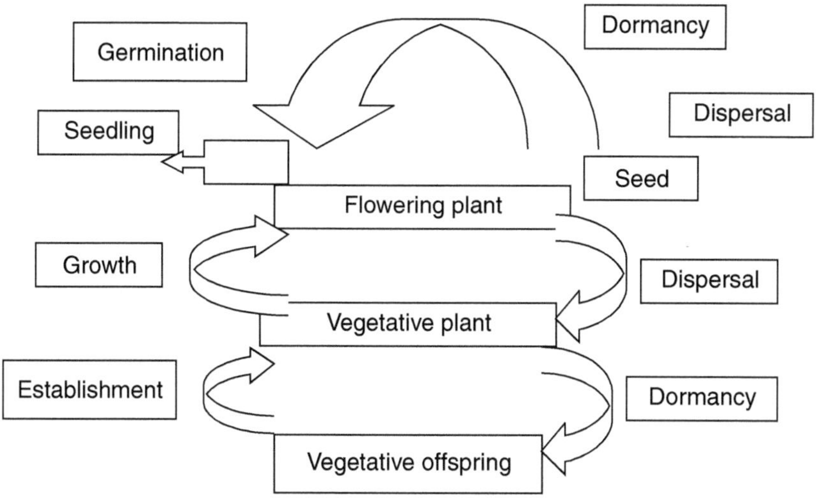

FIGURE 2.2. Cycle of a perennial weed.

Pitelli and Durigan (1984) proposed a classification of periods of competition for practical purposes. These are as follows:

1. *Period for preventing interference* starts from crop emergence or sowing when the crop should be kept weed free. This period is also known as the "weed-free period" (see Figure 2.3).
2. *Period of pre-interference* is the period starting from crop emergence or sowing, when the crop is able to tolerate the presence of some weeds without its productivity being affected.
3. *Initial period of interference* refers to the period when competition becomes important and starts affecting crop productivity.
4. *Critical period for preventing competition* is the period when the presence of weeds is critical, i.e., the maximum period of admissible weed presence.

Knowledge of the "critical periods" would enable farmers to make the most efficient use of limited labor resources. The critical periods vary with the relative competitiveness of the crop and the weed infestation: the lower the crop competitiveness (and/or the higher the weed flora competitiveness), the longer the period the crop must be kept weed free to prevent significant yield losses (Sattin, Berti, and Zanin, 1998).

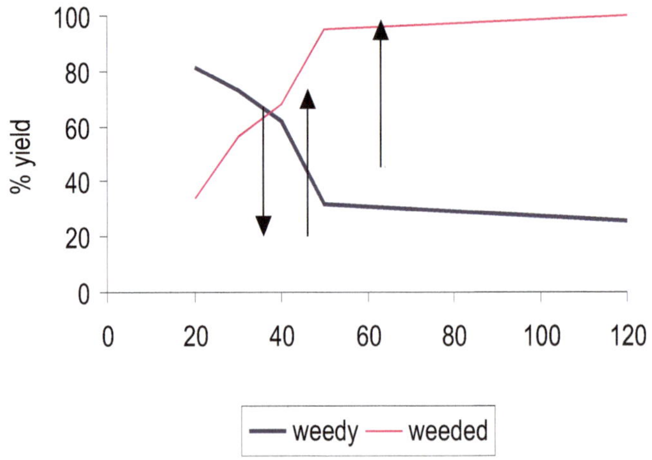

FIGURE 2.3. Critical period of weed competition. Ascending and descending arrows show the critical period of weed competition. The far-right ascending arrow shows the weed-free period, which indicates that there is no further need to weed during the remaining cycle to obtain high crop yields.

Studies have already been conducted on determining weed-free and critical periods of weed competition, but these data, although useful, do not provide advice regarding intervention on weed flushes thriving after the critical periods. It may be that combining knowledge of these periods and weed flushes in a particular site may give a more complete idea of what has to be done to control weeds during the crop cycle in order to prevent weed seed bank buildup in the soil.

CONTROL STRATEGIES

With the advent of herbicides many people thought that the problem of weeding was solved. At that time, the main aim was to identify a selective herbicide that would minimize hand weeding or mechanical cultivation. As a pregraduate student, the author was given to understand that with the use of herbicides, crop rotation would become unnecessary. Experience has demonstrated that these ideas were completely erroneous. Chemical control has been useful and helpful to increase crop yields, but other problems have arisen.

The main weed control strategies that farmers can successfully implement are as follows:

1. Land preparation or tillage systems (minimum till, zero till, and conventional tillage)
2. Cultural: Crop rotation, use of competitive and/or allelopathic cultivars, living or dead cover crops, water management, fertilizer application, mulching, and intercropping
3. Physical: Interrow cultivation and hand weeding
4. Biological control by releasing insects or other animals using beneficial pathogens
5. Chemical control through the use of herbicides
6. Other control methods, e.g., soil solarization

It should be remembered that weed management comprises various operations before and during the crop cycle to reduce weed infestation. From tillage through to harvest, weed control is of major concern to farmers.

Crop Rotation

The key to a successful cropping system is good crop rotation, which helps reduce pest problems. Several weeds can adapt to a specific crop. Rotation with morphologically different crops and with different growing requirements may help to break the cycle of adapted weeds. A good example is when upland crops are rotated with irrigated rice and vice versa. Unfortunately, this is not always possible because either several crops cannot grow in any type of soil or a market for the crop in rotation does not exist and, subsequently, it becomes unprofitable. In fact, monocropping is unsustainable for crop production because of heavy dependence upon agrochemicals, including herbicides.

Cropping sequence has an enormous influence on variations of weed populations as a direct result of implementing various weed management practices associated with crops involved in rotation. The use of densely seeded and competitive plants as preceding crops, such as cereals and legumes, significantly reduces weed density from one season to another. An interesting study has been conducted in Canada by Doucet and colleagues (1999) on the dynamics of standing weed vegetation in *Zea mays* L. (maize), *Glycine max* [L.] Merr. (soybeans), and *Triticum aestivum* L. (wheat). From these studies, maize was shown to be poor preceding crop with respect to weeds, while wheat in the crop rotation reduced weed density. The authors applied an integrated approach, and concluded that the best control was achieved when crop rotation was combined with herbicide use.

In Romania, Petcu and Ionita (1998) examined the effect of some cropping systems on weeds. The systems were wheat monocropping, a two-year wheat-maize rotation, and rotations of wheat-maize-peas or wheat-sugar beet-maize-sunflowers. After 25 years, the authors established that weed infestation decreased within all cropping systems, and it was much lower in the rotations compared when with wheat monocropping.

Tillage Systems

Tillage systems are essential for land preparation, and weed infestation may dictate the type of tillage to be used. Tillage based on plowing and harrowing during land preparation has always been a conventional method of weed control and is still widely used for growing roots, tubers, and vegetables. Minimum- or zero-tillage practices are becoming more widespread in the production of direct-seeded grains. These tillage systems leave the residues and stubble from preceding crop covering the soil, which reduces weed infestation to some extent. However, it is clear that with the switch from conventional tillage to other less disturbing soil methods, weed population changes may take place. For example, a survey conducted in Manitoba (Canada) (Kelner, 1997), in areas with conventional and zero tillage, showed that high densities of perennial weeds, including quackgrass [*Agropyron repens* (L.) Beauv.] and some annual weeds such as *Avena* spp. and volunteer canola, prevail under zero tillage. However, other annual weeds, such as *Setaria* spp., volunteer barley, wild mustard, redroot pigweed (*Amaranthus retroflexus* L.), common lambsquarters (*Chenopodium album* L.) and Kochia [*Kochia scoparia* (L.) Schrad.] had lower densities under zero-tillage crop production. When perennial weeds start prevailing consistently in zero-tillage areas, plowing becomes necessary in order to drag the rhizomes of these weeds to the soil surface to desiccate under the influence of the sun and wind.

Tillage systems alone may reduce the stand of certain species. In conservation agriculture, zero till may succeed in reducing weed infestation in general if combined properly with other control measures, e.g., crop rotation, use of cover crops, increased crop seed densities, and rational use of herbicides.

Competitive and/or Allelopathic Crops

One way to combat weeds is by encouraging the competitive nature of crops at an early stage of growth. Therefore, those plant varieties that are able to compete with weeds and to suppress them are preferable in the context of integrated weed management. Science has demonstrated that plant competitiveness does not necessarily mean tall and robust cultivars. Some studies have been carried out on the competitive ability of rice cultivars with weeds jungle rice [*Echinochola colona* (L.) Link.], late water grass [*E. phyllopogon* (Stapf.) Staph. ex Koss.], early water grass [*E. oryzoides* (Ard.) Fritsch], palisade signal grass [*Brachiaria brizantha* (C. Hoschst ex. A. Rich Stapf], or *Brachiaria decumbens* Stapf. (Fischer and Gibson, 2001), which showed that it is possible to have modern high-yielding, semidwarf cultivars able to compete with weeds efficiently. Competitive rice cultivars, such as semidwarf cv. M-202 effectively suppressed the infestation of *Echinochloa oryzoides* and *E. phyllopogon* emerging 20 days after the crop in California and may help reduce herbicide dependency and decrease selective pressure for resistance (Gibson et al., 2001).

The other area of interest is allelopathic crop varieties. Plants in general are able to produce and exude substances suppressing the growth of other plants in their vicinity. Selection for crop cultivars with allelopathic potential is undergoing research in various crops, with promising results. To date, progress has been made in understanding the genetics of crop allelopathic activity, and successful genetic manipulation of this trait has also been demonstrated (Wu et al., 1999).

Several crop cultivars may exhibit a high level of allelopathic effect over certain weed species. For example, results have shown that there is a wide variation in allelopathy among rice cultivars, some of which are able to suppress both monocot and dicot weed species (Olofsdotter, 2001). Similar results have been obtained in South Korea (Lin, Kim, and Shin, 2000; Chung, Ahn, and Yun, 2001), which represents a potential for improved weed management in the near future. Similar research is being conduced in wheat. Some cultivars such as 'Bezostaya-1', 'Naveed', and 'Niknejad' expressed positive allelopathy against wild oat [*Avena sterilis* ssp. *ludoviciana* (Durieu) Nyman] (Rizvi et al., 2000). Such research has shown that an increase in the wheat sowing density improved the allelopathic inhibition of the weed without any problem of autotoxicity. According to Wu and colleagues (1999), both wheat residue allelopathy and wheat seedling allelopathy can be exploited for managing weeds, including resistant biotypes. Wheat varieties differ in allelopathic potential against weeds, indicating that selection of allelopathic varieties might be a useful strategy in integrated weed management. Several categories of allelochemicals for wheat allelopathy have been identified, namely phenolic acids, hydroxamic acids, and short-chain fatty acids.

Mulch and Cover Crops

Mulches are excellent for weed management. They normally exclude light and serve as a physical barrier against emergence; they also affect soil temperature and moisture. Nonliving mulches can be divided into either *natural materials,* such as plant leaves, stalks, straw, compost, and dry soil, or *synthetic materials,* such as plastic polyethylene, which are widely used in pineapple and strawberry production. Natural mulches add organic matter and nutrients to the soil with their decomposition, and are inexpensive provided that the materials do not have to be transported from distant areas. The major problem is that natural mulches require large quantities to be applied per unit area. It is important to check that these materials are weed seed-free. Experiments show that natural mulches may exert good control of annual weeds but not of perennial weeds.

Before discussing *cover crops,* it is important to clarify that these are crops grown to produce soil cover, regardless of whether they are later incorporated. *Green manures* are plants incorporated into the soil while in an early vegetative stage, or soon after flowering, in order to enrich the soil. The term *green manure* has sometimes been used loosely and may sometimes refer to plants or plant vegetation that may be applied as mulch to the soil.

A cover crop intercropped or interplanted with an annual or perennial crop is called *living mulch,* which is useful to suppress weed infestation and prevent soil erosion, enhance soil fertility, and improve water infiltration. Cover crops also exclude light with their shading, limiting weed emergence. Examples of cover crops in temperate climate countries are seeded hairy vetch, clover, and annual ryegrass. In Brazil, sorghum is planted between rows in citrus with good results. In Colombia, pinto peanut (*Arachis pintoi* Crap. & Greg.) has been identified as an effective cover crop for the control of weeds in grapes (Rojas and Porras, 1999). It was found that 100 percent weed control was reached 240 days after planting, which significantly reduced weed populations without affecting crop yield or quality.

Allelopathic Cover Crops

Allelopathic cover crops are one of the most promising applications for weed control. Fujii (2001) evaluated the allelopathic activity of 53 cover crop plant species, including 26 leguminous, 19 graminaceous, and 8 others. The author found that leguminous cover crops such as hairy vetch (*Vicia villosa* Roth) and velvetbean [*Mucuna pruriens* (L.) DC.] grass cover crops such as oat (*Avena sativa* L.) and rye (*Secale cereale* L.), and certain cultivars of wheat (*Triticum aestivum* L.) and barley (*Hordeum vulgare* L.) were promising. Some of these results were later successfully validated in field experiments, where hairy vetch was found to be the most promising for weed control in abandoned fields. This was due to its ability to die off during the summer season and make thick strawlike mulch.

Cover crops fit well into many different cropping systems during the fallow or noncrop season. Interesting results have been obtained with the use of *Mucuna pruriens* in crop rotation in West Africa. Galiba and colleagues (1998) conducted a survey in farming areas of Benin, where 74 percent of the surveyed farmers were using *Mucuna* for at least three consecutive years, either in crop rotation or intercropped with maize. The farmers acknowledged the contribution of *Mucuna* to soil fertility and its ability to eradicate the stand of perennial weed cogon grass [*Imperata cylindrica* (L.) Beauv.]. The only problem was that farmers did not find any use for *Mucuna*. Because the products of the plant are not used, it is important to identify less persistent varieties of the plant. Carsky and Eteka (1999) are conducting a study to explore these opportunities.

It is obvious that legume cover crops reduce the need for the application of N fertilizers, but indirectly they may also reduce the incidence of some pests including weeds, with subsequent reduction of pesticide inputs. So promoting the use of cover crops is a contribution toward increased sustainability of crop production.

Interrow Cultivation or Weeding During the Crop Cycle

To control weeds by cultivation or manual weeding in the interrows is a traditional practice. These methods have to be used in organic agriculture in several crops to avoiding using herbicides. Organic farming using interrow cultivation to combat weeds in developed economies should carefully consider fuel expenditure and the number of mechanical interventions required. The idea of minimizing or eradicating herbicide applications should be welcome, provided the methods are not high energy and/or labor consuming.

Small-scale farmers in developing countries use hand labor for weed eradication. The problem here is to minimize the work required, which can easily take up to 40 percent of farmers' time. As Mortimer (1998) indicated, a desirable working maxim in developing integrated weed management must be to attempt to minimize the number of interventions per se and, where interventions are necessary, to minimize any negative environmental impacts. In some crop areas of poor small-scale farmers, where root parasitic weeds such as *Striga* and *Orobanche* spp. prevail, hand weeding before flowering of the parasites is advisable.

Biological Weed Control

Control through monophagous/oligophagous natural enemies is a highly sustainable way of managing troublesome weeds, and there are many good examples of effective biological control. One the most important achievements is the control of floating aquatic weeds using various specific insects. Many weeds have been introduced into various countries inadvertently, and the best way to control them is using the classic approach, i.e., introducing specialist natural enemies from the weed's area of origin. Table 2.1 gives some idea of agents that have proved effective in the control of specific weeds.

Biological control is highly specific, and it is impossible for one organism to be able to control several weeds without affecting the crop. However, some herbicide-resistant weeds or other highly invasive plants are extremely difficult to control using mechanical or chemical means. In such situations, the best method of control—if available—is to release biological

TABLE 2.1. List of organisms used effectively for biological control of some weeds.

Weed	Biological control agent
Eichhornia crassipes (Mart.) Solms. (Water hyacinth)	Weevils: *Neochetina eichhorniae, N. bruchi* Moth: *Sameodes albigutalis*
Hydrilla verticillata (Hydrilla)	Grass carp: *Ctenopharyngodon idella* Tuber weevil: *Bagous affinis* Leafmining fly: *Hydrellia pakistanae*
Pistia stratiotes L. (Water lettuce)	Weevil: *Neohydronomus affinis*
Salvinia molesta Mitchell (Water fern)	Weevil: *Cyrtobagous salviniae*
Alternanthera philoxeroides (Mart.) Griseb. (Alligator weed)	Flee beetle: *Agasicles hygrophila* Stem borer: *Vogtia malloi*
Parthenium hysterophorus L. (Ragweed parthenium)	Fungus: *Puccinia abrupta* var. *partheniicola*
Euphorbia esula L. (Leafy spurge)	Flea beetles: *Apthona* spp.
Chromolaena odorata (L.) King & Robins. (Siam weed)	Moth: *Pareuchaetes pseudoinsulata*
Lantana (*Lantana camara* L.)	Lacebug: *Teleonemia scrupulosa*
Cordia curassavica (Jacq.) Roem. & Schult. (Black sage)	Beetle: *Metrogaleruca obscura*
Orobanche spp. (Broomrape)	Stem feeding fly: *Phytomyza orobanchia*

control agents, which are able to spread to areas that cannot be reached, e.g., in large water bodies or forests.

Knowledge of the organism to be used is required as well as methods for rearing and release. It is also essential to know its host range in order to avoid undesirable effects on crops. *The Code of Conduct for the Import and Release of Exotic Biological Control Agents* (FAO, 1996), which is now under revision, gives good guidance on how to proceed in order to introduce new exotic organisms for biological control.

Chemical Weed Control

Herbicides were introduced into agricultural practice more than 50 years ago. At that time 2,4-D was the only chemical available and used solely for the control of annual weeds, mainly broadleaf weeds in cereals. By the mid-1950s selective herbicides started to be used in several crops and, 40 years after the advent of 2,4-D, the list of herbicides has increased (see Table 2.2). In several crops, mixtures of two or three herbicides have been used, giving better results than the application of a single herbicide. Chemical weed control has become cheaper over large areas of crops such as wheat, maize, sugar cane, direct-seeded rice, cotton, soybean, citrus, in vineyards, etc. The use of herbicides has also reduced losses caused by weeds. However, herbicides have been used and handled incorrectly on many occasions, particularly on small farms in developing countries. Accidents and problems affecting human health have often occurred, sometimes involving children; polluted water has been caused mainly by waste cans or herbicide containers; and some carryover effect in subsequent crops due to the high rates of herbicide used. This indicates the need for farmers to be educated in efficient and safe use and handling of herbicides.

The frequent evolution of herbicide resistance is no less serious, particularly with the repeated use of soil-acting herbicides and herbicide groups such as ALS and AACase inhibitors, which have a selection pressure higher than many other herbicide groups. At present there are 296 herbicide-resistant biotypes belonging to 178 crops. Of these 107 are dicots and 71 are monocots, which infest more than 270,000 fields (Heap, 2005).

The evolution of herbicide resistance shows that weed control cannot be solved by using herbicides alone. Herbicides rationally applied are effective tools combined with other control methods (crop rotation, tillage, cover crops, and others). Chemical control should be considered simply as one of a number of alternatives available.

With the repeated use of herbicide, weed monitoring becomes more important in order to detect the shift of weed populations and the development

TABLE 2.2. List of most important herbicide groups and their mode of action.

Herbicide group	Mode of action	Examples of herbicides
Iimidazolinones, sulfonylureas, triazolopyrmidines, pyrimidinylbenzoates, sulfonylaminocarbonyl-triazolinones	Inhibition of acetolactate synthase ALS (acetohydroxyacid synthase AHAS)	Chlorsulfuron, imazaquin, flumetsulam, bispyribac-Na flucarbazone-Na
Aryloxyphenoxy-propionates, cyclohexanediones	Inhibition of acetyl CoA carboxylase (ACCase)	Diclofop, fenoxaprop-ethyl, sethoxydim
S-triazines, trazinones, triazolinone ureas, uracils, amides	Inhibition of photosynthesis at photosystem II	Atrazine, diuron, amicarbazone, propanil, metribuzin, bromacil, desmedipham, pyrazon
Nitriles	Inhibition of photosynthesis at photosystem II	Bromoxynil, ioxynil
Bipyridiliums	Photosystem-I-electron diversion	Paraquat, diquat
Dinitroanilines	Microtubule assembly inhibition	Trifluralin, nitralin, dinitramine, pendimethalin
Mitosis inhibitors	Inhibition of mitosis/ microtubule polymerization inhibitor	Propham
Chloroacetamides, oxyacetamides, and acetamides	Inhibition of cell division (Inhibition of synthesis of very long chain fatty acids)	Butachlor, napropamide, mefenacet
Synthetic auxins	Synthetic auxins (action like indoleacetic acid)	2,4-D, MCPA, MCPB
Thiocarbamates	Inhibition of lipid synthesis—not ACCase inhibition	Triallate, EPTC, butylate
Triazoles	Bleaching: Inhibition of carotenoid biosynthesis (unknown target)	Amitrole
Pyridazinones, pyridinecarboxamides	Bleaching: Inhibition of carotenoid biosynthesis at the phytoene desaturase step (PDS)	Norflurazon, diflufenican
Glycines	Inhibition of EPSP synthase	Glyphosate, sulfosate
PPO inhibitors (diphenylethers, phenylpyrazoles, N-phenylphthalimides, oxadiazoles)	Inhibition of protoporphyrinogen oxidase (PPO)	Oxyfluorfen, fluazolate, oxadiazon
Carbamates	Inhibition of DHP (dihydropteroate) synthase	Asulam
Phosphinic acids	Inhibition of glutamine synthetase	Glufosinate-ammonium bialaphos

of herbicide-resistant weeds. To delay herbicide resistance, it is essential to provide information on the cropping system in place and the number of years and applications for the resistance to evolve. This information may help researchers, extension workers, and farmers to combat problems of resistance.

It is also advisable to avoid continued use of herbicides with the same mode of action unless this is used with other weed control practices and, in the case of probable metabolic resistance, with the use of herbicides that degrade by different metabolic means.

To devise effective herbicide resistance management strategies it is important to have a good understanding of the causes, nature, genetics, mechanisms, and solutions of herbicide-resistant weeds (Heap and LeBaron, 2001). Although this task is extremely difficult, every effort should be made to accomplish it.

Herbicide-Tolerant or Resistant Crops (HRCs)

The use of transgenic crops is a reality in several countries of the western hemisphere. Crops that are either tolerant to herbicides or resistant to insects are increasingly used in the United States (47.6 million ha), Argentina (16.2 million ha), and Canada (5.4 million ha) (James, 2004). Herbicide-resistant crops allow the application to the crops of a broad-spectrum herbicide, e.g., glyphosate or gluphosinate-ammonium, without affecting their growth once the weeds are emerged. Certainly the productivity of the process increases and zero or minimum tillage combined with direct-seeding of several grains becomes easy, with a resulting saving of energy and labor for weeding and prevention of soil erosion problems.

The use of HRCs offers better options for controlling problem weeds. For example, weeds that have become resistant to conventional herbicides can be controlled with the corresponding HRC herbicide. Thus HRCs may provide an additional tool in the management of herbicide-resistant weeds (Hurle, 1998). However, the recent commercialization of crop cultivars containing transgenes for herbicide resistance in North America, Europe, and other parts of the world offers a potential blessing to agriculture as well as posing a potential threat to the well-being of the agroecosystem (Rubin, 1998). There are several concerns about the consequences of development and deployment of transgenic herbicide resistance (HR), including the following:

- The potential transfer of genes from HRCs to wild relatives, thus creating "super" weeds

- The possibility of HRC volunteers becoming weeds in subsequent crops
- Increased selection pressure on weeds due to the repeated use of the herbicide associated to the HRC

The first problem necessitates avoidance of the repeated use of one herbicide mode of action, which may cause a switch in the weed flora toward less sensitive species. Therefore, alternative weed control strategies will have to be adopted to prevent (solve) problems. Farmers should be encouraged to rotate crops and herbicides. For the second problem, again, alternative weed control strategies are required to control HRC volunteers. In addition, certain conditions are necessary for the approval of any new HRC, such as the following:

1. Each country must have a regulatory authority to oversee the introduction of HRCs.
2. Approval should be based on crop, cultivar, HR transformant/gene, and herbicide.
3. The herbicide should already be approved by the pesticide registration scheme in the country concerned or approval should be sought simultaneously with the HR crop approval.

Another important aspect is monitoring the use of HRCs, which requires the following:

1. Postrelease monitoring to assess any significant changes in the weed flora as a result of the use of the HRCs (e.g., presence of surviving weeds and/or volunteers, and overall efficacy).
2. The relevant national agency or institution of the country concerned with the company seeking approval should agree on financial support to this work.
3. Users must keep good records of which HRCs have been used, where planted, and which herbicide has been used. Preferably, regulatory authorities or their local representatives should record this information.

Particular attention should be paid to the use of HRCs in centers of crop origin, where the hazards of gene transfer are higher. Special care should be taken to ensure that native germplasm is not jeopardized by the transfer of herbicide resistance genes.

In order to help national authorities, researchers, and other people involved with the use of HRCs, the FAO (2001) has recently published draft guidelines for risk assessment of the ecological hazards of the use of herbicide- and insect-resistant crops. These draft guidelines provide useful advice for identifying potential hazards of the use of these materials prior to their use in agricultural practice.

As rightly stated by Rubin (1998), too hasty adoption of HRCs may result in irreversible damage to the agroecosystem. It is essential to avoid such damage and to take the necessary precautions before introducing such crops.

Other Control Methods

One of the methods most successfully used during the past 15 years for weed control has been *soil solarization*. This is a safe and effective alternative to toxic, costly soil pesticides and the lengthy crop rotations now needed to control many damaging soil pests. Solarization provides good weed control in some situations, particularly commercial vegetable production, where effective herbicides are unavailable (Hagan and Gazaway, 2000). The method works similarly to a greenhouse. It consists of using a transparent polyethylene sheet (3 to 6 mm) to cover wet soil and trap sun radiation. Soil exposed for four to six weeks reaches the necessary temperature to kill several soilborne pests, including weed seeds. Many countries, particularly those with strong sun radiation (e.g., Mexico, the United States, Israel, Jordan, and others in the Mediterranean basin), use this method effectively to control soilborne pests and also to replace the use of the fumigant methyl bromide. Being nonchemical by definition, solarization has a definite advantage in the framework of nonchemical cropping systems, such as organic and biological farming, and offers potential as an alternative method to chemical soil fumigation. Solarization is compatible with integrated pest management and biological control systems (Grinstein, 1992).

Other control methods are under research, e.g., the use of hot water and hot foam for weed control in western Australia (Collins et al., 2002), band steaming for weed control in Denmark (Melander, Heisl, and Jorgensen, 2002), water-jet cutting for weed control in Sweden (Fogelberg and Blom, 2002), and others. These methods remain to be developed further and economic feasibility, particularly fuel consumption, has yet to be studied for application in developing countries.

Integrated Weed Management

From the various methods described, it is clear that weed management is an integral part of crop and farm systems. Many of the activities carried out by farmers involve weed control. In fact, substantial progress has been made on methods to improve weed management. Crop rotation, tillage, use of cover crops, and flooding (as in rice, mulches, and interrow cultivation) are largely based on the need to prevent weed interference and to avoid losses caused by competition. To regard weed management as a sole method of control is incorrect and will not reduce existing weed problems. Some of the methods listed previously are technically applicable anywhere, but the constraints under which each farmer works vary from one country to another or from one region to another. Farmers in the industrial economies have means that are not available in developing countries, particularly in small farming systems. In the same way that labor can be regarded as a production constraint, sufficient cash or capital may not be available to undertake all the activities that would be profitable if money could be readily borrowed at "market" interest rates (Auld, Menz, and Tisdell, 1987). Therefore, methods applicable in one area may not be applicable elsewhere because farmers may lack the necessary resources (labor and capital) to implement new methods.

However, due to the implications of weed interference in the whole agricultural process, improved crop production (i.e., higher yields and quality) will be attainable only when weed problems and related expenses are minimized. Therefore, methods bringing more productivity and reducing actual cost of weed control will be welcome. In the case of small-scale farming systems, the immediate objective should be to substantially reduce the time farmers and their families spend hand weeding. Accomplishing such an objective will have an enormous socioeconomic impact on the economies of small-scale farmers.

ECONOMIC AREAS AFFECTED BY WEEDS AND THEIR MANAGEMENT

It is worth highlighting briefly the different areas affected by weeds, the most important species in each case, and the causes of incidence. Some sources in the literature give more detailed information on each of the species mentioned, but here the intention is to present only an overview of the problems posed by these troublesome species, together with an explanation of the importance of accepting an integrated management approach for their control.

Weeds affect three major economic areas: agriculture and forestry, industrial and recreational sites, and inland water bodies. Since our work is mainly concerned with agricultural production, industrial and recreational areas are not included in this chapter.

Agricultural Areas

In agricultural areas it is necessary to differentiate between perennial and annual crops. Differentiation of geographic regions is convenient since climatic and soil conditions are important factors affecting the weed flora of any territory or country.

Grass Weeds

Several difficult-to-control grass weeds cause important crop losses worldwide, among them the perennial *Imperata cylindrica,* which is largely found in plantations and fallow areas of Asia and Africa. The weed should be combated with the integration of various methods, including the use of cover crops during fallow periods and the use of broad-spectrum herbicides, particularly to control the cover crop during the process of land preparation for a subsequent crop.

The incidence of johnsongrass [*Sorghum halepense* (L.) Pers.] is no less important in the western hemisphere. This weed is found in annual and perennial crops; its control is difficult due to its extraordinary reproductive capacity, and it is necessary to use systemic herbicides (e.g., glyphosate and asulam) to avoid regrowth and/or the emission of new rhizomes. In heavily infested areas, land preparation by dragging the rhizomes to the soil surface to desiccate is the recommended control method.

The annual grass itchgrass [*Rottboellia cochinchinensis* (Lour.) Clayton] has also been introduced into the western hemisphere; it causes important losses in several annual crops, such as maize, beans, cotton, as well as perennial sugar cane. Its control requires the implementation of cultural methods (including crop rotation) plus chemical control to minimize infestation. Several chlorotriazinic herbicides selectively used in maize and sorghum have not proved effective for its control.

In cereals, *Avena* spp. cause serious problems in North Africa, the Middle East, and Asia. Crop rotation of cereals with other crops, as well as selective chemical control, is required. As with the case of *R. cochinchinensis,* the use of clean seeds is an important preventive control measure.

Broadleaf Weeds

Major problems are those of introduced invasive weeds such as parthenium (*Parthenium hysterophorus* L.) in some parts of Africa, Asia, and Australia. The weed not only affects crop growth but also causes severe skin allergy problems to farmers touching the weed when hand weeding. In highly infested areas, development and implementation of biological control seem to be the best option. In Australia, the weed is largely controlled by the use of biological control agents.

Introduced slamweed [*Chromolaena odorata* (L.) King & Robins] in Africa and Asia, as well as *Prosopis* spp. in Africa and the Middle East, are serious weed problems which are discussed next.

Parasitic Weeds

There are three important complexes of parasitic plants:

1. *Striga* spp., well distributed and causing severe crop losses in Africa south of the Sahara;
2. *Orobanche* spp. in North Africa and the Middle East, with a similar negative effect in legumes, vegetables, sunflower, and tobacco; and
3. *Cuscuta* spp., which are found worldwide and affect a long list of crops.

The first two complexes are root parasitic, while *Cuscuta* spp. are aerial parasites. For the control of these species an integrated approach is required. The idea of combating these species by herbicide use alone is not practical. First, it is important to understand that monocropping is the main cause of the persistent incidence of these weeds, so implementation of crop rotation to substantially reduce the stand of root parasitic weeds should be understood as the key to their control. The use of clean crop seeds, organic manure, and N-fertilizers help to reduce infestation. In the case of *Striga,* intercropping the cereal with a leguminous crop may also reduce infestation. Herbicide use may be helpful if contact compounds are applied over the flowers of emerged parasites to avoid seed-setting of the plants, thus reducing weed seed bank build-up in soil as well. Obviously, the herbicide and its rate of application should be safe to the commercial crop, something that it is not always easy to achieve.

Water Bodies

At present, large dams, lakes, and rivers also become infested by some weeds. The most common species follow:

1. Floating: water hyacinth *(Eichhornia crassipes),* water fern *(Salvinia molesta),* and water lettuce *(Pistia stratiotes);*
2. *Rooting: Typha* spp; and
3. *Submerged: Hydrilla verticillata, Egeria* spp., Canadian waterweed (*Elodea canadensis* Mich.), and others.

Heavily infested water bodies pose a number of serious problems. The weeds limit navigation, reduce fishing, hinder irrigation network, create serious problems for good functioning of hydroelectric power stations, and serve as good habitats for the development of various vectors of human diseases.

The origin of the problem is normally the inappropriate introduction and dissemination of the plant species to the country. These plants always arrive without their natural enemies, so biotic factors able to regulate their reproduction are absent and areas quickly become infested. Another important factor is that these plants become more prolific in water bodies heavily polluted with organic substances. Therefore, a prerequisite for arresting the proliferation of aquatic weeds is to reduce the disposal of organic waste into the water.

Free Trade and Introduced Exotic Plants

As has made clear earlier, several weeds are alien plants in many countries and regions of the world, where they cause serious problems (see Table 2.3). Increased world trade, international travel, and world demographic growth have brought about serious problems of introduced plants, some of them being highly invasive. An active global market is one of the major causes of introduced weeds into new territories. In the United States, losses caused by introduced invasive weeds in the agricultural sector were estimated at US$15.2 billion in 1994 (Bridges, 1994, cited by Westbrooks, 1998).

Table 2.3 contains examples of weed species which have been introduced in various regions, e.g., the Compositae *Parthenium hysterophorus* has been introduced into India, some parts of east and southern Africa and Australia, and presumably in grains coming from the United States. In other cases, species have been introduced by error. Clear examples are *Chromo-*

TABLE 2.3. Examples of introduced weeds in various regions of the world.

Species	Area of introduction
Sorghum halepense	Western hemisphere
Rottboellia cochinchinensis	Western hemisphere
Cyperus rotundus	Western hemisphere, Africa
Mimosa pigra	Africa, Asia, and Oceania
Chromolaena odorata	Africa, Asia, and Oceania
Parthenium hysterophorus	Africa, Asia, and Oceania
Euphorbia esula	North America
Lantana camara	Africa, Asia, and Oceania
Eichhornia crassipes	Central America, Caribbean, Africa, Asia, and Oceania
Hydrilla verticillata	North America
Prosopis spp.	Middle East and Africa
Striga asiatica	North America
Orobanche ramosa	Pacific Coast of Mexico and Cuba

laena odorata (formerly *Eupatorium odoratum*), which was brought from the Caribbean and introduced as a cover crop in plantations of Africa and Asia. This later became an invasive weed that seriously affects crop production and plant biodiversity. Another example is the perennial grass *Sorghum halepense* from the Mediterranean area, introduced into North America as pasture, and at present causing problems in several crops as a highly competitive weed. More recently there is evidence that certain species of *Prosopis,* introduced into arid areas of Africa and the Middle East, are invading large fertile areas of countries in the region.

The invasiveness and economic damage of introduced weeds should be minimized. To this end, some activities are recommended to prevent and/or reduce their effect:

1. Risk assessment of possible adaptation of exotic plants known to be weeds in other countries
2. Provision of knowledge and tools to quarantine inspectors about exotic weed species, their potential for introduction, and methods for their detection

In each country, quarantine services should have a check list of hazardous plants in order to prevent their entry. The list should be prepared on the basis of weeds currently absent and those potentially able to establish themselves in the country. Quarantine inspectors should also have available methodologies for the detection of weed seeds in imported agricultural products. Prevention is an important activity in the context of integrated weed management to avoid future problems and losses caused by exotic plants.

SUGGESTIONS FOR IMPROVED WEED RESEARCH MANAGEMENT

From the present review, the importance of weed management to augment crop production and its quality is evident. Modern agriculture will increasingly depend on the methods used for weed control, and research should regularly provide information and validated weed control methods to achieve such improvement. In conclusion, it is worth emphasizing the studies and actions required to improve weed management. These recommendations are mainly directed to weed scientists in developing countries. The studies include the following:

1. Ecology and biology of prevailing weed species including weed seed bank, and the effect of biotic and abiotic factors on the growth and productivity of the species
2. Weed interference to determine critical periods of competition
3. Tillage systems and their short- and long-term effects on weed flora
4. Use of cover crops and mulches
5. Identification of competitive and/or allelopathic crops for providing more effective competition with weeds
6. Intercropping
7. Preventive methods of weed control, including regulatory measures and weed risk assessment
8. Biological control of prevailing weeds
9. Monitoring of herbicide resistance
10. Hazard assessment of ecological impact of the use of HRCs

No single control measure is perfect for all weed situations. Different actions are required for effective control. IWM is a rational combination of control methods that will reduce a particular weed infestation efficiently, with minimal adverse impacts to the environment.

IWM requires farmers to understand the ecobiology of the main weed species before selecting any control strategy. Knowledge of the life cycle of the species is essential for conducting control at the weeds' most vulnerable stages.

IWM should include practices that prevent the introduction and spread of weeds as well as others that make the crop competitive with weeds. Important preventive methods are the consistent reduction of weed seed bank in the soil and avoiding the invasion of a new weed from one farm or field to another. These measures will make long-term weed management less costly.

Other weed control strategies are available, some of them market-driven such as crop rotation, but governments and growers associations should help farmers to enable them to implement sound crop rotations.

The use of herbicides should be considered as another control method, which must be carefully applied to avoid problems to human health and of environmental contamination, especially in water bodies, and problems of herbicide-resistant weeds. By following the principles of IWM it is possible to reduce the use of herbicides.

For water weed control, it is clear that physical methods combined with biological control provide the best effect, but no less important is to prevent the disposal of organic wastes that serve as nutrients for weeds in water bodies.

REFERENCES

Akobundu, I.O. (1987). *Weed science in the tropics: Principles and practices*. New York: John Wiley and Sons.

Auld, B. (1994). Economic criteria for implementation of weed management. In R. Labrada, J.C. Caseley, and C. Parker (Eds.), *Weed management for developing countries* (pp. 237-246). Plant Production and Protection Paper No. 120. Rome: FAO.

Auld, B., K.M. Menz, and C.A. Tisdell (1987). *Weed control economics*. New York: Academic Press, Inc.

Bridges, D.C. (1994). Impact of weeds on human endeavors. *Weed Technology* 13: 612-625.

Carsky, R.J. and A.C. Eteka (1999). Persistence of *Mucuna pruriens* biomass during the dry season along an agroecological transect in Benin. Publication number IITA/99/20. Available at <http://www.cgiar.org/spipm/news/ccropmtg/results/ccb7/CCBEN7.htm#intro>.

Chung, I.M., J.K. Ahn, and S.J. Yun (2001). Assessment of allelopathic potential on barnyardgrass *(Echinochloa crus-galli)* of rice cultivars. *Crop Protection* 20: 981-928.

Collins, R.M., A. Bertram, J.A. Roche and M.E. Scott (2002). Preliminary studies on the comparison of hot water and hot foam for weed control. In Proceedings of the 5th EWRS Workshop on Physical Weed Control (pp. 207-215), Pisa, Italy, March 11-13.

Doucet, C., S.E. Weaver, A.S. Hamill, Jian-Hua Zhang, and J.H. Zhang (1999). Separating the effects of crop rotation from weed management on weed density and diversity. *Weed Science* 47: 729-735.

Fischer, A. J. and K.D. Gibson (2001). Cultivares competitivos como herramienta para el Manejo Integrado de Malezas. In Resúmenes XV Congreso ALAM (pp. 71-72), Maracaibo, Venezuela, November 26-30.

Fogelberg, F. and A. Blom (2002). Water-jet cutting for weed control. In Proceedings of the 5th EWRS Workshop on Physical Weed Control, Pisa, Italy, March 11-13.

Food and Agriculture Organization of the United Nations (FAO) (1996). *International standards for phytosanitary measures, Part 1: Import regulations, code of conduct for the import and release of exotic biological control agents.* Secretariat of the International Plant Protection Convention Publication No. 3. Rome: FAO.

Food and Agriculture Organization of the United Nations (FAO) (2001). *Draft guidelines for assessment of ecological hazards of herbicide- and insect-resistant crops.* Rome: Plant Production and Protection Division, FAO.

Food and Agriculture Organizaton of the United Nations (FAO) (2003). International code of conduct on the distribution and use of pesticides. Revised version, adopted by the 123rd session of the FAO Council, November 2002. Rome: FAO.

Forcella, F. (1993). Prediction of weed densities from the soil seed reservoir. In *Proceedings of International Symposium on Integrated Weed Management for Sustainable Agriculture* (pp. 53-56). November 18-20. Hisar, India: Indian Society of Weed Science.

Forcella, F. (1998). Application of weed seed bank ecology to weed management. In *Report of FAO Expert Consultation on Weed Ecology and Management* (pp. 23-35), September 22-24, 1997. Rome: FAO Plant Production and Protection Division.

Forcella, F. and O.C. Burnside (1993). Pest management—Weeds. In J.L. Hatfield and D.L. Karlen (Eds.), *Sustainable agriculture systems* (pp. 157-197). Boca Raton, FL: Lewis Publishers.

Fujii, Y. (2001). Screening and future exploitation of allelopathic plants as alternative herbicides with special reference to hairy vetch. *Journal of Crop Production* 4(2): 257-275.

Galiba, M., P. Vissoh, G. Dagbénonbakin, and F. Fagbohoun (1998). Réactions et apprehensions paysannes liées à l'utilisation du pois mascate (*Mucuna pruriens* var. utilis). In D. Buckles, A. Etka, O. Osiname, M. Galiba, and N. Galiano (Eds.), *Cover crops in West Africa.* Canada: IDRC Publications. Available at <http://www.idrc.ca/books/ev-31915-201-1-DO_TOPIC.html>.

Gibson, K.D., J.E. Hill, T.C. Foin, B.P. Caton, and A.J. Fischer (2001). Water-seeded rice cultivars differ in ability to interfere with watergrass. *Agronomy Journal* 93: 326-332.

Grinstein, A. (1992). Introduction of a new agricultural technology—soil solarization—in Israel. *Phytoparasitica* 20(Supplement): 127-131.

Hagan, A.K. and W.S. Gazaway (2000). *Soil solarization for the control of nematodes and soilborne diseases.* The Alabama Cooperative Extension System. Alabama A&M University and Auburn University. Available at <http://www.aces.edu/pubs/docs/A/ANR-0713/>.

Harvey, R.G. (1993). *A simple technique for predicting future weed problems.* Nutrient and Pest Management Program. Madison, WI: Cooperative Extension Division, University of Wisconsin-Madison.

Heap, I. (2005). *International survey of herbicide resistant weeds.* Available at <http://www.weedscience.org/in.asp>.

Heap, I. and H. LeBaron (2001). Introduction and overview of resistance. In S.B. Powles and D.L. Shaner (Eds.), *Herbicide resistance and world grains* (pp. 1-22). Boca Raton, FL: CRC Press LLC.

Hurle, K. (1998). Benefits and risks posed by the use of HRCs. In *Report of FAO Technical Meeting on Benefits and Risks of Transgenic Herbicide Resistant Crops* (pp. 5-6), November 16-18. Rome: FAO.

James, C. (2004). Preview: Global status of commercialized biotech/GM crops: 2004. ISAAA Briefs No. 32. Ithaca, NY: ISAAA.

Kelner, D. (1997). *Weed populations and zero tillage.* Manitoba Agriculture and Food, Soils and Crops Branch. Available at <http://www.gov/mb/ca/agriculture/news/topics/daa03d07.html>.

Labrada, R. and C. Parker (1994). Weed control in the context of integrated pest management. In R. Labrada, J.C. Caseley, and C. Parker (Eds.), *Weed management for developing countries* (pp. 3-8). FAO Plant Production and Protection Paper No. 120. Rome: FAO.

Lin, W.X., K.U. Kim, and D.H. Shin (2000). Rice allelopathic potential and its modes of action on barnyardgrass *(Echinochloa crus-galli). Allelopathy Journal* 7: 215-224.

Melander B., T. Heisel, and M.H. Jorgensen (2002). Band steaming for intra-row weed control. In Proceedings of the 5th EWRS Workshop on Physical Weed Control (pp. 216-219), March 11-13, Pisa, Italy.

Mortensen, D.A., L. Bastiaans, and M. Sattin (2000). The role of ecology in the development of weed management systems: An outlook. *Weed Research* 40: 49-62.

Mortimer M. (1998). The need for studies on weed ecology to improve weed management. In *Report of FAO Expert Consultation on Weed Ecology and Management* (pp. 15-22), Rome, Italy, September 22-24, 1997. Rome: Plant Production and Protection Division, FAO.

Olofsdotter, M. (2001). Rice—A step toward use of allelopathy. *Agronomy Journal* 93: 3-8.

Parker, C. and J. Fryer (1975). Weed control problems causing major reduction in world food supplies. *FAO Plant Protection Bulletin* 23(3/4): 83-95.

Petcu, G. and S. Ionita (1998). Influence of crop rotation on weed infestation, *Fusarium* spp. attack, yield and quality of winter wheat. *Romanian Agricultural Research* 9/10: 83-87.

Pinstrup-Andersen, P. (2001). *The future world food situation and the role of plant diseases.* International Food Policy Research Institute. The Plant Health Instructor. DOI: 10.1094/PHI-I-2001-0425-01. The Glenn Anderson Lecture presented at the joint meeting of the American Phytopathological Society and the Canadian

Phytopathological Society, Montreal, Canada, August 8, 1999. Available at <www.apsnet.org/Education/feature/FoodSecurity>.

Pinstrup-Andersen, P. and M.J. Cohen (1998). *The world food outlook and the role of crop protection.* Prepared for presentation at the 65th Annual Meeting of the American Crop Protection Association, White Sulphur Springs, West Virginia, September 27-29.

Pitelli, R. and J.C. Durigan (1984). Terminologia para períodos de controle e de convivéncia para plantas danhinas em culturas anuais e bianuais. In *Congresso Brasileiro de Herbicidas e Plantas Danhinas,* Belo Horizonte, Resumos.

Rahman, A., T.K. James, and N. Grbavac (2001). Potential of weed seedbanks for managing weeds: A review of recent New Zealand research. *Weed Biology and Management* 1: 89-95.

Rizvi, S.J.H., V. Rizvi, M. Tahir, M.H. Rahiman, P. Shimi, and A. Attri (2000). Genetic variation in allelopathic activity of wheat (*Triticum aestivum* L.) genotypes. *Wheat Information Service* 91: 25-29.

Rojas, P.B. and J.F.C. Porras (1999). *Arachis pintoi* como alternativa sostenible en el manejo integrado de malezas en vid. *Agronomía Colombiana* 16(1-3): 51-59.

Rubin, B. (1998). Herbicide use and transgenic crops resistant to herbicides. In *Report of FAO expert consultation on weed ecology and management* (pp.73-78), Rome, Italy, September 22-24, 1997. Rome: Plant Production and Protection Division, FAO.

Sattin, M., A. Berti, and G. Zanin (1998). Influence of weed time of emergence and removal on crop. In *Report of FAO expert consultation on weed ecology and management* (pp. 53-64), Rome, Italy, September 22-24, 1997. Rome: Plant Production and Protection Division, FAO.

Vleeshouwers, L.M. (1997). Modelling weed emergence pattern. Doctoral thesis, Wageningen Agricultural University, Wageningen, The Netherlands.

Westbrooks, R. (1998). *Invasive plants, changing the landscape of America: Fact book.* Washington, DC: Federal Interagency Committee for the Management of Noxious and Exotic Weeds (FICMNEW).

Wu, H., J. Pratley, D. Lemerle, and T. Haig (1999). Crop cultivars with allelopathic capability. *Weed Research* (Oxford) 39: 171-180.

Zimdahl, R.L. (1995). Weed science in sustainable agriculture. *American Journal of Alternative Agriculture* 10: 138-142.

Zimdahl, R.L. (2004). *Weed-crop competition: A review.* Oxford, UK: Blackwell Publishing.

Chapter 3

Contributions to Weed Suppression from Cover Crops

M. L. Hoffman
Emilie E. Regnier

INTRODUCTION

Sustainable agriculture refers to a strategy of resource management in which the objectives are to satisfy changing human needs while natural resources and the quality of the environment are conserved (Technical Advisory Committee, 1989). Sustainable systems aim to sustain or improve all types of productivity, which may be measured by such diverse indices as crops harvested, soil health, and economic gains. Resources utilized in the process, however, should not be damaged or depleted and should preferably be enhanced. Preferred methods are those that rely on on-farm renewable inputs, although off-farm inputs may be justified if their use results in increased production. Sustainable agriculture should not be considered synonymous with low-input or alternative agriculture (Lal et al., 1991). However, it should be noted that a common strategy shared by all of these crop-management choices involves integrating several management practices rather than depending on a few external inputs, such as herbicides, to control weeds. For growers seeking to manage weeds, utilizing practices that may include but are not limited to herbicides, weed management can be a daunting task. Growing cover crops in association with the cash crop might be a useful weed management alternative for some growers and one that is consistent with the goals of sustainable agriculture. Studies indicate that weed suppression can be achieved through the use of cover crops managed judiciously (Lal et al., 1991). In addition to suppressing weeds, other benefits to agroecosystems from growing cover crops include protecting against soil erosion, increasing water infiltration, enhancing soil structure, contributing to soil fertility, sequestering carbon, and improving soil quality (Lal et al., 1991; Sainju and Singh, 1997). Also, potential exists for cover crops to be utilized in plant disease management and control of crop pests other than weeds (Lal et al., 1991; Grunwald, Hu, and Van Bruggen, 2000b).

Historical Perspective

Growing cover crops in addition to a main crop has been a popular agronomic practice throughout history (Pieters, 1927). Cover crops were originally grown as green manures, which were turned under to improve soil fertility. It is generally believed that the Chinese during the Han Dynasty (1134 B.C.), recognized the fertilizing value of turning under vegetation. By the second century B.C., Theophrastus (372-287 B.C.) of Greece and other writers in Rome advised turning under legumes to improve soil fertility (Pieters, 1927). Utilizing green manures as a primary nitrogen source was a standard practice in North America until the early twentieth century (MacRae and Mehuys, 1985). Because of its universal importance, much attention was paid to green manuring by researchers. In what is generally acknowledged as a treatise on green manuring written at the height of its popularity, Pieters and McKee (1938) reported that over 50 cover crop species were under evaluation at various state experiment stations at the time. Many of the species were nitrogen-supplying legumes, but grasses such as rye (*Secale* spp.) were also included.

Development in 1910 of the Haber-Bosch process for nitrogen fixation increased the supply of synthetic nitrogen fertilizer, which soon replaced green manures as the most economical and preferred nitrogen source. At about this time, when interest in using legume green manures for nitrogen was disappearing (Thomas and Frye, 1984; Power, 1990), the soil conservation movement was gaining attention on a worldwide scale. As an example of the level of concern over soil conservation in the United States, research for control of soil erosion was mandated in legislation appearing in 1928 as an amendment to the Agricultural Appropriation Bill (U.S. Congress, 1930). Formation of the Soil Erosion Service in the 1930s focused even more attention to research on using cover crops to stem soil erosion (Graham, 1941; Carreker, Wilkinson, and Box, 1977). In the opinions of scientists as well as policymakers, the seriousness of the problem of soil erosion demanded action, and the best way to protect soil from eroding was to maintain surface cover such as that provided by cover crops.

During this period, weeds were controlled by mechanical means, which prevented the buildup of residue on the soil surface. Reliance on tillage and cultivation was such that many believed the hallmark of a conscientious farmer was a field with a clean surface, whereas a residue-covered field indicated more slovenly practices. Although tilling the soil was an indispensable practice throughout much of the history of agriculture, its link to soil erosion sent many in search of alternatives. Using the combined efforts of growers, researchers, agricultural engineers, and implement manufacturers, conservation tillage technology was developed (Blevins et al., 1998). A simple definition of conservation tillage, according to the Conservation Tillage Information Center (1983), is any tillage and planting system in which at least 30 percent residue cover is retained on the soil surface after planting. Following the introductions of herbicide formulations such as 2,4-D [(2,4-dichlorophenoxy) acetic acid] that controlled weeds without cultivation, conservation tillage methods became widely accepted as crop management alternatives having numerous benefits including soil erosion control.

The growing popularity of conservation tillage brought about a change in the prevailing view that surface residue somehow indicated negligence on the part of a producer. In addition, because more farmers were choosing to use conservation tillage practices, which by definition must maintain a certain amount of surface residue, manufacturers started to produce a wide range of equipment that operated effectively on residue-covered soil. As a result, some growers became interested in the age-old practice of cover cropping as a means of producing more surface residue. However, because of the newness of conservation tillage, it was generally unknown how cover crops might be best utilized within these systems.

As a result, a number of recent investigations have addressed topics relative to cover cropping and the effects of various types of cover crops on agroecosystem processes. The variety of topics appearing in research papers published during the past two years include the following: impacts of cover crops on soil aggregation (Ball-Coelho, Roy, and Swanton, 2000) and other soil physical properties (Mitchell et al., 2000); cover crop effects on soil nitrogen dynamics (Malpassi et al., 2000; Rasse et al., 2000; Rosecrance et al., 2000; Odhiambo and Bomke, 2001); soil microbe densities in the presence of cover crops (Grunwald, Hu, and Van Bruggen 2000a); mycorrhiza inoculum potential in the presence of a cover crop (Kabir and Koide, 2000); suppression of nematodes by cover crops (Davis, Baird, and McNeill, 2000); plant disease suppression by cover crops (Grunwald, Hu, and Van Bruggen, 2000b); and influences of cover crops on weed community composition (Barberi and Mazzoncini, 2001) and weed control (Akemo, Regnier, and Bennett, 2000; Hutchinson and McGiffen, 2000; Fisk et al., 2001; Herrero et al., 2001).

Managing Cover Crops

Although cover cropping provides many benefits when used within a conservation-tillage crop rotation, a disadvantage of using cover crops is the possibility that they may compete with row crops for light, water, and nutrients or suppress crop seedling growth via allelopathy. Therefore, it is important to understand the advantages and disadvantages of the various techniques that can be utilized to manage cover crops to be able to appropriately match cover crop system with cropping system. Some of the cover crop management approaches that have been evaluated experimentally for weed suppression are allelopathic green manures, smother plants, living mulches, and cover crop mulches.

Green Manuring

As previously mentioned, the cover crop can be treated as a green manure turned into the soil to improve fertility. Although this approach to cover crop management may promote weed seedling emergence (Blum et al., 1997), recent studies using *Brassica* spp. as green manures show promise in reducing weed emergence through allelopathic effects (see Chapter 4 for a review).

Smother Plants

Smother plants are cover crops with short life cycles that are planted in spring with the cash crop and generally die before their presence affects crop growth and yield (DeHaan et al., 1994). The limited number of species having the characteristic traits of a smother plant include mustards (*Brassica* spp.), clover (*Trifolium* spp.), and medics (*Medicago* spp.). In recent studies, it was shown that, if managed correctly, some smother crops have potential to suppress weeds (Buhler, Kohler, and Foster, 1998, 2001). Smother plants should not be confused with smother crops, which have a similar name but are managed differently. Smother crops can be annual or perennial species that are harvested as silage or green chop. At planting, smother crops are seeded thickly at a time when conditions are ideal for vigorous growth such that rapid canopy development is promoted to help crowd and shade out many annual weeds. Examples of smother crops are winter rye (*Secale cereale* L.), buckwheat (*Fagopyrum esculentum* Moench), and sorghum [*Sorghum bicolor* (L.) Moench].

Living Mulch

Live or living mulch refers to an intercropping strategy in which a low-growing cover crop is grown to provide weed control in a summer annual crop such as corn (*Zea mays* L.) or soybean [*Glycine max* (L.) Merr.] (Lal, 1975). Researchers investigated living mulches for their potential to suppress weeds beginning in the 1950s in the midwestern United States (Robinson and Dunham, 1954; Schaller and Larson, 1955). These results lead to other studies carried out in Nigeria and the United States (Hughes and Sweet, 1979; Akobundu, 1980; Cardina and Hartwig, 1982; Echtenkamp and Moomaw, 1989; Enache and Ilnicki, 1990). These early studies showed that living mulches that controlled weeds also suppressed the cash crop and therefore required management by herbicides, cultivation, or mowing to reduce interference with the main crop (Regnier and Stoller, 1987). Such additional inputs may increase crop production costs and reduce net return (see section in this chapter regarding economics). Many recent studies focus on the use of living mulches in vegetable crops (Ateh and Doll, 1996; Masiunas et al., 1997; Martin, Greyson, and Gordon, 1999; Ellis, Guillard, and Adams, 2000; Greenland, 2000; Zemenchik, Albrecht, and Boerboom, 2000), which are generally higher in value than field crops and can be managed more intensively.

Cover Crop Mulch System

Moore, Gillespie, and Swanton (1994) defined the cover crop mulch system as consisting of planting winter-hardy cover crops in the fall and killing the cover crop the following spring to form a potentially weed-suppressive layer of dead residue covering the soil surface before planting the cash crop. Cover crop mulch systems can also be summer cover crops followed by fall crops. Desirable characteristics of cover crop mulches include rapid establishment, excellent winterhardiness (for winter annuals), high biomass accumulation, ease of kill, resistance of dead residue to decay, and, if possible, allelopathic activity toward weeds by shoot and/or root residues. To suppress weeds, cover crops must accumulate large amounts of biomass prior to being killed at the time the cash crop is planted. Because the cover crops are killed, species unsuitable as living mulches or smother plants, due to their competitiveness with the crop, rate of maturity, or time of senescence, may be used.

Types of Cover Crops

Many species function effectively in the role of a cover crop; however, the number with the characteristics necessary to suppress weeds in a cover crop mulch system is limited. Environmental factors such as climate and rainfall may further limit the list of species from which to choose.

Regardless of climatic region, the decision of whether to use a legume or nonlegume cover crop is likely the most fundamental. These two types of cover crops differ in that the nitrogen-fixing ability of legumes, such as hairy vetch (*Vicia villosa* Roth), can add to the supply of nitrogen available for use by the cash crop (or by weeds) (Ebelhar, Frye, and Blevins, 1984). However, legumes frequently accumulate less residue than nonlegumes. Legume residue may decompose more rapidly during the course of the season, and nitrogen provided by decomposing legume residue could enhance emergence and growth of some weed species (Mohler and Teasdale, 1993).

In temperate field-crop production systems, commonly used cover crops include those that could be classified as winter-hardy legumes, clovers, summer annuals, small grains, and pasture grasses (see Table 3.1). Mixtures combining a legume and nonlegume have also been used in both field crops and vegetable systems (Clark, Decker, and Meisinger, 1994; Creamer, Bennett, and Stinner, 1997; Akemo, Bennett, and Regnier, 2000). Among these different types, winter-hardy annuals are the most efficient cover crops for use in temperate regions because they can be grown when fields are otherwise bare, thus providing cover to reduce soil erosion. Winter-hardy species fit into most cropping systems at a time when they are less likely to interfere with other farm operations. Also, soil moisture is usually available for use up by cover crops grown at this time in the season.

The winter annual legume hairy vetch and cereal grains such as winter rye, hereafter referred to as rye, are well suited for use as cover crops to suppress weeds in temperate cropping systems. Winter annual legumes such as crimson clover (*Trifolium incarnatum* L.) and cereals including wheat (*Triticum aestivum* L.), barley (*Hordeum vulgare* L.), and triticale (*Triticosecale rimpaui* Wittm.) likewise provided weed suppression in studies where they were fall-planted and spring-killed (White and Worsham, 1990; Weston, 1990; Nelson, Kahn, and Roberts, 1991; Moore, Gillespie, and Swanton, 1994). However, it has long been known that hairy vetch is one of the most winter hardy of the annual legume species (Piper and McKee, 1923) and that rye is relatively weed suppressive (Barnes and Putnam, 1983). Primarily because of these reasons, hairy vetch and rye were among the first cover crops used in mulch systems for weed suppression and are the cover crops most frequently used today in grain crops. Therefore, they are the focus of the remainder of this chapter.

TABLE 3.1. Planting times for annual cover crop species from several different types of life form categories.

Type	Cover crop	Fall	Winter (frost seeded)	Spring	Mid-summer
Winter-annual legume	Field pea, *Pisum arvense* L.	●		●	
	Austrian winter pea, *P. sativum* L.	●		●	
	Hairy vetch, *Vicia villosa* Roth	●		●	●
Clover	Sweet clovers, *Melilotus* spp.			●	●
	Berseem clover, *Trifolium alexandrinum* L.	●	●	●	●
	Kura clover, *T. ambiguum* L.		●	●	
	Crimson clover, *T. incarnatum* L.	●		●	
	Red clover, *T. pratense* L.	●	●	●	●
	White clover, *T. repens* L.	●	●	●	●
	Subterranean clover, *T. subterraneum* L.	●			●
Summer-annual	Canola/winter rape, *Brassica napus* L.				●
	Black mustard, *B. nigra* (L.) Koch				●
	Rape/turnip, *B. rapa* L.				●
	Buckwheat, *Fagopyrum esculentum* Moench			●	●
	Lupines, *Lupinus* spp.			●	●
	Annual medics, *Medicago* spp.	●		●	●
	Oilseed radish, *Raphanus sativus* L.				●
	Sorghum, *Sorghum bicolor* (L.) Moench				●

TABLE 3.1 *(continued)*

Type	Cover crop	Fall	Winter (frost seeded)	Spring	Mid-summer
	Sudangrass, *S. bicolor* (L.) Moench spp. *drummondii* (Steud.) de Wet				●
	Sudex, *S. bicolor* × *S. bicolor* (L.) Moench spp. *drummondii*				●
	Cowpeas, *Vigna unguiculata* (L.) Walp.			●	
Cereal grain	Oats, *Avena sativa* L.			●	●
	Barley, *Hordeum vulgare* L.	●	●	●	
	Rye, *Secale cereale* L.	●		●	●
	Wheat, *Triticum aestivum* L.	●			●
	Triticale, *Triticosecale rimpaui* Wittm.	●			●
Pasture grass	Annual ryegrass, *Lolium multiflorum* Lam.	●		●	●

MANAGEMENT ISSUES

Whether a cover crop mulch system consisting of hairy vetch or rye contributes to weed control in a conservation tillage system depends primarily on four management issues. The first two issues concerning cover crop establishment are cover crop planting date and seeding rate. The other two issues are date when the cover crop is killed and the kill method, which influence many important processes such as biomass accumulation by the cover crop, nitrogen content of hairy vetch residue, establishment of the cash crop, and interference by the cover crop with the cash crop.

Planting Date and Seeding Rate

Planting date influences the interval for biomass accumulation of cover crops and, as a result, the amount of residue available for weed suppression. In North America, hairy vetch and rye are usually planted in the fall and grow during periods of mild weather in winter and early spring. However, studies conducted in Michigan examining other potential planting dates found that clovers (*Trifolium* spp.) could be successfully frost seeded (late winter or very early spring) and adequate stands of clovers as well as hairy vetch and rye could be achieved by planting in summer (see Table 3.1) (Mutch and Martin, 1998).

Recent studies using hairy vetch in a cover crop mulch system reported seeding rates in the range of 28 to 45 $kg{\cdot}ha^{-1}$ (Hoffman, Regnier, and Cardina, 1993; Clark et al., 1997), and for rye from 62 to 94 $kg{\cdot}ha^{-1}$ (Clark et al., 1997; Kessavalou and Walters, 1999). Seeding rate (56 to 170 $kg{\cdot}ha^{-1}$) of fall-planted rye did not affect rye biomass or weed suppression in Illinois, perhaps due to compensatory tillering of rye at lower seeding rates (Masiunas, Weston, and Weller, 1995). Biomass of fall-planted hairy vetch was not affected by seeding rate in Maryland (14 or 28 $kg{\cdot}ha^{-1}$) (Clark, Decker, and Meisinger, 1994), but weed suppression was not measured. Rye biomass, when spring planted, increased with seeding rate (29 to 114 $kg{\cdot}ha^{-1}$), but weed suppression increased with seeding rate only in one of three years, when weed pressure was severe (Akemo, Regnier, and Bennett, 2000). In Wisconsin, increasing rye seeding rate (56 to 168 $kg{\cdot}ha^{-1}$) of spring-planted rye improved weed suppression but reduced soybean vigor (Ateh and Doll, 1996).

Residual soil N or N additions at cover crop planting may also influence cover crop biomass (Clark, Decker, and Meisinger, 1994) with potential subsequent effects on weed suppression by the killed residue. In rye-hairy vetch mixtures, residual soil N may alter the botanical composition of the mix regardless of initial seeding rates, with high residual soil N favoring the proportion of rye (Clark, Decker, and Meisinger, 1994; Clark, 1993). Changes in rye-hairy vetch proportion may affect weed suppression by the cover crop (Akemo, Regnier, and Bennett, 2000).

Kill Date

Kill date of the cover crop must fall within the time span recommended for planting the cash crop and varies depending on location (and in the case of corn and soybean grown in temperate regions, is generally earlier for corn than soybean). Killing the cover crop too early will result in less accumulated biomass and thus decreased amounts of residue left on the soil surface to suppress weeds (Smeda and Weller, 1996). A later kill date provides more time for the cover crop to accumulate additional biomass but may result in depleted soil moisture content (Clark et al., 1995). Also, it is reasonable to assume that phytotoxicity of cover crop residue may be related to growth stage of the plant when killed: corn yield was reduced the year that rye was killed at an earlier developmental stage, compared with other years when rye was more mature when kill treatments were applied, likely due to increased phytotoxicity associated with less mature rye (Kessavalou and Walters, 1997). In another study, allowing time to elapse between rye kill date and crop planting allowed phytotoxins to dissipate which reduced the negative effects of rye on crop stand and yield (Burgos and Talbert, 1996).

Time of cover crop kill also influences the effectiveness of kill methods. Hoffman, Regnier, and Cardina (1993) evaluated the effect of various kill dates on biomass accumulation and ease of kill of fall-seeded hairy vetch in Ohio. Kill treatments were applied when hairy vetch was at early bud (late April), mid-bloom (mid-May), or late-bloom (early June) growth stages. Generally no differences occurred among kill dates in hairy vetch biomass, which averaged 5,680 $kg{\cdot}ha^{-1}$ over the two years of the study. Kill treatments applied at hairy vetch early-bud growth stage were ineffective; however, hairy vetch was easily killed at mid- or late-bloom growth stages by mechanical or chemical methods (see Table 3.2). Clark and colleagues (1997), in Maryland, found that delaying kill date of rye and hairy vetch from March to April increased biomass accumulated by rye planted alone and by a hairy vetch-rye mixture, but not of hairy vetch when planted alone. The pattern of spring growth differed for rye and hairy vetch: hairy vetch biomass increased rapidly in early March and then slowed, whereas rye biomass increased steadily throughout March and April (Clark et al., 1997). Because rye is actively growing throughout the spring when kill treatments are typically applied, adequate control is often difficult to achieve. Incom-

TABLE 3.2. Effectiveness of kill treatments (in percentages) applied to hairy vetch at early-bud, mid-bloom, and late-bloom growth stages.[a]

Kill treatment	Early bud	Mid-bloom	Late bloom
Herbicide	80	99	100
Mow	47	96	100
Untreated	30	67	96
LSD[b]	16	2	3

Source: Adapted from Hoffman, Regnier, and Cardina, 1993.

[a]Represents control of hairy vetch, which was fall-planted at Columbus, Ohio, by kill treatments consisting of herbicide (glyphosate applied at 2.8 $kg{\cdot}ha^{-1}$), mow (flail chop mower), and untreated. Presented are means of percent hairy vetch that were estimated to be dead by way of visual estimates made on July 1, 1990 and June 17, 1991.

[b]Least significant difference ($p = 0.05$) between kill treatments.

plete control can lead to problems with crop establishment due to physical interference by the rye residue, or crop yield reductions as a result of nitrogen immobilization, soil water depletion, allelopathy, or a combination of these activities (Eckert, 1988; Munawar et al., 1990; Raimbault and Vyn, 1991).

Kill Method

Methods of killing hairy vetch or rye are restricted by the need to leave a weed-suppressive layer of cover crop residue on the soil surface into which the cash crop is planted making use of no-tillage technology. The most advantageous methods should kill effectively, be economical, and result in a layer of residue upon the soil surface.

Few kill methods fit the requirements necessary to be useful in a hairy vetch or rye mulch system. The most popular kill method, based on its use in the vast majority of recently published reports, is to apply glyphosate [N-(phosphonomethyl) glycine]. Variation in application rate is large, ranging from 0.89 to 3.0 $kg{\cdot}ha^{-1}$ (Williams, Mortensen, and Doran, 1998; Kessavalou and Walters, 1999; Swanton et al., 1999; Johnson et al., 2001). Mechanical kill methods such as mowing, rolling, and undercutting have also been used (Dabney, Buehring, and Reginelli, 1991; Hoffman, Regnier, and Cardina, 1993; Creamer et al., 1995; Masiunas, Weston, and Weller, 1995, Creamer and Dabney, 2002). Success of these methods depends primarily on stage of growth of the hairy vetch and rye. At early stages of cover crop growth, chemical methods are more effective than mechanical methods, preventing regrowth of the cover crop and also controlling weeds present within the cover crop, especially if the weeds are too small to be controlled by mechanical methods or can reestablish from underground crowns or propagules. At later stages, mechanical methods may be as effective as chemical methods in controlling cover crop growth (Hoffman, Regnier, and Cardina, 1993; Creamer et al., 1995). Hoffman, Regnier, and Cardina (1993) found that mowing or the mechanical actions associated with corn planting alone were sufficient to kill hairy vetch when hairy vetch was at mid- or late-bloom growth stages (see Table 3.2). Creamer and Dabney (2002) achieved 100 percent kill of a rye/vetch cover crop in Missouri by mowing and over 90 percent kill by a combination of rolling and chopping.

EVIDENCE FOR WEED SUPPRESSION

Early research showed that hairy vetch and rye residues reduced weed emergence but that control was inconsistent unless residue application rates were two to four times those that naturally result from common seeding rates (Teasdale, Beste, and Potts, 1991; Mohler and Teasdale, 1993; Teasdale and Mohler, 1993). Weed control by hairy vetch and rye residues was similar early in the season but was lower in hairy vetch than in rye later in the season due to more rapid residue decomposition rate of vetch. Living hairy vetch was more weed suppressive than killed residue but also reduced stand and yield of the cash crop (Hoffman, Regnier, and Cardina, 1993; Teasdale and Daughtry, 1993). However, when corn was no-till planted into hairy vetch at the mid-bloom stage and hairy vetch was allowed to senesce naturally, weed biomass was reduced by 63 to 75 percent and corn yields were 86 to 100 percent of those in weed-free (and vetch-free) controls (Hoffman, Regnier, and Cardina, 1993). It is generally agreed that rye has characteristics making it highly weed suppressive (Putnam and DeFrank, 1983; Putnam, DeFrank, and Barnes, 1983; Shilling et al., 1986; Weston, 1990; Liebl et al., 1992). Akemo, Regnier, and Bennett (2000) evaluated spring-planted mixes of living rye and pea (*Pisum sativum* L.) on weed suppression prior to cover crop kill and found that weed suppression increased linearly with increasing proportion of rye in the mix. However, rye residues in excessive amounts reduced crop stand and yield by impairing soil-seed contact (Eckert, 1988). Rye may also interfere with crop growth through allelopathy (Kessavalou and Walters, 1997). In more recent studies, efforts to respond to these difficulties are apparent through the use of such methods as additional herbicide applications to control weeds and mechanical techniques for managing residue.

Johnson and colleagues (2001) investigated the effects of tillage and a fall-planted rye cover crop on weed control and crop yield in a peanut (*Arachis hypogaea* L.)-cotton (*Gossypium hirsutum* L.) rotation. In reduced-tillage plots, herbicide was applied in spring to kill the rye, and a seedbed was formed using a strip-tillage implement into which the crop was planted. In conventional-tillage plots, rye was turned under by plowing in spring. Tillage systems were applied to the same plots for five consecutive years. Weed intensity increased with time in all treatments and required additional management inputs in the form of pre- and postemergence herbicide applications to maintain crop yields. Weed densities were greater, and more costly weed control measures were required in reduced tillage than in conventional-tillage systems.

In a similar study conducted for nine years in Ontario, Canada, effects of tillage system, rye cover crop, and nitrogen rate on weed suppression were measured in continuous corn (Swanton et al., 1999). Conventional tillage consisted of spring plowing followed by cultivating for weed control, whereas in no-tillage, trash whippers (planter-mounted notched disks) used to clear residue from the rows prior to planting were the only disturbance other than that caused by planting. Postemergence herbicides were applied in all treatments. Although the effects of tillage on weed density were inconsistent and variable, common lambsquarters (*Chenopodium album* L.) and redroot pigweed (*Amaranthus retroflexus* L.) tended to associate with conventional tillage whereas large crabgrass [*Digitaria sanguinalis* (L.) Scop.] was associated with no-tillage. However, cover crop and nitrogen rate did not affect weed density and community composition, likely due to the effectiveness with which herbicide applications controlled weeds.

The effects of hairy vetch and rye cover crops on weed and soybean growth were monitored in Nebraska under irrigated or rainfed conditions (Williams, Mortensen, and Doran, 1998). The authors concluded that cover crop residues did not effectively suppress all weeds for long enough periods of time. They suggested that a possible management approach might be to apply other integrated weed management tactics such as interrow cultivation or postemergence herbicides. By comparing growth of the crop and weeds, they determined when the crop had a competitive advantage over weeds and concluded that three to five weeks after planting would be an opportune time to affect weeds by using additional control measures in this cover-crop system.

Herbicide applications have been investigated as a means of improving weed control consistency in cover-crop mulch systems by a number of researchers. Weed suppression by hairy vetch or rye in no-tillage sweet corn was evaluated alone and in combination with half- or full-rate herbicides in Arkansas (Burgos and Talbert, 1996). Rye suppressed early-season weed density more than hairy vetch but not as much as half- or full-rate postemergence herbicides in the absence of a cover crop. Rye reduced emergence, height, and yield of sweet corn. Hairy vetch had no detrimental effect on the crop; but it did not control weeds, and full rates of herbicides were required for acceptable weed control. Similarly, in the absence of herbicide treatment, hairy vetch did not suppress weed biomass compared to cover crop-free controls in no-tillage corn, whereas rye did suppress weed biomass (Yenish, Worsham, and York, 1996). The addition of pre- and postemergence herbicide treatments completely controlled weeds in either rye or hairy vetch.

Results of these studies indicate that rye and hairy vetch differ in the amount of weed suppression that they contribute. Although rye is more consistent in suppressing weeds, it does not provide sufficient weed control in all situations so that additional measures are typically necessary. Also, rye tends to suppress the cash crop with apparently little selectivity in its suppressive activity. Both cover crops should be of use in providing additional weed control when included as part of a management program including herbicides, mechanical measures, and manipulation of cover crop kill date.

WEED SUPPRESSION MECHANISMS

After killing the cover crop and planting the cash crop, it is hoped that enough residue remains on the soil surface and has the qualities necessary to interfere with weed germination and growth. Although interference is relatively easy to recognize based on its outcome, the processes (mechanisms) working to produce that outcome can be difficult to separate. Understanding these mechanisms is necessary to predict and improve weed suppression by cover crops. Mechanisms by which cover crop residues interfere with weed seed germination and seedling growth include altered light quantity, light quality, temperature, soil moisture content, and nutrient availability (Creamer et al., 1996; Teasdale and Mohler, 1993). Weed seedlings may also be unable to push through the physical barrier created by the mulch (Teasdale and Mohler, 2000). In addition to these direct physical effects, residues may interfere with weeds by way of allelopathy.

Both cover crops, hairy vetch and rye, release allelochemicals. Allelopathic potential in rye was suspected from early observations and lead to interest in using this species as a weed-suppressive cover crop (Cubbon, 1925). Later studies were successful in characterizing the specific compounds involved. Rye residue inhibited annual forb weeds, in particular, when compared to an inert mulching material used as a control to simulate the physical effects of mulching (Barnes and Putnam, 1983; Putnam and

DeFrank, 1983; Barnes and Putnam, 1986). As further evidence of allelopathy, the same studies determined that rye-root leachates suppressed weed seedling growth. An examination of aqueous extracts of rye-shoot herbage identified 2,4-dihydroxy-1,4(2H)-benzoxazin-3-one (DIBOA) and 2-benzoxazolinone (BOA) as phytotoxic compounds (Barnes and Putnam, 1987; Barnes et al., 1987). In a subsequent series of studies, 2,2'-oxo-1,1'-azobenzene (AZOB) was determined to be a microbially transformed product of DIBOA or BOA metabolism with stronger herbicidal activity than either precursor (Nair, Whitenack, and Putnam, 1990; Chase, Nair, and Putnam, 1991; Chase et al., 1991). Although not as well characterized, volatile emissions and aqueous extracts from hairy vetch residues were shown to elicit phytotoxic responses by indicator species in bioassays (White, Worsham, and Blum, 1989; Bradow and Connick, 1990). A direct comparison of weed suppression by hairy vetch and rye residues in the field showed that more weed seedlings penetrated hairy vetch than rye, an action which the authors thought might be due in part to greater allelopathic toxicity of rye (Teasdale and Mohler, 1993).

To separate the effects of competition from allelopathy in rye interference with barnyardgrass [*Echinochloa crus-galli* (L.) P. Beauv.], Hoffman and colleagues (1996a) developed a greenhouse experiment under non-limiting moisture and nutrient conditions in which partitions were used to prevent shoot and root competition but to allow passage of allelochemicals. Results showed that, averaged over treatments with or without a partition that prevented competition, number of leaves of barnyardgrass was lower when grown for 15 days in the presence of various rye densities than in their absence. However, averaged over rye densities, number of barnyardgrass leaves did not differ between treatments with or without a partition to prevent competition. This similar response by the weed to rye in both partition treatments implicated allelochemicals that could pass through the partition as the interference mechanism. Barnyardgrass shoot dry weight was suppressed by rye at densities greater than six plants per pot, and the primary interference mechanism in this case was likely competition. Barnyardgrass shoot dry weight in treatments where competition was allowed was 20 percent less than in treatments with a partition that prohibited competition but allowed passage of allelochemicals in the soil solution.

Creamer and colleagues (1996) developed a protocol to leach allelochemicals from shoot residue thereby permitting the evaluation of allelopathic against physical effects in weed suppression by cover crop residues. The procedure was used in a field experiment to compare the effects of hairy vetch and rye residues on weed growth. Emergence of eastern black nightshade (*Solanum ptycanthum* Dun.) and yellow foxtail [*Setaria glauca* (L.) Beauv.] was reduced to a similar extent by leached and unleached cover crop shoot residues of both species, so suppression was attributed to physical effects rather than allelopathy. Rye residue suppressed yellow foxtail growth more than did hairy vetch but hairy vetch or rye residues suppressed eastern black nightshade equally. In a greenhouse experiment conducted with surface-applied rye root and shoot tissues under nonlimiting nutrient and moisture conditions, Hoffman and colleagues (1996b) found that rye-root residue inhibited weed growth [barnyardgrass and velvetleaf (*Abutilon theophrasti* Medik.)] more than shoot residue.

Teasdale and Mohler (2000) quantified the relationship between weed emergence and mulch physical characteristics for various dry mulches applied at different rates. Mulch materials differed widely and included hairy vetch shoots, rye shoots, bark chips, corn stalks, and landscape fabric strips. Based on a theoretical framework introduced in the paper, weed suppression by the mulch was defined as a function of mulch area index (area-to-mass ratio multiplied by dry mass of mulch) and fraction of mulch volume that was solid (area index per mulch layer multiplied by height of each layer). Accordingly, weed suppression was related to obstructing elements making up the mulch (which could physically impede seedling growth) and also the degree to which the mulch limited light penetration. This chapter extended earlier research by the same authors demonstrating the relationship between rye or hairy vetch residue mass, soil microclimate (Teasdale and Mohler, 1993), and weed emergence (Mohler and Teasdale, 1993). Dead hairy vetch residues were less weed suppressive than living hairy vetch, in field experiments, due to smaller effects of residues compared to living plants on reducing light quantity, light quality (red:far red light ratio), and diurnal soil temperature amplitude (Teasdale and Daughtry, 1993).

ECONOMICS

Economic considerations associated with cover crops, especially legumes, changed dramatically when technological advances resulted in low-priced, readily available nitrogen fertilizer. Cover crops were no longer necessary to maintain soil fertility. Rather, cover cropping became an option with benefits that included nutrient cycling, erosion control, and weed suppression (Lal et al., 1991). Researchers continued to investigate the economics of hairy vetch relative to its fertilizer replacement value (Ess et al., 1994; Stute and Posner, 1995; Mallory, Posner, and Baldock, 1998). Assigning monetary values to the other benefits is more difficult, and few researchers have studied their effects on economic return in agronomic systems.

The amount of nitrogen contributed by hairy vetch determined if it was economically beneficial in no-tillage corn, according to research conducted in Indiana (Bowen, Jordan, and Biehle, 1991). Frye, Smith, and Williams (1985) compared the effects of hairy vetch and rye combined with nitrogen fertilizer on corn production and soil erosion in the southeastern United States and concluded it was economically prudent to use fall-planted, spring-killed hairy vetch on these sloped lands to control soil erosion while maintaining soil productivity. In the central United States corn belt, Bollero and Bullock (1994) found that the cost of establishing hairy vetch exceeded the benefits of reduced nitrogen fertilizer requirement and increased corn yield but that any profitability associated with rye was eliminated by a large decrease in corn yield. The authors expected that the results would have differed if effects on soil erosion had been included in the analysis. Reddy (2003) found that, although rye cover crops reduced weed biomass by 19 to 38 percent and provided soybean yield comparable to no-cover crop systems, net return was lower with rye cover crops than with no cover crops due to additional costs associated with seeding and killing the rye.

Based on these studies, uncertainty is associated with expectations of economic benefits from using hairy vetch or rye as cover crop mulches in agronomic cropping systems. The most reliable benefit from cover crops was their effect on soil erosion, although an actual dollar amount was not assigned to this benefit. Additional studies are needed to determine the economic benefits of cover crops relative to weed suppression, both on an annual and long-term basis.

CONCLUSIONS

Efforts spent studying ways of enhancing contributions by hairy vetch and rye to weed suppression demonstrate a belief on the part of many that such an idea has merit. However, several limitations associated with using cover crops for weed suppression occur with some regularity. For example, weed suppression by cover crops seems most effective early in the season, so supplemental management may be required for acceptable season-long weed control (Moore, Gillespie, and Swanton, 1994; Masiunas, Weston, and Weller, 1995; Yenish, Worsham, and York, 1996). At other times, effects on weeds ranged from inconsistent to generally lacking (Yenish, Worsham, and York, 1996; Blum et al., 1997; Swanton et al., 1999). Lack of perennial weed control is common (Burgos and Talbert, 1996). Another frequently occurring limitation is interference with establishment of the cash crop by either cover crop residue or regrowth when kill methods have been ineffective (Kessavalou and Walters, 1997).

Reliability could be increased if cover crop cultivars were introduced having improved weed-suppressive ability. The success of attempts to breed hairy vetch and rye cultivars with more reliable weed-suppressive abilities depends on identifying heritable traits that confer the desired characteristics. Research by Teasdale and Mohler (2000) suggests that physical parameters that increase the ability of residue to obstruct weed seedlings and to limit light are critical. Further studies are needed to more fully understand the effect of age and cultivar on shoot tissue amounts of allelochemicals in rye and other allelopathic cover crops (Burgos, Talbert, and Mattice, 1999). Research is also needed to identify characteristics involved in the suppression of weeds observed in the presence of cover crop roots (Hoffman et al, 1996b). It is doubtful whether cover crops will be as economically important as they were when Pieters and McKee (1938) reported that over 50 cover crops species were under evaluation at various state experiment stations. Even so, it is essential that we begin to evaluate cover crops such as hairy vetch and rye more closely to identify lines with improved weed suppressive ability.

Another direction that could be promising is one in which several species of cover crops are planted sequentially over the course of the growing season (Sustainable Agriculture Network, 1998). Research has not been published in this area, although mixtures of two or more cover crops seeded simultaneously have been studied (Creamer, Bennett, and Stinner, 1997; Akemo, Regnier, and Bennett, 2000). The objective of sequential planting is to introduce cover crops in a succession that will make residue continually available to cover the soil surface without interfering with the cash crop. Success depends on identifying a workable schedule comprised of compatible species. If successful, it could supply a means to enhance perennial weed suppression and provide season-long weed control.

Last, researchers should identify easy-to-measure indicators that could be used by farmers to identify best-management practices for use in cover crop systems. It might be possible, for example, to estimate the contribution to late-season weed control, and thus the need for a postemergence herbicide application, from an early-season measure that would indicate cover crop residue thickness and its permeability to light. Such indices could reduce management uncertainties and economic risks associated with growing cover crops, thereby allowing more growers to view cover cropping as an advantageous practice with the potential to contribute to nutrient cycling and erosion control, as well as weed suppression.

REFERENCES

Akemo, M.C., M.A. Bennett, and E.E. Regnier (2000). Tomato growth in spring-sown cover crops. *HortScience* 35:843-848.

Akemo, M.C., E.E. Regnier, and M.A. Bennett (2000). Weed suppression in spring-sown rye *(Secale cereale)*-pea *(Pisum sativum)* cover crop mixes. *Weed Technology* 14: 545-549.

Akobundu, O. (1980). Live mulch: A new approach to weed control and crop production in the tropics. *Brighton Crop Protection Conference—Weeds* 2: 377-382.

Ateh, C.M. and J.D. Doll (1996). Spring-planted winter rye *(Secale cereale)* as a living mulch to control weeds in soybean *(Glycine max). Weed Technology* 10: 347-353.

Ball-Coelho, B.R., R.C. Roy, and C.J. Swanton (2000). Tillage and cover crop impacts on aggregation of a sandy soil. *Canadian Journal of Soil Science* 80: 363-366.

Barberi, P. and M. Mazzoncini (2001). Changes in weed community composition as influenced by cover crop and management system in continuous corn. *Weed Science* 49: 491-499.

Barnes, J.P. and A.R. Putnam (1983). Rye residues contribute weed suppression in no-tillage cropping systems. *Journal of Chemical Ecology* 9: 1045-1057.

Barnes, J.P. and A.R. Putnam (1986). Evidence for allelopathy by residues and aqueous extracts of rye *(Secale cereale). Weed Science* 34: 384-390.

Barnes, J.P. and A.R. Putnam (1987). Role of benzoxazinones in allelopathy by rye (*Secale cereale* L.). *Journal of Chemical Ecology* 13: 889-905.

Barnes, J.P., A.R. Putnam, B.A. Burke, and A.J. Aasen (1987). Isolation and characterization of allelochemicals in rye herbage. *Phytochemistry* 26: 1385-1390.

Blevins, R.L., R. Lal, J.W. Doran, G.W. Langdate, and W.W. Frye (1998). Conservation tillagefor erosion control and soil quality. In F.J. Pierce and W.W. Frye (Eds.), *Advances in soil and water conservation* (pp. 51-54). Chelsea, MI: Sleeping Bear Press.

Blum, U., L.D. King, T.M. Gerig, M.E. Lehman, and A.D. Worsham (1997). Effects of clover and small grain cover crops and tillage techniques on seedling emergence of some dicotyledonous weed species. *American Journal of Alternative Agriculture* 4: 146-161.

Bollero, G.A. and D.G. Bullock (1994). Cover cropping systems for the central corn belt. *Journal of Production Agriculture* 7: 55-58.

Bowen, J., L. Jordan, and D. Biehle (1991). Economics of no-till corn planted into winter cover crops. In W.L. Hargrove (Ed.), *Cover crops for clean water—Proceedings of an international conference* (pp. 181-182). West Tennessee Experiment Station, April 9-11, Jackson, Tennessee. Ankeny, IA: Soil and Water Conservation Society.

Bradow, J.M. and W.J. Connick Jr. (1990). Volatile seed germination inhibitors from plant residues. *Journal of Chemical Ecology* 16: 645-666.

Buhler, D.D., K.A. Kohler, and M.A. Foster (1998). Spring-seeded smother plants for weed control in corn and soybean. *Journal of Soil and Water Conservation* 53: 272-275.

Buhler, D.D., K.A. Kohler, and M.A. Foster (2001). Corn, soybean, and weed responses to spring-seeded smother plants. *Journal of Sustainable Agriculture* 18: 63-79.

Burgos, N.R. and R.E. Talbert (1996). Weed control and sweet corn (*Zea mays* var. *rugosa*) response in a no-till system with cover crops. *Weed Science* 44: 355-361.

Burgos, N.R., R.E. Talbert, and J.D. Mattice (1999). Cultivar and age differences in the production of allelochemicals by *Secale cereale*. *Weed Science* 47: 481-485.

Cardina, J. and N.L. Hartwig (1982). The influence of nitrogen, corn population and crownvetch cover crop on weed-corn competition. *Abstracts of the Weed Science Society of America* 22: 60.

Carreker, J.R., S.W. Wilkinson, and J.E. Box (1977). *Soil and water management systems for sloping land.* ARS-S-160. Washington, DC: Agricultural Research Service, U.S. Department of Agriculture.

Chase, W.R., M.G. Nair, and A.R. Putnam (1991). 2,2'-oxo-1,1'-azobenzene: Selective toxicity of rye (*Secale cereale* L.) allelochemicals to weed and crop species: II. *Journal of Chemical Ecology* 17: 9-19.

Chase, W.R., M.G. Nair, A.R. Putnam, and S.K. Mishra (1991). 2,2'-oxo-1,1'-azobenzene: Microbial transformation of rye (*Secale cereale* L.) allelochemicals in field soils by *Acinetobacter calcoaceticus*: III. *Journal of Chemical Ecology* 17: 1575-1584.

Clark, A.J. (1993). Managing hairy vetch/cereal rye cover crop mixtures for nitrogen and water in no-tillage corn. Doctoral dissertation, University of Maryland, College Park (Diss. Abstr. 9407617).

Clark, A.J., A.M. Decker, and J.J. Meisinger (1994). Seeding rate and kill date effects on hairy vetch-cereal rye cover crop mixtures for corn production. *Agronomy Journal* 86: 1065-1070.

Clark, A.J., A.M. Decker, J.J. Meisinger, and M.S. McIntosh (1997). Kill date of vetch, rye, and a vetch-rye mixture: I. Cover crop and corn nitrogen. *Agronomy Journal* 89: 427-434.

Clark, A.J., A.M. Decker, J.J. Meisinger, F.R. Mulford, and M.S. McIntosh (1995). Hairy vetch kill date effects on soil water and corn production. *Agronomy Journal* 87: 579-585.

Conservation Tillage Information Center (1983). *1982 national survey: Conservation tillage practices.* Fort Wayne, IN: Conservation Tillage Information Center.

Creamer, N.G., M.A. Bennett, and B.R. Stinner (1997). Evaluation of cover crop mixtures for use in vegetable production systems. *HortScience* 32: 866-870.

Creamer, N.G., M.A. Bennett, B.R. Stinner, J. Cardina, and E.E. Regnier (1996). Mechanisms of weed suppression in cover crop-based production systems. *HortScience* 31: 410-413.

Creamer, N.G. and S. M. Dabney (2002). Killing cover crops mechanically: Review of recent literature and assessment of new research results. *Journal of Alternative Agriculture* 17: 32-40.

Creamer, N.G., B. Plassman, M.A. Bennett, R.K. Wood, B.R. Stinner, and J. Cardina (1995). A method for mechanically killing cover crops to optimize weed suppression. *American Journal of Alternative Agriculture* 10: 157-162.

Cubbon, M.H. (1925). Effect of a rye crop on the growth of grapes. *American Society of Agronomy* 17: 568-577.

Dabney, S.M., N.W. Buehring, and D.B. Reginelli (1991). Mechanical control of legume covercrops. In W.L.Hargrove (Ed.), *Cover crops for clean water—Proceedings of an international conference* (pp. 146-147). West Tennessee Experiment Station, April 9-11, Jackson, Tennessee. Ankeny, IA: Soil and Water Conservation Society.

Davis, R.F., R.E. Baird, and R.D. McNeill (2000). Efficacy of cotton root destruction and wintercover crops for suppression of *Hoplolaimus columbus*. *Journal of Nematology* 32: 550-555.

DeHaan, R.L., D.L. Wyse, N.J. Ehlke, B.D. Maxwell, and D.H. Putnam (1994). Simulation of spring-seeded smother plants for weed control in corn *(Zea mays)* *Weed Science* 42: 35-43.

Ebelhar, S.A., W.W. Frye, and R.L. Blevins (1984). Nitrogen from legume cover crops for notillage corn. *Agronomy Journal* 76: 51-55.

Echtenkamp, G.W. and R.S. Moomaw (1989). No-till corn production in a living mulch system. *Weed Technology* 3: 262-266.

Eckert, D.J. (1988). Rye cover crops for no-tillage corn and soybean production. *Journal of Production Agriculture* 1: 207-210.

Ellis, D.R., K. Guillard, and R.G. Adams (2000). Purslane as a living mulch in broccoli production. *American Journal of Alternative Agriculture* 15: 50-59.

Enache, A.J. and R.D. Ilnicki (1990). Weed control by subterranean clover (*Trifolium subterraneum*) used as a living mulch. *Weed Technology* 4: 534-538.

Ess, D.R., D.H. Vaughan, J.M. Luna, and P.G. Sullivan (1994). Energy and economic savings from the use of legume cover crops in Virginia corn production. *American Journal of Alternative Agriculture* 9: 178-185.

Fisk, J.W., O.B. Hesterman, A. Shrestha, J.J. Kells, R.R. Harwood, J.M. Squire, and C.C. Shaeffer (2001). Weed suppression by annual legume cover crops in no-tillage corn. *Agronomy Journal* 93: 319-325.

Frye, W.W., W.G. Smith, and R.J. Williams (1985). Economics of winter cover crops as a source of nitrogen for no-till corn. *Journal of Soil and Water Conservation* 40: 246-249.

Graham, E.H. (1941). *Legumes for Erosion Control and Wildlife*. U.S. Department of Agriculture Publication No. 412. Washington, DC: U.S. Government Printing Office.

Greenland, R.G. (2000). Optimum height at which to kill barley used as a living mulch in onions. *HortScience* 35: 853-855.

Grunwald, N.J., S. Hu, and A.H.C. Van Bruggen (2000a). Short-term cover crop decompositionin organic and conventional soils: Characterization of soil C, N, microbial and plant pathogen dynamics. *European Journal of Plant Pathology* 106: 37-50.

Grunwald, N.J., S. Hu, and A.H.C. Van Bruggen (2000b). Short-term cover crop decompositionin organic and conventional soils: Soil microbial and nutrient cycling indicator variables associated with different levels of soil suppressiveness to *Pythium aphanidermatum.European Journal of Plant Pathology* 106: 51-65.

Herrero, E.V., J.P. Mitchell, W.T. Lanini, S.R. Temple, E.M. Miyao, R.D. Morse, and E. Campiglia (2001). Use of cover crop mulches in a no-till furrow-irrigated processing tomato production system. *HortTechnology* 11: 43-48.

Hoffman, M.L., E.E. Regnier, and J. Cardina (1993). Weed and corn *(Zea mays)* responses to a hairy vetch *(Vicia villosa)* cover crop. *Weed Technology* 7: 594-599.

Hoffman, M.L., L.A. Weston, J.C. Snyder, and E.E. Regnier (1996a). Allelopathic influence of germinating seeds and seedlings of cover crops on weed species. *Weed Science* 44: 579584.

Hoffman, M.L., L.A. Weston, J.C. Snyder, and E.E. Regnier (1996b). Separating the effects of sorghum *(Sorghum bicolor)* and rye *(Secale cereale)* root and shoot residues on weed development. *Weed Science* 44: 402-407.

Hughes, B.J. and R.D. Sweet (1979). Living mulch: A preliminary report on grassy cover crops interplanted with vegetables. *Proceedings of the Northeastern Weed Science Society* 33: 109.

Hutchinson, C.M. and M.E. McGiffen Jr. (2000). Cowpea cover crop mulch for weed control in desert pepper production. *HortScience* 35: 196-198.

Johnson, W.C., III, T.B. Brenneman, S.H. Baker, A.W. Johnson, D.R. Sumner, and B.G. Mullinix Jr. (2001). Tillage and pest management considerations in a peanut-cottonrotation in the southeastern coastal plain. *Agronomy Journal* 93: 570-576.

Kabir, Z. and R.T. Koide (2000). The effect of dandelion or a cover crop on mycorrhiza inoculum potential, soil aggregation and yield of maize. *Agriculture, Ecosystems and Environment* 78: 167-174.

Kessavalou, A. and D.T. Walters (1997). Winter rye as a cover crop following soybean under conservation tillage. *Agronomy Journal* 89: 68-74.

Kessavalou, A. and D.T. Walters (1999). Winter rye cover crop following soybean under conservation tillage: Residual soil nitrate. *Agronomy Journal* 91: 643-649.

Lal, R. (1975). *Role of mulching techniques in tropical soil and water management.* Technical Bulletin 1. Ibadan, Nigeria: International Institute of Tropical Agriculture.

Lal, R., E. Regnier, D.J. Eckert, W.M. Edwards, and R. Hammond (1991). Expectations of cover crops for sustainable agriculture. In W.L. Hargrove (Ed.), *Cover crops for clean water—Proceedings of an international conference* (pp. 1-11). West Tennessee Experiment Station, April 9-11, Jackson, Tennessee. Ankeny, IA: Soil and Water Conservation Society.

Liebl, R., F.W. Simmons, L.M. Wax, and E.W. Stoller (1992). Effect of rye *(Secale cereale)* mulch on weed control and soil moisture in soybean *(Glycine max). Weed Technology* 6: 838-846.

MacRae, R.J. and G.R. Mehuys (1985). The effect of green manuring on the physical properties of temperate-area soils. *Advances in Soil Science* 3: 71-94.

Mallory, E.B., J.L. Posner, and J.O. Baldock (1998). Performance, economics, and adoption of cover crops in Wisconsin cash grain rotations: On-farm trials. *American Journal of Alternative Agriculture* 13: 2-11.

Malpassi, R.N., T.C. Kaspar, T.B. Parkin, C.A. Cambardella, and N.A. Nubel (2000). Oat and rye root decomposition effects on nitrogen mineralization. *Soil Science Society of America Journal* 64: 208-215.

Martin, R.C., P.R. Greyson, and R. Gordon (1999). Competition between corn and a living mulch. *Canadian Journal of Plant Science* 79: 579-586.

Masiunas, J.B., D.M. Eastburn, V.N. Mwaga, and C.E. Eastman (1997). The impact of living and cover crop mulch systems on pests and yields of snap beans and cabbage. *Journal of Sustainable Agriculture* 9: 61-89.

Masiunas, J.B., L.A. Weston, and S.C. Weller (1995). The impact of rye cover crops on weed populations in a tomato cropping system. *Weed Science* 43: 318-323.

Mitchell, J.P., C. Shennan, M.J. Singer, D.W. Peters, R.O. Miller, T. Prichard, S.R. Grattan, J.D. Rhoades, D.M. May, and D.S. Munk (2000). Impacts of gypsum and winter cover crops on soil physical properties and crop productivity when irrigated with saline water. *Agricultural Water Management* 45: 55-71.

Mohler, C.L. and J.R. Teasdale (1993). Response of weed emergence to rate of *Vicia villosa* Roth and *Secale cereale* L. residue. *Weed Research* 33: 487-499.

Moore, M.J., T.J. Gillespie, and C.J. Swanton (1994). Effect of cover crop mulch on weed emergence, weed biomass, and soybean *(Glycine max)* development. *Weed Technology* 8: 512-518.

Munawar, A., R.L. Blevins, W.W. Frye, and M.R. Saul (1990). Tillage and cover crop management for soil water conservation. *Agronomy Journal* 82: 773-777.

Mutch, D.R. and T.E. Martin (1998). Cover crops. In M.A. Cavigelli, S.R. Deming, L.K. Probyn, and R.R. Harwood (Eds.), *Michigan field crop ecology: Managing biological processes for productivity and environmental quality* (pp. 44-53). Michigan State University Extension Bulletin E-2646. East Lansing: Michigan State University.

Nair, M.G., C.J. Whitenack, and A.R. Putnam (1990). 2,2'-oxo-1,1'-azobenzene, a microbially transformed allelochemical from 2,3-benzoxazolinone: I. *Journal of Chemical Ecology* 16: 353-364.

Nelson, W.A., B.A. Kahn, and B.W. Roberts (1991). Screening cover crops for use in conservation tillage systems for vegetables following spring plowing. *HortScience* 26: 860-862.

Odhiambo, J.J.O. and A.A. Bomke (2001). Grass and legume cover crop effects on dry matter and nitrogen accumulation. *Agronomy Journal* 93: 299-307.

Pieters, A.J. (1927). *Green manuring principles and practice.* New York: John Wiley and Sons.

Pieters, A.J. and R. McKee (1938). The use of cover and green-manure crops. *Soils and Men Yearbook of Agriculture* 1938: 431-444.

Piper, C.V. and R. McKee (1923). *Vetches.* U.S. Department of Agriculture Farmers Bulletin Number 515. Washington, DC: United States Department of Agriculture.

Power, J.F. (1990). Legumes and crop rotations. In C.A. Francis, C.B. Flora, and L.D. King (Eds.), *Sustainable agriculture in temperate zones* (pp. 178-204). New York: John Wiley and Sons.

Putnam, A.R. and J. DeFrank (1983). Use of phytotoxic plant residues for selective weed control. *Crop Protection* 2: 173-181.

Putnam, A.R., J. DeFrank, and J.P. Barnes (1983). Exploitation of allelopathy for weed control in annual and perennial cropping systems. *Journal of Chemical Ecology* 9: 1001-1010.

Raimbault, B.A. and T.J. Vyn (1991). Crop rotation and tillage effects on corn growth and soil structural stability. *Agronomy Journal* 83: 979-985.

Rasse, D.P., J.T. Ritchie, W.R. Peterson, J. Wei, and A.J.M. Smucker (2000). Rye cover crop and nitrogen fertilization effects on nitrate leaching in inbred maize fields. *Journal of Environmental Quality* 29: 298-304.

Reddy, K. (2003). Impact of rye cover crop and herbicides on weeds, yield, and net return in narrow-row transgenic, and conventional soybean *(Glycine max). Weed Technology* 17: 28-35.

Regnier, E.E. and E.W. Stoller (1987). Cropping methods affecting the performance of living mulches in soybeans. *Proceedings of the North Central Weed Control Conference* 42: 36-37.

Robinson, R.G. and R.S. Dunham (1954). Companion crops for weed control in soybeans. *Agronomy Journal* 46: 278-281.

Rosecrance, R.C., G.W. McCarty, D.R. Shelton, and J.R. Teasdale (2000). Denitrification and N mineralization from hairy vetch (*Vicia villosa* Roth) and rye (*Secale cereale* L.) cover crop monocultures and bicultures. *Plant and Soil* 227: 283-290.

Sainju, U.M. and B.P. Singh (1997). Winter cover crops for sustainable agricultural systems: Influence on soil properties, water quality, and crop yields. *Hort-Science* 32: 21-28.

Schaller, F.W. and W.E. Larson (1955). Effect of wide spaced corn rows on corn yields and forage establishment. *Agronomy Journal* 47: 271-276.

Shilling, D.G., L.A. Jones, A.D. Worsham, C.E. Parker, and R.F. Wilson (1986). Isolation and identification of some phytotoxic compounds from aqueous extracts of rye. *Journal of Agriculture Food and Chemistry* 34: 633-638.

Smeda, R.J. and S.C. Weller (1996). Potential of rye *(Secale cereale)* for weed management in transplant tomatoes *(Lycopersicon esculentum). Weed Science* 44: 596-602.

Stute, J.K. and J.L. Posner (1995). Legume cover crops as a nitrogen source for corn in an oat-corn rotation. *Journal of Production Agriculture* 8: 385-390.

Sustainable Agriculture Network (1998). *Managing cover crops profitably,* Second edition. Des Moines, IA: Plain Talk Printing Co.

Swanton, C.J., A. Shrestha, R.C. Roy, B.R. Ball-Coelho, and S.Z. Knezevic (1999). Effect of tillage systems, N, and cover crop on the composition of weed flora. *Weed Science* 47: 454-461.

Teasdale, J.R., C.E. Beste, and W.E. Potts (1991). Response of weeds to tillage and cover crop residue. *Weed Science* 39: 195-199.

Teasdale, J.R. and C.S.T. Daughtry (1993). Weed suppression by live and desiccated hairy vetch *(Vicia villosa). Weed Science* 41: 207-212.

Teasdale, J.R. and C.L. Mohler (1993). Light transmittance, soil temperature, and soil moisture under residue of hairy vetch and rye. *Agronomy Journal* 85: 673-680.

Teasdale, J.R. and C.L. Mohler (2000). The quantitative relationship between weed emergence and the physical properties of mulches. *Weed Science* 48: 385-392.

Technical Advisory Committee (1989). *Sustainable agricultural production: Implications for international agricultural research.* Washington, DC: Consultative Group on International Agricultural Research.

Thomas, G.W. and W.W. Frye (1984). Fertilization and liming. In R.E. Phillips and S.H. Phillips (Eds.), *No-tillage agriculture* (pp. 111-126). New York: Van Nostrand Reinhold Co.

U.S. Congress (1930). *House Committee on Appropriation Bill for 1930.* Washington, DC: U.S. (70th) Congress, 2nd Session, pp. 310-330.

Weston, L.A. (1990). Cover crop and herbicide influence on row crop seedling establishment in no-tillage culture. *Weed Science* 38: 166-171.

White, R.H. and A.D. Worsham (1990). Control of legume cover crops in no-till corn *(Zea mays)* and cotton *(Gossypium hirsutum). Weed Technology* 4: 57-62.

White, R.H., A.D. Worsham, and U. Blum (1989). Allelopathic potential of legume debris and aqueous extracts. *Weed Science* 37: 674-679.

Williams, M.M., II, D.A. Mortensen, and J.W. Doran (1998). Assessment of weed and crop fitness in cover crop residues for integrated weed management. *Weed Science* 46: 595-603.

Yenish, J.P., A.D. Worsham, and A.C. York (1996). Cover crops for herbicide replacement in no-tillage corn *(Zea mays). Weed Technology* 10: 815-821.

Zemenchik, R.A., K.A. Albrecht, and C.M Boerboom (2000). Corn production with Kura clover as a living mulch. *Agronomy Journal* 92: 698-705.

Chapter 4

Utilizing *Brassica* Cover Crops for Weed Suppression in Annual Cropping Systems

Rick A. Boydston
Kassim Al-Khatib

INTRODUCTION

Weed management is an essential component of successful crop production. Synthetic herbicides and cultivation are the primary means of managing weeds in modern agriculture. However, there are environmental hazards associated with herbicides, as well as an increase in herbicide-resistant weeds. Erosion and sustainability issues are also associated with cultivation. As a result, there is an increasing interest in developing alternative methods of weed control, including the use of weed-suppressive cover crops. Integrated weed management involves managing weeds through a series of mortality- and fitness-reducing events (Williams, Mortensen, and Doran, 1998). A cover crop is any living ground cover that is planted into or after a main crop and then commonly killed before the next crop is planted (Hartwig and Ammon, 2002) and that provides benefits such as decreased soil erosion, increased soil organic matter, or pest suppression (Sainju and Singh, 1997). Cover crops, as a component of weed management, reduce the fitness of weeds relative to the main crop. Nagabhushana, Worsham, and Yenish (2001) reviewed the use of allelopathic cover crops in cropping systems and concluded that herbicide use can be reduced and top soil conserved with the proper use of cover crops.

Suppressed growth of crops and weeds succeeding *Brassica* species have been noted in numerous reports (Campbell, 1959; Kutacek, 1964; Horricks, 1969; Waddington, 1978; Mason-Sedon, Jessop, and Lovett, 1986; Oleszek, 1987; Vera, McGregor, and Downey, 1987). White mustard (*Brassica hirta* Moench, syn. *Sinapis alba* L.) has been utilized as a cover crop in Europe for over 50 years (Vaughn and Hemingway, 1959). More recently, *Brassica* species have been tested as fall- or spring-planted cover and smother crops to take advantage of their weed-suppressive effects

(DeHaan et al., 1994; Boydston and Hang, 1995; Al-Khatib, Libbey, and Boydston, 1997; Krishnan, Holshouser, and Nissen, 1998; Buhler, Kohler, and Foster, 2001). Other beneficial effects of using *Brassica* species as cover or smother crops include reclaiming soil nutrients (Rogasik, Smukalski, and Obenauf, 1992; Stivers-Young, 1998), decreasing soil erosion, and suppressing nematodes (Mojtahedi et al., 1993), diseases (Papavizas and Lewis, 1971; Muehlchen, 1990; Vaughn, 1993; Angus et al., 1994; Charron and Sams, 1999; Olivier et al., 1999), and insects (Lichtenstein, Morgan, and Mueller, 1964; Williams et al., 1993).

Control of weeds (Al-Khatib and Boydston, 1999) and other soilborne pests (Brown and Morra, 1997) with *Brassica* green manure crops was recently reviewed, as was the pesticidal activity of glucosinolate degradation products (Vaughn, 1999). This chapter summarizes the mechanisms of weed suppression demonstrated with brassicas, and the utilization and integration of brassicas for weed suppression in annual cropping systems.

MECHANISMS OF WEED SUPPRESSION WITH BRASSICA *COVER CROPS*

Weed Suppression Within the Cover Crop

Including cover crops in the rotation at a time when land might otherwise lie uncropped will suppress weed development (Liebman and Davis, 2000; Bond and Grundy, 2001). In general, cover crops suppress weeds by competing for light, water, and soil nutrients. Vigorously growing cover crops shade the soil, reducing light intensity and creating a cooler environment for slower-germinating weed species (Teasdale, 1993). Several studies have noted the rapid smothering action of *Brassica* species (DeHaan et al., 1994; Stivers-Young, 1998; Buhler, Kohler, and Foster, 2001). By 40 days after spring planting, white mustard produced more than 60 percent ground cover compared to less than 35 percent for four legume species tested (Buhler, Kohler, and Foster, 2001).

In a New Zealand study (McLenaghen et al., 1996), white mustard produced 92 percent ground cover and suppressed weeds more than rye (*Secale cereale* L.), Italian ryegrass (*Lolium multiflorum* Lam.), blue lupin (*Lupinus augustifolius* L.), and broad bean (*Vicia faba* L.). Stivers-Young (1998) reported that white mustard and oilseed radish (*Raphanus sativus* L.) planted in late August covered the ground quicker and produced more biomass than oats (*Avena sativa* L.). Likewise, German researchers reported that oilseed radish grew more rapidly than four other cover crops tested (Rogasik, Smukalski, and Obenauf, 1992). Fall-planted cover crops can also dry the soil surface, limiting the moisture essential for many slower-germinating winter annual weed species.

In addition to the effects from shading and plant competition, members of the Brassicaceae family have frequently been cited as potentially allelopathic (Chew, 1988). Black mustard [*Brassica nigra* (L.) Koch] (Muller, 1969; Bell and Muller, 1973) and rapeseed (*B. napus* L.) (Martin and Rademacher, 1960) have been shown to inhibit the germination and growth of other plants.

One class of compounds often cited for suppressing growth of plants growing in proximity to *Brassica* is isothiocyanates (ITCs). Living plants do not actively release high amounts of ITCs (Choesin and Boerner, 1991), although exudation of ITCs from growing roots is known to occur (Elliot and Stowe, 1971). Choesin and Boerner (1991) found no difference in inhibition of alfalfa (*Medicago sativa* L.) growing near rapeseed genotypes that released allyl isothiocyanate at different rates. In replacement series experiments, they reported the competitive ability of a high-glucosinolate-containing wild-type rapeseed was similar to that of a low-glucosinolate-containing mutant. However, other aromatic ITCs released primarily from root tissues may play a greater role in plant suppression, such as, 2-phenylethyl ITC (Martin and Rademacher, 1960; Eberlein et al., 1998; Kirkegaard and Sarwar, 1998a; Kirkegaard and Sarwar, 1999). Suppression of weeds observed during *Brassica* cover crop establishment and growth is probably a combination of strong competition for light, nutrients, and water as well as the allelopathic properties of many *Brassica* species.

Weed Suppression in the Succeeding Crop

Inhibitory Effects of Brassica *Residues*

Numerous crop residues suppress weeds when incorporated in the soil. Barker and Bhowmik (2001) reported that incorporation of various crop residues resulted in increased weed suppression as the amount of residue added to the soil increased from 6 to 24 $Mg{\cdot}ha^{-1}$. In another study, rapeseed, pea (*Pisum sativum* L.), wheat (*Triticum aestivum* L.), and sorghum [*Sorghum bicolor* (L.) Moench] residues added to soil were all equally suppressive to wheat emergence (Jessop and Stewart, 1983). Fujii (2001) screened 53 cover crop species for allelopathic activity using the plant box method. Rapeseed and white mustard residues suppressed weeds more than would be estimated based on their biomass alone, suggesting that allelopathy may be involved. This allelopathy may extend to the succeeding crop, as well as to target weeds. Wheat growth and yield following rapeseed and other brassicas is often reduced (Horricks, 1969; Mason-Sedon, Jessop, and Lovett, 1986; Oleszek, 1987; Vera, McGregor, and Downey, 1987). Vera, McGregor, and Downey (1987) reported that cereals were less sensitive to applied residues of four *Brassica* species than flax (*Linum usitatissimum* L.) or rapeseed. Kirkegaard and colleagues (1994) reported that wheat growth was not inhibited if planting was delayed four weeks after *Brassica* residue incorporation. Germination and growth of cauliflower (*Brassica oleracea* L. var. *botrytis* L.) was inhibited in soil previously cropped with broccoli (*B. oleracea* L. var. *italica* L.), but broccoli and cabbage (*B. oleracea* L. var. *capitata* L.) seedling growth were enhanced (Santos and Leskovar, 1997).

Inhibitory Compounds

The precise mechanism of crop and weed suppression observed following addition of *Brassica* residues is not yet clear. Many brassicas contain high amounts of glucosinolates in their tissues (Fenwick, Heaney, and Mullin, 1983). About 20 different types of glucosinolates are commonly found in *Brassica* tissue (Kirkegaard and Sarwar, 1998a), and over 100 distinct glucosinolates have been identified from brassicas and other plant families (Brown and Morra, 1997; Vaughn, 1999). Upon hydrolysis by myrosinase, glucosinolates are converted to numerous compounds including organic cyanides, ITCs, oxazolidinethiones, nitriles, epithionitriles, and ionic thiocyanates (Fenwick, Heaney, and Mullin, 1983; Borek et al., 1995; Brown and Morra, 1995; Kirkegaard and Sarwar, 1998a). Glucosinolate types and concentrations vary within and among *Brassica* species and in different plant tissues, and the corresponding hydrolysis products also vary (Kirkegaard and Sarwar, 1999).

ITCs released by hydrolysis of glucosinolates during soil degradation of *Brassica* tissues are often cited as potential pesticides. ITCs are known to have broad biocidal activity affecting insects, nematodes, fungi, and weeds (Fenwick, Heaney, and Mullin, 1983; Brown and Morra, 1997; Al-Khatib and Boydston, 1999; Vaughn, 1999). The suppression of soilborne pests and pathogens by release of biocidal compounds, such as ITCs from glucosinolate-containing tissues, has been termed *biofumigation* (Angus et al., 1994).

Inhibition of Germination

The effects of several glucosinolate degradation products on weed germination and growth are well documented. Glucosinolates seldom show biocidal activity with the exception of glucobrassican, which inhibited the germination and growth of wheat (Kutacek, 1964). Although most glucosinolates are relatively nontoxic to plants, volatile glucosinolate breakdown products, such as ITCs, inhibit seed germination and have herbicidal and allelochemical effects (Evenari, 1949; Leblova, Jiracek, and Kostir, 1960; Bell and Muller, 1973; Wolf, Spencer, and Kwolek, 1984; Chew, 1988; Bialy et al., 1990; Brown and Morra, 1995, 1996; Al-Khatib, Libbey, and Boydston, 1997; Petersen et al., 2001). Chopped chou moellier (*Brassica oleracea* L.) tissue placed on filter paper above white clover (*Trifolium repens* L.) seed inhibited germination (Campbell, 1959). Macerated tissue of brown mustard and black mustard inhibited wheat, barnyardgrass [*Echinochloa crus-galli* (L.) Beauv.], and lettuce (*Lactuca sativa* L.) seed germination (Mason-Sedon, Jessop, and Lovett, 1986; Oleszek, 1987). However, ground rapeseed roots had no effect on wheat root growth (Martin and Rademacher, 1960). Macerated leaves of brown mustard, black mustard, and Ethiopian mustard (*Brassica carinata* A. Br.) yielded large amounts of volatiles, in particular, allyl ITC (Tollsten and Bergstrom, 1988). Volatiles (identified mainly as ITCs) from macerated tissue of several *Brassica* inhibited germination of wheat, lettuce, and hemp sesbania [*Sesbania exaltata* (Raf.) Rydb. ex A.W. Hill] (Brown and Morra, 1995; Vaughn and Boydston, 1997).

ITCs inhibit seed germination to varying degrees, and the ability to inhibit germination may be related to volatility. Allyl ITC, the hydrolysis product of allyl glucosinolate, which is found in cabbage, brown mustard, Indian mustard, rapeseed, black mustard, and Crambe (*Crambe abyssinica* Hochst. ex R. E. Fries), inhibited the germination and growth of peas, wheat, and rapeseed in concentrations from 1.1×10^{-4} to 1.1×10^{-2} M (Evenari, 1949; Leblova, Jiracek, and Kostir, 1960; Leblova and Kostir, 1962). Benzyl ITC inhibited sicklepod (*Cassia obtusifolia* L.), sorghum, and velvetleaf (*Abutilon theophrasti* Medik.) germination in separate studies (Wolf, Spencer, and Kwolek, 1984; Dale, 1986). Methyl ITC, the biocidal product of metham sodium, a commercial fumigant, inhibits germination of weed seed at higher concentrations while stimulating germination of dormant seed at lower concentrations (White, 1978; Teasdale and Taylorson, 1986).

Smooth pigweed (*Amaranthus hybridus* L.) germination was inhibited by six ITCs applied in aqueous solutions, with the aryl-ITCs (benzyl and phenylethyl-ITC) being stronger inhibitors than the alkyl-ITCs (allyl, *n*-propyl, and *n*-butyl-ITC) (Petersen et al., 2001). When exposing hydrated seed to volatiles of several ITCs, progressively less crop and weed seed germination occurred with volatiles of phenylethyl, benzyl, 3-butenyl, and allyl-ITC, respectively (Vaughn and Boydston, 1997). These authors attributed the lower activity of phenylethyl and benzyl-ITC to their relatively lower volatility. Bialy and colleagues (1990) reported phenylethyl ITC at 300 to 500 ppm in aqueous solution strongly inhibited wheat germination. Phenylethyl ITC applied in vapor phase at 5 to 20 ppm inhibited germination of redroot pigweed (*Amaranthus retroflexus* L.), but it was not as potent as several other ITCs with greater volatilities (Al-Khatib, Libbey, and Boydston, 1997). Allyl ITC applied in the vapor phase at 1 ppm inhibited germination of barnyardgrass by 45 percent and redroot pigweed by 85 percent. Subsequent shoot growth of both weed species was even more sensitive than germination (Al-Khatib, Libbey, and Boydston, 1997). In studies by Petersen and colleagues (2001), *n*-butyl ITC was more suppressive in the vapor phase than in aqueous solution, while 2-phenylethyl ITC showed the opposite effect. Response of seed germination to ITC volatiles varies with species, and small seeded species tend to be more susceptible to germination inhibition by ITCs than larger seed (Al-Khatib, Libbey, and Boydston, 1997; Vaughn and Boydston, 1997; Petersen et al., 2001).

Other Inhibitory Compounds

Other glucosinolate degradation products, such as thiocyanate ions, also have plant suppressive properties (Stiehl and Bible, 1989). Thiocyanate ions decreased germination, growth, and chlorophyll content of several plant species (Ju, Bible, and Chong, 1983; Stiehl and Bible, 1985, 1989), and may play a role in the allelopathic properties of brassicas. The major degradation products of glucosinolates in soil are organic ITCs and nitriles (Borek et al., 1995; Gardiner et al., 1999). Water-soluble compounds 5-vinyl oxazolidinethione and 5-allyl oxazolidinethione also have been identified (Eberlein et al., 1998). Although present at relatively low levels during *Brassica* tissue decomposition in the soil, ITCs are often cited as the potential herbicidal compound. In studies by Petersen and colleagues (2001), low concentrations (maximum 200 $\mu g \cdot kg^{-1}$) of ITCs were detected in the soil following incorporation of turnip-rape (*Brassica rapa* L.–*Brassica napus* L.), but those authors demonstrated the sensitivity of small-seeded weeds to low levels of ITCs. Other degradation products that may play a role in suppressing weeds were recently reviewed by Brown and Morra (1997).

Root versus Shoot Tissue

If degradation products of glucosinolates play a role in the inhibition of germination and growth of weeds, the total amounts and types of glucosinolates found in various *Brassica* tissues should affect the level of weed control. Several greenhouse bioassays and field studies have found no relationship between weed suppression and total glucosinolate content of the incorporated *Brassica* tissues (Boydston and Al-Khatib, 1994; Krishnan, Holshouser, and Nissen, 1998). However, those authors did not account for glucosinolates in root tissues.

Several authors have stressed the contribution of root tissues when evaluating the potential of pest suppression with brassicas and other cover crops (Eberlein et al., 1998; Kirkegaard and Sarwar, 1998a; Gardiner et al., 1999; Blum, King, and Brownie, 2002). Root tissues of numerous *Brassica* species contain relatively large quantities of 2-phenylethyl glucosinolate (Martin and Rademacher, 1960; Eberlein et al., 1998; Kirkegaard and Sarwar, 1998a; Gardiner et al., 1999). The major glucosinolate found in the root tissue of five fall-planted rapeseed cultivars was phenylethyl glucosinolate, while the main glucosinolates in shoots were 4-pentenyl, 2-hydroxybutenyl, 3-butenyl, and 2-hydroxypentenyl glucosinolate (Eberlein et al., 1998). Total glucosinolate concentrations were 1.5 to 7 times greater in root tissue than shoot tissue, but root tissue biomass was three- to fourfold less than shoot biomass. Roots contributed about half of the total glucosinolates incorporated in the soil when considering both glucosinolate concentrations and biomass of the various tissues. Testing 13 *Brassica* species, Kirkegaard and Sarwar (1998a) reported roots contribute an average of 24 percent of the total plant glucosinolates. Maximum glucosinolates incorporated in the soil over the two-year study was 104 $mol \cdot ha^{-1}$, while the maximum biomass produced was 7,837 $kg \cdot ha^{-1}$.

Decaying roots may release ITCs slower and over a longer period than shoots, and 2-phenylethyl ITC is less volatile and less sensitive to hydrolysis than aliphatic ITCs and may persist for longer periods in the soil (Petersen et al., 2001). Half-life of allyl ITC ranged from 20 to 60 h in six soils tested by Borek and colleagues (1995). Similarly, maximum concentrations of glucosinolate degradation products from rapeseed were found 30 h after plow down (Gardiner et al., 1999). Higher soil temperature, moisture, and organic carbon content may increase losses of ITC from soil (Borek et al., 1995; Petersen et al., 2001).

Maturity of Incorporated Residues

The maturity of the plant at the time of incorporation may also affect the ability of brassicas to suppress weeds. Mature, dried rapeseed residues (4,500 $kg \cdot ha^{-1}$) incorporated into the soil had less suppressive effects on weed emergence than other crop residues tested (Purvis, Jessop, and Lovett, 1985). Mature shoot tissues utilized by Purvis, Jessop, and Lovett (1985) may have contained relatively low levels of glucosinolates and different types of glucosinolates than those found in immature green tissue. Fresh green plant tissues are generally more suppressive than mature, dried stubble and chaff (Jimenez-Orsornio and Gliessman, 1987; Vera, McGregor, and Downey, 1987). Mature rapeseed chaff and straw residues incorporated in greenhouse soil inhibited alfalfa and smooth bromegrass (*Bromus inermis* Leyss.) emergence more than did wheat residues (Waddington, 1978). However, inhibition of barley (*Hordeum vulgare* L.) and bromegrass growth after emergence with rapeseed residues was attributed to a reduction in the availability of nitrogen.

Mojtahedi and colleagues (1991) noted phytotoxicity to tomato (*Lycopersicon esculentum* Mill.) transplants into soil amended with freshly chopped rapeseed residues. Soil bioassays of weed seed germination and growth with white mustard and rapeseed green-tissue-amended soils were conducted by Al-Khatib, Libbey, and Boydston (1997). Fresh white mustard residues suppressed emergence and growth of three weed species more than did equal amounts of fresh rapeseed residues.

Relative Sensitivity of Weed Species to Brassica *Residues*

Weed species composition also affects efficacy of brassicas. Shepherd's-purse [*Capsella bursa-pastoris* (L.) Medik.] emergence and growth were more sensitive to *Brassica* residues than were green foxtail [*Setaria viridis* (L.) Beauv.] and kochia [*Kochia scoparia* (L.) Schrad.]. In similar studies, rapeseed residues had little impact on emergence of five weed species but greatly reduced biomass and vigor of hairy nightshade (*Solanum sarrachoides* Sendtn.), puncturevine (*Tribulus terrestris* L.), and longspine sandbur [*Cenchrus longispinus* (Hack.) Fern.] (Boydston and Al-Khatib, 1994). In general, white mustard residues were more suppressive than rapeseed and reduced emergence and growth of all five weed species, even though total glucosinolate content of the tissues were similar. In similar greenhouse bioassays rapeseed, brown mustard [*Brassica juncea* (L.) Czern.], and white mustard green tissues added to silt loam soil reduced growth and emergence of five weed species to varying degrees (Krishnan, Holshouser, and Nissen, 1998). No relationship between glucosinolate content of the incorporated green manure and suppression of weed growth was found. Only leaf and shoot tissues of the green manure crops were used in these three bioassay studies; however, root tissue may play an important role in the biocidal activity of *Brassica* green manures (Eberlein et al., 1998).

Early spring weed biomass was greatly reduced following fall-planted oilseed radish and white mustard compared to oats or bare ground (Stivers-Young, 1998). Fall- or spring-planted *Brassica* cover crops reduced early-season weed emergence and growth in potato (*Solanum tuberosum* L.), pea, and soybean [*Glycine max* (L.) Merr.] (Boydston and Hang, 1995; Al-Khatib, Libbey, and Boydston, 1997; Krishnan, Holshouser, and Nissen, 1998), although results have not been consistent.

Understanding and optimizing the conditions favoring formation of glucosinolates in brassicas and development of high glucosinolate cultivars could improve the performance and increase the use of *Brassica* cover crops. Glucosinolate production in plants depends on the growth stage of the plant, cultivation and climatic conditions, water and nutrient availability in the soil, and the cultivar (Fenwick, Heaney, and Mullin, 1983; Kirkegaard and Sarwar, 1998b; Al-Khatib and Boydston, 1999). Glucosinolate levels in brassicas can also be increased by mechanical wounding or attack by pathogens (Vaughn, 1999). Manipulation of some of these factors could result in greater efficacy of *Brassica* cover crops on weeds and pests.

INTEGRATING BRASSICA *COVER CROPS INTO CROPPING SYSTEMS*

Fall-Planted Cover Crops

Fall-planted *Brassica* cover crops fit well into annual cropping systems following early harvested crops, such as small grains, sweet corn (*Zea mays* L.), peas, or potatoes. In addition to the weed-suppressive action of *Brassica* cover crops, suppression of nematodes (Mojtahedi et al., 1993) and diseases (Vaughn, 1993; Olivier et al., 1999) common to potato make brassicas attractive cover crops in potato rotations. Potatoes also utilize large amounts of nitrogen and benefit from release of nitrogen from decaying cover crop residues. Brassicas may reclaim residual soil nutrients up to 1 to 2 m deep (Rogasik, Smukalski, and Obenauf, 1992; Laine et al., 1993; Stivers-Young, 1998). In Canadian trials, McKeown and Potter (2001) reported higher yield of potato following a cover crop of brown mustard than following rye plus red clover (*Trifolium pratense* L.). Fall-planted brassicas incorporated in the spring may also improve soil water infiltration (Sarrantonio, 1992) and soil aggregate stability (Thorup-Kristensen, 1993).

In order to accumulate sufficient cover crop biomass, brassicas are fall planted preceding early-planted crops such as potatoes and peas. Biomass production of fall-planted brassicas ranged from 2,800 to 8,000 $kg \cdot ha^{-1}$ and was influenced by species, cultivar, planting date, planting rate, fertility, winter conditions, and growing region (Boydston and Hang, 1995; Eberlein et al., 1998, Stivers-Young, 1998; Boydston and Vaughn, 2002). Planting white mustard earlier than August risks production of viable seed, while planting later than early September limits growth and dry matter production (Stivers-Young, 1998; Boydston, personal observations).

Liebman and Davis (2000) proposed that seedlings arising from heavier seed are more resistant to allelopathy and other stresses. This principle appears true with brassicas as well. Small-seeded plant species are more susceptible to inhibition from *Brassica* cover crop residues than large-seeded species are (Vera, McGregor, and Downey, 1987; Vaughn and Boydston, 1997; Petersen et al., 2001). The large, deeply planted vegetative seed tubers used in potato production are more tolerant to allelopathic cover crop residues than are most small-seeded broadleaf weeds (Boydston and Hang, 1995; Boydston and Vaughn, 2002). Pea is tolerant to white mustard residues when planted two to three weeks after residue incorporation, but rapeseed residues reduced pea stands (Al-Khatib, Libbey, and Boydston, 1997). Delayed planting of many crops, especially small-seeded crops, would likely be required following a plow down of *Brassica* residues to avoid possible phytotoxicity. Pea emergence was not affected following addition of allyl ITC to soil if planting was delayed seven weeks (Papavizas, 1966). Although less sensitive than smaller-seeded weeds, cucumber (*Cucumis sativus* L.), pea, soybean, corn, alfalfa, wheat, and rapeseed germination are inhibited by ITCs (Al-Khatib and Boydston, 1999; Vaughn and Boydston, 1997).

White mustard has become a popular fall-planted cover crop in the potato-producing regions of the Columbia Basin of eastern Washington. Planted in mid to late August, white mustard emerges quickly and produces from 5,000 to 7,000 $kg \cdot ha^{-1}$ dry weight plant residues before succumbing to freezing temperatures in the winter in Washington State. Green residues can be incorporated in late fall, or dead residues incorporated the following spring. Although common dogma is that brassicas should be incorporated while still green to obtain maximum biofumigation activity on nematodes and diseases, studies relating efficacy to the timing of incorporation are lacking. These authors have consistently noted less early season weed emergence and growth following fall-planted, spring-incorporated white mustard that was winter killed several months before incorporation compared to following fallow. Petersen and colleagues (2001) found no difference in ITC levels in the soil following turnip-rape, whether residues were incorporated or left on the surface. Most annual weeds emerge from seed buried at shallow depths and may be susceptible decaying *Brassica* residues left on or near the soil surface.

A delay in weed emergence and suppression of early season weed growth in potato, pea, and soybean has been observed following rapeseed and white mustard cover crops (Boydston and Hang, 1995; Al-Khatib, Libbey, and Boydston, 1997, Krishnan, Holshouser, and Nissen, 1998). In most cases, weed suppression obtained with *Brassica* cover crops must be supplemented with herbicides or cultivation to avoid crop losses from weed competition.

Although weed suppression, nematode control, and disease suppression have been obtained with fall-planted rapeseed, its use has not been adapted extensively by growers. Several disadvantages of rapeseed relative to white mustard include hard seed, potential out crossing to vegetable seed crops, and its ability to over winter requiring an herbicide to kill the cover crop to prevent regrowth after soil incorporation. Poor activity of herbicides has also been noted when applied during cool spring conditions to slow growing rapeseed. Soil moisture depletion by rapeseed can be significant during late winter and spring growth as well. In addition, the incorporation of green rapeseed residues in the spring carries the elevated risk of injury to the succeeding crop and the requirement to delay planting to avoid potential phytotoxicity.

Spring-Planted Smother Crops

Smother crops are cover crops grown simultaneously with the main crop that replace and suppress weeds while having minimal effect on the main crop. The concept of smother crops is not new and has been utilized in establishment of perennial forage crops. Ideally, weed populations are substituted with manageable smother crops. Vigorous early season growth of brassicas can smother and shade-out weeds before they become established (DeHaan et al., 1994). Ideally, the smother crop naturally senesces after providing early weed suppression and after the main crop becomes established.

Brassicas may be useful as spring-planted cover or smother crops, especially on later-planted crops such as sweet corn, beans (*Phaseolus vulgaris* L.), peppers (*Capsicum annuum* L.), or cucurbits. A minimum of four to six weeks of growth under optimum conditions would probably be required to develop enough biomass for biofumigation activity. Under Nebraska conditions, white mustard and brown mustard produced 500 to 1,400 $kg \cdot ha^{-1}$ dry weight when planted in March and allowed to grow 46 to 56 days before incorporating (Krishnan, Holshouser, and Nissen, 1998). Early-season weed biomass was reduced 49 percent at one location, but mustard cover crops had no significant effect on weed biomass in soybeans at a second location.

The relatively low mustard biomass produced in those studies may have accounted for the inconsistent effect on weed suppression. As a green manure crop, incorporated *Brassica* residues have the potential to suppress emergence of the main crop and should be incorporated in the soil at least two to three weeks before planting of the main crop to avoid poor stand establishment, avoid diseases associated with the flourish of microorganisms degrading crop residues, and avoid interference of crop residues with planting equipment.

Weed interference for the first two to eight weeks after corn emergence does not reduce grain yield (Hall, Swanton, and Anderson, 1992; Zimdahl, 1988). Presumably, a smother crop planted just before or soon after planting corn also would not reduce corn yield if naturally senescing after several weeks of early growth, or if it were suppressed by chemical or mechanical means.

Smother crops have not had wide adaptation by growers due to the added seed cost, the need to eliminate the smother crop by herbicides or tillage, soil moisture depletion, or excessive competition with the main crop (Worsham, 1991). Several brassicas naturally senesce after several months, and researchers have selected species with early senescence for use as cover crops (DeHaan et al., 1990). Fast-growing brassicas with early senescence have been utilized in several studies as spring-planted smother crops in corn. DeHaan and colleagues (1994) and Buhler, Kohler, and Foster (2001) suppressed weeds in corn and soybean using white mustard planted in a 25 cm band over the crop row. Delaying planting of white mustard from May 25 to June 6 reduced smother crop density more than 50 percent due to higher temperatures and less soil moisture. *Brassica hirta* smother plants were dead by 72 days after planting at the northernmost location but remained alive longer at locations with shorter day lengths. Ground cover at 28, 40, and 72 days after planting and weed suppression were greater with white mustard than with three legume smother crops tested (Buhler, Kohler, and Foster, 2001). Although weeds were suppressed, crop yields were reduced compared to weed-free plots. The reduction in yield was attributed to possible allelopathic interference by white mustard and allowing the smother crop to compete to maturity. Perhaps killing the smother crop earlier would result in weed suppression while maintaining crop yield (DeHaan et al., 1994). These authors concluded that to develop smother crop systems for annual crops, the effects of various smother crop growth rates and morphologies on crop development and weed control would need to be determined over a wide geographic range.

Further testing of spring-planted *Brassica* smother crops is needed to refine the length of competition required to obtain weed suppression and maintain crop yield. Selection of fast emerging and quicker senescing brassicas could improve the success of these systems. Alternatively, suppressing the smother crop with mowing or herbicides could result in weed suppression while reducing overall herbicide rates and maintaining crop yields.

CONCLUSIONS

The weed suppressive effects associated with growing brassicas are well documented and are beginning to be utilized in cropping systems. Generally, early-season weed suppression from *Brassica* cover crops as green manure incorporated before planting the main crop, or as smother crops growing with the main crop, is not complete enough for a grower to rely totally on the cover crop for weed control and must be combined with other weed control techniques in an integrated weed management system. Increases in the use of *Brassica* cover crops in the near future will likely be driven by benefits other than weed control. Maximizing the utilization of brassicas in cropping systems will require a better understanding of the mechanisms responsible for suppression of weed and other soilborne pests. Screening and breeding of brassicas for maximum biofumigation activity and development of agronomic practices that enhance and compliment the pesticidal activity of brassicas will broaden their use. We feel there are numerous benefits and great potential to using *Brassica* cover crops in crop rotations.

REFERENCES

Al-Khatib, K. and R. A. Boydston (1999). Weed control with *Brassica* green manure crops. In S.S. Narwal (Ed.), *Allelopathy update,* Volume 2, *Basic and applied aspects* (pp. 255-270). New Delhi: Oxford and IBH Publishing Co. Pvt. Ltd.

Al-Khatib, K., C. Libbey, and R. Boydston (1997). Weed suppression with *Brassica* green manure crops in green pea. *Weed Science* 45: 439-445.

Angus, J. F., P. A. Gardner, J. A. Kirkegaard, and J. M. Desmarchelier (1994). Biofumigation: Isothiocyanates released from *Brassica* roots inhibit growth of the take-all fungus. *Plant and Soil* 162: 107-112.

Barker, A. V. and P. C. Bhowmik (2001). Weed control with crop residues in vegetable cropping systems. *Journal of Crop Production* 4: 163-183.

Bell, D. T. and C. H. Muller (1973). Dominance of California annual grasslands by *Brassica nigra. American Midland Naturalist* 90: 277-299.

Bialy, Z., W. Oleszek, J. Lewis, and G. R. Fenwick (1990). Allelopathic potential of glucosinolates (mustard oil glycosides) and their degradation products against wheat. *Plant and Soil* 129: 277-281.

Blum, U., L. D. King, and C. Brownie (2002). Effects of wheat residues on dicotyledonous weed emergence in a simulated no-till system. *Allelopathy Journal* 9: 159-176.

Bond, W. and A. C. Grundy (2001). Non-chemical weed management in organic farming systems. *Weed Research* 41: 383-405.

Borek, V., M. J. Morra, P. D. Brown, and J. P. McCaffrey (1995). Transformation of the glucosinolate-derived allelochemicals allyl isothiocyanate and allylnitrile in soil. *Journal of Agriculture and Food Chemistry* 43: 1935-1940.

Boydston, R. A. and K. Al-Khatib (1994). Brassica green manure crops suppress weeds. *Proceedings Western Society of Weed Science* 47: 24-27.

Boydston, R. A. and A. Hang (1995). Rapeseed *(Brassica napus)* green manure crop suppresses weeds in potato *(Solanum tuberosum). Weed Technology* 9: 669-675.

Boydston, R. A. and S. F. Vaughn (2002). Alternative weed management systems control weeds in potato *(Solanum tuberosum). Weed Technology* 16: 23-28.

Brown, P. D. and M. J. Morra (1995). Glucosinolate-containing plant tissues as bioherbicides. *Journal of Agriculture and Food Chemistry* 43: 3070-3074.

Brown, P. D. and M. J. Morra (1996). Hydrolysis products of glucosinolates in *Brassica napus* tissues as inhibitors of seed germination. *Plant and Soil* 181: 307-316.

Brown, P. D. and M. J. Morra (1997). Control of soil-borne plant pests using glucosinolate-containing plants. *Advances in Agronomy* 61: 167-231.

Buhler, D. D., K. A. Kohler, and M. S. Foster (2001). Corn, soybean, and weed responses to spring-seeded smother plants. *Journal of Sustainable Agriculture* 18: 63-79.

Campbell, A. G. (1959). A germination inhibitor and root-growth retarder in Chou Moellier (*Brassica oleracea* var.). *Nature* 183: 1263-1264.

Charron, C. S. and C. E. Sams (1999). Inhibition of *Pythium ultimum* and *Rhizoctonia solani* by shredded leaves of *Brassica* species. *Journal of the American Society of Horticultural Science* 124: 462-467.

Chew, F. S. (1988). Biological effects of glucosinolates. In H. G. Cutler (Ed.), *Biologically active natural products: Potential use in agriculture* (pp. 155-181). American Chemical Society Symposium Series no. 380. Washington, DC: American Chemical Society.

Choesin, D. N. and R. E. J. Boerner (1991). Allyl isothiocyanate release and the allelopathic potential of *Brassica napus* (Brassicaceae). *American Journal of Botany* 78: 1083-1090.

Dale, J. E. (1986). Decline in phytotoxicity of benzyl isothiocyanate formulated as granules. *Weed Science* 34: 325-327.

DeHaan, R. L., D. L. Wyse, N. J. Ehlke, B. D. Maxwell, and D. H. Putnam (1994). Simulation of spring-seeded smother plants for weed control in corn *(Zea mays). Weed Science* 42: 35-43.

DeHaan, R. L., D. L. Wyse, N. J. Ehlke, and D. H. Putnam (1990). Development of a *Brassica* sp. smother plant for weed control in corn. *Proceedings of the North Central Weed Science Society* 45: 27.

Eberlein, C. V., M. J. Morra, M. J. Guttieri, P. D. Brown, and J. Brown (1998). Glucosinolate production by five field-grown *Brassica napus* cultivars used as green manures. *Weed Technology* 12: 712-718.

Elliot, M. C. and B. B. Stowe (1971). Distribution and variation of indole glucosinolates in woad (*Isatis tinctoria* L.). *Plant Physiology* 48: 498-503.

Evenari, M. (1949). Germination inhibitors. *Botanical Review* 40: 153-194.

Fenwick, G. R., R. K. Heaney and W. J. Mullin (1983). Glucosinolates and their breakdown products in food and food plants. *Critical Reviews in Food Science and Nutrition* 18: 123-201.

Fujii, Y. (2001). Screening and future exploitation of allelopathic plants as alternative herbicides with special reference to hairy vetch. *Journal of Crop Production* 4: 257-275.

Gardiner, J. B., M. J. Morra, C. V. Eberlein, P. D. Brown, and V. Borek (1999). Allelochemicals released in soil following incorporation of rapeseed *(Brassica napus)* green manures. *Journal of Agricultural and Food Chemistry* 47: 3837-3842.

Hall, M. R., C. J. Swanton, and G. W. Anderson (1992). The critical period of weed control in grain corn *(Zea mays). Weed Science* 40: 441-447.

Hartwig, N. L. and H. U. Ammon (2002). Cover crops and living mulches. *Weed Science* 50: 688-699.

Horricks J. S. (1969). Influence of rape residue on cereal production. *Canadian Journal of Plant Science* 49: 632-634.

Jessop, R. S. and L. W. Stewart (1983). Effects of crop residues, soil type and temperature on emergence and early growth of wheat. *Plant and Soil* 74: 101-109.

Jimenez-Orsornio, J. J. and S. R. Gliessman (1987). Allelopathic interference in a wild mustard (*Brassica campestris* L.) and broccoli (*Brassica oleracea* L. var. *italica*) intercrop agroecosystem. In G. R. Waller (Ed.), *Allelochemicals: Role in agriculture and forestry* (pp. 262-288). Washington, DC: American Chemical Society.

Ju, H. Y., B. B. Bible, and C. Chong (1983). Influence of ionic thiocyanate on growth of cabbage, bean and tobacco. *Journal of Chemical Ecology* 9: 1255-1262.

Kirkegaard, J. A., P. A. Gardner, J. F. Angus, and E. Koetz (1994). Effect of *Brassica* break crops on the growth and yield of wheat. *Australian Journal of Agricultural Research* 45: 529-545.

Kirkegaard, J.A. and M. Sarwar (1998a). Biofumigation potential of brassicas. *Plant and Soil* 201: 71-89.

Kirkegaard, J. A. and M. Sarwar (1998b). Biofumigation potential of brassicas II. *Plant and Soil* 201: 91-101.

Kirkegaard, J. A. and M. Sarwar (1999). Glucosinolate profiles of Australian canola (*Brassica napus annua* L.) and Indian mustard (*Brassica juncea* L.) cultivars: Implications for biofumigation. *Australian Journal Agricultural Research* 50: 315-324.

Krishnan, G., D. L. Holshouser, and S. J. Nissen (1998). Weed control in soybean *(Glycine max)* with green manure crops. *Weed Technology* 12: 97-102.

Kutacek, M. (1964) Glucobrassican a potential inhibitor of unusual type affecting the germination and growth of plants: Mechanism of its action. *Biologia Plantarum* 6: 88-98.

Laine, P., A. Ourry, J. Macduff, J. Boucaud, and J. Salette (1993). Kinetic parameters of nitrate uptake by different catch crop species: Effects of low temperatures or previous nitrate starvation. *Physiologia Plantarum* 88: 85-92.

Leblova, S., V. Jiracek, and J. Kostir (1960). The influence of organic isothiocyanates on the growth and metabolism of plants: 1. The effect of allyl isothiocyanate on the germination and growth of rape, wheat and peas. *Rostlinna Vyroba* 6: 651-662.

Leblova, S. and J. Kostir (1962). Action of isothiocyanates on germinating plants. *Experientia* 18: 554-555.

Lichtenstein, E. P., D. G. Morgan, and C. H. Mueller (1964). Naturally occurring insecticides in cruciferous crops. *Journal of Agriculture and Food Chemistry* 12: 158-161.

Liebman, M. and A. S. Davis (2000). Integration of soil, crop and weed management in low-external-input farming systems. *Weed Research* 40: 27-47.

Martin, P. and B. Rademacher (1960). Experimentelle untersuchung zur Frage der Nachwirkung von Rapswurzelruchstanden. *Zeitschrift für Ackerbau und Pflanzenanbau* 111: 105-115.

Mason-Sedon, W., R. S. Jessop, and J. V. Lovett (1986). Differential phytotoxicity among species and cultivars of the genus *Brassica* to wheat: I. Laboratory and field screening of species. *Plant and Soil* 93: 3-16.

McKeown, W. W. and J. W. Potter (2001). Yield of 'Superior' potatoes *(Solanum tuberosum)* and dynamics of root-lesion nematode *(Pratylenchus penetrans)* populations following "nematode suppressive" cover crops and fumigation. *Phytoprotection* 82: 13-23.

McLenaghen, R. D., K. C. Cameron, N. H. Lampkin, M. L. Daly, and B. Deo (1996). Nitrate leaching from ploughed pasture and the effectiveness of winter catch crops in reducing leaching losses. *New Zealand Journal of Agricultural Research* 39: 413-420.

Mojtahedi, H., G. S. Santo, A. N. Hang, and J. H. Wilson (1991). Suppression of root-knot nematode populations with selected rapeseed cultivars as green manure. *Journal of Nematology* 23: 170-174.

Mojtahedi, H., G. S. Santo, J. H. Wilson, and A. N. Hang (1993). Managing *Meloidogyne chitwoodi* on potato with rapeseed as a green manure. *Plant Disease* 77: 42-46.

Muehlchen, A. M. (1990). Evaluation of Crucifer green manures for controlling *Aphanomyces* root rot of peas. *Plant Disease* 74: 651-654.

Muller, C. H. (1969). Allelopathy as a factor in ecological process. *Vegetatio* 18: 348-357.

Nagabhushana, G. G. , A. D. Worsham, and J. P. Yenish (2001). Allelopathic cover crops to reduce herbicide use in sustainable agricultural systems. *Allelopathy Journal* 8: 133-146.

Oleszek, W. (1987). Allelopathic effects of volatiles from some cruciferae species on lettuce, barnyardgrass and wheat growth. *Plant and Soil* 102: 271-273.

Olivier, C., S. F. Vaughn, E. S. G. Mizubuti, and R. Loria (1999). Variation in allyl isothiocyanate production within *Brassica* species and correlation with fungicidal activity. *Journal Chemical Ecology* 25: 2687-2701.

Papavizas, G. C. (1966). Effect of *Aphanomyces* root rot of peas by cruciferous soil amendments. *Phytopathology* 56: 1071-1075.

Papavizas, G. C. and J. A. Lewis (1971). Effect of amendments and fungicides on *Aphanomyces* root rot of peas. *Phytopathology* 61: 215-220.

Petersen, J., R. Belz, F. Walker, and K. Hurle (2001). Weed suppression by release of isothiocyanates from turnip-rape mulch. *Agronomy Journal* 93: 37-43.

Purvis, C. E., R.S. Jessop, and J. V. Lovett (1985). Selective regulation of germination and growth of annual weeds by crop residues. *Weed Research* 25: 415-421.

Rogasik, J., M. Smukalski, and S. Obenauf (1992). Cover crops on sandland in Germany: Husbandry and fate of nitrogen. *Aspects of Applied Biology* 30: 309-316.

Sainju, U. M. and B. P. Singh (1997). Winter cover crops for sustainable agricultural systems: Influence on soil properties, water quality, and crop yields. *HortScience* 32: 21-28.

Santos, J. R. A. and D. I. Leskovar (1997). Interference from broccoli residue on *Brassica* germination and seedling growth. *Journal of the American Society for Horticultural Science* 122: 715-720.

Sarrantonio, M. (1992). Opportunities and challenges for the inclusion of soil-improving crops in vegetable production systems. *HortScience* 27: 754-758.

Stiehl, B. and B. B. Bible (1985). Effect of crucifer toxin on seedling growth and germination of crop species. *HortScience* 20: 185.

Stiehl, B. and B. B. Bible (1989). Reaction of crop species to thiocyanate ion toxicity. *HortScience* 24: 99-101.

Stivers-Young, L. (1998). Growth, nitrogen accumulation, and weed suppression by fall cover crops following early harvest of vegetables. *HortScience* 33: 60-63.

Teasdale, J. R. (1993). Interaction of light, soil moisture, and temperature with weed suppression by hairy vetch residues. *Weed Science* 41: 46-51.

Teasdale, J. R. and R. B. Taylorson (1986). Weed seed response to methyl isothiocyanate and metham. *Weed Science* 34: 520-524.

Thorup-Kristensen, K. (1993). Root development of nitrogen catch crops and of a succeeding crop of broccoli. *Plant and Soil* 43: 58-64.

Tollsten, L. and G. Bergstrom (1988). Headspace volatiles of whole plants and macerated plant parts of *Brassica* and *Sinapsis*. *Phytochemistry* 27: 4013-4018.

Vaughn, J. G. and J. S. Hemingway (1959). The utilization of mustards. *Economic Botany* 13: 196-204.

Vaughn, S. (1993). Inhibition of *Helminthosporium solani* strains by natural isothiocyanates. *American Potato Journal* 70: 852-853.

Vaughn, S. F. (1999). Glucosinolates as natural pesticides. In H. G. Cutler and S. J. Cutler (Eds.), *Biologically active natural products* (pp. 81-92). Boca Raton, FL: CRC Press.

Vaughn, S. F. and R. A. Boydston (1997). Volatile allelochemicals released by crucifer green manures. *Journal of Chemical Ecology* 23: 2107-2116.

Vera, C. L., D. I. McGregor, and R. K. Downey (1987). Detrimental effects of volunteer *Brassica* on production of certain cereal and oilseed crops. *Canadian Journal of Plant Science* 67: 983-995.

Waddington, J. (1978). Growth of barley, bromegrass and alfalfa in the greenhouse in soil containing rapeseed and wheat residues. *Canadian Journal of Plant Science* 58: 241-248.

White, J. G. (1978). Weed control by soil partial sterilant chemicals: A review. In *Proceedings of 1978 British Crop Protection Conference-Weeds,* Volume 3 (pp. 987-995). Farnham, UK: British Crop Protection Council.

Williams, L. III, M. J. Morra, P. D. Brown, and J. P. McCaffrey (1993). Toxicity of allyl isothiocyanate-amended soil to *Limonius californicas* (Mann.) (Coleoptera: Elateridae) wireworms. *Journal of Chemical Ecology* 19: 1033-1046.

Williams, M. M. II, D. A. Mortensen, and J. W. Doran (1998). Assessment of weed and crop fitness in cover crop residues for integrated weed management. *Weed Science* 46: 595-603.

Wolf, R. B., G. F. Spencer, and W. F. Kwolek (1984). Inhibition of velvetleaf *(Abutilon theophrasti)* germination and growth of benzyl isothiocyanate, a natural toxicant. *Weed Science* 32: 612-615.

Worsham, A. D. (1991). Role of cover crops in weed management and water quality. In W. L. Hargrove (Ed.), *Cover crops for clean water* (pp. 82-101). Ankeny, IA: Soil and Water Conservation Society.

Zimdahl, R. L. (1988). The concept and application of critical weed-free period. In M. A. Altieri and M. Liebman (Eds.), *Weed management in agroecosystems: Ecological approaches* (pp. 145-155). Boca Raton, FL: CRC Press.

Chapter 5

Grass-Legume Mixed Cover Crops for Weed Management

Nilda R. Burgos
Ronald E. Talbert
Yong In Kuk

INTRODUCTION

Weed management is a vital component of successful crop production. Crop productivity in modern times relies heavily upon the efficacy of pesticides to manage insects, diseases, and weeds. In U.S. agriculture, herbicides generally comprise the major constituent of pesticide inputs in major field crops, including cotton (*Gossypium hirsutum* L.), corn (*Zea mays* L.), and soybean [*Glycine max* (L.) Merr.] (see Table 5.1). Use of insecticides and fungicides in field crops varies widely from 1 and 0 percent, respectively, in soybean to 93 and 85 percent, respectively, in fall potatoes (*Solanum tuberosum* L.). Insecticides are used more heavily on vegetable crops (76 percent) compared to herbicides (59 percent) (USDA, 2003). Fungicides are also used more intensively on vegetable crops than on field crops. Sustainable agriculture requires good stewardship of pesticide use and development of ways to reduce pesticide input. Weed management with reduced herbicide inputs is possible by integrating complementary technologies. These include reduced tillage and use of allelopathic cover crops and mulches, among other strategies. Many allelopathic cover crops, with high biomass production and good manageability, are grasses such as annual ryegrass (*Lolium multiflorum* Lam.), barley (*Hordeum vulgare* L.), oat (*Avena sativa* L.), rye (*Secale cereale* L.), and wheat (*Triticum aestivum* L.) (Nelson, Kahn, and Roberts, 1991; Fujii, 2001). Rye and wheat are more consistent in terms of cover crop biomass production (Williams, Morten-

---

We acknowledge the support provided by the Arkansas Agricultural Experiment Station, Alternative Pest Control Center, and Gerber's Food, Inc., for various research conducted on this subject matter.

TABLE 5.1. Pesticide use in major field crop production in the United States, 2001.[a]

Crop	Total area (1,000 ha)	Herbicide use %	Herbicide use 1,000 kg	Insecticide use %	Insecticide use 1,000 kg	Fungicide %	Fungicide 1,000 kg	Others[a] %	Others[a] 1,000 kg
Corn	28,642	98	71,472	29	4,093	—	—	—	—
Cotton	5,134	90	9,590	68	10,823	4	96	55	610
Fall potatoes	364	82	618	93	846	95	2,362	61	2,997
Soybean	21,053	96	22,938	1	110	—	—	—	—

Source: Agricultural Statistics Board, National Agricultural Statistics Service, U.S. Department of Agriculture, 2003.

Note: Data were obtained from major crop-producing states: 19 for corn, 7 for cotton, 7 for potato, and 8 for soybean. Where usage was negligible or insufficient information was obtained, data were excluded.

[a]Includes growth regulators and defoliants.

sen, and Doran, 1998) and most reliable for early-season weed suppression (Weston, 1990). Crimson clover (*Trifolium incarnatum* L.) and hairy vetch (*Vicia villosa* Roth) are among the best legume cover crops for ground cover, weed suppression, and manageability. However, crimson clover and hairy vetch produce less biomass and are more difficult to kill with non-selective, burndown herbicides than the grassy species (Nelson, Kahn, and Roberts, 1991). Biomass production is important in choosing a cover crop because, up to a point, biomass is directly related to the degree of soil protection, weed suppression, and nutrient recycling or nutrient contribution to the following crop. Other vetch species, such as Chinese milk vetch (*Astragalus sinicus* L.), have been used as green manure crop in rice (*Oryza sativa* L.) paddies in China, Japan, and Korea (Yasue and Tsutiya, 1982; Seong and Park, 1991; Yasue, 1991; Lee et al., 1999). Chinese milk vetch matures earlier than hairy vetch and narrow leaf vetch [*Vicia angustifolia* (L.) Richard]. This makes Chinese milk vetch a better fit in an alternative rice-based cropping system. In Korea, cultivated areas planted to Chinese milk vetch reached about 37,000 ha, and this is expected to increase in the future (Jeong et al., 1996; Hong et al., 1997). This chapter discusses the utility of temperate small-grain cereal and legume mixtures for weed management and their effect on yield of the subsequent crop.

BENEFITS OF MIXED COVER CROPS

Grass and legume cover crops offer different, but complementary, benefits to the succeeding crop. Grass cover crops produce a higher amount of biomass (Wagger, 1989b; Nelson, Kahn, and Roberts, 1991; Yenish, Worsham, and York, 1996), conserve more residual N during fall and winter (Evanylo, 1991; Meisinger et al., 1991; Shipley, Meisinger, and Decker, 1992), decompose more slowly (Wagger, 1989b; Mohler and Teasdale, 1993;), suppress weeds better (Nelson, Kahn, and Roberts, 1991; Wallace and Bellinder, 1992; Burgos and Talbert, 1996), and are easier to kill with nonselective herbicides compared to legume cover crops. A summary of biomass production of rye, wheat, clover, and vetch planted alone or in a mixture is presented in Table 5.2. This excludes data from studies not involving comparisons of cover crop mixtures with monocultures but shows the general trend of greater biomass production from small-grain cereal than legume cover crops. Research at Fayetteville and Kibler, Arkansas, showed that rye could produce 12 $mg{\cdot}ha^{-1}$ shoot biomass when desiccated at booting to anthesis in late April to early May (Burgos, 1994). Hairy vetch, on the other hand, produced an average of 9 $mg{\cdot}ha^{-1}$ biomass. In Blacksburg, Virginia, similar experiments showed rye produced 7.7 $mg{\cdot}ha^{-1}$ shoot biomass on average when desiccated between the second and third week of May (Vaughan and Evanylo, 1998). In this location, hairy vetch produced almost four times less biomass than did rye. Understandably, cover crop biomass production depends upon soil, seeding rate, stand establishment, environmental conditions during the growing period, as well as timing of cover crop kill.

Cover crop residues or cover crop mulches significantly impact soil moisture in a number of ways, especially when there is no crop canopy closure (Gregory, Simmonds, and Pilbean, 2000). Soil-water evaporation is reduced in mulched soil (Abdul-Baki and Teasdale, 1993). Mulches reduce the evaporation of water from the soil surface by reducing the net radiation absorbed by the soil surface, thereby reducing soil temperature (Gregory, Simmonds, and Pilbean, 2000). Bananas (*Musa sapientum* L.) with organic mulch, such as chopped maize stalks and *Paspalum* spp. grass, took up more water and the soil water recharged more quickly as a result of increased soil porosity than bananas without mulch (McIntyre et al., 2000). In the South Plains of Texas, where cotton (*Gossypium hirsutum* L.) is grown in a water-limited environment, conservation of rain water and reduction of wind erosion is attained by seeding winter wheat into cotton fields (Baumhardt and Lascano, 1996). The wheat is then killed by herbicide in the spring. Wheat mulch increases cumulative infiltration of raindrops into

TABLE 5.2. Seeding rates and biomass production of fall-planted cover crops when used alone or in a mixture at various locations, average of two to three years or two sites at each location.

Location	Species	Seeding rate ($kg \cdot ha^{-1}$)	Time of sampling	Average biomass ($mg \cdot ha^{-1}$)	N content	C/N	Reference
Arkansas	*Secale cereale* L.	123	April 17 to May 4	12.0	nd[a]	nd	Burgos and Talbert, 1996; Burgos, 1994
	Vicia villosa Roth	22		9.2	nd	nd	
	S. cereale + *V. villosa*	39 + 15		14.0	nd	nd	
Maryland	*Secale cereale* L.	94	May 3 to May 15	6.7	74	51	Clark, Decker, and Meisinger, 1994
	Vicia villosa Roth	28		5.0	159	13	
	S. cereale + *V. villosa*	47 + 14		6.9	151	22	
	S. cereale + *V. villosa*	94 + 28		8.7	182	22	
Maryland	*Secale cereale* L.	45	late May	5.7	55	54	Teasdale and Abdul-Baki, 1998
	Trifolium incarnatum L.	22		6.5	149	20	
	Vicia villosa Roth	45		5.3	171	14	
	S. cereale + *T. incarnatum*	45 + 22		8.8	150	27	
	S. cereale + *V. villosa*	45 + 45		7.3	176	20	

Vancouver, British Columbia	*Secale cereale* L.	100	April 27 to May 7	6.0	73	35	Odhiambo and Bomke, 2001
	Triticum aestivum L.	100		8.8	88	39	
	Trifolium incarnatum L.	12		3.6	84	21	
	S. cereale + *T. incarnatum*	80 + 12		6.9	90	32	
	T. aestivum + *T. incarnatum*	80 + 12		8.4	96	36	
Virginia	*Secale cereale* L.	101	April 29 to May 22	6.9	94	36	Vaughan and Evanylo, 1998
	Vicia villosa Roth	28		1.6	61	12	
	S. cereale + *V. villosa*	67 + 19		6.7	105	28	

[a]nd = not determined.

the soil, and in this environment, a minimum of 0.7 $mg \cdot ha^{-1}$ wheat residue is needed to intercept raindrop impact and increase infiltration (Baumhardt and Lascano, 1996). Mulch decreases the impact of raindrops, thereby reducing soil surface disturbance, reduces runoff velocity, and enhances infiltration and percolation. In Nebraska, volumetric soil water content 15 d after planting soybean was about 25 percent without cover crop and about 36 percent with rye residue at 0 to 7.6 cm depth of silty clay loam soil with 2.6 to 2.8 percent organic matter content (Williams, Mortensen, and Doran, 1998). Soil moisture in soil with rye residue remained higher than that of bare soil 22 d after planting, but soil with vetch residue had the same moisture content as bare soil (Williams, Mortensen, and Duran, 1998). When a mixture of grass and legume cover crops are used as surface mulch on tomato (*Lycopersicon esculentum* Mill.) beds, soil moisture is higher relative to the conventional system without mulch (Creamer et al., 1996).

Several producers of winter wheat in the United States would double crop wheat with soybean; however, soybean is planted in early summer which is also the onset of the dry season. To optimize utilization of soil moisture in this system, wheat stubble is cut to about 20 cm and soybean is drilled along the rows of standing wheat stubble. Standing wheat stubble and the cut wheat straw protects the soil from wind and water erosion and reduces evaporation losses of soil water. This also prevents the soybean crop from being sandblasted during windy days.

In semiarid regions such as the central Great Plains of the United States, soil moisture conservation is critical. In these areas, winter wheat is commonly grown followed by a fallow period where stubble mulching with tillage is practiced for weed control (Schlegel and Havlin, 1997). The use of cover crops for weed suppression is highly beneficial in this case because tillage during the fallow months depletes the soil organic matter with time. Dryland forage legumes were tested as cover crops for the fallow months in a semiarid region in Kansas. Hairy vetch and yellow sweetclover [*Melilotus officinalis* (L.) Lam.] produced the most biomass among the 11 leguminous species tested in this environment. However, growing the cover crop during the fallow months depleted the soil moisture enough to cause up to 83 percent yield reduction of the succeeding wheat crop (Schlegel and Havlin, 1997). This illustrates that the use of cover crop needs careful evaluation for site-specific adaptability.

A small-grain cover crop such as rye takes up residual N in the fall and spring, thereby minimizing leaching losses of available N (Meisinger et al., 1991). Rye is more efficient at recycling residual soil N because it is cold tolerant and therefore assimilates more biomass and nutrients than hairy vetch (Shipley, Meisinger, and Decker, 1992) or crimson clover (Ranells and Wagger, 1997) in the fall and early spring. Cereal rye takes up and recycles 21 to 94 kg·ha^{-1} N in its shoot tissue (Ebelhar, Frye, and Blevins, 1984; Wagger, 1989b; Clark, Decker, and Meisinger, 1994; Teasdale and Abdul-Baki, 1998) and can conserve more N than other small grain crops (Brinsfield and Staver, 1989; Blevins, Herbek, and Frye, 1990; Evanylo, 1991). Winter small-grain cover crops such as barley triticale (*Triticum* × *secale*) and wheat can accumulate similar amounts of soil N as rye when grown to maturity (Coale et al., 2001). Small-grain winter cover crops can therefore reduce leaching of NO_3-N to groundwater. Unlike rye, wheat and crimson clover do not increase the retention of residual N in the root zone (Dufault et al., 2000).

Nitrogen taken up by the cover crop becomes available when the cover crop residue decomposes. However, N from rye (or other grass cover crops) is not easily available to the succeeding crop because high C/N ratio in grass residue (Clark, Decker, and Meisinger, 1994; Teasdale and Abdul-Baki, 1998; Vaughan and Evanylo, 1998; Odhiambo and Bomke; 2001) favors initial immobilization of N from the residue and from added fertilizer (Hargrove, 1986). Rye + vetch cover crop mixture may be superior to pure rye cover crops because of intermediate net N-mineralization, which results in reduced N leaching and denitrification losses, and better synchrony of N availability and crop N demands (Rosecrance et al., 2000).

Besides high biomass production, grass cover crops release toxic compounds, which contribute greatly to the inhibition of weed seed germination or seedling growth. Thorough reviews of allelopathic research and its role on weed management are presented by Seigler (1996), Weston (1996), and Einhellig (1996). For a brief review of the allelopathic characteristics of grass cover crops, we use rye and wheat as classic examples. The weed-suppressive activity of rye is attributed to allelochemicals including 2,4-dihydroxy-1,4-(2*H*)-benzoxazine-3-one (DIBOA), 2(3*H*)-benzoxazolinone (BOA), beta-phenyllactic acid (β-PLA), and beta-hydroxybutyric acid (β-HBA) (Shilling, Liebl, and Worsham, 1985; Barnes et al., 1987). BOA is a breakdown product of DIBOA. In the soil, BOA is transformed by microbial action into 2,2'-oxo-1,1'-azobenzene (AZOB), a compound that is reported to be more inhibitory to seedling growth than BOA or DIBOA (Nair, Whitenack, and Putnam, 1990). That rye residue is able to prevent seed germination and suppress weed growth by exudation of chemical inhibitors is proven by bioassays conducted on various species, using aqueous extracts and the purified compounds mentioned previously. DIBOA is most effective against monocots and BOA against dicots (Barnes, Putnam, and Burke, 1986). In seedling bioassays, BOA reduced root length of six vegetable crops and nine weed species. BOA also reduced the germination of tomato, Palmer amaranth (*Amaranthus palmeri* S. Wats.), large crabgrass [*Digitaria sanguinalis* (L.) Scop.], and prickly sida (*Sida spinosa* L.) (Burgos and

Talbert, 2000). DIBOA is more phytotoxic than BOA on species tested, including cucumber (*Cucumis sativus* L.), goosegrass [*Eleusine indica* (L.) Gaertn.], lettuce (*Lactuca sativa* L.), Palmer amaranth, sweet corn (*Zea mays* var. *rugosa* Bonaf), and tomato (Burgos and Talbert, 2000). In general, BOA and DIBOA are more phytotoxic to small-seeded weed species and vegetable crops than large-seeded ones (Burgos and Talbert, 2000). β-PLA inhibits growth of common lambsquarters (*Chenopodium album* L.) and redroot pigweed (*Amaranthus retroflexus* L.) in seedling bioassays (Shilling, Liebl, and Worsham, 1985). Of the four allelopathic compounds identified from rye, BOA and DIBOA are generally more phytotoxic than β-PLA and β-HBA in terms of root and shoot inhibition of large crabgrass, redroot pigweed, barnyardgrass [*Echinochloa crus-galli* (L.) Beauv.], tomato, proso millet (*Panicum miliaceum* L.), and lettuce (Barnes and Putnam, 1987).

For years, rye has been used successfully as a cover crop in conservation tillage crop production and has proven beneficial for weed management in this system. Mowed rye residue provides better weed control than soybean stubble in no-till corn (Johnson, DeFelice, and Helsel, 1993). In no-till soybean, desiccated rye residue provides 90 percent or better control of giant foxtail (*Setaria faberi* Herrm.), smooth pigweed (*Amaranthus hybridus* L.), and velvetleaf (*Abutilon theophrasti* Medik.) (Liebl et al., 1992).

Wheat, on the other hand, exudes phenolic acids and simple acids, which contribute to its allelopathic activity. Ferulic acid is one of the identified phytotoxic compounds (Shilling, Liebl, and Worsham, 1985). Ferulic acid inhibits germination and root growth of common ragweed (*Ambrosia artemisiifolia* L.), crabgrass (*Digitaria* spp.), pitted morning glory (*Ipomoea lacunosa* L.), prickly sida (*Sida spinosa* L.), and tall morning glory [*Ipomoea purpurea* (L.) Roth] (Liebl and Worsham, 1982, 1983; Shilling, Liebl, and Worsham, 1985). Wheat is often double-cropped with soybean where soybean is planted no-till into wheat stubble to conserve soil moisture and prevent weed emergence.

The primary advantage of legume over grass cover crops is the potential of legumes to fix atmospheric N and legumes' readily mineralizable N. Over the years, research on the benefits of legume cover crops has focused on its nitrogen-fixation capability. Legume cover crops add more N to soil than grass cover crops and, therefore, would be more valuable in production systems on marginal land with limited inputs or for reducing the use of inorganic nitrogen fertilizers. Hairy vetch residue may contain 19 to 351 $kg \cdot ha^{-1}$ N depending on amount of biomass produced (Ebelhar, Frye, and Blevins, 1984; Wagger, 1989b; Holderbaum et al., 1990; Clark, Decker, and Meisinger, 1994; Teasdale and Abdul-Baki, 1998; Vaughan and Evanylo, 1998). Crimson clover, another legume cover crop, generally does not contribute as much N (56 to 84 kg·ha^{-1}) as hairy vetch (Ebelhar, Frye, and Blevins, 1984; Wagger, 1989b; Odhiambo and Bomke, 2001). Crimson clover may (Teasdale and Abdul-Baki, 1998) or may not produce as much biomass as hairy vetch (Ebelhar, Frye and Blevins, 1984) and has a lower percentage of N in shoot tissue than hairy vetch (Ebelhar, Frye, and Blevins, 1984; Teasdale and Abdul-Baki, 1998). Chinese milk vetch is widely used as forage and green manure crop in rice fields in China, Japan, and Korea because of its high nitrogen content (Yasue and Tsutiya, 1982; Seong and Park, 1991; Yasue, 1991; Wu and Arima, 1992; Hong et al., 1997; Lee et al., 1999). Besides contributing N to soil, milk vetch also provides some degree of weed control (Hong et al., 1997; Jeong et al., 1996).

Mineralization of N from leguminous cover crops occurs at a faster rate than from grass cover crop (Wagger, 1989b). This is due to a lower C/N ratio in legumes compared to grasses (see Table 5.2). When desiccated around the first week of May, C/N ratios for hairy vetch, crimson clover, and rye were 12:1, 16:1, and 32:1, respectively (Wagger, 1989b). Theoretically, C/N ratios > 25:1 favor immobilization of mineralizable N (Allison, 1966). Accordingly, release of N from cover crop residue is fastest with hairy vetch and slowest with rye (Wagger, 1989b). Because the higher C/N ratio of rye residue immobilizes fertilizer N (Hargrove, 1986), the addition of high amounts of fertilizer N at planting to mitigate N deficiency early in the season is necessary when grass cover crop is used.

The high mineralization rate of legume residue is a double-edged sword. Faster decomposition of legume residue (Corak, Frye, and Smith, 1991; Mohler and Teasdale, 1993) presents the problem of short-duration residue cover on the soil surface, which translates to less soil protection and less weed suppression than grass residue (Nelson, Kahn, and Roberts, 1991; Burgos and Talbert, 1996). Crop response to N from legume residue varies by season because mineralized N may be lost through leaching or denitrification if heavy rains occur early when crop demand for N is low. Nitrogen availability, as measured by corn N content at maturity (after black layer formation), was greatest following vetch, least after rye, and intermediate following the rye + vetch mixture (Clark et al., 1997a). Cover crop residue decomposition and N release in legume and rye mixtures tend to be intermediate between the corresponding monocultures (Ranells and Wagger, 1996).

Combining an allelopathic grass and legume cover crop allows producers to harness the respective strengths of both species, while minimizing their negative traits. A grass-legume mixed cover crop should produce as much, if not more, biomass as the grass cover crop alone; readily available N from the legume should reduce N immobilization by the grass residue;

and the grass residue should increase weed suppression and prolong soil protection afforded by the legume cover crop alone.

MANAGEMENT OF A MIXED COVER CROP

Seeding Rate

For two species to survive in a mixture, each component has to be seeded in a proportion that allows the species to attain optimum growth. Various seeding rates have been used in different locations in North America ranging from 45 to 123 kg·ha^{-1} for rye and 19 to 28 kg·ha^{-1} for hairy vetch (see Table 5.2). Biomass production does not necessarily increase in the same magnitude as seeding rate, which means that environmental factors modify the effect of seeding rate on biomass production. Striking a good balance that works for a particular production situation is important because of economic (seed cost), biological (optimum weed suppression and N fixation), and physical (soil moisture conservation) considerations. In Maryland, the seeding rate for hairy vetch does not have as much impact on total cover crop biomass as the seeding rate of rye (Clark, Decker, and Meisinger, 1994). This is because vetch has a high capacity to compensate for low stand by producing more lateral branches and vining growth. From visual observation vetch seeded at a high density does not produce much shoot biomass per plant because overcrowding occurs quickly and shaded leaves senesce. Vetch, therefore, is expected to have lower optimum seeding rate than rye. Higher total biomass is produced in mixtures planted with a higher rate (94 kg·ha^{-1}) than with a lower rate (47 kg·ha^{-1}) of rye, regardless of seeding rate of vetch (Clark, Decker, and Meisinger, 1994). In a mixture of rye and vetch, rye is the main contributor to biomass. All mixtures of rye and vetch produce more biomass than rye alone (94 kg·ha^{-1}), except for the low seeding rate of both rye and vetch (14 + 47 kg·ha^{-1}). Surprisingly, N content from mixed cover crop, at any seeding rate, was greater than or equal to that of vetch alone (28 kg·ha^{-1}), and two to three times more than from rye alone seeded at 94 kg·ha^{-1} (Clark, Decker, and Meisinger, 1994). If vetch seed is more expensive than rye, then the most economical treatment that would give producers the most agronomic benefit would be a low rate of vetch + a high rate of rye or a medium rate of vetch + a low rate of rye. Optimum seeding rate may differ in latitudes where winters are warmer and crops start growing early in the spring. To illustrate this point, low seeding rates of rye (39 kg·ha^{-1}) and vetch (15 kg·ha^{-1}) in Arkansas produced twice as much biomass as the equivalent seeding rates of rye and vetch (47 + 14 kg·ha^{-1}) in Maryland (Table 5.2).

Cover Crop Control

Cover crops must not compete with the following crop. Thus cover crops are either mowed, plowed under, or desiccated with herbicide(s) before planting the following crop. Mowing does not completely kill hairy vetch (Wallace and Bellinder, 1992); however, vetch can be killed by mechanical undercutting during the mid- to late-flowering stage (Creamer et al., 1995). Red clover (*Trifolium pratense* L.), white sweetclover (*Melilotus alba* Medik.), yellow sweetclover, and ryegrass are difficult to manage or could not be killed by mechanical means (Creamer, Bennett, and Stinner, 1997). If the main objective of cover cropping is to provide soil protection and suppress weeds, desiccation with herbicide and leaving the residue intact on the soil surface is a logical choice. It has been shown that subsequent reduction of weed density is better when rye is desiccated with herbicide and left on the soil surface rather than mowed or incorporated into the soil (Bordelon and Weller, 1997). Glyphosate and paraquat are the most commonly used nonselective herbicides for killing existing vegetation (Wagger, 1989a; Ranells and Wagger, 1992; Clark, Decker, and Meisinger, 1994; Bauer and Busscher, 1996; Burgos and Talbert, 1996; Bordelon and Weller, 1997; Vaughan and Evanylo, 1998). Glyphosate and paraquat kill grass cover crops effectively, but killing legume cover crops with these herbicides can be difficult. This is the case when the legume cover crop is fully grown and has thick foliage.

Glyphosate is translocated within the plant so total coverage of foliage is not required for complete kill; however, it takes longer for glyphosate to kill legume than grass cover crops. In a mixed cover crop system, this would mean a longer waiting period to plant the succeeding crop and reduced time to take advantage of allelopathic weed suppression by the grass cover. On the other hand, a problem of legume regrowth usually occurs when using paraquat as desiccant because this herbicide does not translocate in the plant. For instance, with a thick foliage of hairy vetch, paraquat will desiccate the top foliage but will not kill plant parts that are not directly exposed to the herbicide. The surviving cover crop will then have to be controlled by other means during the growing season. This could be done either by cultivation of middle rows, manual weeding, or application of a broadleaf herbicide that is labeled for the crop grown.

Managing legume cover crops does have some potential difficulties. Research at Cornell University in New York showed that treatment with glyphosate (0.82 $kg{\cdot}ha^{-1}$) + 2,4-D (0.84 $kg{\cdot}ha^{-1}$) did not completely kill red clover (Wallace and Bellinder, 1992). A similar problem was encountered with ladino clover (*Trifolium repens* L.) in Lexington, Kentucky (Galloway and Weston, 1996). Sequential application of glyphosate and 2,4-D was needed to kill legume cover crops in North Carolina (Yenish, Worsham, and York, 1996). A high glyphosate rate (2.2 kg·ha^{-1}) applied two weeks before planting corn, followed by 2,4-D (0.67 kg·ha^{-1}) at corn planting, failed to completely kill crimson clover, subterranean clover (*Trifolium subterraneum* L.), or hairy vetch (Yenish, Worsham, and York, 1996). If a high rate of herbicides, e.g., 2.2 kg·ha^{-1} glyphosate, or more than one herbicide application is needed to kill a cover crop, then that nullifies the goal of reducing herbicide use with cover crops. Other means of killing the cover crop, or alternative herbicide treatments favorable for reduced rates and single applications, need to be explored. Reduced rates of photosynthetic inhibitors (i.e., metribuzin) or other hormone-type herbicides (i.e., triclopyr), which show high activity on leguminous species maybe tested in combination with nonselective herbicides for burndown of grass-legume mixed cover crops. Timing of planting the succeeding crop after burndown needs to be studied because some crops may be sensitive to herbicide residue on the cover crop mulch or soil.

Legume cover crops are susceptible to some photosynthetic inhibitor herbicides such as metribuzin and atrazine, which can be used in a mixture with paraquat to attain complete kill of cover crops (Wagger, 1989a; Burgos et al., 1999). Mixing nonselective herbicides, such as glyphosate and paraquat, with herbicides that control the legume cover crop is better for desiccating a mixed cover crop than the nonselective herbicide alone. Various combinations of nonselective herbicides, auxin-type herbicides, and metribuzin have been tested in Arkansas (see Table 5.3). Glyphosate (0.63 kg·ha^{-1}) + 2,4-D (0.56 kg·ha^{-1}) controlled 98 and 99 percent of rye and hairy vetch, respectively, four weeks after treatment (WAT). Glyphosate is generally used at 0.84 kg·ha^{-1} to kill existing vegetation or for weed control in glyphosate-resistant crops. A lower rate of glyphosate was used to determine if a reduced rate of glyphosate, in combination with another herbicide, will still be effective for burndown of cover crops. This research showed that a reduced glyphosate rate can be used with 2,4-D and still be as effective for grass control as the full rate. Glyphosate (0.32 kg·ha^{-1}) + clopyralid (0.11 kg·ha^{-1}) is also an excellent treatment, providing 96 and 100 percent control of rye and hairy vetch, respectively. Other excellent treatments include two applications of paraquat and a combination of paraquat (0.7 kg·ha^{-1}) and metribuzin (0.35 kg·ha^{-1}). The efficacy of killing hairy vetch can be significantly reduced when a mixture of paraquat and 2,4-D is used. Clopyralid kills hairy vetch, even at a low rate of 0.055 kg·ha^{-1}, but severely antagonizes the efficacy of paraquat in killing rye.

An important factor to consider in choosing herbicide mixtures for cover crop burndown is possible carryover injury to the succeeding crop. Tests

TABLE 5.3. Visual ratings of desiccation of rye and vetch mixed cover crops with herbicides, Fayetteville, Arkansas, 1994 and 1995.

Treatment		Conrol of cover crop[a] (in percent)			
		2 WAT[d]		4 WAT	
Herbicide[b]	Rate[c] ($kg \cdot ha^{-1}$)	Rye	Vetch	Rye	Vetch
Gly + 2,4-D	0.32 + 0.56	78	94	92	100
Gly + 2,4-D	0.63 + 0.56	94	97	98	99
Gly + clopyralid	0.32 + 0.055	88	89	92	99
Gly + clopyralid	0.32 + 0.11	77	92	96	100
Gly + clopyralid	0.42 + 0.22	81	96	92	100
Glu + 2,4-D	0.42 + 0.56	79	100	51	100
Glu + 2,4-D	0.84 + 0.56	88	100	62	100
Glu + clopyralid	0.42 + 0.055	76	95	66	96
Glu + clopyralid	0.42 + 0.11	81	100	59	100
Glu + clopyralid	0.42 + 0.22	78	100	44	100
Paraquat fb paraquat	0.70 + 0.35	92	82	100	98
Paraquat + 2,4-D	0.70 + 0.56	96	90	88	83
Paraquat + clopyralid	0.70 + 0.055	94	100	75	100
Paraquat + clopyralid	0.70 + 0.11	95	100	85	99
Paraquat + metribuzin	0.70 + 0.20	98	88	91	71
Paraquat + metribuzin	0.70 + 0.35	98	97	92	98
$LSD_{0.05}$		6	3	11	8

[a]Values are means of four replications, averaged over two years.

[b]Gly = glyphosate; Glu = glufosinate; fb = followed by.

[c]Rates of glyphosate, 2,4-D, and clopyralid are in terms of acid equivalent (ae). Rates of glufosinate, metribuzin, and paraquat are in terms of active ingredient (ai).

[d]WAT = weeks after treatment application.

conducted at Fayetteville, Arkansas, showed that tomato transplanted into a desiccated rye + hairy vetch cover crop can be injured by 2,4-D and clopyralid (see Table 5.4). Injury from 2,4-D can be minimized by delaying transplanting of tomato to two weeks after desiccation (WAD) of cover crop. Carryover injury from 2,4-D on tomato increased when 2,4-D was applied in combination with glufosinate; but this is inconsequential because glufosinate, in combination with 2,4-D or clopyralid, provides only 44 to 66 percent control of rye. This herbicide combination is therefore not recommended. Clopyralid is more injurious to tomato than 2,4-D, and clopyralid injury can still be observed on tomatoes transplanted 4 WAD. Some exam-

TABLE 5.4. Injury to tomato, from auxinlike herbicides used in cover crop burndown, four weeks after transplanting into desiccated cover crop.

Treatment		Injury on tomato, 4 WAT[a] (in percent)			
Herbicide[b]	**Rate[c] kg·ha^{-1}**	**1 WAD[d]**	**2 WAD**	**3 WAD**	**4 WAD**
Gly + 2,4-D	0.32 + 0.56	8	22	15	0
Gly + 2,4-D	0.63 + 0.56	4	2	9	0
Gly + clopyralid	0.32 + 0.055	57	38	8	0
Gly + clopyralid	0.32 + 0.11	84	59	32	9
Gly + clopyralid	0.32 + 0.22	58	84	41	10
Glu + 2,4-D	0.42 + 0.56	18	33	15	0
Glu + 2,4-D	0.84 + 0.56	52	36	22	0
Glu + clopyralid	0.42 + 0.055	58	35	38	0
Glu + clopyralid	0.42 + 0.11	78	64	10	0
Glu + clopyralid	0.42 + 0.22	87	74	45	0
Paraquat fb paraquat	0.70 + 0.35	0	0	0	0
Paraquat + 2,4-D	0.70 + 0.56	16	2	7	0
Paraquat + clopyralid	0.70 + 0.055	85	26	8	0
Paraquat + clopyralid	0.70 + 0.11	68	70	8	0
Paraquat + metribuzin	0.70 + 0.20	1	0	0	0
Paraquat + metribuzin	0.70 + 0.35	1	0	0	0
$LSD_{0.05}$		20	20	16	7

Note: Tomato was transplanted into desiccated cover at weekly intervals from one to five weeks after desiccation at Fayetteville, Arkansas, in 1994 and 1995.

[a]Visual ratings taken four weeks after transplanting (WAT). Values are means of four replications, averaged over two years.

[b]Gly = glyphosate; Glu = glufosinate; fb = followed by.

[c]Rates of glyphosate, 2,4-D, and clopyralid are in terms of acid equivalent (ae). Rates of glufosinate, metribuzin, and paraquat are in terms of active ingredient (ai).

[d]Timing of transplanting tomato; WAD = weeks after desiccation.

ples of appropriate herbicide mixtures for burndown of mixed cover crops include (1) paraquat or glyphosate + metribuzin for tomato or soybean, (2) glyphosate + 2,4-D for corn or tomato, and (3) paraquat or glyphosate + atrazine for corn.

Timing of Cover Crop Kill

Timing of cover crop kill is the ultimate determinant of biomass produced, amount of N_2 fixed by the legume cover crop, and availability of cover crop N to succeeding crop. Naturally, more biomass is produced if a cover crop is killed late. In Maryland, two weeks of additional growth in May more than doubled the biomass produced by a rye + vetch mixed cover crop from 3,661 to 7,981 $kg \cdot ha^{-1}$ (Clark, Decker, and Meisinger, 1994). The biomass of rye alone almost doubled, but the biomass of vetch alone more than quadrupled from 1,210 to 4,975 $kg \cdot ha^{-1}$ (Clark, Decker, and Meisinger, 1994). This indicates that the increase in biomass of mixed cover desiccated late (first to second week of May) is mostly from additional growth of the legume component, which translates to more N for the succeeding crop. In North Carolina, two weeks of additional growth between mid-April and May resulted in about a 150 percent increase in biomass of crimson clover, hairy vetch, and rye (Wagger, 1989b). The availability of N to the succeeding crop is also influenced by timing of cover crop kill by virtue of the influence of age on the C/N ratio of the grass cover crop. The C/N ratio of rye increases as it matures, starting from 13:1 in the fall to 42:1 in late April (Clark et al., 1997a). Others have also reported rapid increase in C/N ratio of rye as it matures (Sullivan, Parish, and Luna, 1991). The C/N ratio of vetch, on the other hand, remains low (10:1 to 15:1) regardless of age (Clark, Decker, and Meisinger, 1994; Clark et al., 1997a; Teasdale and Abdul-Baki, 1998). Because N immobilization is favored at C/N ratios >25:1 (Allison, 1966), N availability decreases as the grass cover crop matures. This trend is offset in the mixed cover crop system because of N contribution from the legume component, which keeps the C/N ratio below 25:1 (Clark, Decker, and Meisinger, 1994; Clark et al., 1997a; Teasdale and Abdul-baki, 1998). When using mixed cover crops, deciding the time of cover crop kill hinges on the amount of biomass desired and target planting date of the succeeding crop. Generally, giving the legume component more time to grow is most beneficial for the following crop. In the northern United States this entails waiting until mid-April to the first week of May before killing the cover crop. On final analysis, the target window of planting date for the following crop really dictates when the cover crop should be killed. This, for instance, would be earlier in areas along latitudes of the southern United States and later along latitudes of the northern United States and Canada.

EFFICACY OF MIXED COVER CROP FOR WEED CONTROL

Hairy vetch and rye are among the most widely used cover crops in North America not only because these species produce substantial biomass but also because their residues have shown allelopathic inhibition of germination and growth of several weed species (Barnes and Putnam, 1986; Bradow and Connick, 1990; Burgos and Talbert, 2000; Liebl et al., 1992; Shilling, Liebl, and Worsham, 1985; White, Worsham, and Blum, 1989). Hairy vetch is less effective for weed suppression than rye (Burgos and Talbert, 1996; Galloway and Weston, 1996; Mohler and Teasdale, 1993), so mixing the two species should offer added advantage over vetch alone for weed control.

Cover crop residue presents physical and chemical impediments to weed seed germination and growth. Just the mere presence of residue on the soil surface can prevent weed seed germination and seedling growth because of lack of sunlight and space to grow. In addition, chemical compounds released by the cover crop residue during the process of decomposition can inhibit the germination of weed seeds (Shilling, Liebl, and Worsham, 1985; Barnes, Putnam, and Burke, 1986; White, Worsham, and Blum, 1989; Bradow and Connick, 1990; Liebl et al., 1992; Burgos and Talbert, 2000). Logically, the level of weed suppression should increase with amount of cover crop residue, up to a point. Total weed emergence declined as rye residue increased from 0 to 1750 $g{\cdot}m^{-2}$ up to ten weeks after cover crop kill (Mohler and Teasdale, 1993). The weed population was composed of velvetleaf, redroot pigweed, common lambsquarters, witchgrass (*Panicum capillare* L.), curly dock (*Rumex crispus* L.), chickweed [*Stellaria media* (L.) Vill.], and dandelion (*Taraxacum officinale* Weber ex Wiggers). The same trend was observed in plots with hairy vetch residue, but only up to the fifth week. Also, weed emergence was higher in plots with vetch residue than with rye regardless of residue level (Mohler and Teasdale, 1993). This is indicative of the higher allelopathic characteristic of rye and longer duration of weed suppression by rye residue compared to hairy vetch. On the other hand, leguminous cover crops can still reduce the density of winter annual weeds 41 to 78 percent (Fisk et al., 2001).

Allelopathic cover crops, killed before planting the succeeding crop, suppress weeds early in the growing season of the succeeding crop, but additional weed control measures are needed to attain full-season weed control (Burgos and Talbert, 1996; Teasdale, 1996). Because cover crop residues are used in conjunction with reduced or no tillage, either a preemergence or postemergence herbicide application will have to be made during the season. Desiccated cover of rye alone could provide 81 to 97 percent weed suppression six to ten WAD in two locations in Indiana (Smeda and Weller, 1996). Without herbicides, legume + rye mixtures reduced weed biomass (Teasdale and Abdul-Baki, 1998) and provided better control of goosegrass, large crabgrass, Palmer amaranth, redroot pigweed, and yellow nutsedge (*Cyperus esculentus* L.) (Burgos and Talbert, 1996) compared to legume monocultures up to eight WAD, but generally not equal to the level of weed control obtained with herbicides. It is important to note that, like herbicides, allelochemicals from allelopathic cover crops, such as rye, can potentially suppress some weeds more than others (Barnes and Putnam, 1986; Burgos and Talbert, 2000).

In Arkansas, total weed density five WAD in sweet corn plots was 28, 68, and 58 percent less, respectively, when following rye, rye + hairy vetch, and wheat compared to following hairy vetch (Burgos and Talbert, 1996). The highest reduction in weed density was observed in plots following the mixed cover crop. In California, mixed cover crops of rye + woolypod vetch (*Vicia dasycarpa* Ten.) and triticale (*Triticum* × *secale*) + vetch provided the same level of early-season weed suppression in tomato plots as the conventionally tilled fallow treatment with herbicide (Herrero et al., 2001). Due to rapid decomposition of vetch residue relative to rye, weed suppression by vetch alone or mixtures containing vetch may decline faster than weed suppression by rye or wheat alone (see Photo 5.1). This was observed in plots with residues of hairy vetch alone or rye + vetch mixtures, planted to sweet corn, in two locations in Arkansas where weed composition included goosegrass, large crabgrass, Palmer amaranth, redroot pigweed, southwestern cupgrass [*Eriochloa gracilis* (Fourn.) Hitchc.], and yellow nutsedge (Burgos and Talbert, 1996). Atrazine and metolachlor (1.1 + 1.1 $kg \cdot ha^{-1}$) provided > 90 percent weed control for the rest of the season. Cover crop residues control annual weeds better than biennial (Teasdale and Abdul-Baki, 1998) or perennial species (Burgos and Talbert, 1996). Biennial and perennial species are generally naturally tolerant to allelochemicals. In many cases this is because these species have storage organs that support regrowth of the plant after the phytotoxic effect of the residue has dissipated, or sufficient residue decomposition has occurred allowing more light to filter into the soil surface. It is also important to consider that, as discussed in the previous section, many large-seeded annuals are tolerant to allelochemicals from cover crop residues. Hairy vetch, rye, and wheat cover crops did not prevent the proliferation of a perennial species, yellow nutsedge (Burgos and Talbert, 1996). On the other hand, the rye + vetch mixture controlled yellow nutsedge 95 percent 30 d after transplanting tomato (see Photo 5.2, A and B). This level of suppression was better than that of black plastic (71 percent) and wheat + vetch mixture (39 percent) (Burgos et al., 1999) and is largely dependent upon the amount and persis-

PHOTO 5.1. Sweet corn grown without cover crop (A); with hairy vetch (B) or rye + vetch (C) residues. Weed suppression by hairy vetch residue (D) is short-lived compared to rye + vetch (E). (See corresponding plate in the central color section.)

PHOTO 5.2. Tomato and summer squash grown in rye-vetch mixed cover crop, Fayetteville, Arkansas. Yellow nutsedge infestation is apparent in plots without cover crop (A). Thick rye-vetch residue suppresses yellow nutsedge (B). Summer squash in a bare-ground plot (C) has less luxuriant vegetative growth than that in rye-vetch residue (D). (See corresponding plate in the central color section.)

tence of cover crop residue. Postdirected application of metribuzin was needed to suppress yellow nutsedge for the rest of the season (Burgos et al., 1999). In Maryland, rye + vetch mixture without herbicide reduced total weed biomass 34 to 69 percent at the beginning of fruit harvest over three years compared to vetch alone (Teasdale and Abdul-Baki, 1998). Application of metribuzin three to six weeks after transplanting tomato was necessary to kill all emerged weeds and provide residual weed control. Depending upon the geography and climatic conditions from year to year, one postemergence herbicide application may be enough to supplement mixed cover crops for season-long weed control. If no preemergence herbicide is used, a postemergence herbicide application is needed eight WAD or sooner, depending on weed growth. Rye or wheat alone or in mixtures with legumes could provide ≥ 80 percent weed control four to eight WAD (Burgos and Talbert, 1996; Herrero et al., 2001; Smeda and Weller, 1996; Teasdale and Abdul-Baki, 1998). The duration of weed suppression is prolonged during dry years and shortened during wet years. A high amount of rainfall accelerates leaching and decomposition of allelochemicals from the cover crop residue; this shortens duration of weed control by cover crop residue.

Efficacy of cover crop mixture for weed control may be enhanced by using a cereal cover crop with high allelopathic potential. Allelochemical content in shoot tissue was analyzed from eight field-grown rye cultivars in Fayetteville, Arkansas. Concentrations of DIBOA [2,4-dihydroxy-1,4-(2*H*)-benzoxazinone-3-one] and BOA [2-(3*H*)-benzoxazolinone], previously identified as major allelochemicals from rye (Barnes, Putnam, and Burke, 1986), was highest in 'Bonel' shoot tissue compared to seven other cultivars (Burgos, Talbert, and Mattice, 1999). These cultivars include 'Aroostok', 'Bates', 'Elbon', 'Maton', 'Oklon', 'Pastar', and 'Wintercross'. High allelochemical content in 'Bonel' correlated with the highest degree of inhibition of goosegrass germination. Total amount of released allelopathic compounds, however, depends upon biomass production (see Table 5.5). Therefore, 'Bonel', 'Elbon', and 'Maton' cultivars were deemed more beneficial for use in mixed cover crop systems than other cultivars tested.

CROP YIELD FOLLOWING MIXED COVER CROPS

The benefit of added N from a legume to the following crop has been documented for years. Corn and grain sorghum [*Sorghum bicolor* (L.) Moench] produce higher yields following hairy vetch than following rye, regardless of amount of added N (Blevins, Herbek, and Frye, 1990). Like-

TABLE 5.5. Shoot biomass and hydroxamic acid (HA) content in shoot tissues of field-grown *Secale cereale* cultivars, Fayetteville, Arkansas.

	Shoot biomass[a]	HA content of field-grown rye ($g \cdot g^{-1}$)		
Cultivar	**($g \cdot m^{-2}$)**	**BOA**	**DIBOA**	**Total**
Aroostok	589	79	367	443
Bates	820	24	167	191
Bonel	557	229	1,240	1,469
Elbon	741	40	299	339
Maton	820	61	329	390
Oklon	618	23	132	155
Pastar	422	25	112	137
Wintercross	688	27	205	232
$LSD_{0.05}$	21	18	115	127

Source: Adapted from Burgos, Talbert, and Mattice, 1999.

[a]Planted October 22, 1995, and harvested April, 30, 1996. Harvested at booting. Values are the average of three replications.

wise, crimson clover has been shown to increase corn yields 33 to 43 percent compared with no-till fallow treatment in North Carolina (Myers and Wagger, 1991). Mixing a legume with a grass cover crop improves N availability and should increase crop yield. This has been demonstrated in several research studies. Corn yield following legume + wheat (or rye) mixtures was either lower than yield following legumes (Sullivan, Parrish, and Luna, 1991) or intermediate between monoculture legume and monoculture rye or wheat (Holderbaum et al., 1990; Clark, Decker, and Meisinger, 1994; Clark et al., 1997b). The reduction in corn yield following a grass + legume mixture compared to a legume monoculture indicates partial N immobilization by the grass component. To support this argument, N content of a grass + legume mixture is generally equal to or greater than that of a pure legume stand (Clark, Decker, and Meisinger, 1994; Clark et al., 1997a; Teasdale and Abdul-Baki, 1998; Vaughan and Evanylo, 1998; Odhiambo and Bomke, 2001). Sometimes yield of corn following the mixture may not be different from that following rye alone (Mitchell and Teel, 1977). This happens if the legume component in the mixed cover crop has poor stand, is seeded at a low rate (e.g., $< 14\ kg \cdot ha^{-1}$ for vetch), or is not allowed to accumulate enough biomass before cover crop kill.

The amount of N recovered by the following crop is affected greatly by appropriate cover crop seeding rates and time of cover crop kill. In North Carolina, corn grain yields were generally higher following late-killed than early-killed cover crops, most likely because more mulch resulted in greater soil moisture conservation and more available N (Clark, Decker, and Meisinger, 1994). Higher corn grain yield was obtained when vetch or rye-vetch mixture was killed late (Clark et al., 1997b) because of greater biomass and N content from the legume component (Clark et al., 1997a). Across locations and years in North Carolina, the greatest benefit from mixed cover crop was obtained when it was desiccated in late April or early May. When corn was planted following rye + hairy vetch in Blacksburg, Virginia, more N was available to corn, and higher yield was obtained, if the cover crop was desiccated at boot stage of rye, three weeks prior to corn planting, compared to later desiccation (Vaughan and Evanylo, 1998). Mixtures of a high seeding rate of vetch (28 $kg \cdot ha^{-1}$) with a high (94 $kg \cdot ha^{-1}$) or low (47 $kg \cdot ha^{-1}$) seeding rate of rye produced yields equivalent to vetch alone. These treatments produced the best corn yields when killed late (Clark, Decker, and Meisinger, 1994). Low seeding rates of both hairy vetch and rye in a mixture did not improve corn yield versus no cover and rye alone. To reduce inputs without adversely affecting yield of the following crop, a medium seeding rate of vetch ($kg \cdot ha^{-1}$) and a low seeding rate of rye (47 $kg \cdot ha^{-1}$), or low rate of vetch (14 $kg \cdot ha^{-1}$) and high rate of rye (94 $kg \cdot ha^{-1}$) are economically favorable in North Carolina. Cover crop seeding rates and kill dates need to be tested in other geographic locations before appropriate recommendations can be made for that particular region.

In the Coastal Plain of North Carolina, corn N content at black layer formation was 25 $kg \cdot ha^{-1}$ greater when following vetch than that following no cover or rye, and intermediate following the mixture without added fertilizer. Corn N content was 55 to 84 $kg \cdot ha^{-1}$ following rye and was 80 to 99 $kg \cdot ha^{-1}$ following no cover (Clark et al., 1997a). Reduced recovery of N indicates that yield reduction in corn following rye alone is due to partial immobilization of N by rye residue. Without fertilizer N, corn produced 4.4 to 10.6 $mg \cdot ha^{-1}$ grain following vetch or a rye + vetch mixture, but only 1.6 to 5.8 $mg \cdot ha^{-1}$ following rye or no cover (Clark et al., 1997b).

Availability of N from cover crop residue is more important for increasing corn yield than the capacity of the cover crop to conserve moisture (Vaughan and Evanylo, 1998). In Blacksburg, Virginia, corn yields were higher following hairy vetch than following rye and rye + hairy vetch because of N availability from vetch residue (Sullivan, Parrish, and Luna, 1991; Vaughan and Evanylo, 1998). Similarly, lower corn yields were observed following a wheat + vetch mixture than following monoculture legumes (Holderbaum et al., 1990).

Without herbicides, sweet corn yield in Fayetteville and Kibler, Arkansas, was higher following hairy vetch than following rye, rye + vetch, or wheat despite poor weed suppression with hairy vetch residue (Burgos and Talbert, 1996). This occurred despite the application of 56 to 128 $kg \cdot ha^{-1}$ N fertilizer in all plots. Reduced sweet corn yield was due, in part, to stand reduction in plots with rye or wheat residues (Photo 5.1, C and E). Stand loss can be caused by chemical and physical factors associated with the presence of cover crop residue. One consequence of high cover crop biomass can be improper furrow closure, which prevents proper seed soil contact. This happens if the planter used is not equipped with coulters and press wheels appropriate for high-residue situations. If preemergence herbicides are used, the germinating seed may then be damaged by direct contact with herbicide. Crop seed may also fail to germinate or emerge because of chemical inhibitors from the cover crop residue.

Removal of weed competition by handweeding or with herbicides increased weight of sweet corn ears following hairy vetch or fallow, but not when following rye, rye + vetch, and wheat (Burgos and Talbert, 1996). The fact that sweet corn following rye, rye + vetch, or wheat did not respond to weed removal, as did sweet corn following hairy vetch, indicates one or all of four causes. First, weed emergence and growth was lower in the rye, rye + vetch, and wheat than in vetch plots. Second, available N in plots with rye or wheat residue was insufficient to supply crop demand, despite added fertilizer N. Third, sweet corn growth was inhibited by allelochemicals from rye or wheat residue. Vetch residue is not inhibitory to growth of sweet corn. Fourth, loss of crop stand, either by chemical or physical causes, significantly reduced number of harvested ears per unit area regardless of absence of weed competition. In Lane, Oklahoma, a rye cover crop also reduced sweet corn yield compared to hairy vetch and no cover crop regardless of fertilizer N levels (Sojka, Karlen, and Busscher, 1991). Additional N from hairy vetch in the rye + vetch mixture did not compensate for reduction in sweet corn yield by the rye cover crop (Roberts and Cartwright, 1991).

Yields of transplanted tomato in Fayetteville, Arkansas, following rye + vetch and wheat + vetch mixtures tended to be higher than those tomatoes grown without cover crops in flat-bed culture. In bedded culture, tomato yield following rye + vetch was equivalent to yield of tomato grown in black plastic or no cover crop (Burgos et al., 1999). In Beltsville, Maryland, tomato grown in crimson clover, hairy vetch, and rye + vetch produced higher yield than tomato grown in black plastic (Abdul-Baki et al., 1996). Tomato following a rye + vetch mixture produced as much or higher marketable fruit than following monoculture vetch. Tomato following a rye + crimson clover produced 19 to 50 percent more marketable fruit than tomato following monoculture crimson clover (Teasdale and Abdul-Baki, 1998). In California, tomato yield following a triticale + vetch mixture was equal to that of a conventionally tilled fallow control with herbicide, and higher than that following monocultures of snail 'Sava' [*Medicago scutellata*

(L.) Miller] and barrel 'Sephi' (*Medicago truncatulata* Gaertner) medics in one of two years (Herrero et al., 2001). Sometimes tomato yield is unaffected by the presence of cover crop. If difference in available N is equalized by additional fertilizer N, lack of cover crop effect on tomato yield could occur if weed infestation is low enough to be suppressed sufficiently by the cover crop, or if rainfall is limited such that leaching of allelochemicals and flushes of weed seed germination are significantly reduced. In general, application of metribuzin three to six weeks after planting improves tomato yield following a cover crop mixture.

Summer squash (*Cucurbita pepo* L. var. *melopepo* Alef.) grows well following a vetch monoculture or rye + vetch mixed cover crop (Photo 5.2, C, D). It is important to consider, however, that rye + vetch mixture could reduce the stand of summer squash up to 33 percent and monoculture rye or wheat residue 58 percent compared to no residue (see Table 5.6). Hairy vetch residue did not reduce squash stand. Moreover, the presence of cover crop residues may increase leaf feeding by spotted and striped cucumber beetles (*Acalymma vittata* F.) on squash leaves (Burgos, 1994). However, the opposite may be true for other insects in another crop. For instance, cabbages in conventionally tilled plots had higher infestations of diamondback moth (*Plutella xylostella* Linné), imported cabbageworm (*Pieris rapae* Linné), and cabbage looper (*Trichoplusia ni* Hübner) than cabbages grown in cover crop mulch (Mwaja, Masiunas, and Eastman, 1996). The effect of cover crop on insect infestation is noticeable only during years when insect

TABLE 5.6. Effects of cover crop residues on summer squash, averaged over herbicide treatments and years in Arkansas.

Cover crop	Stand reduction (%)	Foliar damage by beetles[b] (%)	Height increase[c] (cm)	Number of harvested fruit (1,000 ha^{-1})	Weight per fruit (g)
No cover	—	3[b]	44[a]	37[a]	306[bc]
Hairy vetch	0[c]	48[a]	46[a]	29[b]	383[a]
Rye	58[a]	65[a]	28[b]	24[bc]	237[c]
Wheat	58[a]	56[a]	34[ab]	21[bc]	256[c]
Rye + vetch	33[b]	48[a]	40[ab]	28[b]	344[ab]

Source: Adapted from Burgos, 1994.

[a]Within the same column, values followed by the same letter are not significantly different based on Fisher's Protected LSD test at the 5 percent level of probability.
[b]Spotted and striped cucumber beetles occurred only in 1992.
[c]Between two and six weeks after planting.

infestation is particularly high (Burgos, 1994; Mwaja, Masiunas, and Eastman, 1996). The incidence of insect pest damage varies from year to year depending upon winter conditions and the presence of overwintering sites around the field. Several studies have shown that cover crops could favor the increase in population of beneficial insects (Tedders, 1983; Yan and Duan, 1988; Zhang et al., 1990) and, therefore, may minimize insect damage (Roberts and Cartwright, 1990). However, cucumber beetles overwinter as adults, and beneficial insects do not impact the adult beetle population. Regardless of high foliar insect damage in one of two years, summer squash following monoculture vetch produced larger fruits than squash with no cover crop, while fruit size following rye + vetch was intermediate between monoculture vetch and no cover crop (Table 5.6). In terms of total fruit weight, squash following monoculture vetch and squash grown on bare ground produced higher yield compared to squash following rye + vetch, monoculture wheat, and monoculture rye. Manually transplanting summer squash into rye + vetch residue eliminates the problem of stand reduction and results in yields higher than or equal to that of no cover crop (unpublished data). This is good for small-scale vegetable production; however, it is not known whether mechanical transplanting of summer squash seedlings can be done successfully or economically.

Cover crop residue maintains lower soil temperature for a longer period of time compared to no cover. At planting time in the spring, soil temperature in plots with cover crops within 5 cm depth are 2 to 7°C cooler than bare-ground plots (Burgos and Talbert, 1996). Similarly, soil temperature 1.3 cm deep remained lower with rye residue compared to bare soil up to around 30 days after planting soybean (Williams, Mortensen, and Doran, 1998). Use of mixed grass and legume cover crop mulch on tomato beds reduced the maximum daily temperatures relative to the conventional system (Creamer et al., 1996). Cooler temperatures in plots with cover crops contribute to slower seedling development and delayed maturity of the succeeding crop. For instance, early-season yield of tomato was lower when grown in rye + vetch that was mowed and left on the soil surface compared to that where the same mulch was incorporated into the soil. Cooler temperature under mulched soil not only could reduce crop growth, but also could delay release of N from the residue (Schonbeck, Elder, and DeGregorio, 1995). Cover crops delayed maturity of tomato by about a week in some years compared to bare soil or soil covered with black polyethylene in Maryland (Abdul-Baki et al., 1996; Teasdale and Abdul-Baki, 1997, 1998). In Arkansas, tomato maturity in plots following a rye + vetch mixture was delayed in one of two years compared to black plastic and bare soil (Burgos et al., 1999). Tomato plants made up for the delay by producing more fruit in later harvests. Despite total yield equivalency, producers using mixed cover crops in some areas would lose the opportunity of selling their product at a premium price for an early tomato crop. This is a trade-off some producers will be confronted with.

CONCLUSIONS

The use of cover crops is highly beneficial in adopting sustainable crop production systems mainly for conservation of soil, soil fertility, and moisture as well as possible reduction of herbicide, fertilizer, and tillage inputs. Mixing of grass and legume cover crops generally increases, or does not affect, biomass production yet offers the advantage of early-season weed control by the grass component and reduced N immobilization because of high N content of the legume component. Cover crop residues alone cannot control weeds for a whole cropping season, so supplemental weed management practices are necessary including herbicide application, hand weeding, or cultivation whichever is appropriate. The success of this system hinges on proper choice and management of cover crops, the availability of equipment and resources to manage the cover crop, and tolerance of the succeeding crop to phytotoxic effects of the cover crop residue. The use of cover crops is not suitable in all environments. Failure to consider any one of these factors may result in yield loss of the following crop.

REFERENCES

Abdul-Baki, A. A. and J. R. Teasdale (1993). A no-tillage tomato production system using vetch and subterranean clover mulches. *HortScience* 28: 106-108.

Abdul-Baki, A. A., A. E. Watada, J. R. Teasdale, and R. D. Morse (1996). Hairy vetch mulch favorably impacts yield of processing tomatoes. *HortScience* 31: 338-340.

Allison, F. E. (1966). The fate of nitrogen applied to soils. *Advances in Agronomy* 18: 219-258.

Barnes, J. P. and A. R. Putnam (1986). Evidence of allelopathy by rye residues and aqueous extracts of rye (*Secale cereale* L.). *Weed Science* 34: 384-390.

Barnes, J. P. and A. R. Putnam (1987). Role of benzoxazinones in allelopathy by rye (*Secale cereale* L.). *Journal of Chemical Ecology* 13: 889-905.

Barnes, J. P., A. R. Putnam, and B. A. Burke (1986). Allelopathic activity of rye (*Secale cereale* L.). In A. R. Putnam and A. S. Tang (Eds.), *The science of allelopathy* (pp. 271-286). New York: John Wiley and Sons, Inc.

Barnes, J. P., A. R. Putnam, B. A. Burke, and A. J. Aasen (1987). Isolation and characterization of allelochemicals in rye herbage. *Phytochemistry* 26: 1385-1390.

Bauer, P. J. and W. J. Busscher (1996). Winter cover and tillage influences on coastal plain cotton production. *Journal of Production Agriculture* 9: 50-54.

Baumhardt, R. L. and R. J. Lascano (1996). Rain infiltration as affected by wheat residue amount and distribution in ridged tillage. *Soil Science Society of America Journal* 60: 1908-1913.

Blevins, R. L., J. H. Herbek, and W. W. Frye (1990). Legume cover crops as a N-source for no-till corn and grain sorghum. *Agronomy Journal* 82: 769-772.

Bordelon, B. P. and S. C. Weller (1997). Preplant cover crops affect weed and vine growth in first-year vineyards. *HortScience* 32: 1040-1043.

Bradow, J. M. and W. J. Connick Jr. (1990). Volatile seed germination inhibitors from plant residues. *Journal of Chemical Ecology* 16: 645-666.

Brinsfield, R. B. and K. W. Staver (1989). Cover crops: A paragon for nitrogen management. In *Ground water issues and solutions in the Potomac River Basin/Chesapeake Bay Region,* March 14-16 (pp. 271-285). Dublin, OH: National Water Well Association.

Burgos, N. R. (1994). Cover crops in weed management systems for vegetable production. Master's thesis, University of Arkansas, Fayetteville.

Burgos, N. R. and R. E. Talbert (1996). Weed control and sweet corn (*Zea mays* var. *rugosa*) in a no-till system with cover crops. *Weed Science* 44: 355-361.

Burgos, N. R. and R. E. Talbert (2000). Differential activity of allelochemicals from *Secale cereale* in seedling bioassays. *Weed Science* 48: 302-310.

Burgos, N. R., R. E. Talbert, and J. D. Mattice (1999). Cultivar and age differences in the production of allelochemicals from *Secale cereale. Weed Science* 47: 481-485.

Burgos, N. R., R. E. Talbert, L. A. Schmidt, and J. J. Wells (1999). Mixed cover crop as an alternative to black plastic mulch for tomato production. In J. R. Clark and M. D. Richardson (Eds.), *Horticultural studies 1998* (pp. 92-95). Arkansas Agricultural Experiment Station Research Series 466. Fayetteville: University of Arkansas, Division of Agriculture.

Clark, A. J., A. M. Decker, and J. J. Meisinger (1994). Seeding rate and kill date effects on hairy vetch-cereal rye cover crop mixtures for corn production. *Agronomy Journal* 86: 1065-1070.

Clark. A. J., A. M. Decker, J. J. Meisinger, and M. S. McIntosh (1997a). Kill date of vetch, rye, and a vetch-rye mixture: I. Cover crop and corn nitrogen. *Agronomy Journal* 89: 427-434.

Clark, A. J., A. M. Decker, J. J. Meisinger, and M. S. McIntosh (1997b). Kill date of vetch, rye, and a vetch-rye mixture: II. Soil moisture and corn yield. *Agronomy Journal* 89: 434-441.

Coale, F. J., J. M. Costa, G. A. Bollero, and S. P. Schlosnagle (2001). Small grain winter cover crops for conservation of residual soil nitrogen in the mid-Atlantic Coastal Plain. *American Journal of Alternative Agriculture* 16: 66-72.

Corak, S. J., W. W. Frye, and M. S. Smith (1991). Legume mulch and nitrogen fertilizer effects on soil water and corn production. *Soil Science Society of America Journal* 55: 1395-1400.

Creamer, N. G., M. A. Bennett, and B. R. Stinner (1997). Evaluation of cover crop mixtures for use in vegetable production systems. *HortScience* 32: 866-870.

Creamer, N. G., M. A. Bennett, B. R. Stinner, and J. Cardina (1996). A comparison of four processing tomato production systems differing in cover crop and chemical inputs. *Journal of the American Society for Horticultural Science* 121: 559-568.

Creamer, N. G., B. Plassman, M. A. Bennett, R. K. Wood, B. R. Stinner, and J. Cardina (1995). A method for mechanically killing cover crops to optimize weed suppression. *American Journal of Alternative Agriculture* 10: 157-162.

Dufault, R. J., D. R. Decoteau, J. T. Garrett, K. D. Batal, D. Granberry, J. M. Davis, G. Hoyt, and D. Sanders (2000). Influence of cover crops an inorganic nitrogen fertilization on tomato and snap bean production and soil nitrate distribution. *Journal of Vegetable Crop Production* 6: 13-25.

Ebelhar, S. A., W. W. Frye, and R. L. Blevins (1984). Nitrogen from legume cover crops for no-tillage corn. *Agronomy Journal* 76: 51-55.

Einhellig, F. A. (1996). Interactions involving allelopathy in cropping systems. *Agronomy Journal* 88: 886-893.

Evanylo, G. K. (1991). Rye nitrogen cycling for corn and potato production. In W. L. Hargrove (Ed.), *Cover crops for clean water* (pp. 101-103). Ankeny, IA: Soil and Water Conservation Society.

Fisk, J. W., O. B. Hesterman, A. Shrestha, J. J. Kells, R. R. Harwood, J. M. Squire, and C. C. Shaeffer (2001). Weed suppression by annual legume crops in no-tillage corn. *Agronomy Journal* 93: 319-325.

Fujii, Y. (2001). Screening and future exploitation of allelopathic plants as alternative herbicides with special reference to hairy vetch. *Journal of Crop Production* 4: 257-275.

Galloway, B. A. and L. A. Weston (1996). Influence of cover crop and herbicide treatment on weed control and yield in no-till sweet corn (*Zea mays* L.) and pumpkin (*Cucurbita maxima* Duch.). *Weed Technology* 10: 341-346.

Gregory, P. J., L. P. Simmonds, and C. J. Pilbean (2000). Soil type, climate regime, and the response of water use efficiency to crop management. *Agronomy Journal* 92: 814-820.

Hargrove, W. L. (1986). Winter legumes as a nitrogen source for no-till grain sorghum. *Agronomy Journal* 78: 70-74.

Herrero, E. V., J. P. Mitchell, W. T. Lanini, S. R. Temple, E. M. Miyao, R. D. Morse, and E. Campiglia (2001). Use of cover crop mulches in a no-till furrow-irrigated processing tomato production system. *Horticulture Technology* 11: 43-48.

Holderbaum, J. F., A. M. Decker, J. J. Meisinger, F. R. Mulford, and L. R. Vough (1990). Fall-seeded legume cover crops for no-tillage corn in the humid East. *Agronomy Journal* 82: 117-124.

Hong, K. R., J. Y. Kim, D. J. Kang, and Z. R. Choe (1997). Effects of different vetch sward treatments on soil and rice growth in no-till direct-sown rice-vetch interrelaying cropping systems. *Korean Journal of Crop Science* 42: 564-570.

Jeong, J. H., S. Y. Choi, B. W. Shin, and J. D. So (1996). Effect of milk vetch (*Astragalus sinicus* L.) cultivation in reduction of nitrogen fertilizer application rate in paddy soil. *RDA Journal of Agricultural Science* 38: 299-303.

Johnson, G. A., M. S. DeFelice, and Z. R. Helsel (1993). Cover crop management and weed control in corn. *Weed Technology* 7: 425-430.

Lee, Y., H. R. Shin, S. W. Kim, O. D. Kwon, H. G. Park, and Y. J. Kim (1999). Growth of rice plant and chemical properties of soil as affected by nitrogen fertilization level in milk vetch *(Astragalus sinicus* L.) cultivation in paddy field. *Korean Journal of Crop Science* 44: 1-3.

Liebl, R. A., F. W. Williams, L. M. Wax, and E. W. Stoller (1992). Effect of rye (*Secale cereale*) mulch on weed control and soil moisture in soybean *(Glycine max). Weed Technology* 6: 838-846.

Liebl, R. A. and A. D. Worsham (1982). Allelopathic suppression of broadleaf weeds in no-till cropping systems. *Proceedings Southern Weed Science Society* 35: 316.

Liebl, R. A. and A. D. Worsham (1983). Allelopathic suppression of pitted morningglory and certain other weed spceis by phytotoxic components of wheat (*Triticum aestivum* L.) straw. *Journal of Chemical Ecology* 9: 1027-1043.

McIntyre, B. D., P. R. Speijer, S. J. Riha, and F. Kizito (2000). Effects of mulching on biomass, nutrients, and soil water in banana inoculated with nematodes. *Agronomy Journal* 92: 1081-1085.

Meisinger, J. J., W. L. Hargrove, R. L. Mikkelsen, J. R. Williams, and V. W. Benson (1991). Effects of cover crops on groundwater quality. In W. L. Hargrove (Ed.), *Cover crops for clean water* (pp. 57-68). Ankeny, IA: Soil and Water Conservation Society.

Mitchell, W. H. and M. R. Teel (1977). Winter annual cover crops for no-tillage corn production. *Agronomy Journal* 69: 569-573.

Mohler, C. L. and J. R. Teasdale (1993). Response of weed emergence to rate of *Vicia villosa* Roth. and *Secale cereale* L. residue. *Weed Research* 33: 487-499.

Mwaja, V. N., J. B. Masiunas, and C. E. Eastman (1996). Rye (*Secale cereale* L.) and hairy vetch (*Vicia villosa* Roth.) intercrop management in fresh-market vegetables. *Journal of American Society of Horticultural Science* 121: 586-591.

Myers, J. L. and M. G. Wagger (1991). Reseeding potential of crimson clover as a cover crop for no-tillage corn. *Agronomy Journal* 83: 985-991.

Nair, M., C. J. Whitenack, and A. R. Putnam (1990). 2,2'-oxo-1,1'-azobenzene: A microbially-transformed allelochemical from 2,3-benzoxazolinone. *Journal of Chemical Ecology* 16: 353-364.

Nelson, W. A., B. A. Kahn, and B. W. Roberts (1991). Screening cover crops for use in conservation tillage systems for vegetables following spring plowing. *HortScience* 860-862.

Odhiambo, J. J. O. and A. A. Bomke (2001). Grass and legume cover crop effects on dry matter and nitrogen accumulation. *Agronomy Journal* 93: 299-307.

Ranells, N. N. and M. G. Wagger (1992). Crimson clover management to enhance reseeding and no-till corn grain production. *Agronomy Journal* 85: 62-67.

Ranells, N. N. and M. G. Wagger (1996). Nitrogen release from grass and legume cover crop monocultures and bicultures. *Agronomy Journal* 88: 777-782.

Ranells, N. N. and M. G. Wagger (1997). Nitrogen-15 recovery and release by rye and crimson clover cover crops. *Soil Science Society of America Journal* 61: 943-948.

Roberts, B. W. and B. Cartwright (1990). *A reduced input approach to cabbage production: Managing erosion, fertility, and pests.* Research Reports Agricultural Experiment Station, Oklahoma State University, Division of Agriculture.

Roberts, B. W. and B. Cartwright (1991). Sweet corn yield and insect damage as affected by cover crops and nitrogen. *HortScience* 26: 489.

Rosecrance, R. C., G. W. McCarty, D. R. Shelton, and J. R. Teasdale (2000). Denitrification and N mineralization from hairy vetch (*Vicia villosa* Roth) and rye (*Secale cereale* L.) cover crop monocultures and bicultures. *Plant and Soil* 227: 283-290.

Schlegel, A. J. and J. L. Havlin (1997). Green fallow for the central Great Plains. *Agronomy Journal* 89: 762-767.

Schonbeck, M., P. Elder, and R. DeGregorio (1995). Winter annual cover crops for the home food garden. *Journal of Sustainable Agriculture* 6: 29-53.

Seigler, D. S. (1996). Chemistry and mechanisms of allelopathic interactions. *Agronomy Journal* 88: 876-885.

Seong, R. C. and K. Y. Park (1991). Forage productivity of collected Chinese milk vetch varieties. *Korean Journal of Crop Science* 436: 7-11.

Shilling, D. G., R. A. Liebl, and A. D. Worsham (1985). Rye (*Secale cereale* L.) and wheat (*Triticum aestivum* L.) mulch: The suppression of certain broadleaf weeds and the isolation and identification of phytotoxins. In A. C. Thompson (Ed.), *The chemistry of allelopathy* (pp. 243-271). Washington, DC: American Chemical Society.

Shipley, P. R., J. J. Meisinger, and A. M. Decker (1992). Conserving residual corn fertilizer N with winter cover crops. *Agronomy Journal* 84: 869-876.

Smeda, R. J. and S. C. Weller (1996). Potential of rye (*Secale cereale*) for weed management in transplant tomatoes (*Lycopersicon esculentum*). *Weed Science* 44: 596-602.

Sojka, R. E., D. L. Karlen, and W. J. Busscher (1991). A conservation tillage research update from the Coastal Plain Soil and Water Conservation Research Center of South Carolina: A review of previous research. *Soil and Tillage Research* 21: 361-376.

Sullivan, P. G., D. J. Parrish, and J. M. Luna (1991). Cover crop contributions to N supply and water conservation in corn production. *American Journal of Alternative Agriculture* 6: 106-113.

Teasdale, J. R. (1996). Contribution of cover crops to weed management in sustainable agricultural systems. *Journal of Production Agriculture* 9: 475-479.

Teasdale, J. R. and A. A. Abdul-Baki (1997). Growth analysis of tomatoes in black polyethylene and hairy vetch production systems. *HortScience* 32: 659-663.

Teasdale, J. R. and A. A. Abdul-Baki (1998). Comparison of mixtures vs. monocultures of cover crops for fresh-market tomato production with and without herbicide. *HortScience* 33: 1163-1166.

Tedders, W. L. (1983). Insect management in deciduous orchard ecosystems: Habitat manipulation. *Environmental Management* 7: 29-34.

U.S. Department of Agriculture (2003). National Agricultural Statistics Service reports. Available at <usda.mannlib.cornell.edu/reports/nassr/>.

Vaughan, J. D. and G. K. Evanylo (1998). Corn response to cover crop species, spring desiccation time, and residue management. *Agronomy Journal* 90: 536-544.

Wagger, M. G. (1989a). Cover crop management and nitrogen rate in relation to growth and yield of no-till corn. *Agronomy Journal* 81: 533-538.

Wagger, M. G. (1989b). Time of desiccation effects on plant composition and subsequent nitrogen release from several winter annual cover crops. *Agronomy Journal* 81: 236-241.

Wallace, R. W. and R. R. Bellinder (1992). Alternative tillage and herbicide options for successful weed control in vegetables. *HortScience* 27: 745-749.

Weston, L. A. (1990). Cover crop and herbicide influence on row crop seedling establishment in no-tillage culture. *Weed Science* 38: 166-171.

Weston, L. A. (1996). Utilization of allelopathy for weed management in agroecosystems. *Agronomy Journal* 88: 860-866.

White, R. H., A. D. Worsham, and U. Blum (1989). Allelopathic potential of legume debris and aqueous extracts. *Weed Science* 37: 674-679.

Williams, M. M., D. A. Mortensen, and J. W. Doran (1998). Assessment of weed and crop fitness in cover crop residues for integrated weed management. *Weed Science* 46: 595-603.

Wu, J. Z. and Y. Arima (1992). Effect of rhizobium inoculation and application of N, P, K fertilizer on the growth and nitrogen fixation of field-grown Chinese milk vetch. *Soil Science and Plant Nutrition* 38: 75-84.

Yan, Y. H. and J. J. Duan (1988). Effects of cover crops in apple orchards on predator community on the trees. *Acta Phytophylactica Sinica* 15: 23-27.

Yasue, T. (1991). The change of cultivation and utilization of Chinese milk vetch (*Astragalus sinicus* L.) and the effect of fertilizer and soil fertility on paddy fields as a green manure. *Japanese Journal of Crop Science* 60: 583-592.

Yasue, T. and U. Tsutiya (1982). Chinese milk vetch and its history of cultivation. *Kyoiku Shupan, Gifu* 52: 107.

Yenish, J. P., A. D. Worsham, and A. C. York (1996). Cover crops for herbicide replacement in no-tillage corn (*Zea mays*). *Weed Technology* 10: 815-821.

Zhang, W. H., H. Chang, Y. Y. Zhu, and W. H. Xu (1990). Influence of cultivation methods on the natural enemy populations in cotton-barley intercropping systems in Jiangsu. *Chinese Journal of Biological Control* 6: 1-4.

Chapter 6

Rye As a Weed Management Tool in Vegetable Cropping Systems

John B. Masiunas

INTRODUCTION

Current weed management systems for vegetable crops are not sustainable. They rely on extensive inputs of herbicides, tillage, and hand weeding. Many of the herbicides used in vegetable production are older products that cause environmental or health concerns. Some vegetable cropping systems have only a few registered herbicides, making resistant weeds a concern and causing a desperate need for alternative weed management techniques. Herbicides account for most of the pesticide active ingredients applied in many vegetable cropping systems (Clark et al., 1998), and alternative weed management systems are needed to reduce use of pesticides. Hand weeding is the most costly input used in vegetable production and must be minimized if alternative weed management systems are to be economical (Clark et al., 1998; Masiunas, 1999). Thus more environmentally sound and economical approaches are needed to manage weeds in vegetable cropping systems.

If sustainable weed management systems are to be developed for vegetable crops, our reliance must be shifted from external resources such as herbicides to internal, farm-derived, renewable resources (Francis and King, 1988). One way to accomplish this goal is to increase surface residues and organic matter to create agroecosystems controlled more by biological processes than are our current vegetable production systems (Coolman and Hoyt, 1993). Using cover crops for weed management meets the requirement that more sustainable vegetable cropping systems should have greater surface residues and organic matter.

Cover crops have been used for thousands of years to enhance soil fertility and as "smother crops" to suppress weeds (Hill, 1926; Putnam, 1986; Burket, Hemphill, and Dick, 1997), but generally they have been incorporated into the soil prior to planting vegetables. Recently, cover crop systems have been developed that leave plant residues on the soil surface to suppress weeds. In the northeast and midwest United States, cereal rye (*Secale cereale* L.) is one of the most promising cover crops for these systems (Putnam, 1986; Schonbeck et al., 1993; Masiunas, 1999).

THE ADVANTAGES OF RYE COVER CROPS

Rye cover crops can suppress weeds in vegetables, but they are often used for the other advantages they provide in cropping systems. Cover crops reduce or eliminate tillage, controlling soil erosion compared to conventional tillage systems (Skarphol, Corey, and Meisinger, 1987; Coolman and Hoyt, 1993; Morse, 1993; Hoyt, Monks, and Monaco, 1994; Hoyt, Bonanno, and Parker, 1996). Reduced-tillage systems also conserve soil water (Coolman and Hoyt, 1993; Morse, 1993; Sainju et al., 2000) by increasing the infiltration capacity and reducing crusting and evaporative loss of water compared to conventional tillage (Skarphol, Corey, and Meisinger, 1987; Coolman and Hoyt, 1993). The more efficient water use in reduced-tillage systems lessens irrigation requirements for producing vegetables (Morse, 1993; Hoyt, Monks, and Monaco, 1994).

Continued use of reduced-tillage systems results in soil improvement and enhanced soil productivity (Morse 1993; Hoyt, Monks, and Monaco, 1994). Reduced-tillage systems make more efficient use of fossil fuel–based nonrenewable resources and have reduced labor requirements (Skarphol, Corey, and Meisinger, 1987; Coolman and Hoyt, 1993; Hoyt, Monks, and Monaco, 1994; Hoyt and Walgenbach, 1995). But many of the advantages of reduced-tillage systems are dependent on the presence of a surface mulch (Skarphol, Corey, and Meisinger, 1987).

Cereal rye is ideal for creating a surface mulch for sustainable vegetable production systems (Nelson, Kahn, and Roberts, 1991; Abdul-Baki et al., 1996). Rye does well in heavy clayey loam soils (Knott, 1996; Wyland, Jackson, and Schulbach, 1995; Wyland et al., 1996). It survives harsh winters without lodging and grows rapidly the following spring (Weston, 1990; Roberts and Cartwright, 1991; Masiunas, 1999). Vegetable growers have used rye extensively to protect against soil erosion, and some growers leave 50 to 150 cm-wide strips of rye to protect fruit and flowers of vegetable plants against wind damage (Putnam, 1986; Kahn, Monke, and Foster, 1988; Phatak, 1992; Hoyt, Monks, and Monaco, 1994; Lal, 1995; Paine and Harrison 1996; Smeda and Weller, 1996). Rye also produces a large quantity of biomass and has a great tillering capacity that makes it an effective surface mulch (Weston, 1990; Abdul-Baki et al., 1996; Smeda and Weller,

1996; Knott, 1996; Bottenberg, Masiunas, and Eastman, 1999; Masiunas, 1999).

Rye improves soil's physical and chemical properties (Eckert, 1991; Masiunas, Weston, and Weller, 1995). Mwaja, Masiunas, and Eastman (1996) found that a rye cover crop increased soil organic matter and pH compared to conventional tillage. The increased soil pH was caused by an assimilation of N by the rye. Large continuous soil biopores are produced when the root system of the rye decays (Stirzaker and White, 1995). This increases the water and air entry into the soil (Stirzaker and White, 1995).

Rye planted after a summer crop can recover residual fertilizer N in the soil that otherwise might leach to the groundwater (Shennan, 1992; Wyland, Jackson, and Schulbach, 1995; Knott, 1996; Sheperd and Lord, 1996; Brandi-Dohrn et al., 1997; Burket, Hemphill, and Dick, 1997; Kuo and Sainju, 1998; Masiunas, 1999). Because of its extensive root system and early establishment in the fall, rye is better able to scavenge residual NO_3 from the soil and reduce NO_3 leaching than are other cover crops such as red clover, oats, or hairy vetch (Sainju, Singh, and Whitehead, 1998). As rye residues decompose they can be a source of essential nutrients for the next season's vegetables (Eckert, 1991; Lal, 1995; Abdul-Baki et al., 1996; Masiunas, 1999) and weeds.

THE DISADVANTAGES OF RYE COVER CROPS

Rye cover crop mulch systems also have disadvantages that must be overcome if they are to be used for weed management in vegetables. On heavier soils, soil compaction is one of the most severe problems caused by reduced-tillage systems with rye residues (Morse, 1993; NeSmith and McCracken, 1994; Bottenberg, Masiunas, and Eastman, 1999; Amisi, 2000; Sainju et al., 2000). Bottenberg, Masiunas, and Eastman (1999) reported that soil bulk density was higher in the rye mulch treatments than in the conventionally tilled plots, and the greater bulk density reduced snap bean growth (see Table 6.1). The soil compaction was likely caused by fall disking and natural soil consolidation during the winter and lack of spring tillage to disrupt the compacted layer (NeSmith and McCracken, 1994).

A cereal rye cover crop may result in net microbial N immobilization and deplete the soil of available N for the subsequent vegetable crops and weeds (Weston, 1990; Schonbeck et al., 1993; Hoyt, Monks, and Monaco, 1994; Mwaja, Masiunas, and Eastman, 1996; Burket, Hemphill, and Dick, 1997). Fall-planted rye has a significant lignin content and a high C:N ratio (>60), resulting in net immobilization of soil N (Wyland, Jackson, and Schulbach, 1995; Kuo and Sainju, 1998). Kuo and Sainju (1998) found that rye surface

TABLE 6.1. The effect of a rye cover crop mulch and strip tillage on soil bulk density of a Flanagan silt loam (fine montimorrellontic, mesic, Aquic Arguidoll; pH 6.3 to 6.6 and organic matter 3.1 to 3.3%).

	Soil bulk density ($g \cdot cm^{-3}$)			
Management system[a]	**June 4**	**June 18**	**July 17**	**August 11**
Conventional tillage	1.43[a,b]	1.36[b]	1.42[b]	1.30[b]
Solid rye cover crop mulch	1.52[a]	1.50[a]	1.56[a]	1.58[a]
Strip-tilled rye cover crop	—[c]	1.37[b]	1.61[a]	1.52[a]

Source: Modified by permission of the American Society for Horticultural Science from Table 1 of Bottenberg, Masiunas, and Eastman, 1999.

[a]The cropping systems were conventional tillage, mold-board plowed in the fall followed by spring disked and trifluralin incorporated at 0.56 $kg \cdot ha^{-1}$; solid rye cover crop mulch, fall-planted cereal rye killed with glyphosate at 1.1 $kg \cdot ha^{-1}$ and mowed; and strip-tilled rye cover crop, fall-planted cereal rye killed with glyphosate and a 45 cm wide strip rotatilled into the mulch.

[b]Mean separation within columns by Fisher's Protected LSD, $p = 0.05$.

[c]Strips were not yet established.

residues required much longer than 20 weeks of incubation for net N mineralization.

Rye cover crop mulch systems are more management intensive, more risky, and less consistent than conventional vegetable cropping systems (Masiunas, 1999; Amisi, 2000). Cover crops can inhibit emergence of vegetable seedlings, reducing stand, and can affect subsequent crop growth and yield (Weston, 1990; Amisi, 2000; see Table 6.2). These risks have caused vegetable farmers to be reluctant to adopt rye cover crop mulch systems and must be overcome if their use is to become more widespread (Morrison et al., 1973; Putnam, 1986; Coolman and Hoyt, 1993; Mundy et al., 1999).

Rye cover crops have a variable impact on insect pests of vegetables depending on the species and the management of the cover crop (Smith, Hammond, and Stinner, 1988). They can affect pest dispersal, colonization, or reproduction on vegetable crops (House and Del Rosario Alzaugary, 1989; Bugg, 1992; Creamer, Bennett, Stinner, Cardina, and Regnier, 1996). Cover crops can reduce insect pest colonization of vegetable crops, have nutritional effects on host plants, cause microclimate changes that reduce pest success, or enhance performance of natural enemies (Zehnder and Linduska, 1987; Bugg and Ellis, 1990; Bugg, Phatak, and Dutcher, 1990; Bugg, 1992).

Potato leafhopper [*Empoasca fabae* (Harris)] populations on snap beans (*Phaseolus vulgaris* L.) have been found to be significantly higher in con-

TABLE 6.2. Influence of plastic and rye cover crop mulch on 'Howden' pumpkin yields.

	Total number of fruit/plot		Number of green fruit/plot		Yield ($Mg \cdot hg^{-1}$)	
Treatment[a]	**1998**	**1999**	**1998**	**1999**	**1998**	**1999**
Conventional	23[a,b]	44[a]	0	8	8.3[a]	17.2[a]
Conv + plastic	23[a]	44[a]	5	4	6.9[a]	17.5[a]
Solid rye	1[b]	25[b]	1	3	0.1[b]	8.2[b]
Strip-till + plast	8[b]	23[b]	3	3	1.4[b]	8.2[b]

Source: Adapted by permission of Karen Amisi from Table 2.7 of Amisi, 2000.

[a]The treatments included conventional: conventionally tilled with clomazone and ethalfluran at 0.28 and 1.68 $kg \cdot ha^{-1}$, respectively; conv + plastic: conventionally tilled with black plastic mulch within the rows and clomazone and ethalfluran at 0.28 and 1.68 $kg \cdot ha^{-1}$, respectively between the rows; solid rye: fall-seeded cereal rye killed with glyphosate and mowed prior to no-till planting pumpkins; and strip-till + plast: fall-seeded cereal rye killed with glyphosate, mowed, and strip tilled, black plastic mulch laid in the crop row prior to planting pumpkins.

[b]Mean separation within columns by Fisher's Protected LSD, $p = 0.05$.

ventional tillage than in rye cover crop mulch (Bottenberg, Masiunas, and Eastman, 1999; Table 6.3). Snap bean plant growth in rye plots was retarded, resulting in smaller plants with fewer leaves that supported smaller insect populations than in conventional tillage. Also, grasses are the least suitable host for potato leafhopper, and the presence of rye might serve to alter the colonization and dispersal of potato leafhopper (Smith, Hammond, and Stinner, 1988).

Diseases in rye cover crops can be reduced, enhanced, or unaffected depending on type of inoculum (Creamer, Bennett, Stinner, Cardina, and Regnier, 1996). If the inoculum survives on the rye surface residue until the vegetable is planted, the disease will be more severe (Creamer, Bennett, Stinner, Cardina, and Regnier, 1996). The additional organic matter in rye cover crops may increase populations of saprophytic fungi and bacteria that antagonize and parasitize root pathogens (Creamer, Bennett, Stinner, Cardina, and Regnier, 1996; Clark et al. 1998; Bottenberg, Masiunas, and Eastman, 1999). Rye cover crops can affect the potential for disease development by modifying soil moisture and temperature or affecting vegetable crop susceptibility to diseases (Creamer, Bennett, Stinner, Cardina, and Regnier, 1996; Bottenberg, Masiunas, and Eastman, 1999). For example, we found that the incidence of white mold [*Sclerotina sclerotiorum* (Lib.)

TABLE 6.3. Potato leafhopper populations on snap beans.

		Number per plant	
Management system[a]	**Insecticide[b]**	**July 12**	**August 6**
Conventional tillage	+	—	14.2 bc[c]
Conventional tillage	–	8.1 a	32.3 a
Solid rye cover crop mulch	+	—	8.0 c
Solid rye cover crop mulch	–	3.5 b	8.0 c
Strip-tilled rye cover crop	+	—	6.7 c
Strip-tilled rye cover crop	–	5.2 ab	21.8 ab

Source: Adapted by permission of the American Society for Horticultural Science from Table 5 of Bottenberg, Masiunas, and Eastman, 1999.

[a]The cropping systems were conventional tillage, mold-board plowed in the fall followed by spring disked and trifluralin incorporated at 0.56 $kg \cdot ha^{-1}$; solid rye cover crop mulch, fall-planted cereal rye killed with glyphosate at 1.1 $kg \cdot ha^{-1}$ and mowed; and strip-tilled rye cover crop, fall-planted cereal rye killed with glyphosate and a 45 cm wide strip rotatilled into the mulch.

[b]With (+) or without (–) dimethoate applied on July 16.

[c]Mean separation within columns by Fisher's Protected LSD, $p = 0.05$.

deBary] was reduced by rye treatments (Bottenberg et al., 1997a; Bottenberg, Masiunas, and Eastman, 1999). The lower incidence was because the mulch may have acted as a barrier between soil and pods, minimizing splashing and spore dispersal.

EFFECTS OF RYE COVER CROPS ON WEED POPULATIONS

Numerous studies have found that rye can suppress weeds. Putnam (1986) evaluated spring-planted rye that was killed after planting the vegetable crop. The spring-planted living rye reduced weed biomass by 93 percent over plots with no rye (Barnes and Putnam, 1983, 1986). Ateh and Doll (1996) evaluated a spring-planted rye system for soybeans [*Glycine max* (L.) Merr.] and found that it reduced weed biomass by 60 to 90 percent relative to the bare ground control. The living cover of spring-planted winter rye reduced the early-season biomass of common lambsquarter (*Chenopodium album* L.) by 80 to 98 percent, large crabgrass [*Digitaria sanguinalis* (L.) Scop.] by 42 percent, and common ragweed (*Ambrosia artemisiifolia* L.) by 90 percent (Barnes and Putnam, 1983, 1986).

The residue decays rapidly after the spring-planted rye is killed, resulting in inconsistent weed control (see Table 6.4; Figure 6.1). Ateh and Doll (1996) found that killing the rye 45 days after planting soybean favored the establishment of late-season broadleaf weeds by allowing weed seedlings that were hitherto suppressed by rye and the soybean canopy to grow. Residues from fall-planted rye suppressed weeds more than did residues of spring-planted rye (Masiunas, 1999). These differences were primarily related to the greater amounts of biomass and their greater persistence. The spring-planted rye residues did not persist long enough to suppress weeds through the critical weed-free periods for most vegetables (Masiunas, 1999).

Most research on cover crop mulch systems has therefore concentrated on fall-planted, spring-killed rye. A number of studies have found that fall-planted, spring-killed rye suppressed early-season weeds as well as pre-emergent herbicides (Barnes and Putnam, 1983, 1986; Creamer, Bennett, Stinner, Cardina, and Regnier, 1996; Bottenberg et al., 1997a,b). Residues from fall-planted, spring-killed rye reduced total weed biomass by over 30 percent (Barnes and Putnam, 1986; Burgos and Talbert, 1996) and over-wintering weeds by 93 percent compared to bare ground (Smeda and Weller, 1996).

Weeds differ in their susceptibility to rye cover crop mulches. Residues of rye are generally more toxic to annual dicotyledonous weeds, moderately toxic to annual monocotyledonous weeds, and have little or no effect

TABLE 6.4. Weed control from spring-planted rye.

Cropping system[c]	Broadleaf weeds (m^{-2})		Grass weeds (m^{-2})	
	1994	1995	1994	1995
Conventional till	160 b[d]	660 b	144 a	310
Rye cover crop	130 a	577 a	229 b	345

Source: Modified by permission of Elizabeth Wahle from Table 3.5 of Wahle, 1996.

[a]The dominant broadleaf weeds were common purslane and pigweed species [redroot pigweed (*Amaranthus retroflexus* L.), smooth pigweed (*A. hybridus* L.), tumble pigweed (*A. albus* L.), and prostate pigweed (*A. blitoides* S. Wats.)].

[b]The grass weeds were mainly large crabgrass and giant foxtail.

[c]The cropping systems were conventional till, mold-board plowed in the fall followed by spring disked and trifluralin incorporated at 0.56 kg·ha^{-1} and rye cover crop, cereal rye killed with glyphosate at 1.1 kg·ha^{-1} and mowed.

[d]Mean separation within columns by Fisher's Protected LSD, $p = 0.05$.

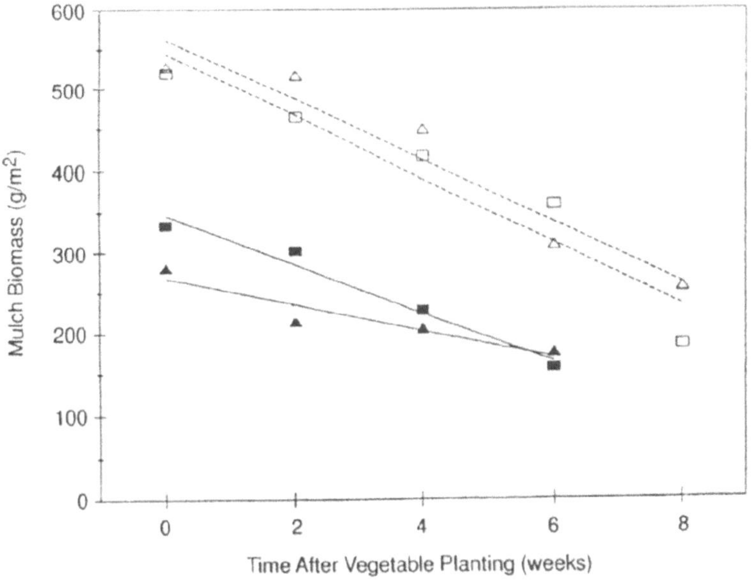

FIGURE 6.1. The amount of residue remaining at various times after killing a rye cover crop mulch. The results are filled points (actual) or solid lines (predicted), while the 1993 results are open points or dashed lines. The regression equations for rye residues in cabbage are, for 1992, $Y = 344.8–30.1X$ (solid squares; Y = rye residue biomass in g·m^{-2}; X = weeks after vegetable planting; $r^2 = 0.88$), and for 1993, $Y = 544–38.8X$ (open squares; $r^2 = 0.88$). The regression equations for rye residues in snap beans are, for 1992, $Y = 268–16.3X$ (solid triangles; $r^2 = 0.84$) and for 1993, $Y = 562–37.5X$ (open triangles; $r^2 = 0.92$) (*Source:* Reproduced by permission of The Haworth Press from Figure 3 of Masiunas et al., 1997, p. 71).

on perennial weeds (Putnam, 1986, 1990; Phatak, 1992; Masiunas, 1999). But actual data do not always support this trend. For example, barnyardgrass [*Echinochloa crus-galli* (L.) Beauv.] biomass was reduced up to 74 percent, and redroot pigweed (*Amaranthus retroflexus* L.) was reduced up to 55 percent compared to the bare ground control (Barnes and Putnam, 1983). Cereal rye equally suppressed eastern black nightshade (*Solanum ptycanthum* Dun.) and yellow foxtail [*Setaria glaucia* (L.) Beauv.] (Creamer, Bennett, Stinner, Cardina, and Regnier, 1996). In other studies, rye cover crop mulch reduced the population of yellow nutsedge and gave >65 percent control of giant foxtail (*Setaria faberi* Herrm.), velvetleaf (*Abutilon theophrasti* Medikus), common lambsquarters, large crabgrass, redroot, and smooth pigweed (*Amaranthus hybridus* L.) (Burgos and Talbert, 1996; Masiunas, 1999; see Table 6.5).

TABLE 6.5. The impact of cropping system on the total number of weeds in cabbage and snap beans. The numbers are the sum of the counts taken every two weeks between 2 and 10 weeks after planting the crop.

	Broadleaf weeds[b] (m^{-2})		Grasses[c] (m^{-2})	
Management system[a]	**1992**	**1993**	**1992**	**1993**
Cabbage				
Conventional till	32	126 b[d]	0	7
Rye cover crop	23	42 a	1	9
Snap bean				
Conventional till	18	85	0	8 a
Rye cover crop	14	80	0	27 b

Source: Modified by permission of The Haworth Press from Table 1 of Masiunas et al., 1997, p. 73.

[a]The cropping systems were conventional till, mold-board plowed in the fall followed by spring disked and trifluralin incorporated at 0.56 kg·ha^{-1}, and rye cover crop, cereal rye killed with glyphosate at 1.1 kg·ha^{-1} and mowed.

[b]The dominant broadleaf weed was redroot pigweed. Other broadleaf weeds included velvetleaf, ivyleaf morning glory, and dandelion.

[c]The grass weeds were mainly large crabgrass and giant foxtail.

[d]Mean separation within columns by Fisher's Protected LSD, $p = 0.05$.

Weed suppression with rye has been correlated with the amount of residue present, but the actual shape of the response varied among weed species. Emergence of some species, including redroot pigweed, common lambsquarters, fall panicum (*Panicum capillare* L.), and barnyardgrass, showed a montonic decline in response to increasing rates of residue (Mohler and Teasdale, 1993). Germination of these species is promoted by light and warm temperatures. Emergence of other weed species such as curly dock (*Rumex crispus* L.), chickweed [*Stellaria media* (L.) Vill.], and dandelion (*Taraxacum officinale* Weber in Wiggers) increased at low amounts of rye residue before declining at higher amounts (Mohler and Teasdale, 1993).

The length of weed suppression with fall-seeded rye cover crops is dependent on the weed species, amount of residue, whether the rye has been mowed, the location, and the environmental conditions. In New York, 4000 kg·ha^{-1} of rye residues suppressed weed emergence and growth for up to 12 weeks (Bellinder, Wallace, and Wilkins, 1996; Wilkins and Bellinder, 1996). In Michigan, weed control from rye residues lasts for 4 to 11 weeks (Putnam, 1986). Masiunas, Weston, and Weller (1995) found that rye suppressed weeds for four to eight weeks, and the length of weed suppression was longer in Illinois and Indiana than Kentucky. This was probably due to the more rapid decomposition of cover crop residues and allelochemicals in Kentucky (Masiunas, Weston, and Weller, 1995; Masiunas, 1999). Weed suppression also lasts longer if the vegetables are planted soon after killing the rye cover crop (Masiunas, 1999). Weed suppression persists longer when rainfall occurs soon after the rye is killed and is followed by relatively dry conditions (Putnam, 1986; Weston, 1996).

There are problems in obtaining consistent weed control with fall-planted rye cover crops. In rye cover crop mulch systems, populations of weeds with wind-blown seed, such as common dandelion, are increased (Bottenberg et al., 1997a; Masiunas, 1999). Many species with wind-blown seed are adapted for growth in grass sods, and they may be less susceptible to allelochemicals. Also, wind is an effective dissemination agent for moving weed seed into no-till systems. If the cropping system requires soil disturbance for sidedressing fertilizers or for hilling potatoes (*Solanum tuberosum* L.), weed control can be reduced (Bellinder, Wallace, and Wilkins, 1996). Supplemental weed control is often necessary in rye cover crop mulch systems later in the season (Masiunas, Weston, and Weller, 1995; Bottenberg et al., 1997a; Masiunas, 1999).

MANAGING RYE COVER CROPS

How rye is managed is critical to obtaining good weed suppression without reducing vegetable crop yield. Spring-planted rye generally will not tiller, but its seeding rate can be varied to affect biomass production. In spring-seeded rye, a higher seeding rate provided more ground cover and better weed control than the lower seeding rate, but the rye also reduced soybean vigor (Ateh and Doll, 1996). Ateh and Doll (1996) found the optimum seeding rate for spring-planted rye to be 112 $kg \cdot ha^{-1}$. Spring-planted cereal rye is characterized by leafy shoots with tender stems and remains vegetative; thus, it may be easier to manage than fall-planted rye (Ateh and Doll, 1996).

The seeding rate has less of an effect on biomass production of fall-planted rye because it can compensate for low seeding rates by tillering. Masiunas, Weston, and Weller (1995) evaluated rye fall-seeding rates between 56 and 170 $kg \cdot ha^{-1}$. Rye biomass varied from 320 to 1150 $g \cdot m^{-2}$ and was not affected by seeding rate (Masiunas, Weston, and Weller, 1995). Smeda and Weller (1996) found that optimum weed control and crop yields can be obtained in Indiana if the rye is seeded at 157 $kg \cdot ha^{-1}$ in early to mid-September (Smeda and Weller, 1996). This seeding rate produced an initial rye residue biomass of approximately 12,000 to 20,000 kg·ha^{-1} (Smeda and Weller, 1996). High amounts of rye biomass are important to obtain acceptable weed suppression.

Rye planting date is important to allow biomass production for spring-seeded rye and to ensure winter survival for fall-seeded rye. Generally, the earlier that rye can be seeded in the spring, the more biomass will be produced. It is important with fall-planted rye to select a seeding date that provides consistent survival regardless of winter conditions (Burket, Hemphill, and Dick, 1997). Burket, Hemphill, and Dick (1997) found that late plantings (after October 1), a dry fall, excessive rainfall and/or a cold spell (<0°C for several days) in November or December severely reduced rye growth and survival. We found similar results where late plantings (after October 1) or low-vigor rye stands were much more susceptible to winter injury than were early plantings (before September 20) or more vigorous stands (Masiunas, Weston, and Weller, 1995).

The timing and method used to kill rye will also influence rye regrowth, biomass production, weed suppression, and vegetable crop yield. Spring-planted rye is generally allowed to grow until the vegetable crop is established and then the rye is killed with a postemergence graminicide (i.e., fluazifop, sethyoxydim). The goal is to maximize weed suppression while minimizing competition and maximizing yield of the vegetable crop. Ateh and Doll (1996) reported that for spring-planted rye, killing the cover crop 45 days after planting soybean gave optimum weed control (Ateh and Doll, 1996).

Much more research has been conducted on timing and method of killing fall-seeded rye. Rye's use of soil moisture during the spring and immobilization of N depend on when it is killed (Raimbault, Vyn, and Tollenaar, 1991). Low soil moisture or N will reduce vegetable crop establishment and yield. Killing the rye earlier in the spring allows more time for the soil profile to be recharged (Doss, Turner, and Evans, 1981; Wagner-Riddle, Gillespie, and Swanton, 1994). The problem with a late rye killing date is that there will be a greater reduction in crop yield than with an early rye killing date if either May or July were dry (Wagner-Riddle, Gillespie, and Swanton, 1994; Wagner-Riddle, Gillespie, and Hunt, 1997).

The time interval between killing the rye and planting the vegetable is important in determining weed suppression, since allelochemicals are produced mainly during the early periods of rye residue decomposition and the compounds have a short period of survival (<30 days) in the soil (Raimbault, Vyn, and Tollenaar, 1991). In Michigan, Putnam (1986) reported that for best weed control and vegetable growth the rye should be desiccated between May 1 and 10 when it reaches a height of 0.5 m. Smeda and Weller (1996) evaluated killing the rye between one and four weeks before planting tomatoes. Killing rye at four weeks before planting the tomatoes resulted in no weed suppression, whereas killing the rye two weeks before planting the tomatoes resulted in up to 97 percent weed suppression at five weeks after tomato planting (Smeda and Weller, 1996). The optimum killing time for rye was one week before planting the crop (Wagner-Riddle, Gillespie, and Swanton, 1994; Smeda and Weller, 1996; Wagner-Riddle, Gillespie, and Hunt, 1997).

Several methods to kill rye have been evaluated. Tillage can be used to kill the rye. Rye is usually incorporated with a disk or moldboard plow two weeks before spring planting to conserve soil water and to initiate decomposition of the cover crop so planting operations are not impaired (Creamer, et al., 1995). Incorporation of residues into the soil greatly reduces weed control (Putnam, 1986; Creamer, et al. 1995; Masiunas, 1999). When the rye is incorporated, there is less surface coverage, the tillage brings new weed seed to the soil's surface, and there is more rapid decomposition of incorporated residues (Masiunas, 1999). Also, the rye cover crop must be incorporated early in the spring when growth is rapid, forfeiting significant amounts of dry matter (Creamer et al., 1995).

Because of these disadvantages, rye residues are generally left on the surface as a mulch, and the subsequent vegetable crop is direct seeded or transplanted with minimum soil disturbance into the surface mulch. This best exploits the cover crop's weed control potential (Creamer et al., 1995). Herbicides have been the most successful method used to kill the rye cover crop (Putnam, 1986, 1990; Creamer et al., 1995). Contact herbicides such as paraquat or glufosinate, nonselective and translocated herbicides such as glyphosate, or selective graminicides such as fluazifop or sethyoxydim have all been used to kill rye cover crops (Putnam, 1986, 1990; Weston, 1990). Paraquat or glufosinate rapidly desiccate rye, but regrowth can be a problem. Sethoxydim and fluazifop controlled rye but required 20 to 30 days to achieve pronounced desiccation (Weston, 1990). Glyphosate at 1.1 or 2.2 $kg \cdot ha^{-1}$ completely killed rye within 10 to 14 days without any regrowth (Weston, 1990; NeSmith and McCracken, 1994; Masiunas, Weston, and Weller, 1995; Masiunas, 1999). When glyphosate was used to kill the rye, vegetable crop seedling establishment and weed suppression were better than when other herbicides were used to kill rye.

Some cropping systems do not allow the use of herbicides. Instead mowing, undercutting, or rolling can be used to kill rye. The success of using mowing to kill rye is dependent on the type of mower used and the cover crop growth stage (Masiunas, 1999). Sickle bar and flail mowers deposit residue evenly across the mowed width which will evenly suppress weeds and the regrowth of rye (Wilkins and Bellinder, 1996). Mowing with a flail-chop or rotary mower produces small pieces of rye that rapidly decompose because they are in close contact with the soil and decomposing microorganisms (Creamer et al., 1995).

Rye will regrow if it is mowed before maturity. Schonbeck and colleagues (1993) and Wilkins and Bellinder (1996) found that rye regrew when mowed at the boot stage, but mowing after flowering resulted in no regrowth of rye. When rye was cut at the first or second nodes, the apices responsible for tillering were below the 10 cm mowing height and regrowth was from undamaged tiller elongation, while rye mowed after the boot stage had regrowth originating from secondary apices (Wilkins and Bellinder, 1996). Two factors must be balanced in deciding when to mow rye. One factor is the need to wait until rye reaches the optimum stage for biomass production while having minimal regrowth. The other factor is the need to mow rye early enough so vegetable crop establishment is not reduced by low soil moisture (Wilkins and Bellinder, 1996).

Rye can be managed to improve its control by mowing. Late fall rye seeding will favor late spring flowering and less regrowth (Wilkins and Bellinder, 1996). High seeding rates of rye cause fewer tillers per plant and make regrowth less likely. Consistent planting depth, resulting in uniform rye growth, will make mowing a more consistent method of control. Nutrient-poor soil during the spring, which reduces rye's ability to recover and renew vegetative growth, will minimize regrowth but must be balanced against optimizing nutrients for vegetable crop growth (Wilkins and Bellinder, 1996).

Undercutting and rolling have also been evaluated for killing rye. Rolling at a proper growth stage will kill cover crops, but it has not been very consistent with rye and has not been widely adopted. Creamer and colleagues (1995) developed and evaluated an undercutter that killed the rye by severing their roots while flattening the aboveground biomass on the soil surface. When rye was in mid- to late-bloom stage or beyond, it was easily killed by the undercutter, and the resulting mulch provided weed control similar to the conventional tillage treatment (Creamer et al., 1995).

MECHANISMS EXPLAINING RYE EFFECTS ON WEEDS

Rye cover crop mulches suppress weeds through physical, environmental, and chemical factors (Lanfranconi, Bellinder, and Wallace, 1993; Bellinder, Wallace, and Wilkins, 1996; Weston, 1996). Rye residues are a physical barrier restricting emergence of some seedlings (Masiunas, Weston, and Weller, 1995). For example, Creamer, Bennett, Stinner, Cardina, and Regnier (1996) reported that suppression of yellow foxtail by rye could be attributed to physical suppression alone. But in another field study, rye biomass (up to 1,650 or 2,300 kg·ha^{-1}) was not correlated with weed emergence (Lanfranconi, Bellinder, and Wallace, 1993). Mohler and Teasdale (1993) felt that feeding by larger numbers of insects, slugs, snails, and voles in the high-residue treatments also tends to decrease the number of weed seedlings.

Rye cover crop mulches modify the soil microenvironment, thereby affecting moisture, temperature, and radiation, which influences the growth of weeds (Schonbeck et al., 1993; Masiunas, Weston, and Weller, 1995; Masiunas, 1999). Changes in soil microclimate may create suboptimal conditions and alter the environmental cues needed for weed seed germination (Masiunas, Weston, and Weller, 1995). The surface coverage of the rye cover crop is correlated with light transmittance and weed suppression (Mohler and Teasdale, 1993; Teasdale, 1993a,b; Masiunas, Weston, and Weller, 1995; Bottenberg et al., 1997a; Masiunas, 1999). The reduced radiation reaching the soil surface with rye residues is important because many small-seeded weed species require a light stimulus for germination (Bottenberg et al., 1997a; Masiunas, 1999). For example, Creamer, Bennett, Stinner, and Cardina (1996) reported that the inhibition of eastern black nightshade emergence in rye cover crop was likely due to sensitivity of nightshade to light exclusion.

Soils under rye cover crop mulches are more moist and subject to less moisture fluctuations than soils in conventional tillage systems (Kahn, Monke, and Foster, 1988; Steiner, 1989; Coolman and Hoyt, 1993; Lal, 1995; Masiunas, Weston, and Weller, 1995; Abdul-Baki et al., 1996; Wilkins and Bellinder, 1996; Wagner-Riddle, Gillespie, and Hunt, 1997; Smeda and Weller, 1996; Mundy et al., 1999). The greater soil moisture and reduced moisture fluctuations under rye residues decrease the cues triggering weed germination (Mohler and Teasdale, 1993; Bottenberg et al., 1997a; Masiunas, 1999). The excess moisture in rye cover crop mulch systems can cause low oxygen levels in the seed germination zone, inhibiting growth or killing seedlings before emergence.

Soils with cover crops have lower daytime temperatures (Coolman and Hoyt, 1993; Schonbeck et al. 1993; Wagner-Riddle, Gillespie, and Swanton, 1994; Stirzaker and White, 1995; Smeda and Weller, 1996; Wilkins and Bellinder, 1996; Hoyt, 1999; Masiunas, 1999; Mundy et al., 1999; Figure 6.2). The lower daytime soil temperatures under residue reduce weed emergence (Mohler and Teasdale, 1993; Creamer, Bennett, Stinner, Cardina, and Regnier, 1996) and delay growth of weeds, affecting weed-crop competition (Masiunas, 1999). The greatest differences in soil temperatures are early in the season before crop canopy closure and when weed seed germination occurs (Wagner-Riddle, Gillespie, and Swanton, 1994; Masiunas, 1999).

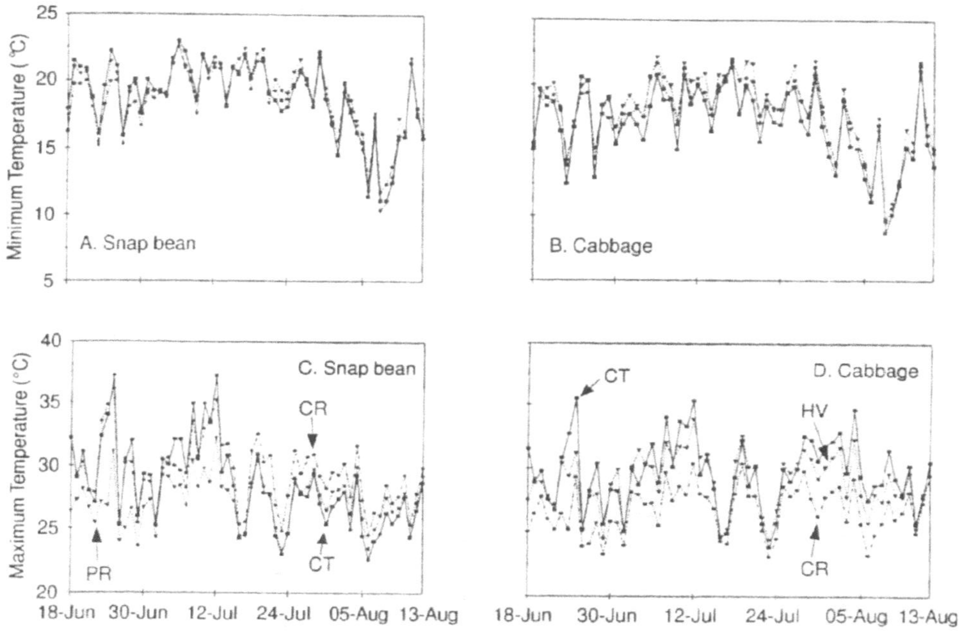

FIGURE 6.2. The effect of rye cover crop mulch on the soil temperature. The graphs are minimum soil temperature for snap bean (A), minimum soil temperature for cabbage (B), maximum soil temperature for snap bean (C), and maximum soil temperature for cabbage (D). The cropping systems are CT, mold-board plowed in the fall followed by spring discing and trifluralin incorporated at 0.56 kg·ha^{-1}; CR, cereal rye killed with glyphosate at 1.1 kg·ha^{-1} and mowed; HV, hairy vetch killed with 2,4-D at 0.56 kg·ha^{-1} and mowed; PR, perennial ryegrass interseeded between rows of snap beans (*Source:* Reproduced by permission of The Haworth Press from Figure 1 of Masiunas et al., 1997, p. 69).

Rye cover crop mulches also cause less variable diurnal soil temperatures (Coolman and Hoyt, 1993; Bistow et al., 1996; Smeda and Weller, 1996; Wilkins and Bellinder, 1996; Wagner-Riddle, Gillespie, and Hunt, 1997; Mundy et al., 1999). The reduced soil temperature fluctuations under rye residues decrease weed seed germination for species that require diurnal temperature fluctuations to break dormancy (Mohler and Teasdale, 1993; Creamer, Bennett, Stinner, Cardina, and Regnier, 1996; Masiunas, 1999). Creamer, Bennett, Stinner, and Cardina (1996) reported that temperature fluctuations greater than 10°C are generally required to break weed seed dormancy, and that cover crop mulch reduced diurnal fluctuations to less than 10°C.

The microenviromental changes do not explain all the weed suppression from rye residues. Rye residue reduced total weed biomass by 63 percent more than popular excelsior, an inert mulch with similar microclimate effects as rye residues (Barnes and Putnam, 1983). This suggests that allelopathy, in addition to physical effects of the mulch, contributed to weed control (Barnes and Putnam, 1983, 1986; Anaya, 1999). Allelopathic inhibition of rye typically results from the combined action of a group of allelochemicals, as allelochemicals, in general, are collectively known to interfere with several physiological processes (Einhellig, 1996). The joint action of allelochemicals is important because the concentrations of single chemicals in field situations are below their inhibition thresholds (Einhellig, 1996).

In rye, the allelochemicals are primarily organic acids and include vanillic, ferulic, phenylacetic, 4-phenylbutyric, *p*-hydroxybenzoic, salicylic, *o*-coumaric, β-phenyllactic (β-PLA), and β-hydroxybutyric (β-MHA) acids, along with salicylaldehyde (Perez and Ormeno-Nunez, 1993; Mwaja, Masiunas, and Weston, 1995; Creamer, Bennett, Stinner, Cardina, and Regnier, 1996). Most research has concentrated on two hydroxamic acids, 2,4-dihydroxy-1,4 (*2H*)-benzoxazin-3-one (DIBOA) and its metabolite 2(*3H*)-benzoxazolinone (BOA) (Barnes et al., 1987; Mwaja, Masiunas, and Weston, 1995; Creamer, Bennett, Stinner, Cardina, and Regnier, 1996). DIBOA and BOA are more toxic to seedling growth than are other allelochemicals from rye residues (Barnes et al. 1987; Mwaja, Masiunas, and Weston, 1995).

In intact rye tissue, the glycoside, 2-O-β-glucopyranosyl-4-hydroxyl-1,4-benzoxazin-3-one (DIBOA-glc), was the main hydroxamic acid (Barnes et al., 1987; Perez and Ormeno-Nunez, 1993; Mwaja, Masiunas, and Weston, 1995). β-glucosidases are released upon injury to the rye and rapidly hydrolyze the glucoside (DIBOA-glc) to release DIBOA (Barnes et al., 1987). DIBOA decomposes in water to release BOA (Mwaja, Masiunas, and Weston, 1995). Additional allelochemicals arise by microbial transformation of compounds from rye residues (Barnes and Putnam, 1986; Mwaja, Masiunas, and Weston, 1995). For example, *Acinetobacter calcoaceticus,* a gram-negative bacterium, was responsible for the biotransformation of BOA to 2,2'-oxo-1,1'-azobenzene (AZOB) (Chase et al., 1991). AZOB is more toxic to curly cress (*Lepidium sativum* L.) and barnyardgrass than either DIBOA or BOA (Chase et al. 1991; Mwaja, Masiunas, and Weston, 1995).

In general, the extent and duration of weed suppression by allelopathy is dependent upon the concentration of allelochemicals in the zone of weed emergence and the rate of breakdown of those chemicals (Smeda and Weller, 1996). Mwaja, Masiunas, and Weston (1995) found the concentration of DIBOA ranged between 128 to 423 $\mu g \cdot g^{-1}$ and BOA ranged between 2.5 and 31 $\mu g \cdot g^{-1}$ in freshly killed rye shoots. The amounts of DIBOA and BOA correlated with the amount of rye herbage and suppression of weeds (Masiunas, 1999). Smeda and Weller (1996) also found that higher amounts of rye herbage would produce greater quantities of allelochemicals, causing greater weed suppression than lower amounts.

Allelochemical content and weed suppression also vary depending on the plant part, its age, and the rye genotype. Rye shoot herbage is twice as inhibitory as root herbage (Barnes et al., 1987). Young rye tissue contains higher levels of allelochemicals but also decays more rapidly than older tissue (Barnes et al., 1987; Wahle and Masiunas, 2002). We found the greatest concentrations of allelochemicals in spring-grown rye (Mwaja, Masiunas, and Weston, 1995). Rye genotypes differ in their allelochemical content and ability to suppress weeds (Barnes et al., 1987; Wahle and Masiunas, 2002; see Table 6.6).

After the rye is killed, its tissue and allelochemical content will rapidly decline. Yenish, Worsham, and Chilton (1995) found that the biomass of killed rye shoots declined linearly over time, with 50 percent of residue disappearing by 15 weeks. Because of the delay in residue decomposition, the maximum allelochemical content occurs within 20 days after rye decomposition begins (Masiunas, Weston, and Weller, 1995). Once in the soil environment, allelochemicals are rapidly decomposed, with 50 percent of the allelochemicals disappearing within 10 to 12 days (Bottenberg et al., 1997a). The duration of weed suppression by cover crops more closely follows the disappearance of allelochemicals from rye residue than the disappearance of the residue (Yenish, Worsham, and Chilton, 1995).

Environmental factors influence allelochemical production in rye and their weed-suppressive activity (Mwaja, Masiunas, and Weston, 1995). Stressful conditions often enhance allelochemical production, and plants are more sensitive to allelopathy when moisture, temperature, or nutrient levels are less than optimal (Einhellig, 1996). Increased air temperatures

TABLE 6.6. The effect of genotype on weed suppression of fall-seeded, spring-killed rye. Weed counts (m^{-2}) were averaged across vegetable crops and totaled over sampling dates.

Cultivar	Broadleaf	Grass
Conventional tillage[a]	808 e[b]	17.0
Aroostook	252 ab	25.8
Bonel	230 a	19.4
Dacol	602 d	47.3
Florida 401	898 f	40.6
Prima	498 c	61.9
Wheeler	295 ab	26.1
Wrens Abruzzi	330 b	35.9

Source: Adapted by permission of Elizabeth Wahle from Table 3.1 of Wahle, 1996.

[a]Conventional tillage with no cover crop and trifluralin preplant incorporated at 0.8 $kg{\cdot}ha^{-1}$.

[b]Mean separation within columns using Fisher's protected LSD test, $p = 0.05$.

enhance allelochemical concentration and consequent inhibition of weeds (Mwaja, Masiunas, and Weston, 1995). Heavy rains will reduce allelochemical content of rye tissues and will move water-soluble allelochemicals out of the zone where seeds are germinating (Smeda and Weller, 1996). Plants produced under high light intensity generally produce greater levels of extractable allelochemicals because of greater exposure to ultra-violet radiation, which damages the tissue (Mwaja, Masiunas, and Weston, 1995).

Mwaja, Masiunas, and Weston (1995) found that fast-growing plants in fertile environments contained lower allelochemical concentrations than slow-growing species in areas of limited fertility. Also, some allelochemicals interfere with mineral uptake, causing a subsequent reduction in the nutrient content of weed tissues (Einhellig, 1996), and may impact on weed crop competition. Amisi (2000) and Wahle and Masiunas (2002) found that increasing fertilization did not overcome the negative action of allelochemicals (see Tables 6.7 and 6.8).

THE EFFECTS OF WEED CONTROL AND RYE ON VEGETABLES

Rye cover crop mulch systems not only provide weed control but also affect vegetable crop establishment and growth, which influences crop com-

TABLE 6.7. The effect of added nitrogen on 'NK 530' squash N content and yield. Urea (46-0-0) was surface applied three days after planting the squash.

Cropping system	Nitrogen level ($kg \cdot ha^{-1}$)	Total N[a] (%)	Yield	
			Number/plot	Weight ($kg \cdot plot^{-1}$)
Conventional[b]	0	6.0 b[c]	16.8 b	92.2 c
Conventional	45	6.6 d	16.5 b	103.1 de
Conventional	90	6.4 cd	15.8 b	96.2 cd
Conventional	135	6.1 b	16.5 b	108.7 e
Conventional	270	6.4 cd	17.3 b	107.1 e
Rye	0	4.4 a	10.5 a	39.3 a
Rye	45	6.2 bc	9.5 a	43.2 a
Rye	90	6.1 b	9.0 a	60.3 b
Rye	135	6.5 d	10.3 a	58.3 b
Rye	270	6.5 d	9.3 a	51.2 ab

[a]Total N is the nitrogen content of the youngest fully expanded leaf on August 28 during fruit enlargement.

[b]Conventional was moldboard plowed and disked in the fall. In the spring the plot was tilled with a roterra and clomazone applied at 0.56 $kg \cdot ha^{-1}$. Rye was fall planted; in the spring glyphosate was applied on May 13 and mowed on May 22.

[c]Mean separation within columns using Fisher's protected LSD test, $p = 0.05$.

petitiveness against weeds and their subsequent yield. In general, cover crop mulches are more management intensive and vegetable yields more variable than in systems using conventional tillage with herbicides (Masiunas, Weston, and Weller, 1995; Masiunas, 1999). Large-seeded, quickly germinating vegetable crops, particularly legumes, preform better in rye cover crop mulch systems than do small-seeded or slowly germinating species such as carrots (*Daucus carota* L.) and onions (*Allium cepa* L.) (Putnam, 1986; Weston, 1990; Galloway and Weston, 1996; Hoyt, 1999).

Rye cover crops can reduce the emergence and plant stand of some vegetables. Cereal rye residues are phytotoxic to direct-seeded tomato (*Lycopersicon esculentum* Mill.); however, using transplants overcomes the phytotoxicity (Smeda and Weller, 1996). Potato populations in rye systems were lower than in conventional tillage (Bellinder, Wallace, and Wilkins, 1996; Mundy et al., 1999). Potatoes in reduced-tillage systems emerged more slowly than potatoes in conventional tillage (Mundy et al., 1999). Rye reduced the stand of corn and caused less plant growth (Eckert, 1991; Burgos and Talbert, 1996). There was no effect of rye on snap bean stand, and when

TABLE 6.8. The effect of rye residues and nitrate on pumpkin mass and leaf N.

Treatment	Nitrogen (ppm)	Fresh mass (g/plant)	Dry mass (g/plant)	Leaf nitrogen (%)
Sand only	20	7.4 gh[a]	1.3 e	1.8 g
	100	20.2 de	3.4 d	2.8 c-g
	300	48.4 c	6.7 c	3.9 a-c
Unleached shoots	20	0.4 h	0.5 e	3.8 a-d
	100	0.8 h	0.7 e	3.6 a-e
	300	0.5 h	0.4 e	2.7 d-g
Leached shoots[b]	20	6.9 gh	1.3 e	1.9 fg
	100	17.8 ef	2.7 d	3.1 c-f
	300	66.1 d	8.5 a	4.1 ab
Rye leachate[c]	20	1.4 h	0.7 e	2.9 c-g
	100	0.5 h	0.9 e	2.9 c-g
	300	0.3 h	0.9 e	3.4 b-e

Source: Modified by permission from Karen Amisi from Table 3.15 of Amisi, 2000.

[a]Mean separation within columns using Fisher's protected LSD test, $p = 0.05$.

[b]The leached shoots were shaken in deionized water for 24 hours to remove water-soluble allelochemicals. The rye leachate was the water solution resulting from this shaking.

[c]The leachate was used for daily watering of the pots in the rye leachate treatment.

adequate stands and weed control were obtained, there was no effect on yield (NeSmith and McCracken, 1994).

Vegetable crop growth and maturity can be affected by rye cover crops. For example, potato plants were smaller in reduced-tillage systems (Mundy et al., 1999), while there were no differences in the size of tomatoes grown in conventional tillage or rye mulch systems (Drost and Price, 1991). Galloway and Weston (1996) found that sweet corn and pumpkins produced in rye residues had delayed maturity in comparison to conventional tillage. In other instances, crop maturity was not affected. For example, maturity of snap beans, tomatoes, and peas (*Pisum sativum* L.) was not delayed by rye cover crops compared to conventional systems (Bellinder, Wilson, and Hines, 1987; Knott, 1996; Smeda and Weller, 1996).

The effect of rye cover crop mulch on vegetable crop yields vary. Yields and fruit weight of tomato plants grown in rye were higher or comparable to those grown in the conventional black polyethylene mulch systems (Zehnder and Linduska, 1987; Drost and Price, 1991; Masiunas, Weston, and Weller, 1995; Abdul-Baki et al., 1996; Bottenberg, Masiunas, and Eastman, 1999; Akemo, Bennett, and Regnier, 2000). In other instances, tomato yields in rye cover crops were reduced compared to those in conventional tillage (Doss, Turner, and Evans, 1981; Mwaja, Masiunas, and Eastman, 1996; Masiunas, 1999). For example, in one study, marketable yields averaged 2.2 $Mt{\cdot}ha^{-1}$ greater in convention tillage than in rye treatments (Doss, Turner and Evans, 1981).

Total potato yields were similar or increased for rye cover crop systems when compared to conventional tillage (Bellinder, Wallace, and Wilkins, 1996; Mundy et al., 1999). Even though total yields were similar, Bellinder, Wallace, and Wilkins (1996) found that in the rye system marketable yields were about 14 percent lower due to an increased number of predominately green culls (Bellinder, Wallace, and Wilkins, 1996). The increased greening is likely caused by exposure of tubers to more light. Potato fields with rye cover crops often have problems in obtaining adequate planting depth and soil coverage.

Studies have also evaluated the effect of cover crop mulches on yields of transplanted cabbage (*Brassica oleracea* L.) and broccoli (*Brassica oleracea* L.). Burket, Hemphill, and Dick (1997) reported that the yield of the broccoli in a cereal rye cover crop was 81 percent of the yield in the conventional tillage. Knavel and Herron (1981) and Drost and Price (1991) found higher marketable yields of spring cabbage due to higher soil moisture throughout the growing season in cover crops than in conventional tillage (Drost and Price, 1991). Other studies found that the greatest cabbage yields were in conventional tillage and not the rye cover crop (Roberts and Cartwright, 1991; Mwaja, Masiunas, and Eastman, 1996; Bottenberg et al., 1997b).

In no-till rye mulch systems, yields of pumpkin (*Cucurbita pepo* L.), melons (*Cucumis melo* L.), cucumbers (*Cucumis sativus* L.), and squash [*Cucurbita pepo* L. and *C. moschata* (Duchesne ex Lam.) Duchesne ex Poir.] were comparable to or less than those in conventional tillage (Drost and Price, 1991; Singogo, Lamont, and Marr, 1996; Bottenberg, Masiunas, and Eastman, 1999; Amisi, 2000; see Table 6.2). The similar marketable yields in rye cover crops were probably due to higher soil moisture compensating for other adverse factors (Drost and Price, 1991), while lower yields were caused by soil compaction and allelochemicals (Amisi, 2000).

Yields of direct-seeded vegetables, such as snap beans and sweet corn, were also determined in rye cover crop mulch systems. Snap bean yield in plots with rye mulch was reduced or comparable to those in conventional tillage (NeSmith and McCracken, 1994; Mwaja, Masiunas, and Eastman, 1996; Bottenberg et al., 1997a; Bottenberg, Masiunas, and Eastman, 1999).

For example, Mwaja, Masiunas, and Eastman (1996) found that total snap bean yields were greater by 2 to 3 $Mg \cdot ha^{-1}$ in the conventional tillage treatment than in the rye cover crop treatments. There was also a consistent trend for rye cover crops to cause a small decrease in sweet corn yields compared to those in conventional tillage (Burket, Hemphill, and Dick, 1997). In general, transplanted vegetables grow and yield better in rye cover crop systems than do direct-seeded vegetables.

The reduced or variable yields for direct-seeded vegetables were likely caused by a variety of factors. Vegetable yields are correlated with weed control obtained from rye cover crop systems (Smeda and Weller, 1996; Masiunas, 1999). Bottenberg and colleagues (1997a,b) found that yield loss and poor growth of snap beans and cabbage in plots with rye mulch was partly attributable to poor killing of the rye (Bottenberg et al., 1997a). Potato yields were higher in reduced tillage than in conventional tillage when early-emerging weeds were suppressed (Lanfranconi, Bellinder, and Wallace, 1993). Weed abundance had a significant negative correlation with corn and tomato yields (Clark et al., 1998).

The microenvironmental effects of cover crop mulch systems also impact vegetable crop establishment, growth, and yield. Rye cover crop mulch systems reduce soil temperatures which are correlated with reduced yields in a variety of vegetable cropping systems (Drost and Price, 1991; Masiunas, Weston, and Weller, 1995; Bottenberg et al., 1997a; Bottenberg, Masiunas, and Eastman, 1999; Hoyt, 1999; see Figure 6.2). Yield reductions have been reported for short-season vegetable crops or crops planted in early spring when large amounts of cover crop residue slows soil warming (Hoyt and Walgenbach, 1995). For example, crucifer crops grown with cover crop mulch systems in early spring or late fall have lower soil temperatures and delayed growth compared to conventional tillage systems (Hoyt, 1999). Even though maximum temperatures in rye cover crops were lowered, the growth and yield reductions of snap beans were not due to suboptimum temperatures in the rye but were likely caused by soil compaction and allelochemicals (Bottenberg, Masiunas, and Eastman, 1999).

Soil moisture is either decreased by rye growth in the early spring or increased by the surface mulching effects of killed rye cover crops. The reduced vegetable yield, because of lower soil moisture, can probably be eliminated by killing the rye earlier in the spring, allowing more time for soil recharge by rainfall (Doss, Turner, and Evans, 1981), but earlier killing of rye will reduce weed suppression. Rye cover crop mulches also increase water infiltration and reduce runoff and evaporation, resulting in greater available soil moisture and crop yields (Drost and Price, 1991; Morse, 1993). But the wetter soils can cause cooler soil temperatures and greater problems with slugs, which limit sweet corn production in rye cover crop systems (Drost and Price, 1991).

Rye cover mulch systems and the wetter condition can result in soil compaction and reduce crop yield (Morse, 1993; Bottenberg et al., 1997a; Bottenberg, Masiunas, and Eastman, 1999; Mundy et al., 1999). The soil compaction is caused by increased wheel traffic during fall planting and spring kill of rye, and a subsequent lack of tillage (Bottenberg, Masiunas, and Eastman, 1999). The more dense soil may restrict vegetable crop root growth or it may inhibit microbial activity and reduce nitrate uptake (Knavel and Herron, 1981; Bottenberg, Masiunas, and Eastman, 1999).

Nitrogen immobilization commonly occurs during decomposition of rye mulch (Schonbeck et al., 1993; Masiunas, Weston, and Weller, 1995; Wyland, Jackson, and Schulbach, 1995; Bottenberg et al., 1997a; Bottenberg, Masiunas, and Eastman, 1999; Hoyt, 1999). For example, the chlorotic snap bean growth and lower yields were partly due to N immobilization during mulch decomposition (Mwaja, Masiunas, and Eastman, 1996; Bottenberg et al., 1997a). The yields of snap beans, tomatoes, and cabbage grown in rye cover crops were all increased by the addition of N fertilizer (Skarphol, Corey, and Meisinger, 1987; Roberts and Cartwright, 1991; Stivers and Shennan, 1991), but N fertilization alone could not overcome the lower yields (Knavel and Herron, 1981; Amisi, 2000; see Tables 6.7 and 6.8).

The reduced vegetable growth and yields may be due to the effects of allelochemicals on the crop. Rye cover crop allelopathy may inhibit the emergence or early growth of small-seeded crops such as lettuce (*Lactuca sativa* L.) or direct-seeded tomato (Schonbeck et al., 1991). The stand reduction of sweet corn was partly due to allelopathy of the rye (Burgos and Talbert, 1996).

METHODS TO MAKE RYE A FEASIBLE WEED MANAGEMENT TACTIC

Rye cover crop mulch systems can be combined with either preemergent or postemergent herbicides to improve weed control. An ideal surface-applied preemergence herbicide for use in rye cover crop systems must not bind to the rye residues and must be selective and registered on the crop to be grown (Hoyt, Monks, and Monaco, 1994). Also, more irrigation or rainfall is needed to move the herbicide through the rye residues into the soil where it is active (Hoyt, Monks, and Monaco, 1994). In sweet corn, snap beans, and potatoes, rye cover crop mulch allowed rates of preemergent herbicides to be reduced by one-third or one-half of the standard rate without loss of weed control (Lanfranconi, Bellinder, and Wallace, 1993; Burgos and Talbert, 1996).

Rye cover crop mulches can be used to provide early-season weed control, and postemergent herbicides can be applied as necessary for later-season weed control. Burgos and Talbert (1996) found that cover crops allowed postemergence herbicide applications to be delayed for four to six weeks after transplanting tomatoes with no loss of weed control. In another study with tomatoes, excellent results were obtained by using a rye cover crop for early-season weed control followed by one application of metribuzin at 0.28 $kg \cdot ha^{-1}$ for later-season weed control (Putnam, 1990).

Strip tillage and combinations of rye with other cover crop species have been evaluated as methods to improve vegetable growth and yield while maintaining weed control. Strip tillage combines in-row tillage for improved planting efficiency and soil physical properties with between-row cover for weed suppression and moisture and soil conservation (Wilhoit, Morse, and Vaughan, 1990; Putnam, 1990; Morse, 1993; Hoyt, Bonanno, and Parker, 1996; Bottenberg, Masiunas, and Eastman, 1999). The soil disturbance during tillage of the strip and loss of the rye surface residues aggravates weed problems (Putnam, 1990) but also allows incorporation of herbicides, cultivation, or the laying of plastic within the crop row to control weeds (Wilhoit, Morse, and Vaughan, 1990; Masiunas, 1999).

The tillage within the crop row breaks up the compacted top soil layer before planting, but subsequent soil consolidation through rainfall and machinery traffic increases soil bulk density compared to conventional tillage (Bottenberg, Masiunas, and Eastman, 1999; see Table 6.1). Strip tillage may reduce allelochemical effects on vegetable crops by diluting allelochemicals within the row and by keeping rye residues away from close contact with the plants (Hoyt, Bonanno, and Parker, 1996; Bottenberg, Masiunas, and Eastman, 1999). But Amisi (2000) found that even when the rye is incorporated into the tilled strip, it still has allelopathic effects on pumpkins (see Table 6.2).

Even though strip tillage may not alleviate soil compaction or allelochemical effects on vegetables, it can result in vegetable crop yields similar to conventional tillage. Strip tillage resulted in higher corn, tomato, squash, and pepper yields than those in no-till treatments (Loy et al., 1987; Raimbault, Vyn, and Tollenaar, 1991; Hoyt, 1999). Hoyt and Monks (1996) reported that yields of Irish potato and sweet potato [*Ipomoea batatas* (L.) Lam.] were equal in strip tillage or conventional tillage treatments with herbicides. Cabbage yields varied; in some instances, yields in strip tillage were the same (Wilhoit, Morse, and Vaughan, 1990; Hoyt, Bonanno, and Parker, 1996), while in other instances, cabbage yields were lower than in conventional tillage (Hoyt and Walgenbach, 1995). The strip tillage system caused cabbage roots to become established in a shallow cultivated area, and on soils prone to compaction, cabbage yields were reduced (Hoyt and Walgenbach, 1995).

Using polycultures of cover crops can combine the weed suppression of rye with the vegetable crop yield enhancement of legumes (Creamer, Bennett, and Stinner, 1997). Legumes grown with cereal rye may produce more organic matter, protect the soil, and suppress weeds better than either cover crop grown alone (Schonbeck et al., 1993; Creamer, Bennett, Stinner, Cardina, and Regnier, 1996; Akemo, Bennett, and Regnier, 2000). Cereal rye offsets the rapid decomposition of legumes, aids in their overwintering, and increases their above-ground biomass (Creamer, Bennett, Stinner, Cardina, and Regnier, 1996; Mwaja, Masiunas, and Eastman, 1996; Akemo, Bennett, and Regnier, 2000). The greater biomass and slower decomposition improves the weed suppression of rye-legume polycultures compared to legumes alone.

Combining cereal rye with legumes may improve the nitrogen advantages of cover crop systems. Cereal rye is able to capture more residual soil N during the fall and winter than are legumes alone (Clark, Decker, and Meisinger, 1994; Burket, Hemphill, and Dick, 1997; Ranells and Wagger, 1997; Akemo, Bennett, and Regnier, 2000), and also reduces fertilizer use and groundwater contamination (Ranells and Wagger, 1997). Combining a legume with rye takes advantage of the legume's ability to fix N and may result in the legume accumulating more N than if it were growing by itself (Creamer, Bennett, Stinner, Cardina, and Regnier, 1996; Creamer, Bennett, and Stinner, 1997; Kuo and Sainju, 1998; Akemo, Bennett, and Regnier, 2000). By combining plants whose C:N ratio is high (mature cereals) with those that are low (legumes), immobilization can be reduced, and the release of N can theoretically be timed with crop uptake (Creamer, Bennett, and Stinner, 1997). The positive effects on nitrogen availability cause increased vegetable yields compared to cropping systems with rye alone (Schonbeck et al., 1993; Clark, Decker, and Meisinger, 1994; Akemo, Bennett, and Regnier, 2000). But further research is necessary to determine the best legume combinations with rye and to obtain consistent weed control and vegetable yields.

CONCLUSIONS AND RECOMMENDATIONS

Rye cover crop systems can reduce soil erosion and nitrogen leaching, protecting the environment. Rye residues can suppress annual weeds for up to twelve weeks and can result in vegetable crop yields similar to conventional tillage with herbicides. The challenge is to improve the consistency of weed control and increase vegetable crop yield. Until the management intensity and the risks associated with the potential of poorer vegetable quality and yields in rye cover crop systems can be overcome, vegetable grower adoption of this system will be limited.

The difficulty for researchers is to overcome the site- and environment specific aspects of rye cover crop mulch systems and develop systems that have similar or better vegetable crop yields and lower productions costs than conventional tillage systems using herbicides. Researchers need to concentrate on hybrid systems such as those that use polyethylene mulch and trickle irrigation within rows of transplanted vegetables and rye cover crop mulch between rows. Fertilization recommendations and irrigation schedules will need to be developed specifically for these systems. Reduced rate or total postemergence herbicide programs will need to be developed for vegetable crops to supplement the weed suppression by rye cover crop mulches.

For direct-seeded vegetables, more research is needed on strip-tillage systems and on polycultures of rye and legumes. Preemergent herbicides need to be registered for many vegetable crops that can be effectively used in the tilled strips containing rye residue. Fertilization requirements of vegetables grown in tilled strips between rye or polycultures of rye and legumes are unknown. Breeding of rye specifically for use in cover crop mulch systems should be conducted. Critically, economic and cost-benefit analysis needs to be conducted with rye cover crop mulch systems. None of this additional research will be useful unless participatory and on-farm research involving key vegetables is conducted with rye cover crop mulch systems. Vegetable grower acceptance of rye cover crop systems will be the ultimate gauge of the success of this additional research.

REFERENCES

Abdul-Baki, A. A., J. R. Teasdale, R. Korcak, D. J. Chitwood, and R. N. Huettel (1996). Fresh-market tomato production in a low input alternative system using cover-crop mulch. *HortScience* 31: 65-69.

Akemo, M. C., M. A. Bennett, and E. E. Regnier (2000). Tomato growth in spring-sown cover crops. *HortScience* 35: 843-848.

Amisi, K. (2000). Alternative weed management systems for squash. Master's thesis, University of Illinois, Urbana.

Anaya, A. L. (1999). Allelopathy as a tool in the management of biotic resources in agroecosystems. *Critical Reviews in Plant Sciences* 18(6): 697-739.

Andow, D. A. (1991). Vegetational diversity and arthropod response. *Annual Reviews in Entomology* 36: 561-586.

Ateh, C. M. and J. D. Doll (1996). Spring-planted winter rye *(Secale cereale)* as a living mulch to control weeds in soybeans *(Glycine max)*. *Weed Technology* 10: 347-353.

Barnes, J. P. and A. P. Putnam (1983). Rye residuals contribute to weed suppression in no-tillage cropping systems. *Journal of Chemical Ecology* 9: 1045-1057.

Barnes, J. P. and A. P. Putnam (1986). Evidence for allelopathy by residues and aqueous extracts of rye (*Secale cereale* L.). *Weed Science* 34: 384-390.

Barnes, J. P., A. P. Putnam, B. A. Burke, and A. J. Aasen (1987). Isolation and characterization of allelochemicals in rye herbage. *Phytochemistry* 26: 1385-1390.

Bellinder, R. R., R. W. Wallace, and E. D. Wilkins (1996). Reduced rates of herbicides following hilling controlled weeds in conventional and reduce tillage potato *(Solanum tuberosum)* production. *Weed Technology* 10: 311-316.

Bellinder, R. R., H. P. Wilson, and T. E. Hines (1987). Comparative studies of conventional and no-tillage systems for snap bean production. HortScience 22: 159-162.

Bistow, K. L., G. S. Campbell, R. I. Papendick, and L. F. Elliott (1996). Simulation of heat and moisture through a surface residue soil system. *Agricultural and Forest Meteorology* 36: 193-214.

Bottenberg, H., J. Masiunas, and C. Eastman (1999). Strip tillage reduces yield loss in snapbean planted in rye mulch. *HortTechnology* 9: 235-239.

Bottenberg, H., J. Masiunas, C. Eastman, and D. Eastburn (1997a). The impact of rye cover crops on weeds, insects, and diseases in snapbean cropping systems. *Journal of Sustainable Agriculture* 9: 133-155.

Bottenberg, H., J. Masiunas, C. Eastman, and D. Eastburn (1997b). Yield and quality constraints of cabbage planted in rye mulch. *Biological, Agriculture and Horticulture* 14: 323-342.

Brandi-Dohrn, F., R. P. Dick, M. Hess, S. M. Kaufman, D. D. Hemphill, and J. S. Selker (1997). Nitrate leaching under a cereal rye cover crop. *Journal of Environmental Quality* 26: 181-188.

Bugg, R. L. (1992). Using cover crops to manage arthropods on truck farms. *HortScience* 27: 741-745.

Bugg, R. L. and R. T. Ellis (1990). Insects associated with cover crops in Massachusetts. *Biological, Agriculture and Horticulture* 7: 47-68.

Bugg, R. L., S. C. Phatak, and J. D. Dutcher (1990). Insects associated with cool-season cover crops in southern Georgia: Implications for pest control in truck-farm and pecan agroecosystems. *Biological, Agriculture and Horticulture* 7: 17-45.

Burgos, N. R. and R. E. Talbert (1996). Weed control and sweet corn (*Zea mays* var. *rugosa*) response in no-till system with cover crops. *Weed Science* 44: 355-361.

Burket, J. Z., D. D. Hemphill, and R. P. Dick (1997). Winter cover crops and nitrogen management in sweet corn and broccoli rotations. *HortScience* 32: 664-668.

Chase, W. R., M. G. Nair, A. R. Putnam, and S. K. Mishra (1991). 2,2'-oxo-1,1'-azobenzene: Microbial transformation of rye (*Secale cereale* L.) allelochemical in field soils by *Acinetobacter calcoaceticus* III. *Journal of Chemical Ecology* 17: 1575-1584.

Clark, A. J., A. M. Decker, and J. J. Meisinger (1994). Seedling rate and kill date effects on hairy vetch, cereal rye cover crop mixtures for corn production. *Agronomy Journal* 86: 1065-1070.

Clark, M. S., H. Ferris, K. Klonsky, W. T. Lanini, A. H. C. van Bruggen, and F. G. Zalom (1998). Agronomic, economic, and environmental comparison of pest in conventional and alternative tomato and corn systems in northern California. *Agriculture, Ecosystems and Environment* 68: 51-71.

Coolman, R. M. and G. D. Hoyt (1993). The effects of reduced tillage on the soil environment. *HortTechnology* 3: 143-145.

Creamer, N. G., M. A. Bennett, and B. R. Stinner (1997). Evaluation of cover crop mixtures for use in vegetable production systems. *HortScience* 32: 866-870.

Creamer, N. G., M. A. Bennett, B. R. Stinner, and J. Cardina (1996). A comparison of four processing tomato production systems differing in cover crop and chemical inputs. *Journal of the American Society for Horticultural Sciences* 121: 559-568.

Creamer, N. G., M. A. Bennett, B. R. Stinner, J. Cardina, and E. E. Regnier (1996). Mechanisms of weed suppression in cover-crop based production systems. *HortScience* 31: 410-413.

Creamer, N. G., B. Plassman, M. A. Bennett, R. K. Wood, B. R. Stinner, and J. Cardina (1995). A method for mechanically killing cover crops to optimize weed suppression. *American Journal of Alternative Agriculture* 10: 156-161.

Doss, B. D., J. L. Turner, and C. E. Evans (1981). Influence of tillage, nitrogen, and rye cover crop on growth and yield of tomatoes. *Journal of the American Society for Horticultural Science* 106: 95-97.

Drost, D. T. and H. C. Price (1991). Effect of tillage system and planting date on growth and yield of transplanted tomato. *HortScience* 26: 1478-1480.

Eckert, D. J. (1991). Chemical attributes of soils subjected to no-till cropping with rye cover crops. *Soil Science Society of America Journal* 55: 405-409.

Einhellig, F. A. (1996). Interactions involving allelopathy in cropping systems. *Agronomy Journal* 88: 886-893.

Francis, C. A. and J. W. King (1988). Cropping systems based on farm-derived renewable resources. *Agricultural Systems* 27: 67-77.

Galloway, B. A. and L. A. Weston (1996). Influence of cover crop and herbicide treatment on weed control and yield in no-till sweet corn (*Zea mays* L.) and pumpkin (*Cucurbita maxima* Duch.). *Weed Technology* 10: 341-346.

Hill, H. H. (1926). Decomposition of organic matter in soil. *Journal of Agricultural Research* 33: 77-99.

House, G. J. and M. Del Rosario Alzaugary (1989). Influence of cover cropping and no-till practices on community composition of soil arthopods in a North Carolina agroecosystems. *Environmental Entomology* 18: 302-307.

Hoyt, G. D. (1999). Tillage and cover crop residue affects on vegetable yields. *HortTechnology* 9: 351-365.

Hoyt, G. D., A. R. Bonanno, and G. C. Parker (1996). Influence of herbicides and tillage on weed control, yield, and quality of cabbage (*Brassica oleracea* L. var. *capitata*). *Weed Technology* 10: 50-54.

Hoyt, G. D. and D. W. Monks (1996). Weed management in strip-tilled Irish potato and sweetpotato systems. *HortTechnology* 6: 238-240.

Hoyt, G. D., D. W. Monks, and T. J. Monaco (1994). Conservation tillage for vegetable production. *HortTechnology* 4: 129-135.

Hoyt, G. D. and J. F. Walgenbach (1995). Pest evaluation in sustainable cabbage production systems. *HortScience* 30: 1046-1048.

Kahn, M. J., E. J. Monke, and G. R. Foster (1988). Mulch cover and canopy effect on soil loss. *Transactions of the American Society for Agricultural Engineering* 31: 706-711.

Knavel, D. E. and J. W. Herron (1981). Influence of tillage system, plant spacing, and nitrogen on head weight, yield, and nutrient concentration of spring cabbage. *Journal of the American Society for Horticultural Science* 106: 540-545.

Knott, C. M. (1996). The effect of cover crops on the establishment and yield of vining peas and nitrate leaching. *Journal of Agricultural Science* 126: 471-479.

Kuo, S. and U. M. Sainju (1998). Nitrogen mineralization and availability of mixed leguminous and non-leguminous cover crop residues in soil. *Biology and Fertility of Soils* 26: 346-353.

Lal, R. (1995). The role of residue management in sustainable agricultural systems. *Journal of Sustainable Agriculture* 5: 51-78.

Lanfranconi, L. E., R. R. Bellinder, and R. W. Wallace (1993). Grain rye residue and weed control strategies in reduced tillage potatoes. *Weed Technology* 6: 1021-1026.

Loy, S. J. W., L. C. Peirce, G. O. Estes, and O. S. Wells (1987). Productivity in a strip tillage vegetable production system. *HortScience* 22: 415-417.

Masiunas, J. B. (1999). Production of vegetables using cover crops and living mulches: A review. *Journal of Vegetable Crop Production* 4: 11-31.

Masiunas, J. B., D. M. Eastburn, V. N. Mwaja, and C. E. Eastman (1997). The impact of living and cover crop mulch systems on pests and yields of snap beans and cabbage. *Journal of Sustainable Agriculture* 9(2/3): 61-89.

Masiunas, J. B., L. A. Weston, and S. C. Weller (1995). The impact of rye cover crops on weed populations in a tomato cropping system. *Weed Science* 43: 318-323.

Mohler, C. L. and J. R. Teasdale (1993). Response of weed emergence to rate of *Vicia villosa* Roth. and *Secale cereale* L. residues. *Weed Research* 33: 487-499.

Morrison, J. E. Jr., D. C. Milbocker, W. O. Atkinson, and J. H. Smiley (1973). Transplanter modification and survival of transplants under no-tillage conditions. *HortScience* 8: 483-485.

Morse, R. D. (1993). Components of sustainable production systems for vegetables: Conserving soil moisture. *HortTechnology* 3: 211-214.

Mundy, C., N. G. Creamer, C. R. Crozier, L. G. Wilson, and R. D. Morse (1999). Soil physical properties and potato yield in no-till, subsurface-till, and conventional-till systems. *HortTechnology* 9: 240-247.

Mwaja, V. N., J. B. Masiunas, and C. E. Eastman (1996). Rye and hairy vetch intercrop management in fresh market vegetables. *Journal of the American Society for Horticultural Science* 121: 586-591.

Mwaja, V. N., J. B. Masiunas, and L. A. Weston (1995). The effect of fertility on biomass, phytotoxicity, and allelochemical content of cereal rye. *Journal of Chemical Ecology* 21: 81-96.

Nelson, W. A., B. A. Kahn, and B. W. Roberts (1991). Screening cover crops for use in conservation tillage systems for vegetables following spring plowing. *HortScience* 26: 860-862.

NeSmith, D. S. and D. V. McCracken (1994). Snapbean response to soil tillage management and cover crops. *Communications in Soil Science and Plant Analysis* 25(13-14): 2501-2512.

Paine, L. K. and H. Harrison (1996). The historic roots of living mulch and related practices. *HortTechnology* 3: 137-143.

Perez, F. J. and J. Ormeno-Nunez (1993). Weed growth interference from temperate cereals: The effect of a hydroxamic-acid-exuding rye (*Secale cereale* L.) cultivar. *Weed Research* 33: 115-119.

Phatak, S. C. (1992). An integrated sustainable vegetable production system. *HortScience* 27: 738-741.

Putnam, A. R. (1986). Allelopathy: Can it be managed to benefit horticulture? *HortScience* 21: 411-412.

Putnam, A. R. (1990). Vegetable weed control with minimal herbicide inputs. *HortScience* 25: 165-169.

Raimbault, B. A., T. J. Vyn, and M. Tollenaar (1991). Corn response to cereal rye cover crop, tillage methods, and planter options. *Agronomy Journal* 83: 287-290.

Ranells, N. N. and M. G. Wagger (1997). Nitrogen-15 recovery and release by rye and crimson clover cover crops. *Soil Science Society of America Journal* 61: 943-948.

Roberts, B. W. and B. Cartwright (1991). Alternative soil and pest management practices for sustainable production of fresh-market cabbage. *Journal of Sustainable Agriculture* 1: 21-35.

Sainju, U. M., B. P. Singh, S. Rahman, and V. R. Reddy (2000). Tomato root growth is influenced by tillage, cover cropping, and nitrogen fertilization. *HortScience* 35: 78-82.

Sainju, U. M., B. P. Singh, and W. F. Whitehead (1998). Cover crop root distribution and its effect on soil nitrogen cycling. *Agronomy Journal* 90: 511-518.

Schonbeck, M., J. Browne, G. Deziel, and R. DeGregorio (1991). Comparison of weed biomass and flora in four cover crops and a subsequent lettuce crop on three New England organic farms. *Biological, Agriculture and Horticulture* 8: 123-143.

Schonbeck, M., S. Herbert, R. DeGregorio, F. Mangan, K. Guillard, E. Sideman, J. Herbst, and R. Jaye (1993). Cover cropping systems in the Northeastern United States: I. Cover crop and vegetable yields, nutrients, and soil conditions. *Journal of Sustainable Agriculture* 3: 105-132.

Shennan, C. (1992). Cover crops, nitrogen recycling, and soil properties in semi-irrigated vegetable production systems. *HortScience* 27: 749-753.

Sheperd, M. A. and E. I. Lord (1996). Nitrate leaching from a sandy soil: The effect of previous crop and post-harvest soil management in an arable rotation. *Journal of Agricultural Science* 127: 215-229.

Singogo, W., W. J. Lamont Jr., and C. W. Marr (1996). Fall-planted cover crops support good yields of muskmelons. *HortScience* 31: 62-64.

Skarphol, B. J., K. A. Corey, and J. J. Meisinger (1987). Response of snapbean to tillage and cover crop combinations. *Journal of the American Society for Horticultural Science* 112: 936-941.

Smeda, R. J. and S. C. Weller (1996). Potential of rye *(Secale cereale)* for weed management in transplanted tomatoes *(Lycopersicon esculentum). Weed Science* 44: 596-602.

Smith, A. W., R. B. Hammond, and B. R. Stinner (1988). Influence of rye-cover crop management on soybean foliage arthropods. *Environmental Entomology* 17: 109-114.

Steiner, J. L. (1989). Tillage and surface residue effects on evaporation from soils. *Soil Science Society of America Journal* 53: 911-916.

Stirzaker, R. J. and I. White (1995). Amelioration of soil compaction by a cover-crop for no-tillage lettuce production. *Australian Journal of Agricultural Research* 46: 553-568.

Stivers, L. J. and C. Shennan (1991). Meeting the nitrogen needs of processing tomatoes through winter cover cropping. *Journal of Production Agriculture* 4: 330-335.

Teasdale, J. R. (1993a). Interaction of light, soil moisture, and temperature with weed suppression by hairy vetch residue. *Weed Science* 41: 46-52.

Teasdale, J. R. (1993b). Reduced-herbicide weed management systems for no-tillage corn *(Zea mays)* in a hairy vetch cover crop *(Vicia villosa)* cover crop. *Weed Technology* 7: 879-883.

Wagner-Riddle, C., T. J. Gillespie, and L. A. Hunt (1997). Modeling a rye cover crop and subsequent soybean yield. *Agronomy Journal* 89: 208-218.

Wagner-Riddle, C., T. J. Gillespie, and C. J. Swanton (1994). Rye cover crop management impact on soil water content, soil temperature, and soybean growth. *Canadian Journal of Plant Science* 74: 485-495.

Wahle, E. A. (1996). Evaluation of different cereal rye cultivars for weed management in vegetables. Master's thesis, University of Illinois, Urbana.

Wahle, E. A. and J. B. Masiunas (2002). The effect of spring planted cereal rye on the yield of fresh market vegetables. *Journal of Vegetable Crop Production* 8: 123-132.

Weston, L. A. (1990). Cover crop and herbicide influence on row crop seedling establishment in no-tillage culture. *Weed Science* 166-171.

Weston, L. A. (1996). Utilization of allelopathy for weed management in agroecosystems. *Agronomy Journal* 88: 860-866.

Wilhoit, J. H., R. D. Morse, and D. H. Vaughan (1990). Strip-tillage production of summer cabbage using high residue levels. *Applied Agricultural Research* 5: 338-342.

Wilkins, E. D. and R. R. Bellinder (1996). Mow-kill regulation of winter cereals for spring no-till production. *Weed Technology* 10: 247-252.

Wyland, L. J., L. E. Jackson, W. E. Chaney, K. Klonsky, S. T. Koike, and B. Kimple (1996). Winter cover crops in a vegetable cropping system—Impacts on nitrate leaching, soil water, crop yield, pests, and management costs. *Agriculture Ecosystems and Environment* 59: 1-17.

Wyland, L. J., L. E. Jackson, and K. F. Schulbach (1995). Soil-plant dynamics following incorporation of a mature cereal rye cover crop in a lettuce production system. *Journal of Agricultural Science* 124: 17-25.

Yenish, J. P., A. D. Worsham, and W. S. Chilton (1995). Disappearance of DIBOA-glucoside, DIBOA, and BOA from rye (*Secale cereale* L.) cover crop residues. *Weed Science* 43: 18-20.

Zehnder, G. W. and J. J. Linduska (1987). Influence of conservation tillage practices on populations of Colorado potato beetle (Coleoptera: Chrysomelidae) in rotated and nonrotated tomato fields. *Environmental Entomology* 16: 135-139.

Chapter 7

A Rotation Design That Aids Annual Weed Management in a Semiarid Region

Randy L. Anderson

CROP PRODUCTION IN THE GREAT PLAINS OF THE UNITED STATES

The predominate rotation in the Great Plains is winter wheat (*Triticum aestivum* L.)-fallow. Producers developed this rotation in response to the region's semiarid climate where yearly precipitation ranges from 250 to 450 mm per year and evaporation potential approaches 1,500 mm. To compensate for these climatic limitations, producers include fallow in the rotation. During fallow, no crop is grown and weeds are controlled, thus precipitation is stored in soil. The reservoir of soil water gained during fallow improves growth of the following winter wheat, subsequently reducing yield variability and crop loss due to drought stress. In addition, producers grow sorghum [*Sorghum bicolor* (L.) Moench], proso millet (*Panicum miliaceum* L.), corn (*Zea mays* L.), and sunflower (*Helianthus annuus* L.), but presently, winter wheat is the major crop of the region.

Status of Current Weed Management

Producers in the Great Plains rely on a multitude of strategies to control weeds, such as tillage or cultural practices, but especially emphasize herbicides. Initially, herbicides were so effective that producers perceived them to be the "silver bullet," controlling weeds with one management tactic. A startling consequence of this single-tactic approach, however, is the development of weeds resistant to herbicides (Holt and LeBaron, 1990). In the Great Plains, herbicide-resistant weeds such as downy brome (*Bromus tectorum* L.), Russian thistle (*Salsola iberica* Sennen & Pau), and kochia [*Kochia scoparia* (L.) Schrad.] are now common (Lyon, Miller, and Wicks, 1996).

A second trend with current production systems is proliferation of weed-crop associations (also known as crop mimics). In the Great Plains, brome species (*Bromus* spp.) and jointed goatgrass (*Aegilops cylindrica* Host), winter annual grasses, have plagued winter wheat producers for decades (Wicks, 1984; Donald and Zimdahl, 1987). Crop mimics also occur with summer annual crops such as proso millet or corn; green foxtail [*Setaria viridis* (L.) Beauv.] and longspine sandbur [*Cenchrus longispinus* (Hack.) Fern.], summer annual grasses, are also becoming prominent (Holtzer et al., 1996). A further trend of Great Plains cropping systems is that production costs continue to increase, yet prices for crop products remain unchanged or even decrease. If weeds are not managed, yields are significantly reduced. To ensure economic survival, producers increase crop inputs with the goal of increasing yields to offset these costs; the final consequence, however, is a crop input treadmill.

Because of these trends, herbicide-based weed management in the Great Plains is often ineffective and uneconomical (Holtzer et al., 1996). Producers still lose considerable crop yield due to weeds, whereas herbicide resistance is forcing producers to use more expensive management tactics (Beckie et al., 1999). Furthermore, this approach has consistently failed to achieve long-term weed management (Mortensen, Bastiaans, and Sattin, 2000).

An Alternative: Ecologically Based Weed Management

One possible solution is to devise ecologically based weed management systems, an approach that emphasizes weed prevention and population management (Pedigo, 1995). Cropping systems are designed to lower average density of weeds and improve crop tolerance to weeds. With the ecologically based approach, herbicides supplement preventive tactics rather than serve as the sole tactic. Knowledge of weed ecology is integrated with multiple control tactics, with initial emphasis placed on system design. A key component of this approach is to use rotations comprised of a diversity of crops, which increases the number of control opportunities for producers in devising management strategies (Lewis et al., 1997).

Historically, crop rotations were an integral component of weed control; however, herbicides led producers to rely less on crop rotations for weed management (Holt and LeBaron, 1990). With the ecologically based approach, rotations again are the foundation of weed management; well-designed rotations strengthen effectiveness of other weed management strategies, thus improving overall weed management (Holtzer et al., 1996).

Great Plains Cropping Systems Are Changing

Two factors are stimulating producers in the central Great Plains to change their crop rotations. First, no-till systems enable producers to preserve crop residue on the soil surface, which improves precipitation storage during noncrop periods. With this extra soil water, producers can grow more crops in succession before fallow is needed (Peterson et al., 1996). Because of no-till systems, land productivity is doubled with rotations such as winter wheat-corn-proso millet or winter wheat-corn-sunflower-fallow (Peterson, Westfall, and Cole, 1993; Anderson et al., 1999). Diverse rotations not only improve net returns compared to winter wheat-fallow but also reduce financial risks for producers (Dhyuvetter et al., 1996).

Second, severe infestations of downy brome, jointed goatgrass, and feral rye (*Secale cereale* L.) are forcing producers to change their rotation (Holtzer et al., 1996). Because no effective herbicides control these grasses in winter wheat, producers insert summer annual crops such as corn or proso millet in the rotation to lengthen the time between wheat crops. This strategy is effective because the time interval before the next wheat crop favors the natural decline of weed seed viability in soil, consequently reducing future weed densities. Seedlings of jointed goatgrass or downy brome, both winter annual grasses, are easily controlled during noncrop periods between summer annual crops, thus preventing seed production and entry into the seed bank.

Renewed interest among producers in exploring diverse rotations provides us with an opportunity to restructure cropping systems by integrating crop diversity and sequencing with our greater understanding of weed ecology. Therefore, in a series of studies, we examined weed dynamics as affected by rotations and cultural control practices. Our long-term goal is to develop guidelines for producers to integrate rotation design with weed control tactics; ultimately, we hope to reduce reliance on herbicides, minimize development of herbicide resistance and crop mimics, and reduce input costs for weed control.

IMPACT OF ROTATION DESIGN ON WEED DYNAMICS

A long-term cropping systems study was established at the Central Great Plains Research Station, Akron, Colorado, in 1990 to examine alternative rotations for this semiarid region. Within this study, a series of rotations with winter wheat, corn, proso millet, and fallow contained a range of winter and summer crops combinations, enabling us to evaluate impact of rotation design on weed dynamics.

Weed biomass was measured in four rotations: continuous proso millet (M-M), wheat-proso millet-fallow (W-M-F), wheat-corn-proso millet (W-C-M), and wheat-corn-proso millet-fallow (W-C-M-F). Herbicides used in proso millet controlled only broadleaf weeds, which favored proliferation of green foxtail and longspine sandbur. Conventional herbicides effectively controlled these grasses in winter wheat and fallow, thus preventing seed production. Grass control in corn was approximately 90 percent, which allowed isolated plants to produce weed seed during the growing season.

After eight years, biomass of these grasses was 180 $g \cdot m^{-2}$ in continuous proso millet (see Figure 7.1). In contrast, weed biomass was 85 percent less in W-M-F, a rotation with only one summer crop in three years. Weed biomass in W-C-M was one half of continuous proso millet (M-M), whereas biomass in W-C-M-F was only 30 percent of weed biomass in W-C-M. An intriguing finding was that weed biomass did not differ between W-M-F and W-C-M-F, even though W-C-M-F had two summer crops in a row. One similarity between W-M-F and W-C-M-F is the two-year period without summer crops (winter wheat and fallow). Apparently, this two-year period affected weed dynamics more than did two years of consecutive summer crops.

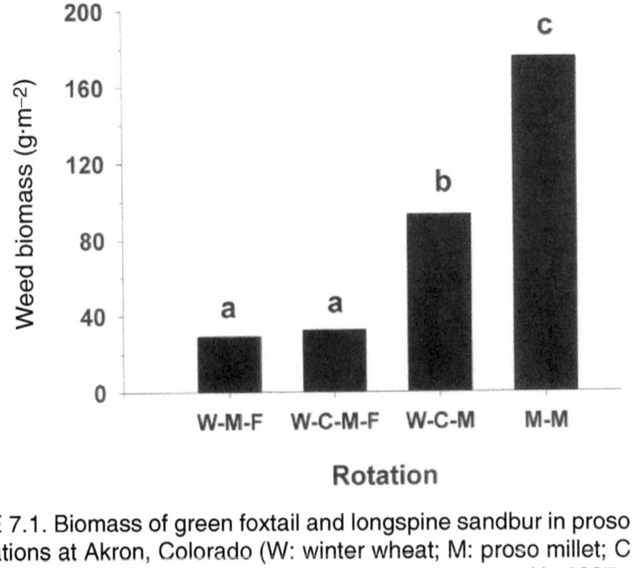

FIGURE 7.1. Biomass of green foxtail and longspine sandbur in proso millet for four rotations at Akron, Colorado (W: winter wheat; M: proso millet; C: corn; F: fallow). Study initiated in 1990, with weed biomass measured in 1997 and 1999. Bars with the same letter are not significantly different based on Fisher's LSD (0.05). (*Source:* Adapted from Anderson, 1998a.)

A contributing factor to this trend is length of weed seed survival in soil. The reservoir of weed seeds in soil, referred to as the seed bank, is the main source of weed infestation in crops (Roberts, 1981). Seeds in soil either germinate, die, or are consumed by microorganisms and fauna (Harper, 1977). Decline of seed density in soil over time is exponential, with a dramatic decline occurring during the first two years after seeds enter the seed bank (Roberts, 1981; Egley and Williams, 1990). For example, seed survival of green foxtail and longspine sandbur is less than 10 percent after two years in soil (see Figure 7.2). Both W-M-F and W-C-M-F have two years in the rotation when green foxtail and longspine sandbur can be easily controlled before seed is produced, thus favoring rapid decline of seed density in soil (see Figure 7.2). In other words, the two-year interval reduces potential seedling densities by 90 percent.

The two-year interval in summer crops also aids control of winter annual grasses in winter wheat. Seed decline of downy brome is similar to the summer annual grasses, with less than 5 percent of seed alive after two years in soil (see Figure 7.3). Jointed goatgrass persists somewhat longer because its seed is encased within glumes of the spikelet (its dispersal unit). The spikelet protects seed from environmental conditions, yet, within three years, the number of viable seeds still declined to less than 5 percent.

The two-year interval has shown similar effects on weed densities in other agroecosystems. In the Corn Belt region of the United States, a four-

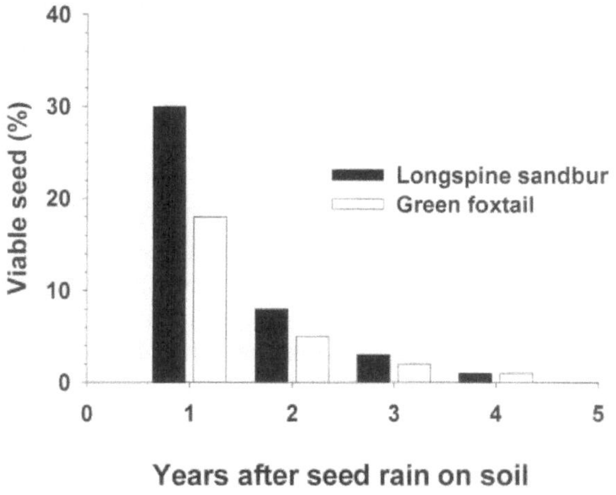

FIGURE 7.2. Persistence of seed viability for green foxtail and longspine sandbur when seed remains within the top 3 cm of soil. (*Source:* Adapted from Dawson and Burns, 1975; Donald and Zimdahl, 1987; Boydston, 1990; and Anderson, 1998a.)

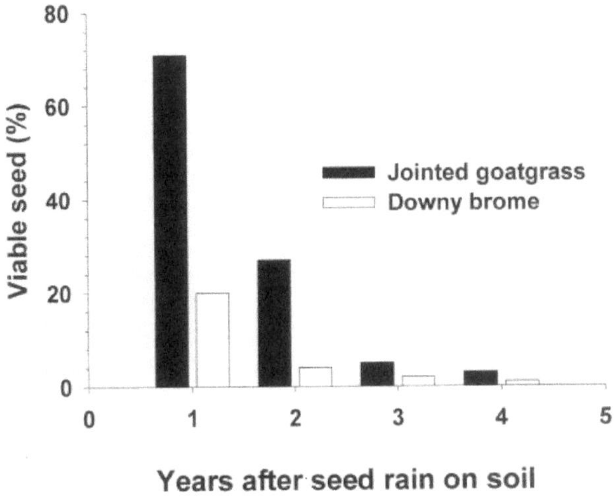

FIGURE 7.3. Persistence of seed viability for downy brome and jointed goatgrass, when seed remains within the top 3 cm of soil. (*Source:* Adapted from Dawson and Burns, 1975; Donald and Zimdahl, 1987; Boydston, 1990; and Anderson, 1998a.)

year rotation of corn-soybean [*Glycine max* (L.) Merr.]-hay-hay was designed to prevent green foxtail seed production during the two years of hay (Jordan et al., 1995). Subsequently, green foxtail densities were drastically reduced in corn with this rotation compared to continuous corn or corn-soybean rotations. A second example occurs in the Northern Great Plains (Anderson et al., 1998). Fairy candelabra (*Androsace occidentalis* Pursh), a native cool-season species in rangelands, proliferates in no-till spring wheat-fallow. However, a no-till rotation comprised of spring wheat-winter wheat-sunflower eliminates fairy candelabra. Fairy candelabra emerges before sunflower is planted, thus seedlings are easily controlled, whereas winter wheat growth in the spring is more competitive than spring wheat and prevents fairy candelabra seedlings from establishing. The two-year interval out of spring wheat essentially eliminates fairy candelabra in this rotation.

Can We Enhance the Natural Decline of the Weed Seed Bank?

Scientists have explored various strategies to stimulate weed seed germination, with the goal of reducing weed seed densities in soil (Egley, 1986). Research has focused primarily on tillage; however, placing weed seeds into soil may prolong persistence by protecting seeds from environmental extremes. For example, the number of viable seeds of green foxtail was four times greater after two years in soil when seed was buried 10 cm compared to 2.5 cm (Dawson and Burns, 1975). Sagar and Mortimer (1976) found that wild oat (*Avena fatua* L.) viability loss was five times greater if seed remained on the soil surface over winter compared to burial 5 cm deep. Similar results occurred with green foxtail in Canada (Thomas, Banting, and Bowes, 1986) and with several weed species in the United States (Egley and Williams, 1990); seeds remain viable longer if buried 5 cm or deeper in soil compared to remaining on or near the soil surface.

Froud-Williams, Chancellor, and Drennan (1984) hypothesized that shallow tillage to bury weed seed less than 4 cm deep would enhance seed-bank decline due to greater weed germination. We tested this hypothesis with jointed goatgrass by tilling with a sweep plow* (Anderson, 1998b). The sweep plow places jointed goatgrass seeds 1 to 3 cm deep in soil. During the first year, tillage increased seedling emergence twofold compared to seeds remaining on the soil surface; however, seedling emergence did not differ between these treatments during the next four years. Thus, tillage did not influence long-term seed-bank dynamics. Lack of differences between tillage treatments in later years probably reflected greater winter kill of seeds on the soil surface.

Other tactics to reduce weed seed longevity in soil also have not been successful (Egley, 1986). Producers can accrue more weed management benefits by emphasizing diverse crop rotations and cultural strategies within crops that reduce weed seed production (Anderson, 1998b).

Density Growth of Green Foxtail

Because jointed goatgrass seed persists for several years in soil, some producers use rotations that include three consecutive summer annual crops. With this practice, we were concerned that summer annual weeds may proliferate. Therefore, we estimated population growth of green foxtail with the following assumptions: (1) A green foxtail plant growing in corn will produce approximately 2,300 seeds per plant (Anderson, 1998a); (2) 8 percent of green foxtail seed in soil emerges each year (Forcella et al., 1997); and (3) weed control in the summer crop is 90 percent.

Based on these assumptions, one green foxtail plant establishing in the first year will result in 18 plants the second year and 324 plants in the third

*A sweep plow is composed of V-shaped blades that sever weed roots with minimum soil disturbance, tilling to a depth of 5 to 8 cm.

year; an 18-fold increase per year. This scenario suggests extensive green foxtail population growth will occur in the third year of a summer annual crop. If proso millet is included in the rotation, density of green foxtail will increase even more rapidly, since currently registered herbicides for proso millet control green foxtail only marginally (< 50 percent).

A ROTATION DESIGN THAT AIDS WEED MANAGEMENT

Our research demonstrates that rotation design can have a striking impact on weed dynamics; note the sixfold difference among rotations in Figure 7.1. Both winter annual and summer annual crops are well adapted in the Great Plains, which helps weed management because growth periods differ considerably between the crop types. For example, winter wheat is usually planted in late September and harvested in early July; in contrast, proso millet is planted in early June and harvested in early September. Prevalent weeds in each crop differ because of the lack of similarity in growth periods for weeds. Growing two winter annual crops followed by two summer annual crops will further enhance this growing season difference; the cycle-of-four design favors natural seed decline in soil yet avoids proliferation of weed densities that may occur if three crops with similar growing seasons are grown consecutively.

Furthermore, the cycle-of-four design can improve land productivity. In the Akron, Colorado, rotation study W-C-M-F and W-C-S-F ranked among the most productive rotations, almost doubling land productivity and improving gross returns compared to winter wheat-fallow (Anderson et al., 1999). From a weed perspective, fallow favors the impact of both summer and winter crop cycles, as weeds are easily prevented from producing seed during fallow. However, fallow is detrimental to long-term soil productivity because it enhances soil erosion and loss of organic matter (Peterson, Westfall, and Cole, 1993). Therefore, producers are seeking rotations with continuous cropping. One promising rotation in our study is winter wheat-corn-proso millet-dry pea (*Pisum sativum* L.), where dry pea is managed either for green fallow, hay, or seed, depending on growing season precipitation. Dry pea is planted in March and harvested before summer annual weeds can produce seed, thus dry pea fits the winter annual crop cycle in managing weeds. Also, different planting dates of dry pea and winter wheat will help producers in managing winter annual grasses.

Rotations based on the cycle-of-four will enable producers to counter herbicide resistance, crop mimics, and rising input costs. Crop diversity provides opportunities for producers to rotate herbicides with different modes of action, whereas cultural systems may allow producers to eliminate herbicide use in some crops of a rotation. Both strategies delay development of herbicide resistance by eliminating selection pressure in some years (Holt and LeBaron, 1990). Crop diversity, because of differences in planting dates, also minimizes proliferation of crop mimics (Froud-Williams, 1988).

Crop diversity also may reduce input costs. A study (Jordan et al., 1997) in England compared a system based on diverse rotations and ecologically based pest management to commonly used pesticide-based production systems. Input costs with the ecologically based system were reduced 32 percent because of substantial reductions in applied herbicides (26 percent), fungicides (79 percent), and insecticides (78 percent). Lower weed densities and increased crop competitiveness also will improve herbicide efficacy (Hoffman and Lavy, 1978; Dieleman, Mortensen, and Martin, 1999).

Rotation Design Can Help Manage Plant Diseases

Other rotations are being used in the Great Plains, such as W-M-F or W-C-F. These rotations will help manage weeds, as shown in Figure 7.1, but they minimize other potential benefits of longer rotations. For example, some plant diseases are related to how frequently the crop is grown in rotation. The term *crop interval* refers to number of years before the same crop is grown again and was developed to guide management strategies for root diseases (Cook and Veseth, 1991). With winter wheat-fallow, the crop interval is two, whereas in winter wheat-corn-fallow, the crop interval is three for each crop.

Grain yield in some crops is highly related to its crop interval. In the central Great Plains, sunflower yields are maximum when grown once every four years (see Figure 7.4); if grown more frequently, the soilborne disease phoma (*Phoma macdonaldii* Boerma) severely reduces yield (Anderson et al., 1999). Yields of corn and winter wheat are reduced when grown every two years compared to a four-year crop interval, whereas proso millet is not affected by crop interval (see Figure 7.5). Corn and winter wheat respond similarly to crop interval in other regions of the United States. In Minnesota, corn yield is highest when grown once every three years (Porter et al., 1997), whereas winter wheat yield in the Pacific Northwest is highest with a crop interval of four years (Cook and Veseth, 1991). Furthermore, a cycle-of-four rotation enables producers to include oilseed crops such as sunflower or canola (*Brassica napus* L.) in rotations, as a four-year crop interval is necessary to prevent serious plant diseases with oilseeds (Bailey, 1996). Crop interval also affects plant-*rhizobia* interaction; N fixation by

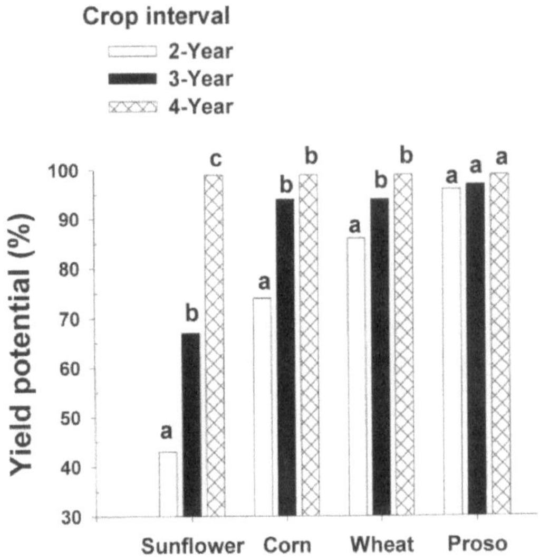

FIGURE 7.4. Sunflower, corn, proso millet, and winter wheat response to crop interval at Akron, Colorado, 1993-1996. An asterisk signifies that the two-year or three-year crop interval mean differed from the four-year mean within each crop. Rotations for sunflower: sunflower-proso, wheat-sunflower-fallow, and wheat-corn-sunflower-fallow; for corn: proso-corn, wheat-corn-fallow, and wheat-corn-proso-fallow; for proso: wheat-proso, wheat-proso-fallow, and wheat-corn-proso-fallow; and for wheat: wheat-fallow, wheat-corn-fallow, and wheat-corn-proso-fallow. (*Source:* Adapted from Anderson, 1998a; Anderson et al., 1999.)

legumes is improved in a rotation comprised of four different crops compared to rotations with fewer crops (Matus et al., 1997).

Rotating Crops Can Increase Grain Yield

Crop diversity favors the *rotation effect,* where rotating crops generally increases yield compared to monoculture (Kurtz et al., 1984; Porter et al., 1997). For example, wheat grain yield was increased 15 to 30 percent if rotated with broadleaf crops such as dry pea or flax (*Linum usitatissimum* L.) compared to continuous wheat (Wright, 1990). Rotating corn and soybean [*Glycine max* (L.) Merr.] also increased grain yield of both crops, but this effect interacted with crop interval (Crookston et al., 1991). Yield of both crops increased approximately 10 percent in a corn-soybean rotation compared to monocultures of each crop, whereas yields increased 15 percent if

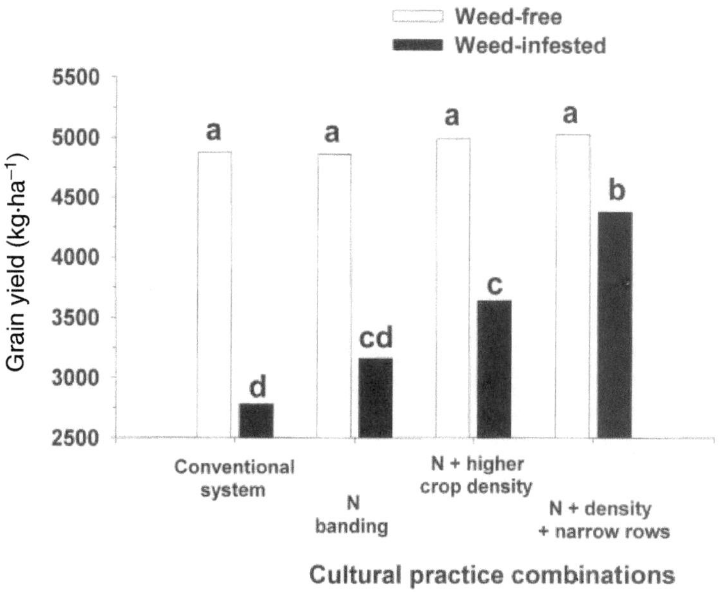

FIGURE 7.5. Corn grain yield in weed-free and weed-infested conditions as affected by cultural practice combinations. Conventional system was 37,000 plants/ha at a row spacing of 76 cm, with N fertilizer broadcast at planting. Cultural practices were banding N by the seed, increasing crop density to 47,000 plants/ha, and reducing row spacing to 38 cm. Study conducted at Akron, Colorado, with data averaged across three years. Bars with the same letter are not significantly different based on Fisher's LSD (0.05). (*Source:* Adapted from Anderson, 2000b.)

either crop was grown only once in three years. The authors concluded that a three-crop rotation would be most favorable for rotational yield benefits with corn and soybean.

Rotating crops within a crop type, such as grasses, is usually not as favorable as rotating between grass and broadleaf crops, but anomalies do occur. For example, the rotation study at Akron, Colorado, showed an intriguing rotation effect: Corn as a previous crop increased grain yield of proso millet 15 to 20 percent, compared to winter wheat as the previous crop (Anderson et al., 1999). This yield trend resulted from improved water use efficiency of proso millet; if corn is the previous crop, proso millet is more efficient in converting water into grain. Also, corn improved water use efficiency of winter wheat. These trends indicate that corn exerted some unknown effect that is favorable for other crops in semiarid rotations.

The cause of the rotation effect has not been identified and most likely represents a complex interaction of multiple soil factors (Wright, 1990). However, the rotation effect is enhanced by diversity of crops, longer crop intervals, and no-till systems (Pierce and Rice, 1988; Crookston et al., 1991); thus, a no-till rotation comprised of four different crops should accentuate rotational yield benefits in semiarid regions.

CULTURAL PRACTICES THAT IMPROVE CROP COMPETITIVENESS

Ecologically based weed management focuses on two key goals: reducing density of weeds infesting the crop, which can be achieved with diverse crop rotations, and improving the crop's ability to compete with and tolerate weeds. To strengthen crop tolerance to weeds, cultural practices such as N fertilizer placement, narrow row spacing, higher plant densities, or delayed planting, either alone or in combinations, were evaluated to determine their impact on weed growth and crop yields.

Winter Wheat

Jointed goatgrass proliferates in winter wheat-fallow because its seeds persists for several years in the soil. To aid control of jointed goatgrass, we evaluated cultural practices such as N fertilizer placement, seeding rate, and row spacing for their impact on jointed goatgrass seed production in winter wheat (Anderson, 1997). Common practices used by producers to grow winter wheat include planting cultivars at 1.2 million seeds/ha in rows spaced 30 cm apart, with N fertilizer applied broadcast. A treatment composed of three cultural practices, seeding rate of 1.8 million seeds/ha, row spacing of 18 cm, and N applied in a band near the seed row, reduced jointed goatgrass seed production 45 percent. We observed that synergism occurred among cultural practices if three cultural practices were combined; when two cultural practices were combined, jointed goatgrass seed production was reduced less than 10 percent, a fivefold difference compared to the combination of three practices.

Cultural practices will enable producers to reduce seed entry into the soil seed bank by almost half. Integrating this approach with diverse rotations that include summer annual crops can reduce weed density in future winter wheat 85 to 95 percent (Anderson, 1998b). A similar approach is successful in Australia, where wild oat (*Avena fatua* L. and *A. ludoviciana* Dur.) is a serious weed of winter wheat-fallow rotations. Pandley and Medd (1990) developed a *seed kill* strategy, where postemergence herbicides applied to wild oat at flowering drastically reduce weed seed production in winter wheat. Combining this strategy with rotations that included summer annual crops not only controlled wild oats more effectively, but yielded a threefold economic benefit compared to the conventional system of winter wheat-fallow, when projected over a 15-year period (Jones and Medd, 1997).

Corn

Common practices for growing corn in the Great Plains include a target population of 37,000 plants/ha, row spacing of 76 cm, and N fertilizer applied broadcast. With this system, corn is not competitive with weeds because of its open canopy and availability of N to weeds. Therefore, we evaluated a series of cultural practices to improve corn's competitiveness and tolerance to weeds (Anderson, 2000b). One treatment consisted of N banded near the seed; a second treatment combined N banding with higher corn density (47,000 plants/ha); and for the third treatment, corn was planted in rows spaced 38 cm apart in addition to N banding and higher corn density. Foxtail millet (*Setaria* spp.) was planted as an indicator species to measure effects of cultural practices.

Foxtail millet biomass was reduced 60 percent with the treatment composed of three cultural practices; biomass reduction was 10 percent with N banding alone and 25 percent with N banding plus higher crop density. Yield loss due to grass interference was 43 percent with the conventional system (see Figure 7.5). In contrast, yield loss was only 13 percent with the system composed of three cultural practices. As found with cultural practices in winter wheat, synergism among cultural practices occurred; impact on weed biomass and yield loss was four- to sixfold greater when three cultural practices were combined compared to one practice alone.

Proso Millet

Proso millet is well adapted to semiarid climates because of its low water use (Anderson et al., 1999). Various aspects of proso millet production suggest that weeds possibly could be controlled with cultural practices, thus eliminating the need for herbicides. First, proso millet usually is planted after the majority of weeds emerge in this region. Second, erratic precipitation can result in extensive dry periods with minimal weed seedling emergence. A third aspect is the synergistic effect of combining cultural practices; a crop's competitiveness to weeds can be increased four- to sixfold (Anderson, 1997, 2000b). Therefore, we hypothesized that the interaction among weed emergence patterns, erratic precipitation, and synergism of cultural systems would enable producers to control weeds in proso millet without herbicides.

To test this hypothesis, we compared two systems of proso millet production, a common system used by producers versus an ecologically based system, for impact on proso millet's interaction with redroot pigweed (*Amaranthus retroflexus* L.) (Anderson, 2000a). The ecologically based system was designed to (1) strengthen proso millet's canopy by increasing seeding rate 50 percent and banding N fertilizer near the seed and (2) reduce weed densities by delaying planting (allowing the first cohort of redroot pigweed seedlings to be controlled before planting). With the conventional system, N was applied broadcast. Two tillage treatments also were compared: no-till versus minimum-till (tillage with the sweep plow before planting).

Redroot pigweed density and biomass were reduced severalfold with the ecologically based system compared to the conventional system in both tillage treatments (see Table 7.1). An intriguing result was that pigweed density was sevenfold greater in the minimum-till treatment compared to the no-till treatment; tilling with the sweep plow apparently placed pigweed seed in more favorable germination sites in soil. However, the ecologically based system was still effective in the minimum-till treatment. Yield loss

TABLE 7.1. Weed management system effect on weed growth and competition in proso millet at Akron, Colorado.

	Redroot pigweed			Weed-infested proso millet
Management system	Density (plants/m²)	Biomass (g/m²)	Seed production (seeds/m²)	Yield loss (%)
Minimum-till				
Conventional[a]	42 a[b]	475 a	88,400 a	29 a
Ecologically based	8 b	56 b	8,100 b	2 bc
No-till				
Conventional	6 b	72 b	9,700 b	4 b
Ecologically based	1 c	7 c	600 c	0 c

Source: Adapted from Anderson, 2000a.

Notes: Data averaged across three years. Minimum-till consisted of tillage with a sweep plow. Proso millet yields in the weed-free treatments ranged from 1,500 to 2,800 kg·ha^{-1} across four years.

[a]Conventional system was composed of a seeding rate of 11 kg·ha^{-1}, N broadcast, and early planting; ecologically-based system was proso planted at 17, kg·ha^{-1}, N banded near the seed, and planting delayed two weeks.

[b]Means within columns followed by the same letter do not differ as determined with the Fisher's LSD ($p = 0.05$).

due to redroot pigweed was eliminated with the ecologically based system in both tillage treatments. Proso millet grain yield in weed-free conditions did not differ among systems within a tillage treatment. However, yield was 25 percent greater in no-till than in minimum-till, which was attributed to improved precipitation storage in soil during the noncrop period before planting.

With an ecologically based system, proso millet can be grown without using herbicides in-crop, which will accrue additional benefits for producers. First, weed densities infesting proso millet are variable in the field, with areas of no weeds. Second, proso millet can be injured by herbicides, especially in dry years, with yield losses approaching 20 percent (Lyon and Miller, 1999). Thus, the ecologically based system avoids unnecessary herbicide use and possible yield loss due to herbicide injury. A further benefit is that the ecologically based approach, by eliminating need for herbicides in-crop, removes selection pressure in one year of the rotation, thus delaying development of herbicide resistance (Holt and LeBaron, 1990).

Producers believe that weed management without herbicides will lead to extensive proliferation of weed densities in future years. A surprising result of our study, however, was that seed production by redroot pigweed was reduced 10- to 16-fold in the ecologically based system (Table 7.1). Population dynamics studies have shown that weed populations in rotations remain relatively stable if control efficacy is approximately 90 percent (Cousens et al., 1987). The ecologically based system achieved this level of control in both tillage systems, suggesting that weed populations will not increase over time.

ECOLOGICALLY BASED WEED MANAGEMENT IN THE SEMIARID GREAT PLAINS

Cropping systems are changing in the Great Plains, with producers exploring a multitude of rotations. Rotations designed to accentuate ecological trends of weed dynamics, such as the natural decline of weed seed in soil, will help producers develop weed management systems that are not so reliant on herbicides. Also, synergism among cultural practices improves crop competitiveness with weeds; we speculate that a cycle-of-four rotation, which reduces weed densities and improves crop vigor due to the rotation effect and disease management, will further strengthen cultural weed management. In addition, diverse crop rotations provide other opportunities to manage weeds. For example, weed emergence and establishment is reduced by winter wheat residue (3,000 kg·ha^{-1} or more) lying on the soil surface (Crutchfield, Wicks, and Burnside, 1986). Cultural systems of winter wheat that reduce jointed goatgrass seed production (Anderson, 1997) produce 2,000 to 3,000 kg/ha more crop residue than conventional practices; consequently, weed emergence was 30 percent less in the following summer annual crop (Anderson, 1999).

Producers are concerned that adding less-profitable crops to the rotation will reduce farm economics; however, diverse rotations can maintain profitability because of reduced input costs. A study in the Netherlands compared conventional and ecologically based systems across 15 years (Lewis et al., 1997). Pesticide use was reduced eightfold with the ecologically based system; yields were somewhat lower with this system, yet reduced input costs led to favorable net returns compared to the conventional system. Furthermore, the ecologically based system improved financial risk management.

The goal of producers and research scientists in the Central Great Plains is to develop ecologically sound rotations that improve farm economics, conserve soil health, and enhance pest management yet minimize inputs. The cycle-of-four rotation design enhances the opportunities to achieve this goal.

REFERENCES

Anderson, R. L. (1997). Cultural systems can reduce reproductive potential of winter annual grasses. *Weed Technology* 11: 608-613.

Anderson, R. L. (1998a). Designing rotations for a semiarid region. In *Proceedings, 10th Annual Meeting, Colorado Conservation Tillage Association* (pp. 4-15). Akron, CO: Colorado Conservation Tillage Association.

Anderson, R. L. (1998b). Seedling emergence of winter annual grasses as affected by limited tillage and crop canopy. *Weed Technology* 12: 262-267.

Anderson, R. L. (1999). Cultural strategies reduce weed densities in summer annual crops. *Weed Technology* 13: 314-319.

Anderson, R. L. (2000a). A cultural systems approach eliminates the need for herbicides in semiarid proso millet. *Weed Technology* 14: 602-607.

Anderson, R. L. (2000b). Cultural systems to aid weed management in semiarid corn (*Zea mays*). *Weed Technology* 14: 630-634.

Anderson, R. L., R. A. Bowman, D. C. Nielsen, M. F. Vigil, R. M. Aiken, and J. G. Benjamin (1999). Alternative crop rotations for the central Great Plains. *Journal of Production Agriculture* 12: 95-99.

Anderson, R. L., D. L. Tanaka, A. L. Black, and E. E. Schweizer (1998). Weed community and species response to crop rotation, tillage, and nitrogen fertility. *Weed Technology* 12: 531-536.

Bailey, K. L. (1996). Diseases under conservation tillage systems. *Canadian Journal of Plant Science* 76: 635-639.

Beckie, H. J., A. G. Thomas, A. Legere, D. J. Kelner, R. C. van Acker, and S. Meers (1999). Nature, occurrence, and cost of herbicide-resistant wild oat *(Avena fatua)* in small-grain production area. *Weed Technology* 13: 612-625.

Boydston, R. A. (1990). Time of emergence and seed production of longspine sandbur *(Cenchrus longispinus)* and puncturevine *(Tribulus terrestris). Weed Science* 38: 16-21.

Cook, R. J. and R. J. Veseth (1991). *Wheat health management.* St. Paul, MN: American Phytopathological Society Press.

Cousens, R., S. R. Moss, G. W. Cussans, and B. J. Wilson (1987). Modeling weed populations in cereals. *Reviews of Weed Science* 3: 93-112.

Crookston, R. K., J. E. Kurle, P. J. Copeland, J. H. Ford, and W. E. Lueschen (1991). Rotational cropping sequence affects yield of corn and soybean. *Agronomy Journal* 83: 108-113.

Crutchfield, D. A., G. A. Wicks, and O. C. Burnside (1986). Effect of winter wheat *(Triticum aestivum)* straw mulch level on weed control. *Weed Science* 34: 110-114.

Dawson, J. H. and V. F. Burns (1975). Longevity of barnyardgrass, green foxtail, and yellow foxtail seeds in soil. *Weed Science* 23: 437-440.

Dhuyvetter, K. C., C. R. Thompson, C. A. Norwood, and A. D. Halvorson (1996). Economics of dryland cropping systems in the Great Plains: A review. *Journal of Production Agriculture* 9: 212-216.

Dieleman, J. A., D. A. Mortensen, and A. R. Martin (1999). Influence of velvetleaf *(Abutilon theophrasti)* and common sunflower *(Helianthus annuus)* density variation on weed management outcomes. *Weed Science* 47: 81-87.

Donald, W. W. and R. L. Zimdahl (1987). Persistence, germinability, and distribution of jointed goatgrass *(Aegilops cylindrica)* seed in soil. *Weed Science* 35: 149-154.

Egley, G. H. (1986). Stimulation of weed seed germination in soil. *Reviews of Weed Science* 2: 67-89.

Egley, G. H. and R. D. Williams (1990). Decline of weed seeds and seedling emergence over five years as affected by soil disturbance. *Weed Science* 38: 504-510.

Forcella, F., R. Wilson, J. Dekker, R. Kremer, J. Cardina, R. Anderson, D. Alm, K. Renner, R., Harvey, S. Clay, and D. Buhler (1997). Weed seedbank emergence across the Corn Belt. *Weed Science* 45: 67-76.

Froud-Williams, R. J. (1988). Changes in weed flora with different tillage and agronomic management systems. In M. A. Altieri and M. Liebman (Eds.), *Weed management in agroecosystems: Ecological approaches* (pp. 215-236). Boca Raton, FL: CRC Press, Inc.

Froud-Williams, R. J., R. J. Chancellor, and D. S. H. Drennan (1984). The effects of seed burial and soil disturbance on emergence and survival of arable weeds in relation to minimal cultivation. *Journal of Applied Ecology* 21: 629-641.

Harper, J. L. (1977). The seed bank. In J. L. Harper (Ed.), *Population biology of plants* (pp. 83-110). New York: Academic Press.

Hoffman, D. W. and T. L. Lavy (1978). Plant competition for atrazine. *Weed Science* 26: 94-99.

Holt, J. S. and H. M. LeBaron (1990). Significance and distribution of herbicide resistance. *Weed Technology* 4: 141-149.

Holtzer, T. O., R. L. Anderson, M. P. McMullen, and F. B. Peairs (1996). Integrated pest management for insects, plant pathogens, and weeds in dryland cropping systems of the Great Plains. *Journal of Production Agriculture* 9: 200-208.

Jones, R. and R. Medd (1997). Economic analysis of integrated management of wild oats involving fallow, herbicide and crop rotational options. *Australian Journal of Experimental Agriculture* 37: 683-691.

Jordan, N., D. A. Mortensen, D. M. Prenzlow, and K. C. Curtis (1995). Simulation analysis of crop rotation effects on weed seedbanks. *American Journal of Botany* 82: 390-398.

Jordan, V. W. L., J. A. Hutcheon, G. V. Donaldson, and D. P. Farmer (1997). Research into and development of integrated farming systems for less-intensive arable crop production: Experimental progress (1989-1994) and commercial implementation. *Agriculture Ecosystems and Environment* 64: 141-148.

Kurtz, L. T., L. V. Boone, T. R. Peck, and R. G. Hoeft (1984). Crop rotations for efficient nitrogen use. In R. D. Hauck (Ed.), *Nitrogen in crop production* (pp. 295-306). Madison WI: American Society of Agronomy.

Lewis, W. J., J. C. van Lenteren, S. C. Phatak, and J. H. Tumlinson (1997). A total system approach to sustainable pest management. *Proceedings, National Academy of Sciences, USA,* 94: 12243-12248.

Lyon, D. J. and S. D. Miller (1999). Herbicide injury in proso and foxtail millets. *Proceedings, Western Society Weed Science Annual Meeting* 52: 24.

Lyon, D. J., S. D. Miller, and G. A. Wicks (1996). The future of herbicides in weed control systems of the Great Plains. *Journal of Production Agriculture* 9: 209-215.

Matus, A., D. A. Derksen, F. L. Walley, H. A. Loeppky, and C. van Kessel (1997). The influence of tillage and crop rotation on nitrogen fixation in lentil and pea. *Canadian Journal of Plant Science* 77: 197-200.

Mortensen, D. A., L. Bastiaans, and M. Sattin (2000). The role of ecology in the development of weed management systems: An outlook. *Weed Research* 40: 49-62.

Pandley, S. and R. W. Medd (1990). Integration of seed and plant kill tactics for control of wild oats: An economic evaluation. *Agricultural Systems* 34: 65-76.

Pedigo, L. P. (1995). Closing the gap between IPM theory and practice. *Journal of Agricultural Entomology* 12: 171-181.

Peterson, G. A., D. G. Westfall, and C. V. Cole (1993). Agroecosystern approach to soil and crop management research. *Soil Science Society of America Journal* 57: 1354-1360.

Peterson, G. A., A. J. Schegel, D. L. Tanaka, and O. R. Jones (1996). Precipitation use efficiency as affected by cropping and tillage systems. *Journal of Production Agriculture* 9: 180-186.

Pierce, F. J. and C. W. Rice (1988). Crop rotation and its impact on efficiency of water and nitrogen use. In W. L. Hargrove (Ed.), *Cropping strategies for efficient use of water and nitrogen* (pp. 21-42). Madison, WI: American Society of Agronomy.

Porter, P. M., R. K. Crookston, J. H. Ford, D. R. Huggins, and W. E. Lueschen (1997). Interrupting yield depression in monoculture corn: Comparative effectiveness of grasses and dicots. *Agronomy Journal* 89: 247-250.
Roberts, H. A. (1981). Seed banks in soil. *Advances in Applied Biology* 6: 1-55.
Sagar, G. R. and A. M. Mortimer (1976). An approach to the study fo the population dynamics of plants with special reference to weeds. *Advances in Applied Biology* 1: 1-47.
Thomas, A. G., J. D. Banting, and G. Bowes (1986). Longevity of green foxtail seeds in a Canadian prairie soil. *Canadian Journal of Plant Science* 66: 189-192.
Wicks, G. A. (1984). Integrated systems for control and management of downy brome (*Bromus tectorum*) in cropland. *Weed Science* 32(Supplement 1): 26-31.
Wright, A. T. (1990). Yield effect of pulses on subsequent cereal crops in the northern plains. *Canadian Journal of Plant Science* 70: 1023-1032.

Chapter 8

Examining Tillage and Crop Rotation Effects on Weed Populations in the Canadian Prairies

R. E. Blackshaw
A. G. Thomas
D. A. Derksen
J. R. Moyer
P. R. Watson
A. Légère
G. C. Turnbull

INTRODUCTION

The Canadian Prairies occur at the northern portion of the Great Plains of North America (see Figure 8.1). This large area of 35 million ha of cropped land has been subdivided into eight ecoregions (Ecological Stratification Working Group, 1995) that are areas of similar landforms, climate, natural vegetation, soils, and land use. Soils are commonly Aridic and Typic Haplustolls in the Mixed Grassland, Moist Mixed Grassland and Aspen Parkland Ecoregions and Molic Haplocryalfs in the Boreal Transition, Mid-Boreal Upland, and Peace Lowland Ecoregions (Clayton et al., 1977). Despite covering a vast geographic area with a range of grassland and forest soils, crop production on the Canadian Prairies has not been highly diversified. Historically, the area has been dominated by spring cereal production with fallow being a common part of the rotation. On the Canadian Prairies in 2001, there were 10.9 million ha of spring wheat (*Triticum aestivum* L.), 4.7 million ha of spring barley (*Hordeum vulgare* L.), and 3.9 million ha of spring canola (*Brassica napus* L. and *B. rapa* L.) (Anonymous, 2001).

We thank S. A. Brandt, A. M. Johnston, G. P. Lafond, J. Y. Leeson, B. G. McConkey, and R. C. Van Acker for their helpful comments and for their technical assistance as site managers at the various study locations.

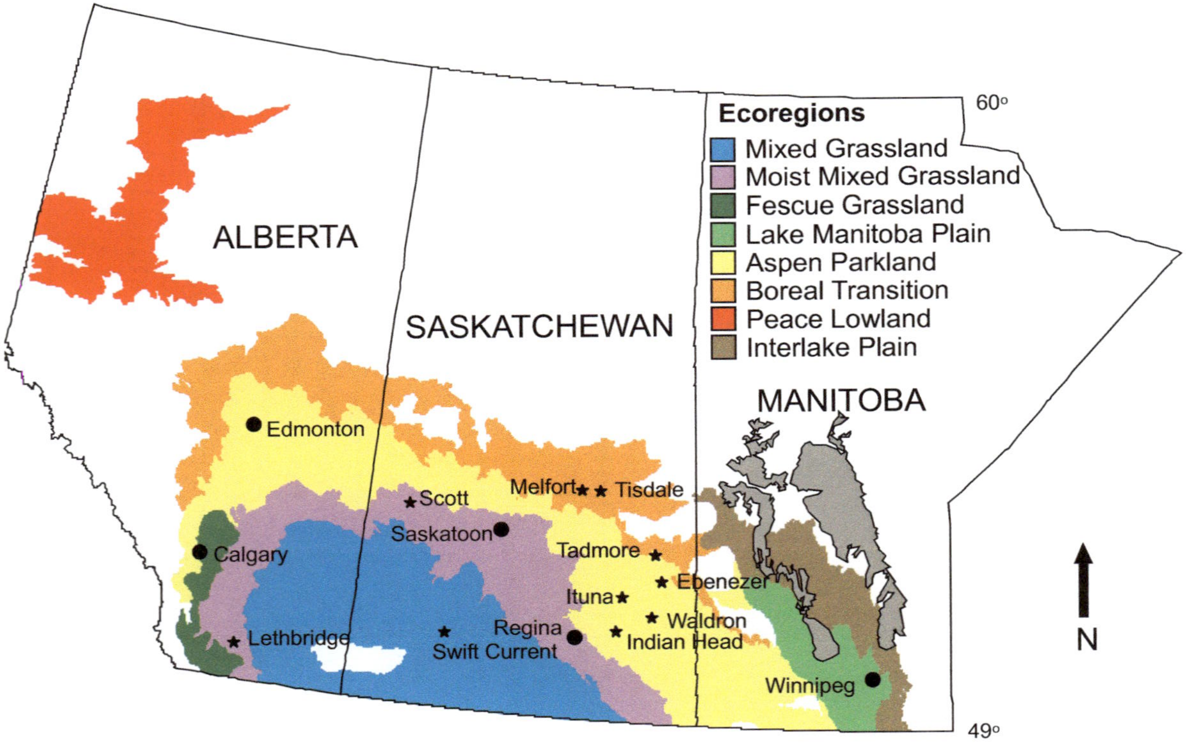

FIGURE 8.1. Major ecoregions of the Canadian Prairies. Study site locations are indicated by ★.

Spring oats (*Avena sativa* L.), dry pea (*Pisum sativum* L.), lentil (*Lens culinaris* Medik.), flax (*Linum usitatissimum* L.), chickpea (*Cicer arietinum* L.), and mustard (*Brassica juncea* L.) were each grown on 1.8 to 0.5 million ha, respectively. Winter wheat (0.2 million ha) has not been a significant crop in our rotations because of survival problems during harsh winters. The remaining cultivated land area is used for fallow (4.7 million ha), hay (4.7 million ha), and pasture (16.9 million ha).

The primary limiting factors for crop production are limited annual precipitation and a short growing season. An annual precipitation deficit exists in nearly all areas of the Canadian Prairies (see Figure 8.2) and the frost-free growing season ranges from 88 days in the northern regions to 127 days in the southern regions (Hunter, Morrison, and Rourke, 1990; Agriculture and Agri-Food Canada, 1997). The high frequency of spring crops in rotation has selected for spring-germinating annual weeds that dominate the weed community throughout the cropping region (Thomas et al., 1996, 1998; Thomas, Frick, and Hall, 1998). Some facultative winter annual and herbaceous perennial species also are problems on cropland. Most weed species have been introduced from Europe with some native species, such as foxtail barley (*Hordeum jubatum* L.), recently increasing in abundance (Blackshaw et al., 2000).

The adoption of conservation tillage systems, such as minimum or zero tillage, as a means of increasing soil moisture availability and subsequent crop yield, has been widespread in recent years (Anonymous 1997a,b,c). In 1996, conservation tillage was practiced on 48 percent of cropland on the Canadian Prairies. Conservation tillage has led to a reduction in fallow and increased frequency of broadleaf crops in rotations (Lafond and Derksen, 1996). Questions have arisen regarding adverse changes in weed communities due to the adoption of these practices. For example, will Canada thistle [*Cirsium arvense* (L.) Scop.] and dandelion (*Taraxacum officinale* Weber in Wiggers) increase in abundance in zero-tillage systems, thereby increasing costs of production and negating the benefits of these tillage systems?

In the mid-1990s, a number of weed researchers involved in independent cropping system studies from several ecoregions on the Canadian Prairies recognized the value that could be derived from pooling their data of 56 site-years and testing hypotheses about the relationship between management practices and weed communities. It was recognized that a team approach would be needed to achieve this objective. Dow AgroSciences Canada researchers recognized that important shifts in the weed community were occurring due to changes in prairie crop production practices and consequently decided to join and support the team. All participants wanted to gain a greater understanding of the ecology and dynamics of prairie weed communities and to be better able to predict future weed problems. During

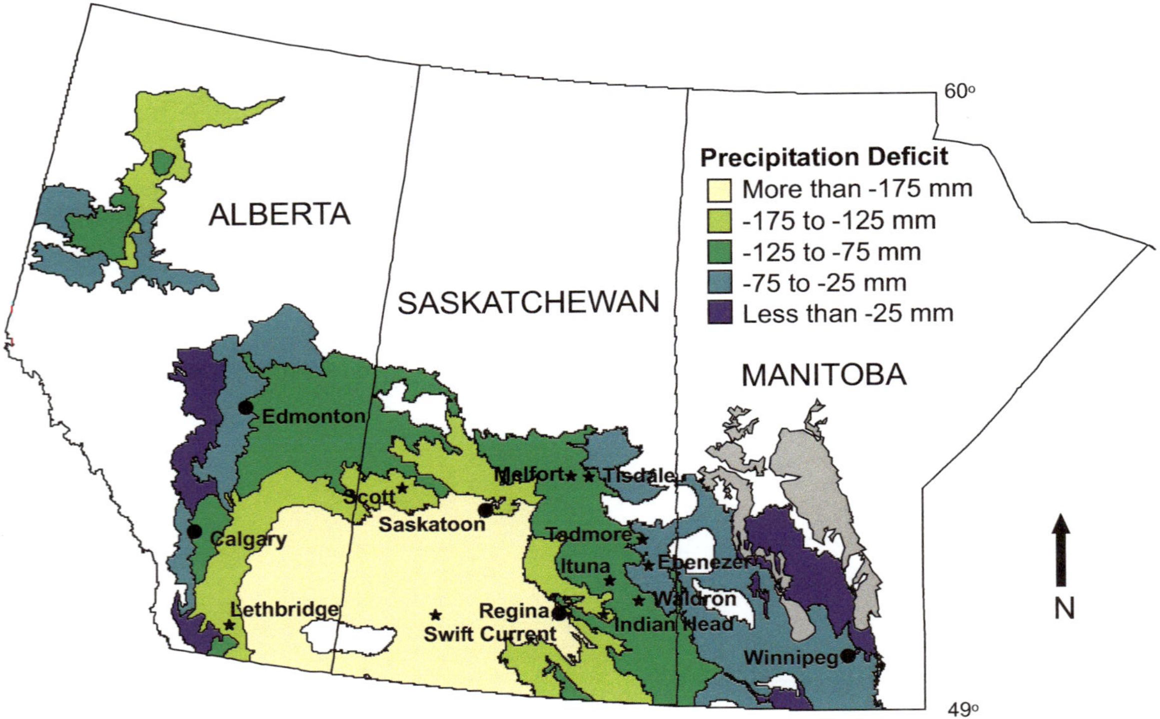

FIGURE 8.2. Precipitation surplus/deficit estimates on the Canadian Prairies based on 30-year average annual precipitation values and potential evapotranspiration calculated by the Thornthwaite method. Study site locations are indicated by ★. (*Source:* Agriculture and Agri-Food Canada, 1997.)

the initial planning meeting, it was recognized that the analysis and synthesis of weed community data in cropping systems necessitated a multivariate approach and therefore an expert with knowledge of classification and ordination methodology in vegetation ecology was recruited. The final team was composed of weed scientists from Agriculture and Agri-Food Canada, University of Manitoba, and Dow AgroSciences Canada, Inc. Additional advisors to the team were the agronomists who managed the individual study sites. The team met every three to six months for four years to interpret data in relation to agricultural and ecological selection pressures.

The broad objective of this chapter is to demonstrate the use of a multisite data set from a large geographic area as a case study of how to understand weed community dynamics in relation to the selection pressures of conservation tillage and crop rotation on the Canadian Prairies. The specific research objectives were to determine (1) the relative effect of year, tillage system, and crop rotation on weed populations; (2) species associations with specific tillage systems; (3) species associations with certain crop rotations compared with results of previous rotation studies in this region; (4) if species respond individually or as functional groups to cropping systems selection pressures.

MULTISITE PROJECT

Study Site Selection

The initial step for the research team was the identification of appropriate studies that represented conservation tillage and crop rotation systems for the various ecoregions of the Canadian Prairies. Although studies were not available for each ecoregion, the selected studies were located in areas where the adoption of conservation tillage systems has been prevalent (see Figure 8.1). Studies had to meet certain criteria to be included in the project (Table 8.1).

Criteria were that cropping systems have

1. completed at least one rotational cycle;
2. more than one crop rotation so that tillage system and crop rotation were not confounded;
3. all phases of each rotation present each year so that year and phase were not confounded;
4. at least two tillage systems for comparative purposes;
5. a minimum of four replicates;

6. used typical farm machinery to ensure selection pressures are adequately represented;
7. herbicide selection based on weed flora present (not necessarily the same for each tillage system);
8. plots large enough for representative weed communities to establish.

Weed assessment criteria were as follows:

9. must be quantitative (density);
10. collected at least once per growing season;
11. have at least one assessment time in common with other studies;
12. have all species present counted;
13. have enough quadrats (12 or greater) assessed per plot to accurately represent the density and diversity of the weed community present;
14. weeds did not have to be counted in fallow years.

Definition of Tillage Systems

Tillage systems were defined by their respective level of soil disturbance. In the *zero-tillage* system, crops were planted directly into the previous crop's stubble with <30 percent soil disturbance, and herbicides alone were used to control weeds in fallow. *Conventional tillage* referred to the typical practices where the study was located. Conventional systems used a fall and spring tillage prior to seeding spring crops, and tillage was the only tool used for weed control in fallow. *Minimum tillage* in our study was defined as intermediate between zero and conventional. It was usually represented by a single tillage pass in the spring and a combination of herbicides and tillage to control weeds in fallow. Due to the typically dry conditions at Lethbridge, tillage frequency and intensity are further reduced in this ecoregion. Therefore, conventional tillage was defined as one pass prior to seeding with a disc, and minimum tillage was one pass with a wide-blade cultivator.

Due to moisture deficits, the short growing season, and harsh winters on the Canadian Prairies, tillage systems have been adapted to address these limitations. Generally, the most intensive tillage operations used in these studies were two passes with a field cultivator. Thus conventional tillage systems would be considered minimal in many parts of the world. Regardless of system, tillage did not occur before seeding winter wheat to leave standing stubble to trap snow and enhance winter survival. In all cases, fallow was the absence of a crop for one growing season, which left approximately 24 months between the seeding of subsequent spring crops.

TABLE 8.1. Cropping system and weed assessment details for the 11 studies selected for the project.

Study	Tillage systems	Crop rotations	Rotation phases	Replicates	Plots	Site years	Number of weed assessments	Quadrats per plot	Species
CMS	3	3	4	4	144	12	1	20	42
SCMS	2	6	4	4	192	4	3	20	46
MTRS1	3	3	4	4	144	4	1	20	32
MTRS2	3	3	4	4	144	4	1	20	26
STSS	2	2	3	4	32	6	1	20	34
LLTS	3	4	2	4	96	8	3	12	22
SCAC	2	5	3	4	120	3	1	20	28
ECSC1	3	2	3	4	72	2	1	20	30
ECSC2	3	2	3	4	72	5	1	20	37
ECSC3	3	2	3	6	108	3	1	20	40
ECSC4	3	2	3	4	72	5	1	20	35
TOTAL					1,196	56			82

CMS = Crop Management Study at Indian Head, Saskatchewan; SCMS = Special Crop Management Study at Indian Head, Saskatchewan; MTRS = Melfort Tillage and Rotation Study at (1) Melfort and (2) Tisdale, Saskatchewan; STSS = Tillage System Study at Scott, Saskatchewan; LLTS = Long-Term Rotation and Tillage Study at Lethbridge, Alberta; SCAC = Agronomy of Alternate Crops Study at Swift Current, Saskatchewan; ECSC = East Central Saskatchewan Soil Conservation Project at (1) Ebenezer, (2) Ituna, (3) Tadmore, and (4) Waldron, Saskatchewan.

Multivariate Analysis

Klecka (1980) outlines the advantages and disadvantages of several approaches to multivariate data analysis. Numerous methods were evaluated for their appropriate data representation and statistical testing as described in Kenkel and colleagues (2002). Canonical discriminant analysis (CDA) was chosen to analyze the data (SAS Institute Inc., 1999) from our individual studies because the method allowed comparison of the relative importance of tillage system, crop rotation, and time.

CDA represents multidimensional data structures in one or two dimensions, while maximally preserving the trends in variation that are present in the data. Ordination axes are extracted in order, with each axis summarizing the maximum amount of information not yet accounted for by previous axes. Thus the first axis is always the "most important," followed by the second, the third, and so on. Higher-dimensional axes are then discarded on the assumption that most of the intrinsic structure of the data is preserved in the first few axes. Biplots of weed species and tillage system, crop rotation and time are formed, and species associations can be ascertained by the direction of vectors. Vector length indicates the relative strength of the association; the longer the vector the stronger the association.

Although 82 species were documented in the 11 studies (see Table 8.1), only 71 were retained for analyses. Discarded species were either very rare weeds or minor speciality crops for which insufficient data were available. Species data were log-transformed prior to analysis. Examples of the use of this method for weed community analysis in cropping systems have previously been published (Derksen et al., 1994; Derksen, Watson and Loeppky, 1998; Blackshaw, Larney, et al., 2001).

Indicator Species Analysis

Once the relative effect of tillage system was understood in relation to crop rotation and time, the association of individual species with tillage system across studies was further explored using Indicator Species Analysis (Dufrêsne and Legendre, 1997). Each species indicator value (IV) was calculated by multiplying its within-species relative abundance by its within-treatment frequency using PC-ORD (McCune and Mefford, 1999). The calculation of IVs does not incorporate the relative abundance of other species and is therefore not a community measure. Indicator values were calculated for each species within each site on nontransformed density data. Because IVs are affected by the number of tillage systems within a study, an IV for a species across all studies could not be calculated. The higher the IV the greater the association with a tillage system. An example of indicator values for the most abundant annual grass, facultative winter annual, and herbaceous perennial species is presented in Table 8.2.

Response Groups

Response groups were defined as groups of species that were associated with a specific tillage system. Each of the 71 indicator value tables was interpreted by the team, and each species was assigned to one or more tillage response groups. Many of the decisions were not straightforward and the specific management of the site had to be considered. For example, while wild oat (*Avena fatua* L.) was clearly associated with zero tillage at the SCMS and STSS sites (see Table 8.2), the reverse was true at the MTRS site, with no clear indication from the ECSC sites. The SCMS site had atypically high numbers of noncompetitive crops in rotation and/or reduced herbicide inputs, and the STSS study was established prior to the advent of postemergence wild oat herbicides. Therefore, the team assigned wild oats to the ubiquitous response group. A similar process was used to assign field pennycress (*Thlaspi arvense* L.) to a conventional tillage response group and Canada thistle to both zero- and minimum-tillage response groups.

Once all species had been assigned to groups, the team tried to find common functional traits among the species within a response group. Traditionally, weed scientists have grouped species by morphological and/or taxonomic similarities. However, the team found that these simple groupings were inadequate. The team then used a suite of functional characteristics to explain the responses to tillage systems. Functional characteristics considered included such factors as life cycle, seed dormancy, seed bank persistence, recruitment depth, and dispersal mechanisms.

WEED COMMUNITY ASSOCIATIONS

Crop Rotation, Tillage, and Environmental Effects on Weeds

Changes in weed community composition are the result of selection pressures imposed by agronomic practices in conjunction with the modifying effects of the prevailing environmental conditions within a cropping region. Thus the relative importance of any one factor (climate, rotation, tillage) is dependent on the variability within other factors. For instance, tillage system changes may be expected to have larger effects on weed communities in simple than in diverse crop rotations.

TABLE 8.2. Indicator values for an annual grass, facultative winter annual, and herbaceous perennial species.

		Wild oat			Field pennycress			Canada thistle		
Study	Assessment	ZT	MT	CT	ZT	MT	CT	ZT	MT	CT
CMS	Residual	<	<	9	4	17	31	19	15	4
SCMS	Seedbank	60		22	25		71	14		7
SCMS	Prespray	62		35	11		83	42		23
SCMS	Residual	70		25	13		66	45		28
MTRS1	Residual	18	21	34	24	21	14	28	29	26
MTRS2	Residual	11	25	17	15	13	18	32	38	15
STSS	Residual	30		1	25		50	3		4
LLTS	October	—	—	—	2	2	3	—	3	1
LLTS	May	—	—	—	4	8	8	—	1	<
LLTS	June	—	—	—	1	17	28	—	—	2
SCAC	Residual	—	—		6	30	—	1	—	
ECSC1	Residual	3	7	2	<	1	—	20	12	13
ECSC2	Residual	22	10	12	5	1	<	17	14	13
ECSC3	Residual	<	<	4	10	5	4	<	2	<
ECSC4	Residual	2	2	2	<	<	13	10	16	9

ZT = zero tillage systems, MT = minimum tillage systems; CT = conventional tillage systems; CMS = Crop Management Study at Indian Head, Saskatchewan; SCMS = Special Crop Management Study at Indian Head, Saskatchewan; MTRS = Melfort Tillage and Rotation Study at (1) Melfort and (2) Tisdale, Saskatchewan; STSS = Tillage System Study at Scott, Saskatchewan; LLTS = Long-Term Rotation and Tillage Study at Lethbridge, Alberta; SCAC = Agronomy of Alternate Crops Study at Swift Current, Saskatchewan; ECSC = East Central Saskatchewan Soil Conservation Project at (1) Ebenezer, (2) Itiuna, (3) Tadmore, and (4) Waldron, Saskatchewan. *Legend:* blank line = tillage system not present in study; — species absent in tillage system; < = an indicator value of less than 0.5.

Canonical discriminant analysis of weed density data from 56 site-years of western Canadian data indicated that the overall ranking of effects on weed communities was climate > crop rotation > tillage. This finding is well demonstrated by a study conducted at Melfort, Saskatchewan, consisting of three crop rotations (canola-wheat-barley-barley, canola-barley-pea-wheat, and canola-pea-flax-barley) and three tillage intensities (cultivated fall and spring, cultivated in spring only, or zero till). Weed species were most affected by year-to-year climatic differences (separated on axis one of the biplot), followed by tillage and rotation effects (see Figure 8.3). The Canadian Prairies have a continental climate with short, hot summers and long, cold winters and highly variable precipitation, thus it should not be surprising that weed populations were markedly affected by year-to-year variation in climate.

Weed densities in zero tillage were often greater in years in which above-normal rainfall was received in the spring (see Figure 8.4A) or the previous fall (see Figure 8.4B) because seeds trapped in crop residue were exposed to sufficient moisture for the radicle to reach the soil and become rooted. A dry spring preceded by a dry fall resulted in greater weed densities in conventional than in zero tillage (see Figure 8.4C) because tillage moved weed seeds below the surface into moister soil. Weed densities usually increased in wet years following dry years, likely due to a combination of seed-bank expansion in dry years and increased seedling recruitment in wet years.

Weed species respond differently to climate. In years of low rainfall and high temperatures, Russian thistle (*Salsola iberica* Sennen & Pau) and kochia [*Kochia scoparia* (L.) Schrad.] increased regardless of crop rotation or tillage system (Blackshaw, Larney, et al., 2001). These weeds are well adapted to drought conditions and may compete more vigorously with crops under dry than moist soil conditions (Wiese and Vandiver, 1970; Dwyer and Wolde-Yohannis, 1972; Young, 1986). Redroot pigweed (*Amaranthus retroflexus* L.), a C_4 species, became more prevalent in years with above-average summer temperatures. Winter annual weeds, such as downy brome (*Bromus tectorum* L.), flixweed [*Descurainia sophia* (L.) Webb ex Prantl], and narrowleaf hawksbeard (*Crepis tectorum* L.), increased in density when adequate rainfall for germination was received in the fall, especially if they were not controlled by tillage or herbicides. Dandelion and quackgrass [*Elytriga repens* (L.) Nevski] thrived in wet years, whereas deep-rooted perennials such as Canada thistle were less affected by year-to-year variation in rainfall.

Two contrasting studies conducted at Indian Head, Saskatchewan, illustrate that the influence of crop rotation and tillage on weed populations can be complex. Study 1 consisted of various tillage treatments (cultivated or zero till) and crop rotations (primarily spring cereal-oilseed-spring cereal-

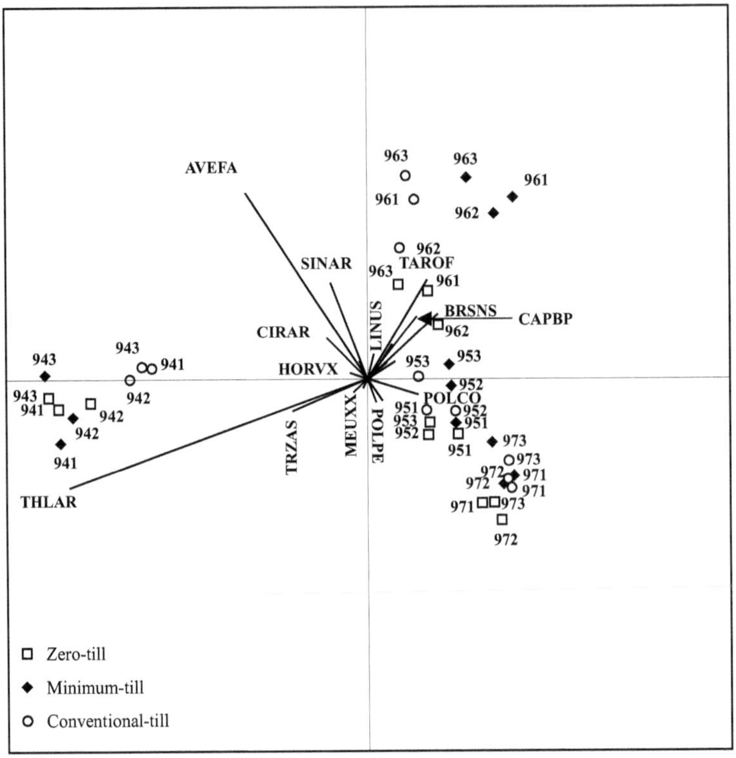

FIGURE 8.3. Canonical discriminant analysis ordination biplot of treatment (year by tillage by rotation) centroids and weed species present in residual assessments of 1994 to 1997 in the MTRS1 study at Melfort, Saskatchewan. Weed species codes are given in Table 8.3. Treatments are coded with a three-digit code; the first two digits are year and the third digit is rotation (1: canola-spring wheat-barley-barley; 2: canola-barley-pea-spring wheat; and 3: canola-pea-flax-barley). Conventional tillage consisted of one fall plus one spring cultivation before crop seeding in May. Minimum tillage consisted of one cultivation before spring seeding without any prior fall tillage. Zero tillage consisted of a glyphosate application to control weeds before planting directly into crop residues with a seed drill that caused <30 percent soil disturbance.

pulse). Canonical discriminant analysis of weed density data taken before application of in-crop herbicides (Figure 8.5A) and before crop harvest (Figure 8.5B) indicated that weed communities over a four-year period were more affected by tillage intensity (separated on axis one in Figure 8.5A and axis two in Figure 8.5B) than crop rotation. Crop rotation treat-

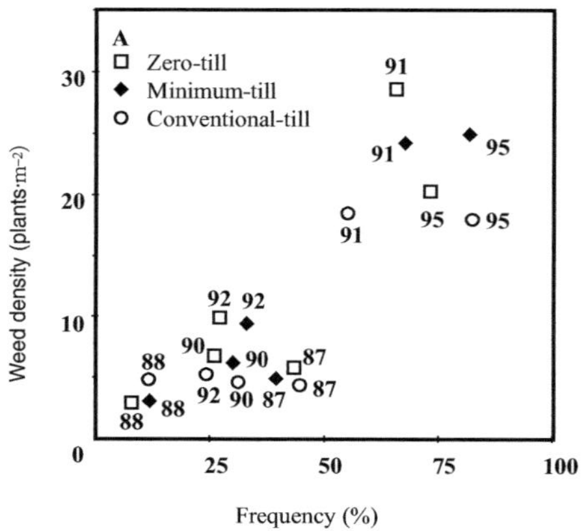

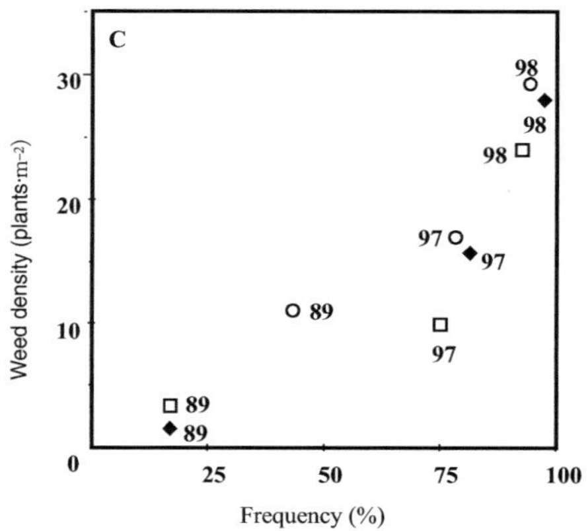

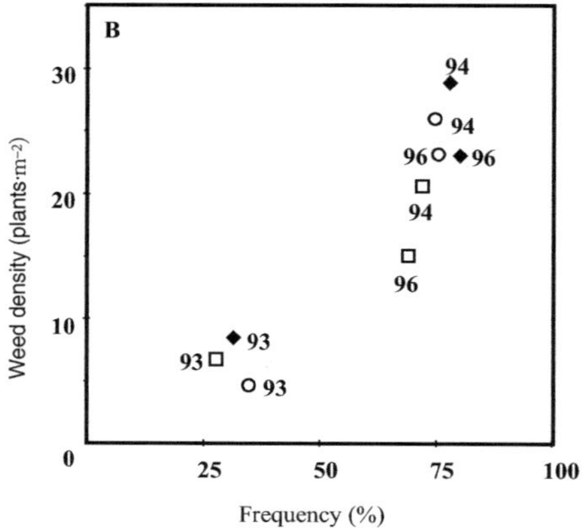

FIGURE 8.4. Residual weed assessments for 12 years (numbers indicate year) in the CMS study at Indian Head, Saskatchewan. Scatter plots of total mean weed density versus frequency (percent quadrats with weeds present) as influenced by tillage intensity in (A) a wet spring, (B) a dry spring preceded by a wet fall, and (C) a dry spring preceded by a dry fall. Dry and wet were defined as <75 percent and >125 percent, respectively, of mean duration-of-study precipitation for the designated time period (fall or spring). Conventional tillage consisted of one fall plus one spring cultivation before crop seeding in May. Minimum tillage consisted of one cultivation before spring seeding. Zero tillage consisted of a glyphosate application to control weeds before planting directly into crop residues with a seed drill that caused <30 percent soil disturbance.

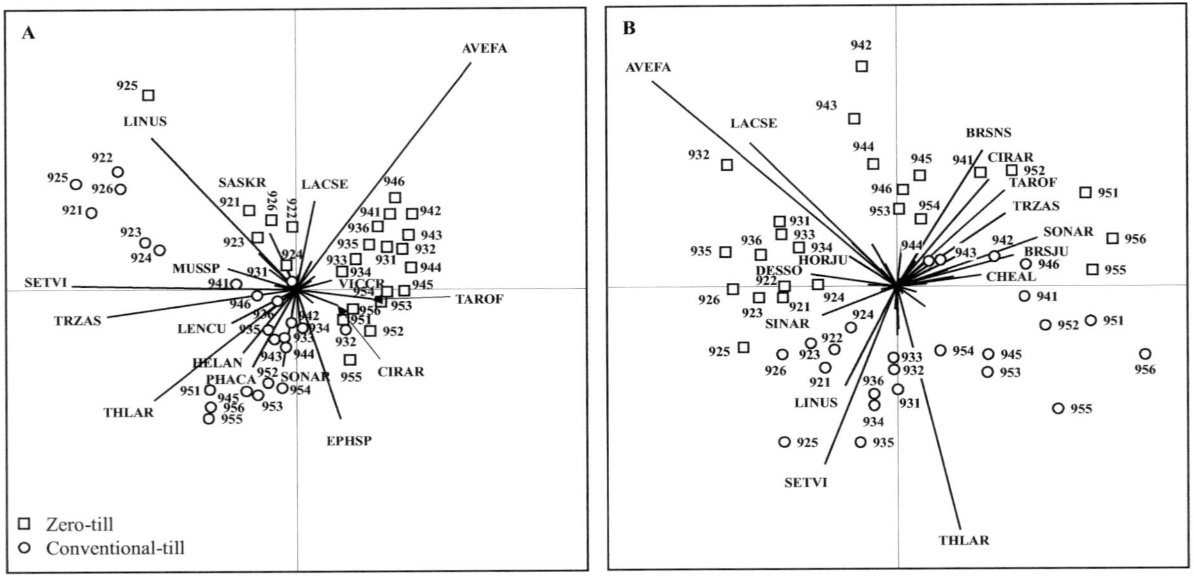

FIGURE 8.5. Canonical discriminant analysis ordination biplot of treatment (year by tillage by rotation) centroids and weed species present in (A) prespray and (B) residual assessments of 1992 to 1995 in the CSMS study at Indian Head, Saskatchewan. Weed species codes are given in Table 8.3. Treatments are coded with a three-digit code; the first two digits are year and the third is rotation (1: spring wheat-canola-spring wheat-lentil [only postemergence herbicides]; 2: spring wheat-canola-spring wheat-lentil [only preemergence herbicides]; 3: spring wheat-canola-spring wheat-lentil [low-input herbicides]; 4: spring wheat-canola-spring wheat-lentil [low-input herbicides and fertilizer]; 5: canaryseed-sunflower-spring wheat-lentil; and 6: spring wheat-mustard-canaryseed-lentil). Conventional tillage consisted of one fall plus one spring cultivation before crop seeding in May. Zero tillage consisted of a glyphosate application to control weeds before planting directly into crop residues with a seed drill that caused <30 percent soil disturbance.

ments in this study included cereals, oilseeds, and pulses, all spring-planted annual crops.

Study 2 at Indian Head, Saskatchewan, consisted of various tillage treatments (cultivated in fall and spring, cultivated in spring only, or zero tillage) and crop rotations (spring wheat-spring wheat-winter wheat-summerfallow; spring wheat-flax-winter wheat-spring wheat; spring wheat-flax-winter wheat-pea). Weed populations in the last four years of this 12-year study were more affected by crop rotation than tillage intensity (see Figure 8.6). Year effects were separated on axis one and rotation effects on axis two, indicating that rotation influenced weed populations to a greater de-

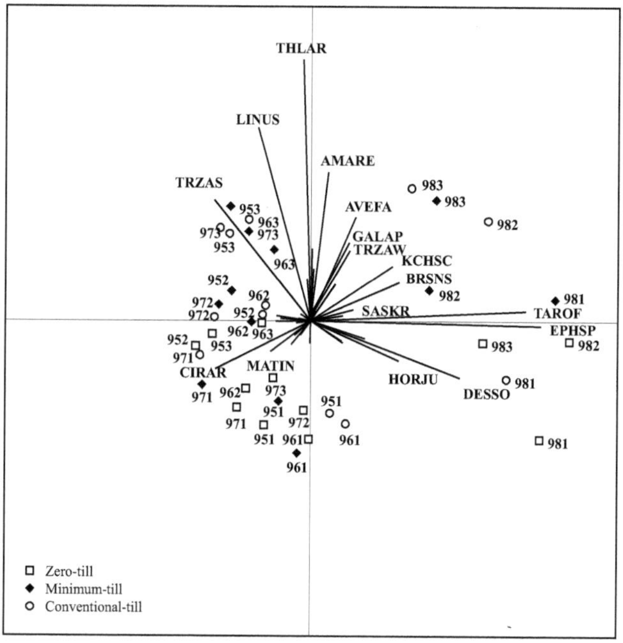

FIGURE 8.6. Canonical discriminant analysis ordination biplot of treatment (year by tillage by rotation) centroids and weed species present in postherbicide assessments of 1995 to 1998 in the CMS study at Indian Head, Saskatchewan. Weed species codes are given in Table 8.3. Treatments are coded with a three-digit code; the first two digits are year and the third is rotation (1: spring wheat-spring wheat-winter wheat-summerfallow; 2: spring wheat-flax-winter wheat-spring wheat; and 3: spring wheat-flax-winter wheat-pea). Conventional tillage consisted of one fall plus one spring cultivation before crop seeding in May. Minimum tillage consisted of one cultivation before spring seeding. Zero tillage consisted of a glyphosate application to control weeds before planting directly into crop residues with seed drill that caused <30 percent soil disturbance.

gree than tillage. The larger effect of crop rotation was attributed to the inclusion of a fall-planted crop (winter wheat) and summer fallow in this study compared with study 1 which contained only spring-planted crops. An eight-year study at Lethbridge, Alberta, similarly found that weed communities were more affected by crop rotation than tillage (Blackshaw, Larney, et al., 2001).

Tillage and Weed Associations

Previous studies have indicated that perennials, annual grasses, plants with wind-dispersed seeds, and some native species often increase with conservation tillage (Froud-Williams, Chancellor, and Drennan, 1981; Froud-Williams, 1988; Fay, 1990; Moyer et al., 1994). Results from 56 site-years of data in western Canada both support and refute these generalized observations.

Analysis of weed and volunteer crop density data allowed determination of species associations with conventional, minimum, or zero tillage. Of all weed species enumerated, 56 percent were associated with minimum-zero tillage and 27 percent were associated with conventional tillage (see Table 8.3). Species ubiquitous across tillage systems represented the remaining 17 percent. Despite more than one-half of all weeds being associated with minimum-zero tillage, most of these weeds were not new weeds to our agricultural systems, but they tended to increase in density and distribution with the adoption of conservation-tillage cropping systems.

All perennial species were more strongly associated with minimum-zero tillage than conventional tillage (see Table 8.3). Canada thistle and perennial sowthistle (*Sonchus arvensis* L.) were present in all tillage systems, but their densities often increased with zero tillage. Others, such as dandelion and foxtail barley, exhibited large increases in density with zero tillage. Native species, such as Canada goldenrod (*Solidago canadensis* L.) and rose (*Rosa* spp.), rarely occurred with conventional tillage but became increasingly prevalent with time in zero-tillage systems.

Quackgrass densities were sometimes greater in minimum than zero tillage, likely because of rhizome fragmentation and spread caused by tillage. Grower experience indicates that quackgrass control often improved with zero tillage if efficacious in-crop herbicides or preharvest glyphosate were used as compared to control levels attained with similar herbicide use in either minimum- or conventional-tillage systems.

Biennial weeds have been predicted to increase with conservation tillage, and this occurred in our study. Biennial wormwood (*Artemisia biennis* Willd.), yellow sweetclover [*Melilotus officinalis* (L.) Lam.], and western

TABLE 8.3. Species association with tillage system as determined from 56 site-years of data from the Canadian Prairies.

Common name	Scientific name	Five-letter code
Minimum-zero tillage		
Annuals		
Annual sowthistle	*Sonchus oleraceus* L.	SONOL
Common groundsel	*Senecio vulgaris* L.	SENVU
Kochia	*Kochia scoparia* (L.) Schrad.	KCHSC
Pineapple-weed	*Matricaria matricariodes* (Less.) C.L. Porter	MATMT
Prickly lettuce	*Lactuca serriola* L.	LACSE
Russian thistle	*Salsola iberica* Sennen & Pau	SASKR
Winter annuals		
European sticktight	*Lappula echinata* Gilib.	LPLSQ
Catchweed bedstraw	*Galium aparine* L.	GALAP
Downy brome	*Bromus tectorum* L.	BROTE
Greenflower pepperweed	*Lepidium densiflorum* Schrad.	LEPDE
Horseweed	*Conyza canadensis* (L.) Cronq.	ERICA
Narrowleaf hawksbeard	*Crepis tectorum* L.	CVPTE
Pigmyflower	*Androsace septentrionalis* L.	ANSSE
Redstem filaree	*Erodium cicutarium* (L.) L'Her. ex Ait.	EROCI
Scentless chamomile	*Matricaria perforata* Merat	MATIN
Shepherd's-purse	*Capsella bursa-pastoris* (L.) Medikus	CAPBP
Wood whitlowgrass	*Draba nemorosa* L.	DRBNE
Biennials		
Biennial wormwood	*Artemisia biennis* Willd.	ARTBI
Yellow sweetclover	*Melilotus officinalis* (L.) Lam.	MEUOF
Western salsify	*Tragopogon dubius* Scop.	TRODM
Perennials		
Absinth wormwood	*Artemisia absinthium* L.	ARTAB
Canada goldenrod	*Solidago canadensis* L.	SOOCA
Canada thistle	*Cirsium arvense* (L.) Scop.	CIRAR
Dandelion	*Taraxacum officinale* Weber in Wiggers	TAROF
Field horsetail	*Equisetum arvense* L.	EQUAR
Foxtail barley	*Hordeum jubatum* L.	HORJU
Perennial sowthistle	*Sonchus arvensis* L.	SONAR
Rose	*Rosa* spp.	ROSXX
Quackgrass	*Elytriga repens* (L.) Nevski	AGRRE
Smooth brome	*Bromus inermis* Leyss.	BROIN

TABLE 8.3 *(continued)*

Common name	Scientific name	Five-letter code
Conventional tillage		
Annuals		
Ball mustard	*Neslia paniculata* (L.) Desv.	NEAPA
Common lambsquarters	*Chenopodium album* L.	CHEAL
Cutleaf nightshade	*Solanum triflorum* Nutt.	SOLTR
Dog mustard	*Erucastrum gallicum* (Willd.) O.E. Schulz	ERWGA
Green foxtail	*Setaria viridis* (L.) Beauv.	SETVI
Green smartweed	*Polygonum scabrum* Moench	POLSC
Oakleaf goosefoot	*Chenopodium glaucum* L.	CHEGL
Prostrate pigweed	*Amaranthus blitoides* S. Wats.	AMABL
Redroot pigweed	*Amaranthus retroflexus* L.	AMARE
Thyme-leaved spurge	*Euphorbia serpyllifolia* Pers.	EPHSP
Wild buckwheat	*Polygonum convolvulus* L.	POLCO
Wild mustard	*Brassica kaber* (DC.) L.C. Wheeler	SINAR
Winter annuals		
Field pennycress	*Thlaspi arvense* L.	THLAR
Volunteer crops		
Barley	*Hordeum vulgare* L.	HORVX
Lentil	*Lens culinaris* Medik.	LENCU
Pea	*Pisum sativum* L.	PISSA
Sunflower	*Helianthus annuus* L.	HELAN
Rye	*Secale cereale* L.	SECCE
Ubiquitous across tillage systems		
Annuals		
Barnyardgrass	*Echinochloa crus-galli* (L.) Beauv.	ECHCG
Common chickweed	*Stellaria media* (L.) Vill.	STEME
Cowcockle	*Vaccaria pyramidata* Medikus	VAAPY
European sticktight	*Lappula echinata* Gilib.	LPLSQ
Nightflowering catchfly	*Silene noctiflora* L.	MELNO
Prostrate knotweed	*Polygonum aviculare* L.	POLAV
Round-leaved mallow	*Malva pusilla* Sm.	MALPU
Wild oat	*Avena fatua* L.	AVEFA
Winter annual		
Flixweed	*Descurainia sophia* (L.) Webb ex Prantl	DESSO
Volunteer crops		
Canola	*Brassica napus* L.	BRANP
Flax	*Linum usitatissimum* L.	LINUS
Mustard	*Brassica juncea* L.	BRAJU
Wheat	*Triticum aestivum* L.	TRZAS

Note: Species scientific names and five-digit codes are approved by the Weed Science Society of America and are available on computer disk from WSSA, 810 East 10th Street, Lawrence, KS 66044-8897.

salsify (*Tragopogon dubius* Scop.) were associated with minimum-zero tillage and rarely occurred with conventional tillage (Table 8.3).

The association of annual weed species with tillage system was more complex than that noted with biennial or perennial weeds. Of the 39 annual weed species present in these studies, 44 percent were associated with minimum-zero tillage, 33 percent with conventional tillage, and 23 percent were ubiquitous across tillage systems (Table 8.3). Many of the species associated with minimum-zero tillage have wind-disseminated seed capable of germinating on the soil surface; examples include annual sowthistle (*Sonchus oleraceus* L.), common groundsel (*Senecio vulgaris* L.), prickly lettuce (*Lactuca serriola* L.), and narrowleaf hawksbeard. Many of the hard-seeded annuals, such as mustards, pigweeds, wild buckwheat (*Polygonum convolvulus* L.), and common lambsquarters (*Chenopodium album* L.), were associated with conventional tillage. The prediction that annual grasses will proliferate with conservation tillage was not supported in our studies. While downy brome was associated with minimum-zero tillage, green foxtail [*Setaria viridis* (L.) Beauv.] was associated with conventional tillage, and barnyardgrass [*Echinochloa crus-galli* (L.) Beauv.] and wild oat were ubiquitous across tillage systems.

There was a strong association of winter annual weeds with minimum-zero tillage; 10 of 12 winter annual species increased in density with these tillage systems (Table 8.3). Lack of fall tillage and greater "snow trapping" with standing stubble in conservation tillage systems has likely contributed to the proliferation of winter annual weeds. Redstem filaree [*Erodium cicutarium* (L.) L'Her. ex Ait.] and catchweed bedstraw (*Galium aparine* L.) now commonly survive as winter annuals in zero-tillage systems because of the insulating effect of snow cover where previously they were killed by the cold winter temperatures commonly experienced in western Canada.

In contrast to previous reports (Izaurralde et al., 1993; Moyer et al., 1994), volunteer crops in our study were not associated with minimum-zero tillage (Table 8.3). The large-seeded crops such as lentil, pea, and sunflower (*Helianthus annuus* L.) exhibited a strong association with conventional tillage, indicating the need for greater soil-seed contact for germination of these species. Barley and rye (*Secale cereale* L.) also were associated with conventional tillage, but canola, flax, mustard, and wheat were present at similar levels across tillage systems.

Crop Rotation and Weed Associations

As there was greater variability in crop rotations than tillage systems among our individual studies and because they did not represent the full range of crop diversity (e.g., forages) currently being practiced by farmers, the following discussion relates to findings of previous crop rotation studies as well as those of our present study.

Spring Wheat Rotations

Farmers on the Canadian Prairies have historically selected crop rotations that include high frequencies of spring wheat and fallow (Campbell et al., 1990). The weed that most often presented the greatest problem in dryland grain-fallow rotations was wild oat (Siemens, 1963; Dryden, Bailey, and Grant, 1983). Freyman and colleagues (1982) indicated that Russian thistle, common lambsquarters, and wild mustard [*Brassica kaber* (DC.) L.C. Wheeler] had been the most common weeds until 2,4-D was widely used and then wild oat became the predominant weed. Similarly, wild oat was found to be the most serious weed in an irrigated spring wheat-oat rotation in southern Alberta. After seven years, wild oat densities were >300 plants·m^{-2} in the oat phase of the rotation under zero tillage and when nitrogen fertilizer was surface broadcast at 100 kg·ha^{-1} (unpublished data, Lethbridge, Alberta). A study conducted from 1995 through 1998 in Saskatchewan found that wild oat, volunteer wheat, and foxtail barley were more strongly associated with continuous crop than crop-fallow systems and with rotations that included wheat in three of four years than in rotations that included wheat in only two of four years (see Figure 8.7). Broadleaf weeds, such as wild mustard and field pennycress, were associated with the rotation that included the highest frequency of broadleaf crops.

A rotation experiment initiated in 1958 at Indian Head, Saskatchewan, had 2,4-D applied annually for broadleaf weeds and, since 1966, triallate was applied annually for wild oat. Weed assessments made in 1979 and 1980 indicated that green foxtail densities were >1000 plants (plants·m^{-2}) in continuous spring wheat but <5 plants (plants·m^{-2}) in spring wheat-spring wheat-fallow and spring wheat-fallow rotations (Hume, 1982). Another study similarly found that green foxtail densities were >2000 plants (plants·m^{-2}) in continuous spring wheat but virtually absent in a spring wheat-fallow rotation (Hume, Tessier, and Dyck, 1991). In the absence of tillage, foxtail barley replaced green foxtail as the predominant weed in continuous spring wheat production systems.

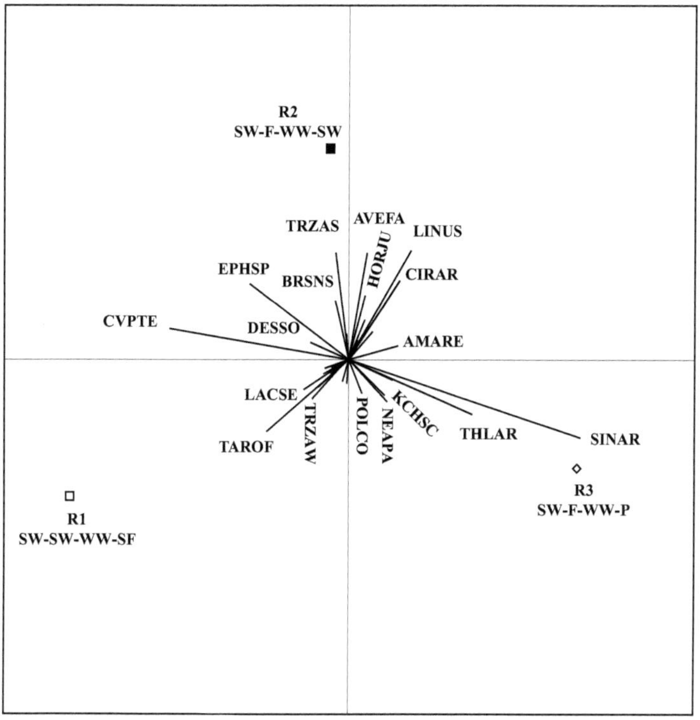

FIGURE 8.7. Canonical discriminant analysis ordination biplot of rotation centroids and weed species presenting postherbicide assessments taken in the final year of the 12-year CMS study at Indian Head, Saskatchewan. Weed species codes are given in Table 8.3. Crop designations are SW = spring wheat; WW = winter wheat; SF = summer fallow; F = flax; and P = field pea.

The winter annual weed downy brome, rather than wild oat, dominated in continuous winter wheat (Blackshaw, Larney, et al., 2001). Downy brome densities were greatly reduced when winter wheat was grown in rotation with spring-planted crops or with fallow.

Diverse Annual Crop Rotations

Low prices for cereal crops, production problems in wheat monoculture systems, and adoption of conservation tillage systems that allow increased soil water conservation has recently led to reduced fallow acreage and more diverse crop rotations (Miller et al., 2001). Canola, flax, mustard, pea, lentil, and chickpea are now commonly grown in rotation with cereals. Increased densities of volunteer crops have been associated with continuous cropping compared with crop-fallow systems (Derksen et al., 1994). However, many weed species including perennials such as Canada thistle, perennial sow-thistle, and quackgrass were not associated with the presence or absence of fallow, and it was concluded that fallow may not be necessary to manage weeds in diverse crop rotations.

A rotation study that compared continuous winter wheat production with winter wheat grown in rotation with canola, flax, or fallow found that the major difference in weed populations was between continuous winter wheat and the more diverse rotations (Blackshaw, Larney, et al., 2001). Mean weed densities in May in the winter wheat phase of the rotation were 49 plants plants·m^{-2} in continuous winter wheat but less than 10 plants plants·m^{-2} in the other rotations.

In high broadleaf crop frequency rotations that included canola, pea, or flax in at least three of four years, perennial sowthistle and dandelion were found to increase in density (Stevenson and Johnston, 1999). The frequent production of crops where there was no herbicide option for controlling perennial sowthistle and dandelion contributed to their increase over years.

Forage Crops and Cover Crops in Rotations

The use of forage crops to manage weed infestations was commonly recommended before herbicides were available (Wheeler, 1919; Pavlychenko, 1942). Leighty (1938) stated that alfalfa (*Medicago sativa* L.) may not kill all weeds but they were subdued sufficiently so that other crops can be grown for several years without excessive interference.

The ability of forages, either alfalfa or perennial grasses, to reduce wild oat populations is well documented in western Canada (Pavlychenko, 1942; Siemens, 1963; Dryden, Bailey, and Grant, 1983). After 24 years, a rotation of fallow-wheat-wheat-oats was heavily infested with wild oat while a rotation of corn-wheat-hay-hay-wheat-oats was almost free of wild oat (Table 8.4). Pavlychenko (1942) established crested wheatgrass [*Agropyron cristatum* (L.) Gaertn.] and smooth brome (*Bromus inermis* Leyss.) on land heavily infested with wild oat and, after seven years of production, very few wild oats persisted in the succeeding spring wheat crop grown in the eighth year.

Alfalfa that is managed to maintain a vigorous stand will provide sufficient shade to inhibit weed germination and growth, but winter annuals or some perennials that establish when alfalfa is dormant can sometimes flourish (Peters and Linscott, 1988). In the first year of alfalfa establishment wild

Table 8.4. Percent wild oat in threshed grain from different rotations at Brandon, Manitoba, 24 years after initiation of the rotations.

Rotation	Percent wild oat by weight in grain
A	
Summer fallow	—
Wheat	2.6
Wheat	13
Oats	9
B	
Corn	—
Wheat	0.3
Hay (alfalfa/brome)	—
Hay	—
Wheat	0
Oats	0
C	
Summer fallow	—
Wheat	0
Hay (alfalfa/brome)	—
Hay	—
Wheat	0.2
Oats	0.2

oat and Russian thistle were predominant, but neither weed was present in the second year of alfalfa production (Moyer, 1992). In an experiment where alfalfa was mowed twice each year, the number of Canada thistle shoots remaining after three years was less than 1 percent of the original population (Hodgson, 1958). Derscheid, Nash, and Wicks (1961) similarly reported that alfalfa or smooth brome nearly eliminated Canada thistle and perennial sowthistle infestations.

In a recent survey of 253 forage producers on the Canadian Prairies, 83 percent indicated fewer weeds were present in grain crops following forages than when following annual crops (Entz, Bullied, and Katepa, 1995). Respondents indicated there were reduced populations of wild oat, Canada thistle, green foxtail, and wild mustard in crops grown after forages. Ominski, Entz, and Kenkel (1999) similarly documented reduced populations of wild oat, wild mustard, Canada thistle, and catchweed bedstraw in cereals following alfalfa than in continuous cereal production systems.

In addition to perennial forages, annual or biennial crops grown as short-term forage or cover crops can also contribute to improved weed management when grown in rotation with annual cereals or oilseeds. Schoofs and

Entz (2000) found that winter triticale (*Triticosecale* spp.) grown for grazing or hay effectively competed with weeds. Biennial sweetclover grown as an alternative to fallow suppressed weeds during the fallow year, and populations of kochia, Russian thistle, downy brome, dandelion, and perennial sowthistle were reduced in the succeeding wheat crop (Blackshaw, Moyer, et al., 2001). Similarly, fall rye grown as a fallow alternative effectively reduced Canada thistle and dandelion populations (Moyer et al., 2000). Growers rely heavily on cover crops such as sweetclover, red clover (*Trifolium pratense* L.), and fall rye for weed management in organic production systems on the Canadian Prairies.

CONCLUSIONS

The 56 site-year data set consisting of 11 separate cropping studies conducted on the Canadian Prairies was very valuable in examining the effects of changing tillage and crop rotation practices on weed community dynamics. Studies were conducted on different soil types and in different ecoregions, but there was sufficient similarity in tillage and rotation treatments and in weed data collection to allow comparison of results across studies and over years.

Numerous statistical approaches were evaluated and canonical discriminant analysis was the most suitable to discern weed associations with variables such as tillage, rotation, and year. The overall finding that weed populations were more affected by year-to-year differences in environmental conditions, followed by crop rotation, and then tillage intensity has large implications for farmers. Growers concerned about greater weed problems if they move to conservation tillage systems can be reassured that other factors such as crop rotation can be more important, and indeed crop rotation may be manipulated to reduce potential adverse effects on weed populations when adopting conservation-tillage practices. Growers in western Canada that have made concurrent changes in crop rotation and tillage practices have experienced fewer weed problems than those that adopted ony conservation tillage.

Year-to-year changes in weed populations as influenced by the highly variable rainfall and temperature conditions on the Canadian Prairies indicates that farmers should be actively scouting their fields on a yearly basis and be willing to make appropriate adjustments in herbicide use and cropping systems. It is important to recognize that in other parts of the world, where climate may be more uniform and crop rotations less diversified, the effect of adopting conservation-tillage cropping systems on weed populations could be larger than noted in our study.

Indicator species analysis combined with expert knowledge allowed determination of weed species associations with specific tillage systems. Most biennial and perennial weeds were associated with minimum and zero tillage. In contrast, the effect of tillage on annual weeds was much more complex. Winter annual weeds that emerge in fall and survive cold Canadian winters often became more prevalent with conservation tillage, perhaps due to the combined effect of less fall tillage and the insulating effect of increased snow cover facilitated by standing crop stubble. Summer annual weeds with wind-disseminated seed capable of germinating on or near the soil surface also were associated with conservation tillage in our study. However, many other annual weeds were more strongly associated with conventional than conservation tillage and yet others were ubiquitous across tillage systems.

The goal of defining common functional traits for species response groups to specific tillage practices was not realized. Although numerous characteristics such as growth habit, seed dormancy, seed-bank persistence, recruitment depth, and dispersal mode were considered, there were always exceptions to each guideline or association proposed. For many weed species insufficient information on weed biology and weed ecophysiology was applicable to crop production on the Canadian Prairies to allow these types of associations to be discerned. This highlights the need for increased weed biology research in future years.

Crop rotation was found to strongly affect weed populations. The predominance of spring-planted annual crops on the Canadian Prairies has resulted in a weed community dominated by summer annual weeds. Pavlychenko (1942) stated that continuous grain farming was creating a weed problem in Canada that was national in scope. Our study and those of others indicates that summer annual weed densities can be markedly reduced by including winter crops, annual and perennial forages, and cover or green manure crops in the rotation. A diverse crop rotation should be a key component in any integrated weed management program because it facilitates varied selection pressure on weed populations. Andersson and Milberg (1998) concluded that a diverse rotation is important for weed management, but the specific type of rotation is less important when herbicides are used. Others have concluded that a major benefit of diverse crop rotations is that these rotations facilitate the rotation of herbicides with the ability to control different weed species (Ball, 1992; Liebman and Dyck, 1993).

This study on weed population changes in medium- and long-term agronomic studies proved useful in several ways. It facilitated identification and evaluation of appropriate statistical approaches for weed population data in these types of cropping systems studies. Results confirmed some of our hypotheses and rejected others. Some questions were answered and many more were generated. Additional long-term cropping studies are required throughout the world in various climatic zones and with contrasting cropping systems to further understand weed species' response to changing agricultural crop production practices.

REFERENCES

Agriculture and Agri-Food Canada (1997). Canadian ecodistrict climate normals 1961-1990. Available online at <http://sas.agr.gc.ca/cansis/nsdb/ecostrat/district/climate.html>.

Andersson, T.N. and P. Milberg (1998). Weed flora and the relative importance of site, crop, crop rotation, and nitrogen. *Weed Science* 46: 30-38.

Anonymous (1997a). *Agricultural profile of Alberta.* Catalog no. 95-180-XPB. Ottawa, Canada: Agriculture Division, Statistics Canada.

Anonymous (1997b). *Agricultural profile of Manitoba.* Catalog no. 95-178-XPB. Ottawa, Canada: Agriculture Division, Statistics Canada.

Anonymous (1997c). *Agricultural profile of Saskatchewan.* Catalog no. 95-179-XPB. Ottawa, Canada: Agriculture Division, Statistics Canada.

Anonymous (2001). *November estimate of production of principal field crops, Canada, 2001.* Field Crop Reporting Series No. 8. Ottawa, Canada: Statistics Canada.

Ball, D.A. (1992). Weed seed bank response to tillage herbicides and crop rotation sequence. *Weed Science* 40: 654-659.

Blackshaw, R.E., F.J. Larney, C.W. Lindwall, P.R.Watson, and D.A. Derksen (2001). Tillage intensity and crop rotation affects weed community dynamics in a winter wheat cropping system. *Canadian Journal of Plant Science* 81: 805-813.

Blackshaw, R.E., J.R. Moyer, R.C. Doram, and A.L. Boswall (2001). Yellow sweetclover, green manure, and its residues effectively suppress weeds during fallow. *Weed Science* 49: 406-413.

Blackshaw, R.E., G. Semach, X. Li, J.T. O'Donovan, and K.N. Harker (2000). Tillage, fertiliser and glyphosate timing effects on foxtail barley *(Hordeum jubatum)* management in wheat. *Canadian Journal of Plant Science* 80: 655-660.

Campbell, C.A., R.P. Zentner, H.H. Janzen, and K.E. Bowren (1990). *Crop rotation studies on the Canadian Prairies.* Publ. no. 1841/E. Ottawa, Canada: Agriculture Canada.

Clayton, J.S., W.A. Ehrlich, D.B. Cann, J.H. Day, and I.B. Marshall (1977). *Soils of Canada.* Volume 1: *Soil report.* Ottawa, Canada: Canada Department of Agriculture.

Derksen, D.A., A.G. Thomas, G.P. Lafond, H.A. Loeppky, and C.J. Swanton (1994). Impact of agronomic practices on weed communities: Fallow within tillage systems. *Weed Science* 42: 184-194.

Derksen, D.A., P.R. Watson, and H.A. Loeppky (1998). Weed community composition in seedbanks, seedling, and mature plant communities in a multi-year trial in western Canada. *Aspects of Applied Biology* 51: 43-50.

Derscheid, L.A., R.L. Nash, and G.A. Wicks (1961). Thistle control with cultivation cropping and chemicals. *Weeds* 9: 90-102.

Dryden, R.D., L.D Bailey, and C.A. Grant (1983). Crop rotations at the Brandon Research Station: 1893 to 1982. In *Proceedings of the 26th Annual Manitoba Society of Soil Science Meeting,* January 10-11 (pp. 198-210). Winnipeg, Canada: University of Manitoba.

Dufrêsne, M. and P. Legendre (1997). Species assemblages and indicator species: The need for a flexible asymmetrical approach. *Ecological Monographs* 67: 345-366.

Dwyer, D.D. and K. Wolde-Yohannis (1972). Germination, emergence, water use and production of Russian thistle. *Agronomy Journal* 64: 52-55.

Ecological Stratification Working Group (1995). *A national ecological framework for Canada.* Agriculture and Agri-Food Canada, Research Branch, Centre for Land and Biological Resources Research and Environment Canada, State of the Environment Directorate, Ecozone Analysis Branch, Ottawa/Hull. [Report and national map at 1:7 5000 000 scale.]

Entz, M.H., W.J. Bullied, and F. Katepa (1995). Rotational benefits of forage crops in Canadian prairie cropping systems. *Journal of Production Agriculture* 8: 521-529.

Fay, P.K. (1990). A brief overview of the biology and distribution of weeds of wheat. In W.W. Donald (Ed.), *Systems of weed control in wheat in North America* (pp. 33-50). Champaign, IL: Weed Science Society of America.

Freyman, S., C.J. Palmer, E.H. Hobbs, J.F. Dormaar, G.B. Schaalje, and J.R. Moyer (1982). Yield trends on long-term dryland wheat rotations at Lethbridge. *Canadian Journal of Plant Science* 62: 609-619.

Froud-Williams, R.J. (1988). Changes in weed flora with different tillage and agronomic management systems. In M.A. Altieri and M. Liebman (Eds.), *Weed management in agroecosystems: Ecological approaches* (pp. 213-236). Boca Raton, FL: CRC Press.

Froud-Williams, R.J., R.J. Chancellor, and D.S.H. Drennan (1981). Potential changes in weed floras associated with reduced-cultivation systems for cereal production in temperate regions. *Weed Research* 21: 99-109.

Hodgson, J.M. (1958). Canada thistle (*Cirsium arvense* Scop.) control with cultivation cropping and chemical sprays. *Weeds* 6: 1-11.

Hume, L. (1982). The long-term effects of fertilizer application and three rotations on weed communities in wheat after 21-22 years at Indian Head, Saskatchewan. *Canadian Journal of Plant Science* 62: 741-750.

Hume, L., S. Tessier, and F.B. Dyck (1991). Tillage and rotation influences on weed community composition in wheat (*Triticum aestivum* L.) in southwestern Saskatchewan. *Canadian Journal of Plant Science* 71: 783-789.

Hunter, J.H., I.N. Morrison, and R.S. Rourke (1990). The Canadian prairie provinces. In W.W. Donald (Ed.), *Systems of weed control in wheat in North America* (pp. 51-89). Champaign, IL: Weed Science Society of America.

Izaurralde, R.C., N.G. Juma, W.B. McGill, D.S. Chanasyk, S. Pawluk, and M.J. Dudas (1993). Performance of conventional and alternative cropping systems in cryoboreal subhumid central Alberta. *Journal of Agricultural Science* 120: 33-41.

Kenkel, N.C., D.A. Derksen, A.G. Thomas and P.R. Watson (2002). Multivariate analysis in weed science research. *Weed Science* 50: 281-292.

Klecka, W. R. (1980). *Discriminant analysis*. Beverly Hills, CA: Sage Publications.

Lafond, G. P. and D.A. Derksen (1996). Long-term potential of conservation tillage systems on the Canadian Prairies. *Canadian Journal of Plant Pathology* 18: 151-158.

Leighty, C.E. (1938). Crop rotation. In H. G. Knight (Ed.), *Soils and men* (pp. 406-431). Washington, DC: United States Department of Agriculture.

Liebman, M. and E. Dyck (1993). Crop rotation and intercropping strategies for weed management. *Ecological Applications* 3: 92-122.

McCune, B. and M.J. Mefford (1999). *PC-ORD: Multivariate analysis of ecological data,* Version 4. Gleneden Beach, OR: MjM Software Design.

Miller, P.R., C.L. McDonald, D.A. Derksen, and J. Waddington (2001). The adaptation of seven broadleaf crops to the dry semiarid prairie. *Canadian Journal of Plant Science* 81: 29-43.

Moyer, J.R. (1992). Alfalfa yield in establishment and subsequent years after herbicides and phosphorus applications during establishment. *Canadian Journal of Plant Science* 72: 619-625.

Moyer, J.R., R.E. Blackshaw, E.G., Smith, and S.M. McGinn (2000). Cereal cover crops for weed suppression in a summer fallow-wheat cropping sequence. *Canadian Journal of Plant Science* 80: 441-449.

Moyer, J. R., R.S. Roman, C.W. Lindwall, and R.E. Blackshaw (1994). Weed management in conservation tillage systems for wheat production in North and South America. *Crop Protection* 13: 243-259.

Ominski, P.D., M.H. Entz, and N. Kenkel (1999). Weed suppression by *Medicago sativa* in subsequent cereal crops: A comparative survey. *Weed Science* 47: 282-290.

Pavlychenko, T.V. (1942). *Root systems of certain forage crops in relation to their management of agricultural soils.* National Research Council Publication no. 1088. Ottawa, Canada: National Research Council of Canada and Dominion Department of Agriculture.

Peters, E.J. and D.L. Linscott (1988). Weed control in alfalfa and alfalfa improvement. In A.A. Hanson, D.K. Barnes, and R.R. Hill (Eds.), *Alfalfa and alfalfa improvement* (pp. 705-735). Agronomy Monograph no. 29. Madison, WI: American Society of Agronomy.

SAS Institute Inc. (1999). *SAS/STAT user's guide,* Version 8. Cary, NC: SAS Institute Inc.

Schoofs, A. and M.H. Entz (2000). Influence of annual forages on weed dynamics in a cropping system. *Canadian Journal of Plant Science* 80: 189-198.

Siemens, L.B. (1963). *Cropping systems: An evaluative review of literature*. Technical Bulletin no. 1. Winnipeg, Canada: University of Manitoba.

Stevenson, F.C. and A.M. Johnston (1999). Annual broadleaf crop frequency and residual weed populations in Saskatchewan Parkland. *Weed Science* 47: 208-214.

Thomas, A.G., B.L. Frick, and L.M. Hall (1998). *Alberta weed survey of cereal and oilseed crops in 1997.* Weed Survey Series Publication 98-2. Saskatoon, Canada: Saskatoon Research Centre, Agriculture and Agri-Food Canada.

Thomas, A.G., B.L. Frick, R.C. Van Acker, S.Z. Knezevic, and D. Joosse (1998). *Manitoba weed survey of cereal and oilseed crops in 1997.* Weed Survey Series Publication 98-1. Saskatoon, Canada: Saskatoon Research Centre, Agriculture and Agri-Food Canada.

Thomas, A.G., R.F. Wise, B.L. Frick, and L.T. Juras (1996). *Saskatchewan weed survey of cereal, oilseed and pulse crops.* Weed Survey Series Publication 96-1. Saskatoon, Canada: Saskatoon Research Centre, Agriculture and Agri-Food Canada.

Wheeler, S. (1919). *Seager Wheeler's book on profitable grain growing.* Winnipeg, Canada: The Grain Growers Guide, Limited.

Wiese, A.F. and C.W. Vandiver (1970). Soil moisture effects on competitive ability of weeds. *Weed Science* 18: 518-519.

Young, F. L. (1986). Russian thistle *(Salsola iberica)* growth and development in wheat *(Triticum aestivum). Weed Science* 34: 901-905.

Chapter 9

Potential of Allelopathy and Allelochemicals for Weed Management

Daizy R. Batish
H. P. Singh
R. K. Kohli
G. P. Dawra

INTRODUCTION

Effective management of various weeds competing with agricultural crops greatly enhances the crop yield. Interference of troublesome weeds with crops is an age-old problem occurring from the advent of agriculture. Alhough crops and weeds have coevolved, recently various anthropogenic factors have made weeds more aggressive and problematic. A huge amount of money is spent worldwide to control weeds. In the United States alone weeds cause about 12 percent loss in crop yield, and it costs nearly US$35 billion to control them (Piementel et al., 2001). The costs are even more in developing countries. In underdeveloped countries, unmechanized traditional agricultural practices further increase the loss of yield due to weeds.

Weeds can be managed through various means, including mechanical, chemical, cultural, and biological. However, the use of synthetic herbicides is more common, especially since 1950. In fact, the use of herbicides for weed control is a twentieth-century technology; in 1932, DNOC (4,6-dinitro-*o*-cresol)—the first modern synthetic herbicide—was patented in France for the control of annual weeds in cereals. The discovery of synthetic growth regulators such as 2,4-D (2,4-dichlorophenoxyacetic acid) and MCPA (2-methyl-4-chlorophenoxyacetic acid) in the 1950s led to their widespread popularity as selective herbicides (Kirby, 1980). Since then agriculture has become totally dependent upon them. As a result, a variety of herbicides have been synthesized and are utilized for this purpose.

Unfortunately, excessive and indiscriminate use of herbicides has led to several environmental and health hazards. This fact was realized in the 1970s through the writings of Rachel Carson, whose 1962 book *Silent*

Spring generated much public concern. The increased awareness about their ill effects has put substantial pressure on the herbicide industry to reassess the toxicological and environmental impacts of various herbicide classes. It is further aggravated by a dramatic increase in the herbicide-resistant weed biotypes posing a serious threat to agriculture. To date, nearly 296 weed-resistant biotypes belonging to 178 species (107 dicots and 71 monocots) have been developed (Heap, 2005). Owing to these problems, efforts are being made to find alternatives to synthetic herbicides. Among the alternatives, use of natural plant products through the phenomenon of allelopathy has generated a greater interest among scientists worldwide. The reason for this could be attributed to following factors:

- It is an environmentally friendly and sustainable way of managing weeds, as it involves natural plant products which do not leave any residual effect.
- Allelopathic plants are a source of a wide variety of chemicals that could be explored for discovery of new herbicides.
- Allelochemicals possess novel target sites of action that could help in overcoming the problem of rapidly increasing herbicidal resistance.
- Use of allelopathic crops improved for allelopathic properties could provide a cost-effective, sustainable, and innovative approach for weed management.

ALLELOPATHY, ALLELOCHEMICALS: A BRIEF BACKGROUND

Plants synthesize a variety of chemicals as a result of secondary metabolism, and this forms the basis of allelopathy—a type of plant-plant interaction. Though the historical data indicate that knowledge of plant interactions mediated through chemicals is very old and dates back to 300 B.C., the term *allelopathy* originated only in the twentieth century (Singh, Batish, and Kohli, 2001). The genesis of the term *allelopathy* dates back to 1937, when Hans Molisch combined two Greek words, *Allelo* and *Pathos,* literally meaning "mutual sufferings." According to Molisch, all plants, including microbes, mutually interact through the release of chemicals in the environment. Later, Rice (1984) defined allelopathy as any direct or indirect positive or negative effect of one plant (including microbe) on the other through the release of chemicals into the environment. Since the time of Molisch, the progress in the field has been rather slow, but after the pioneering work of Muller and co-workers in the 1960s (Willis, 1995, and references therein), research has progressed steadily. It has now become an interdisciplinary science with a multifaceted approach and is one of the exciting fields offering challenging studies in the fields of botany, chemistry, agriculture, ecology, agronomy, and soil sciences. In spite of criticism from a section of scientists, allelopathy is now an integral component of chemical ecology (Anaya, 1999). The current focus within allelopathic research is its applications in solving various ecological and agricultural problems such as its utility for weed and pest management, role in soil sickness, and regeneration problems in forest areas. However, greater emphasis is on the use of allelopathy and allelochemicals for the management of weeds due to the environmental hazards of synthetic herbicides. Various allelopathic plants are being screened for the purpose of exploring them for selective weed management. For example, Solymosi (1994) explored 430 Hungarian species for their weed-suppressing ability; Hong and colleagues (2003) researched several allelopathic plants from Southeast Asia for allelopathic potential.

Allelochemicals (the chemicals that bring about allelopathic effects) are synthesized within plants as secondary metabolites that serve as a defensive adaptation, besides playing an important role between plants and their environment (Harborne, 1989; Seigler, 1996). Unlike the earlier belief that secondary metabolites or allelochemicals are useless, they are now considered as compounds of great potential owing to their role in protecting plants against the attack of pathogens, insects, and pests, in addition to providing reproductive fitness for the plants (Rice, 1984). Allelochemicals belong to a variety of chemical types and are released from the plant through leachation, volatilization, root exudation, and death and decay of plant parts, aided by various biotic and abiotic factors (Rice, 1984). Thus the biological activity of the allelochemicals depends upon concentration at a given time, flux rate, cultivar, age and metabolic stage of the donor plant, their contact with the other propagules, and environmental conditions (Einhellig, 1996).

ALLELOPATHIC INTERACTIONS AND THEIR SCOPE FOR WEED MANAGEMENT IN AGROECOSYSTEMS

In agroecosystems various crops, weeds, trees, and microbes, constituting the biotic components, are known to interact with the abiotic environment. The allelopathic interactions among various biotic components have great potential for improving crop production through management of weeds and pests (Weston, 1996; Kohli, Batish, and Singh, 1998; Singh, Batish, and Kohli, 2001, 2003). Weeds are generally known to exert negative effects on crops through their allelopathic potential of greater competitive ability. A number of weeds have been found to be allelopathic (Qasem and Foy, 2001). Their use for weed management is of little interest because their presence in agroecosystems further aggravates the problem. However, they may provide a source of some potential allelochemicals. For example, parthenin from ragweed parthenium (*Parthenium hysterophorus* L.) (Batish et al., 1997) and artemisinin from *Artemisia* spp. (Duke et al., 1987). Even trees grown in agroecosystems as a component of agroforestry can also be considered as potential candidates for selective weed management programs (Rizvi et al., 1999). However, like weeds they can serve as only the source of allelochemicals that can be exploited for weed control, for example, mimosine from lead tree (*Leucaena leucocephala* Lam. De Wit.) (Chou and Kuo, 1986; Chou, 1993); volatile oils of *Eucalyptus* spp. (Kohli, Batish, and Singh, 1998; Singh, Batish, Kaur, et al., 2002b); and ailanthone from tree of heaven, *Ailanthus altissima (*Miller) Sw. (Heisey, 1996).

However, crops, unlike weeds and trees, have great potential for managing weeds in a number of ways. They can directly interfere with the weedy species by releasing chemicals into the environment, or they can serve as a source of potential allelochemicals. Further, allelopathic crops can be manipulated under various cropping patterns. Batish and colleagues (2001) listed 35 crops affecting weeds, a property that can be capitalized on for sustainable weed management. Even the residues of allelopathic crops such as rye, sunflower, wheat, and barley could be extremely useful in suppressing weeds (Leather, 1983b; Liebl and Worsham, 1983; Barnes and Putnam, 1986; Weston, 1996; Batish et al., 2001). The allelopathic property of cover, smother, and green manure crops or crops grown in rotation could form another strategy worth exploiting for weed management (Leather, 1987; Jordan, 1993; Weston, 1996; Anaya, 1999; Singh, Batish, and Kohli, 2001; Batish, Singh, and Kohli, 2002). Moreover, crops can be improved for allelopathic properties. For example, the genotypes with lesser allelochemical content can be improved by transferring traits encoding the synthesis of allelochemicals. Some cultivars of crops such as wheat, oat, and barley are known to contain more allelochemicals and are competitively better than others with lesser amounts (Baghestani et al., 1999). These cultivars with greater competitive ability can thus be selected and used for controlling weeds (Olofsdotter, Jensen, and Courtois, 2002). Further, the diverse type of allelochemicals found in crops can also be utilized directly as natural herbicides, for example sorgoleone from *Sorghum* spp. (Weston and Czarnota, 2001).

CULTURAL PRACTICES, ALLELOPATHIC CROPS, AND WEED MANAGEMENT

Crop Rotation and Weed Management

Crop rotation is an important strategy for weed management in which some crops, particularly legumes, are grown in short rotation with the main crops. Continuous sole cropping on the same land results in buildup of a particular combination of crop-weed species that have similar growth requirements. However, with diverse types of crops grown in rotation, germination and growth cycles of weeds are disrupted by variations in cultural practices (such as tillage, planting dates, and competition) associated with each crop and thus the weeds can be controlled. It is one of the very old and traditional practices that declined with the use of synthetic herbicides. However, it has resurfaced recently in order to maintain the sustainability of agroecosystems. The manipulation of allelopathic rotational crops can provide an effective means of weed management. Leather (1983a) reported that crop rotation with allelopathic sunflower reduces the density of both broadleaf and grassy weeds. Soybean [*Glycine max* (L.) Merr.] is another crop that has the potential to decrease weed incidence when grown in rotation. Jones, Jessop, and Sindel (2001) observed that the weed incidence was least in the soybean crop harvested after eight weeks compared to other crops and opined that inclusion of soybean in crop rotation can substantially reduce weeds in the next season. Further, under reduced or no till, a considerable decrease in population of giant foxtail (*Setaria faberi* Herrm.) was observed when allelopathic soybean-corn-wheat rotation was followed as compared to corn alone, and a significant decrease in herbicide input was observed (Schreiber, 1992). A similar observation was made earlier by Forcella and Lindstorm (1988) that weed density was reduced when corn was rotated with soybeans with ridge tillage compared to corn alone. Liebman and Dyck (1993) summarized various combinations of crop rotation that have the potential for suppressing weeds.

Inclusion of alfalfa (*Medicago sativa* L.), a known allelopathic species, (Miller, 1996) in the crop rotation sequence significantly reduces the incidence of weeds in the next crop (Entz, Bullied, and Katepa-Mupondwa, 1995). Ominski, Entz, and Kenkel (1999), on the basis of a survey conducted in 117 fields in Manitoba, Canada, found that rotation with alfalfa can effectively reduce the incidence of wild oat (*Avena fatua* L.), Canada thistle [*Cirsium arvense* (L.) Scop.], wild mustard [*Brasscia kaber* (DC.) L. C. Wheeler], and catchweed bedstraw (*Galium aparine* L.) in the succeeding cereal crops. These researchers concluded that including alfalfa in crop rotation could be an effective part of integrated weed management. The success of crop rotation for suppressing weeds, however, depends upon several factors such as crop sequences that create varying patterns of resource competition, allelopathic interference, soil disturbance, and mechanical damage so as to provide an unstable environment and thereby prevent the growth and establishment of a particular weed species (Liebman and Dyck, 1993; Liebman and Davis, 2000).

Allelopathic Cover Crops and Weed Management

Use of cover crops (also known as smother crops because of their overshadowing effect) is also one of the cultural practices of reducing weed incidence (Foley, 1999). Cover crops are grown for several motives such as conservation of soil and moisture, improvement of nutrient cycling, lowering of temperature, supplying forages in emergency, protection of cash crops from winds, suppression of weeds, and increase in crop production (Barnes and Putnam, 1983; Weston, 1990; Teasdale, 1996; Reicosky and Forcella, 1998; Gallandt, Liebman, and Huggins, 1999; Masiunas, 1999), and they can reduce herbicide input (Worsham and Blum, 1992). In addition to other mechanisms of weed suppression by cover crops, a number of reports indicate the role of allelochemicals released by cover crops in bringing about weed reduction (Eliott, Cochran, and Papendick, 1981; Leather, 1983b; Liebel and Worsham, 1983; Putnam and DeFrank, 1983; Weston, Harmon, and Mueller, 1989; Blum et al., 1991; Yenish, Worsham, and Chilton, 1995; Liebman and Davis, 2000). A number of cover crops are known to suppress weeds, owing to their allelopathic property (see Table 9.1). The allelochemicals identified from such cover crops are listed in Table 9.2.

Hairy vetch is a well-known nitrogen-fixing cover crop grown extensively in Europe, the United States, and Japan. Besides preventing soil erosion, promoting soil porosity, ameliorating microclimates, and providing biomass addition, it suppresses weeds through its allelopathic effect (Fujii, 2001). Its mulch has been observed to reduce the incidence of weeds in grasslands, orchards, and paddy (Fujii, 2001) and corn (Johnson, Defelice, and Helsel, 1993) fields. Even its residues reduce the emergence of common lambsquarters (*Chenopodium album* L.) (Teasdale, 1993), Cogon grass [*Imperata cylindrica* (L.) Räusch.] (Chikoye et al., 2002), pitted morning glory (*Ipomoea lacunosa* L.), and solanceous weeds (White, Worsham, and Blum, 1989). Fujii (2001) demonstrated that hairy vetch is even better than Chinese milkvetch (*Astragalus sinicus* L.)—another commonly used cover crop in Japan. Its use completely removes weeds, and the action was quite similar to that of synthetic herbicides (Fujii, 2001).

TABLE 9.1. Cover crops with allelopathic property and weed-suppressing ability.

Cover crop	Common name	Reference(s)
Avena sativa L.	Oat	Guenzi and McCalla, 1966
Brassica spp.	Turnip, rape, mustard, radish, etc.	Narwal, 1999; Al-Khatib, Libbey, and Boydston, 1997
Fagopyrum esculentum Moench	Buckwheat	Tsuzuki, Yamamoto, and Shimizu, 1987; Eskelsen and Crabtree, 1995; Iqbal et al., 2002
Hordeum vulgare L.	Barley	Putnam, DeFrank, and Barnes, 1983
Ipomoea tricolor Cav.	Pearly Gates Morning glory	Anaya et al., 1990; Anaya, 1999
Medicago polymorpha L. *M. lupilina* L.	Medics	Sheaffer et al., 2002.
Medicago sativa L.	Alfalfa	Miller, 1983, 1996
Mucuna pruriens (L.) DC.	Velvetbean	Fujii, 1999a
Ophiopogon japonicus K.	Dwarf lilyturf	Lin et al., 2003
Secale cereale L.	Rye	Barnes and Putnam, 1983, 1987; Yenish, Worsham, and Chilton, 1995
Sorghum spp.	Sorghum	Forney, Foy, and Wolf, 1985; Weston, 1996
Trifolium spp.	Clover	Rice, 1995, Breland, 1996
Triticum aestivum L.	Wheat	Shilling, Liebl, and Worsham, 1985
Vicia villosa Roth	Hairy vetch	Fujii, 1999a

Cover crops of velvetbean [*Mucuna cochinchinensis* (Lour.) A. Chev.] and tropical kudzu [*Pueraria phaseoloides* (Roxb.) Benth.] can effectively control speargrass (*Imperata* spp.) in fallows and maize fields (Udensi et al., 1999; Akobundu, Udensi, and Chikoye, 2000; Chikoye et al., 2002). Holt (1995) reported that velvetbean suppresses weeds by reducing photosynthesis, thereby causing rapid aging and dieback. Casini and Olivero (2001) observed that pod cultures, seed leachates, and water extracts, residues, and roots of kudzu and velvetbean affect the germination and seedling growth of cogon grass (*Imperata brasiliensis* Tring.). Likewise, the use of leguminous cover crops such as jumbi bean (*Leucaena leucocephala*), tamarind (*Tamarindus indica* L.), and velvetbean [*Mucuna deeringiana* (Bort.) Merr.] reduces incidence (in terms of types, number, and density), growth,

TABLE 9.2. Allelochemicals identified from the cover crops.

Cover Crop	Alleochemical(s)	Reference(s)
Avena sativa	Scopoletin, phenolic acids	Guenzi and McCalla, 1966; Rice, 1984
Brassica spp.	Glucosinolates and their breakdown products including isothiocyanates, nitriles, epithinitriles, ionic thiocyanates	Fenwick, Heaney, and Mullin, 1983; Brown and Morra, 1999
Fagopyrum esculentum	Fatty acids (palmitic, stearic, arachidic, and behenic acid), fagomine, 4-Pipedone and 2-piperdinemethanol	Tsuzuki, Yamamoto, and Shimizu, 1987; Iqbal et al., 2002
Hordeum vulgare	Gramine, hordenine	Liu and Lovett, 1993a,b; Hanson et al., 1981
Ipomoea tricolor	Tricolorin A	Pereda-Miranda et al., 1993
Medicago sativa	Medicarpin, 4-methoxy medicarpin, sativan, 5-methoxy sativan, canavanine, saponins, and phenolic acids such as ferulic, chlorogenic, and salicylic acid	Miller, 1983; Miller et al., 1988; Dornbos, Spencer, and Miller, 1990; Gorski, Miersch, and Ploszynski, 1991; Wyman-Simpson et al., 1991; Waller, Jurzysta, and Thorne, 1995
Mucuna pruriens	L-DOPA (L-3,4-dihydroxyphenylalanine)	Fujii, 1999b
Secale cereale	2,4-dihydroxy-1,4 (2H)-benzoxazin-3-one (DIBOA) and 2(3H)-benzoxazolinone (BOA), b-hydroxybutyric acid, b-phenyllactic acid, phenolics acids (ferulic, p-hydroxybenzoic, salicylic, o-coumaric, and succinic)	Patrick, 1971; Chou and Patrick, 1976; Barnes et al., 1987; Shilling, Liebl, and Worsham, 1985; Wojcik-Wojtkowiak et al., 1990
Sorghum spp.	Sorgoleone (p-benzoquinone), dhurrin (cyanogenic glycoside), breakdown products such as p-hydroxybenzoic acid and p-hydroxybenzaldehyde	Guenzi and McCalla, 1966; Nicollier, Pope, and Thompson, 1983; Netzley et al., 1988; Weston, 1996
Trifolium spp.	Phenolic acids	Ohno and Doolan, 2001; cf. Rice, 1995

Triticum aestivum	DIMBOA (2,4-dihydroxy-7-methoxy-1,4-benzoxazin-3-one) and phenolic acids such as syringic, vanillic, *trans*-ferulic, *cis*-ferulic, *trans*-p-coumaric, *cis*-p-coumaric, and p-hydroxybenzoic acid	Guenzi and McCalla, 1966; Shilling, Liebl, and Worsham, 1985; Lodhi, Bilal, and Malik, 1987; Wu, Haig, et al., 2000
Vicia villosa	Cyanamide	Fujii, 1999a

and biomass of weeds in corn fields (Caamal-Maldolando et al., 2001). The aqueous extracts of velvetbean significantly reduced the germination and growth of barnyardgrass and prince's feather amaranth (*Amaranthus hypochondriacus* L.) (Caamal-Maldolando et al., 2001).

Putnam, DeFrank, and Barnes (1983) reported that cover crops of rye, wheat, sorghum, and barley are very effective in suppressing redroot pigweed (*Amaranthus retroflexus* L.), common purslane (*Portulaca oleracea* L.), and common ragweed (*Ambrosia artemisiifolia* L.). It was attributed to various physical and chemical attributes including shading, reduced temperature, and allelopathic interference of these crops. Worsham and Blum (1992) reported that cover crops of rye, subterranean clover (*Trifolium subterraneum* L.), crimson clover (*T. incarnatum* L.), and hairy vetch reduced the incidence of redroot pigweed, spiny amaranth (*Amaranthus spinosus* L.), and smooth pigweed (*A. hybridus* L.). The effect persisted up to four weeks after planting into killed cover crops and can greatly reduce the need for herbicide inputs (Worsham and Blum, 1992).

Rye is a widely used cover crop grown under several cropping systems and is an ideal cover crop for vegetable production systems (Masiunas, 1999). It is known to suppress weeds by allelopathic interference coupled with its physical attributes (Barnes and Putnam, 1986, 1987). Living cover of spring-planted, winter-killed rye inhibited broadleaf weeds such as common lambsquarters, large crabgrass [*Digitaria sanguinalis* (L.) Scop.], and common ragweed (Barnes and Putnam, 1983) and reduced weed biomass in soybeans by 60 to 90 percent (Ateh and Doll, 1996). Masiunas (1999), however, reported that residues from fall-planted rye are more effective in suppressing weeds as compared to spring-planted rye. The degree and length of weed suppression by rye depends upon the type of weed species, amount of rye biomass, prevailing environmental conditions, and the cultural practices (Smeda and Weller, 1996; Masiunas, 1999). In general, residues of rye are more toxic to annual dicots, moderately toxic to annual grasses, and only slightly toxic to perennial ones (Putnam, 1986, 1990; Masiunas, 1999). Although rye cover crop is an effective weed management tool, supplemental weed control with herbicides is required for complete weed management (Smeda and Weller, 1996). Residues of rye have also been observed to suppress the germination of triazine-resistant weeds (Prezepiorkowski and Gorski, 1994). A number of allelochemicals including cyclic hydroxamic acids such as DIBOA and BOA and some phenolic acids have been identified from rye (Table 9.2). Among these, BOA is more inhibitory than DIBOA. Upon activity of gram-positive bacterium *Acinetobacter calcoaceticus,* it is biotransformed into an even more toxic form, AZOB (2,2'-oxo-1,1'-azobenzone), in soil (Chase, Nair, and Putnam, 1991; Chase et al., 1991). However, with the degradation of allelochemicals, the toxicity of rye residues diminishes (Yenish, Worsham, and Chilton, 1995). Weed suppression by rye is negatively correlated with its biomass vis-à-vis the amount of allelochemicals (Masiunas, 1999). Further, the inhibitory activity of rye herbage and the content of hydroxamic acids depends upon plant part, age, and genotype of rye (Barnes et al., 1987; Hoffman et al., 1996b) and is affected by the fertility regimes and production environment (Mwaja, Masiunas, and Weston, 1995).

Recently, Lin and colleagues (2003) reported that dwarf lilyturf (*Ophiopogon japonicus* K.), a common cover crop in Japan, applied at 50 to 150 $g{\cdot}m^{-2}$ significantly suppresses paddy weeds such as arrow-leaved monochoria [*Monochoria vaginalis* (Burm. f.) Presl. ex Kunth.], small flower umbrella (*Cyperus difformis* L.), and bur-Marigold (*Bidens biternata* L.) without any effect on rice. Even the aqueous extracts of the dried plants inhibited the germination and seedling growth of these weeds, which was attributed to allelopathic potential (Lin et al., 2003).

Cover crops can be used as living plant tissue (Lehman and Blum, 1997) or as living mulch (Enache and Ilnicki, 1990) or residues prepared either through decomposition or after desiccation with contact herbicides or freezing (Putnam and DeFrank, 1983; Teasdale, 1996; Teasdale and Mohler, 2000). However, Lehman and Blum (1997) reported that cover crops of rye and subterranean clover, in some cases, lose weed-suppressing ability after desiccation with the herbicide glyphosate. Although cover crops offer a great option in sustainable farming systems by controlling weeds, they need to be supplemented with the other cultural practices such as herbicides. However, the use of herbicide is greatly minimized (Enache and Ilnicki, 1990; Mohler and Teasdale, 1993; Jordan et al., 1999). Moreover, weed control by a cover crop depends on its type, residue management tillage, and prevailing environmental conditions, and is species specific (Teasdale, 1996; Teasdale and Mohler, 2000).

Green Manure Crops for Weed Management

Some cover crops such as legumes (soybeans, clovers, and *Mucuna* spp.) and cruciferous plants (*Brassica* spp.) not only improve soil through addition of organic matter but are also an effective tool for weed management (Al-Khatib and Boydston, 1999). Such crops are known as *green manure crops*. The weed-suppressing effect of these green manure crops is attributed to their allelopathic effects. Among these, legume green manure crops are not only effective nitrogen sources but also manage weeds depending upon residue quality and quantity, soil moisture, temperature, and other factors (Dyck and Liebman, 1994; Al-Khatib and Boydston, 1999; Ohno et al., 2000). Allelochemicals released from the residues of crimson clover and red clover (*T. pratense* L.) suppress the growth of common lambsquarters (Dyck and Liebman, 1994), pitted morning glory (White, Worsham, and Blum, 1989), and wild mustard (*Sinapis alba* L.) (Ohno et al., 2000).

Several members of family Brassicaceae, namely, field mustard (*B. campestris* L.), white or yellow mustard (*Brassica hirta* Moench), brown/Indian mustard [*B. juncea* (L.) Czern.], white mustard (*B. hirta*), rapeseed/oilseed rape/canola (*B. napus* L.), black mustard [*B. nigra* (L.) Koch.), and garden cress (*Lepidium sativum* L.), are excellent sources of green manure and effectively control weeds (Boydston et al., 1994; Al-Khatib and Boydston, 1999; Narwal, 1999). Leaves from rapeseed incorporated into soil reduce the incidence of a number of broadleaf weeds such as common lambsquarters, redroot pigweed, hairy nightshade (*Solanum sarrachoides* Sendtner) and longspine sandbur [*Cenchrus longispinus* (Hack.) Fern], shepherd's purse [*Capsella bursa-pastoris* (L.) Medik.], kochia [*Kochia scoparia* (L.) Schrad.], and grassy weeds such as barnyard grass [*Echinochloa crus-galli* ([L.) P. Beauv.] and green foxtail [*Setaria viridis* (L.) Beauv.] in potato and pea (Boydston and Hang, 1995; Al-Khatib, Libbey, and Boydston, 1997). Krishnan, Holshouser, and Nissen (1998) reported that mulch from white and brown mustard and rapeseed effectively controls the incidence of shepherd's purse, kochia, redroot pigweed, and green foxtail in soybean crops. Likewise, turnip-rape mulch incorporated into soil suppresses the growth of weeds including scentless mayweed (*Matricaria inodora* L.) and spiny sowthistle [*Sonchus asper* (L.) Hill.] (Petersen et al., 2001). The effect was attributed to the release of volatile isothiocyantes (Petersen et al., 2001). The weed-suppressing ability of cruciferous green manure crops is attributed to the presence of volatile glucosinolates in them (Vaughn and Boydston, 1997) and their breakdown products (Fenwick, Heaney, and Mullin, 1983; Brown and Morra, 1996, 1999) (see Table 9.2). However, the extent of weed suppression by Crucifers depends upon species, cultivar, and the seed size of the target species (Waddington, 1978; Al-Khatib and Boydston, 1999). Even the glucosinolate content varies with the species and cultivar type (Vaughn and Boydston, 1997; Eberlein et al., 1998), with wild types being more inhibitory and phytotoxic owing to greater content of glucosinolates (Choesin and Boener, 1991). Isothiocynates are potent germination inhibitors (Brown and Morra, 1996) and can be used as promising bioherbicides (Brown and Morra, 1995).

Allelopathic Crop Residues and Weed Control

Residues from a number of crops (wheat, rice, sorghum, barley, sunflower, etc.) are known to suppress weeds through their allelopathic effects. The presence of crop residues on the soil surface as mulch suppresses weeds through allelopathy and reduces reliance on herbicides (White, Worsham, and Blum, 1989; Worsham and Blum, 1992; Weston, 1996; Batish et al., 2001). Generally, the effect of crop residues for weed management declines after four to six weeks owing to loss of residue mass and breakdown of allelochemicals (Patrick, Tousson, and Snyder, 1963; Kimber, 1973; Smeda and Weller, 1996).

Wheat

Under no-till systems, wheat residues mulched into the soil suppress the growth of redroot pigweed, spiny amaranth, tall morning glory [*Ipomoea purpurea* (L.) Roth], and barnyard grass in the next season's crop (Banks and Robinson, 1980; Thilsted and Murray, 1980; AlSaadawi, 2001). The inhibitory effect was similar to the treatment of herbicide in the nonmulched field (Banks and Robinson, 1980). Both root and shoot residues are effective in controlling weeds such as ivyleaf morning glory (*Ipomoea hederacea* Jacq.), redroot pigweed, and prickly sida (*Sida spinosa* L.), although the root residues are more effective (Blum, King, and Browne, 2002). Wheat straw mulch can also be used for the control of forest weeds such as red raspberry (*Rubus idaeus* L.) (Jobidon, Thibault, and Fortin, 1989). The inhibitory activity of the straw was attributed to the release of phenolics from them (Jobidon, Thibault, and Fortin, 1989). Weed control by wheat residues depends upon the amount of soil covered by the residues. A soil cover of 60 percent with wheat residues was found to be optimum for suppressing the next season's weeds (Bilalis et al., 2003). Inhibitory activity of wheat mulch is due to allelopathic interference (Worsham, 1984). and the weed-suppressing effect persists up to six weeks (AlSaadawi, 2001).

Aqueous extracts from wheat residues inhibit the growth of a number of weed species such as ivyleaf morning glory, velvetleaf (*Abutilon theophrasti* Medik.), barnyardgrass and redroot pigweed, pitted morning glory, sicklepod (*Cassia obtusifolia* L.), wild oat (*Avena ludoviciana* Dur.), turnip weed [*Rapistrum rugosum* (L.) All.], paradoxa grass (*Phalaris paradoxa* L.) and common or annual sowthistle (*Sonchus oleraceus* L.), annual ryegrass (*Lolium rigidum* Gaud.) (Steinsiek, Oliver, and Collins, 1982; Liebl and Worsham, 1983; Rambakudzibga, 1991; Lehman and Blum, 1997; Jones et al., 1999; AlSaadawi, 2001; Wu et al., 2001; Wu, Pratley, and Haig, 2003). A number of allelochemicals such as DIMBOA and phenolic acids have been identified from the wheat (see Table 9.2).

Rice

Like wheat, rice residues (stubbles and straw) when incorporated into the soil reduce the incidence of both grassy and broadleaf weeds and thus could be exploited as a possible weed management strategy (Narwal, 2000). Rice residues when added to the soil at 5 ton·ha^{-1} reduced biomass of jungle rice [*Echinochloa colona* (L.) Link.], monarch redstem (*Ammannia baccifera* L.), ammannia (*A. multiflora* Roxb.) and gulf leaf flower (*Phyllanthus fraternus* Webster) (Khan and Vaishya, 1992), and flatsedge (*Cyperus iria* L.) (Lin, Smith, and Dilday, 1992b). The weed-suppressing effect is similar to that of herbicides (Lin, Smith, and Dilday, 1992b). The inhibitory effect of rice residues is due to the release of allelochemicals from them. A number of allelochemicals have been identified from rice residues under flooded (3-hydroxybenzoic, 3,4-dihydroxycinnamic, and 4-hydroxyphenylacetic acid), upland (4-hydroxy benzaldehyde, 4-hydroxy benzoic acid, 3-hydroxy-4-methoxy benzoic acid, 4-hydroxy cinnamic acid) (Mattice et al., 1998), and waterlogged (*p*-hydroxybenzoic, vanillic, ferulic, *o*-hydroxy phenylacetic, and syringic acid) conditions (Chou and Lin, 1976; Chou, Chiang, and Cheng, 1981), and momilactone from the root exudates (Kato-Noguchi et al., 2002).

Sorghum

Even the residues from *Sorghum* spp. reduce the incidence and growth of a number of weed species such as common purslane and smooth crabgrass [*Digitaria ischaemum* (Schr.) Schr. ex Muhl.] (Putnam and DeFrank, 1983), green foxtail, velvetleaf, and smooth pigweed (Panasiuk, Bills, and Leather, 1986; Hoffman et al., 1996a). Cheema and Khaliq (2000) found that incorporation of herbage of mature sorghum plants into soil at 2 to 6 ton/ha and spray of water extracts reduces the incidence of weeds and increases the yield of irrigated wheat under the semiarid regions of Pakistan. The inhibitory activity of the sorghum residues is dependent on their age, with two- to four-week-old residues being more effective compared to six- or more-week older residue (Putnam and DeFrank, 1983). Even the interplanting of the sorghum is observed to reduce the growth and biomass of weeds such as barnyard grass, redroot pigweed, and red sorrel (*Rumex acetosella* L.) (Panasiuk, Bills, and Leather, 1986). Sorghum residues release a number of allelochemicals such as sorgoleone, cyanogenic glycosides-dhurrin, and a number of breakdown products that bring about weed suppression (Table 9.2).

Alfalfa

Likewise, residues from alfalfa and their aqueous extracts inhibit the germination and growth of a number of broadleaf and grassy weed species such as common lambsquarters, redroot pigweed, velvetleaf, giant foxtail, cheatgrass (*Bromus secalinus* L.), and crabgrass (*Digitaria* spp.) (Chung and Miller, 1995). Aerial residues of alfalfa when used as pellets at 1 to 3 ton/ha inhibit the growth of lowland weeds such as barnyardgrass [*Echinochloa oryzicola* (Vas.) Vas.], Henry's crabgrass [*Digitaria ciliaris* (Retz.) Koeler], variable flatsedge (*Cyperus difformis* L.), and arrowleaved monochoria, abunome [*Dopatrium junceum* (Roxb). Buch.-Ham. ex Benth.], false pimpernel (*Lindernia pyxidaria* sensu Pennll. L.), long-stemmed waterwort (*Elatine tiandra* Schk. var. *pedicellata* Krylov), needle or slender spikerush (*Eleocharis acicularis* Roem et. Schult. var. *longiseta* Sven.), and spike-flowered rotala (*Rotala indica* Koehne var. *uliginosa* Koehne) and reduced emergence and dry weight of barnyardgrass (Tsuzuki et al., 1999; Xuan and Tsuzuki, 2001; Xuan et al., 2001, 2003). Researchers concluded that it could be used as a natural herbicide (Chung and Miller, 1995) in the paddy fields (Xuan et al., 2001). Even decomposed roots of alfalfa incorporated into the soil inhibit the germination and growth of cogon grass and the production of new plants from rhizomes (Abdul-Rahman and Habib, 1989) and thus could be used to control cogon grass. A number of allelochemicals have been identified in alfalfa (Table 9.2).

Jones and colleagues (1999) reported that residues of crops such as barley, wheat, and canola reduce the germination, growth, and biomass of sowthistle, paradoxa grass, wild oat, and turnip weed; among them barley was most effective. Extracts from the residues of barley and canola have been found to be more inhibitory to flixweed [*Descurainia sophia* (L.) Webb ex Prantl], stinkweed (*Thlaspi arvense* L.), and downy brome

(*Bromus tectorum* L.) than to wheat and could be used for the selective management of weeds in wheat (Moyer and Huang, 1997). Both barley and wheat are known to contain a variety of allelochemicals (Table 9.2).

From this discussion it is clear that residues from a number of crops such as wheat, rice, sorghum, barley, and alfalfa have weed-suppressing ability. In the light of this, cultivars and accessions from different regions of the world are being screened for this activity and correlated with the allelochemical content in them (discussed in the next section). Although crop residues hold great potential as a weed management strategy, there have been reports that allelochemicals released from them also affect the growing crops (Rice, 1984, 1995; Batish et al., 2001).

SCREENING OF CULTIVARS WITH GREATER ALLELOPATHIC POTENTIAL AND THEIR ROLE IN CROP IMPROVEMENT

It is well-known that various accessions of crops differ in their allelochemical/phytochemical content, imparting to them various levels of resistance against pests and diseases. In addition, crop varieties and genotypes may possess varying levels of allelochemicals (Christensen, 1995; Wu et al., 1999, Olofsdotter, Jensen, and Courtois, 2002). Hoult and Lovett (1993) indicated that wild accessions of a number of crop plants were more resistant to pests and weeds owing to the presence of allelopathic traits in them. These, however, got lost during the process of cultivation because of selection pressure favoring high yield and fast growth. Nevertheless, current crop cultivars can be improved by reincorporating the lost genes into them. But this requires a comprehensive screening effort to search for superior accessions/genotypes/varieties possessing higher levels of allelochemicals. The genes encoding the synthesis of allelochemicals can be incorporated through either conventional breeding methods or molecular genetic techniques. A number of reports indicate the efforts made by different scientists in searching for accessions, particularly wild ones, possessing allelopathic traits (Table 9.3). Rice (1984), in his comprehensive compilation, has cited a number of examples indicating the importance of this search for allelopathic accessions. Among various examples cited in Table 9.3, rice and wheat have been extensively studied for weed-suppressing varieties. The work in rice allelopathy has been conducted in Taiwan, the United States, Japan, Korea, Egypt, and the Philippines. Dilday and colleagues (1998, 2001) evaluated 17,279 rice accessions collected from 110 countries for allelopathic activity at USDA-ARS. Of these, 5,000 were evaluated for their inhibitory effect toward redstem (*Ammania coccinea* Rottb.) and barnyardgrass, and 12,000 were evaluated against ducksalad [*Heteranthera limosa*

TABLE 9.3. List of crop plants where accessions/varieties have been screened for allelopathic traits.

Crop plant	Number of accessions/varieties/ cultivars screened	Reference(s)
Avena spp.	3,000	Fay and Duke, 1977
Cucumis sativus L.	526	Putnam and Duke, 1974
Glycine max	141	Massantini, Caporali, Zellini, 1977
	12	James, Banks, and Karnok, 1988
Helianthus annuus L.	13	cf. Hilton, 1979
Hordeum spp. (*H. vulgare,* and *H. spontaneum* Koch and *H. agriocrithon* Aberg)	43	Hanson et al., 1981; Lovett and Hoult, 1992
Medicago sativa	8	Xuan and Tsuzuki, 2002
Oryza perennis Moench	24	Chou, Chang, and Oka, 1991
O. sativa	189	Fujii, 1992
	38	Park, 1996
	1000	Hassan et al., 1998
	134	Kim and Shin, 1998
	111 (under field trials)	Olofsdotter and Navarez, 1996; Olofsdotter et al., 1999
	91	Ahn and Chung, 2000
	44	Chung, Ahn, and Yun, 2001; Chung et al., 2002
	17279	Dilday et al., 2001
Pennisetum typhoides (Burm. f.) Stapf. & Hubbard (pearlmillet)	13	Narwal, 2000
Secale cereale	9	Reberg-Horton, Creamer, and Mohler, 2000; Reberg-Horton, Creamer, Burton, Ranells, and Mohler, 2001
Sorghum spp.	25, 7	Nimbal et al., 1996; Czarnota et al., 2001; Czarnota, Rimando, and Weston, 2003
S. bicolor	100	AlSaadawi et al., 1986
Triticum spp. [*T. speltoides* (Tausch) Gren. ex K. Richt., *T. aestivum, T. tauschii* (Coss.) Schmalh.]	55	Niemeyer, 1988

T. aestivum	286	Spruell, 1984
	47	Nicol et al., 1992
	38	Wu et al., 1998
	453	Wu, Pratley, et al., 2000
	4	Wu et al., 2001
	58	Wu et al., 2002
T. aestivum and *T. durum* Desf.	52	Copaja, Niemeyer, and Wratten, 1991
T. speltoides	19	Hashem and Adkins, 1996
	26	Quader et al., 2001

Source: Adapted from Singh, Batish, and Kohli, 2001.

(Sw.) Willd.]. Of these, 145 accessions of rice exhibited allelopathy against redstem, 412 against ducksalad, and 94 against barnyardgrass (Dilday et al., 2001). Accessions possessing strong allelopathy can even suppress a mixture of weeds (Lin, Smith, and Dilday, 1992a,b). At the International Rice Research Institute (IRRI, Philippines), Olofsdotter and Navarez (1996) selected 11 varieties allelopathic toward barnyardgrass both in dry and wet seasons. Rice varieties possessing higher allelopathic potential can be used for breeding purposes (Olofsdotter, 1998), although a concerted effort is required. Likewise, in wheat, several researchers have screened accessions possessing allelopathic effects against weeds and the allelochemical content (see Table 9.3 and 9.4). Wu and colleagues (2002) have selected 58 wheat accessions strongly allelopathic toward annual ryegrass, and the allelopathic activity correlated with the allelochemical content. Likewise, in *Sorghum,* germplasm consisting of 25 genotypes was screened for sorgoleone (a known allelochemical) production in five-day-old seedling roots (Czarnota et al., 2001). Considerable variability in sorgoleone content was found among the genotypes. Maximum content was found in genotype 'B Redlan' followed by 'IS 1318C', possessing 17.8 and 14.2 mg/g sorgoleone, respectively (Czarnota et al., 2001).

Besides screening, crop improvement programs have been initiated, though to a limited extent. Probably the first attempts were made by Panchuk and Prutenskeya (1973) and Grodzinsky and Panchuk (1974). They reported that residue extracts of wheatgrass [*Agropyron glaucum* (Desf.) Roem.] were more inhibitory toward radish (*Raphanus sativus* L.) compared to wheat, and if a cross is made between them, F_1 hybrids containing more characters from wheatgrass were also very inhibitory. Based on the study, it was concluded that a possibility exists of transferring traits from one cultivar to another. After these studies, little attention was paid toward crop improvement programs. This may be attributed to the difficulties

TABLE 9.4. List of crop plants where accessions/varieties have been screened for allelochemicals.

Crop plant	Allelochemical screened	Reference
Avena spp.	Scopoletin	Fay and Duke, 1977
Hordeum spp. (*H. vulgare,* and *H. spontaneum, H. agriocrithon*)	Gramine	Hanson et al., 1981; Lovett and Hoult, 1992
Oryza sativa	Phenolic acids	Chung et al., 2002
Sorghum spp.	Sorgoleone, 5-ethoxysorgoleone, 2,5-dimethoxysorgoleone	Nimbal et al., 1996; Czarnota et al., 2001; Czarnota, Rimando, and Weston, 2003
Triticum spp. (*T. speltoides, T. aestivum, T. tauschii*)	DIMBOA	Niemeyer, 1988
T. aestivum and *T. durum*	DIMBOA, Phenolic acids	Copaja, Niemeyer, and Wratten, 1991; Wu et al., 1998
T. aestivum	Phenolic acids	Wu et al., 2001, 2002; Wu, Haig, et al., 2000
T. speltoides	DIMBOA	Quader et al., 2001

in breeding crops because of the quantitative characters of allelopathic traits and polygenic nature of their inheritance. However, work in this direction began with the advancement of molecular genetic techniques. Nevertheless, attempts are being made to improve rye varieties for increased allelopathy by employing breeding techniques (Reberg-Horton, Creamer, Burton, Ranells, and Murphy, 2001).

Duke and colleagues (2001) proposed two approaches for improving the crops with allelopathic traits. These are (1) to enhance existing allelopathic potential of a particular crop and (2) to produce transgenics by inserting genes encoding for a particular allelochemical. The first approach is assisted by the use of several DNA markers such as PCR, RFLP, RAPD, and AFLP. With the help of these markers, QTLs are located on the plant genome, which help in selection of a particular genotype. RFLP markers are quite useful in locating QTLs associated with allelopathic effects. In rice, this technique has been utilized to analyze QTLs in the F_2 population arising from a cross between highly inhibitory and less inhibitory germplasm (Rimando et al., 2001; Ebana et al., 2001). Likewise, QTL mapping has also been performed on 142 recombinant inbred rice lines using relay-seeding technique for phenotype (Jensen et al., 2001). Four main effect QTLs associated with allelopathic effect against barnyardgrass have been identified on three chromosomes (2, 3, and 8). Based on these studies, Olofsdotter (2001) and Ebana and colleagues (2001) have suggested that future research on rice should focus on understanding the mechanism of allelopathic expressions from both agronomic and environmental viewpoints, identification of allelochemicals responsible for inhibitory effects, and assessment of environmental and toxicological effects, as well as continuing the work on understanding the genetics of rice in order to have complete mapping of allelopathic genes.

Based on DNA marker-assisted selection, RFLP techniques have also been used to identify QTLs for the production of glucosinolates in rapeseed (Toroser et al., 1995). In *Sorghum,* efforts are also being made to explore the biosynthetic pathway for the production of sorgoleone in root hair cells (Duke et al., 2001). One completely sequenced novel gene with expressions localized in root hair cells has also been identified (Weston, Yang, and Scheffler, 2002). Improvement in the content of the hydroxamic acid 2,4-dihydroxy-7-methoxy-2H-1,4-benzoxazin-3 (4H)-one (DIMBOA) found in cereals and responsible for pest and herbicidal resistance has also been made. DIMBOA is synthesized in plants via two pathways: from methoxylated lactam (Tipton et al., 1973) or from DIBOA (Zuniga et al., 1990). Niemeyer and Jerez (1997) investigated the genes responsible for accumulation of DIMBOA in wheat and identified the chromosomes containing genes for transformation of DIBOA into DIMBOA or for the transfer of methoxylated lactam to DIMBOA and inhibitor of accumulation of DIMBOA. This identification of genes would help in improving cereals (wheat, maize, and rye) for enhancing the amount of DIMBOA in them. With the identification of genes responsible for synthesis and release of allelochemicals, our current high-yielding cultivars can be improved for allelochemical production. Even various plant genomic and proteomic techniques can be utilized for enhancing their weed-suppressing ability (Birkett et al., 2001). However, the major limitations and problems associated with this approach are autotoxicity of the allelochemicals and the metabolic costs involved in production of an adequate amount of allelochemicals for desired weed control (Scheffler et al., 2001).

ALLELOCHEMICALS AS HERBICIDES

Allelochemicals from Higher Plants with Herbicidal Potential

Allelochemicals can also be used in either crude or pure form for the purpose of weed management (Duke, 1986; Duke et al., 2000, 2002; Macias et al., 2001; Vyvyan, 2002; Singh, Batish, and Kohli, 2003). The reasons for this are due to several characteristics of allelochemicals:

- They exhibit structural diversity and thus offer greater scope for exploring novel compounds for herbicidal properties.
- They may serve as invaluable sources of lead compounds.
- They have no halogen or heavy elements; rather, they contain more oxygen atoms.
- They are environmentally friendly and have, in general, comparatively less residual activity.
- They have novel target sites of action different from synthetic herbicides.

Among these, the last two, i.e., their rapid degradation and novel target sites of action, are most sought after and are gaining the attention of the agrochemical industry throughout the world. Moreover, the use of such compounds can also help in restoring the deteriorating natural balance of agroecosystems. Various allelochemicals in pure or crude form isolated from higher plants with weed-suppressing ability are listed in Table 9.5.

In spite of many compounds being tested for herbicidal activity, only a few have been marketed. For example, cinmethylin (Cinch) and Mesotrione (Callisto). Cinch is based on the chemistry of cineole (Grayson et al., 1987) and Mesotrione is based on the allelochemical leptospermone (Mitchell et al., 2001). The low marketing potential of natural phytotoxins/allelochemicals from higher plants is due to their poor performance under field conditions compared to laboratory studies.

Phytotoxins from Microbes with Weed-Suppressing Ability

Besides allelochemicals from higher plants, a number of phytotoxins of microbial (fungi, bacteria, and actinomycetes) origin are known to possess herbicidal activity. In fact, microbes offer a great choice for potential bioactive molecules with weed-suppressing ability. A number of excellent reviews providing insight into this property of microbial metabolites are

TABLE 9.5. Allelochemicals from plants with potential herbicidal activity.

Category of compound	**Source**	**Weeds affected**	**Reference(s)**
Volatile essential oils	*Eucalyptus* spp.	Ragweed parthenium	Kohli, Batish, and Singh, 1998
	Lemon-scented eucalypt (*E. citriodora* Hook.)	Hairy beggarticks (*Bidens pilosa* L.), wild oat, coffee weed (*Cassia occidentalis* L.)	Singh, Batish, Kaur, et al., 2002b
	Lemon grass [*Cymbopogon citratus* (DC. ex Nees) Stapf], Bible hyssop (*Origanum syriacum* L.), tea hyssop [*Micromeria fruticosa* (L.) Druce]	Black mustard (*Sinapis nigra* L.), palmer amaranth (*Amaranthus palmeri* S. Wats.) mat amaranth (*A. blitoides* S. Wats.), garden spurge (*Euphorbia hirta* L.)	Dudai et al., 1999
	Red thyme (*Thymus vulgaris* L.), summer savory (*Satureja hortensis* L.), cinnamon (*Cinnamomum zeylanicum* Blume), clove [*Syzygium aromaticum* (L.) Merr. & Perry]	Dandelion (*Taraxacum officinale* Weber in Wiggers), common lambsquarters, common ragweed, and johnsongrass [*Sorghum halepense* (L.) Pers.]	Tworkoski, 2002
	Japanese beauty berry (*Callicarpa japonica* Thunb.)	Bent grass	Kobaisy et al., 2002
Volatile monoterpenes			
Geraniol α-terpeniol	Pure compound	Italian ryegrass (*Lolium multiflorum* Lam.), large crabgrass, redroot pigweed, velvetleaf	Vaughn and Spencer, 1993
1,4- and 1,8-cineole	Pure compound	Sickle pod (*Cassia obtusifolia* L.) Barnyardgrass	Romagni, Allen, and Dayan, 2000
Cineole and citronellol	Pure compound	Billy goat weed (*Ageratum conyzoides* L.)	Singh, Batish, and Kohli, 2002
Citronellol, citronellal, linalool, cineole	Pure compound	Coffee weed	Singh, Batish, Kaur, et al., 2002a

TABLE 9.5 *(continued)*

Category of compound		**Weeds affected**	**Reference(s)**
Sesquiterpene lactones			
Artemisinin	Wormweed	Redroot pigweed, common purslane, velvetleaf	Duke et al., 1987; Chen and Leather, 1990; Lydon, Teasdale, and Chen, 1997
Parthenin	Ragweed pathenium	Billy goat weed, hairy beggarticks, wild oat, sicklepod	Batish et al., 1997; Datta and Saxena, 2001; Batish et al., 2002; Singh, Batish, Kohli, et al., 2002
Heliannuols	Sunflower	Dicot weeds	Macias et al., 2001
Arturin, ovatifolin, and their derivatives	*Podanthus* spp.	Italian ryegrass, red clover, lettuce	Cespedes et al., 2001
Benzoxazinones			
BOA, DIMBOA, DIBOA	Rye, maize, wheat	Barnyardgrass, crabgrass, prosomillet (*Panicum miliaceum* L.), velvetleaf, wild oat	Wolf, Spencer, and Plattner, 1985; Whitenack, Nair, and Putnam, 1988; Pérez, 1990; Pérez and Ormeño Nuñez, 1991, 1993
Glucosinolates			
Isothiocyanates (methyl-, phenylethyl-, benzyl-, allyl-, *n*-propyl, *n*-butyl)	*Brassica* spp.	Barnyardgrass, velvetleaf, spiny sowthistle, scentless mayweed, smooth pigweed, black grass (*Alopecurus myosuroides* Huds.)	Wolf, Spencer, and Kwolek, 1984; Oleszek, 1987; Petersen et al., 2001
Hirsutin	Creeping yellowcress [*Rorippa sylvestris* (L) Bess.]	Lettuce	Kawabata et al., 1989
Quassinoids			
Ailanthone	Tree of heaven	Redroot pigweed, garden cress, yellow foxtail [*Setaria glauca* (L.) P. Beauv.], barnyardgrass	Lin et al., 1995; Heisey, 1996

Hailanthone	Pure compound	Green foxtail, sicklepod, tef [*Eragrostis tef* (Zucc.) Trotter]	Ashton and Craft, 1981; Lin et al., 1995
Chaparrinone, glaucarubolone, holacanthone	Pure compound	Creeping bent grass (*Agrostis stolonifera* L.), green foxtail, sicklepod	Dayan, Watson, et al., 1999
Saponins			
Medicagenic acid, hederogenin, lecernic acid, zhanic acid, soyasapogenol B	Alfalfa	Barnyardgrass, cheat grass (*Bromus secalinus* L.)	Waller et al., 1987; Waller, Jurzysta, and Thorne, 1993, 1995
Canavanine, Medicagenic acid	Alfalfa	Pigweed and cress	Gorski, Miersch, and Ploszynski, 1991
Quinones			
Sorgoleone (*p*-benzoquinone)	*Sorghum* spp.	Large crabgrass, Black nightshade (Solanum nigrum L.), redroot pigweed, sicklepod, common purslane, ragweed	Nimbal et al., 1996; Weston and Czarnota, 2001
Juglone (5-hydroxy-1,4-naphthoquinone)	*Juglans* spp.	Velvetleaf, redroot pigweed, prickly sida, garden cress, alfalfa	Williams and Hoagland, 1982; Shettel and Balke, 1983; Kocacalsikan and Terzi, 2001
Other compounds			
Caffeine	Coffee (*Coffea arabica* L.)	Barnyardgrass, jungle rice, spiny amaranth	Rizvi, Mukerji, and Mathur, 1981; Rizvi et al., 1987
α-terthienyl	*Tagetes* spp.	Redroot pigweed, common lambsquarters, milkweed [*Morrenia odorata* (H & A.) Lindley], timothy grass (*Phleum pratense* L.)	Campbell et al., 1982
Podophyllotoxins	May apple (*Podophyllum peltatum* L.)	Italian ryegrass, lettuce	Oliva et al., 2002
Lignans	Texas sage [*Leucophyllum frutescens* (Berl.) I.M. Johnst.]	Bent grass and lettuce	Rimando et al., 1999

TABLE 9.5 *(continued)*

Category of compound		**Weeds affected**	**Reference(s)**
Umbelliferone	Pure compound	Velvetleaf, pigweed, prosomillet	Shettel and Balke, 1983
L-DOPA	Velvetbean	Broadleaf weeds	Fujii, 1999b
Withanolides	*Iochroma australe* Griseb.	Johnsongrass, common lambsquarters	Vaccarini and Bornetto, 2000
Clerodane	*Viguiera tucumanensis* (Hook et. Am.) Griseb.	Tall morning glory, Johnsongrass	Vaccarini et al., 1999

available (Fischer and Bellus, 1983; Misato and Yamaguchi, 1984; Cutler, 1988; Rice, 1995; Duke et al., 1996; Kremer and Kennedy, 1996; Hoagland, 1999; Mallik, 2001; Saxena and Pandey, 2001; Li et al., 2003; Mohan Babu et al., 2003). Anisomycin was perhaps the first phytotoxin isolated form the culture filtrates of *Streptomyces toyocaensis,* which exhibited selective toxicity against barnyard grass and crabgrass without any effect on broadleaved horticultural plants (Yamada et al., 1972). Its synthetic analog methoxyphenone (or NK049) was later commercialized as a herbicide for paddy fields (Yamada et al., 1974). Since then, a number of phytotoxins have been isolated from other microbes and tested against weeds; this number is increasing day by day. Singh, Batish, and Kohli (2003) have listed over 100 phytotoxins of microbial origin possessing weed-suppressing potential. Some of the microbes providing excellent sources of phytotoxins are *Streptomyces, Alternaria, Fusarium,* and *Dreschlera.* In fact, several potential phytotoxins with herbicidal activity have been isolated from a single species of *Streptomyces,* i.e., *S. hygroscopicus.* Some of the phytotoxins isolated from *Streptomyces* spp. are Phosphinothricin, anisomycin, bialaphos, blasticidins, coaristeromycin, coformycin, herbicidins A & B, geldanamycin, herbimycins, herboxidiene, homoalanosine, hydantocidin, monensin, nigericin, naramycin B, resormycin, toyocamycin, tubercidin SF2454, and vulgamycin.

In spite of so much diversity of microbial products, only a few have been commercially exploited. Among these, bialaphos, a product from *Streptomyces viridiogenes* and *S. hygroscopicus* and its active form L-phosphinothricin (L-PPT) have been widely exploited. Phosphinothricin is a unique amino acid linked to 2-L-alanyl moieties. The biological activity of bialaphos is related to L-PPT—a natural form. Wild and Ziegler (1989) pointed out that bialaphos is a proherbicide that has to be converted into L-PPT so as to bring about the action. L-PPT has been commercialized as glufosinate, which is its ammonium salt, and is at present being marketed under the trade names Basta, Herbiace, Ignite, Challenge, and Harvest. It is a nonselective postemergent herbicide that can control both broadleaf and grassy weeds from orchards, vineyards, and fallows (Vasil, 1996). The production of transgenic crops with resistance to PPT and glufosinate has further expanded its use. Moreover, bialaphos has a novel target site of action and inhibits glutamine synthetase (Lydon and Duke, 1999). Its activity is higher than glyphosate. The low rate of commercialization of microbial phytotoxins is due to the problems associated with their extraction, purification, culturing, cost of synthesis, and mammalian toxicity (Dayan, Romagni, et al., 1999; Duke et al., 2002). Some of the phytotoxins possessing toxicity toward other life forms, including mammals, are cercosporin, isocercosporin, fumonisin B1, AAL-toxin, and australifungin (Hoagland, 2001).

Not only the microbial phytotoxins but also some microbes (both pathogenic and nonpathogenic) can be used directly for the control of weeds. In this context, the term *bioherbicide* or *mycoherbicide* (especially for fungal pathogens) is generally applied. Several reviews are available on this topic (Watson, 1991; Cardina, 1995; Boyetchko, 1999; Charudattan, 2001). Charudattan (1991, 2001) listed over 100 pathogens with the potential for direct use as weed-suppressing agents. Of these, only eight have been approved for use as bioherbicides (Charudattan, 2001). These are Collego, Biochon, DeVine, BioMal, Dr. BioSedge, Hakatak, StumpOut, and Camperico. However, use of phytotoxins is preferred over the direct use of pathogens because of logistics of storage, difficulties in application and formulation, compatibility and chance of spread to nontargets, difficulties in potential marketing and registration, etc. (Charudattan, 2001; Hoagland, 2001).

Lichens also can serve as source of secondary metabolites, particularly anthraquinones and usnic acid, possessing new molecular target sites and herbicidal activity (Dayan and Romagni, 2001). Among anthraquinones, emodin and rhodocladonic acid are highly active against seedlings of higher plants, particularly grasses. Moreover, these are simple chemicals with easy synthesis which can serve as excellent candidates for synthesis of new herbicides. Usnic acid is another unique chemical that inhibits *p*-hydroxyphenylpyruvate dioxygenase—an enzyme of carotenoid biosynthesis.

POTENTIAL OF ALLELOPATHIC INTERACTIONS FOR PARASITIC WEED MANAGEMENT

Parasitic weeds, especially *Striga* and *Orobanche* species, pose a serious threat to agriculture as they are difficult to manage. However, allelopathy can be successfully exploited to control them (1) by using allelopathic crops as catch and trap crops (Chittapur, Hunshal, and Shenoy, 2001) or as intercrops (Khan et al., 2002; Oswald et al., 2002), (2) through allelochemicals (Sugimoto et al., 1998; Welzel, Röhrig, and Milkova, 1999), and (3) by the use of microbes or their phytotoxins (Zonno and Vurro, 1999; Amsellem et al., 2001). Parasitic species generally germinate in response to germination-stimulant compounds such as strigol, alectrol, sorgoleone, and sorgolactone, which are released from the host roots (Wegmann, 1998; Yokota et al., 1998). These compounds are effective at very low concentrations. For example, at 10^{-4} to 10^{-6} M concentration of strigol, 85 to 100 percent germination of witchweed seed occurs. These compounds also act as signals between the host and the weed. Sugimoto and colleagues (1998) and Welzel, Röhrig, and Milkova (1999) have made extensive studies in this regard, especially the regulatory signals between the host and the *Striga* species. The stimulation of germination of parasitic weeds by such compounds can be exploited for their management. For example, use of stimulants before the growth of the host plant would lead to their premature germination and the seedlings will die in the absence of the host. Such a suicidal germination can be stimulated by the use of a number of stimulating allelochemicals such as terpenoids, sesquiterpene lactones, sesquilactone, and coumarin-type compounds (Johnson, Rosebery, and Parker, 1976; Rice, 1984; Fischer et al., 1990; de Luque et al., 2000). Qasem and Foy (2001) have pointed that even the residues of plant species can be used for controlling both *Striga* and *Orobanche*. They concluded that allelopathy can be utilized in controlling these weeds either by preventing their germination or by destroying the seed bank.

Parasitic weeds can also be controlled by employing cultural practices through the use of allelopathic plants as intercrops or in crop rotations. Use of various leguminous crops such as red leucaena [*Leucaena diversifolia* (Schltdl.) Benth.], sesban [*Sesbania sesban* (L.) Merr.], sweet potato [*Ipomoea batatas* (L.) Lam.], peanut (*Arachis hypogaea* L.), yellow gram (*Cicer arietinum* L.), and soybean is extremely useful as intercrops with the crops being affected by parasitic weeds (Oswald et al., 1996, 2002). Further, the use of silver leaf [*Desmodium uncinatum* (Jacq.) DC.] and green leaf [*D. intortum* (Mill.) Urb.] desmodiums have also been found to be very useful intercrops in maize for controlling witchweed infestation (Khan et al., 1997; Pickett, 1999). It was attributed to the allelopathic interference of desmodiums (Khan et al., 2002). Recently, Tsanuo and colleagues (2003) reported that four isoflavonones (uncinanone A, B, C, and genistein) are released from the root exudates of silver leaf desmodium, and of these unicinanone C inhibits the radicle growth of *Striga*. It is the first report of an allelochemical inhibiting (not stimulating) the growth of *Striga* spp. Intercropping has also been useful in controlling *Orobanche* species. In tomato and faba bean fields, incidence of branched broomrape (*O. ramosa* L.)

is significantly reduced if intercropped with fenugreek, Egyptian lupine, and turnip (Al-Menoufi, 1991, 1992). Thijssen (1995) reported that green manure of red calliandra (*Calliandra calothyrsus* Meissn.) can effectively reduce the infestation of *Striga,* and this was attributed to the allelopathic interference of the tree. Even the trees of musine (*Croton megalocarpus* Hutch.) induce suicidal germination of *Striga* spp. and can thus help in reducing its incidence (Thijssen, 1995).

Like allelopathic crops, microbes and their phytotoxins can also provide an effective control of parasitic weeds (for example, mycoherbicides from *Fusarium* spp. [Citola, Watson, and Hallet, 1995; Abbasher and Sauerborn, 1992; Thomas et al., 1999; Ahmed et al., 2000; Amsellem et al., 2001] and fungal toxins such as tricothecenes [Zonno and Vurro, 1999].

ALLELOPATHY AND MANAGEMENT OF AQUATIC WEEDS

A number of aquatic weeds choke rivers, streams, ponds, and lakes and pose serious problems for other useful aquatic flora and fauna. Some of these difficult to control weeds are water hyacinth [*Eichhornia crassipes* (Mart.) Solms.], hydrilla (*Hydrilla* spp.), *Salvinia* spp., and *Potamogeton* spp. Their control through the use of synthetic chemicals is fraught with dangers of contaminating water and causing adverse effect on the biota. Allelopathy can offer an environmentally safe way of controlling these weeds. Several reports have indicated that use of allelopathic aquatic plants with greater competitive ability and their allelochemicals can control serious aquatic weeds (Yeo and Fisher, 1970; Szezepanski, 1977; Putnam, 1988; Elakovich and Wooten, 1989). Needle spikerush [*Eleocharis acicularis* (L.) Roem. & Schult.] can eliminate Canadian pondweed (*Elodea canadensis* Michx.) and curly pondweed (*Potamogeton crispus* L.) through its allelopathy (Frank and Dechoretz, 1980). Allelochemicals and aqueous extracts from spike rush species inhibited the growth of duckweed (Sutton and Portier, 1989). Saxena (2000) reported that aqueous extracts of twigs and flowers, and residues of *Lantana* can be effectively used for controlling water hyacinth. Likewise, dried leaf powder and residues of parthenium have the potential to inhibit the growth of water hyacinth and giant salvinia (*Salvinia molesta* Mitch.) (Pandey, Kauraw, and Bhan, 1993; Pandey, 1994). Later, Pandey (1996) reported that parthenin and phenolic acids such as *p*-hydroxybenzoic acid are toxic to a number of aquatic weeds, including water hyacinth, giant salvinia, floating aquatic fern (*Azolla nilotica* Decne. ex Mett.), hydrilla, and water lettuce (*Pistia stratiotes* L.), and thus can be used for their control.

Even microbes and their metabolites show great potential in controlling aquatic weeds (Barreto et al., 2000), e.g., use of Mexican fungi and *Alternaria* spp. (Shabana, Charudattan, and Elwakil, 1995; Jimenez and Charudattan, 1998; Mohan Babu, Sajeena, and Seetharaman, 2003) for the control of water hyacinth; *Fusarium culmorum* (W. G. Sm.) Sacc. (Smither-Kopperl, Charudattan, and Berger, 1998) and *Macrophomina phaseolina (Tossi) Goid.* (Joye, 1990) for the control of hydrilla.

CONCLUSIONS

From this discussion it is clear that allelopathy could be a useful tool for weed management under sustainable farming systems. Worth mentioning in this regard are the allelopathic crops with better competitive ability that can overtake weeds. Allelopathic cover/smother crops and green manure crops (including their residues) grown either as intercrops or in rotation fit well in this type of model and hold great promise for the future. Care, however, should be taken to avoid autotoxic effects of crops as well as their effects on other crops. Thus, it is essential to determine the useful combination of main crops and those grown as intercrops or in rotation so as to have minimal weed incidence. Allelopathic crops with weak allelopathic traits can also be improved genetically through DNA recombinant technology to obtain cultivars with more allelochemical content or transgenes. Encouraging results have been obtained, particularly in rice, wheat, and sorghum, and in the future improved genotypes with greater competitive ability (vis-à-vis allelopathy) can be expected. Likewise, development of transgenics with allelopathic traits from other sources has also been proposed, but a cautious effort is desired in this direction as transgenics could themselves lead to several environmental problems.

Besides allelopathic plants, microbes could also provide a useful source of novel herbicides possessing no or little environmental hazards. These chemicals possess unique target sites of action besides many other benign properties and could thus replace synthetic herbicides that are environmental and toxicological hazards. In addition, microbes themselves can be utilized as bioherbicides for the control of weeds, especially aquatic and parasitic ones. Despite a large number of reports on allelochemicals/phytotoxins of higher plant or microbial origin, only a few have been utilized commercially, for example, cinmethylin and glufosinate. This limited use of natural products could be attributed to several factors including difficulties in extraction, purification, or culturing (in the case of microbes), huge costs involved in their discovery, mammalian toxicity, toxicity to nontarget organisms, and rapid degradation in the environment, as well as the difficulties associated with their patenting, intellectual property rights, and marketing. In spite of these limitations, allelopathy and allelochemicals can serve as important components of integrated weed management for future sustainable weed management programs.

REFERENCES

Abbasher, A.A. and J. Sauerborn (1992). *Fusarium nygamai,* a potential bioherbicide for *Striga hermonthica* control in sorghum. *Biological Control* 2: 291-296.

Abdul-Rahman, A.A and S.A. Habib (1989). Allelopathic effect of alfalfa *(Medicago sativa)* on bladygrass *(Imperata cylindrica). Journal of Chemical Ecology* 15: 2289-2300.

Ahmed, N.E., Y. Sugimoto, A.G.T. Babiker, O.E. Mohamed, Y. Ma, S. Inanga, and H. Nakajima (2000). Effects of *Fusarium solani* isolates and metabolites on *Striga* germination. *Weed Science* 49: 354-358.

Ahn, J.K. and I.M. Chung (2000). Allelopathic potential of rice hulls on germination and seedling growth of barnyard grass. *Agronomy Journal* 92: 1162-1167.

Akobundu, I.O., E.U. Udensi, and D. Chikoye (2000). Velvetbean (Mucuna spp.) suppresses speargrass (Imperata cylindrica [L.] Raeuschel) and increases maize yield. *International Journal of Pest Management* 46: 103-108.

Al-Khatib, K. and R. Boydston (1999). Weed control with *Brassica* green manure crops. In S.S. Narwal (Ed.), *Allelopathy update,* Volume 2, *Basic and applied aspects* (pp. 255-270). New Delhi, India: Oxford and IBH Publishing Co. Pvt. Ltd.

Al-Khatib, K., C. Libbey, and R. Boydston (1997). Weed suppression with *Brassica* green manure and cover crops in green pea. *Weed Science* 45: 439-445.

Al-Menoufi, A.O. (1991). Crop rotation as a control measure of *Orobanche crenata* in *Vicia faba* fields. In K. Wegmann, and L.J. Musselman (Eds.), *Progress in* Orobanche *Research: Proceedings of International Workshop on* Orobanche *Research* (pp. 241-247). Tubingen, Germany: Eberhard-Karls-Universität.

Al-Menoufi, A.O. (1992). *Cultural practices to control* Orobanche crenata *in faba bean field in Egypt.* Final report. [Project Financed by Alexandria University, Egypt (1991-1992).]

AlSaadawi, I.S. (2001). Allelopathic influence of decomposing wheat residues in agroecosystems. *Journal of Crop Production* 4(2): 185-196.

AlSaadawi, I.S., J.K. Al-Uqaili, A. Al-Rubeaa, and S.M. Al-Hadithy (1986). Allelopathic suppression of weeds and nitrification with selected cultivars of *Sorghum bicolor* (L.) Moench. *Journal of Chemical Ecology* 12: 209-219.

Amsellem, Z., Y. Kleifeld, Z. Kerenyi, L. Hornok, Y. Goldwasser, and J. Gressel (2001). Isolation, identification, and activity of mycoherbicidal pathogens from juvenile broomrape plants. *Biological Control* 21: 274-284.

Anaya, A.L. (1999). Allelopathy as a tool in the management of biotic resource in agroecosystems. *Critical Reviews in Plant Sciences* 18: 697-739.

Anaya, A.L., M.R. Calera, R. Mata, and R. Pereda-Miranda (1990). Allelopathic potential of compounds isolated from *Ipomoea tricolor* Cav. (Convolvulaceae). *Journal of Chemical Ecology* 16: 2145-2152.

Ashton, F.M. and A.S. Craft (1981). *Mode of action of herbicides.* New York: Wiley.

Ateh, C.M., and J.D. Doll (1996). Spring-planted winter rye *(Secale cereale)* as a living mulch to control weeds in soybeans *(Glycine max). Weed Technology* 10: 347-353.

Baghestani, A., C. Lemieux, G.D. Leroux, R. Baziramakenga, and R.R. Simard (1999). Determination of allelochemicals in spring cereal cultivars of different competitiveness. *Weed Science* 47: 498-504.

Banks, P.A. and E.L. Robinson (1980). Effect of straw mulch on pre-emergence herbicides. *Proceedings of Southern Weed Science Society* 33: 286.

Barnes, J.P. and A.R. Putnam (1983). Rye residues contribute weed suppression in no-tillage cropping systems. *Journal of Chemical Ecology* 9: 1045-1057.

Barnes, J.P. and A.R. Putnam (1986). Evidence for allelopathy by residues and aqueous extracts of rye. *Weed Science* 34: 384-390.

Barnes, J.P. and A.R. Putnam (1987). Role of benzoxazinones in allelopathy by rye (*Secale cereale* L.). *Journal of Chemical Ecology* 13: 889-906.

Barnes, J.P., A.R. Putnam, B.A. Burke, and A.J. Aasen (1987). Isolation and characterization of allelochemicals in rye herbage. *Phytochemistry* 26: 1385-1390.

Barreto, R., R. Charudattan, A. Pomella, and R. Hanada (2000). Biological control of neotropical aquatic weeds with fungi. *Crop Protection* 19: 697-703.

Batish, D.R., R.K. Kohli, H.P. Singh, and D.B. Saxena (1997). Studies on herbicidal activity of parthenin—A constituent of *Parthenium hysterophorus* toward billgoat weed. *Current Science* 73: 369-371.

Batish, D.R., H.P. Singh, and R.K. Kohli (2002). Utilization of allelopathic interactions for weed management. *Journal of Plant Disease and Protection* 18: 589-596.

Batish, D.R., H.P. Singh, R.K. Kohli, and S. Kaur (2001). Crop allelopathy and its role in ecological agriculture. *Journal of Crop Production* 4: 121-161.

Batish, D.R., H.P. Singh, R.K. Kohli, D.B. Saxena, and S. Kaur (2002). Allelopathic effects of parthenin against two weedy species, *Avena fatua* and *Bidens pilosa. Environmental and Experimental Botany* 47: 149-155.

Bilalis, D., N. Sidiras, G. Economou, and C. Vakali (2003). Effect of different levels of wheat straw soil surface coverage on weed flora in *Vicia faba* crops. *Journal of Agronomy and Crop Science* 189: 233-241.

Birkett, M.A., K. Chamberlain, A.M. Hooper, and J.A. Pickett (2001). Does allelopathy offer real promise for practical weed management and for explaining rhizosphere interactions involving higher plants? *Plant and Soil* 232: 31-39.

Blum, U., L.D. King, and C. Browne (2002). Effect of wheat residues on dicotyledonous weed emergence in a simulated no-till systems. *Allelopathy Journal* 9: 159-176.

Blum, U., T.R. Wentworth, K. Klein, A.D. Worsham, L.D. King, T.M. Gerig, and S.W. Lyu (1991). Phenolic acid content of soils of wheat-no till, wheat-conven-

tional till, and fallow-conventional till soybean cropping systems. *Journal of Chemical Ecology* 17: 1045-1068.
Boydston, R.A., K. Al-Khatib, A. Hang, G. Krishnan, and S. Nissen (1994). Weed control with rapeseed *(Brassica napus)* and white mustard *(Brassica hirta)* as green manure crops. *Weed Science Society America, Abstracts* 34: 89.
Boydston, R.A. and A. Hang (1995). Rapeseed *(Brassica napus)* green manure crop suppresses weeds in potato *(Solanum tuberosum). Weed Technology* 9: 669-675.
Boyetchko, S. (1999). Innovative applications of microbial agents for biological weed control. In K.G. Mukerji, B.P. Chamola, and K. Upadhyay (Eds.), *Biotechnological approaches in biocontrol of plant pathogens* (pp. 73-97). New York: Kluwer Academic Publishers.
Breland, T.A. (1996). Phytotoxic effects of fresh and decomposing cover crop residues. *Norwegian Journal of Agricultural Science* 10: 355-362.
Brown, P.D. and M.J. Morra (1995). Glucosinolate-containing plant tissues as bioherbicides. *Journal of Agricultural and Food Chemistry* 43: 3070-3074.
Brown, P.D. and M.J. Morra (1996). Hydrolysis products of glucosinolates in Brassica napus tissue as inhibitors of seed germination. *Plant and Soil* 181: 307-316.
Brown, P.D. and M.J. Morra (1999). Glucosinolate-derived allelochemicals in the soil environment. In S.S. Narwal (Ed.), *Allelopathy update,* Volume 2, *Basic and applied aspects* (pp. 65-89). New Delhi, India: Oxford and IBH Publishing Co. Pvt. Ltd.
Caamal-Maldonaldo, J.A., J.J. Jiménez-Osornio, A. Torres-Barragán, and A.L. Anaya (2001). The use of allelopathic legume cover and mulch species for weed control in cropping systems. *Agronomy Journal* 93: 27-36.
Campbell, G., J.D.H. Lambert, J.T. Arnason, and G.H.N. Towers (1982). Allelopathic properties of μ-terthienyl and phenylheptatriyne, naturally occurring compounds from species of Asteraceae. *Journal of Chemical Ecology* 8: 961-972.
Cardina, J. (1995). Biological weed management. In A.E. Smith (Ed.), *Handbook of weed management systems* (pp. 279-341). New York: Marcel Dekker.
Casini, V. and L. Olivero (2001). Allelopathic effects of legume cover crops on cogongrass (*Imperata brasiliensis* Tring.). *Allelopathy Journal* 8: 189-200.
Cespedes, C.L., M. Hoeneisen, M. Bittner, J. Becerra, and M. Silva (2001). Comparative study of ovatifolin antioxidant and growth inhibition activities. *Journal of Agricultural and Food Chemistry* 49: 4243-4251.
Charudattan, R. (1991). *The mycoherbicide approach with plant pathogens.* New York: Chapman and Hall.
Charudattan, R. (2001). Biological control of weeds by means of plant pathogens: Significance for integrated weed management in modern agroecology. *Biocontrol* 46: 229-260.
Chase, W.R., M.G. Nair, and A.R. Putnam (1991). 2,2' oxo-1,1'-azobenzene selective toxicity of rye (*Secale cereale* L.) allelochemicals to weed and crop species, II. *Journal of Chemical Ecology* 17: 9-19.
Chase, W.R., M.G. Nair, A.R. Putnam, and S.K. Mishra (1991). 2,2'-oxo-1,1' azobenzene: Microbial transformation of rye (*Secale cereale* L.) allelochemical in

field soils by *Acinetobacter calcoaceticus:* III. *Journal of Chemical Ecology* 17: 1575-1584.

Cheema, Z.A. and A. Khaliq (2000). Use of sorghum allelopathic properties to control weeds in irrigated wheat in semi-arid region of Punjab. *Agriculture, Ecosystems and Environment* 79: 105-112.

Chen, P.K. and G.R. Leather (1990). Plant growth regulatory activities of artemisinin and its related compounds. *Journal of Chemical Ecology* 16: 18667-18876.

Chikoye, D., V.M. Manyong, R.J. Carsky, F. Ekeleme, G. Gbehounou, and A. Ahanchede (2002). Response of speargrass *(Imperata cylindrica)* to cover crops integrated with hand weeding and chemical control in maize and cassava. *Crop Protection* 21: 145-156.

Chittapur, B.M., C.S. Hunshal, and H. Shenoy (2001). Allelopathy in parasitic weed management: Role of catch and trap crops. *Allelopathy Journal* 8: 147-160.

Choesin, D.N. and R.E. Boerner (1991). Allyl isothiocyanate release and allelopathic potential of *Brassica napus* (Brassicaceae). *American Journal of Botany* 78: 1083-1090.

Chou, C.H. (1993). The role of allelopathy in the diversity of plant-communities in Taiwan. *Botanical Bulletin of Academia Sinica* 34: 211-221.

Chou, C.H., F.J. Chang, and F.J. Oka (1991). Allelopathic potential of wild rice, *Oryza perennis. Taiwania* 36: 201-210.

Chou, C.H., Y.C. Chiang, and H.H. Cheng (1981). Autointoxication mechanism of *Oryza sativa:* III. Effect of temperature on phytotoxin production during rice straw decomposition in soil. *Journal of Chemical Ecology* 7: 741-752.

Chou, C.H. and Y.L. Kuo (1986). Allelopathic research of subtropical vegetation in Taiwan: III. Allelopathic exclusion of understorey by *Leucaena leucocephala* (Lam.) de Wit. *Journal of Chemical Ecology* 12: 1431-1438.

Chou, C.H. and H.J. Lin (1976). Autointoxication mechanism of *Oryza sativa:* I. Phytotoxic effects of decomposing rice residues in soil. *Journal of Chemical Ecology* 2: 353-367.

Chou, C.H. and Z.A. Patrick (1976). Identification and phytotoxicity of compounds produced during the decomposition of corn and rye residue in soil. *Journal of Chemical Ecology* 2: 369-387.

Christensen, S. (1995). Weed suppression ability of spring barley varieties. *Weed Research (Oxford)* 35: 241-247.

Chung, I.M., J.K. Ahn, and S.J. Yun (2001). Assessment of allelopathic potential on barnyard grass *(Echinochloa crus-galli)* of rice cultivars. *Crop Protection* 20: 921-928.

Chung, I.M., K.H. Kim, J.K. Ahn, S.C. Chun, C.S. Kim, J.T. Kim, and S.H. Kim (2002). Screening of allelochemicals on barnyard grass *(Echinochloa crus-galli)* and the identification of potentially allelopathic compounds from rice *(Oryza sativa)* variety hull extracts. *Crop Protection* 21: 913-920.

Chung, I.M. and D.A. Miller (1995). Natural herbicide potential of alfalfa residue on selected weed seed species. *Agronomy Journal* 87: 920-925.

Citola, M., A.K. Watson, and S.G. Hallet (1995). Discovery of an isolate of *Fusarium oxysporum* with potential to control *Striga hermonthica* in Africa. *Weed Research (Oxford)* 35: 303-309.

Copaja, S.V., H.M. Niemeyer, and S.D. Wratten (1991). Hydroxamic acid levels in Chilean and British wheat seedlings. *Annals of Applied Biology* 118: 223-227.

Cutler, H.G. (1988). Perspectives on discovery of microbial phytotoxins with herbicidal activity. *Weed Technology* 2: 525-532.

Czarnota, M.A., R.N. Paul, F.E. Dayan, C.I. Nimbal, and L.A. Weston (2001). Mode of action, localization of production, chemical nature and activity of sorgoleone: A potent PS II inhibitor in *Sorghum* spp. root exudates. *Weed Technology* 15: 813-825.

Czarnota, M.A., A.M. Rimando, and L.A. Weston (2003). Evaluation of root exudates of seven sorghum accessions. *Journal of Chemical Ecology* 29: 2073-2083.

Datta, S. and D.B. Saxena (2001). Pesticidal properties of parthenin (from *Parthenium hysterophorus*) and related compounds. *Pest Management Science* 57: 95-101.

Dayan, F. and J. Romagni (2001). Lichens as a potential source of pesticides. *Pesticide Outlook* 12: 229-232.

Dayan, F., J. Romagni, M. Tellez, A. Rimando, and S. Duke (1999). Managing weeds with natural products. *Pesticide Outlook* 10: 185-188.

Dayan, F., S.B. Watson, J.C.G. Galindo, A. Hernández, J. Dou, J.D. McChesney, and S.O. Duke (1999). Phytotoxicity of quassinoids: Physiological responses and structural requirements. *Pesticide Biochemistry and Physiology* 65: 15-24.

de Luque, P.A., J.C.G. Galindo, F.A. Macías, and J. Jorrín (2000). Sunflower sesquiterpene lactone model induce Orobanche cumana seed germination. *Phytochemistry* 53: 45-50.

Dilday, R.H., J.D. Mattice, K.A. Moldenhauer, and W. Yan (2001). Allelopathic potential in rice germplasm against ducksalad, redstem and barnyard grass. *Journal of Crop Production* 4(2): 287-301.

Dilday, R.H., W.G. Yan, K.A. Moldenhauer, and K.A. Gravois (1998). Allelopathic activity in rice for controlling major aquatic weeds. In M. Olofsdotter (Ed.), *Allelopathy in rice: Proceedings of the workshop on allelopathy in rice* (pp. 7-26). Manila, Philippines: International Rice Research Institute.

Dornbos, D.L. Jr., G.F. Spencer, and R. Miller (1990). Medicarpin delays alfalfa seed germination and seedling growth. *Crop Science* 30: 162-166.

Dudai, N., A. Poljakoff-Mayber, A.M. Mayer, E. Putievsky, and H.R. Lerner (1999). Essential oils as allelochemicals and their potential use as bioherbicides. *Journal of Chemical Ecology* 25: 1079-1089.

Duke, S.O. (1986). Naturally occurring chemical compounds as herbicides. *Reviews of Weed Science* 2: 15-44.

Duke, S.O., H.K. Abbas, M.V. Duke, H.J. Lee, K.C. Vaughn, T. Amagasa, and T. Tanaka (1996). Microbial phytotoxins as potential herbicides. *Journal of Environmental Science and Health* B31: 427-434.

Duke, S.O., F.E. Dayan, A.M. Rimando, K.K. Schrader, G. Aliotta, A. Oliva, and J.G. Romagni (2002). Chemicals from nature for weed management. *Weed Science* 50: 138-151.

Duke, S.O., F.E. Dayan, J.G. Romagni, and A.M. Rimando (2000). Natural products as sources of herbicides: Current status and future trends. *Weed Research* (Oxford) 40: 99-111.

Duke, S.O., B.E. Scheffler, F.E. Dayan, L.A. Weston, and E. Ota (2001). Strategies for using transgenes to produce allelopathic crops. *Weed Technology* 15: 826-834.

Duke, S.O., K.C. Vaughn, E.M. Groom, and H.N. Elsholy (1987). Artemisinin, a constituent of annual wormwood *(Artemisia annua)* is a selective phytotoxin. *Weed Science* 35: 499-505.

Dyck, E. and M. Liebman (1994). Soil fertility management as a factor in weed control: The effect of crimson clover residue, synthetic nitrogen fertilizer, and their interactions on emergence and early growth of lambsquarters and sweet corn. *Plant and Soil* 167: 227-237.

Ebana, K., W. Yan, R.H. Dilday, H. Namai, and K. Okuno (2001). Analysis of QTL associated with the allelopathic effect of rice using water-soluble extracts. *Breeding Science* 51: 47-51.

Eberlein, C.V., M.J. Morra, M.J. Guttieri, P.D. Brown, and J. Brown (1998). Glucosinolate production by five field-grown *Brassica napus* cultivars used as green manures. *Weed Technology* 12: 712-718.

Einhellig, F.A. (1996). Interactions involving allelopathy in cropping systems. *Agronomy Journal* 88: 886-893.

Elakovich, S.D. and J.W. Wooten (1989). Allelopathic potential of sixteen aquatic and wetland plants. *Journal of Aquatic Plant Management* 27: 78-84.

Eliott, L.F., V.L. Cochran, and R.I. Papendick (1981). Wheat residue and nitrogen placement effect on wheat growth in the greenhouse. *Soil Science* 131: 48-52.

Enache, A.J. and R.D. Ilnicki (1990). Weed control by subterranean clover *(Trifolium subterraneum)* used as a living mulch. Weed Technology 4: 534-538.

Entz, M.H., W.J. Bullied, and F. Katepa-Mupondwa (1995). Rotational benefits of forage crops in Canadian thistle cropping systems. *Journal of Production Agriculture* 8: 521-529.

Eskelsen, S.R. and G.D. Crabtree (1995). The role of allelopathy in buckwheat *(Fagopyrum sagittatum)* inhibition of Canada thistle *(Cirsium arvense)*. *Weed Science* 43: 70-74.

Fay, P.K. and W.B. Duke (1977). An assessment of allelopathic potential in Avena germplasm. *Weed Science* 25: 224-228.

Fenwick, G.R., R.K. Heaney, and W.J. Mullin (1983). Glucosinolates and their breakdown products in food and plants. *Critical Reviews in Food Science* 18: 123-201.

Fischer, H.-P. and D. Bellus (1983). Phytotoxicants from microorganisms and related compounds. *Pesticide Science* 14: 334-346.

Fischer, N.H., J.D. Weidenhamer, J.L. Riopel, L. Quijano, and M.A. Menelaou (1990). Stimulation of witchweed germination by sesquiterpene lactones: A structure-activity study. *Phytochemistry* 29: 2479-2483.

Foley, M.C. (1999). Genetic approaches to the development of cover crops for weed management. *Journal of Crop Production* 2: 77-93.

Forcella, F. and M.J. Lindstorm (1988). Movement and germination of weed seeds in ridge-till crop production systems. *Weed Science* 36: 56-59.

Forney, D.R., C.L. Foy, and D.D. Wolf (1985). Weed suppression in no-till alfalfa *(Medicago sativa)* by prior cropping of summer-annual grasses. *Weed Science* 33: 490-497.

Frank, P.A. and N. Dechoretz (1980). Allelopathy in dwarf spikerush *(Eleocharis coloradoensis). Weed Science* 28: 499-505.

Fujii, Y. (1992). The potential biological control of paddy weeds with allelopathy: Allelopathic effect of some rice varieties. In *Proceedings of the International Symposium on Biological Control and Integrated Management of Paddy and Aquatic Weeds in Asia,* October 19-25 (pp. 305-320). Tsukuba, Japan: National Agriculture Research Centre.

Fujii, Y. (1999a). Allelopathy of hairy vetch and Mucuna: Their application for sustainable agriculture. In C.H. Chou, G.R. Waller, and C. Reinhardt (Eds.), Biodiversity and allelopathy: From organisms to ecosystems in the Pacific (pp. 289-300). Taipei, Taiwan, Republic of China: Academia Sinica.

Fujii, Y. (1999b). Allelopathy of velvetbean: Determination and identification of L-DOPA as a candidate of allelopathic substances. In H.G. Cutler and S.J. Cutler (Eds.), *Biologically active natural products* (pp. 33-48). Boca Raton, FL: CRC Press.

Fujii, Y. (2001). Screening and future exploitation of allelopathic plants as alternative herbicides with special reference to hairy vetch. *Journal of Crop Production* 4(2): 257-275.

Gallandt, E.R., M. Liebman, and D.R. Huggins (1999). Improving soil quality: Implications for weed management. *Journal of Crop Production* 2: 95-121.

Gorski, P.M., J. Miersch, and M. Ploszynski (1991). Production and biological activity of saponins and canavanine in alfalfa seedlings. *Journal of Chemical Ecology* 17: 1135-1143.

Grayson, B.T., K.S. Williams, P.A. Freehauf, R.R. Pease, W.T. Ziesel, R.L. Sereno, and R.E. Reinsfelder (1987). The physical and chemical-properties of the herbicide cinmethylin (SD-95481). *Pesticide Science* 21: 143-153.

Grodzinsky, A.M. and M.A. Panchuk (1974). Allelopathic properties of crop residues of wheat-wheat grass hybrids. In A.M. Grodzinsky (Ed.), *Physiological-biochemical basis of plant interactions in phytocenoses* (pp. 51-55). Kiev: Naukova Dumka. [In Russian with English summary.]

Guenzi, W.D. and T.M. McCalla (1966). Phenolic acids in oats, wheat, sorghum and corn residues and their phytotoxicity. *Agronomy Journal* 58: 303-304.

Hanson, A.D., P.L. Traynor, K.M. Ditz, and D.A. Reicosky (1981). Gramine in barley forage—Effects of genotype and environment. *Crop Science* 21: 726-730.

Harborne, J.B. (1989). Introduction to biochemical ecology. New York: Academic Press.

Hashem, A. and S.W. Adkins (1996). Introduction of allelopathy to wheat: Effect of Triticum speltoides on two weeds of wheat. In *Proceedings of 11th Australian*

Weeds Conference (pp. 45-48). Melbourne, Australia: Australian Society of Agronomy.

Hassan, S.M., I.R. Aidy, A.O. Bastawisi, and A.E. Draz (1998). Weed management in rice using allelopathic rice varieties in Egypt. In M. Olofsdotter (Ed.), *Allelopathy in rice—Proceedings of the Workshop on Allelopathy in Rice* (pp. 27-37). Philippines: International Rice Research Institute.

Heap, I. (2005). International Survey of Herbicide Resistant Weeds. Available online at <www.weedscience.org/in.asp>.

Heisey, R.M. (1996). Identification of an allelopathic compound from *Ailanthus altissima* (Simaroubaceae) and characterization of its herbicidal activity. *American Journal of Botany* 83: 192-200.

Hilton, J.L. (1979). Research on the physiology and biology of weeds. *Weeds Today* 10: 5-6.

Hoagland, R.E. (1999). Allelopathic interactions of plants and pathogens. In F.A. Macías, J.C.G. Galindo, J.M.G. Molinillo, and H.G. Cutler (Eds.), *Recent advances in allelopathy,* Volume 1, *A science for the future* (pp. 423-450). Universidad de Cádiz, Spain: Servicio de Publicaciones.

Hoagland, R.E. (2001). Microbial allelochemicals and pathogens as bioherbicidal agents. *Weed Technology* 15: 835-857.

Hoffman, M.L., L.A. Weston, J.C. Snyder, and E.E. Regnier (1996a). Allelopathic influence of germinating seeds and seedlings of cover crops on weed species. *Weed Science* 44: 579-584.

Hoffman, M.L., L.A. Weston, J.C. Snyder, and E.E. Regnier (1996b). Separating the effects of sorghum *(Sorghum bicolor)* and rye *(Secale cereale)* root and shoot residues on weed development. *Weed Science* 44: 402-407.

Holt, J.S. (1995). Plant response to light: A potential tool for weed management. *Weed Science* 43: 472-482.

Hong, N.H., T.D. Xuan, E. Tsuzuki, T. Hiroyuki, M. Mitsuhiro, and T.D. Khanh (2003). Screening for allelopathic potential of higher plants from Southeast Asia. Crop Protection 22: 829-836.

Hoult, A.H.C. and J.V. Lovett (1993). Biologically active secondary metabolites of barley: III. A method for identification and quantification of hordenine and gramine in barley by high-performance liquid chromatography. *Journal of Chemical Ecology* 19: 2245-2254.

Iqbal, Z., S. Hiradate, A. Noda, S.-I. Isojima, and Y. Fujii (2002). Allelopathy of buckwheat: Assessment of allelopathic potential of extracts of aerial parts of buckwheat and identification of fagomine and other related alkaloids. *Weed Biology and Management* 2: 110-115.

James, K.L., P.A. Banks, and K.J. Karnok (1988). Interference of soybean, *Glycine max,* cultivars with sicklepod, *Cassia obtusifolia. Weed Technology* 2: 404-409.

Jensen, L.B., B. Courtois, L. Shen, Z. Li, M. Olofsdotter, and R.P. Mauleon (2001). Locating genes controlling allelopathic effects against barnyard grass in upland rice. *Agronomy Journal* 93: 21-26.

Jimenez, M. and R. Charudattan (1998). Survey and evaluation of Mexican native fungi for potential biocontrol of water hyacinth. *Journal of Aquatic Plant Management* 36: 145-148.

Jobidon, R., J.R. Thibault, and J.A. Fortin (1989). Effect of straw residues on black spruce seedling growth and mineral nutrition, under greenhouse conditions. *Canadian Journal of Forest Research* 19: 1291-1293.

Johnson, A.W., G. Rosebery, and C. Parker (1976). A novel approach to *Striga* and *Orobanche* control using synthetic germination stimulants. *Weed Research (Oxford)* 16: 223-227.

Johnson, G.A., M.S. Defelice, and Z.R. Helsel (1993). Cover crop management and weed control in corn. *Weed Technology* 7: 425-430.

Jones, E., R.S. Jessop, and B. Sindel (2001). The potential of summer crops to affect weed growth. In *Proceedings of the 10th Australian Agronomy Conference, Hobart.* Available online at <http://www.regional.org.au/au/asa/2001/p/12/jones.htm>.

Jones, E., R.S. Jessop, B.M. Sindel, and A. Hoult (1999). Utilising crop residues to control weeds. In A. Bishop, M. Boersma, and C.D. Bames (Eds.), *Proceedings 12th Australian Weed Conference* (pp. 373-376). Devonport, Australia: Tasmanian Weed Science Society.

Jordan, D.L., P.K. Bollich, M.P. Braverman, and D.E. Sanders (1999). Influence of tillage and *Triticum aestivum* cover crop on herbicide efficacy in *Oryza sativa. Weed Science* 47: 332-337.

Jordan, N. (1993). Prospects for weed control through crop interference. *Ecological Applications* 3: 84-91.

Joye, G.F. (1990). Biocontrol of *Hydrilla verticillata* with an endemic fungus *Macrophomina phaseolina. Plant Disease* 74: 1035-1036.

Kato-Noguchi, H., T. Ino, N. Sata, and S. Yamamura (2002). Isolation and identification of a potent allelopathic substance in rice. *Physiologia Plantarum* 115: 401-405.

Kawabata, J., Y. Fukushi, R. Hayashi, K. Suzuki, Y. Mishima, A. Yamane, and J. Mizutani (1989). 8-methylsulfinyloctyl isothiocyanate as allelochemical candidate from *Rorippa sylvestris* Besser. *Agricultural and Biological Chemistry* 53: 3361-3362.

Khan, A.H. and R.D. Vaishaya (1992). Allelopathic effects of different crop residues on germination and growth of weeds. In P. Tauro, and S.S. Narwal (Eds.), *Proceedings of National Symposium Allelopathy in Agroecosystems* (pp. 59-60). Haryana Agricultural University, Hisar, India: Indian Society of Allelopathy.

Khan, Z.R., K. Ampong-Nyarko, P. Chiliswa, A. Hassanali, S. Kimani, W. Lwande, W.A. Overholt, J.A. Pickett, L.E. Smart, L.J. Wadhams, and C.M. Woodcock (1997). Intercropping increases parasitism of pests. *Nature* 388: 631-632.

Khan, Z.R., A. Hassanali, W. Overholt, T.M. Khamis, A.M., Hooper, J.A. Pickett, L.J. Wadhams, and C.M. Woodcock (2002). Control of witchweed *Striga hermonthica* by intercropping with *Desmodium* spp., and the mechanism defined as allelopathic. *Journal of Chemical Ecology* 28: 1871-1885.

Kim, K.U. and D.H. Shin (1998). Rice allelopathy research in Korea. In M. Olofsdotter (Ed.), *Allelopathy in Rice—Proceedings of the Workshop on Allelopathy in Rice* (pp. 39-44). Philippines: International Rice Research Institute.

Kimber, R.W.L. (1973). Phytotoxicity from plant residues: II. The effect of time of rotting straw from some grasses and legumes on the growth of wheat seedlings. *Plant and Soil* 38: 347-361.

Kirby, C. (1980). *The hormone weed killers: A short history of their discovery and development.* Suffolk, UK: Lavenham Press, BCPC Publications.

Kobaisy, M., M.R. Tellez, F.E. Dayan, and S.O. Duke. (2002). Phytotoxicity and volatile constituents from leaves of *Callicarpa japonica* Thunb. *Phytochemistry* 61: 37-40.

Kocacaliskan, I. and I. Terzi (2001). Allelopathic effects of walnut leaf extracts and juglone on seed germination and seedling growth. *Journal of Horticultural Science and Biotechnology* 76: 436-440.

Kohli, R.K., D.R. Batish, and H.P. Singh (1998). Eucalypt oils for the control of parthenium (*Parthenium hysterophorus* L.). *Crop Protection* 17: 119-122.

Kremer, R.J. and A.C. Kennedy (1996). Rhizobacteria as biocontrol agents of weeds. *Weed Technology* 10: 601-609.

Krishnan, G., D.L. Holshouser, and S.J. Nissen (1998). Weed control in soybean (Soybean max) with green manure crops. *Weed Technology* 12: 97-102.

Leather, G.R. (1983a). Sunflowers *(Helianthus annuus)* are allelopathic to weeds. *Weed Science* 31: 37-42.

Leather, G.R. (1983b). Weed control using allelopathic crop plants. *Journal of Chemical Ecology* 9: 983-989.

Leather, G.R. (1987). Weed control using allelopathic sunflowers and herbicides. *Plant and Soil* 8: 17-23.

Lehman, M.E. and U. Blum (1997). Cover crop debris effects on weed emergence as modified by environmental factors. *Allelopathy Journal* 4: 69-88.

Li, Y., Z. Sun, X. Zhuang, L. Xu, S. Chen, and M. Li (2003). Research progress on microbial herbicides. *Crop Protection* 22: 247-252.

Liebl, R.A. and A.D. Worsham (1983). Inhibition of pitted morningglory (*Ipomoea lacunosa* L.) and certain other weed species by phytotoxic components of wheat (*Triticum aestivum* L.) straw. *Journal of Chemical Ecology* 9: 1027-1043.

Liebman, M. and A.S. Davis (2000). Integration of soil, crop and weed management in low-external-input farming system. *Weed Research* (Oxford) 40: 27-47.

Liebman, M. and E. Dyck (1993). Crop rotation and intercropping strategies for weed management. *Ecological Applications* 3: 92-122.

Lin, D., E. Tsuzuki, Y. Sugimoto, Y. Dong, M. Matsuo, and H. Terao (2003). Assessment of dwarf lilyturf (*Ophiopogon japonicus* K.) dried powders for weed control in transplanted rice. *Crop Protection* 22: 431-435.

Lin, J., R.J. Smith Jr., and R.H. Dilday (1992a). Allelopathic activity of rice germplasm on weeds. *Proceedings of Southern Weed Science Society* 45: 99.

Lin, J., R.J. Smith Jr., and R.H. Dilday (1992b). Comparison of allelopathy and herbicides for weed control in rice. In *Proceedings of 24th Rice Technical Working Group* (p. 127). Little Rock: Cooperative Extension Service, Division of Agriculture, University of Arkansas.

Lin, L.-J., G. Peiser, B.-P. Ying, F. Mathias, F. Karasina, Z. Wang, J. Itatani, L. Green, and Y.S. Hwang (1995). Identification of plant growth inhibitory principles in *Ailanthus altissima* and *Castela tortuosa*. *Journal of Agricultural and Food Chemistry* 43: 1708-1711.

Liu, D.L. and J.V. Lovett (1993a). Biologically active secondary metabolites of barley: I. Developing techniques and assessing allelopathy in barley. *Journal of Chemical Ecology* 19: 2217-2230.

Liu, D.L. and J.V. Lovett (1993b). Biologically active secondary metabolites of barley: II. Phytotoxicity of barley allelochemicals. *Journal of Chemical Ecology* 19: 2231-2244.

Lodhi, M.A.K., R. Bilal, and K.A. Malik (1987). Allelopathy in agroecosystems: Wheat phytotoxicity and its possible role in crop rotation. *Journal of Chemical Ecology* 13: 1881-1891.

Lovett, J.V. and A.H.C. Hoult (1992). Gramine: The occurrence of a self-defence chemical in barley, *Hordeum vulgare*. In *Proceedings of 6th Australian Agronomy Conference* (pp. 426-429). Armidale, Australia: Agronomy Society of Australia.

Lydon, J., and S.O. Duke (1999). Inhibitors of glutamine synthetase. In B.K. Singh (Ed.), *Plant amino acids: Biochemistry and biotechnology* (pp. 445-464). New York: Marcel and Dekker.

Lydon, J., J.R. Teasdale, and P.K. Chen (1997). Allelopathic activity of wormwood *(Artemisia annua)* and the role of artemisinin. *Weed Science* 45: 807-811.

Macias, F.A., J.M.G. Molinillo, J.C.G. Galindo, R.M. Varela, A.M. Simonet, and D. Castellano (2001). The use of allelopathic studies in the search for natural herbicides. *Journal of Crop Production* 4(2): 237-255.

Mallik, M.A.B. (2001). Selective isolation and screening of soil microorganisms for metabolites with herbicidal potential. *Journal of Crop Production* 4(2): 219-236.

Masiunas, J.B. (1999). Production of vegetables using cover crops and living mulches—A review. *Journal of Vegetable Crop Production* 4: 11-31.

Massantini, F., F. Caporali, and G. Zellini (1977). Evidence for allelopathic control of weeds in lines of soybean. In *Proceedings of the 2nd EWRS Symposium on the Different Methods of Weed Control and Their Integration,* Volume 1 (pp. 23-28). Uppsala, Sweden: European Weed Research Society.

Mattice, J., T. Lavy, B. Skulman, and R. Dilday (1998). Searching for allelochemicals in rice that control ducksalad. In M. Olofsdotter (Ed.), *Allelopathy in rice: Proceedings of the workshop on allelopathy in rice* (pp. 81-98). Manila, Philippines: International Rice Research Institute.

Miller, D.A. (1983). Allelopathic effects of alfalfa. *Journal of Chemical Ecology* 9: 1059-1072.

Miller, D.A. (1996). Allelopathy in forage crop systems. *Agronomy Journal* 88: 854-859.

Miller, R.W., R. Kleiman, R.G. Powell, and A.R. Putnam (1988). Germination and growth inhibitors of alfalfa. *Journal of Natural Products* 51: 328-330.

Misato, T. and I. Yamaguchi (1984). Pesticides of microbial origin. *Outlook on Agriculture* 13: 117-120.

Mitchell, G., D.W. Bartlett, T.E.M. Fraser, T.R. Hawkes, D.C. Holt, J.K. Townson, and R.A. Wichert (2001). Mesotrione: A new selective herbicide for use in maize. *Pest Management Science* 57: 120-128.

Mohan Babu, R., A. Sajeena, and K. Seetharaman (2003). Bioassay of the potentiality of *Alternaria alternata* (Fr.) Keissler as a bioherbicide to control waterhyacinth and other aquatic weeds. *Crop Protection* 22: 1005-1013.

Mohan Babu, R., A. Sajeena, K. Seetharaman, P. Vidhyasekaran, P. Rangasamy, M. Som Prakash, A. Senthil Raja, and K.R. Biji (2003). Advances in bioherbicides development—An overview. *Crop Protection* 22: 253-260.

Mohler, C.L. and J.R. Teasdale (1993). Response of weed emergence to rate of *Vicia villosa* Roth and *Secale cereale* L. residue. *Weed Research (Oxford)* 33: 487-499.

Molisch, H. (1937). *Der Einfluss einer Pflanze auf die andere—Allelopathie.* Jena: Fischer.

Moyer, J.R. and H.C. Huang (1997). Effect of aqueous extracts of crop residues on germination and seedling growth of ten weed species. *Botanical Bulletin of Academia Sinica* 38: 131-139.

Mwaja, V.N., J.B. Masiunas, and L.A. Weston (1995). Effect of fertility on biomass, phytotoxicity, and allelochemical content of cereal rye. *Journal of Chemical Ecology* 21: 81-96.

Narwal, S.S. (1999). Smothering crops as in weed management: Preliminary studies. In F.A. Macias, J.C.G. Galindo, J.M.G. Molinillo, and H.G. Cutler (Eds.), *Recent advances in allelopathy,* Volume I, *A science for the future* (pp. 365-382). Cádiz, Spain: Servicio de Publicaciones de la Universidad de Cádiz.

Narwal, S.S. (2000). Weed management in rice: Wheat rotation by allelopathy. *Critical Reviews in Plant Sciences* 19: 249-266.

Netzley, D.H., J.L. Reopel, G. Ejeta, and L. Butler (1988). Germination stimulants of witchweed *(Striga asiatica)* from hydrophobic root exudate of sorghum *(Sorghum bicolor). Weed Science* 36: 441-446.

Nicol, D., S.V. Copaja, S.D. Wratten, and H.M. Niemeyer (1992). A screen of worldwide wheat cultivars for hydroxamic acid levels and aphid antixenosis. *Annals of Applied Biology* 121: 11-18.

Nicollier, J.F., D.F. Pope, and A.C. Thompson (1983). Biological activity of dhurrin and other compounds from johnsongrass *(Sorghum halepense). Journal of Agricultural and Food Chemistry* 31: 744-748.

Niemeyer, H.M. (1988). Hydroxamic acid content of *Triticum* species. *Euphytica* 37: 289-293.

Niemeyer, H.M. and J.H. Jerez (1997). Chromosomal location of genes for hydroxamic acid accumulation in *Triticum aestivum* L. (wheat) using wheat aneuploids and wheat substitution line. *Heredity* 79: 10-14.

Nimbal, C.I., J. Pedersen, C.N. Yerkes, L.A. Weston, and S.C. Weller (1996). Phytotoxicity and distribution of sorgoleone in grain sorghum germplasm. *Journal of Agricultural and Food Chemistry* 44: 1343-1347.

Ohno, T. and K. Doolan (2001). Effects of red clover decomposition on phytotoxicity to wild mustard seedling growth. *Applied Soil Ecology* 16: 187-192.

Ohno, T., K. Doolan, L.M. Zibilske, M. Liebman, E.R. Gallandt, and C. Berube (2000). Phytotoxic effects of red clover amended soils on wild mustard seedling growth. *Agriculture, Ecosystem and Environment* 78: 187-192.

Oleszek, W. (1987). Allelopathic effects of volatiles from some cruciferae species on lettuce, barnyard grass and wheat growth. *Plant and Soil* 102: 271-273.

Oliva, A., R.M. Moraes, S.B. Watson, S.O. Duke, and F.E. Dayan (2002). Aryltetralin lignans inhibit plant growth by affecting microtubular organizing centers. *Pesticide Biochemistry and Physiology* 72: 45-54.

Olofsdotter, M. (1998). Allelopathy for weed control in organic farming. In N. El-Bassam, R.K. Behl, and B. Prochnow (Eds.), Sustainable agriculture for food, energy and industry: Strategies toward achievement (pp. 453-457). London, UK: James and James (Science Publishers) Ltd.

Olofsdotter, M. (2001). Rice—A step toward use of allelopathy. *Agronomy Journal* 93: 1-8.

Olofsdotter, M., L.B. Jensen, and B. Courtois (2002). Improving crop competitive ability using allelopathy—An example from rice. *Plant Breeding* 121: 1-9.

Olofsdotter, M. and D. Navarez (1996). Allelopathic rice for *Echinochloa crus-galli* control. In H. Brown, G.W. Cussans, M.D. Devine, S.O. Duke, C. Fernandez-Quintanilla, A. Helweg, R.E. Labrada, M. Landes, P. Kudsk, and J.C. Streibig (Eds.), *Proceedings of the 2nd International Weed Control Congress,* June 25-28 (pp. 1175-1181). Copenhagen, Denmark: DJF, Flakkebjerg.

Olofsdotter, M., D. Navarez, M. Rebulanan, and J.C. Streibig (1999). Weed suppressing rice cultivars—Does allelopathy play a role? *Weed Research (Oxford)* 39: 441-454.

Ominski, P.D., M.H. Entz, and N. Kenkel (1999). Weed suppression by *Medicago sativa* in subsequent cereal crops: A comparative survey. *Weed Science* 47: 282-290.

Oswald, A., H. Frost, J.K. Ransom, J. Kroschel, K. Shephard, and J. Sauerborn (1996). Studies on the potential for improved fallow using tree and shrubs to reduce *Striga* infestations in Kenya. In M.T. Moreno, J. Cubero, D. Berner, D. Joel, L.J. Musselman, and C. Parker (Eds.), *Proceedings of 6th International Symposium on Parasitic Weeds* (pp. 795-800). Cordoba, Spain: Dirección General de Investigación Agraria, Servicio de Publicaciones y Divulgación.

Oswald, A., J.K. Ransom, J. Kroschel, and J. Sauerborn (2002). Intercropping controls *Striga* in maize based farming systems. Crop Protection 21: 367-374.

Panasiuk, O., D.D. Bills, and G.R. Leather (1986). Allelopathic influence of *Sorghum bicolor* on weeds during germination and early development of seedling. *Journal of Chemical Ecology* 12: 1533-1543.

Panchuk, M.A. and N.L. Prutenskaya (1973). On the problem of the presence of allelopathic properties in wheat-wheat grass hybrids and their initial forms. In A.M. Grodzinsky (Ed.), *Physiological-biochemical basis of plant interactions in phytocenoses* (pp. 44-47). Kiev: Naukova Dumka. [In Russian with English Summary.]

Pandey, D.K. (1994). Inhibition of salvinia (*Salvinia molesta* Mitchell) by parthenium (*Parthenium hysterophorus* L.): I. Effect of leaf residues and allelochemicals. *Journal of Chemical Ecology* 20: 3111-3122.

Pandey, D.K. (1996). Phytotoxicity of sesquiterpene lactone parthenin on aquatic weeds. *Journal of Chemical Ecology* 22: 151-160.

Pandey, D.K., I.P. Kauraw, and V.M. Bhan (1993). Inhibitory effect of parthenium (*Parthenium hysterophorus* L.) residues on growth of water hyacinth (*Eichhornia crassipes* Mart. Solms.): I. Effect of leaf residues. *Journal of Chemical Ecology* 19: 2651-2662.

Park, K.H. (1996). A potential weed management by allelopathic rice germplasm in Korea. In *Abstracts First World Congress on Allelopathy* (p. 269). Cadiz, Spain: International Allelopathy Society.

Patrick, Z.A. (1971). Phytotoxic substances associated with decomposition in soil of plant residues. *Soil Science* 111: 13-18.

Patrick, Z.A., T.A. Tousson, and W.C. Snyder (1963). Phytotoxic substances in arable soils associated with decomposition of plant residues. *Phytopathology* 53: 152-161.

Pereda-Miranda, R.R., A.L. Mata, A.L. Anaya, J.M. Pezzuto, D.B.M. Wickramaratne, and A.D. Kinghorn (1993). Structure and biological activities of tricolorin A, major phytogrowth-inhibitor from *Ipomoea tricolor. Journal of Natural Products* 56: 571-582.

Pérez, F.J. (1990). Allelopathic effects of hydroxamic acids from cereals on *Avena sativa* and *A. fatua. Phytochemistry* 29: 773-776.

Pérez, F.J. and J. Ormeño-Nuñez (1991). Differences in hydroxamic acids content in roots and root exudates of wheat (*Triticum aestivum* L.) and rye (*Secale cereale* L.): Possible role in allelopathy. *Journal of Chemical Ecology* 17: 1037-1043.

Pérez, F.J. and J. Ormeño-Nuñez (1993). Weed growth interference from temperate cereals: The effect of a hydroxamic-acids exuding rye (*Secale cereale* L.) cultivar. *Weed Research (Oxford)* 33: 115-119.

Petersen, J., R. Belz, F. Walker, and K. Hurle (2001). Weed suppression by release of isothiocyanates from turnip-rape mulch. *Agronomy Journal* 93: 37-43.

Pickett, J.A. (1999). Pest control that helps control weeds at the same time. Biotechnology and Biological Sciences Research Council (BBSRC) Business, April 16-17.

Pimentel, D., S. McNair, J. Janecka, J. Wightman, C. Simmonds, C. O'Connell, E. Wong, L. Russel, J. Zern, T. Aquino, and T. Tsomondo (2001). Economic and environmental threats of alien plant, animal, and microbe invasions. *Agriculture, Ecosystems and Environment* 84: 1-20.

Prezepiorkowski, T. and S.F. Gorski (1994). Influence of rye *(Secale cereale)* plant residue on germination and growth of three triazine-resistant and susceptible weeds. *Weed Technology* 8: 744-747.

Putnam, A.R. (1986). Allelopathy: Can it be managed to benefit horticulture? *HortScience* 21: 411-412.

Putnam, A.R. (1988). Allelochemicals from plants as herbicides. *Weed Technology* 2: 510-518.

Putnam, A.R. (1990). Vegetable weed control with minimal herbicide inputs. *HortScience* 25: 165-169.

Putnam, A. R. and J. DeFrank (1983). Use of phytotoxic plant residues for selective weed control. *Crop Protection* 2: 173-181.

Putnam, A.R., J. DeFrank, and J.P. Barnes (1983). Exploration of allelopathy for weed control in annual and perennial cropping systems. *Journal of Chemical Ecology* 9: 1001-1010.

Putnam, A.R. and W.B. Duke (1974). Biological suppression of weeds: Evidence for allelopathy in accessions of cucumber. *Science* 185: 370-372.

Qasem. J.R. and C.L. Foy (2001). Weed allelopathy, its ecological impacts and future prospects: A review. *Journal of Crop Production* 4: 43-119.

Quader, M., G. Daggard, R. Barrow, S. Walker, and M.W. Sutherland (2001). Allelopathy, DIMBOA production and genetic variability in accessions of *Triticum speltoides*. *Journal of Chemical Ecology* 27: 747-760.

Rambakudzibga, K.M. (1991). Allelopathic effects of aqueous wheat (*Triticum aestivum* L.) straw extracts on eight arable weeds commonly found in Zimbabwe. *Zimbabwe Journal of Agricultural Science* 29: 77-79.

Reberg-Horton, S.C., N.G. Creamer, J. Burton, N. Ranells, and C.L. Mohler (2001). Bioassay and field evaluation of rye cultivars for allelopathy. *Weed Science Society of America, Abstracts* 41: 109.

Reberg-Horton, S.C., N.G. Creamer, J. Burton, N. Ranells, and P.L. Murphy (2001). Breeding rye *(Secale cereale)* for increased allelopathy. *HortScience* 36: 561.

Reberg-Horton, S.C., N.G. Creamer, and C.L. Mohler (2000). The suppression of weeds by nine varieties of rye *(Secale cereale)*. *Weed Science Society of America, Abstracts* 40: 27-28.

Reicosky, D.C. and F. Forcella (1998). Cover crop and soil quality interactions in agroecosystems. *Journal of Soil and Water Conservation* 53: 224-229.

Rice, E.L. (1984). *Allelopathy,* Second edition. Orlando, FL: Academic Press, Inc.

Rice, E.L. (1995). *Biological control of weeds and plant diseases: Advances in applied allelopathy.* Norman: University of Oklahoma Press.

Rimando, A.M., F.E. Dayan, J.R. Mikell, and R.M. Moraes (1999). Phytotoxic lignans of *Leucophyllum frutescens*. *Natural Toxins* 7: 39-43.

Rimando, A.M., M. Olofsdotter, F.E. Dayan, and S.O. Duke (2001). Searching for rice allelochemicals: An example of bioassay guided isolation. *Agronomy Journal* 93: 16-20.

Rizvi, S.J.H., D. Mukerji, and S.N. Mathur (1981). 1,3,7-Trimethylxanthine—A new natural herbicide. *Agricultural and Biological Chemistry* 54: 1255-1256.

Rizvi, S.J.H., V. Rizvi, D. Mukerji, and S.N. Mathur (1987). 1,3,7-trimethylxanthine, an allelochemical from seeds of *Coffea arabica,* some aspects of its mode of action as a natural herbicide. *Plant and Soil* 98: 81-91.

Rizvi, S.J.H., M. Tahir, V. Rizvi, R.K. Kohli, and A. Ansari (1999). Allelopathic interactions in agroforestry systems. *Critical Reviews in Plant Sciences* 18: 773-796.

Romagni, J.G., S.N. Allen, and F.E. Dayan (2000). Allelopathic effects of volatile cineoles on two weedy plant species. *Journal of Chemical Ecology* 26: 303-313.

Saxena, M.K. (2000). Aqueous leachates of *Lantana camara* kills water hyacinth. *Journal of Chemical Ecology* 26: 2435-2447.

Saxena, S. and A.K. Pandey (2001). Microbial metabolites as eco-friendly agrochemicals. *Applied Microbiology and Biotechnology* 55: 395-403.

Scheffler, B.E., S.O. Duke, F.E. Dayan, and E. Ota (2001): Crop allelopathy: Enhancement through biotechnology. *Recent Advances in Phytochemistry* 35: 257-274.

Schreiber, M.M. (1992). Influence of tillage, crop rotation, and weed management on giant foxtail *(Setaria faberii)* population dynamics and corn yield. *Weed Science* 40: 645-653.

Seigler, D.S. (1996). Chemistry and mechanism of allelopathic interactions. *Agronomy Journal* 88: 876-885.

Shabana, Y.M., R. Charudattan, and M.A. Elwakil (1995). Evaluation of *Alternaria eichhorniae* as a bioherbicide for water hyacinth *(Eichhornia crassipes)* in greenhouse trials. *Biological Control* 5: 136-144.

Sheaffer C.C., J.L. Gunsolus, J.G. Jewett, and S.H. Lee (2002). Annual *Medicago* as a smother crop in soybean. *Journal of Agronomy and Crop Science* 188: 408-416.

Shettel, N.L. and N.E. Balke (1983). Plant growth response to several allelopathic chemicals. *Weed Science* 31: 293-298.

Shilling, D.G., R.A. Liebl, and A.D. Worsham (1985). Rye (*Secale cereale* L.) and wheat (*Triticum aestivum* L.) mulch: The suppression of certain broad-leaved weeds and the isolation and identification of phytotoxins. In A.C. Thompson (Ed.), *The chemistry of allelopathy: Biochemical interactions among plants* (pp. 243-271). ACS Symposium Series 268. Washington, DC: American Chemical Society.

Singh, H.P., D.R. Batish, S. Kaur, H. Ramezani, and R.K. Kohli (2002a). Comparative phytotoxicity of four monoterpenes against *Cassia occidentalis. Annals of Applied Biology* 141: 111-116.

Singh, H.P., D.R. Batish, S. Kaur, H. Ramezani, and R.K. Kohli (2002b). Potential utilization of volatile oils from *Eucalyptus citriodora* Hook. for weed management. *Journal of Plant Disease and Protection* 18: 607-614.

Singh, H.P., D.R. Batish, and R.K. Kohli (2001). Allelopathy in agroecosystems: An overview. *Journal of Crop Production* 4(2): 1-41.

Singh, H.P., D.R. Batish, and R.K. Kohli (2002). Allelopathic effects of two volatile monoterpenes against bill goat weed (*Ageratum conyzoides* L). *Crop Protection* 21: 347-350.

Singh, H.P., D.R. Batish, and R.K. Kohli (2003). Allelopathic interactions and allelochemicals: New possibilities for sustainable weed management. *Critical Reviews in Plant Sciences* 22: 239-311.

Singh, H.P., D.R. Batish, R.K. Kohli, D.B. Saxena, and V. Arora (2002). Effect of parthenin—A sesquiterpene lactone from *Parthenium hysterophorus,* on early growth and physiology of *Ageratum conyzoides. Journal of Chemical Ecology* 28: 2169-2179.

Smeda, R.J. and S.C. Weller (1996). Potential of rye *(Secale cereale)* for weed management in transplant tomatoes *(Lycopersicon esculentum). Weed Science* 44: 596-602.

Smither-Kopperl, M.L., R. Charudattan, and R.D. Berger (1998). Dispersal of spores of *Fusarium culmorum* in aquatic systems. *Phytopathology* 88: 382-388.

Solymosi, P. (1994). Crude plant extracts as weed biocontrol agents. *Acta Phytopathologica et Entomologica Hungarica* 29: 361-370.

Spruell, J.A. (1984). Allelopathic potential of wheat accessions. *Dissertation Abstracts B (Science and Engineering)* 45: 1102B.

Steinsiek, J.W., L.R. Oliver, and F.C. Collins (1982). Allelopathic potential of wheat *(Triticum aestivum)* straw on selected weed species. *Weed Science* 30: 495-497.

Sugimoto, Y., S.C.M. Wigchert, J.W.J.F. Thuring, and B. Zwanenburg (1998). Synthesis of all eight stereoisomers of germination stimulant sorgolactone. *Journal of Organic Chemistry* 63: 1259-1267.

Sutton, D.L. and K.M. Portier (1989). Influence of allelochemicals and aqueous plant extracts on growth of duckweed. *Journal of Aquatic Plant Management* 27: 90-95.

Szezpanski, A.J. (1977). Allelopathy as a means of biological control of waterweeds. *Aquatic Botany* 3: 193-197.

Teasdale, J.R. (1993). Interaction of light, soil moisture, and temperature with weed suppression by hairy vetch residues. *Weed Science* 41: 46-51.

Teasdale, J.R. (1996). Contribution of cover crops to weed management in sustainable agricultural systems. *Journal of Production Agriculture* 9: 457-479.

Teasdale, J.R. and C.L. Mohler (2000). The quantitative relationship between weed emergence and the physical properties of mulches. *Weed Science* 48: 385-392.

Thijssen, R. (1995). Weeds and trees. *ILEIA Newsletter* 11: 20.

Thilsted, D.E. and B.S. Murray (1980). Effect of wheat straw on weed control in no-tillage soybeans. *Proceedings of Southern Weed Science Society* 33: 42.

Thomas, H., A. Heller, J. Sauerborn, and D. Müller-Stöver (1999). *Fusarium oxysporum* f. sp. *orthoceras,* a potential mycoherbicide, parasitizes seeds of *Orobanche cumana* (sunflower broomrape): A cytological study. *Annals of Botany (London)* 83: 453-458.

Tipton, C.I., M.C. Wang, F.H.C. Tsao, T.C.C. Lin, and R.R. Husted (1973). Biosynthesis of 1,4-benzoxazin-3-one in *Zea mays*. *Phytochemistry* 12: 347-352.

Toroser, D., C.E. Thormann, T.C. Osborn, and R. Mithen (1995). RFLP mapping of quantitative trait loci controlling seed aliphatic-glucosinolate content in oil seed rape (*Brassica napus* L.). *Theoretical and Applied Genetics* 91: 802-808.

Tsanuo, M.K., A. Hassanali, A.M. Hooper, Z. Khan, F. Kaberia, J.A. Pickett, and L.J. Wadhams (2003). Isoflavonones isolated from the allelopathic root exudates of *Desmodium uncinatum*. *Phytochemistry* 64: 265-273.

Tsuzuki, E., M. Miura, N. Sakaki, and T. Yoshino (1999). Study on the control of weeds by using higher plants. *Report of Kyushu Branch of the Crop Science Society of Japan* 65: 39-40.

Tsuzuki, E., Y. Yamamoto, and T. Shimizu (1987). Fatty acids in buckwheat are growth inhibitors. *Annals of Botany* 60: 69-70.

Tworkoski, T. (2002). Herbicide activity of essential oils. *Weed Science* 50: 425-431.

Udensi, E.A., I.O. Akobundu, A.O. Ayeni, and D. Chikoye (1999). Management of cogongrass *(Imperata cylindrica)* with velvetbean (*Mucuna pruriens* var. *utilis*) and herbicides. *Weed Technology* 13: 201-208.

Vaccarini, C.E. and G.M. Bonetto (2000). Selective phytotoxic activity of withanolides from *Iochroma australe* to crop and weed species. *Journal of Chemical Ecology* 26: 2187-2196.

Vaccarini, C.E., S.M. Palacios, K.M. Meragelman, and V.E. Sosa (1999). Phytogrowth-inhibitory activities of a clerodane from *Viguiera tucumanensis. Phytochemistry* 50: 227-230.

Vasil, I.K. (1996). Phosphinothricin-resistant crops. In S.O. Duke (Ed.), *Herbicide resistant crops* (pp. 85-91). Boca Raton, FL: CRC Press.

Vaughn, S.F. and R.A. Boydston (1997). Volatile allelochemicals released by crucifer green manures. *Journal of Chemical Ecology* 23: 2107-2116.

Vaughn, S.F. and G.F. Spencer (1993). Volatile monoterpenes as potential parent structures for new herbicides. *Weed Science* 41: 114-119.

Vyvyan, J.R. (2002). Allelochemicals as leads for new herbicides and agrochemicals. *Tetrahedron* 58: 1631-1646.

Waddington, J. (1978). Growth of barley, bromegrass and alfalfa in the greenhouse in the soil containing rapeseed and wheat residues. *Canadian Journal of Plant Science* 58: 241-248.

Waller, G.R., M. Jurzysta, and R.L.Z. Thorne (1993). Allelopathic activity of root saponins from alfalfa *(Medicago sativa)* against weeds and wheat. *Botanical Bulletin of Academia Sinica* 34: 1-11.

Waller, G.R., M. Jurzysta, and R.L.Z. Thorne (1995). Root saponins from alfalfa (*Medicago sativa* L.) and their allelopathic activity on weeds and wheat. *Allelopathy Journal* 2: 21-30.

Waller, G.R., E.G., Krenzer, Jr., J.K. McPherson, and S.R. McGown (1987). Allelopathic compounds in soil from no-tillage vs. conventional tillage in wheat production. *Plant and Soil* 98: 5-15.

Watson, A.K. (1991). The classical approach with plant pathogens. In D.O. TeBeest (Ed.), *Microbial control of weeds* (pp. 3-23). New York: Chapman and Hall.

Wegmann, K. (1998). Progress in *Orobanche* research during the past decade. In K. Wegmann, L.J. Musselman, and D.M. Joel (Eds.), *Current problems of Orobanche research: Proceedings of the 4th International* Orobanche *Workshop* (pp. 13-17). Albena, General Toshevo, Bulgaria: Institute for Wheat and Sunflower "Dobroudja."

Welzel, P., S. Röhrig, and Z. Milkova (1999). Strigol-type germination stimulants: The C-2' configuration problems. *Chemical Communications Issue* 20: 2017-2022.

Weston, L.A. (1990). Cover crops and herbicide influence on row crop seeds establishment in no-tillage culture. *Weed Science* 38: 166-171.

Weston, L.A. (1996). Utilization of allelopathy for weed management in agroecosystems. *Agronomy Journal* 88: 860-866.

Weston, L.A. and M.A. Czarnota (2001). Activity and persistence of sorgoleone, a long-chain hydroquinone produced by *Sorghum bicolor. Journal of Crop Production* 4(2): 363-377.

Weston, L.A., R. Harmon, and S. Mueller (1989). Allelopathic potential of sorghum-sudangrass hybrid (Sudex). *Journal of Chemical Ecology* 15: 1855-1865.

Weston, L.A., X. Yang, and B.E. Scheffler (2002). Gene regulation by bioactive root exudates produced by *Sorghum* sp. In *Proceedings of 3rd World Allelopathy Congress*. Tsukuba, Japan: International Allelopathy Society. Abstract no. 76.

White, R.H., A.D. Worsham, and U. Blum (1989). Allelopathic potential of legume debris and aqueous extracts. *Weed Science* 37: 674-679.

Whitenack, C.J., M.G. Nair, and A.R. Putnam (1988). 2,2 -oxo-1,1 azobenzene: A potential allelopathic compound from breakdown of rye (*Secale cereale* L.) residues. *Weed Science Society of America, Abstracts* 28: 144.

Wild, A. and C. Ziegler (1989). The effects of bialaphos on ammonium-assimilation and photosynthesis: I. Effect on the enzymes of ammonium assimilation. *Zeitschrift für Naturforschung* C 44: 97-102.

Williams, R.D. and R.E. Hoagland (1982). The effects of naturally occurring phenolic compounds on seed germination. *Weed Science* 30: 206-212.

Willis, R.J. (1995). Pioneers of allelopathy: IV. Cornelius H. Muller. *Allelopathy Journal* 2: 109-116.

Wojcik-Wojtkowiak, D., B. Politycka, M. Schneider, and J. Perkowski (1990). Phenolic substances as allelopathic agents arising during the degradation of rye (Secale cereale) tissues. *Plant and Soil* 124: 143-147.

Wolf, R.B., G.F. Spencer, and W.F. Kwolek (1984). Inhibition of velvetleaf (Abutilon theophrasti) germination and growth by benzyl isothiocyanate, a natural toxicant. *Weed Science* 32: 612-615.

Wolf, R.B., G.F. Spencer, and R.D. Plattner (1985). Benzoxazolinone-2-4-dihydroxy-1,4-benzoxazin-3-one, and its glucoside from *Acantus mollis* seeds inhibit velvetleaf germination and growth. *Journal of Natural Products* 48: 59-63.

Worsham, A.D. (1984). Crop residues kill weeds: Allelopathy at work with wheat and rye. *Crops Soils* 37: 18-20.

Worsham, A.D. and U. Blum (1992). Allelopathic cover crops to reduce herbicide inputs in cropping systems. In R.G. Richardson (Ed.), *Proceedings of the First International Weed Control Congress* (pp. 577-579). Melbourne, Australia: Weed Science Society of Victoria.

Wu, H., T. Haig, J. Pratley, D. Lemerle, and M. An (2000). Distribution and exudation of allelochemicals in wheat *Triticum aestivum. Journal of Chemical Ecology* 26: 2141-2154.

Wu, H., T. Haig, J. Pratley, D. Lemerle, and M. An (2002). Chemical basis for wheat seedling allelopathy on the suppression of annual ryegrass *(Lolium rigidum). Journal of Agricultural and Food Chemistry* 50: 4567-4571.

Wu, H., J. Pratley, and T. Haig (2003). Phytotoxic effects of wheat extracts on herbicide resistant biotypes of annual ryegrass *(Lolium rigidum). Journal of Agricultural and Food Chemistry* 51: 4610-4616.

Wu, H., J. Pratley, D. Lemerle, and T. Haig (1999). Crop cultivars with allelopathic capability. *Weed Research* (Oxford) 39: 171-180.

Wu, H., J. Pratley, D. Lemerle, and T. Haig (2000). Evaluation of seedling allelopathy in 453 wheat *(Triticum aestivum)* accessions by Equal-Compartment-Agar-Method. *Australian Journal of Agricultural Research* 51: 937-944.

Wu, H., J. Pratley, D. Lemerle, and T. Haig (2001). Allelopathy in wheat *(Triticum aestivum). Annals of Applied Biology* 139: 1-9.

Wu, H., J. Pratley, D. Lemerle, T. Haig, and B. Verbeek (1998). Differential allelopathic potential among wheat accessions to annual ryegrass. In D.L. Michalk and J.E. Pratley (Eds.), *Proceedings of the 9th Australian Agronomy Conference* (pp. 567-571). Wagga Wagga, New South Wales: Australian Agronomy Society.

Wyman-Simpson, C.L., G.R. Waller, M. Jurzysta, J.K. McPherson, and C.C. Young (1991). Biological activity and chemical isolation of root saponins of six cultivars of alfalfa (*Medicago sativa* L.). *Plant and Soil* 135: 83-94.

Xuan, T.D. and E. Tsuzuki (2001). Effect of application of alfalfa pellet on the germination and growth of weeds. *Journal of Crop Production* 4: 303-312.

Xuan, T.D. and E. Tsuzuki (2002). Varietal differences in allelopathic potential of alfalfa. *Journal of Agronomy and Crop Science* 188: 2-7.

Xuan, T.D., E. Tsuzuki, H. Terao, M. Mitsuo, T.D. Khanh, S. Murayama, and N.H. Hong (2003). Alfalfa, rice by-products and their incorporation for weed control in rice. *Weed Biology and Management* 3: 137-144.

Xuan, T.D., E. Tsuzuki, H. Uematsu, and H. Terao (2001). Weed control with alfalfa pellets in transplanting rice. *Weed Biology and Management* 1: 231-235.

Yamada, O., S. Ishida, F. Futatsuya, K. Ito, H. Yamamoto, and K. Munakata (1974). Plant growth regulating activities of 4-methoxydiphenylmethams and their related compounds. *Agricultural and Biological Chemistry* 38: 1235-1237.

Yamada, O., Y. Kaise, F. Futatsuya, S. Ishida, K. Ito, H. Yamamoto, and K. Munakata (1972). Studies on plant growth regulating activities of anisomycin and toyocamycin. *Agricultural and Biological Chemistry* 36: 2013-2015.

Yang, X., B.E. Scheffler, and L.A. Weston (2001). Analysis of gene expression related to sorgoleone production using mRNA differential display. *Weed Science Society of America, Abstracts* 41: 37.

Yenish, J.P., A.D. Worsham, and W.S. Chilton (1995). Disappearance of DIBOA-glucoside, DIBOA, and BOA from rye (*Secale cereale* L.) cover crop residue. *Weed Science* 43: 18-20.

Yeo, R.R. and T.W. Fisher (1970). Progress and potential for biological weed control with fish, pathogens, competitive plants, and snails. In J.T. Holstun (Ed.), *Proceedings of FAO International Conference on Weed Control* (pp. 450-463). Champaign, IL: Weed Science Society of America.

Yokota, T., H. Sakai, K. Okuno, K. Yoneyama, and Y. Takeuchi (1998). Alectrol and orobanchol, germination stimulants for *Orobanche minor,* from its host red clover. *Phytochemistry* 49: 1967-1973.

Zonno, M.C. and M. Vurro (1999). Effect of fungal toxins on germination of *Striga hermonthica. Weed Research* (Oxford) 39: 15-20.

Zuniga, G.E., S.V. Copaja, H.R. Bravo, and V.H. Argandona (1990). Hydroxamic acids accumulation by wheat callus. *Phytochemistry* 29: 2139-2141.

Chapter 10

Progress in Developing Weed-Suppressive Rice Cultivars for the Southern United States

David R. Gealy
Karen A. Moldenhauer

RICE IN THE UNITED STATES

For thousands of years, rice (*Oryza sativa* L.) has been among the most important crop species. Currently, it is the main food source for more than half of the earth's population, with Asia producing and consuming about 90 percent of all rice (Singh and Khush, 2000). Approximately 596 million metric tons of rough rice is produced annually on 155 million ha worldwide (IRRI, 2002). In the United States 9.5 metric tons of rice are now harvested from about 1.3 million ha in two major regions: the southern Mississippi valley/Gulf Coast (Missouri ~6 percent of total ha; Arkansas ~49 percent; Mississippi ~8 percent; Louisiana ~17 percent; and Texas ~6 percent), and California (~14 percent) (NASS, 2002). The United States exports more than 40 percent of its rice production, ranking fourth in the world among rice exporters according to recent estimates (Childs and Burdett, 2000).

Rice production systems in the United States fall into two main groups, both being irrigated "lowland" systems as distinct from rainfed "upland" systems in many other parts of the world. California systems produce primarily medium grain japonica rice, and to a much lesser extent, short grain rice, using water-seeded, aerial planting systems. Medium grain and short grain rice accounts for about 24 percent and 1 percent, respectively, of the total U.S. rice production Tropical japonica rice is typically grown in the southern United States. It is predominantly long grain, which accounts for about 75 percent of all U.S. rice (Street and Bollich, 2003). In Missouri, Arkansas, Mississippi, north Louisiana, and Texas rice is typically produced in drill-seeded (dry-seeded) systems. However, much of the Gulf Coast area in southwest Louisiana is predominantly water seeded. Rice in Texas, southwest Louisiana, and Florida is typically ratoon cropped (new panicles allowed to grow after first harvest) late in the summer due to the longer growing seasons in these sections of the rice-producing area (Street and Bollich, 2003). Most rice in the South is produced using conventional tillage systems, but conservation systems such as no-till or stale seedbed are becoming more prevalent (Street and Bollich, 2003).

More than 70 species of weeds infest rice fields in the United States (Smith, 1989). In the southern United States, the most common and economically important of these species in direct-seeded rice is barnyardgrass [*Echinochloa crus-galli* (L.) P. Beauv.] (Smith, Flinchum, and Seaman, 1977; Chandler, 1981; Smith, 1988). In California, where rice is predominantly water seeded, the *Echinochloa* complex of early watergrass [*E. oryzoides* (Ard.) Fritsch] and late watergrass [*E. phyllopogon* (Stapf.) Stapf. ex Koss] are the most problematic weeds (Fischer et al., 2000). *Echinochloa* spp. will be a primary focus of discussion in this chapter because they are important in rice production in the United States and throughout the world.

Red rice (*Oryza sativa* L.) is a highly aggressive competitor and crop mimic of cultivated rice (*Oryza sativa* L.), generally producing tall plants with large numbers of tillers. Although red rice infests fewer hectares in the South than barnyardgrass, it is even more troublesome when present (Bridges and Baumann, 1992; Dowler, 1997; Webster, 2000), especially in drill-seeded systems. This weed is such a severe problem in parts of southwest Louisiana that preplant tillage operations are performed in puddled, muddy water ("mudded in") to control emerged red rice before rice is planted (Street and Bollich, 2003). Red rice is not present in California, largely because of water-seeding practices. Pantone and Baker (1991) have shown that the competitive ability of one red rice plant is equivalent to that of three rice plants.

Essentially all U.S. rice is produced in high-input, high-yielding systems that produce rice suitable for the world export market. Thus, government subsidies for U.S. rice are high by world standards, exceeding 40 percent of farmers' total receipts through direct or indirect payments; however, these subsidy levels are only about half of those for rice in Japan and Korea (Sumner, 2003). These high-quality, high-yield, high-subsidy conditions place U.S. rice production in a substantially different setting than that found in many other areas of the world. Although the U.S. industry can direct substantial resources into rice production, the final product must be of very high quality and produced in great abundance in order to satisfy the economic needs of the complex system involving the growers, rice mills, marketers, and consumers.

Growers and agricultural experts in the United States and throughout the world consistently indicate that weeds are one of the most economically important pests of rice. It has been estimated that total annual world losses in rice due to weeds and their control amount to 15 percent of the crop's value (Smith, 1983). Annual losses due to all weeds in rice fields in the southern United States have been estimated at $45 million when growers use current "best management practices," as recommended by extension and industry specialists, along with herbicides (Bridges and Baumann, 1992). However, if herbicides were to be omitted entirely from rice production in the southern United States, the estimated loss balloons to $640 million, amounting to more than two-thirds of the $900 million average total annual value of southern U.S. rice from 1998 to 2002 (NASS, 2002). If this large benefit from herbicides could be partly substituted with low-cost alternatives such as growing weed-suppressive cultivars, rice could be produced more economically and with greater flexibility. Further, herbicide use can produce unintended adverse consequences such as toxicity to off-site or nontarget flora and fauna. Repeated herbicide use can select for resistant weed populations (Valverde and Itoh, 2001; Valverde, Riches, and Caseley, 2000), and recent trends in the agricultural chemical industry have led to the existence of fewer companies developing and marketing fewer herbicides with new and different modes of action.

Herbicide use is a mainstay in modern U.S. rice production, with about 20 different herbicides presently available or under development for use (Kendig, Williams, and Smith, 2003). These include acifluorofen (Blazer), bensulfuron (Londax), bentazon (Basagran), bispyribac (experimental; Regiment proposed), carfentrazone (Aim), clefoxydim (experimental; Aura proposed), clomazone (Command), cyhalofop (experimental; Clincher proposed), fenoxaprop (Whip 360), halosulfuron (Permit), molinate (Molinate), pendimethalin (Prowl), propanil (Stam and others), quinclorac (Facet), thiobencarb (Bolero), triclopyr (Grandstand), and 2,4-D. Propanil has long been the main herbicide used in drill-seeded systems of the South, but clomazone use has become increasingly widespread because of its low cost and effectiveness against major grass weeds. Rice cultivars resistant to imazethapyr are now being marketed as 'Clearfield' cultivars in the United States, and cultivars with resistance to glufosinate and glyphosate are in the experimental stage of development (Gealy, Mitten, and Rutger, 2003; Kendig, Williams, and Smith, 2003). From the 1980s to the end of the century, the numbers of herbicide-resistant weed species increased nearly linearly (Heap, 1997) despite the introduction of many new herbicide products with new chemistries and/or modes of action during this time. Thus, continuation of similar herbicide use practices may well result in additional resistant weed populations.

Although herbicides are crucial, the flood that is maintained during most or all of the growing season is thought to account for 40 to 60 percent of the overall weed control in the U.S. rice crop (Kendig, Williams, and Smith,

2003). Continuous flooding (e.g., water seeding) generally provides greater weed control than delayed flooding systems (e.g., drill seeding), but with the disadvantage of higher water costs.

Presently, farmers select cultivars for a variety of reasons including yield potential, disease susceptibility, length of growing season, grain type, processing characteristics, and market demand (Street and Bollich, 2003), but not for their weed-suppression capabilities. Rice cultivars that could help control the infestation of weeds without the application of herbicides would be an economic and environmental benefit to farmers.

Developing suppressive rice cultivars to reduce the impact of weeds places added emphasis on genetic solutions to this ever-present, intractable agricultural problem. Genetic approaches can be inherently proactive and forward looking, and may be especially applicable for alternative or organic-based cropping systems. Weed-suppressive cultivars may be advantageous because they do not require the manipulation of a complex environment. Thus, only the genetics of a single crop organism need be altered; and our ability to manipulate genes for crop improvement has increased markedly in the past few decades.

The notion of weed-suppressive rice cultivars is appealing in its simplicity. Ideally, once the grower chooses a suppressive cultivar, few or no additional inputs would be required and additional management decisions would be minimal compared to those for traditional nonsuppressive cultivars. Although the current suppressive rice cultivar technology falls far short of this ideal, the potential for eventual successes has encouraged the development and maintenance of research programs that have generally concentrated on competition systems (rice depriving weeds of necessary resources) or on allelopathic systems (rice producing and releasing substances that inhibit weeds). The most effective initial use of weed-suppressive rice cultivars is likely to be in altered versions of presently successful weed control systems in which herbicide applications are reduced in rate or in number, delayed, or omitted entirely depending on the initial effectiveness of natural weed suppression.

THE CONCEPT OF WEED-SUPPRESSIVE RICE

Competition

Before the modern agricultural era, rice cultivars typically were aggressive weed competitors (De Datta, 1981). However, many of the genetic improvements incorporated into modern, high-yielding rice cultivars (erect, short, low-tillering plant types) have decreased their competitive ability and increased dependence on herbicides and nitrogen fertilizer (De Datta, 1981). The fact that modern rice cultivars differ in their ability to suppress weeds has been of interest for some time. Thus, manipulation of crop architecture or seedling vigor (such as emergence rates, tiller production, leaf growth, shape, angle, etc.) in order to deny resources to competing weeds has been a focal point of interest. Genetic control and physiological understanding of crop vigor/architecture characteristics as a means to deny resources to weeds have been evaluated in low-input systems of Africa and Asia (Jennings and Aquino, 1968; Garrity, Movillon, and Moody, 1992; Jones et al., 1997; Dingkuhn et al., 1999; Fofana and Rauber, 2000). Researchers in the United States have recently conducted similar investigations into suppression of *Echinochloa* spp. in water-seeded rice (Gibson, Foin, and Hill, 1999; Gibson et al., 2001). In other work with cereal crops, Ogg and Seefeldt (1999) used path analysis to suggest that wheat cultivars with the greatest early-season height-development rates would likely be more competitive against jointed goatgrass (*Aegilops cylindrica* Host), especially in dry years. Donald and Khan (1996) also employed path analysis to show that Canada thistle [*Cirsium arvense* (L.) Scop.] interactions with spring wheat yield components (*Triticum aestivum* L.) were most pronounced early in the crop season. Jordan (1993) and Jannink and colleagues (2000) have similarly shown that the greatest effects on soybean [*Glycine max* (L.) Merr.] competitiveness against broadleaf weed species occurred in the early season.

Allelopathy

The prospect of achieving weed control with limited or no herbicide application has sparked interest in the allelopathic potential of rice (Olofsdotter, Navarez, and Moody; 1995; Ebana et al., 2001a,b; Mattice et al., 2001; Olofsdotter, 2001a,b; Rimando et al., 2001; Olofsdotter, Jensen, and Courtois, 2002). The term *allelopathy* was coined by Molisch (1937) and generally refers to the detrimental effects, including those from microorganisms, of higher plants of one species on the germination, growth, or development of plants of another (Rice, 1984). A related phytotoxic phenomenon is autotoxicity in which crop residues of a particular species inhibit growth of the following crop of the same species. Rice and other crop species can become autotoxic in continuous cropping systems (Singh, Batish, and Kohli, 1999). A broad view of weed-suppression mechanisms includes any combination of whole plant competition and/or allelochemicals. Thus, the most useful approaches to breeding suppressive commercial cultivars may prove to be those that address both competition and allelopathy when feasible. Although complete control of weeds through natural suppression is probably unrealistic, moderate improvements in weed suppression by rice cultivars might facilitate reduced use of herbicides with the accompanying economic and environmental benefits. A better understanding of the mechanism(s) of interference and their relative contributions in agroecosystems would be useful in optimizing breeding objectives for enhanced weed suppression.

The distinction between simple competition and allelopathy in rice systems often is not clear. Competition, the concept of one plant denying resources to another, is relatively simple to understand because it is found in everyone's common life experience. Certainly, it is a major factor in most crop-weed interactions. Allelopathy, the concept of one plant producing and dispersing phytochemicals in concentrations and toxicities sufficient to adversely affect another plant, is more complex, is less readily understood, and is usually not part of one's common life experience. Its suppressive effects are likely to be relatively small compared to those due to competition over the entire growing season. However, small allelopathic benefits early may lead to significant increases in crop yield and/or reductions in weed growth by the end of the crop season. In their review, Weston and Duke (2003) reported that several important crop species including mustards, barley, fescues, rice, rye, sorghum, sudangrass, sunflower, and wheat exhibit allelopathic potential and have been screened aggressively for superior weed-suppression activity. To date, however, we are aware of no cultivars that were developed and grown primarily or exclusively for their allelopathic or competitive characteristics.

Beginning in the late 1980s, mass screenings of rice accessions in the USDA-ARS National Small Grains Collection for natural suppression (termed allelopathy in most of these studies) of ducksalad [*Heteranthera limosa* (Sw.) Willd.], red stem (*Ammannia coccinea* Rottb.), and barnyardgrass were conducted in Arkansas (Dilday, Nastasi, and Smith, 1989; Dilday et al., 1991; Dilday, Lin, and Yan, 1994; Dilday, Yan, et al., 1998; Dilday, Mattice, et al., 2001) (see Table 10.1). Allelochemical involvement was initially suspected in these studies because a weed-suppressed zone was visible near the base of some cultivars and not others. The possibility that weed-suppressive rice cultivars might be exploited to control the more economically significant weed *E. crus-galli* was further investigated in Arkansas (Dilday, Mattice, et al., 2001; Gealy and Dilday, 2001; Mattice et al., 2001; Gealy et al., 2003) (see Table 10.1) and elsewhere (Olofsdotter, Jensen, and Courtois, 2002) with modest success.

TABLE 10.1. Selected entries from the USDA-ARS world rice collection showing substantial suppression of barnyardgrass, ducksalad, or redstem in field tests without herbicide at Stuttgart, Arkansas, as compared to U.S. commercial standards.[a]

Cultivar, variety, or pedigree	Place of origin	PI or CI number in U.S. rice collection	Weed species tested	Field plot type	Reference
Suppressive entries					
Donduni Kunluz	Afghanistan	406061	Ducksalad	Hill plot, inhibition zone	Dilday, Nastasi, and Smith, 1989
Red Khosha Cerma	Afghanistan	277414	Ducksalad	Hill plot, inhibition zone	Dilday, Nastasi, and Smith, 1989
CICA 4	Brazil	uncertain	Ducksalad	Hill plot, inhibition zone	Dilday et al., 1991
Guichao no. 2[b]	China	615013	Barnyardgrass	Multiple-row plots	Dilday, Moldenhauer, et al., 1998
Mon Z Wuan	China	389680	Ducksalad	Hill plot, inhibition zone	Dilday, Nastasi, and Smith, 1989
Pan Ju	China	160530	Barnyardgrass	Two-row plots	Dilday et al., 1995
Teqing	China	536047	Barnyardgrass	Multiple-row plots	Dilday, Moldenhauer, et al., 1998
			Barnyardgrass	Multiple-row plots	Gealy, Dilday, and Rutger, 1998
			Barnyardgrass	Multiple-row plots	Gealy and Dilday, 2001
Tono Brea 439	Dominican Republic	430254	Ducksalad	Hill plot, inhibition zone	Dilday, Nastasi, and Smith, 1989
AC 1423	India	400723	Ducksalad	Hill plot, inhibition zone	Dilday, Nastasi, and Smith, 1989
Arc 11611	India	376253	Barnyardgrass	Two-row plots	Dilday et al., 1995

TABLE 10.1 *(continued)*

Cultivar, variety, or pedigree	Place of origin	PI or CI number in U.S. rice collection	Weed species tested	Field plot type	Reference
CR52-3	India	373020	Redstem	Hill plot, inhibition zone	Dilday, Mattice, et al., 2001
			Ducksalad	Hill plot, inhibition zone	Dilday, Mattice, et al., 2001
Johna 349 (early)	India	247882	Redstem	Hill plot, inhibition zone	Dilday, Mattice, et al., 2001
	Ducksalad	Hill plot, inhibition zone			Dilda, Mattice, et al., 2001
Mahlar 346	India	247883	Redstem	Hill plot, inhibition zone	Dilday, Mattice, et al., 2001
			Ducksalad	Hill plot, inhibition zone	Dilday, Mattice, et al., 2001
Masrai	Malaysia	372764	Redstem	Hill plot, inhibition zone	Dilday, Mattice, et al., 2001
			Ducksalad	Hill plot, inhibition zone	Dilday, Mattice, et al., 2001
Reata	Mexico	230105	Barnyardgrass	Two-row plots	Dilday et al., 1995
S 12 DZK	Pakistan[c]	247887	Redstem	Hill plot, inhibition zone	Dilday, Mattice, et al., 2001
			Ducksalad	Hill plot, inhibition zone	Dilday, Mattice, et al., 2001
Lambaygue no. 1	Peru	180177	Barnyardgrass	Two-row plots	Dilday et al., 1995
CP231/SL017/ Nanng mon	Philippines	321184	Redstem	Hill plot, inhibition zone	Dilday, Mattice, et al., 2001
			Ducksalad	Hill plot, inhibition zone	Dilday, Mattice, et al., 2001
IR 36-27-1-1	Philippines	312524	Barnyardgrass	Two-row plots	Dilday et al., 1995
IR 4-90-3-2	Philippines	312709	Barnyardgrass	Two-row plots	Dilday et al., 1995

IR 644-1-63-1-1	Philippines	338046	Redstem	Water-seeded plots	Dilday et al., 1995
			Ducksalad	Water-seeded plots	Dilday et al., 1995
			Ducksalad	Hill plot, inhibition zone	Dilday, Nastasi, and Smith, 1989
			Barnyardgrass	Multiple-row plots	Dilday, Yan, et al., 2001
IR 753-175	Philippines	338060	Barnyardgrass	Two-row plots	Dilday et al., 1995
IR 781-92-1-2-1-2-2	Philippines	350468	Barnyardgrass	Multiple-row plots	Dilday, Moldenhauer, et al., 1998
IR 788-16-1-1-1	Philippines	373026	Redstem	Water-seeded plots	Dilday, et al., 1995
			Ducksalad	Water-seeded plots	Dilday et al., 1995
			Redstem	Hill plot, inhibition zone	Dilday, Mattice, et al., 2001
			Ducksalad	Hill plot, inhibition zone	Dilday,Mattice, et al., 2001
IR 1317-385	Philippines	348835	Redstem	Hill plot, inhibition zone	Dilday, Mattice, et al., 2001
			Ducksalad	Hill plot, inhibition zone	Dilday, Mattice, et al., 2001
IR 1321-12	Philippines	348842	Redstem	Hill plot, inhibition zone	Dilday, Mattice, et al., 2001
			Ducksalad	Hill plot, inhibition zone	Dilday, Mattice, et al., 2001
IR 238	Philippines	321183	Redstem	Hill plot, inhibition zone	Dilday, Mattice, et al., 2001
			Ducksalad	Hill plot, inhibition zone	Dilday, Mattice, et al., 2001
IR 52 16 7 3	Philippines	312576	Redstem	Hill plot, inhibition zone	Dilday, Mattice, et al., 2001
			Ducksalad	Hill plot, inhibition zone	Dilday, Mattice, et al., 2001

TABLE 10.1 *(continued)*

Cultivar, variety, or pedigree	Place of origin	PI or CI number in U.S. rice collection	Weed species tested	Field plot type	Reference
IR 52 30 6 2	Philippines	312580	Redstem	Hill plot, inhibition zone	Dilday, Mattice, et al., 2001
			Ducksalad	Hill plot, inhibition zone	Dilday, Mattice, et al., 2001
IR 781-497-2-3	Philippines	348798	Redstem	Hill plot, inhibition zone	Dilday, Mattice, et al., 2001
			Ducksalad	Hill plot, inhibition zone	Dilday, Mattice, et al., 2001
Paraga Selection	Philippines	12108	Barnyardgrass	Two-row plots	Dilday et al., 1995
Phil MGVR	Philippines	Uncertain	Ducksalad	Hill plot, inhibition zone	Dilday, Nastasi, and Smith, 1989
T65/2X-TN-1	Philippines	312777	Redstem	Water-seeded plots	Dilday et al., 1995
			Ducksalad	Water-seeded plots	Dilday et al., 1995
			Barnyardgrass	Multiple-row plots	Dilday, Yan, et al., 2001
			Barnyardgrass	Multiple-row plots	Gealy, Dilday, and Rutger, 1998
			Barnyardgrass	Multiple-row plots	Gealy and Dilday, 2001
Lui chou 25 108 30	Taiwan	303683	Redstem	Hill plot, inhibition zone	Dilday, Mattice, et al., 2001
			Ducksalad	Hill plot, inhibition zone	Dilday, Mattice, et al., 2001
SH 30-21	Taiwan	366150	Redstem	Water-seeded plots	Dilday et al., 1995
			Ducksalad	Water-seeded plots	Dilday et al., 1995
			Barnyardgrass	Multiple-row plots	Dilday, Moldenhauer, et al., 1998

Taichung Native 1 (= TN-1)	Taiwan	271672	Redstem	Hill plot, inhibition zone	Dilday, Mattice, et al., 2001
			Ducksalad	Hill plot, inhibition zone	Dilday, Mattice, et al., 2001
			Barnyardgrass	Multiple-row plots	Gealy and Dilday, 2001
Tsai Yuan Chon	Taiwan	294400	Ducksalad	Hill plot, inhibition zone	Dilday, Nastasi, and Smith, 1989
Woo Co Chin Yu	Taiwan	279171	Ducksalad	Hill plot, inhibition zone	Dilday et al., 1991
Drew	United States	596758	Barnyardgrass	Multiple-row plots	Gealy and Dilday, 2001
LaGrue	United States	568891	Barnyardgrass	Multiple-row plots	Gealy and Dilday, 2001
NATO/9209/sel// AROS/3/NROS	United States	Not available	Ducksalad	Hill plot, inhibition zone	Dilday, Nastasi, and Smith, 1989
NSSL 10/28 STP 8	United States	Not available	Ducksalad	Hill plot, inhibition zone	Dilday, Nastasi, and Smith, 1989
Stg 95L05-039 (PI 338046/Alan)	United States	Not available	Barnyardgrass	Multiple-row plots	Dilday, Mattice, et al., 2001
Stg 95L05-054 (PI 338046//Lemont/ RA 73)	United States	Not available	Barnyardgrass	Multiple-row plots	Dilday, Mattice, et al., 2001
B850/Cros 1-7-18-3-2 sein (298965-2)	Unknown	12496	Redstem	Hill plot, inhibition zone	Dilday, Mattice, et al., 2001
			Ducksalad	Hill plot, inhibition zone	Dilda, Mattice, et al., 2001
Nonsuppressive or poorly suppressive cultivar standards					
Katy	United States	527707	Barnyardgrass	Multiple-row plots	Gealy and Dilday, 2001
			Barnyardgrass	Multiple-row plots	Dilday, Mattice, et al., 2001

TABLE 10.1 *(continued)*

Cultivar, variety, or pedigree	Place of origin	PI or CI number in U.S. rice collection	Weed species tested	Field plot type	Reference
Kaybonnet	United States	583278	Barnyardgrass	Multiple-row plots	Gealy, Dilday, and Rutger, 1998
Lemont	United States	475833	Barnyardgrass	Multiple-row plots	Gealy and Dilday, 2001
			Barnyardgrass	Multiple-row plots	Dilday, Moldenhauer, et al., 1998
Rexmont[d]	United States	502968	Redstem and Ducksalad	Various	Dilday, Mattice, et al., 2001; Dilday et al., 1995
			Barnyardgrass	Multiple-row plots	Dilday, Moldenhauer, et al., 1998
			Barnyardgrass	Multiple-row plots	Dilday, Yan, et al., 2001
Starbonnet[e]	United States	388582	Barnyardgrass	Multiple-row plots	Gealy and Dilday, 2001

[a]Most rice entries, places of origin, and PI or CI numbers can be referenced online at the GRIN (2002) Web site of the USDA-ARS Germplasm Resources Information Network <http://www.ars-grin.gov/npgs/acc/acc_queries.html>. Plots were dry seeded unless otherwise stated.

[b]Also spelled 'Guichou' or 'Guichow' in some publications.

[c]India was listed as 'Origin' of S 12 DZK (PI 247887) in Dilday, Mattice, et al. (2001).

[d]'Rexmont' is an early-maturing, long grain, semi-dwarf, tropical japonica cultivar (Bollich et al., 1990).

[e]'Starbonnet' is a very late maturing, long grain, tall, tropical japonica cultivar that was grown extensively in the southern United States from the late 1960s to the mid-1980s (Moldenhauer, Gibbons, and McKenzie, 2004).

OPTIMUM PERIOD FOR MAXIMUM WEED SUPPRESSION

Regardless of the mode of weed control employed, rice, like most other field crop species, responds most favorably when weeds are controlled or removed very early in the crop season. It is well established that weed competition during early growth stages reduces rice yield more than competition during later growth stages (De Datta, 1981; Olofsdotter, 2001a). Weeds will usually not affect final rice yield when rice is maintained weed-free during the initial stages of production, especially before permanent flood in drill-seeded systems. If the crop is maintained weed-free for a limited initial period (critical weed-free period), later-emerging weeds need not be treated or removed. Achieving this critical weed-free or weed-suppressed period during crop establishment is crucial for minimizing the impact of weeds on rice yield. The duration of the critical weed-free period depends on factors such as planting method, cropping system, weed species, and rice cultivar. This critical period can be as brief as ten days (De Datta, 1981). However, in Colombia, rice yields increased with weed-free periods up to 70 days after emergence (Fischer et al., 1993).

Preventing or delaying germination of *E. crus-galli* even for brief periods can be advantageous to rice. In Arkansas field tests, both the highly suppressive cultivar PI 312777 and 'Kaybonnet' (Gravois et al., 1995), a less suppressive, tall, long grain, cultivar, suppressed *E. crus-galli* nearly twice as much when initial irrigation or rainfall occurred at 11 days compared to 2 days after planting (Gealy, Dilday, and Rutger, 1998; see Table 10.2). Similarly, delaying *E. crus-galli* germination for brief periods reduced its growth and its ability to reduce yields of water-seeded rice in California (Gibson and Fischer, 2002). In these studies, rice yields fell markedly when weed emergence was delayed beyond 30 days.

TABLE 10.2. Effect of flush irrigation timing on rice yield and natural weed suppression without grass herbicide.[a]

Rice cultivar	Time between planting and first irrigation or rainfall (days)	Rough rice yield (% of weed free)	Barnyardgrass visual control rating (%)
Kaybonnet	2	12	33
	11	35	58
PI 312777	2	35	41
	11	67	78

[a]Reconstructed with permission from Gealy, Dilday, and Rutger, 1998.

Weeds that compete with the crop for only a limited period during the seedling stage will not reduce final crop yields. Final yields of three tall tropical japonica rice cultivars in Arkansas were usually not affected when the period of competition with *E. crus-galli* was 40 days or less (Smith, 1974).

Aggressive weed suppression early in the season is most desirable because it is unlikely that rice cultivars would become rapidly suppressive to weeds at midseason. Olofsdotter, Navarez, and Moody (1995) have suggested that research on allelopathic components of weed control would be most effectively targeted in the early developmental stages of rice.

*DYNAMICS OF RICE-*ECHINOCHLOA *INTERFERENCE*

Interactions of growth and yield of *Echinochloa* spp. and rice have been investigated extensively. High-tillering, wet-seeded rice genotypes were more competitive against a weed population dominated by *E. glabrescens* Munro ex Hook. f. (no common name was given in the paper) than were low-tillering ones (Tuong, 2000). Increasing early-season leaf area expansion rates of rice is also a potential means to improve competitiveness against *Echinochloa* species (Lindquist and Kropff, 1996). Competitive/suppressive rice cultivars were economically viable with reduced herbicide rates (Chavez, Gealy, and Black, 1999; Gibson et al., 2001). Competition for light and nitrogen between rice and barnyardgrass was simulated with a model that accounted for tiller production, and mass of leaves, culm, roots, and grain on each tiller. The model predicted general trends of growth of rice and barnyardgrass in monoculture as well as in competition with each other (Graf and Hill, 1992). Kropff and colleagues (1995) used early-season relative leaf area measurements of *E. crus-galli* and several other species in a model to predict crop loss from competition. The model took into account realistic field situations in which seedling emergence dates and maximum plant heights differed between weeds and rice. Simulations of *E. crus-galli* competition in rice accurately accounted for the lower competitiveness of direct seeded rice in comparison with transplanted rice. It also accounted for greater crop losses under higher weed densities and when weeds emerged with rice (Kropff and Lotz, 1993; Kropff et al., 1993).

Differences in plant type have long been known to influence weed competition. For instance, vigorous, high-tillering, tall, leafy tropical indica cultivars were much more competitive than cultivars with small, erect, sturdy plant types (including 'Taichung Native 1' from Taiwan) when grown in mixtures. However, pure stands of the small, erect types outyielded the tall types (Jennings and DeJesus, 1968). Yield and competitive ability were negatively associated, and the latter was attributable to a single major locus (Jennings and Herrera, 1968). These researchers further showed that although tall types were more competitive than short types under normal conditions, the short types were more competitive under conditions of high nitrogen fertilizer and high plant densities.

Rice growth patterns in mixtures of strongly and weakly competitive cultivars showed that strong competitors tended to intercept light at the expense of weak ones, suggesting that light was a major factor for which rice competes in mixed rice populations (Jennings and Aquino, 1968). Tiller and leaf number, leaf length, spreading growth habit, leaf area development, dry weight, and height were greater in strong competitors. Thus, a genetic disposition toward increased plant size and vigor in early stages appears to confer greater competitive ability. Such characteristics may be desirable under conditions of limited weed control but viewed as undesirable in the case where weeds could be controlled adequately and as a limitation to cultivar improvement in tropical breeding programs (Jennings and Aquino, 1968).

Studies in Sri Lanka have shown that root competition is more important than shoot (light) competition against barnyardgrass (Perera, Ayres, and Gunasena, 1992). In a 50:50 species mixture, barnyardgrass growth was greater (than rice), especially if fertilized, since this promoted growth of the weed relative to rice. A lower relative crowding coefficient for rice in fertilized soil than in nonfertilized soil demonstrated this. They concluded that barnyardgrass reduced rice growth primarily by inhibiting root growth of rice, which reduces its ability to obtain resources from the soil. However, recent reports on pregerminated, broadcast seeded IR-72 rice from the Philippines indicated that above-ground factors were the primary source of interference against *E. crus-galli* under normal nitrogen fertilization, whereas below-ground competition became predominant without added nitrogen (Caton, Navarez, and Mortimer, 2002). Thus, under high fertility management problems, weed species that compete most vigorously for light may be the most problematic.

In Colombia, indica cultivars CICA 8 and ORYZICA 1 were similarly competitive. Weeds that emerged with the crop caused the greatest yield losses. After a 70-day weed-free period, both cultivars suppressed weeds similarly, even though the leaf area index of 'CICA 8' was much larger than 'ORYZICA' under weed-free conditions. Both cultivars required a critical height of 44 to 50 cm to suppress weeds. Economic analysis showed that three postemergence herbicide applications were justified under heavy weed infestations in the absence of continuous flooding (Fischer et al., 1993).

Planting density can also influence economics of weed control in rice. In Colombia, high planting densities that led to high tiller densities were a more economical way to control mixed weed populations, which included *E. crus-galli* and *Leptochloa* spp., than were late postemergence herbicides. This provided the added benefit of reducing herbicide use (Fischer and Ramirez, 1993).

Weed control can alter the response of rice yield to nitrogen fertilizer. In India, nitrogen application increased rice yields when plots were cultivated three times. However, nitrogen application reduced yields when plots were cultivated only once, due to increased weed growth (Sharma, 1994).

RED RICE

We have observed that some rice cultivars provide modest levels of suppression against certain red rice ecotypes in field trials at Stuttgart, Arkansas. The rice cultivar PI 312777 suppressed growth and seed production of red rice to a greater degree than did commercial cultivars. A short-statured rice-red rice cross was less competitive against rice than were taller red rice types (Estorninos, 2000; Estorninos, Gealy, and Talbert, 2002). 'PI 312777' is a T65/2X-TN-1 cross from the Philippines with both parents originating from Taiwan. Similarly, a tall rice cultivar suppressed red rice growth more than did a semi-dwarf cultivar (Kwon, Smith, and Talbert, 1991). Notwithstanding these examples of moderate red rice suppression, it seems unlikely that rice cultivars capable of reducing the economic impacts of red rice could be developed given the competitive superiority of most red rice types over rice.

INTEGRATING SUPPRESSIVE CULTIVARS WITH REDUCED HERBICIDE RATES IN ARKANSAS

For many decades, propanil has controlled a wide array of rice weed species in the southern United States. It has been estimated that propanil is applied one or more times to 98 percent of the rice in Arkansas (Carey, Hoagland, and Talbert, 1995). However, propanil-resistant barnyardgrass populations have become increasingly common in Arkansas (Baltazar and Smith, 1994; Carey, Hoagland, and Talbert, 1995). Although quinclorac has been used to control propanil-resistant barnyardgrass, some of these populations also have become resistant to quinclorac (R. Talbert, personal communication, 2002; Valverde, Riches, and Caseley, 2000). Therefore, alternative control methods, including natural suppression systems, may be needed in future weed management systems. Weed control programs using reduced herbicide rates have been recommended by extension specialists for some rice cropping systems in Arkansas as a means of reducing input costs (Baldwin, Boyd, and Smith, 2000). Applying herbicides at reduced rates or in combinations with other herbicides can increase economic returns in rice production in the South (Jordan, 1997a,b).

Weed-suppressive rice germplasm lines have reduced the density and growth of barnyardgrass while producing greater yields compared to non-suppressive rice cultivars (Dilday, Moldenhauer, et al., 1998). This opens the possibility that these rice lines may be grown with greatly reduced rates of propanil. In California, a competitive rice cultivar required lower herbicide rates than did a less competitive cultivar to achieve similar levels of weed suppression (Gibson et al., 2001).

In a study evaluating the economics and efficacy of reducing herbicide rates in combination with weed-suppressive cultivars, high-yielding indica cultivars reduced *E. crus-galli* biomass more and produced higher yields than did tropical japonica U.S. cultivars (Gealy et al., 2003). Projected net economic returns for the indica cultivar PI 312777 were more than double that of the tropical japonica, early-maturing, long grain, semi-dwarf cultivar Lemont (Bollich et al., 1985), apparently because 'PI 312777' produced greater yields and suppressed weeds better than 'Lemont' (see Table 10.3) (Gealy et al., 2003). The net return for 'PI 312777' with no propanil was greater than returns for 'Lemont' at any herbicide rate, and did not respond to propanil application. Conversely, maximum net returns for 'Lemont' were obtained only with propanil application. Responses of 'Kaybonnet', a tall long grain U.S. cultivar, were intermediate between those of 'PI 312777' and 'Lemont'. It must be noted that certain grain quality characteristics of 'PI 312777' such as milling quality (see Table 10.3) are presently inferior to those of commercial U.S. cultivars and would require substantial improvements before a completely viable economic analysis can be made. A related study compared the indica cultivar Teqing with 'Lemont'. 'Teqing' provided better control of propanil-susceptible barnyardgrass (Lovelace et al., 2001) and propanil-resistant barnyardgrass (M. Lovelace, 2001, personal communication) alone or with reduced rates of thiobencarb herbicide. These results confirm the economic benefit of herbicide use on commercial tropical japonica cultivars and highlight the potential of enhancing natural weed suppression in U.S. rice.

TABLE 10.3. Effect of propanil rate on weed biomass, rough rice yield, and net returns of weed-suppressive and nonsuppressive rice cultivars in drill-seeded field plots at Stuttgart, Arkansas.

Rice cultivar	Propanil rate ($kg \cdot ha^{-1}$)	Weed biomass (% of no crop, weedy check)	Rough rice yield ($kg \cdot ha^{-1}$)	Estimate net return ($/ha)
Lemont (nonsuppressive)	0.0	39	4,800	90
	1.1	27	5,800	260
	2.2	23	5,900	270
	4.5	13	6,200	400
Kaybonnet (nonsuppressive)	0.0	29	5,400	200
	1.1	23	5,700	220
	2.2	18	6,600	400
	4.5	10	6,400	360
PI 312777 (weed suppressive)	0.0	8	7,300	650
	1.1	7	7,300	590
	2.2	6	7,100	560
	4.5	1	7,600	700
LSD (0.05)		2	500	none given

Source: Data were reconstructed with permission (from Gealy et al., 2003; copyright held by Weed Science Society of America).
Notes: Weed biomass and rough rice yield data represent means from 1995, 1996, and 1998 studies. Estimated net returns were not corrected for milling yield and represent means from 1995 and 1996 studies only. Historic milling yields for these cultivars were as follows: 'Lemont', 54 percent head rice and 16.0 percent broken rice; 'Kaybonnet', 56 percent head rice and 12 percent broken rice; and 'PI 312777', 45 percent head rice and 25 percent broken rice. The milling yields of 'Lemont' and 'Kaybonnet' are commercially acceptable, while those of 'PI 3122777' are not. In 1995 and 1996 heights of 'Lemont', 'Kaybonnet', and 'PI 312777' treated with 4.5 $kg \cdot ha^{-1}$ propanil averaged 90, 124, and 115 cm, respectively. Propanil is a postemergence, contact-type herbicide that selectively controls a broad spectrum of weed species in rice and has been the

RICE GENETICS AND CROSSING CONSIDERATIONS

Rice (*Oryza sativa* L.), in addition to being a major world food crop, has become a model cereal crop for molecular genetic research (Izawa and Shimamoto, 1996). Rice is ideal because of its small genome size (430 Mb) relative to other crop species and the low proportion of repetitive sequences in its genome (Kurata et al., 1997). The dramatic similarities in location and function of genes in the rice genome compared to other cereal crop genomes (synteny) (Devos and Gale, 1997), especially corn (Ahn and Tanksley, 1993), at both the map and megabase levels allow researchers to predict gene locations in other cereals based on the rice genetic map. Moore and colleagues (1997) looked at the possibility of a holocentric chromosome ancestor existing between the cereal genomes. This possible level of homogeneity has served to further promote research interest in rice, particularly the application of functional genomics. Many of the resources and tools that are developed by rice researchers will be applicable to the study of other cereals such as wheat and maize.

Oryza sativa is a diploid ($2n = 24$) in the 'AA' genome group of *Oryza*. *Oryza sativa* plant types are capable of intercrossing with other species in the 'AA' group such as African rice (*O. glaberrima* Steud.), brown beard rice (*O. rufipogon* Griff.), and Indian wild rice (*O. nivara* Sharma et Shastra) (Sitch, 1990; Singh and Khush, 2000). Rice cultivars are primarily self-pollinated, averaging about 0.5 percent outcrossing in the United States (Beachell et al., 1938). Typically, rice florets open only once at the time of fertilization (Tseng, 1972). Pollen is shed just before or at the same time that the floret opens (Adair, 1934). Once exposed to the external environment, rice pollen becomes mostly nonviable within 5 to 20 minutes of shedding (Khatun and Flowers, 1995; Moldenhauer and Gibbons, 2003). The stigma typically remains viable for several days and, compared to pollen, is quite tolerant to temperature extremes (Jodon, 1938; Yoshida, 1981). Rice floret sterility and particularly rice pollen viability can be affected by temperature extremes, drought, nutrient deficiencies, or toxicities during spikelet formation and microsporogenesis (pollen formation) which occurs in late boot when the penultimate leaf and flag leaves are emerging (Stansel, 1975; Yoshida, 1981; Hoshikawa, 1989). Rice florets typically are most receptive to cross-pollination near midday under warm temperatures, bright sunlight, and high humidity that maximize the duration and degree of floret opening and pollen survival (Adair, 1934; Moldenhauer and Gibbons, 2003). Time and duration of blooming of individual florets and panicles are affected by cultivar and environmental conditions (Adair, 1934). Greatest blooming generally occurred near midday, and cool temperatures and cloudy skies delay or slowed flowering.

Most U.S. cultivars are of tropical japonica (temperate climate) background, and several of the promising weed-suppressive lines tested in Arkansas have been indica (tropical) types. Wide genetic crosses between these two rice types can lead to infertility. *Indica*×*japonica* hybrids have produced more than 50 percent abortive pollen and similarly low fertility

(Kuang and Tu, 1949). *Indica*×*tropical japonica* crosses also exhibit sterility.

RICE BREEDING EFFORTS IN ARKANSAS AND SOME POTENTIAL SCREENING METHODS

Introduction of natural tolerance mechanisms into crop species has been a relatively safe and inexpensive approach to protect crop plants from nonweed pests. Similar approaches might also be useful in weed suppression and management. Thus, several highly weed-suppressive and high-yielding cultivars that are adapted to southern U.S. conditions have been identified at the Dale Bumpers National Rice Research Center (DB NRRC) and University of Arkansas Rice Research and Extension Center (RREC) at Stuttgart, Arkansas (Dilday, Mattice, et al., 2001). These include the indica rices 'PI 338046' and 'PI 312777', which are typical of the promising weed-suppressive cultivars tested under southern U.S. conditions. By contrast, japonica types comprised a larger proportion of suppressive types than did indica types in research conducted in Asia (Olofsdotter, 2001a). 'Taichung Native 1' (TN 1) is present in the parentage of both 'PI 338046' and 'PI 312777'. This line has significant weed-suppressive activity against several grass and broadleaf weed species in field tests in several climates and in laboratory assays (Dilday, Lin, and Yan, 1994; Dilday, Mattice, et al., 2001; Gealy and Dilday, 2001; Olofsdotter, 2001a,b; Rimando and Duke, 2003). The degree to which allelopathy may be involved in the weed suppression mechanism(s) of 'PI 312777' and 'PI 338046' is uncertain, but both cultivars can aggressively suppress *E. crus-galli* in drill-seeded rice before application of the permanent flood. 'PI 312777' produces greater numbers of tillers and a greater root mass (Dilday, Mattice, et al., 2001; Gealy, unpublished data), and can deplete soil nitrogen (see Table 10.4) and reduce photosynthesis rates of barnyardgrass (Gealy, Dilday, and Rutger, 1998) to a greater extent than do commercial cultivars, which may account for a portion of its weed-suppression activity. In Arkansas, yields of indica lines such as 'Guichao' and 'PI 312777' have typically been greater than those of common U.S. long grain cultivars under recommended (Dilday, Yan, et al., 2001) and less-than-recommended nitrogen fertilization rates (Gealy and Dilday, 2001; Gealy et al., 2003). Thus, only modest rates of nitrogen fertilizer may be necessary for these indica rice lines to achieve a yield advantage or weed suppression (Gealy et al., 2003) with these indica rice lines. The lower nitrogen rates may in turn help prevent the crop from producing excess vegetative growth at the expense of seed production.

TABLE 10.4. Depletion of soil nitrogen by rice and barnyardgrass before application of permanent flood after three planting dates at Stuttgart, Arkansas, in 2000.[a]

	Total N concentration (ppm)					
	Planting date					
	April		May		June	
	Soil depth					
Rice cultivar	**0-8 cm**	**8-15 cm**	**0-8 cm**	**8-15 cm**	**0-8 cm**	**8-15 cm**
PI 31277 only	9.3	8.8	13.3	9.3	8.0	9.8
Teqing only	11.8	10.3	13.5	17.0	10.5	14.8
Kaybonnet only	10.0	12.8	14.3	20.0	13.3	11.0
RU9701151 only (cross of PI 338046 / Katy)	10.8	12.5	16.8	20.5	14.3	17.0
Barnyardgrass only	13.8	10.3	13.5	8.0	11.0	11.8
No vegetation	29.3	19.5	25.0	18.5	19.8	20.3
LSD (0.05)	3.7	3.3	6.1	6.3	5.2	4.7

Source: Gealy, unpublished data.

[a]April planting date April 20, 2000, with soil sample taken on May 31, 2000; May planting date May 16, 2000, with soil sample taken on June 23, 2000; June planting date June 7, 2000, with soil sample take on July 5, 2000.

Grain quality and agronomic characteristics of suppressive lines discovered at Stuttgart, Arkansas, have not been commercially acceptable to the U.S. rice industry. The combination of high yields and weak straw renders 'PI 312777' prone to lodging. Further, its head rice yields are approximately 45 percent, well below the typical commercial levels of 55 to 60 percent. 'PI 338046' has produced commercially acceptable head rice yields in Arkansas (Robert Dilday, 2001, personal communication) and thus may be a useful source of germplasm for weed suppression (Dilday, Mattice, et al., 2001). 'PI 338046' has suppressed aquatic weeds effectively in Arkansas field trials (Dilday et al., 1995; see Table 10.1), but usually has been less effective than 'PI 312777' against *E. crus-galli* (Dilday, Mattice, et al., 2001).

Discovery by Dilday, Nastasi, and Smith (1989) of potential allelopathic properties in the rice line 'PI 338046' was instrumental in the initiation of a breeding effort for weed-suppressive rice in Arkansas. In 1991 Karen Moldenhauer made crosses with 'PI 338046' as the female with the best breeding material adapted to Arkansas, with the objective of improving natural weed suppression in future commercial cultivars. The male parents included the long grain tropical japonica cultivars Katy (Moldenhauer et al.,

1990), Alan (Moldenhauer et al., 1991), Newbonnet (Johnston et al., 1984), and five other advanced breeding lines. Lines from these crosses with 'PI 338046' have been advanced through the traditional rice breeding program at the RREC at Stuttgart, Arkansas. The initial selection technique entailed discarding all of the plant types that lacked good agronomic characteristics. This selection resulted in 14 lines with acceptable plant types that were screened further for weed-suppression activity (R. H. Dilday, 2002, personal communication). One of the lines, 'RU9701151', is a long grain rice with standard U.S. cooking quality, high yield potential, and good plant type, but is not tolerant to blast disease. In the 1997 and 1998 Arkansas Rice Performance Trials, it had an average yield of 8,225 $kg \cdot ha^{-1}$ compared to 'LaGrue' at 8,228 $kg \cdot ha^{-1}$. 'LaGrue' is a tall, long grain, tropical japonica (Moldenhauer et al., 1994) and is among the highest-yielding cultivars in Arkansas. 'RU9701151' was derived from the cross 'PI 338046'/'Katy' and had acceptable milling yields (64 percent head rice and 71 percent total rice) (Moldenhauer et al., 1999) with standard U.S. cooking quality and good plant type.

In Arkansas the breeding program is currently continuing to look at these lines in an attempt to capture this weed-suppressive activity. In the second stage of introgressing weed-suppression activity into adapted rice cultivars, the line 'RU9701151' was selected from the original set of crosses and was crossed to 'PI 312777'. 'RU9701151' was one of the lines utilized in these crosses. F_2 to F_4 lines from these crosses, when grown at the RREC, segregated for all phenotypic characteristics. F_4 lines and above from these plants are currently being evaluated for weed suppression in various tests. To date we have successfully maintained adequate grain quality in these advanced lines, but levels of weed suppression appear to be lower than those of both of the weed-suppressive parents we introduced into the pedigree (i.e., 'PI 338046' and 'PI 312777'). In tests over several years, 'RU9701151' has exhibited moderately high barnyardgrass suppression capability, but it does not consistently achieve levels of suppression as high as those from 'PI 312777' (see Table 10.5).

Even if desirable grain quality, agronomic characteristics, and weed suppression can be developed from this work, none of these lines is likely to control weeds totally, but they may be economically beneficial to the rice producers if used in combination with reduced rates of herbicide (Gealy et al., 2003). Meaningful progress toward the development of weed-suppressive rice cultivars will require sustained team collaboration among numerous specialists, including plant breeders, chemists, plant physiologists, molecular geneticists, microbiologists, ecologists, and weed scientists. The fact that no commercial U.S. rice cultivar with enhanced weed-suppression capacity has been developed to date is a testament to the

TABLE 10.5. Barnyardgrass suppression and rice yields from the 'PI 338046'/'Katy' cross ('RU9701151') and selected rice cultivars in 2000, 2001, and 2002.[a]

	Visual weed suppression rating (% control)			**Rough rice yield in weedy plots ($t \cdot ha^{-1}$)**		
Rice cultivar	**2000**	**2001**	**2002**	**2000**	**2001**	**2002**
Kaybonnet	67	42	47	4.4	5.1	2.2
Katy	53	55	—	2.5	3.3	—
Lemont	48	35	48	2.8	1.1	2.9
PI 312777	81	72	78	6.6	6.0	5.6
PI 338046	66	68	60	5.1	9.2	4.3
RU9701151 (cross of PI 338046 / Katy)	67	48	67	5.2	2.6	4.6
LSD (0.05)	15	—	13	—	—	1.6

Source: Reconstructed with permission from Gealy and Dilday, 2001; Gealy, unpublished data.

[a]No grass control herbicides were applied to these plots. Visual ratings were taken mid- to late season. Dashed line (—) indicates data were missing or not available. In separate Arkansas Rice Performance Trials conducted at five locations in 1997 and four locations in 1998, 'RU9701151' ('PI 338046'/'Katy' cross; designated 'Line E') produced average grain yields as high or higher, and milling yields similar to those of commercial cultivars Drew, Cypress, and Madison (Moldenhauer et al., 1999).

complexity of the biological systems involved and the quality demands of the U.S. rice industry and the world export market. None of the other southern U.S. rice breeding programs is working with these traits at this time. California scientists are presently investigating genetic improvement of weed-competitive characteristics in water-seeded rice systems.

'PI 338046'/'Lemont' ('Stg95L05-058' and 'Stg95L05-054') and 'PI 338046'/'Alan' ('Stg95L05-039') crosses have reduced *E. crus-galli* populations to the same level as those in 'PI 338046' (Dilday, Mattice, et al., 2001; see Table 10.1). Although encouraging, the practical significance of these data is uncertain because the crosses and nonsuppressive checks in the study were not compared directly to more competitive commercial cultivars such as 'Drew' or 'LaGrue' (Gealy and Dilday, 2001; Moldenhauer et al., 1998). U.S. rice growers are not likely to accept new cultivars unless they produce yields and grain quality similar to or greater than those of existing commercial cultivars and also exhibit qualitatively greater weed suppression.

Scientists in West Africa have attempted to utilize desirable yield and weed-competitiveness characteristics of upland *O. glaberrima* rice types to improve *O. sativa* cultivars (Jones et al., 1997; Dingkuhn et al., 1999). Several *O. sativa* x *O. glaberrima* crosses from the West Africa Rice Development Association have been evaluated for barnyardgrass suppressiveness in Arkansas. These included 'WAB450-11-1-2-P1-HB' ('PI 602630' or 'QF 36192'); 'WAB450-4-1-1-P23-1-1' ('PI 602631' or 'QF 36194'); 'WAB450-I-B-P-148-2-1' ('PI 602632' or 'QF 36195'); 'WAB450-I-B-P-23-HB' ('PI 602633' or 'QF 36197'); and 'WAB450-24-2-3-P33-HB' ('PI 602634' or 'QF 36199'). However, these crosses have generally been less barnyardgrass suppressive than most of the commercial and foreign cultivars tested, probably because they are poorly adapted to Arkansas conditions (Gealy and Dilday, 2001). Scientists at DB NRRC have been using similar wide crossing approaches with *Oryza* relatives to incorporate disease resistance into U.S. rice cultivars (Eizenga and Rutger, 2003). Future efforts may reveal other useful competitive genes from these *Oryza* relatives.

Effective breeding for suppressive rice phenotypes requires accurate, rapid, inexpensive screening methods that can identify the desired traits as early as possible in the growth cycle. Without such tools a breeder's progress will be limited. Thus, developing improved techniques for screening rice lines in the early generations after crossing could greatly increase progress toward the goal of improved weed suppression and maintenance of agronomically acceptable plant types. Even if weed-suppressive lines are not agronomically superior, they potentially can be improved through future crossing. The initial successes with lines such as 'PI 312777' suggest that through breeding and biotechnology weed-suppression capabilities of rice cultivars can be improved, thereby decreasing our reliance on synthetic weed killers.

Several potential methods of screening for weed-suppressive rice that have been attempted in Arkansas are presented next. Generally, these methods are not meant to differentiate between competitive and allelopathic mechanisms, but rather to assist us in identifying rice lines or crosses with substantially elevated levels of weed-suppressive activity under field conditions regardless of the relative contributions from each mechanism. From our perspective, a highly suppressive cultivar that derived 95 percent of its suppressive activity from competitive factors and 5 percent from allelopathy could be just as useful as an equally suppressive cultivar in which the relative contributions were 60 percent competition and 40 percent allelopathy. In either case, adequate performance under field conditions is paramount, and aggressive early-season suppression is preferred over mid- or late-season suppression.

The simplest field assay involves visual evaluation of inhibition zones in which weed densities and/or weed growth are reduced near the stems of individual rice hills or rows. These assay systems have been used with great success, especially for suppression of aquatic weeds such as red stem and duck salad (see Table 10.1). Generally, this approach has not worked as well for evaluation of barnyardgrass suppression for which larger multiple-row plots (at least 2 × 3 m) are preferable.

Measuring early-season leaf area and tiller production of rice and density and growth of weed seedlings has been useful. Indica cultivars such as 'PI 312777' and 'Teqing', which produce large numbers of tillers in the seedling stage and have dramatically reduced emergence and growth of weeds before permanent flood, have consistently been more weed suppressive than low-tillering commercial cultivars in Arkansas field tests at this growth stage as well as at harvest (Gealy, unpublished data). The indica cultivars have an erect growth habit in the seedling stage and, therefore, do not generally shade the inter-row areas excessively at this time. Similarly, in a growth analysis of nine diverse *Oryza* spp. cultivars in competition with *E. crus-galli,* two indica cultivars produced the greatest dry matter (Caton, Eusebio, and Mortimer, 2002). They showed that leaf area and other shoot-related traits were primarily responsible for total dry matter production, which supports the hypothesis that nondestructively screening for such traits is a viable approach to select lines with increased competitiveness against weeds.

The weed-suppressive cultivars PI 312777 and PI 338046 produced greater root mass under field conditions than did the nonsuppressive cultivars Lemont and M-201 (Dilday, Mattice, et al., 2001). The relative root masses of C-3 and C-4 plants in mixtures can be determined by ^{13}C isotope depletion analysis of roots extracted in bulk from soil (Svejcar and Boutton, 1985; Gealy, 2001). Conveniently, rice is a C-3 plant and *E. crus-galli* is a C-4 plant. This method may have limited potential as a plant breeding tool because ^{13}C analysis is time consuming and expensive. However, it may be useful when below-ground root interactions are a major component of the rice-weed interference system and could readily be employed for a limited number of advanced breeding lines.

Sampling soil nitrogen depletion levels in the field before application of nitrogen fertilizer and the permanent flood (see Table 10.4), or in small pot assays (John Mattice, 2002, personal communication) may be useful as a screening method for barnyardgrass suppression. In these studies 'PI 312777' has often depleted nitrogen levels more than commercial rice standards, suggesting that nitrogen deprivation may contribute to barnyardgrass suppression. Indirect sampling has been done by using density-dependent phytotoxicity experiments to determine whether allelochemicals (direct toxic effect) or merely resource removal (nutrient depletion) are responsible for plant suppression. When allelochemicals are responsible for phytotoxicity, decreased plant density in a constant soil volume will result in proportionately greater inhibition of the target plants, whereas ordinary competition for nutrients would not (Weidenhamer, 1996). Dilday and colleagues (1995) have shown that in a water-seeded system, the rice lines 'PI 312777', 'PI 338046', 'PI 366150', and 'PI 373026' became increasingly suppressive against ducksalad with increasing nitrogen fertilizer rates. These results suggest that nitrogen depletion by rice was not involved directly in the suppression of ducksalad, and the possibility exists that higher nitrogen rates may have resulted in greater above-ground growth, and thus more shading by the crop.

Soils in Arkansas can experience brief periods of drought between rice emergence and permanent flood if fields do not receive timely rainfall or flush irrigation. In some instances, aggressive cultivars such as 'PI 312777' and 'Teqing' appear to have depleted soil moisture levels more quickly than less aggressive cultivars such as 'Kaybonnet' (unpublished data). We have readily measured moisture depletion in Arkansas rice field soils with commercially available instruments such as thermocouple psychrometers (e.g., Tru-Psi* from Campbell Scientific Corp., Logan, Utah) that are capable of measuring soil water potential, or transmission line oscillator (TLO) instruments (e.g., HydroSense from Decagon Corp., Pullman, Washington) that are capable of measuring volumetric water content (dielectric permittivity). To date we have not observed consistent differences in soil moisture depletion between suppressive and nonsuppressive rice cultivars, but this potential environmental interaction under moisture-stressed conditions in drill-seeded rice fields early in the season cannot be ruled out.

Delaying planting of *E. crus-galli* seeds for brief periods following rice planting into petri dishes or small pots in small-volume laboratory assays may be useful to identify cultivars with weed-suppressive or allelopathic activity in the seedling stage (Jensen et al., 2001; Mattice et al., 2001).

Rice leaves or roots may be analyzed for the presence of allelochemicals or chemicals correlated with suppression (Mattice et al., 2001; Rimando et al., 2001; Kong et al., 2002). Rimando and Duke (2003) have recently reviewed several laboratory bioassays that discriminate between rice allelopathy and competition. These include agar assays in which phytotoxic chemicals exuded by a large mass of well-developed rice roots diffuse a short distance to inhibit root growth of a target species (Fujii, 1995, as cited

*All tradenames and company names are included for the benefit of the reader and do not imply endorsement or preferential treatment of the product by the U.S. Department of Agriculture.

in Rimando and Duke, 2003) "relay seeding" assays in which roots of a target species planted between adjacent rows of seven-day-old rice plants on perlite in a petri dish are inhibited (Navarez and Olofsdotter, 1996, as cited in Rimando and Duke, 2003) and continuous rice root exudate trapping systems employing various resins and solid phase materials to remove allelochemicals from rice roots growing in aqueous solutions (Mattice et al., 1998, as cited in Rimando and Duke, 2003).

INDUCED MUTATION: AN ALTERNATIVE TO TRADITIONAL BREEDING METHODS

Induced mutation has been a highly successful breeding tool in rice. These efforts have resulted in more than 300 mutant cultivar accessions worldwide, by far more than for any other crop species, with most being direct releases of mutants (Maluszynski, 1999). Primarily through gamma irradiation of seeds with ^{60}Co, this approach to creating genetic variability has been used since the 1940s. It has led directly or indirectly to improvements in rice such as semi-dwarfism, lodging resistance, early maturity, cold tolerance, disease resistance, yield, cooking quality, and breeding assistance tools (Bhatia, 1991; Kawai and Amano, 1991; Rutger, 1991; Wang, 1991; Maluszynski, 1999; Eizenga and Rutger, 2003).

USDA scientists in Arkansas are using induced-mutation techniques in a genetic improvement program to import and adapt high-yielding foreign indica germplasm to meet the production and marketing needs in the United States (Rutger, Bryant and Yan, 2003). Several of these indica lines can produce up to 20 percent greater yields than the best-yielding local cultivars (Yan et al., 2002). Other traits of interest in this program include earlier maturity, increased straw strength (lodging resistance), reduced plant height, and reduced amylose content (to acceptable levels of 21 to 22 percent).

Similar approaches to improving the acceptability of plant type and grain quality of 'PI 312777' also may be useful (J. N. Rutger, 2002, personal communication). Although it is high yielding, high tillering, highly weed suppressive, and of medium plant height, 'PI 312777' is prone to lodging and lacks commercially acceptable milling and cooking quality (Gealy et al., 2003, unpublished data). Induced mutation may be a relatively efficient method to improve a few quality characteristics of cultivars such as 'PI 312777' that already are suppressive to weeds when compared to the alternative of manipulating the transfer of complex weed-suppressive traits from suppressive cultivars into commercial cultivars. Weed-suppressive traits of 'PI 312777' are thought to be quantitatively inherited (Dilday, Mattice, et al., 2001; J. N. Rutger, 2002, personal communication).

QUANTITATIVE TRAIT LOCI AND MARKER-ASSISTED SELECTION AS BREEDING TOOLS

Numerous factors affect the suppression of weeds by crops, including rapid vegetative growth, high capacity to acquire and utilize nutrients, tolerance to pests and abiotic stress, and chemical defense mechanisms (e.g., allelopathy) (Olofsdotter, 2001a). However, determining the relative contributions of each of these components is complex. The genetically driven components, breeding considerations, and the potential for an allelopathy component in the improvement of weed-suppressive rice systems have been summarized recently (Olofsdotter, 2001a,b; Olofsdotter, Jensen, and Courtois, 2002). QTL mapping can be useful for complex genetic systems such as quantitatively inherited traits. QTL mapping involves correlation of phenotype (e.g., allelopathy or weed suppression in our example) and molecular marker data from a mapping population of known parentage (McCouch and Tanksley, 1991). Where strong correlations exist, marker-assisted selection can be used to further manipulate appropriate phenotypic traits through breeding.

QTLs for allelopathic activity have been identified in studies in the Philippines. A mapping population consisting of recombinant inbred lines was evaluated for allelopathy against barnyardgrass root growth using the relay seeding technique in which rice seedlings were seeded prior to weed seedlings and weed seedlings experienced no competition for light, water, or nutrients (Jensen et al., 2001). Inbred lines were derived from single seed decent from the cross 'IAC 165' (japonica upland cultivar; strongly allelopathic to *E. crus-galli*) / 'CO 39' (indica irrigated cultivar; weakly allelopathic). In these studies, four QTL on three rice chromosomes explained about 35 percent of the variability in allelopathy observed, and the genetic control of allelopathic activity appeared to be independent from that of root morphology. Alleles on chromosomes 2 and 3 from the 'IAC 165' parent appeared to be particularly responsible for the allelopathic effect against barnyardgrass. These authors also have shown that allelopathic rice 'IR 64' inhibited secondary root growth of barnyardgrass, whereas nonallelopathic rice 'Aus 196' did not (Olofsdotter, Jensen, and Courtois, 2002). They are also conducting similar QTL studies with another mapping population ('AC 1423'/'Aus 196'). Thus, even though allelopathy did not account for the majority of the weed-suppressive activity, it appeared to be under genetic control and its potential contribution was sufficiently biologically significant to be considered in future breeding and genetics programs.

Other investigators developed a mapping population ('PI 312777'/'Rexmont') to test allelopathic effects of rice leaf extracts against lettuce root growth in petri dish studies (Ebana et al., 2001a). They determined that a total of seven QTL associated with the allelopathic effect were located on chromosomes 1, 3, 5, 7, 11, and 12. The QTL on chromosome 6, and five QTL overall, explained 16 and 37 percent, respectively, of the total variation in allelopathy.

Because both of these QTL studies identified chromosome 3 as being associated with allelopathy, this chromosome may be a useful target for further investigations. Although the results from these studies may not be directly relevant to rice production under field conditions, they demonstrate the power of using marker-assisted selection to investigate quantitative traits such as weed suppression.

Potential contributions of allelochemicals to weed suppression in 'PI 312777' and other rice cultivars also have been studied recently (Ebana et al., 2001b; Mattice et al., 2001). Putative allelochemicals have been extracted from roots of 'Taichung Native 1' (Rimando et al., 2001), which is in the parentage of 'PI 312777' and 'PI 338046'. Identified chemicals included the known allelochemical *p*-coumaric acid, but this was not active against barnyardgrass at concentrations less than 3 mM (Rimando et al., 2001). Chemical identities and herbicidal modes of action of putative and proven allelochemicals and other natural products have been extensively reviewed elsewhere (Abbas and Duke, 1995; Duke et al., 2002). In their reviews, Rimando and Duke (2003) and Weston and Duke (2003) indicate that secondary metabolites produced by rice typically are limited to fatty acids, long-chain hydrocarbons, sterols, tricyclic diterpenes, cinnamic and benzoic acids, and associated derivatives. Momilactones, a class of tricarboxylic diterpenes, are interesting candidate allelochemicals in rice because they appear unique to *Oryza* species and can be induced by stress.

FUTURE PROSPECTS

Modern breeding approaches, including use of QTLs and marker assisted selection, potentially can combine weed-suppressive traits such as production of allelochemicals (Olofsdotter, Jensen, and Courtois, 2002) and competitive plant characteristics (Jennings and Aquino, 1968; Jennings and Herrera, 1968;) along with an agronomically acceptable plant type. Genomics approaches may be useful in identifying the production and modes of action of active allelochemicals and maximizing their activity. New DNA microchip analysis, which can simultaneously evaluate biological, environmental, or temporal influences on expression of thousands of rice genes, may prove to be a useful means to identify and exploit the genes involved in allelochemical production in rice (Weston and Duke, 2003).

Sorghum serves as a convincing model system of allelochemical involvement in natural weed suppression in grass grain crops. Sorgoleone is a phytotoxin produced in root hairs of grain sorghum and weedy *Sorghum* spp. relatives that reduces growth of a number of weed species via photosystem II inhibition (Czarnota et al., 1999, 2001; Gonzalez et al., 1997; Weston and Czarnota, 2001). In recent studies quantitative PCR analysis confirmed that *SOR9,* a gene thought to be involved in the biosynthesis of sorgoleone, was expressed to a greater degree in *Sorghum* spp. that produced high levels of sorgoleone, and was not detectable in wheat, corn, or rice (Yang et al., 2002). Genes conferring allelopathic activity in sorghum or other grass species such as barley (*Hordeum vulgare* L.) or winter rye (*Secale cereale* L.) (Weston and Duke, 2003) may eventually be suitable for the transformation of rice given the high proportion of shared genes among grass species. However, any such transgenic approach is likely to bring with it significant constraints inherent in identifying the appropriate target genes in one species, successfully moving and expressing these genes, and safely and consistently exploiting the allelopathic products at the proper concentration and location in the soil environment (Duke et al., 2001; Rimando and Duke, 2003; Scheffler et al., 2001).

A potential benefit from the basic research of weed-suppressive mechanisms is an improved understanding and approach to dealing with environmental concerns and increasingly limited economic resources in agriculture. Increasing awareness of conditions under which certain crop residues can inhibit subsequent crops should improve our ability to manage crop rotations. Through breeding and biotechnology, future rice cultivars may be equipped with competitive mechanisms and other genetically based interference capabilities to suppress and perhaps control substantially a broad spectrum of the troublesome weed species in reduced-input systems.

REFERENCES

Abbas, H.K. and S.O. Duke (1995). Phytotoxins from plant pathogens as potential herbicides. *Journal of Toxicology—Toxin Reviews* 14: 523-543.

Adair, C.R. (1934). Studies on blooming in rice. *Journal of the American Society of Agronomy* 26: 965-973.

Ahn, S.N. and S.D. Tanksley (1993). Comparative linkage maps of the rice and maize genomes. *Proceedings of the National Academy of Sciences, USA* 90: 7980-7984.

Baldwin, F.L., J.W. Boyd, and K.L. Smith (2000). *Recommended chemicals for weed and brush control* (MP-44). Little Rock, AR: University of Arkansas Extension Service and U.S. Department of Agriculture.

Baltazar, A.M. and R.J. Smith Jr. (1994). Propanil-resistant barnyardgrass *(Echinochloa crus-galli)* control in rice *(Oryza sativa). Weed Technology* 8: 576-581.

Beachell, H.M., C.R. Adair, N.E. Jodon, L.L. Davis, and J. Jones (1938). Extent of natural crossing in rice. *Journal of the American Society of Agronomy* 30: 743-753.

Bhatia, C.R. (1991). Economic impact of mutant varieties in India. In Editorial Staff of International Atomic Energy Agency (Eds.), *Proceedings of a Symposium of the International Atomic Energy Agency: Plant Mutation Breeding for Crop Improvement,* Volume 1 (pp. 33-46). Vienna: International Atomic Energy Agency.

Bollich, C.N., B.D. Webb, M.A. Marchetti, and J.E. Scott (1985). Registration of 'Lemont' rice. *Crop Science* 25: 883-885.

Bollich, C.N., B.D. Webb, M.A. Marchetti, and J.E. Scott (1990). Registration of 'Rexmont' rice. *Crop Science* 30: 1160.

Bridges, D.C and P.A. Baumann (1992). Weeds causing losses in the United States. In D.C. Bridges (Ed.), *Crop losses due to weeds in the United States* (pp. 75-147). Champaign, IL: Weed Science Society of America.

Carey, V.F. III, R.E. Hoagland, and R.E. Talbert (1995). Verification and distribution of propanil-resistant barnyardgrass *(Echinochloa crus-galli)* in Arkansas. *Weed Technology* 9: 366-372.

Caton, B.P., A.A. Eusebio, and A.M. Mortimer (2002). Growth analysis of nine diverse rice cultivars in competition. *Weed Science Society of America, Abstracts* 42: 33.

Caton, B.P., D. Navarez, and A.M. Mortimer (2002). Quantifying the relative effects of above- and below-ground competition on irrigated, direct-seeded rice growth in the tropics. *Weed Science Society of America, Abstracts* 42: 53.

Chandler, J.M. (1981). Estimated losses of crops to weeds. In D. Pimentel (Ed.), *Handbook of pest management in agriculture,* Volume 1 (pp. 95-109). Boca Raton, FL: CRC Press.

Chavez, R.S.C., D.R. Gealy, and H.L. Black (1999). Reduced propanil rates and naturally suppressive cultivars for barnyardgrass control in drill-seeded rice. In R.J. Norman and T.H. Johnson (Eds.), *Bobby R. Wells Rice Research Studies—1998* (pp. 43-50). Arkansas Agricultural Experiment Station Series 468. Fayetteville: University of Arkansas.

Childs, N. and A. Burdett (2000). The U.S. rice export market. In M.R. Evans (Ed.), *Rice situation and outlook yearbook* (pp. 48-54). U.S. Department of Agriculture, Market and Trade Economics Division, Economic Research Service, November, RCS-2000. Available online at <http://www.ers.usda.gov/Briefing/Rice/SpecialArticle/USriceMarket.pdf>.

Czarnota, M.A., F. Dayan, A. Rimando, and L.A. Weston (1999). Sorgoleone: Production, localization, and mode of action of a potent inhibitor produced by *Sorghum* spp. In A. Mallik, B. Barnes, D. Berube, J. Graham, K. Price, and L. Ruxton (Eds.), *Abstracts of the Second World Congress on Allelopathy—Critical analysis and future prospects.* Thunder Bay, Ontario, Canada: Lakehead University.

Czarnota, M.A., R.N. Paul, F.E. Dayan, C.I. Nimbal, and L.A. Weston (2001). Mode of action, localization of production, chemical nature, and activity of sorgoleone: A potent PSII inhibitor in *Sorghum* spp. root exudates. *Weed Technology* 15: 813-825.

De Datta, S.K. (1981). Weeds and weed control in rice. In S.K. De Datta (Ed.), *Principles and practices of rice production* (pp. 460-512). New York: John Wiley and Sons.

Devos, K.M. and M.D. Gale (1997). Comparative genetics in the grasses. *Plant Molecular Biology* 35: 3-5.

Dilday, R.H., R.S. Helms, B.R. Wells, K.A. Gravois, W. Yan, J. Lin, M. Price, M. Jalaluddin, C. Guy, and K.A. Moldenhauer (1995). Genetic, physiological and biochemical enhancement of exotic rice germplasm. In B.R. Wells (Ed.), *Bobby R. Wells rice research studies—1994* (pp. 9-15). Arkansas Agricultural Experiment Station, Series 446. Fayetteville: University of Arkansas.

Dilday, R.H., J. Lin, and W. Yan (1994). Identification of allelopathy in the USDA-ARS rice germplasm collection. *Australian Journal of Experimental Agriculture* 34: 907-910.

Dilday, R.H., J.D. Mattice, K.A. Moldenhauer, and W. Yan (2001). Allelopathic potential in rice germplasm against ducksalad, redstem and barnyardgrass. *Journal of Crop Production* 4(2): 287-301.

Dilday, R.H., K.A. Moldenhauer, W.G. Yan, and D.R. Gealy (1998). Allelopathic activities to barnyardgrass in rice and yield reduction due to barnyardgrass infestation. In R.J. Norman and T.H. Johnson (Eds.), *Bobby R. Wells Rice Research Studies—1997* (pp. 27-31). Arkansas Agricultural Experiment Station, Series 460. Fayetteville: University of Arkansas.

Dilday, R.H., P. Nastasi, J. Lin, and R.J. Smith (1991). Allelopathic activity in rice (*Oryza sativa* L.) against ducksalad (*Heteranthera limosa* (Sw.) Willd.). In J.D. Hanson, M.J. Schaffer, D.A. Ball, and C.V. Cole (Eds.), *Sustainable agriculture for the Great Plains* (pp. 193-202). Washington, DC: USDA, ARS, ARS-89.

Dilday, R.H., P. Nastasi, and R.J. Smith (1989). Allelopathic observation in rice (*Oryza sativa* L.) to ducksalad *(Heteranthera limosa). Proceedings of the Arkansas Academy of Science* 43: 21-22.

Dilday, R.H., W.G. Yan, K.A. Moldenhauer, J.W. Gibbons, F.N. Lee, and R.J. Bryant (2001). Chinese and other foreign germplasm evaluation. In R.J. Norman and J.-F. Meullenet (Eds.), *Bobby R. Wells Rice Research Studies—2000* (pp. 1-12). Arkansas Agricultural Experiment Station, Research Series 485. Fayetteville: Division of Agriculture, University of Arkansas.

Dilday, R.H., W.G. Yan, K.A.K. Moldenhauer, and K.A. Gravois (1998). Allelopathic activity in rice for controlling major aquatic weeds. In M. Olofsdotter (Ed.), *Allelopathy in rice: Proceedings of the Workshop on Allelopathy in Rice* (pp. 7-26). Manila, Philippines: International Rice Research Institute.

Dingkuhn, M., D.E. Johnson, A. Sow, and A.Y. Audebert (1999). Relationships between upland rice canopy characteristics and weed competitiveness. *Field Crops Research* 61: 79-95.

Donald, W.W. and M. Khan (1996). Canada thistle *(Cirsium arvense)* effects on yield components of spring wheat *(Triticum aestivum). Weed Science* 44: 114-121.

Dowler, C.C. (1997). Weed survey-southern states: Grass crops subsection. *Proceedings Southern Weed Science Society* 50: 227-246.

Duke, S.O., F.E. Dayan, A.M. Rimando, K.K. Schrader, G. Aliotta, A. Oliva, and J.G. Romagni (2002). Chemicals from nature for weed management. *Weed Science* 50: 138-151.

Duke, S.O., B.E. Scheffler, F.E. Dayan, L.A. Weston, and E. Ota (2001). Strategies for using transgenes to produce allelopathic crops. *Weed Technology* 15: 826-834.

Ebana, K., W. Yan, R.H. Dilday, H. Namai, and K. Okuno (2001a). Analysis of QTL associated with the allelopathic effect of rice using water-soluble extracts. *Breeding Science* 51: 47-51.

Ebana, K., W. Yan, R.H. Dilday, H. Namai, and K. Okuno (2001b). Variation in the allelopathic effect of rice with water soluble extracts. *Agronomy Journal* 93: 12-16.

Eizenga, G.C. and J.N. Rutger (2003). Genetics, cytogenetics, mutation and beyond. In C. Wayne Smith and Robert H. Dilday (Eds.), *Rice: Origin, history, technology, and production* (pp. 153-175). Crop Production Series #6149. New York: John Wiley and Sons, Inc.

Estorninos, L.E. Jr. (2000). Interference and genetic characterization of red rice (*Oryza sativa* L.) and rice *(O. sativa)*. Doctoral dissertation, University of Arkansas, Fayetteville.

Estorninos, L.E. Jr., D.R. Gealy, and R.E. Talbert (2002). Growth response of rice (*Oryza sativa)* and red rice (*O. sativa*) in a replacement series study. *Weed Technology* 16: 401-406.

Fischer, A.J., C.M. Ateh, D.E. Bayer, and J.E. Hill (2000). Herbicide-resistant *Echinochloa oryzoides* and *E. phyllopogon* in California *Oryza sativa* fields. *Weed Science* 48: 225-230.

Fischer, A.J., J. Lozano, A. Ramirez, and L.R. Sanint (1993). Yield loss prediction for integrated weed management in direct-seeded rice. *International Journal of Pest Management* 39: 175-180.

Fischer, A.J. and A. Ramirez (1993). Mixed-weed infestations: Prediction of crop losses for economic weed management in rice. *International Journal of Pest Management* 39: 354-357.

Fofana, B. and R. Rauber (2000). Weed suppression ability of upland rice under low-input conditions in West Africa. *Weed Research* 40: 271-280.

Fujii, Y. (1995). The use of allelopathic chemicals in sustainable agriculture. Proceedings of the Plant Growth Regulators Society of America 22nd National Conference, Minneapolis, MN, July 14-18.

Garrity, D.P., M. Movillon, and K. Moody (1992). Differential weed suppression ability in upland rice cultivars. *Agronomy Journal* 84: 586-591.

Gealy, D.R. (2001). Determination of root distribution of weed-suppressive rice *(Oryza sativa)* and barnyardgrass *(Echinochloa crus-galli)* in field soil using ^{13}C isotope analysis. In R.J. Norman and J.-F. Meullenet (Eds.), *Bobby R. Wells rice research studies—2000* (pp. 69-74). Arkansas Agricultural Experiment Station, Series 485. Fayetteville: University of Arkansas.

Gealy, D.R. and R.H. Dilday (2001). Progress in suppressing barnyardgrass *(Echinochloa crus-galli)* with foreign and domestic rice cultivars. In R.J. Norman and J.-F. Meullenet (Eds.), *Bobby R. Wells rice research studies—2000* (pp. 62-68). Arkansas Agricultural Experiment Station, Series 485. Fayetteville: University of Arkansas.

Gealy, D.R., R.H. Dilday, and J.N. Rutger (1998). Interaction of flush irrigation timing and suppression of barnyardgrass with potentially allelopathic rice lines. In R.J. Norman and J.-F. Meullenet (Eds.), *Bobby R. Wells rice research studies—1997* (pp. 49-55). Arkansas Agricultural Experiment Station, Series 460. Fayetteville: University of Arkansas.

Gealy, D.R., D.H. Mitten, and J.N. Rutger (2003). Geneflow between red rice *(Oryza sativa)* and herbicide-resistant rice *(O. sativa):* Implications for weed management. *Weed Technology* 17: 627-645.

Gealy, D.R., E.J. Wailes, L.E. Estorninos Jr., and R.S.C. Chavez (2003). Rice cultivar differences in suppression of barnyardgrass *(Echinochloa crus-galli)* and economics of reduced propanil rates. *Weed Science* 51: 601-609.

Gibson, K.D. and A.J. Fischer (2002). Implications of *Echinochloa* germination and duration of competition for weed management in water-seeded rice. *Weed Science Society of America Abstracts* 42: 84.

Gibson, K.D., T.C. Foin, and J.E. Hill (1999). The relative importance of root and shoot competition between water-seeded rice and *Echinochloa phyllopogon. Weed Research* 39: 181-190.

Gibson, K.D., J.E. Hill, T.C. Foin, B.P. Caton, and A.J. Fischer (2001). Water-seeded rice cultivars differ in ability to interfere with watergrass. *Agronomy Journal* 93: 181-190.

Gonzalez, V.M., J. Kazimir, C. Nimbal, L.A. Weston, and G.M. Cheinae (1997). Inhibition of a photosystem II electron transfer reaction by the natural product sorgoleone. *Journal of Agricultural and Food Chemistry* 45: 1415-1421.

Graf, B. and J.E. Hill (1992). Modeling the competition for light and nitrogen between rice and *Echinochloa crus-galli. Agricultural Systems* 40: 345-359.

Gravois, K.A., K.A.K. Moldenhauer, F.N. Lee, R.J. Norman, R.S. Helms, J.L. Bernhardt, B.R. Wells, R.H. Dilday, P.C. Rohman, and M.M. Blocker (1995). Registration of 'Kaybonnet' rice. *Crop Science* 35: 587-588.

GRIN (Germplasm Resources Information Network) (2002). Beltsville Agricultural Research Center, Agricultural Research Service, U.S. Department of Agriculture, Beltsville, MD. Available online at <http://www.ars-grin.gov/npgs/acc/acc_queries.html>.

Heap, I.M. (1997). The occurrence of herbicide-resistant weeds world-wide. *Pesticide Science* 51: 235-243.

Hoshikawa, K. (1989). *The growing rice plant: An anatomical monograph.* Tokyo, Japan: Nobunkyo.

IRRI (International Rice Research Institute) (2002). Rice science for a better world—Rice facts (1999 estimates). Available online at <http://www.riceweb.org/Ricetabindex.htm>.

Izawa, T. and K. Shimamoto (1996). Becoming a model plant: The importance of rice to plant science. *Trends in Plant Science* 1: 95-99.

Jannink, J.L., J.H. Orf, N.R. Jordan, and R.G. Shaw (2000). Index selection for weed suppressive ability in soybean. *Crop Science* 40: 1087-1094.

Jennings, P.R. and R.C. Aquino (1968). Studies on competition in rice: III. The mechanism of competition among phenotypes. *Evolution* 22: 529-542.

Jennings, P.R. and J. DeJesus Jr. (1968). Studies on competition in rice: I. Competition in mixtures of varieties. *Evolution* 22: 119-124.

Jennings, P.R. and R.M. Herrera (1968). Studies on competition in rice: II. Competition in segregating populations. *Evolution* 22: 332-336.

Jensen, L.B., B. Courtois, L. Shen, Z. Li, M. Olofsdotter, and R.P. Mauleon (2001). Locating genes controlling allelopathic effects against barnyardgrass in upland rice. *Agronomy Journal* 93: 21-26.

Jodon, N.E. (1938). Experiments on artificial hybridization of rice. *Journal of the American Society of Agronomy* 30: 294-305.

Johnston, T.H., K.A. Kuenzel, F.N. Lee, B.R. Wells, S.E. Henry, and R.H. Dilday (1984). Registration of 'Newbonnet' rice. *Crop Science* 24: 209-210.

Jones, M.P., M. Dingkuhn, G.K. Aluko, and M. Semon (1997). Interspecific *Oryza sativa* L. × *O. glaberrima* Steud. progenies in upland rice improvement. *Euphytica* 92: 237-246.

Jordan, D.L. (1997a). Efficacy of reduced-rate herbicide combinations in dry-seeded rice *(Oryza sativa)* on alluvial clay soil. *Weed Science* 45: 151-157.

Jordan, D.L. (1997b). Efficacy of reduced rates of quinclorac applied with propanil or propanil plus molinate in dry-seeded rice *(Oryza sativa). Weed Science* 45: 824-828.

Jordan, N. (1993). Prospects for weed control through crop interference. *Ecological Applications* 3: 84-91.

Kawai, T. and E. Amano (1991). Mutation breeding in Japan. In Editorial Staff of International Atomic Energy Agency (Eds.), *Proceedings of a symposium of the International Atomic Energy Agency plant mutation breeding for crop improvement,* Volume 1 (pp. 47-66). Vienna: International Atomic Energy Agency.

Kendig, A., B. Williams, and C.W. Smith (2003). Rice weed control. In C. Wayne Smith and Robert H. Dilday (Eds.), *Rice: Origin, history, technology, and production* (pp. 457-472). Crop Production Series #6149. New York: John Wiley and Sons, Inc.

Khatun, S. and T.J. Flowers (1995). The estimation of pollen viability in rice. *Journal of Experimental Botany* 46(282): 151-154.

Kong, C., X. Xu, F. Hu, X. Chen, B. Ling, and Z. Tan (2002). Using specific secondary metabolites as markers to evaluate allelopathic potentials of rice varieties and individual plants. *Chinese Science Bulletin* 47(10): 839-843.

Kropff, M.J. and L.A.P. Lotz (1993). Eco-physiological characterization of the species. In M.J. Kropff and H.H. van Laar (Eds.), *Modeling crop-weed interactions* (pp. 83-104). Wallingford, UK: CAB International.

Kropff, M.J., L.A.P. Lotz, S.E. Weaver, H.J. Bos, J. Wallinga, and T. Migo (1995). A two parameter model for prediction of crop loss by weed competition from early observations of relative leaf area of the weeds. *Annals of Applied Biology* 126: 329-346.

Kropff, M.J., S.E. Weaver, L.A.P. Lotz, J.L. Lindquist, W. Joenje, B.J. Schnieders, N.C. van Keulen, T.R. Migo, and F.F. Fajardo (1993). Understanding crop-weed interaction in field situations. In M.J. Kropff and H.H. van Laar (Eds.), *Modeling crop-weed interactions* (pp. 105-136). Wallingford, UK: CAB International.

Kuang, H.H. and D.S. Tu (1949). Studies on the fertile percentage in varietal crosses of rice hybrids. *Agronomy Journal* 41: 195-199.

Kurata, N., Y. Umehara, H. Tanoue, and T. Sasaki (1997). Physical mapping of the rice genome with YAC clones. *Plant Molecular Biology* 35: 101-113.

Kwon, S.L., R.J. Smith Jr., and R.E. Talbert (1991). Interference of red rice *(Oryza sativa)* densities in rice *(O. sativa)*. *Weed Science* 39: 169-174.

Lindquist, J.L. and M.J. Kropff (1996). Applications of an ecophysiological model for irrigated rice *(Oryza sativa)-Echinochloa* competition. *Weed Science* 44: 52-56.

Lovelace, M.L., R.E. Talbert, R.H. Dilday, E.F. Scherder, and N.W. Buerhing (2001). Use of allelopathic rice with reduced herbicide rates for control of barnyardgrass *(Echinochloa crus-galli)*. In R.J. Norman and J.-F. Muellenet (eds.), *Bobby R. Wells rice research studies—2000* (pp. 75-79). Arkansas Agricultural Experiment Station Series 485. Fayetteville: University of Arkansas.

Maluszynski, M. (1999). Crop germplasm enhancement through mutation techniques. In J.N. Rutger, J.F. Robinson, and R.H. Dilday (Eds.), *Proceedings of the International Symposium on Rice Germplasm Evaluation and Enhancement* (pp. 74-82). University of Arkansas Special Report 195. Stuttgart: University of Arkansas.

Mattice, J.D., R.H. Dilday, E.E. Gbur, and B.W. Skulman (2001). Barnyardgrass growth inhibition with rice using high-performance liquid chromatography to identify rice accession activity. *Agronomy Journal* 93: 8-11.

Mattice, J.D., T. Lavy, B. Skulman, and R.H. Dilday (1998). Searching for allelochemicals in rice that control ducksalad. In M. Olofsdotter (Ed.), *Allelopathy in rice: Proceedintgs of the Workshop on Allelopathy in Rice* (pp. 81-98). November 25-27, 1996. Manila, The Philippines: International Rice Research Institute.

McCouch, S.R. and S.D. Tanksley (1991). Development and use of restriction fragment length polymorphism in rice breeding and genetics. In G.S. Khush and G.H. Toenniessen (Eds.), *Rice biotechnology* (pp. 109-133). Wallingford, UK: CAB International.

Moldenhauer, K., J. Gibbons, and K. McKenzie (2004). Rice varieties. In E.T. Champagne (Ed.), *Rice: Chemistry and technology,* Third edition (pp. 49-76). St. Paul, MN: The American Association of Cereal Chemists, Inc.

Moldenhauer, K.A.K. and J.H. Gibbons (2003). Rice morphology and development. In C. Wayne Smith and Robert H. Dilday (Eds.), *Rice: Origin, history, technology, and production* (pp. 103-127). Crop Production Series #6149. New York: John Wiley and Sons, Inc.

Moldenhauer, K.A.K., J.W. Gibbons, F.N. Lee, R.J. Norman, J.L. Bernhardt, R.H. Dilday, J.N. Rutger, M.M. Blocker, and A.C. Tolbert (1999). Breeding and evaluation for improved rice varieties—The Arkansas rice breeding and development program. In R.J. Norman and T.H. Johnson (Eds.), *Bobby R. Wells rice research studies—1998* (pp. 20-27). Arkansas Agricultural Experiment Station, Series 468. Fayetteville: University of Arkansas.

Moldenhauer, K.A.K., K.A. Gravois, F.N. Lee, R.J. Norman, J.L. Bernhardt, B.R. Wells, R.H. Dilday, M.M. Blocker, P.C. Rohman, and T.A. McMinn (1998). Registration of 'Drew' rice. *Crop Science* 38: 896-897.

Moldenhauer, K.A.K., K.A. Gravois, F.N. Lee, R.J. Norman, J.L. Bernhardt, B.R. Wells, R.S. Helms, R.H. Dilday, P.C. Rohman, and M.M. Blocker (1994). Registration of 'LaGrue' rice. *Crop Science* 34:1123-1124.

Moldenhauer, K.A.K., F.N. Lee, R.J. Norman, R.S. Helms, B.R. Wells, R.H. Dilday, P.C. Rohman, and M.A. Marchetti (1990). Registration of 'Katy' rice. *Crop Science* 30: 747-748.

Moldenhauer, K.A.K., R.J. Norman, K.A. Gravois, F.N. Lee, R.S. Helms, B.R. Wells, R.H. Dilday, and P.C. Rohman (1991). Registration of 'Alan' rice. *Crop Science* 31: 229.

Molisch, H. (1937). *Der Einfluss einer Pflanze auf die andere-Allelopathie*. Jena: Fischer.

Moore, G., L. Aragon-Alcaide, M. Roberts, S. Reader, T. Miller, and T. Foote (1997). Are rice chromosomes components of a holocentric chromosome ancestor? *Plant Molecular Biology* 35: 17-23.

NASS (National Agricultural Statistics Service) (2002). U.S. Department of Agriculture. Available online at <http://www.usda.gov/nass>.

Navarez, D. and M. Olofsdotter (1996). Seeding technique for screening allelopathic rice (*Oryza sativa* L.). In H. Brown, G.W. Cussans, M.D. Devine, S.O. Duke, C. Fernandez-Quintanilla, A. Helweg, R.E. Labrada, M. Landes, P. Kudsk, and J.C. Streibig (Eds.), *Proceedings of the 2nd International Weed Control Conference,* June 25-28 (pp. 1285-1290). Copenhagen, Denmark: DJF, Flakkebjerg.

Ogg, A.G., and S.S. Seefeldt (1999). Characterizing traits that enhance the competitiveness of winter wheat *(Triticum aestivum)* against jointed goatgrass *(Aegilops cylindrica). Weed Science* 47: 74-80.

Olofsdotter, M. (2001a). Getting closer to breeding for competitive ability and the role of allelopathy—An example from rice *(Oryza sativa). Weed Technology* 15: 798-806.

Olofsdotter, M. (2001b). Rice—A step toward use of allelopathy. *Agronomy Journal* 93: 3-8.

Olofsdotter, M., L.B. Jensen, and B. Courtois (2002). Improving crop competitive ability using allelopathy—An example from rice. *Plant Breeding* 121: 1-9.

Olofsdotter, M., D. Navarez, and K. Moody (1995). Allelopathic potential in rice (*Oryza sativa* L.) germplasm. *Annals of Applied Biology* 127: 543-560.

Pantone, D.J. and J.B. Baker (1991). Weed-crop competition models and response-surface analysis of red rice competition in cultivated rice: A review. Crop *Science* 31: 1105-1110.

Perera, K.K., P.G. Ayres, and H.P.M. Gunasena (1992). Root growth and the relative importance of root and shoot competition in interactions between rice *(Oryza sativa)* and *Echinochloa crus-galli. Weed Research* 32: 67-76.

Rice, E.L. (1984). *Allelopathy,* Second edition. New York: Academic Press.

Rimando, A.M. and S.O. Duke (2003). Studies on rice allelochemicals. In C. Wayne Smith and Robert H. Dilday (Eds.), *Rice: Origin, history, technology, and production* (pp. 221-244). Crop Production Series #6149. New York: John Wiley and Sons, Inc.

Rimando, A.M., M. Olofsdotter, F.E. Dayan, and S.O. Duke (2001). Searching for rice allelochemicals: An example of bioassay-guided isolation. *Agronomy Journal* 93: 16-20.

Rutger, J.N. (1991). Mutation breeding in California and the United States of America. In Editorial Staff of International Atomic Energy Agency (Eds.), *Proceedings of a symposium of the International Atomic Energy Agency: Plant mutation breeding for crop improvement,* Volume 1 (pp. 155-165). Vienna: International Atomic Energy Agency.

Rutger, J.N., R.J. Bryant, and W. Yan (2003). Indica base-broadening for temperate rice. In *Third International Temperate Rice Conference, abstracts* (p. 36). Punta del Este, Uruguay: National Agricultural Research Institute.

Scheffler, B.E., S.O. Duke, F.E. Dayan, and E. Ota (2001). Crop allelopathy: Enhancement through biotechnology. *Recent Advances in Phytochemistry* 35: 257-274.

Sharma, A.R. (1994). Effect of tillage, weed-control practices and nitrogen fertilization on performance of rice *(Oryza sativa)* under intermediate deep-water condition. *Indian Journal of Agricultural Sciences* 64: 829-835.

Singh, H.P., D.R. Batish, and R.K. Kohli (1999). Autotoxicity: Concept, organisms, and ecological significance. *Critical Reviews in Plant Sciences* 18: 757-772.

Singh, R.J. and G.S. Khush (2000). Cytogenetics of rice. In J.S. Nanda (Ed.), *Rice breeding and genetics—Research priorities and challenges* (pp. 287-311). Enfield, NH: Science Publishers, Inc.

Sitch, L.A. (1990). Incompatibility barriers operating in crosses of *Oryza sativa* with related species and genera. In J. P. Gustafson (Ed.), *Gene manipulation in plant improvement II: 19th Stadler Genetics Symposium* (pp. 77-93). New York: Plenum Press.

Smith, R.J. Jr. (1974). Competition of barnyardgrass with rice cultivars. *Weed Science* 22: 423-426.

Smith, R.J. Jr. (1983). Weeds of major economic importance in rice and yield losses due to weed competition. In *Weed control in rice: Proceedings of the conference jointly sponsored by the International Rice Research Institute and the International Weed Science Society* (pp. 19-36). Manila, Philippines: International Rice Research Institute.

Smith, R.J. Jr. (1988). Weed thresholds in southern U.S. rice *(Oryza sativa). Weed Technology* 2: 232-241.

Smith, R.J. Jr. (1989). Economics of weed control in U.S. rice. In J.O. Guh, J.Y. Pyon, S.C. Kim, G.C. Chung, S.L. Kwon, and J.C. Kim (Eds.), *Weed problems and their economic management: A symposium in the twelfth conference of the Asian Pacific Weed Science Society* (pp. 39-51). Seoul, Korea: Asia-Pacific Weed Science Society and the Korean Society of Weed Science.

Smith, R.J. Jr., W.T. Flinchum, and D.E. Seaman (1977). *Weed control in U.S. rice production.* Agriculture Handbook No. 497. Washington, DC: U.S. Government Printing Office.

Stansel, J.W. (1975). The rice plant—Its development and yield. In J.E. Miller (Director), *Six decades of rice research in Texas* (pp. 9-21). Research Monograph No. 4. College Station: Texas Agricultural Experiment Station, Texas A&M University System, and U.S. Department of Agriculture.

Street, J.E. and P.K. Bollich (2003). Rice production. In C. Wayne Smith and Robert H. Dilday (Eds.), *Rice: Origin, history, technology, and production* (pp. 271-296). Crop Production Series #6149. New York: John Wiley and Sons, Inc.

Sumner, D.A. (2003). Rice subsidies and the economics of rice trade and trade negotiations. In *Third International Temperate Rice Conference, abstracts* (p. 22). Punta del Este, Uruguay: National Agricultural Research Institute.

Svejcar, T.J. and T.W. Boutton (1985). The use of stable carbon isotope analysis in rooting studies. *Oecologia* 67: 205-208.

Tseng, S.T. (1972). Some factors that contribute to poor cross-pollination in rice. In *Proceedings of the 14th Rice Technical Working Group* (p. 21). College Station: Texas Agricultural Experiment Station, Texas A&M University.

Tuong, T.P. (2000). Increasing water productivity and weed suppression of wet seeded rice: Effect of water management and rice genotypes. *Experimental Agriculture* 36: 71-89.

Valverde, B.E. and K. Itoh (2001). World rice and herbicide resistance. In S.B. Powles and D.L. Shaner (Eds.), *Herbicide resistance and world grains* (pp. 195-249). New York: CRC Press.

Valverde, B.E., C.R. Riches, and J.C. Caseley (2000). Prevention and management of herbicide resistant weeds in rice: Experiences from Central America with *Echinochloa colona.* San Jose, Costa Rica: Camara Insumos Agropecuarios. Available online at <http://www.weedscience.org/Bernal/RiceRW.html>.

Wang, L.Q. (1991). Induced mutation for crop improvement in China. In Editorial Staff of International Atomic Energy Agency (Eds.), *Proceedings of a symposium of the International Atomic Energy Agency: Plant mutation breeding for crop improvement,* Volume 1 (pp. 9-32). Vienna: International Atomic Energy Agency.

Webster, T.M. (2000). Weed survey—Southern states: Grass crops subsection. *Proceedings Southern Weed Science Society* 53: 247-274.

Weidenhamer, J.D. (1996). Distinguishing resource competition and chemical interference: Overcoming the methodological impasse. *Agronomy Journal* 88: 866-875.

Weston, L.A. and M.A. Czarnota (2001). Activity and persistence of sorgoleone, a long-chain hydroquinone produced by *Sorghum bicolor. Journal of Crop Production* 4: 363-377.

Weston, L.A. and S.O. Duke (2003). Weed and crop allelopathy. *Critical Reviews in Plant Sciences* 22: 367-389.

Yan, W., J.N. Rutger, F.N. Lee, and R.J. Bryant (2002). High-yielding indica germplasm from China. In *Proceedings of the 29th Rice Technical Working*

Group (p. 57). Little Rock: Division of Agriculture, Cooperative Extension Service, University of Arkansas.

Yang, X., R. Harmon, B. Scheffler, and L. Weston (2002). SOR9, A gene involved in sorgoleone biosynthesis? *Weed Science Society of America Abstracts* 42: 41.

Yoshida, S. (1981). *Fundamentals of rice crop science* (pp. 55-58). Los Banos, Philippines: International Rice Research Institute.

Chapter 11

The Ecology of Weed Seed Predation in Herbaceous Crop Systems

Fabián D. Menalled
Matt Liebman
Karen A. Renner

INTRODUCTION

Could weed seed predation (i.e., the consumption and destruction of weed seeds by granivorous animals) be a component of management strategies that reduce reliance on herbicide applications and cultivation? Plant ecologists have long recognized the important role that seed predation plays in the evolution, and population and community dynamics of native plants in nonagricultural systems (see Janzen, 1971; Harper, 1977; Crawley, 1989, 1992; Louda, 1989, 2001; Zhang et al., 1997a; Hulme, 2002; and references therein). As we show in this review, researchers are now devoting a substantial amount of effort to better understanding the role of weed seed predation in agroecosystems, since this process appears to be important for regulating weed density and community composition.

The study of weed seed predation in agroecosystems is part of a broader shift in the weed science research agenda that has arisen in response to increasing public support for environmentally sound agricultural systems that are more reliant on ecological processes and less reliant on synthetic chemicals (Hall et al., 2000; Kropff and Walter, 2000). The development of such ecologically based weed management strategies requires a clear understanding of how crop-weed interactions might be managed through suites of tactics that delay weed emergence, favor crop resource consumption, subject weeds to toxic allelochemicals, promote weed diseases, and encourage weed mortality (Liebman and Gallandt, 1997).

We thank Paula Westerman, Douglas Landis, Francis Drummond, and three anonymous reviewers for their helpful comments on the manuscript.

In this chapter, we evaluate the last of these processes by exploring the role that seed predators might have as weed biological control agents in herbaceous crop systems. We start by reviewing several key concepts of seed predation developed in nonagricultural systems. Second, we evaluate the magnitude and impact of weed seed predation in herbaceous crops. Third, we assess the relationship between crop and noncrop habitats in determining weed seed predation rates. Fourth, we explore habitat management strategies to conserve weed seed predators in agricultural systems. Finally, we describe potentially fruitful avenues for future research. We must acknowledge that in comparison with available data concerning insect pest regulation through predation, the information available on the magnitude and impact of weed seed predation in herbaceous crops is sparse and sometimes contradictory. It is hoped that this review will help to fill that knowledge gap by stimulating future studies addressing the role of seed predation in agroecosystems.

OVERVIEW OF KEY CONCEPTS

Seed predation, also referred to as granivory, is a particular type of predation in which a mobile predator consumes a sessile prey. This prey has high food quality but is usually too small to allow the full development of the predator with the consumption of only a single prey item (Janzen, 1971; Crawley, 1992; Hulme, 1993). Conceptually, and due to differences in mechanisms and ecological consequences, it is helpful to distinguish between predispersal seed predation and postdispersal seed predation. These two types of seed mortality differ with regard to the guilds of animals involved, the relative importance of various plant traits affecting the impacts of seed predators, and their influence on plant population and community dynamics (Hulme, 2002).

Predispersal seed predation involves the consumption of seeds on the parent plant before they are shed. These seeds represent a spatially and temporally aggregated resource, and are often defended physically and chemically through mechanisms provided by the parent plant. Small and sedentary specialist feeders belonging to the insect orders Diptera, Lepidoptera, Coleoptera, and Hymenoptera are mostly responsible for predispersal seed predation (Crawley, 1992). Numerous studies have documented the existence of predispersal seed predation in nonagricultural habitats such as prairies (Platt, Hill, and Clark, 1974; Louda, Potvin, and Collinge, 1990), tropical forests (Peres, 1991; Greig, 1993; Levey and Byrne, 1993; Calvo-Irabien and Islas-Luna, 1999), temperate forests (Holm, 1994; Kaye, 1999), wetlands (Kudoh and Whigham, 1998), mangroves (Smith, 1987; Smith et al., 1989; Clarke, 1992), old-fields (Franson and Willson, 1983), and deserts (Boyd and Brum, 1983; Keeley et al., 1984; Mull and MacMahon, 1997).

Conversely, large and mobile generalist herbivores such as rodents, birds, and insects, including Coleoptera and Hymenoptera, consume seeds after they have been dispersed from the parent plant, i.e., postdispersal seed predation (Crawley, 1992). Seeds may be partially protected from seed predators by physical mechanisms (e.g., thorns, hard seed covers), chemical components, or mimicry strategies. As with predispersal seed predation, postdispersal seed predation has been assessed in many nonagricultural habitats including old-fields (Mittelbach and Gross, 1984; Myster and Pickett, 1993; Reader 1993), grasslands (Hoffmann, Redente, and McEwen, 1995; Howe and Brown, 2000), mangroves (Robertson, Giddins, and Smith, 1990), and deserts (Abramsky, 1983; Brown et al., 1986; Davidson, 1993).

Janzen (1971), Louda (1989), Crawley (1992), and Hulme (1998, 2002) have provided thorough reviews of the magnitude and impact of seed predation in nonagricultural settings. Tooley and Brust (2002) reviewed the importance of carabid beetles as weed seed predators in agroecosystems. Several generalizations emerge from these reviews:

1. Seed predation can act at many different organizational levels within ecosystems ranging from internal plant energy budgets to plant community composition.
2. Due to several biotic and abiotic factors, seed predation is highly variable in time and space.
3. Seed predators usually feed selectively and have a strong potential to affect the balance between plant species.
4. The cumulative effects of seed predation and other biotic sources of plant mortality, such as seedling herbivory and seed decay, can determine spatial patterns of plant abundance.
5. Seed predators interact with seed dispersing agents and herbivores in determining plant demographic patterns.
6. Seed predation involves plant-insect interactions that have coevolved in chemical, spatial, and temporal dimensions.

Based on these points, critical questions for agroecologists interested in weed seed predation include the following:

1. Do weed seed predators eat crop seeds?
2. What is the relationship between management practices and seed predator abundance and activity?

3. To what extent does weed seed predation influence weed plant demography and community dynamics?
4. How should seed predation be integrated with other sources of plant mortality and stress in the development of sustainable weed management systems?

In the following sections we address these and other issues related to weed seed predation in agricultural settings.

WEED SEED PREDATION IN HERBACEOUS CROP SYSTEMS

Do Weed Seed Predators Eat Crop Seeds?

Weed seed predation, as with any biological control method, has inherent risks. Gurr and Wratten (2000) comprehensively reviewed off-target effects of pest animal, weed, and pathogen biological control programs and concluded that ecology could provide the necessary framework to minimize risks and enhance chances of achieving success. In herbaceous crop systems, a key issue in the implementation of practices that increase weed seed consumption is whether seed predators eat crop seeds. Vertebrate and invertebrate seed predators differ in the temporal and spatial scale of their effects and plant preferences (Hulme, 2002). Therefore, it is helpful to assess them separately with regard to their potential to remove crop seeds.

Brust and House (1988) reported that invertebrate seed predators, particularly ground beetles, consumed wheat (*Triticum aestivum* L.) seeds located on the soil surface of tilled and no-tilled fields. In another study, Bassett (1978) showed that larvae of coleoptera *Zabrus tenebrioides* (Goeze) damage wheat seeds. We conducted a series of greenhouse and laboratory trials to test if seed predation by invertebrates represents a threat to crops (Menalled et al., 2001). Specifically, we assessed the feeding preference for dry and germinated corn (*Zea mays* L.), soybean [*Glycine max* (L.) Merr.], and wheat seeds by three ground beetles commonly found in Michigan agroecosystems: *Amara aenea* (DeGeer), *Anisodactylus sanctaecrusis* (Fabricius), and *Harpalus pensylvanicus* (DeGeer). Although these three species are known to eat large numbers of weed seeds (Johnson and Cameron, 1969; Best and Beegle, 1977; Lund and Turpin, 1977; Brust and House, 1988; Brust, 1994), crop seed predation was very low in all conditions included in the experiment (see Table 11.1).

The red imported fire ant (*Solenopsis invicta* Buren) is an opportunistic omnivore that colonizes no-till fields of herbaceous crops including soy-

TABLE 11.1. Mean number (1 SEM) of 50 crop seeds eaten by three common ground beetles (sample size = 10 trials).

	Corn		Soybean		Wheat	
Seed predator	**Dry**	**Germinated**	**Dry**	**Germinated**	**Dry**	**Germinated**
Amara aenea	0	1.00 (0.21)	0.10 (0.10)	0.40 (0.16)	0.10 (0.10)	1.41 (0.37)
Anisodactylus sanctaecrusis	0	0.55 (0.22)	0	0.79 (0.25)	0	0.22 (0.13)
Harpalus pensylvanicus	*	*	*	*	1.56 (0.76)	1.21 (0.44)

Source: Adapted from Menalled et al., 2001.
*Not tested because seed predators are not active at the time of corn and soybean planting.

bean, corn, and sorghum [*Sorghum bicolor* (L.) Moench]. Seaman and Marino (2003) observed *S. invicta* consuming seeds of several agricultural weeds, including redroot pigweed (*Amaranthus retroflexus* L.), giant ragweed (*Ambrosia trifida* L.), common lambsquarter (*Chenopodium album* L.), and goldenrod (*Solidago altissima* L.). However, several investigators have reported that fire ants consume crop seeds and seedlings in both field and laboratory conditions (Adams et al., 1983; Morrison et al., 1997; Morrison, Williams, and Oi, 1999; Shatters and Vander Meer, 2000) with associated yield reductions (Banks et al., 1990). Drees et al. (1991) suggested a series of cultural practices to reduce the period of crop seed susceptibility to fire ant. These practices aim to avoid leaving crop seeds close to the soil surface, close furrow openings, and ensure that furrows remain closed.

Certain vertebrates have also been cited as potential agricultural pests capable of damaging or consuming crop seeds. For example, wood mice (*Apodemus sylvaticus* L.) have been observed eating sugar beet seeds in Europe (Green, 1979). Interestingly, this mouse is also considered a beneficial organism because it consumes large quantities of weed seeds; its damaging impact on crop seeds can be reduced by providing alternative food resources or by changing crop sowing times (Pelz, 1986).

Most birds that feed on weed seeds in agroecosystems prefer stubble fields and avoid growing cereals (Wilson et al., 1999). However, in European crop fields, the skylark (*Alauda arvensis* L.) has been reported to eat seeds of sugar beet and various cereals (Edgar and Isaacson, 1974; Robinson and Sutherland, 1996). Red-winged blackbirds (*Agelaius phoeniceus* L.), common grackles (*Quiscalus quiscula* L.), and starlings (*Sturnus vulgaris* L.) eat corn seeds (Dolbeer, Woronecki, and Stehn, 1986; Bernhardt, van Allsburg, and Dolbeer, 1987; Blackwell and Dolbeer, 2001). The use of management practices that include resistant cultivars and bird-scaring techniques can reduce bird damage on seeds (Dolbeer, 1990). Also, the provision of alternative food resources has been postulated as a strategy to reduce bird pressure toward crop seeds (Wilson et al., 1999). This objective can be reached through the conservation of uncultivated field margins, hedgerows, riparian margins, and road verges. As discussed later in this chapter, provision of these habitats within agricultural landscapes represents a viable tool to enhance the survival of several groups of weed seed predators.

Because there is little evidence that native granivores consume large quantities of crop seeds, favoring management practices that promote high levels of seed predation appears to be a relatively safe strategy for weed control in many herbaceous crop systems. The specialized host-feeding range of predispersal weed seed predators and their short-distance dispersal habits appear to be the main factors responsible for the reduced risk of crop attack (Crawley, 1992). At least four factors may be responsible for a low-risk crop seed consumption by postdispersal weed seed predators. First, seeds of most annual agronomic crops are one to three orders of magnitude larger than weed seeds (Mohler, 1996). It has been observed that invertebrate seed predators, including carabid beetles such as *Harpalus rufipes* De Geer and small birds such as chaffinches (*Fringilla coelebs* L.), feed selectively, preferring smaller seeds (Hartke, Drummond, and Liebman, 1998; Holmes and Froud-Williams, 2001). For example, Tooley and Brust (2002) reported that because carabids feed by holding prey items between their tarsal spurs, there is a maximum seed size they can handle. Thus, large crop seeds should be less prone to be eaten than small weed seeds. Second, the activity density of postdispersal weed seed predators, such as the field cricket (*Gryllus pennsylvanicus* Burmeister) and several ground beetles species, peaks at the time that many weed seeds are shed (Zhang et al., 1997b; Carmona and Landis, 1999; Carmona, Menalled, and Landis, 1999).

Synchronization between predator activity density and weed seed shed at a time when crop seeds are not present at the soil surface suggests that field crickets, ground beetles, and other invertebrates may play an important role in weed population and community dynamics with no risk to crop seeds. Third, postdispersal seed predation occurs mostly on the soil surface and to a much lesser extent after seeds have been incorporated into the seedbank (Crawley, 1992; White, 2000). Planting buries crop seeds several centimeters below ground, and therefore crop seed predation should be very low. Finally, crop seedlings emerge a few days after they are planted, reducing their temporal window of susceptibility to vertebrate and invertebrate seed predators. A caveat should be noted: the extent to which these assertions represent all pre- and postdispersal seed predation agents, rodents and birds in particular, is unknown. Future studies should clearly address this issue.

Predispersal Weed Seed Predation

Relatively little is known about the extent and impact of predispersal seed predation in herbaceous crop systems. In this section we summarize the major outcomes of the few studies that have addressed this issue and highlight areas where more research is needed.

Forsyth and Watson (1985) determined that 20 to 85 percent of Canada thistle [*Cirsium arvense* (L.) Scop.] plants were infested by *Orellia ruficauda* (Fabr.), with 20 to 80 percent of seeds on the plants damaged by insects. Canada thistle is a perennial weed found in habitats such as abandoned fields, ditch banks, roadsides, pastures, rangelands, and crop fields (Uva, Neal, and DiTomaso, 1997). Because Forsyth and Watson (1985) conducted this study along field edges, roadsides, and marginal areas near Montreal, Canada (A. K. Watson, personal communication, 2001), the implications of their results for the biological control of Canada thistle in herbaceous crop systems are unclear.

Giant ragweed is a serious herbaceous weed of corn, soybean, and other crops. Vitolo and Stiles (1987) observed different types of larval insects eating giant ragweed seeds prior to dispersal. In corn fields, Regnier, Harrison, and Schmoll (1999), and Harrison and colleagues (2001) observed that insect larvae such as fruit flies, weevils, and moths accounted for 13 to 19 percent of the loses in viability of giant ragweed seeds before dispersal.

Although management practices may influence the biology and ecology of predispersal seed predators, few studies have specifically attempted to analyze this relationship. Exceptions are the investigations of Griffiths and Swanton (1999), Swanton, Griffiths, and colleagues (1999), and Nurse and Swanton (2000), who evaluated the effect of crop row width and tillage on common lambsquarters, redroot pigweed, and green pigweed (*Amaranthus powellii* S.Wats.) predispersal seed predation. In corn fields, predispersal seed predation increased by 12 percent with a 38 cm row width as compared with the traditional 76 cm row width (Griffiths and Swanton, 1999). In another study conducted in corn fields, Swanton, Griffiths, and colleagues (1999) reported that predispersal seed predation reduced redroot and green pigweed seed production by 3 to 40 percent. Predispersal predation of the two *Amaranthus* species was highest in wide-row, no-till soybean fields, but no effect of row width or tillage regime was detected on the predation of common lambsquarters seeds (Nurse and Swanton, 2000).

Very little is known regarding synergistic associations between weed seed predators and weed seed pathogens. However, the integration of these mortality sources represents a potentially valuable tool for enhancing the weed biological control (Kremer 2000). In two soybean fields in Missouri, Kremer and Spencer (1989a,b) evaluated the integration of *Niesthrea louisianica* Sailer, a scentless plant bug that feeds on seeds, with fungi of the genera *Fusarium* and *Alternaria* for velvetleaf (*Abutilon theophrasti* Medik.) control. They observed that the combined effect of seed predation and fungal infection reduced velvetleaf seed viability 15 to 17.5 percent compared with 87.5 to 95.5 percent at insect-free sites. The authors speculated that if no more seeds were incorporated into the seed bank, the integrated use of these biological control agents could reduce velvetleaf abundance by 98 percent after 24 months. Future studies could evaluate the accuracy of this prediction.

The fact that the importance of predispersal weed seed predation on weed demography, weed community dynamics, and crop-weed interactions has not been fully evaluated in herbaceous agroecosystems opens many new research opportunities. As stated by Louda and Potvin (1995), critical demographic questions include the extent to which inflorescence damage reduces seed output, seedling establishment, and mature progeny. These studies should also consider the existence of compensatory flowering (McNaughton, 1986) and seed development (Hendrix, 1979) in response to damage. Also, it is important to understand how predispersal seed predation may affect seed dispersal (Hulme, 2002).

Postdispersal Weed Seed Predation

Postdispersal seed predation acts as a filter on the seed rain, and thus affects seed densities and distributions (Hulme, 2002). Mice and birds have been identified as vertebrate postdispersal weed seed predators in agroecosystems (Getz and Brighty, 1986; Brust and House, 1988; Díaz, 1994; Díaz and Tellería, 1994; Cardina et al., 1996; Cromar, Murphy, and Swanton, 1999; Wilson et al., 1999; White, 2000). Invertebrate postdispersal weed seed predators in crop fields include several species of ground beetles (Briggs, 1965; Kirk, 1972; Best and Beegle, 1977; Brust and House, 1988; Povey, Smith, and Watt, 1993; Sunderland, Lövei, and Fenlon, 1995; Cardina et al., 1996; Hartke, Drummond, and Liebman, 1998), ants (Carroll and Risch, 1984; Risch and Carroll, 1986; Brust and House, 1988; Tooley and Brust, 2002), slugs (Cardina et al., 1996), and crickets (Brust and House, 1988; Carmona, Menalled, and Landis, 1999). Tooley and Brust (2002) have provided a comprehensive review of weed species attacked by granivorous carabid beetles. In the following sections we discuss several ecological aspects of postdispersal weed seed predation in herbaceous crops.

Invertebrate Postdispersal Seed Predator Behavior

Understanding the response of seed predators to variations in seed abundance, location, and plant species composition lies at the heart of the development of a weed biological control program. Feeding-behavior studies have shown that carabid beetle larvae of several species, including *Harpalus rufipes* and *Harpalus erraticus* Say, consume weed seeds on the soil surface and accumulate them in burrows 8 to 20 cm deep (Kirk, 1972; Luff, 1980; Zhang, 1993; Hartke, Drummond, and Liebman, 1998). The implications of this behavior for weed management are multiple. First, weed seedling germination and emergence rates are negatively related to depth of burial. That is, deep-buried seeds have fewer chances of establishing seedlings than those located close to the soil surface. Second, since weed seeds germinating deep in the soil take more time to emerge and consume a greater amount of seed resources during growth to the surface, the resulting seedlings are weaker than those generated by seeds located closer to the soil surface (Herr and Straub, 1970; Mester and Buhler, 1990). Finally, tillage could bring buried seeds back to the surface, thereby increasing their chances of arriving in sites favorable for germination. However, as discussed later in this chapter, tillage also reduces the abundance of carabid beetles.

Ants are highly selective seed collectors with the capability of gathering large numbers of weed seeds and storing them in their nests (Kelrick et al., 1986; Risch and Carroll, 1986; Crist and MacMahon, 1992). Despite the large number of seeds attacked by ants, the implications of this behavior for weed population and community dynamics are not straightforward. On one hand, ant nests may provide microsites rich in essential nutrients such as phosphorus and nitrogen that may boost weed seedling establishment and growth (Culver and Beattie, 1983). On the other hand, weed seeds captured by ants suffer high rates of destruction. An important issue that future studies may evaluate is the combined effect of ant seed transport and ant seed predation on weed germination, abundance, and dispersion. Future studies could also evaluate the existence of switching patterns of seed predator behavior under the presence of a more diverse weed seed community and the existence of inter- and intraspecific interference among weed seed predators.

Tillage Regime and Postdispersal Seed Predation

Tillage represents a major disturbance in agricultural fields that buries weed seeds and affects the abundance and diversity of seed predators. Studies have shown that the increased ground cover and reduced soil disturbance associated with no-tillage systems provide a favorable habitat for seed predators (House and All, 1981; Brust, Stinner, and McCartney, 1986). Because the abundance of invertebrate generalist predators correlates with prey removal rates in corn and soybean fields (Menalled, Lee, and Landis, 1999, 2001), it follows that weed seed predation should be higher in no-tillage fields compared to conventionally tilled fields. Different investigators have tested this hypothesis, but results are far from conclusive.

In a one-year study, Brust and House (1988) evaluated the magnitude of postdispersal seed predation of four broadleaf weed species [common ragweed (*Ambrosia artemisiifolia* L.), redroot pigweed, sicklepod (*Cassia obtusifolia,* L.) and jimsonweed (*Datura stramonium* L.)] and one grain species (wheat). Tests included 100 replicates in two 0.5 ha soybean fields planted in a wheat-soybean-corn rotation in North Carolina. In accordance with the proposed hypothesis, they determined that seed removal was 2.3 times greater in no-till and low-input soybean fields (68 percent seed removal) than in conventionally farmed fields (27 percent seed removal). However, in a four-year study conducted in continuous corn fields in Ohio, Cardina and colleagues (1996) found similar rates of velvetleaf seed predation between no-till and conventionally tilled systems. In a third study, Cromar, Murphy, and Swanton (1999) evaluated postdispersal seed predation of common lambsquarters and barnyardgrass [*Echinochloa crus-galli* (L.) Beauv.] in a corn-soybean rotation located in Ontario, Canada. Their results showed that seed consumption was higher in no-till and moldboard plowed plots (averaging 32 percent in both) than in chisel-plowed plots (averaging 24 percent).

The differential response of vertebrates and invertebrates to tillage regimes, and their foraging behavior and dispersal ability, may provide a possible explanation for the lack of consistency among studies. For example, radar-tracking and mark and recapture studies showed that carabid beetles living in agricultural fields have a net displacement ranging from 5.3 $m \cdot day^{-1}$ (Wallin and Ekbom, 1988) to 11.2 $m \cdot day^{-1}$ (Zhang et al., 1997b). Radiotelemetry studies indicated that the total distance traveled per night by an adult wood mouse ranges between 130 ± 26.1 m and 1201 ± 112.5 m, depending on habitat characteristics and season (Corp, Gorman, and Speakman, 1997). It is possible that the small size of the plots used by Cromar, Murphy, and Swanton (1999) (6 m × 12 m) and Cardina et al. (1996) (30 m × 12 m) affected the investigators' ability to determine the influence of tillage on seed predation.

To overcome difficulties associated with high granivore mobility and small plot size, we evaluated the influence of different crop management systems on carabid beetle activity-density and weed seed predation by invertebrates in relatively large soybean fields (Dauer, Fox, and Menalled, 2001). Studies were conducted in six replicated 1 ha plots established in a corn-soybean-wheat rotation in Michigan. Activity density of carabid beetle species known to eat weed seeds was higher in the no-till plots that in moldboard plowed plots. In accordance, invertebrates removed more fall panicum (*Panicum dichotomiflorum* Michx.) and common lambsquarters seeds in the no-till system than in the moldboard plowed systems.

Influence of Crop Residue on Seed Predation

Crop residue management influences the survivorship, behavior, and abundance of vertebrate and invertebrate seed predators (Cárcamo, Neimalä, and Spence, 1995; Clark, Gage, and Spence, 1997). Ground cover provides suitable microhabitats for seed predators but could also reduce their mobility. This, in turn, could translate into spatially aggregated patterns of seed predation (Reader, 1993; Cromar, Murphy, and Swanton, 1999). Cromar, Murphy, and Swanton (1999) used the experimental plots described in the previous section to evaluate the influence of different kinds of crop residue on postdispersal seed predation. In no-till systems, weed seed predation rates were highest in corn residue (31 percent) and lowest in soybean and wheat residue (24 and 21 percent, respectively). Certainly, more work is needed to understand the relationship between crop residue management and weed seed consumption.

Crop Type and Postdispersal Seed Predation

Different crops have different planting and harvest dates, growth habits, and management practices. These differences may affect seed predator abundance and behavior. For example, crop type significantly affected abundance and movement patterns of seed predators including wood mice (*Apodemus sylvaticus* L.) (Macdonald et al., 2000) and carabid beetles (Kromp, 1999). Therefore, it is logical to expect different values of weed seed removal among different crops. Despite the potential importance that this hypothesis has for weed management, to our knowledge only two studies evaluated this aspect of seed predation.

Anderson (1998) estimated seed predation rates of common lambsquarters, scentless chamomile (*Matricaria perforata* Merat), black bindweed [*Bilderdykia* (*Polygonum*) *convolvulus* (L.) Dumort], shepherd's-purse [*Capsella bursa-pastoris* (L.) Medikus], field pennycress (*Thlaspi arvense* L.), and pale smartweed (*Polygonum lapathifolium* L.) in two crops: oat (*Avena sativa* L.) and a forage mixture of red clover (*Trifolium pratense* L.), timothy (*Phleum pratense* L.), and meadow fescue (*Festuca pratensis* Huds.). Although not all weed species were tested in replicated plots, Anderson (1998) reported the highest rates of black bindweed seed predation in oat fields, whereas seed consumption of field pennycress and pale smartweed were greater in the forage mixture than in oat fields. Potential causes for the observed differences include changes in seed predator fauna as a function of crop type and/or variation in seed predator behavior in different habitats. Future studies could assess the validity of these hypotheses.

In another study, Zhang, Drummond, and Liebman (1998) observed that the activity density of *Harpalus rufipes* was higher in barley (*Hordeum vulgare* L.) than in potato (*Solanum tuberosum* L.) fields in northern Maine. *Harpalus rufipes* is the dominant carabid beetle in arable crop systems in the region and feeds on various types of seeds as well as soft-bodied insects (Sunderland et al., 1987). Accordingly, Zhang (1993) observed that seed predation of barnyardgrass and wild mustard [*Brassica kaber* (DC.) L.C. Wheeler] was greater in barley fields than in potato fields. Wire mesh cages that excluded vertebrate seed predators were used in experiments, and Zhang (1993) attributed the majority of seed losses to *H. rufipes.*

IMPACT OF SEED PREDATION ON WEED POPULATION AND COMMUNITY DYNAMICS

The implementation of management practices that encourage pre- and/or postdispersal seed predation requires a clear understanding of how these forms of mortality influence weed population and community structure. The scarcity of information available precludes us from making definitive statements, but several trends can be identified.

Seed Predation and Weed Population Dynamics

Although seed-feeding organisms may reduce the density of seeds in and on the soil, the relationship between seed abundance and seedling recruitment may not be linear. Thus, high seed loses may have little impact on the density of established plants. Andersen (1989) and Crawley (1989, 1992) proposed that when seedbank density was high, an increase in the abundance of "safe sites" for germination and establishment would translate into higher seedling densities. In agroecosystems, management practices maintain weed density at low values and competition for suitable microsites for recruitment probably plays a minor role in weed establishment. Consequently, an increase in seed predation large enough to reduce seed-bank density might have a large impact on weed seedling abundance. Confirming this hypothesis, modeling analysis has shown that postdispersal seed mortality is one of the most sensitive parameters regulating the abundance of wild oat (*Avena sterilis* L.) (Gonzales-Andujar and Fernandez-Quintanilla, 1991), velvetleaf, and green foxtail [*Setaria viridis* (L.) Beauv.] (Jordan et al., 1995). In an empirical study conducted within corn fields in Michigan, White (2000) assessed the importance of seed removal on weed emergence by establishing a series of cages to exclude predators and observed that seed predation by vertebrates decreased emergence of velvetleaf by 9 percent. In the same study, the joint predation by vertebrates and invertebrates decreased giant foxtail (*Setaria faberi* Herrm.) emergence by 15 percent.

Impact of Seed Predation on Weed Communities

Several studies conducted in herbaceous crops have demonstrated that seed predators exhibit preferences for particular species of weeds (Lund and Turpin, 1977; Brust and House, 1988; Zhang, 1993; Marino, Gross, and Landis, 1997; White, 2000). Also, weed predator behavior is affected by weed seed density (Zhang, 1993; Cardina et al., 1996; Cromar, Murphy, and Swanton, 1999). Differential seed consumption coupled with selective herbivory and limited seedling survival have been shown to be important factors associated with changes in plant communities of natural habitats (Louda, 1982a,b; Davidson, Samson, and Inouye, 1985; Smith, 1987; Smith et al., 1989; Samson, Philippi, and Davidson, 1992). The extent to which changes in pre- and postdispersal seed predation may affect weed community structure in herbaceous crop systems is still poorly understood. However, Tooley and Brust (2002) argue that the selective feeding of carabid beetles on seeds of different weed species could result in changes in weed community composition.

In a greenhouse experiment, Brust (1994) observed that the different predation habits of invertebrates influenced broadleaf weed growth and competitive ability. He further suggested that these types of studies should be conducted under more realistic conditions. Carroll and Risch (1984) conducted a field experiment in a tropical agroecosystem of Mexico to evaluate how predator feeding preference influenced weed community composition. They determined that the seed-feeding ant, *Solenopsis geminata* Fabricius, is the most abundant ant species consuming large amounts of small weed seeds, particularly grasses. Although the authors did not specifically test the impact of *S. geminata* on weed community dynamics, they argued that the differential ant feeding preference might modify weed species composition.

A manipulative field trial consisting of a "pulse/chase" experiment combined with seed predator exclosures could be used to test the combined effect of seed-bank abundance, seed consumption, and weed community dynamics. The "pulse/chase" protocol has been used by several investigators (Teo-Sherrell, Mortensen, and Keaton, 1996; Buhler and Hartzler, 2001) as an approach to alleviate difficulties associated with the heterogeneous and patchy nature of native weed seed banks. In such experiments, a large and uniform source of weed seed is added to the seed bank at a sufficient density to minimize "background noise" attributed to the existent seed bank. Exclusion provides a simple approach to determine the effect of seed predation on seedling establishment and cohort demography (Louda and Potvin, 1995).

HERBACEOUS CROP FIELDS, NONCROP HABITATS, AND WEED SEED PREDATION

Many trophic interactions occurring in agricultural systems act on large spatial scales (Thies and Tscharntke, 1999; Tscharntke, 2000). Several studies have shown that the presence of noncrop habitats within agricultural landscapes represents a critical factor affecting the abundance, diversity, and fecundity of natural enemies of pests (Gut et al., 1982; Bommarco, 1998; Menalled et al., 1999; Thies and Tscharntke, 1999). Specifically, noncrop habitats can provide shelter and food for many potential weed seed predators including mammals (Pollard and Relton, 1970; Angelstam, Hansson, and Pehrsson, 1987; Montgomery and Dowie, 1993, Kotzageorgis and Mason, 1997; Todd, Tew, and Macdonald, 2000), birds (Best, 1983; Sykes and Hannon, 2001), and insects (Lys and Nentwig, 1992; Kiss et al., 1997; Lee, Menalled, and Landis, 2001; Thomas et al., 2001). Consequently, the study of weed seed predation should go beyond the within-field level of analysis.

Agricultural Landscape Structure and Weed Seed Predation

The current tendency of removing hedgerows, fencerows, and woodlots generates simplified agricultural landscapes composed of intensively disturbed early successional systems (the crop fields) interspersed with fragments of more stable habitats (Auclair, 1976; Pogue and Schnell, 2001). This tendency probably affects both predispersal and postdispersal seed removal.

To our knowledge, only one study assessed the influence of agricultural landscape structure on predispersal seed predators (Steffan-Dewenter, Münzenberg, and Tscharntke, 2001). Indeed, the authors quantified not just predispersal seed predation, but the influence of landscape structure on pollinator biology, and the joint effects of seed predation and pollination. Flowering plants of brown knapweed (*Centaurea jacea* L.) were placed along a gradient of increasing agricultural landscape complexity. The number of flower-visiting bees and the proportion of flower heads damaged by predispersal seed predators increased as landscape complexity increased. Interestingly, the counterbalancing effects of these two processes resulted in the absence of a correlation between the abundance of nonagricultural areas and the number of damaged seeds per flower. The authors speculated that this result was a consequence of the contrasting responses of insects to habitat complexity, with changes in pollinator numbers explained by landscape variables occurring at small scales (up to 1,000 m) and seed predator numbers responding to large-scale variables (up to 2,500 m).

In a comparison of a complex agricultural landscape composed of a mixture of crop and noncrop habitats with a simple landscape consisting of a homogenous area with low crop diversity and few noncrop habitats, Menalled and colleagues (2000) found a trend toward higher weed seed removal rates in the complex landscape. Because Menalled and colleagues (2000) did not assess the abundance of seed predators in these landscapes, a complete analysis of the mechanisms responsible for the observed results is missing.

Díaz (1992a,b, 1994) and Díaz and Tellería (1994) evaluated the distribution patterns of granivorous birds, rodents, and ants, as well as rates of postdispersal weed seed predation in cereal crop landscapes of central Spain. Overall, these studies demonstrated that the magnitude of postdispersal weed seed predation was severely limited by the availability of microsites suitable for nest building and lack of refuges from adverse conditions. Interestingly, the distribution pattern of seed predators was determined mainly by vegetation structure and patterns of human use of the land, with each seed predator group responding differently to these variables (Díaz, 1994). These results sharply contrast with the complex network of direct and indirect relationships observed in natural systems among different groups of seed predators. For example, in desert systems of the United States, seed predation has been postulated as one mechanism responsible for the coexistence of diversified plant and animal communities (Brown et al., 1986).

Field Margins and Weed Seed Predation

In an ideal situation, seed predators overwinter in refuges, colonize crop fields, eat weed seeds, and return to refuges following a "cyclic colonization pattern" (Wissinger, 1997). Therefore, it would seem that weed seed removal should be higher closer to field margins than in the center of crop fields. Marino, Gross, and Landis (1997) tested this hypothesis in four conventional-tillage corn fields and one no-tillage corn field having a hedgerow of unmanaged trees or shrubs on one side. The authors placed seeds of velvetleaf, redroot pigweed, common lambsquarters, fall panicum, and yellow foxtail [*Setaria glauca* (L.) Beauv.] at 5 m and 100 m from hedgerows, but no differences were detected in rates of seed removal near hedgerows compared with interior areas of corn fields. The researchers speculated that their inability to detect differences in seed removal in edge versus interior areas might be a consequence of the small size of the selected fields. They further suggested that evaluation of the impact of noncrop habitats on weed seed removal should be conducted at a larger scale of analysis.

HOW CAN WE CONSERVE SEED PREDATORS IN AGROECOSYSTEMS?

Weed seed predation represents a form of biological control in which insects, rodents, and birds help farmers to control weeds by reducing seed abundance. In this context, we argue that to maximize the beneficial impact of seed predators, agricultural systems should provide for the basic biological requirements needed by seed predators. These requirements include food, overwintering sites, presence of mates, and shelter from adverse conditions. However, in extensively tilled annual monocultures the detrimental effects of management practices, such as pesticide applications, plowing, and the removal of noncrop habitats, impede the establishment of large and effective numbers of natural enemies (Menalled, Landis, and Dyer, 2004). Also, the adoption of newly developed technologies, such as herbicide-tolerant crops paired with highly effective herbicides, may reduce weed abundance to low levels, creating cascade effects that could decrease the abundance of weed seed predators such as birds (Watkinson et al., 2000).

Habitat management, defined as a series of practices meant to "alter habitats to improve availability of resources required by natural enemies for optimal performance," represents a viable approach to provide the resources needed by weed seed predators (Landis, Wratten, and Gurr, 2000, p. 176). A key element in habitat management is to understand the spatial scale at which practices should be implemented (within-field level, farm level, and landscape level). It is also important to know the time when resources are needed most.

Certainly, a big challenge in the design of a habitat management program in herbaceous crop systems is to find approaches compatible with modern production practices. In many instances, a change in management practices such as the adoption of cover crops or reduced tillage could enhance seed predator abundance. In the case where refuge habitats are created in close spatial association with crop fields, the most obvious problems are the increased costs of production, the need to take some land out of production, and the potential that these refuge habitats may serve as a corridor for pest distribution. Nentwig, Frank, and Lethmayer (1998), Landis, Wratten, and Gurr (2000), and Lee and Landis (2002) discuss how noncrop habitats could be established within agricultural landscapes, while minimizing environmental and economic costs. In this context, several habitat management practices aimed at increasing the abundance of weed seed predators can be integrated with other soil, water, and wildlife conservation practices. For example, grass waterways, riparian buffers, and cross wind traps established to reduce surface chemical runoff, soil erosion, and water loss from agricultural fields could enhance weed seed predator abundance (Povey, Smith, and Watt 1993; Landis and Dyer, 1998; Carmona, Menalled, and Landis, 1999; Kollmann and Bassin, 2001; Menalled, Lee, and Landis, 2001).

CONCLUSIONS AND DIRECTIONS FOR FUTURE STUDIES

Most agricultural weeds are prolific seed producers. However, if the correct ecological infrastructure is present, seed predation might help farmers manage weed abundance and community composition. Three concepts appear particularly worthwhile to achieve this goal. First, the use of seed predators should be integrated with other management practices aimed at reducing weed emergence, survivorship, and/or competitive ability. Possible complementary practices include cover crops, reduced tillage, soil fertility management (Liebman and Gallandt, 1997), use of pathogens (Bürki et al., 1997; Kremer and Spencer, 1989a,b; Müller-Schärer, Scheepens, and Greaves, 2000), and herbicides (Messersmith and Adkins, 1995).

Second, it is necessary to identify management practices that promote the survival of weed seed predators in agricultural settings. Simple habitat management practices, such as disturbance reduction, conservation of noncrop habitats in close spatial association with highly disturbed crop fields, and delaying postharvest tillage, represent viable ways to achieve this goal.

Finally, the implementation of integrated weed management requires a clear understanding of the ecological principles governing the abundance and dynamics of weeds and beneficial organisms. In this sense, weed seed predation in herbaceous crop fields represents an area of agricultural ecology where our current knowledge opens opportunities for future research. There is a need for multiple-year studies evaluating the magnitude and impact of seed predation under different management scenarios. Future studies should also evaluate how pre- and postdispersal seed predators differ between weed species, the joint impact of pre- and postdispersal seed predation on plant demography, the existence of mast seeding as a mechanism to enhance the reproductive success of participating weeds, and the response of seed predators to variations in weed density and weed community composition. Other possible avenues to explore are as follows.

Weed Seed Predation and Weed Threshold Level

Although the concept of a threshold level represents a core component of integrated arthropod management, there is a strong debate among weed scientists about its applicability to weed management. Several authors argue that management decisions based on threshold weed densities and critical periods of interference can help producers reduce herbicide applications (Swanton, Weaver, et al., 1999, and references therein). Others believe that, because of the high reproductive potential of agricultural weeds, all weeds that might set seeds should be eliminated and a no-seed threshold level should be implemented (Norris, 1999, and references therein). The economic cost associated with a no-seed threshold level policy may make this idea impractical in low-value crops. Hopefully, future research will confirm or reject the suitability of threshold levels in weed management. If threshold levels are adopted, maximizing weed seed predation might represent a viable approach for helping farmers reach low seed-bank density levels. In this context, research on weed seed predation should be integrated with an evaluation of the relationship between seed-bank density, weed density, and yield reduction.

Field Experiments and Mathematical Model

Combining field experiments with analytical models represents a promising approach to fully understand the impact of seed predation on weed population and community dynamics. Two approaches appear to be especially useful: path analysis and matrix population models. Path analysis is a general form of multiple regression that allows the consideration of multiple dependent variables and the effects of dependent variables on one another (Mitchell, 1993). This technique was proposed by Louda (1989) as a tool to calculate the relative contribution of seed consumption to plant population dynamics. Price and Jenkins (1986) used path analysis to model the importance of vertebrate seed predation in nonagricultural settings, and future studies done in herbaceous crop systems could benefit from it.

Matrix population models are used extensively in conservation biology and wildlife ecology to guide management strategies (Caswell, 2001) and have been used to study agricultural weeds (e.g, Gonzales-Andujar and Fernandez-Quintanilla, 1991; Jordan et al., 1995). Such models include population vectors composed of numbers of individuals within different life-stage classes (e.g., seeds within the soil seed bank, seedlings, reproductive adults) and projection matrices specifying the probability of making transitions between those life stages. The transition probabilities can be adjusted to reflect impacts of different management practices on weed survivorship (such as a change in pre- and/or postdispersal seed predation due to a change in tillage or row spacing). Future studies could use this technique to evaluate (1) the fate of weed seeds with contrasting life history strategies and (2) the potential shifts in weed community composition due to changes in seed predation pressures.

Trophic Relationships Between Weed Seeds and Their Predators

Weed seed consumption can be conceptualized as a simple predator-prey linkage. In this framework, it is necessary to understand the biology of resource acquisition and conversion and determine the functional and numerical responses that relate trophic levels (Holling, 1959). The functional response provides the per capita rate of resource acquisition by the consumers, i.e., the number of weed seeds that are acquired by each weed seed predator at different weed seed densities. The numerical response is the relationship between number of prey items (i.e., weed seeds) and number of predators (i.e., seed predators) and represents the conversion and allocation efficiency of resources within the consumer trophic level.

Although considerable work has been done on the importance of functional and numerical responses in the biological control of insect pests (Mills and Gutierrez, 1999), few studies have attempted to quantify the relationship between agricultural weed seeds and weed seed predators. Zhang, Drummond, and Liebman (1998) determined that *Harpalus rufipes* adults do not appear to respond to a change in the density of barnyard grass and wild mustard seeds. In contrast, Holmes, Norris, and Froud-Williams (personal communication, 2001) observed that chaffinches, granivorous birds commonly found in Europe, exhibit a Type II functional response to changes in seed density of common lambsquarters, common chickweed [*Stellaria media* (L.) Vill.], and prostrate knotweed (*Polygonum aviculare* L.). A Type II response is characterized by seed intake rate being positively related to seed density up to a threshold value, beyond which seed handling time impedes further increases. Future studies could evaluate the extent to which these preliminary observations represent generalizations about herbaceous crop systems, the existence of numerical responses in weed seed predation, and changes in functional and numerical responses in relation to cropping systems, seed burial depth, and seed characteristics.

Integrative Research and New Partnerships

The nature of weed seed predation requires that experimental studies take into account the real complexity of weed-predator interactions. This requires that scientists interested in weed seed predation consider the benefits of linking their research with others such as land-use planners, landscape ecologists, economists, mathematicians, and wildlife ecologists. Finally, to maximize adoption of these ideas, it is necessary to find effective approaches to transfer the information through extension or other outreach sources (Menalled, Landis, and Dyer, 2004).

Undoubtedly, the design of weed management programs aimed at reducing reliance on herbicides and cultivation represents one of the biggest challenges of industrialized agriculture. To achieve this goal it is necessary to incorporate a whole-system perspective. Within this framework, maximizing weed seed predation represents a tool that should be integrated with other management practices to increase reliance on ecological processes while maintaining or increasing crop productivity, farm revenues, and environmental quality. We believe that such an integrative approach represents the foundation of sustainable farming systems.

REFERENCES

Abramsky, Z. (1983). Experiments on seed predation by rodents and ants in the Israeli desert. *Oecologia* 57: 328-332.

Adams, C.T., W.A. Banks, C.S. Lofgren, B.J. Smittle, and D.P. Harlan (1983). Impact of the red imported fire ant, *Solenopsis invicta* (Hymenoptera: Formicidae) on the growth and yield of soybeans. *Journal of Economic Entomology* 76: 1129-1132.

Andersen, A.N. (1989). How important is seed predation to recruitment in stable populations of long-lived perennials? *Oecologia* 81: 310-315.

Anderson, L. (1998). Post-dispersal seed removal in some agricultural weeds. *Aspects of Applied Biology* 51: 159-164.

Angelstam, P., L. Hansson, and S. Pehrsson (1987). Distribution borders of field mice *Apodemus:* The importance of seed abundance and landscape composition. *Oikos* 50: 123-130.

Auclair, A.U. (1976). Ecological factors in the development of intensive-management ecosystems in the midwestern United States. *Ecology* 57: 431-444.

Banks, W.A., C.T. Adams, C.S. Lofgren, and D.P. Wojcik (1990). Imported fire ant infestation of soybean fields in the southern United States. *Florida Entomologist* 73: 503-504

Bassett, P. (1978). Damage to winter cereals by *Zabrus tenebrioides* (Goeze) (Coleoptera: Carabidae). *Plant Pathology* 27: 48.

Bernhardt, G.E., L. van Allsburg, and R.A. Dolbeer (1987). Blackbird and starling feeding behavior on ripening corn ears. *Ohio Journal of Science* 87: 125-125.

Best, L.B. (1983). Bird use of fencerows: Implications of contemporary fencerow management practices. *Wildlife Society Bulletin* 11: 343-347.

Best, R.L. and C.C. Beegle (1977). Food preferences of five species of carabids commonly found in Iowa cornfields. *Environmental Entomology* 6: 9-12.

Blackwell, B.F. and R.A. Dolbeer (2001). Decline of the red-winged blackbird population in Ohio correlated to changes in agriculture (1965-1996). *Journal of Wildlife Management* 65: 661-667.

Bommarco, R. (1998). Reproduction and energy reserves of a predatory carabid beetle relative to agroecosystem complexity. *Ecological Applications* 8: 846-853.

Boyd, R.S. and G.D. Brum (1983). Predispersal reproductive attrition in a Mojave Desert population of *Larrea tridentata* (Zygophyllaceae). *American Midland Naturalist* 110: 14-24.

Briggs, J.B. (1965). Biology of some ground beetles (Coleoptera: Carabidae) injurious to strawberries. *Bulletin of Entomological Research* 56: 79-93.

Brown, J.H., D.W. Davidson, J.C. Munger, and R.S. Inouye (1986). Experimental community ecology: The desert granivore system. In J. Diamond and T.J. Case (Eds.), *Community ecology* (pp. 41-61). New York: Harper and Row Publishers.

Brust, G.E. (1994). Seed-predators reduce broadleaf weed growth and competitive ability. *Agriculture, Ecosystems and Environment* 48: 27-34.

Brust, G.E. and G.J. House (1988). Weed seed destruction by arthropods and rodents in low-input soybean agroecosystems. *American Journal of Alternative Agriculture* 3: 19-25.

Brust, G.E., B.J. Stinner, and D.A. McCartney (1986). Predator activity and predation in corn agroecosystems. *Environmental Entomology* 15: 1017-1021.

Buhler, D.D. and R.G. Hartzler (2001). Emergence and persistence of seeds of velvetleaf, common waterhemp, woolly cupgrass, and giant foxtail. *Weed Science* 49: 230-235.

Bürki, H.M., D. Schroeder, J. Lawrie, L. Cagan, M. Vrablova, M. El Aydam, F. Szentkiralyi, R. Ghorbani, B. Jüttersonke, and H.U. Ammon (1997). Biological control of pigweeds (*Amaranthus retroflexus* L., *A. powellii* S. Watson and *A. buchonii* Thell.) with phytophagous insects, fungal pathogens and crop management. *Integrated Pest Control* 2: 51-59.

Calvo-Irabien, L.M. and A. Islas-Luna (1999). Predispersal predation of an understory rainforest herb *Aphelandra aurantiaca* (Acanthaceae) in gaps and mature forest. *American Journal of Botany* 86: 1108-1113.

Cárcamo, H.A., J.K. Neimalä, and J.R. Spence (1995). Farming and ground beetles: Effects of agronomic practice on populations and community structure. *Canadian Entomologist* 127: 123-140.

Cardina, J., H.M. Norquay, B.R. Stinner, and D.A. McCartney (1996). Postdispersal predation of velvetleaf *(Abutilon theophrasti)* seeds. *Weed Science* 44: 534-539.

Carmona, D.M. and D.A. Landis (1999). Influence of habitat and cover crops on seasonal activity-density of ground beetles (Coleoptera: Carabidae) in field crops. *Environmental Entomology* 28: 1145-1153.

Carmona, D.M., F.D. Menalled, and D.A. Landis (1999). Northern field cricket, *Gryllus pennsylvanicus* Burmeister (Orthoptera: Gryllidae): Laboratory weed seed predation and within field activity-density. *Journal of Economical Entomology* 92: 825-829.

Carroll, C.R. and S.J. Risch (1984). The dynamics of seed harvesting in early successional communities by a tropical ant, *Solenopsis geminata. Oecologia* 61: 388-392.

Caswell, H. (2001). *Matrix population models: Construction, analysis, and interpretation,* Second edition. Sunderland, MA: Sinauer Associates, Inc.

Clark, M.S., S. Gage, and J. Spence (1997). Habitats and management associated with common ground beetles (Coleoptera: Carabidae) in a Michigan agricultural landscape. *Environmental Entomology* 26: 519-527.

Clarke, P.J. (1992). Predispersal mortality and fecundity in the grey mangrove *(Avicennia marina)* in Southern Australia. *Australian Journal of Ecology* 17: 161-168.

Corp, N., M.L. Gorman, and J.R. Speakman (1997). Ranging behaviour and time budgets of male wood mice *Apodemus sylvaticus* in different habitats and seasons. *Oecologia* 109: 242-250.

Crawley, M.J. (1989). Insect herbivores and plant population dynamics. *Annual Review of Entomology* 34: 531-564.

Crawley, M.J. (1992). Seed predators and plant population dynamics. In M. Fenner (Ed.), *Seeds: The ecology of regeneration in plant communities* (pp. 157-191). Wallingford, Oxon: CAB International.

Crist, T.O. and J.A. MacMahon (1992). Harvester ant foraging and shrub-steppe seeds: Interactions of seed resources and seed use. *Ecology* 73: 1768-1779.

Cromar, H.E., S.D. Murphy, and C.J. Swanton (1999). Influence of tillage and crop residue on postdispersal predation of weed seeds. *Weed Science* 47: 184-194.

Culver, D.C. and A.J. Beattie (1983). Effects of ant mounds on soil chemistry and vegetation patters in a Colorado montane meadow. *Ecology* 64: 485-492.

Dauer, J., T. Fox, and F.D. Menalled (2001). Impact of agricultural management systems on carabid beetle communities and post-dispersal weed seed predation. *86th Annual Meeting of Ecological Society of America,* August 5-10, Madison, Wisconsin, p. 78 (Abstracts). Available online at <http://abstracts.co.allenpress.com/pwed/esa2001/document/?id=24223>.

Davidson, D.W. (1993). The effects of herbivory and granivory on terrestrial plant succession. *Oikos* 68: 23-35.

Davidson, D.W., D.A. Samson, and R.S. Inouye (1985). Granivory in the Chihuahuan desert: Interactions within and between trophic levels. *Ecology* 66: 486-502.

Díaz, M. (1992a). Rodent seed predation in cereal crop areas of central Spain: Effects of physiognomy, food availability, and predation risk. *Ecography* 15: 77-85.

Díaz, M. (1992b). Spatial and temporal patterns of granivorous ant seed predation in patchy cereal crop areas of central Spain. *Oecologia* 91: 561-568.

Díaz, M. (1994). Granivory in cereal crop landscapes of central Spain: Environmental correlates of rodents, birds, and ants. *Acta Oecologica* 15: 739-752.

Díaz, M. and J.L. Tellería (1994). Predicting the effects of agricultural changes in central Spain croplands on seed-eating overwintering birds. *Agriculture, Ecosystems and Environment* 49: 289-298.

Dolbeer, R.A. (1990). Ornithology and integrated pest management. *Ibis* 132: 309-322.

Dolbeer, R.A., P.P. Woronecki, and R.A. Stehn (1986). Resistance of sweet corn to damage by blackbirds and starlings. *Journal of the American Society of Horticultural Sciences* 111: 306-311.

Drees, B.M., L.A. Berger, R. Cavazos, and B. Vinson (1991). Factors affecting sorghum and corn seed predation by foraging red imported fire ants (Hymenoptera: Formicidae). *Journal of Economic Entomology* 84: 285-289.

Edgar, W.H. and A.J. Isaacson (1974). Observations on skylark damage to sugar-beet and lettuce seedlings in East Anglia. *Annals of Applied Biology* 76: 335-337.

Forsyth, S.F. and A.K. Watson (1985). Predispersal seed predation of Canada thistle. *Canadian Entomologist* 117: 1075-1081.

Franson, S.E. and M.F. Willson (1983). Seed predation and patterns of fruit production in *Asclepias syriaca* L. *Oecologia* 59: 370-376.

Getz, L.L. and E. Brighty (1986). Potential effects of small mammals in high-intensity agricultural systems in East-Central Illinois USA. *Agriculture Ecosystems and Enironment* 15: 39-50.

Gonzales-Andujar, J.L. and C. Fernandez-Quintanilla (1991). Modeling the population dynamics of *Avena sterilis* under dry-land cereal cropping systems. *Journal of Applied Ecology* 28: 16-27.

Green, R. (1979). The ecology of wood mice *(Apodemus sylvaticus)* on arable farmland. *Journal of Zoology* 188: 357-377.

Greig, N. (1993). Predispersal seed predation on five *Piper* species in tropical rainforest. *Oecologia* 93: 412-420.

Griffiths J.T. and C.J. Swanton (1999). Predispersal seed predation in corn. *Proceedings of the Weed Science Society of America* 39: 76.

Gurr G. and S. Wratten (2000). *Biological control: Measures of success*. Dordrecht, the Netherlands: Kluwer Academic Publishers.

Gut, L.J., C.E. Jochums, P.H. Westigard, and W.J. Liss (1982). Variations in pear psylla (*Psylla pyricola* Foerster) densities in southern Oregon orchards and its implications. *Acta Horticulterae* 124: 101-111.

Hall, J.C., L.L. Van Eerd, S.D. Miller, M.D.K. Owen, T.S. Prather, D.L. Shaner, M. Singh, K.C. Vaughn, and S.C. Weller (2000). Future research directions for weed science. *Weed Technology* 14: 647-658.

Harper, J.L. (1977). The predation of seeds and fruits. In J.L. Harper (Ed.), *Population biology of plants* (pp. 457-482). New York: Academic Press.

Harrison, S.K., E.E. Regnier, J.T. Schmoll, and J.E. Webb (2001). Competition and fecundity of giant ragweed in corn. *Weed Science* 49: 224-229.

Hartke, A., F.A. Drummond, and M. Liebman (1998). Seed feeding, seed caching, and burrowing behaviors of *Harpalus rufipes* De Geer larvae (Coleoptera: Carabidae) in the Maine potato agroecosystem. *Biological Control* 13: 91-100.

Hendrix, S.D. (1979). Compensatory reproduction in a biennial herb following defloration. *Oecologia* 42: 107-118.

Herr, D.E. and E.W. Straub (1970). Velvetleaf control as influenced by herbicide placement and seed depth. *Weed Science* 18: 459-461.

Hoffmann, L.A., E.F. Redente, and L.C. McEwen (1995). Effects of selective seed predation by rodents on shortgrass establishment. *Ecological Applications* 5: 200-208.

Holling, C.S. (1959). Some characteristics of simple types of predation and parasitism. *Canadian Entomologist* 91: 385-398.

Holm, S.O. (1994). Reproductive variability and pollen limitation in three *Betula* taxa in northern Sweden. *Ecography* 17: 73-81.

Holmes, R.J. and R.J. Froud-Williams (2001). The predation and dispersal of weed seeds by birds. In *Proceedings of the British Crop Protection Conference—Weeds* (pp. 333-336). Brighton, UK: British Crop Protection Council.

House, G.J. and J.N. All (1981). Carabid beetles in soybean agroecosystems. *Environmental Entomology* 10: 194-196.

Howe, H.F. and J.S. Brown (2000). Early effects of rodent granivory on experimental forb communities. *Ecological Applications* 10: 917-924.

Hulme, P.E. (1993). Post-dispersal seed predation by small mammals. *Symposium of the Zoological Society of London* 65: 269-287.

Hulme, P.E. (1998). Post-dispersal seed predation: Consequences for plant demography and evolution. *Perspectives in Plant Ecology, Evolution and Systematics* 1: 32-46.

Hulme, P.E. (2002). Seed-eaters: Seed dispersal, destruction, and demography. In D.J. Levey, W.R. Silva, and M. Galetti (Eds.), *Seed dispersal and frugivory: Ecology, evolution and conservation* (pp. 257-273). Wallingford, Oxon: CAB International.

Janzen, D.H. (1971). Seed predation by animals. *Annual Review of Ecology and Systematics* 2: 465-492.

Johnson, N.E. and R.S. Cameron (1969). Phytophagous ground beetles. *Annals of the Entomological Society of America* 62: 909-914.

Jordan, N., D.A. Mortensen, D.M. Prenzlow, and K.C. Cox (1995). Simulation analysis of crop rotation effects on weed seedbank. *American Journal of Botany* 82: 390-398.

Kaye, T.N. (1999). From flowering to dispersal: Reproductive ecology of an endemic plant, *Astragalus australis* var. *olympicus* (Fabaceae). *American Journal of Botany* 86: 1248-1256.

Keeley, J.E., S.C. Keeley, C.C. Swift, and J. Lee (1984). Seed predation due to the Yucca-moth symbiosis. *American Midland Naturalist* 112: 187-191.

Kelrick, M.I., J.A. MacMahon, R.R. Parmenter, and D.V. Sisson (1986). Native seed preferences of shrub-steppe rodents, birds and ants: The relationship of seed attributes and seed use. *Oecologia* 68: 327-337.

Kirk, V.M. (1972). Seed-caching by larvae of two ground beetles, *Harpalus pensylvanicus* and *H. erraticus*. *Annals of the Entomological Society of America* 65: 1426-1428.

Kiss, J., K. Penksza, F. Tóth, and F. Kádár (1997). Evaluation of field margins in nature production capacity with special regard to plant protection. *Agriculture, Ecosystems and Environment* 63: 227-232.

Kollmann, J. and S. Bassin (2001). Effect of management on seed predation in wildflower strips in northern Switzerland. *Agriculture, Ecosystems, and Environment* 83: 285-296.

Kotzageorgis, G.C. and C.F. Mason (1997). Small mammal populations in relation to hedgerow structure in an arable landscape. *Journal of Zoology London* 242: 425-434.

Kremer, R.J. (2000). Combinations of microbial and insect biocontrol agents for management of weed seeds. In N.R. Spencer (Ed.), *Proceedings of the 10th International Symposium on Biological Control of Weeds* (pp. 799-806). July 4-14, 1999, Montana State University, Bozeman, Montana. Bozeman: Montana State University.

Kremer, R.J. and N.R. Spencer (1989a). Impact of a seed-feeding insect and microorganisms on velvetleaf *(Abutilon theophrasti)* seed viability. *Weed Science* 37: 211-216.

Kremer, R.J. and N.R. Spencer (1989b). Interactions of insects, fungi, and burial on velvetleaf *(Abutilon theophrasti)* seed viability. *Weed Technology* 3: 322-328.

Kromp, B. (1999). Carabid beetles in sustainable agriculture: A review on pest control efficacy, cultivation impacts and enhancement. *Agriculture Ecosystems and Environment* 74: 187-228.

Kropff, M.J. and H. Walter (2000). EWRS and the challenges for weed research at the start of a new millennium. *Weed Research* 40: 7-10.

Kudoh, H. and D.F. Whigham (1998). The effect of petal size manipulation on pollinator/seed predator mediated female reproductive success of *Hibiscus moscheutos*. *Oecologia* 117: 70-79.

Landis, D.A. and L.E. Dyer (1998). *Conservation buffers and beneficial insects, mites, and spiders*. Conservation information sheet, agronomy series. Lansing, MI: Natural Resource Conservation Service, U.S. Department of Agriculture.

Landis, D.A., S.D. Wratten, and G.M. Gurr (2000). Habitat management to conserve natural enemies of arthropod pests in agriculture. *Annual Review of Entomology* 45: 175-201.

Lee, J.C. and D.A. Landis (2002). Non-crop habitat management for carabid beetles. In J.M. Holland (Ed.), *The agroecology of carabid beetles* (pp. 279-303). Andover, Hampshire: Intercept Ltd.

Lee, J.C., F.D. Menalled, and D.A. Landis (2001). Refuge habitats modify impact of insecticide disturbance on carabid beetle communities. *Journal of Applied Ecology* 38: 472-483.

Levey, D.J. and M.M. Byrne (1993). Complex ant-plant interactions: rain-forest ants as secondary dispersers and post-dispersal seed predators. *Ecology* 74: 1802-1812.

Liebman, M. and E.R. Gallandt (1997). Many little hammers: Ecological approaches for management of crop-weed interactions. In L.E. Jackson (Ed.), *Ecology in agriculture* (pp. 291-343). San Diego, CA: Academic Press.

Louda, S.M. (1982a). Distribution ecology: Variation in plant recruitment in relation to insect seed predation. *Ecological Monograph* 52: 25-41.

Louda, S.M. (1982b). Limitation of the recruitment of the shrub *Haplopappus squarrosus* (Asteracea) by flower- and seed-feeding insects. *Journal of Ecology* 70: 43-53.

Louda, S.M. (1989). Predation in the dynamics of seed regeneration. In M.A. Leck, V.T. Parker, and R.L. Simpson (Eds.), *Ecology of soil seed banks* (pp. 25-51). New York: Academic Press.

Louda. S.M. (2001). Discovering an effect of insect floral herbivory on plant population density and distribution in a "Green World." *Bulletin of the Ecological Society of America* 82: 229-231.

Louda, S.M. and M.A. Potvin (1995). Effect of inflorescence-feeding insects on the demography and lifetime fitness of a native plant. *Ecology* 76: 229-245.

Louda, S.M., M.A. Potvin, and S.K. Collinge (1990). Predispersal seed predation, postdispersal seed predation and competition in the recruitment of seedlings of a native thistle in Sandhills Prarie (Nebraska, USA). *American Midland Naturalist* 124: 105-113.

Luff, M.L. (1980). The biology of the ground beetle *Harpalus rufipes* in a strawberry field in Northumberland. *Annals of Applied Biology* 101: 164-170.

Lund, R.D. and F.T. Turpin (1977). Carabid damage to weed seeds found in Indiana cornfields. *Environmental Entomology* 6: 695-698.

Lys, J.A. and W. Nentwig (1992). Augmentation of beneficial arthropods by strip-management: 4. Surface activity, movements, and activity density of abundant carabid beetles in cereal fields. *Oecologia* 92: 373-382.

Macdonald, D.W., T.E. Tew, I.A. Todd, J.P. Garner, and P.J. Johnson (2000). Arable habitat use by wood mice *(Apodemus sylvaticus):* 3. A farm-scale experiment on the effect of crop rotation. *Journal of Zoology London* 250: 313-320.

Marino, P.C., K.L. Gross, and D.A. Landis (1997). Weed seed loss due to predation in Michigan maize fields. *Agriculture, Ecosystems and Environment* 66: 189-196.

McNaughton, S.J. (1986). On plants and herbivores. *American Naturalist* 128: 765-770.

Menalled F., J. Dauer, T. Fox, and K. Renner (2001). *Managing your farm to increase weed seed predation.* Extension Bulletin E-2749. Michigan State University.

Menalled, F.D., D.A. Landis, and L. Dyer (2004). Research and extension supporting ecologically based IPM systems. *Journal of Crop Improvement* 11: 153-174.

Menalled, F.D., J. Lee, and D. Landis (1999). Manipulating carabid beetle abundance alters prey removal rates in corn fields. *Biocontrol* 43: 441-456.

Menalled, F.D., J. Lee, and D. Landis (2001). Herbaceous filter strips in agroecosystems: Implications for ground beetle (Coleoptera: Carabidae) conservation and invertebrate weed seed predation. *The Great Lakes Entomologist* 34: 77-91.

Menalled, F.D., P.C. Marino, S.H. Gage, and D.A. Landis (1999). Does agricultural landscape structure affect parasitism and parasitoid diversity? *Ecological Applications* 9: 634-641.

Menalled, F.D., P.C. Marino, K. Renner, and D.A. Landis (2000). Post-dispersal weed seed predation in Michigan crop fields as a function of agricultural landscape structure. *Agriculture, Ecosystem and Environment* 77: 193-202.

Messersmith, C.G. and S.W. Adkins (1995). Integrating weed-feeding insects and herbicides for weed control. *Weed Technology* 9: 199-208.

Mester, T.C. and D.D. Buhler (1990). Effect of seedling depth on velvetleaf *(Abutilon theophrasti)* seedling development and response to cyanazine. *Weed Science* 38: 34-38.

Mills, N.J. and A.P. Gutierrez (1999). Biological control of insect pests: A tritrophic perspective. In B.A. Hawkins and H.V. Cornell (Eds.), *Theoretical approaches to biological control* (pp. 89-102). Cambridge, UK: Cambridge University Press.

Mitchell, R.J. (1993). Path analysis: Pollination. In S.M. Scheiner and J. Gurevithch (Eds.), *Design and analysis of ecological experiments* (pp. 211-231). New York: Chapman and Hall.

Mittelbach, G.G. and K.L. Gross (1984). Experimental studies of seed predation in old-fields. *Oecologia* 65: 7-13.

Mohler, C.L. (1996). Ecological bases for cultural control of weeds. *Journal of Production Agriculture* 9: 468-474.

Montgomery, W.I. and M. Dowie (1993). The distribution and population regulation of the wood mouse *Apodemus sylvaticus* on field boundaries of pastoral farmland. *Journal of Applied Ecology* 30: 783-791.

Morrison, J.E., D.F. Williams, and D.H. Oi (1999). Effect of crop seed water content on the rate of seed damage by red imported fire ants (Hymenoptera: Formicidae). *Journal of Economic Entomology* 92: 215-219.

Morrison, J.E., D.F. Williams, D.H. Oi, and K.N. Potter (1997). Damage of dry crop seed by red imported fire ants (Hymenoptera: Formicidae). *Journal of Economic Entomology* 90: 218-222.

Mull, J.F. and J.A. MacMahon (1997). Spatial variation in rates of seed removal by harvester ants *(Pogonomyrmex occidentalis)* in a shrub-steppe ecosystem. *American Midland Naturalist* 138: 1-13.

Müller-Schärer, H., P.C. Scheepens, and M. P. Greaves (2000). Biological control of weeds in European crops: Recent achievements and future work. *Weed Research* 40: 83-98.

Myster R.W. and S.T.A. Pickett (1993). Effects of litter, distance and vegetation patch type on postdispersal tree seed predation in old fields. *Oikos* 66: 381-388.

Nentwig, W., T. Frank, and C. Lethmayer (1998). Sown weed strips: Artificial ecological compensation areas as an important tool in conservation biological control. In P. Barbosa (Ed.), *Conservation biological control* (pp. 133-153). San Diego, CA: Academic Press.

Norris, R.F. (1999). Ecological implications of using thresholds for weed management. *Journal of Crop Production* 2: 31-58.

Nurse, R.E., and C.J. Swanton (2000). The effect of row width and soil tillage on pre-dispersal weed seed predation in soybeans. *Proceedings of the Weed Science Society of America* 40: 100.

Pelz, H.J. (1986). Ecological aspects of damage to sugar beet seeds by *Apodemus sylvaticus*. In R.J. Putman (Ed.), *Mammals as pests* (pp. 34-50). New York: Chapman and Hall, Inc.

Peres, C.A. (1991). Seed predation of *Cariniana micrantha* (Lecythidaceae) by brown capuchin monkeys in Central Amazonia. *Biotropica* 23: 262-270.

Platt, W.J., G.R. Hill, and S. Clark (1974). Seed production in a prairie legume (*Astragalus canadensis* L.): Interactions between pollination, predispersal seed predation, and plant density. *Oecologia* 17: 55-63.

Pogue, D.W. and G.D. Schnell (2001). Effects of agriculture on habitat complexity in a prairie-forest ecotone in the Southern Great Plains of North America. *Agriculture, Ecosystems and Environment* 87: 287-298.

Pollard, E. and J. Relton (1970). Hedges: V. A study of small mammals in hedges and uncultivated fields. *Journal of Applied Ecology* 7: 549-557.

Povey, F.D., H. Smith, and T.A. Watt (1993). Predation of annual grass weed seeds in arable field margins. *Annals of Applied Biology* 122: 323-328.

Price, M.V. and S.H. Jenkins (1986). Rodents as seed consumers and dispersers. In D.R. Murray (Ed.), *Seed dispersal* (pp. 191-235). Orlando, FL: Academic Press, Inc.

Reader, R.J. (1993). Control of seedling emergence by ground cover and seed predation in relation to seed size for some old-field species. *Journal of Ecology* 8: 169-175.

Regnier, E.E., S.K. Harrison, and J.T. Schmoll (1999). Identification and characterization of seed-borne insect pests of giant ragweed. *Proceedings of the North Central Weed Science Society* 54: 173.

Risch, S. and C.R. Carroll (1986). Effects of seed predation by a tropical and on competition among weeds. *Ecology* 67: 1319-1327.

Robertson, A.I., R. Giddins, and T.J. Smith (1990). Seed predation by insects in tropical mangrove forests: Extent and effects on seed viability and the growth of seedlings. *Oecologia* 83: 213-219.

Robinson, R.A. and W.J. Sutherland (1996). The winter distribution of seed-eating birds: Habitat structure, seed density, and seasonal depletion. *Ecography* 22: 447.

Samson, D.A., T.E. Philippi, and D.W. Davidson (1992). Granivory and competition as determinants of annual plant diversity in the Chihuahuan desert. *Oikos* 65: 61-80.

Seaman, R.E. and P.C. Marino (2003). Influence of mound building and selective seed predation by the red imported fire ant *(Solenopsis invicta)* on an old-field plant assemblage. *Journal of the Torrey Botanical Society* 130: 193-201.

Shatters, R.G. and R.K. Vander Meer (2000). Characterizing the interaction between fire ants (Hymenoptera: Formicidae) and developing soybean. *Journal of Economic Entomology* 93: 1680-1687.

Smith, T.J. (1987). Seed predation in relation to tree dominance and distribution in mangrove forests. *Ecology* 68: 266-273.

Smith, T.J., H.T. Chan, C.C. McIvor, and M.B. Robblee (1989). Comparisons of seed predation in tropical, tidal forests from three continents. *Ecology* 70: 146-151.

Steffan-Dewenter, I., U. Münzenberg, and T. Tscharntke (2001). Pollination, seed set and seed predation on a landscape scale. *Proceedings of the Royal Society London, Series B* 268: 1685-1690.

Sunderland, K.D., N.E. Crook, D.L. Stacey, and B.J. Fuller (1987). A study of feeding by polyphagous predators on cereal aphids using ELISA and gut dissection. *Journal of Applied Ecology* 24: 907-933.

Sunderland, K.D., G.L. Lövei, and J. Fenlon (1995). Diets and reproductive phenologies of the introduced ground beetles *Harpalus affinis* and *Clivina australasiae* (Coleoptera: Carabidae) in New Zealand. *Australian Journal of Zoology* 43: 39-50.

Swanton, C.J., J.T. Griffiths, H.E. Cromar, and B.D. Booth (1999). Pre- and post-dispersal weed seed predation and its implications to agriculture. *The 1999 Brighton Conference—Weeds: Proceedings of an international conference held at the Brighton Metropole Hotel, Brighton, UK, 15-18 November 1999,* Volume 3 (pp. 829-834). British Crop Protection Council.

Swanton, C.J., S. Weaver, P. Cowan, R.R. Van Acker, W. Deen, and A. Shreshta (1999). Weed thresholds: Theory and applicability. *Journal of Crop Production* 2: 9-29.

Sykes, A.K. and S.J. Hannon (2001). Fencerows as habitats for birds in an agricultural landscape. *EcoScience* 8 : 441-449.

Teo-Sherrell, C.P.A., D.A Mortensen, and M.E. Keaton (1996). Fates of weed seeds in soils: A seeded core method of study. *Journal of Applied Ecology* 33: 1107-1113.

Thies, C. and T. Tscharntke (1999). Landscape structure and biological control in agroecosystems. *Science* 285: 893-895.

Thomas, C.F.G., L. Parkinson, G.J.K. Griffiths, A. Fernández García, and E.J.P. Marshall (2001). Aggregation and temporal stability of carabid beetle distribution in field and hedgerow habitats. *Journal of Applied Ecology* 38: 100-116.

Todd, I.A., T.E. Tew, and D.W. Macdonald (2000). Arable habitat use by wood mice *(Apodemus sylvaticus):* 1. Macrohabitat. *Journal of Zoology London* 250: 299-303.

Tooley, J. and G. Brust (2002). Weed seed predation by carabid beetles. In J.M. Holland (Ed.), *The agroecology of carabid beetles* (pp. 215-229). Andover, Hampshire: Intercept Ltd.

Tscharntke, T. (2000). Parasitoid populations in the agricultural landscape. In M.E. Hochberg and A.R. Ives (Eds.), *Parasitoid population biology* (pp. 235-253). Princeton, NJ: Princeton University Press.

Uva, R.H., J.C. Neal, and J.M. DiTomaso (1997). *Weeds of the Northeast.* Ithaca, NY: Cornell University Press.

Vitolo, D.B. and E.W. Stiles (1987). The effect of density of *Ambrosia trifida* L. on seed predation by *Euaresta festiva* (Loew) (Diptera: Tephritidae). *Journal of the New York Entomological Society* 95: 491-494.

Wallin, H. and B.S. Ekbom (1988). Movements of carabid beetles (Coleoptera: Carabidae) inhabiting cereal fields: A field tracing study. *Oecologia* 17: 39-43.

Watkinson, A.R., R.P. Freckleton, R.A. Robinson, and W.J. Sutherland (2000). Predictions of biodiversity response to genetically modified herbicide-tolerant crops. *Science* 289: 1554-1557.

White, S.S. (2000). Weed seed predation in agroecosystems. Master's of science thesis, East Lansing, Michigan State University.

Wilson, J.D., A.J. Morris, B.E. Arroyo, S.C. Clark, and R.B. Bradbury (1999). A review of the abundance and diversity of invertebrate and plant foods of granivorous birds in northern Europe in relation to agricultural change. *Agriculture, Ecosystems and Environment* 75: 13-30.

Wissinger, S.A. (1997). Cyclic colonization in predictably ephemeral habitats: A template for biological control. *Biological Control:* 10: 4-15.

Zhang, J. (1993). Biology of *Harpalus rufipes* DeGeer (Coleoptera: Carabidae) in Maine and dynamics of seed predation. Master's of science thesis, Orono, University of Maine.

Zhang, J., F.A. Drummond, and M. Liebman (1998). Effect of crop habitat and potato management practices on the population abundance of *Harpalus rufipes* (Coleoptera: Carabidae). *Journal of Agricultural Entomology* 15: 63-74.

Zhang, J., F.A. Drummond, M. Liebman, and A. Hartke (1997a). *Insect predation of seeds and plant population dynamics.* Maine Agricultural and Forest Experiment Station Technical Bulletin 163. Orono: University of Maine.

Zhang, J., F.A. Drummond, M. Liebman, and A. Hartke (1997b). Phenology and dispersal of *Harpalus rufipes* De Geer (Coleoptera: Carabidae) in agroecosystems in Maine. *Journal of Agricultural Entomology* 14: 171-186.

Chapter 12

Mowing for Weed Management

William W. Donald

INTRODUCTION

Introductory weed science texts dismiss mowing as a method for controlling weeds (Anderson, 1983; Anonymous, 1985; Zimdahl, 1993). For example, Anderson (1983) remarked,

> Mowing has limited value as a means of weed control. It is primarily used to reduce seed production and to restrict unsightly or rank weed growth. Mowing commonly is used for these purposes in meadows and pastures, along roadsides, and in waste places. (p. 72)

Some recent weed science texts do not index or mention mowing as a weed management practice (Aldrich and Kremer, 1997). Others acknowledge that mowing contributes to integrated weed management systems but cite few references or examples (Monaco, Weller, and Ashton, 2002; Ross and Lembi, 1999). Although mowing often does not adequately control weeds when used alone, it contributes to weed management systems when used in combination or sequence with other weed control methods.

Most past research on mowing was reductionist and focused on using mowing alone, independently of other weed control practices. Past research was also largely descriptive and exploratory, rather than analytical and predictive. Negative preconceptions about the effectiveness of mowing for weed control probably have discouraged weed scientists from more creatively exploiting mowing in integrated weed management systems.

Research publications on mowing for weed management are scattered among several applied scientific disciplines, including agricultural engineering, agronomy, horticulture, forage science, forestry, landscape man-

---

I thank Drs. Mark Renz, Joe DiTomaso, Harminder Pal Singh, John Teasdale, and several anonymous reviewers for critical evaluation of the manuscript and helpful suggestions.

agement, range science, restoration ecology, and turf science. Often weed control was of only secondary interest to the authors' published research objectives. Mowing mechanics and mowing machinery management have been reviewed by agricultural engineers (Persson, 1987). Mowing mechanics deals with issues such as critical speeds for efficient mowing, mower power requirements, mechanical wear on equipment, and the dynamics of different types of mowers on different plant species (Kanafojski and Karwowski, 1972; O'Dogherty, 1982; O'Dogherty and Gale, 1991a; Tuck et al., 1991). As far as the author is aware, mowing for weed management has not been reviewed before from an agronomic perspective. Because the literature on mowing is vast, old, and scattered, only selected references are presented in this brief overview.

REPORTING MOWING PRACTICES IN WEED SCIENCE

To fully describe a mowing "regime," three basic features of mowing must be summarized:

1. crop and weed growth in relation to mowing (see Figures 12.1 and 12.2);
2. mown residue handling after mowing (Figure 12.2); and
3. mowing timing in relation to crop and weed growth stages (see Figure 12.3).

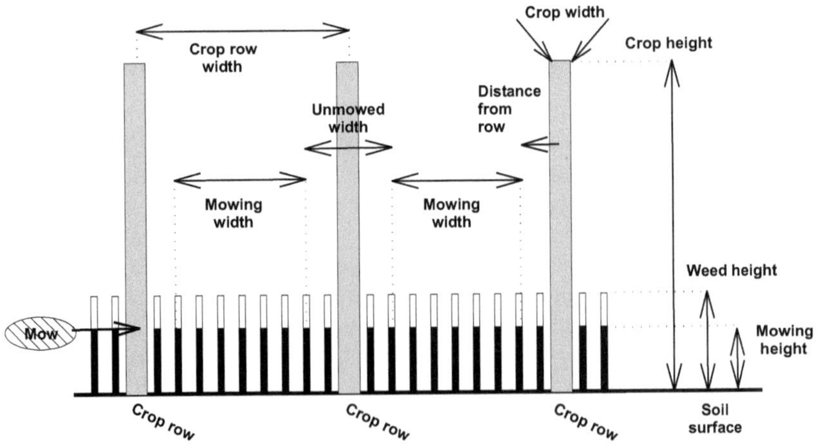

FIGURE 12.1. The vertical and horizontal relationships between crop and weed (or ground cover) height and spacing in relationship to mowing width and height.

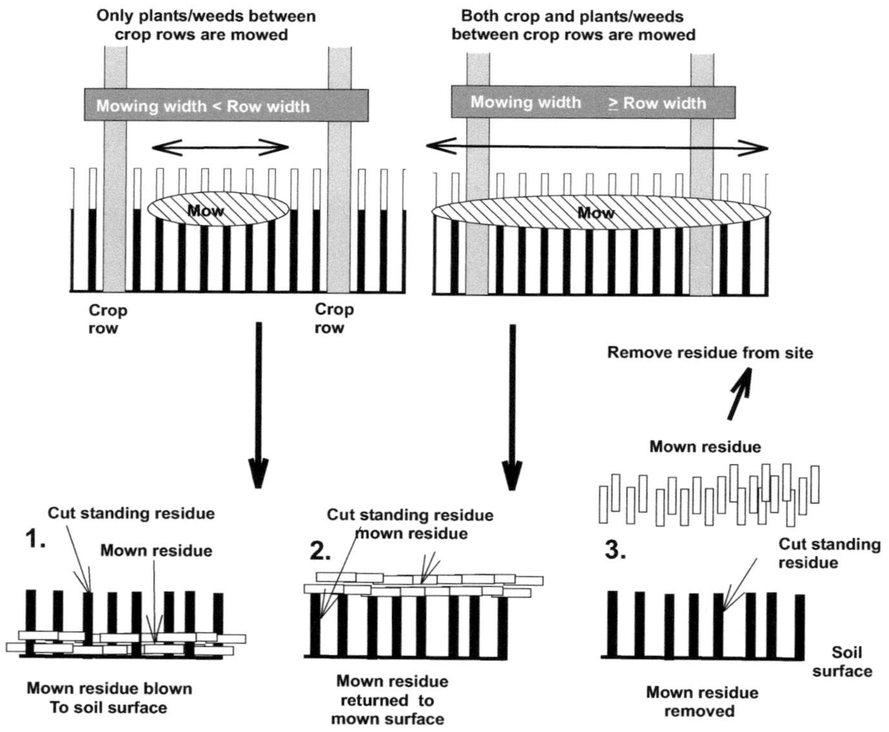

FIGURE 12.2. Mowing is either performed between crop rows or both the crop and weeds (or ground cover) between crop rows are mown together. After mowing, mown residue may be (1) blown into the cut standing residue, (2) returned to the surface of the cut standing residue, or (3) removed from the site.

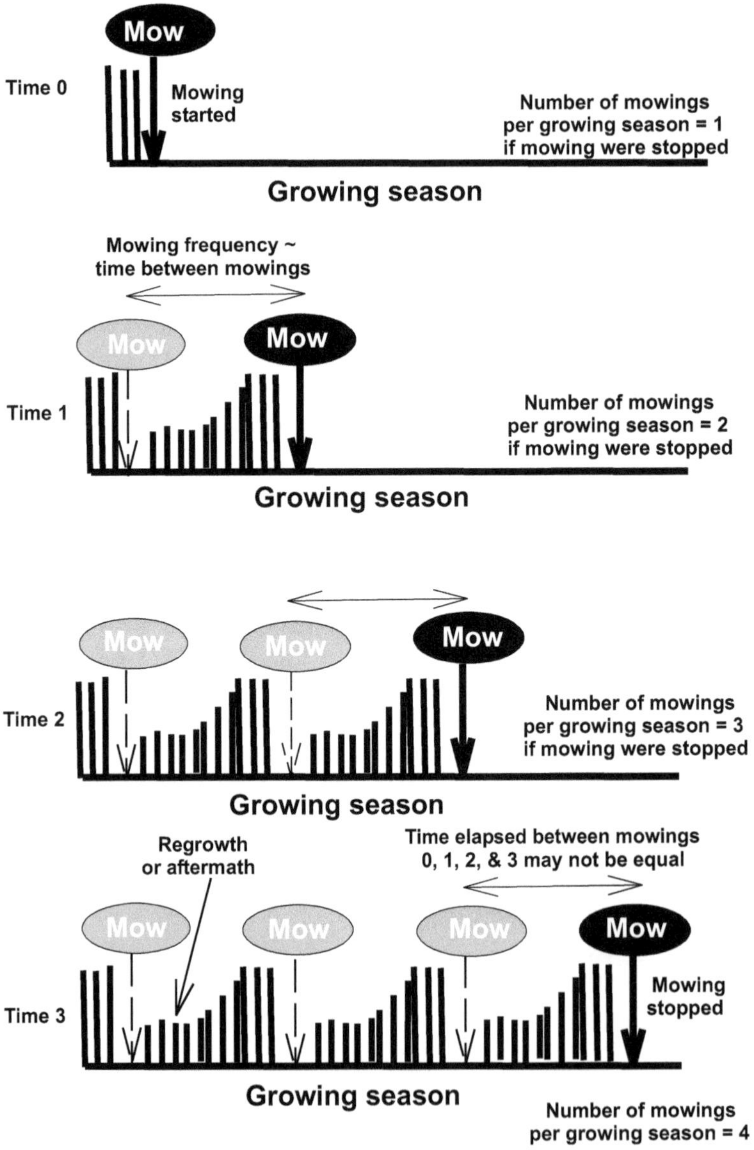

FIGURE 12.3. Mowing timing during the growing season is characterized by (1) start times, (2) mowing frequency, (3) number of mowings per growing season, and (4) stop times. Generally, time intervals between mowings are not equal.

No universally accepted reporting standards for mowing have been established in weed science. The following data are suggested to thoroughly describe mowing treatments (or mowing regimes):

1. a description of the mowing equipment and mowing mechanism used, including manufacturer and implement model number;
2. mowing height above the ground;
3. number of mowings per growing season;
4. mowing timing (start and stop times) and frequency (i.e., time interval between mowings);
5. mowing width in relationship to crop row width;
6. the amount of residue mown off per unit area;
7. whether mown residue was returned to the soil surface, to the surface of standing cut residue, or removed from the site; and
8. the weed and crop height and growth stage at mowing.

A measure of total production on an area basis subdivided into mown residue plus cut standing residue remaining attached to the soil surface (e.g., density, ground cover, fresh weight, or dry weight, etc.) before and after mowing would be desirable. To date, most research on mowing for weed control has concerned varying only one or two of these factors at a time for a given weed(s) and crop.

MOWER EQUIPMENT

Commercial mowers use either impact cutting or shear cutting (O'Dogherty and Gale, 1991b). Impact cutting severs plants using blades spun at high speed and moves or deflects stems or shoots during mowing. In shear cutting, two close, sharp-edged blades slide over each other like scissors to cut plant material between them. Experimental CO_2 lasers have been used to defoliate dicot weeds under controlled conditions (Heisel et al., 2001), but this approach has yet to be field-tested.

Several types of commercial mowers are available:

1. rotary or disc mowers (impact cutting);
2. flail mowers (impact cutting);
3. reciprocating sicklebar mowers (shear cutting); and
4. reel mowers (shear cutting) (Persson, 1987).

The first three types of mowers are used for field and roadside mowing and windrowing or swathing hay and forage for harvest (Smith and Wilkes,

1976; Ross and Lembi, 1985; Hunt, 2001). Sicklebar mowers are also used on some grain combines (Hunt, 2001; Smith and Wilkes, 1976). Rotary or disc mowers and reel mowers are used routinely on lawns, turf, and golf greens. Additional specialized types of mowers have been described (Kanafojski and Karwowski, 1972; Persson, 1987). These mowers are generally adapted for herbaceous rather than woody plants (Vallentine, 1989). Heavy-duty rotary mowers are available for cutting brush and woody stems up to 10 cm in diameter. Persson (1987) provides line drawings and photographs of mower equipment, as well as the agricultural engineering classification of mowers.

Mowers can be mounted (rear, mid-, or forward), semi-mounted, or trailed behind tractors (Hunt, 2001; Smith and Wilkes, 1976) or all-terrain vehicles. Often mowers are mounted on tractors with three-point hitches and powered by the power takeoff (PTO) drive train. Gangs of rotary or reel mowers can be mounted on tool bars to increase mowing width. Hydraulic power is often used to raise and lower mower units to avoid obstructions. The construction, dimensions, adjustment, and operation of sicklebar and rotary mowers have been summarized (Hunt, 2001; Persson, 1987).

The height and angle of cut of sicklebar mowers can be varied relative to the soil surface to either pick up and mow lodged foliage or to skip over heavy residue or rough, rocky terrain (Hunt, 2001; Smith and Wilkes, 1976). Sicklebar mowers can be used to cut tall weeds because plant height and volume do not greatly affect sicklebar operation (Ross and Lembi, 1985). But weeds with vining growth habits or lodged weeds can clog sicklebar mowers. Flail mowers process this material better.

Rotary or disc mowers cut plant tissue using mowing elements rotated at high speed in a horizontal plane (Hunt, 2001). Typical rotary mowing elements include single blades, arms with knives on their ends (flat, suction, pickup), or plastic cord of various thicknesses. Rotary mowers are usually operated tipped forward at an angle relative to the ground to reduce remowing cut foliage. If remowing is desired, rotary mowers can be operated horizontally or tipped backward. Mulching rotary lawn mowers remow leaf blades into smaller pieces than traditional rotary lawn mowers and return or "blow" leaf fragments back into lawns to decompose.

PRECONDITIONS FOR MOWING

Mowing is not adapted for all farming or agricultural systems. A smooth soil surface, which is free of rocks or other obstructions, is a prerequisite for mowing. Consequently, it is not used in ridge tillage or furrow and bed irrigation systems. Because an even soil surface is required for mowing close to soil surfaces (e.g., 4 to 7 cm cutting height in turf or golf greens), rocks may need to be pulled and land may need to be leveled. Soil surface evenness becomes less critical as mowing height is increased in forage harvesting or windrowing and/or combine harvesting cereal, oilseed, or legume grain.

Usually, mowing is used to control weeds and manage vegetation in untilled, short- and long-term, herbaceous and woody perennial crops. Mowing is an integral part of managing pastures, rangeland, conservation reserve land, cover crops, grassed waterways, stream buffer zones, field margins, roadsides, orchards, tree plantations, vineyards, horticultural ornamentals, lawns, turf, golf greens, and wild land areas. Although these perennial crops and areas are planted into tilled or no-tillage seedbeds, they are usually maintained without further soil disturbance after establishment.

GOALS OF MOWING IN DIFFERENT FARMING SYSTEMS

In many crops, mowing is used at several landscape scales to achieve several goals:

1. to prevent or reduce losses in yield or quality due to weed growth and competition with crops,
2. to minimize viable weed seed production,
3. to assist in crop harvest,
4. to maintain ground cover for erosion control,
5. to enhance aesthetic appeal, and
6. to direct plant succession.

In conservation or restoration ecology, mowing is one of several methods that are sometimes used to shift plant succession (Wilson and Clark, 2001). Mowing is used to encourage native plant reestablishment and discourage undesirable, exotic, invasive weeds on prairies, rangeland, wetlands, and wild lands.

Often mowing is used to achieve several goals simultaneously. For example, both weeds and crops are mown simultaneously in herbaceous perennial crops, such as grass and legume forages, home lawns, and golf greens. The goals are to suppress weed competition and weed seed production while harvesting and managing crop biomass quality and quantity, or aesthetics. In noncropland areas, such as along roadsides, waterways, fence rows, or field borders, herbaceous ground cover species, usually perennial grasses, legumes, and weeds, are mown to maintain ground cover and prevent soil erosion, but also can be harvested for forage.

Mowing goals may conflict. Because some goals may be less important than others, certain goals may be favored. For example, in California the optimum time to control yellow starthistle (*Centaurea solstitialis* L.) with mowing is in spring, but road crews favor mowing in early summer to minimize the chance of fire (Mark Renz, personal communication, 2002; Joe DiTomaso, personal communication, 2002).

Goal 1: Mowing to Prevent or Reduce Losses in Yield or Quality Due to Weed Growth and Competition with Crops

When not planted in wide rows, both crops and weeds can be mown together at the same height and time (see Figure 12.2). The mowing regime is chosen to enhance crop stand persistence and competition with weeds and to minimize weed survival, growth, competition, and viable weed seed production. When crops are planted in rows (e.g., orchards, fruit trees, forest tree plantations, horticultural ornamentals), only ground cover plants and weeds between crop rows are mown. The chief goal of mowing between rows of perennial crops is to minimize competition from both weeds and the undesirable excessive growth of ground cover species. Because ground cover plants help prevent soil erosion (Dabney, 1998; Schwab et al., 1981), they should be mown in such a way as to maintain stand persistence.

Initial mowing for weed control must wait until weeds exceed a threshold height which depends upon the available mowing equipment. Mowing also can be repeated over a longer, later "window of opportunity" than most herbicide treatments or cultivation and can control weeds that become too large to be killed with either herbicides or cultivation (Muenscher, 1980). But mowing alone seldom provides season-long weed control. Repeated mowing is often required because annual and perennial weeds emerge in "flushes" over an extended period, especially in perennial crops, and environmental conditions may favor regrowth (aftermath) of mown weeds and ground cover vegetation.

Direct and Indirect Mowing Effects on Weeds

Even when used for purposes other than weed control, mowing directly and indirectly impacts weed biology. Mowing directly defoliates plants and, in turn, can reduce growth, decrease plant survival, and reduce or prevent seed production. Mowing directly and indirectly changes the abiotic and biotic environment of mown plant communities. Abiotic factors include light, temperature, and soil moisture, among others. Mowing creates gaps by removing part of the canopy shading the ground. Light intensity increases at soil surfaces and light quality changes in gaps, resulting in increased red/far-red ratios at soil surfaces. In turn, increased light energy at soil surfaces increases the average daily soil temperature and diurnal range in soil temperature down the soil profile. For example, mowing removed bunch grass cover and increased subsequent emergence of Lehmann lovegrass (*Eragrostis lehmanniana* Nees) (Sumrall, Roundy, and Winkel, 1991) by increasing light intensity and red light at soil surfaces which stimulated the biologically active form of phytochrome in dormant seed (Roundy, Taylorson, and Sumrall, 1992; Ballare and Casal, 2000). It also increased the diurnal range in temperature at the soil surface (Roundy, Taylorson, and Sumrall, 1992). Both abiotic changes stimulated Lehmann lovegrass germination. In addition, mown residue itself can change abiotic factors, as well at the chemical environment (e.g., allelochemicals, nutrients, pH). Mowing changes competitive relationships between neighboring plants because different species die or regrow at different rates following mowing (Vila and Terradas, 1998). Unintended, indirect effects of mowing on weed biology deserve study because they can be manipulated to improve crop competitiveness and weed control.

Direct Effects: Weed Susceptibility, Tolerance, and Resistance to Mowing

I define mowing-susceptible and -tolerant weeds as those that are killed by one and more than one properly timed mowing, respectively. In contrast, I define mowing-resistant weeds as those that survive several mowings to set seed. Because mowing regimes differ among different farming systems, a weed species that is mowing susceptible in one farming system may be mowing tolerant or mowing resistant in another. Consequently, a mowing regime must be described in order to better understand weed response to mowing (see Figures 12.1 to 12.3). Mowing height limits the vertical space above soil surfaces available for plant growth, but it is not the only component of mowing regime that impacts weed response to mowing.

Once a mowing regime is defined for a given farming system, the mowing susceptibility, tolerance, or resistance of weeds may best be understood in terms of (1) weed life cycle length (e.g., annual versus perennial), (2) plant growth stage when mown, (3) plant phenology, and (4) regenerative capacity based on the number and height of plant meristems (shoot apical, lateral, tiller, and adventitious meristem buds) relative to the mowing height (growth habit) (McIntyre, Diaz, et al., 1999; McIntyre, Lavorel, et al., 1999). Raunkiaer's (1934) plant classification system based on life cycle length and the position of meristem buds above or below the soil surface may help group weeds by likely mowing susceptibility (e.g., Therophytes, Cryptophytes, Geophytes, Hemicryptophytes, Chamaephytes, and Phanerophytes) (Hakansson, 1995 a,b,c; McIntyre, Diaz, et al., 1999). After mowing at a given height, regeneration of new shoots from meristem buds is controlled by meristem bud numbers, meristem bud display relative to soil surfaces, and growth rates (Rosenthal and Kotanen, 1994). However, Room, Maillette, and Hanan (1994) remarked that the mere presence of numerous meristems below the disturbance height does not ensure resistance to disturbance. Meristems must be stimulated to grow following disturbance, and sprouting must increase plant fitness (Room, Maillette, and Hanan, 1994).

By my definition, mowing-tolerant weeds escape a single mowing, but not more than one mowing, to later set seed. For annual weeds, possible mechanisms for mowing tolerance include (1) emergence over a protracted period of time with some emergence after the final mowing of the growing season or (2) growth below the mowing height followed by later set seed (i.e., escape in time or space, respectively). If emergence and subsequent flowering are delayed until after mowing ends, seedling fitness may be reduced because of crop shading (i.e., fitness is the ability to vegetatively or sexually reproduce and persist). Although partial to total defoliation by mowing stresses mowing-resistant weeds, they recover to set seed, although total seed production may be reduced.

Both mowing height and plant growth stage when mown can determine annual dicot susceptibility or tolerance to mowing because these factors define whether apical and lateral bud meristems of annual dicot seedlings extend above the mowing height. For example, one mowing killed seedlings of the annual dicots common lambsquarters (*Chenopodium album* L.) (Bassett and Crompton, 1978; Heisel et al., 2001) and wild mustard [*Brassica kaber* (DC.) L. C. Wheeler] (Heisel et al., 2001) best when it decapitated shoot apical meristems extending above the mowing height. Otherwise, these weeds regrow from undamaged buds on lateral branches below the mowing height. In this way, normally erect growing, relatively unbranched annual dicot plants can become prostrate and highly branched following repeated mowings (Ross and Lembi, 1985). Seedlings of some perennial dicot weeds may respond to mowing like annual dicots if they are mown before they form regenerative meristematic buds below the mowing height. For example, seedlings of honey mesquite [*Prosopis juliflora* (Swartz) DC], a woody rangeland shrub, died when shoots were cut below the cotyledons, but many seedlings survived when mown above the cotyledons between seven and 56 days after emergence (Scifres and Hahn, 1971) or at the soil surface between one and two weeks after emergence (Bovey and Meyer, 1974). Likewise, yellow starthistle was killed by mowing at the soil surface but survived mowing above the soil surface (Benefield et al., 1999). However, obstructions, such as rocks, plant debris, or uneven soil surfaces, make it impractical to mow low enough to decapitate honey mesquite or yellow starthistle seedling shoots at the soil surface.

Timing mowing to precede adventitious bud formation on some perennial weed seedlings kills more seedlings than does mowing started after adventitious buds form. Adventitious buds first form on different plant parts of perennial weed seedlings at different growth stages soon after emergence. For example, buds formed on hypocotyls of leafy spurge (*Euphorbia esula* L.) seedlings between 15 and 27 days after emergence, but buds formed on roots 27 days after emergence (Raju, 1975). Common milkweed (*Asclepias syriaca* L.) shoots emerged from adventitious buds when seedlings shoots were clipped at the soil surface at the four-leaf stage, 21 days after planting (Evetts and Burnside, 1972). As common milkweed seedlings became older after the four-leaf stage, shoot numbers from adventitious buds increased and the time required to emerge after clipping decreased (Jeffery and Robison, 1971; Evetts and Burnside, 1972).

Many herbaceous perennial broadleaf weeds become resistant to mowing after establishment because they form adventitious buds below the mowing height on underground stolons, bulbs, rhizomes, or thickened roots. They regrow new shoots from these adventitious buds, and their underground storage organs supply carbohydrate reserves to support bud growth. It takes repeated mowing over an extended period to deplete this adventitious bud bank and carbohydrate storage reserves. Mowing when seasonal carbohydrate reserves are at a minimum both delays stand recovery and improves perennial weed control (Aldrich and Kremer, 1997; Monaco, Weller, and Ashton, 2002; Ross and Lembi, 1999).

Mown pasture species and perennial weeds recover from mowing at different rates depending upon species and environmental conditions. Bellingham and Sparrow (2000) suggested that mowing severity (e.g., proportion of aboveground biomass removed by mowing) influenced shoot emergence from bud meristems in a hierarchical fashion. Because many perennial herbaceous weeds are much less susceptible to mowing than are annual weeds, mowing used alone must usually be repeated, often over several years, for managing perennial weeds in pastures. For example, mowing Canada thistle [*Cirsium arvense* (L.) Scop.] once or twice per year for several years was required to reduce infestations in established alfalfa (*Medicago sativa* L.) or mixed grass-legume pasture (Table 12 in Donald, 1990) (Welton, Morris, and Hartzler, 1929; Hodgson, 1958; Derscheid, Nash, and Wicks, 1961; Schreiber, 1967; Amor and Harris, 1977). Often mowing had to be sustained for at least three years to adequately reduce Canada thistle stands. Less frequent or unsustained mowing was ineffective

(Amor and Harris, 1977). In Ohio pastures, three repeated mowings starting in June (June 1, July 1, and August 1) controlled Canada thistle better than three repeated mowings starting in July (July 1, August 1, and September 1) (Welton, Morris, and Hartzler, 1929). Although changes in Canada thistle stand were not documented, Welton, Morris, and Hartzler (1929) asserted that frequent early mowing (June 1, July 1, August 1, and September 1; or June 1, July 1, and August 1; or July 1, August 1, and September 1) nearly eliminated Canada thistle after three years. Such frequent mowing might limit forage stand persistence and production in arid regions. In other research, mowing Canada thistle to a cutting height of 5 cm, one to three times per growing season, either alone or in combination with MCPB [4-(4-chloro-2-methylphenoxy)butanoic acid], did not eradicate Canada thistle from a pasture but did reduce shoot density and the vigor of surviving shoots (Amor and Harris, 1977). Mowing Canada thistle three times per year controlled it better in a subirrigated pasture, but not in an unirrigated upland pasture in Colorado (Beck and Sebastian, 2000). Beck and Sebastian (2000) suggested that the high water table in the subirrigated pasture had restricted Canada thistle's perennial roots, making it easier to control. Mowing did not enhance the phytotoxicity of fall-applied picloram, picloram plus 2,4-D, chlorosulfuron, clopyralid plus 2,4-D or dicamba on Canada thistle in these pastures.

Perennial broadleaf weed response to mowing in pastures depends on the number of mowings per growing season and the interval between mowings. For example, when established multiflora rose (*Rosa multiflora* Thunb. ex Murr.) was repeatedly mown to a height of 7.5 cm every two weeks during the growing season, it took nearly two years to kill plants in pastures (Bryan and Mills, 1988). Likewise, clipping purple loosestrife (*Lythrum salicaria* L.) shoots at two- to four-week intervals for two years was required to severely reduce root dry weight in Minnesota (Katovich, Becker, and Ragsdale, 1999). When tropical soda apple (*Solanum viarum* Dunal) was mown to a 7.5 cm height starting in mid-June, three mowings at 60-day intervals controlled it 90 percent in bahiagrass (*Paspalum notatum* Flugge) or pangolagrass (*Digitaria decumbens* Stent) pastures in Florida (Mislevy, Mullahey, and Martin, 1999).

Annual and perennial grasses respond to mowing differently (Garay, Matthew, and Hodgson, 1999) because they differ in life cycle and meristem bud display (Jewiss, 1972). Preformed tiller buds on both types of grasses reinitiate growth following one mowing. However, fewer tiller buds form in annual grasses than in perennial grasses, and tillering often ends after flowering, because flowering signals plant death in annual grasses (Harper, 1989; Garay, Matthew, and Hodgson, 1999). Compared to annual grasses, perennial grasses have both larger carbohydrate storage reserves and adventitious bud banks, allowing more regrowth after mowing.

Growth stage when mown influenced annual grass regrowth from tiller buds. For example, barnyardgrass [*Echinochloa crus-galli* (L.) Beauv.] regrew better when clipped to a 5 or 10 cm height at the vegetative stage compared to later reproductive stages (i.e., internode elongation, heading, or maturity) (Vengris, Hill, and Field, 1966). Proso millet (*Panicum miliaceum* L.) regrew following complete defoliation at the one- and three-leaf stage, but was killed when completely defoliated at the six-leaf stage. Partial defoliation did not kill proso millet at any of these stages. Proso millet was resistant to mowing to a 3 cm height in the field when it was mown either once 25 days after planting or once per week from 25 days after planting until maturity (Cavers and Kane, 1990). Apparently, annual grasses must grow beyond a threshold size or growth stage in order to be killed by mowing.

The presence of undamaged foliage and lateral branch or tiller buds below the mowing height influenced regrowth and mowing resistance (Sachs and Hassidim, 1996; Vengris, Hill, and Field, 1966). Observations on plant physiological response following grazing may be helpful in understanding plant recovery following defoliation by mowing. However, grazing differs from mowing in severity. Grazed plants which were not completely defoliated mobilized carbohydrate reserves and changed carbon allocation patterns for shoot regrowth. The photosynthetic rates of foliage remaining after defoliation from grazing increased to compensate for lost foliage, and roots increased nutrient uptake rates (Rosenthal and Kotanen, 1994). Defoliation reduced short-term plant productivity and root growth because plants increased assimilate reallocation to shoots until the root/shoot ratio present before defoliation was reestablished (Paez et al., 1995). After perennial ryegrass (*Lolium perenne* L.) was defoliated, it utilized remobilized carbohydrate reserves from plant parts which survived defoliation (i.e., fructans in leaf sheaths) to support shoot growth (Morvan-Bertrand, Boucaud, and Prud'homme, 1999a,b). But new and elongating leaves also contributed photosynthate used during shoot regrowth following mowing. In fact, at two days after defoliation, half of the sucrose used in the growth zone of recovering perennial ryegrass leaves was newly fixed photosynthate by elongating, previously cut leaves (Morvan-Bertrand, Boucaud, and Prud'homme, 1999a,b).

Although weed functional traits which determine mowing susceptibility, tolerance, or resistance have been emphasized in this discussion, weed response to mowing is not simply a set of species-specific characteristics. Weed response to mowing is modified by (1) the mowing regime, (2) the response of neighboring plants in the community to mowing (i.e., the biotic environment), and (3) changes in the abiotic environment following mowing which are either independent or dependent on mowing. These interrelationships are poorly defined and unpredictable at present.

Mowing Effects on Weed Ecotypes

Repeated mowing is a selection pressure which permits some weed species and genotypes within a species to set seed more than others, shifting the relative proportions of weeds or weed genotypes to those that can reproduce even when repeatedly mown. Intraspecific genetic variation and differentiation partially explains mowing responses of some weeds (Warwick, 1990). Mowing was correlated with differences in plant size and degree of erectness due to genetic differences in annual bluegrass (*Poa annua* L.) and broadleaf plantain (*Plantago major* L.) from mown habitats. In contrast, common yarrow (*Achillea millefolium* L.), English daisy (*Bellis perennis* L.), buckhorn plantain (*Plantago lanceolata* L.), and healall (*Prunella vulgaris* L.) from mown habitats responded to mowing using both genetic differentiation and phenotypic plasticity.

Genetic differentiation and local adaptation between plant populations can be demonstrated by planting seed (or reciprocal transplants) of populations from different habitats (e.g., unmown, mown, or grazed) in common gardens and observing subsequent phenotypic differences in growth habit and reproduction (Warwick, 1990). However, nongenetic maternal effects, such as initial seed size, may partially contribute to observed differences between populations (McKinney and Fowler, 1991). In common garden experiments, the annual grass field sandbur (*Cenchrus incertus* M. A.Curtis) from mown habitats was shorter and had more leaves, tillers, panicles, and burs (seed dispersal units) per plant and fewer burs per panicle than plants from either unmown or grazed habitats (McKinney and Fowler, 1991). Prostrate and erect ecotypes of buckhorn plantain, a rosette-forming perennial dicot, from mown and grazed grassland habitats, respectively, maintained their growth habit and seed production in reciprocal garden experiments using either seed or vegetative transplants (Van Tienderen and Van Der Toorn, 1991a,b). Seed production was greatest for each buckhorn plantain genotype in its native habitat and was correlated with the height of nearby vegetation (Van Tienderen and Van Der Toorn, 1991b). In addition, flowering phenology remained early or late for populations from early- and late-mown habitats, respectively, when grown as reciprocal transplants (Van Tienderen and Van Der Toorn, 1991a).

Indirect Effects of Mown Residue on Weed Control

Mown residue is either harvested and removed from the field or returned to the surface of the cut standing residue or blown into it (see Figure 12.2). Forages may be windrowed before being harvested for hay or silage, depending upon moisture content and available storage facilities. Lawn clippings are also often harvested. In some regions, grain or oilseed crops are windrowed before being harvested with pickup combines. Whether pickup or direct combining is used, the grain is removed from the field and the remaining cut residue is returned to the field. Postharvest residue handling redistributes weed seed (see discussion of dispersal).

The type of mower influences mown residue distribution and fragment size, which influence residue decomposition rate after being returned to the field. Sicklebar mowers create large plant fragments which rest above soil surfaces on cut standing plant residue, delaying decay (Creamer et al., 1995). Flail mowers recut mown residue, which is usually removed from fields. In contrast, mulching lawn mowers recut mown residue into smaller fragments before blowing it back into the lawn surface. Small residue fragments readily decay when they come in contact with soil microorganisms in moist soil (Creamer et al., 1995). Evenly spreading mown clippings helps recycle nutrients during residue decomposition, reduces fertilizer requirements for turf, and provides some residual weed control (Heckman et al., 2000). For example, clippings returned to Kentucky bluegrass (*Poa pratensis* L.) turf controlled weeds, gave the turf a darker green color, and halved nitrogen fertilizer requirements (Heckman et al., 2000).

Differences in mown residue fragment size and decomposition rate influenced the duration of residual weed control. For example, sicklebar mowing crop residues [e.g., barley (*Hordeum vulgare* L.), crimson clover (*Trifolium incarnatum* L.), hairy vetch (*Vicia villosa* Roth.), rye, subterranean clover (*Trifolium subterraneum* L.)] at or after the mid- to late-bloom stages suppressed weeds [yellow foxtail (*Setaria glauca* (L.) Beauv.), common lambsquarters, and redroot pigweed (*Amaranthus retroflexus* L.)] better than flail mowing in the absence of crop competition (Creamer et al., 1995). This may be because sicklebar mowing leaves larger residue fragments on the soil surface than does flail mowing.

In Maryland, flail mowing fall-sown cover crops of hairy vetch and subterranean clover in spring killed them and provided weed-suppressive mulch before transplanting no-till tomato (*Lycopersicon esculentum* Mill.) (Abdul-Baki and Teasdale, 1993) or no-till broccoli (*Brassica oleracea* L.) (Morse, 1999). In the northeastern United States, mowing fall-sown mixtures of hairy vetch plus rye (*Secale cereale* L.) or rye cover crops with a sicklebar mower at a height of 5 cm in spring before planting broccoli killed the cover crop and controlled summer annual weeds better than incorporating the cover crop into the soil (Mangan et al., 1995).

The duration of weed control achieved by mown cover crop residues has been inconsistent between sites and years and often does not control weeds all season. Although flail-mown hairy vetch suppressed weeds {common lambsquarters, eastern black nightshade (*Solanum ptycanthum* Dun.), fall panicum (*Panicum dichotomiflorum* Michx.), green foxtail [*Setaria viridis* (L.) Beauv.], horseweed [*Conyza canadensis* (L.) Cronq.]} for about six weeks, early in the growing season, in corn (*Zea mays* L.) in Maryland, late-season weed control was often unacceptable, compared to preemergence herbicides (Teasdale, 1993). In other research, foxtail millet [*Setaria italica* (L.) Beauv.] was used as a cover crop before transplanting broccoli into it (Abdul-Baki et al., 1997). Although herbicides suppressed foxtail millet better than flail mowing to a height of 6 cm, mown foxtail millet controlled weeds and prevented yield loss. However, large crabgrass [*Digitaria sanguinalis* (L.) Scop.] regrew and perennial weeds [broadleaf dock (*Rumex obtusifolius* L.) and red sorrel (*Rumex acetosella* L.)] became established later in the growing season. Several other transplanted crops performed well in mown residue (Galloway and Weston, 1996). Mowing with a sicklebar mower at 5 cm height controlled rye alone or hairy vetch plus rye cover crops and weeds [horseweed, red sorrel, white clover (*Trifolium repens* L.)] in no-till transplanted broccoli as well as did herbicides (Mangan et al., 1995). However, the cover crop regrew to a limited extent after mowing. Because mown rye alone or hairy vetch plus rye did not control weeds consistently between sites, herbicide or cultivation would be needed to achieve acceptable yields at some locations.

Goal 2: Mowing to Minimize Weed Seed Production

Mowing to reduce weed seed production is usually started well after mowing to minimize weed competition and crop yield loss. If only one mowing per growing season is possible, then it should be timed to match weed flowering (Muenscher, 1980). Mowing must also be properly timed to minimize both total weed seed production and seed viability. To achieve these goals, developing seedheads should be mown before viable seed first form. For example, when giant foxtail (*Setaria faberi* Herrm.), an annual grass, was mown during early vegetative growth, it flowered and set seed (Schreiber, 1965). But mowing seedheads at maturity reduced total seed production and panicle size. Timing is critical and must precede anthesis, pollination, and fertilization (Ross and Lembi, 1985). In fact, some viable seed can form less than a week after anthesis. For instance, mowing musk thistle (*Carduus nutans* L.), a biennial dicot, within two days after anthesis totally prevented viable seed production, but mowing became ineffective if mowing was delayed four or more days after anthesis (McCarty and Hatting, 1975). Nevertheless, several mowings were needed to prevent musk thistle seed production because plants in the population flowered and set seed over a protracted period in summer. Besides reducing total seed production, mowing also can delay flowering of surviving plants [e.g., plumeless thistle (*Carduus acanthoides* L.)] (Feldman and Lewis, 1990).

Although mowing height and timing likely interact to affect weed seed production, few studies quantitatively document this interaction. Mowing height must be low enough to cut off developing seedheads. If plants are initially mown too low, however, seeds later form so close to the soil surface that a second mowing misses them (Muenscher, 1980). In California, yellow starthistle competing with grass grew more erect and branched less than did isolated plants (Benefield et al., 1999). When mown to a height of 10 cm or less at early flowering, these erect-growing plants did not recover as well and produced fewer seed than did isolated plants,

By preventing viable seed formation, properly timed mowing helps minimize weed reinfestation, population increase, and seed dispersal of new weed species or herbicide-resistant weeds within fields or from field borders into fields (Thill, O'Donovan, and Mallory-Smith, 1993). If mowing is poorly timed, it can spread viable weed seed beyond the borders of current infestations. For example, mowing chervil (*Chaerophyllum aureum* L.), a dicot infesting pastures, "early" reduced seed production by decreasing both shoot density and seed set, whereas mowing "later" reduced apparent seed production by spreading seed outside the study site (Magda and Jarry, 2000). Mowing yellow starthistle at early flowering (2 to 5 percent) reduced total seed production more than mowing earlier at bolting or spine formation (Benefield et al., 1999). Properly timed mowing may not be optimal and may involve tradeoffs when several weeds with different flowering phenologies co-infest the same site. Mowing timed to prevent viable seed production of one weed may inadvertently be timed to spread viable mature seed of others.

Goal 3: Mowing to Assist in Crop Harvest

Forage Harvesting

The mowing regime for harvest impacts total forage production, forage quality, and stand persistence. It also directly and indirectly affects weed population biology, survival, and seed production, weed invasion, and weed shifts in pastures and rangeland. Indirect effects of mowing on weeds result from the regrowth rate and competitive ability of mown forage aftermath. However, forage harvest research generally focuses on maximizing economic returns, total forage production, and forage quality. The effect of forage harvest mowing on weed biology is usually of secondary interest.

Because some weeds cannot survive even one mowing per growing season, mowing prevents mowing-susceptible weeds from invading pastures. However, these weeds can invade pastures after mowing is stopped. For example, Japanese knotweed (*Reynoutria japonica* Houtt.), Sakhalin or giant knotweed [*R. sachalinensis* (F. Schmidt ex Maxim.) Nakai], and their hybrid invaded pastures in Czechoslovakia when mowing twice per year was stopped (Brabec and Pysek, 2000). If improperly timed, mowing may allow mowing-tolerant weeds to invade pastures by reducing the competitive ability of pasture forages. For example, diffuse knapweed (*Centaurea diffusa* Lam.) was better able to establish in mown than unmown bluebunch wheatgrass [*Pseudoroegneria spicata* (Pursh.) Schrb.] range in eastern Washington (Sheley, Olson, and Larson, 1997).

Both mowing height and frequency interact to influence weed invasion of pastures (Aiken, Sladden, and Bransby, 1995). Less frequent mowing reduced stress on desirable forage species and helped competitive forage crops resist weed encroachment. For example, large crabgrass control by mowing in bermudagrass [*Cynodon dactylon* (L.) Pers.] pasture was reduced as mowing height was progressively increased from 2.5 to 17.5 cm, but became better as mowing frequency was decreased (i.e., from one- to six-week intervals between successive mowings) (Aiken, Sladden, and Bransby, 1995). However, because these factors interacted, when a mowing height of 2.5 cm was combined with a six-week mowing interval, control of large crabgrass was best in bermudagrass. This management combination likely helped bermudagrass shade soil surfaces, depriving large crabgrass of the light stimulus needed for germination (Turgeon, 1994). In other research, progressively increasing the mowing frequency by decreasing the interval between mowing from 37 to 25 days encouraged yellow foxtail invasion in established, irrigated alfalfa (*Medicago sativa* L.) in California (Norris and Ayres, 1991). If alfalfa was mown with a sicklebar mower at a 5 to 7 cm height every 37 days, yellow foxtail did not invade alfalfa. Other research verified that mowing alfalfa every two weeks stressed it, delayed regrowth, and encouraged weed encroachment, which, in turn, reduced alfalfa production and shortened stand persistence (Hoveland, Durham, and Bouton, 1996). Bermudagrass or large crabgrass did not invade mowing-tolerant alfalfa cultivars if alfalfa was mown every four to six weeks instead of every two weeks (Hoveland, Durham, and Bouton, 1996).

The time of initial mowing can influence forage yield and weed competition, as it did between alfalfa and cheat (*Bromus secalinus* L.), an annual grass, in Oklahoma (Pike and Stritzke, 1984). Following fall seeding of alfalfa, cheat yield increased as initial forage harvest date was delayed in the following spring. Mowing at a 6 cm height between mid-April and mid-May corresponded to early bud and late bloom of alfalfa. But the total yield of cheat-infested alfalfa was not consistently improved by mowing early and never yielded as much as cheat-free alfalfa. Early mowing suppressed cheat only partially and alfalfa competing with cheat did not achieve its full weed-free yield potential under irrigation.

Dispersal of Weed Seed by Mowing Pastures

Mowing can spread hundreds of thousands of weed seed throughout pastures (Strykstra, Verweij, and Bakker, 1997). Weed seed dispersal depends upon weed flowering phenology and seedhead height, as well as mower construction and height (Strykstra Verweij, and Bakker, 1997). Mowers can spread weed seed only if weeds are setting seed when mown and seedheads extend above the mowing height (Strykstra, Bekker, and Verweij, 1996). In this study, seed dispersal was independent of seed size, but different mower parts selectively dispersed seed of different species, depending upon seedhead height (Strykstra, Verweij, and Bakker, 1997). For example, seed of tall (> 15 cm) species were selectively transported on mower safety skirts. Seeds which accumulated on disk mower skids and upright support beams were transported and deposited both within and between fields, but seed accumulated on safety skirts were moved only within fields. Seed accumulations on mower skids were deposited when raising and lowering the disk mower, whereas those seed accumulated on safety skirts were dislodged when the mower hit bumps or uneven ground. The starting point, direction, and pattern of repeated mowing or combine harvesting should be consciously planned to minimize spreading weed seed across fields. Cleaning mowers (or combines) between fields helps prevent weed seed spread.

Weed seed dispersal by mowing is greatest when mowing and seed set coincide (Coulson et al., 2001). In English grasslands, dispersal of yellow-rattle (*Rhinanthus minor* L.) and oxeye daisy (*Leucanthemum vulgare* Lam.) seed were different when mown either once in July or twice (July followed by September). Yellow-rattle flowered and set seed from May to August, whereas oxeye daisy completed seed set by August and September. Mowing distributed seed of yellow-rattle further than oxeye daisy because yellow-rattle had set some seed when mown in July, but oxeye daisy had not yet begun flowering. Moreover, oxeye daisy had completed natural seed dispersal before the second mowing (September). Mowing once (July) or twice (July followed by September) also stressed oxeye daisy, reducing total seed production. As expected, seed were dispersed further in the direction of mower travel (i.e., dispersal was directional). Seed dispersal by mowing was modeled as an inverse power function of distance.

Information is limited on dispersal of vegetative propagules of weeds by mowing, but it does occur. Rotary mowers did not kill Macartney rose (*Rosa bracteata* J. C. Wendl.) in Texas pastures, but mowing did spread cut canes, which rerooted in moist soil (cited in Haas, Lehman, and Morton, 1970).

Windrowing and Combine-Harvesting Effects on Weed Seed Production and Dispersal

Grain harvesting is not normally grouped with mowing. However, sicklebar cutters are used when direct combining grain crops or when windrowing to dry oilseed, legume, and cereal grain crops in the field before pickup combine harvesting, and in windrowers for drying forages before baling or silage (Persson, 1987).

Harvest height and timing in relation to weed seedhead height and flowering phenology determine whether harvesting reduces weed seed production or, conversely, contributes to redistributing and spreading viable weed seed, as noted for grassland mowing. Some viable weed seed may be removed from fields with harvested grain or baled hay (Wilson, 1970), and some viable seed are returned to the field concentrated in wide bands behind combines or windrowers in the direction of travel. Chaff spreaders on combines can redistribute weed seed more widely and laterally than with other types of mowing. Combining and mowing disperse weed seed further than natural dispersal mechanisms.

In one of the few quantitative studies of weed seed dispersal by combine harvesting, combines collected more than 90 percent of the capsules of Chinese thornapple (*Datura ferox* L.) growing in soybean [*Glycine max* (L.) Merr.] in Argentina because preharvest seed shed was limited (about 1 percent) (Ballare et al., 1987a). Between 7 and 40 percent of seed were returned to the field and were deposited between 21 and 98 m from the dispersal site by different combines. Chinese thornapple seed viability was unaffected by combining. In a second study, seed of two biotypes (black-seeded and gold-seeded) of proso millet were distributed almost uniformly 50 m behind a combine in corn in Ontario, Canada (McCanny and Cavers, 1988). Differences between biotypes in seed density within this 50 m swath were related to initial differences in source patch seed density. Mark and recapture methods were used to study poverty brome (*Bromus sterilis* L.) seed dispersal during winter wheat (*Triticum aestivum* L.) combining (Rew, Froud-Williams, and Boatman, 1996). Using this research approach, total seed recovery was 31 percent for marked seed in dyed seedheads. Forty-six percent of poverty brome seed were recovered within 1 m of the source, but combining moved 43.3 percent of the seed as far as 7 and 53 m behind and forward from the dispersal site. Unlike the previous study, combine-mediated seed distribution was markedly skewed. Combine-mediated dispersal was directional in the forward direction with little lateral spread in either study.

Combine harvesting can modify seed germinability. For example, the percent germination of weed seed [curly dock, *Rumex crispus* L.; slimleaf lambsquarters, *Chenopodium leptophyllum* (Moq.) Nutt ex S. Wats.; and Venice mallow, *Hibiscus trionum* L.] increased after combine harvesting winter wheat, compared to hand-harvested seed (Currie and Peeper, 1988). Enhanced germination was attributed to mechanical scarification of seed as it passed through the mowing parts of the combine. Conventional mowers are unlikely to scarify weed seed to influence germination, but this remains to be tested. A wider range of species with different dormancy mechanisms and different mowing or harvesting equipment need to be studied.

Recently, the factors controlling weed distribution in fields, including harvest combining, were reviewed (Ghersa et al., 2000; Rew and Cousens, 2001). Explanations and models for combine-assisted weed dispersal and distribution during harvest were also summarized. The impact of several hypothetical dispersal patterns on population biology and later yield loss have also been modeled (Ballare et al., 1987b; Gonzalez-Andujar, Perry, and Moss, 1999; Maxwell and Ghersa, 1992). Unfortunately, models without experimental verification are no more than hypotheses. After modification, such models may have application for refocusing research questions and better designing field studies to test and verify more quantitative descriptions of mowing-assisted weed seed dispersal. The model prediction that the proportion of weed seed removed by combine harvesting determines how quickly weed populations can be reduced below economic thresholds warrants study in the field (Gonzalez-Andujar, Perry, and Moss, 1999).

Weed control during crop growth can influence whether harvest combining reduces weed seed production and dispersal. Because seedlings of common lambsquarters and field pennycress (*Thlaspi arvense* L.), which either emerged late or survived herbicide treatment were shorter than normal, they survived below the combining height during spring wheat harvest and later set seed in fall (Hume, 1988).

Windrow or combine mowing height can impact postharvest weed regrowth and weed seed production. For example, cutter bar height during spring wheat windrowing influenced subsequent postharvest growth and seed production of green foxtail in fall in North Dakota (Black and Siddoway, 1977). After harvest, green foxtail produced more seed in short (15 cm) than tall (38 cm) spring wheat straw.

Goal 4: Mowing to Maintain Ground Cover for Erosion Control

Mowing is used to manage ground cover kept for both soil erosion prevention and weed control along roadsides, grassed waterways, and field borders, as well as in pastures, old fields, and Conservation Reserve Program set-aside land (Anonymous, 1985; Fleischer et al., 1989). Mowing also helps improve visibility for vehicle traffic along roads and is used on field borders to help manage insects and diseases of field crops by killing weeds which act as alternative hosts (Fleischer et al., 1989). In addition, no-till farming practices contribute to controlling erosion on highly erodible land. Mowing has a place in several no-till systems for managing ground cover and weeds in perennial and annual crop production.

Mowing for Pasture Renovation

Generally, herbicides are used for no-till pasture renovation on highly erodible land and to improve pasture quality (i.e., protein) by increasing legumes to between 30 and 50 percent of grass-legume mixtures (Sequin et al., 2001). However, under some conditions, herbicides can reduce total forage production, lower legume-grass ratios, and encourage weed encroachment (Sequin et al., 2001).

Mowing has been studied as a substitute for tillage or herbicides in conventional tillage or no-till pasture renovation. After pasture grasses grew 30 cm tall in spring, rotary mowing to a height of 5 cm suppressed them enough to allow red clover (*Trifolium pratense* L.) or white clover to be no-till seeded. Establishment by this method was as good as glyphosate at 2.6 kg a.i.·ha^{-1} (Sequin et al., 2001). In contrast to glyphosate treatment, mowing at 5 cm prevented weed encroachment and allowed maximum forage production in the establishment year. However, forage quality was not greatly improved because percent clover remained low and renovated legumes did not persist into the following growing season. Additional research is needed on mowing as a means of sod suppression and pasture renovation.

Ground Cover Maintenance in Orchards

In humid regions of North America, ground cover in orchards usually consists of

1. between-row alleys of planted perennial grasses and naturally occurring herbaceous annual and perennial weeds, which are repeatedly mown, and
2. in-row regions in which the same ground cover vegetation is managed using either (a) herbicides, (b) repeated mowing, (c) mulches, or (d) repeated cultivation (Oliveira and Merwin, 2001). Between-row vegetation minimizes erosion and allows machinery access for orchard operations, such as pruning, when the soil is wet (Anonymous, 1985; Connell et al., 2001).

Between-row vegetation also keeps orchards cooler and freer of dust, although it can increase water and nitrogen requirements of orchards. In-row vegetation competes with orchard trees for water and nutrients, especially young trees. If in-row vegetation is not suppressed, it can reduce fruit yields and quality (Oliveira and Merwin, 2001). Alley mowing three times per growing season (May, July, and October) has also been used to help successfully establish Corsican pine [*Pinus nigra* var. *maritima* (Ait.) Melville] and Norway maple (*Acer platanoides* L.) over a four-year period in English tree plantations (Willoughby and McDonald, 1999).

Sod alleys in orchards are usually maintained by repeated mowing between trees beyond the drip line. Other methods have been studied for managing ground cover vegetation and weeds around individual trees. In New England in the past, peach [*Prunus persica* (L.) Batsch] orchards were either mulched or cultivated around trees to suppress weeds, but this led to soil erosion on sloping land and increased production costs (Lord and Vlach, 1973). Near trees, herbicides (either paraquat plus simazine once a year or two annual applications of paraquat), hay mulching, or rototilling in May and July controlled weeds better than mowing four to five times per growing season with an offset rotary mower (Lord and Vlach, 1973). Mowing was repeated whenever grass became 20 to 30 cm tall. Over six years, nitrogen uptake by trees, tree growth, and fruit yield were greater with herbicides, mulching, or cultivation than with mowing.

Orchard floor management is different in arid regions where orchards require supplemental irrigation (Connell et al., 2001). Commercial practices for almond [*Prunus dulcis* (Mill.) Webb in Heywood] in California varied during the growing season in response to different weeds and crop production concerns. In conventional practice, cover crops and winter annual weeds are usually mown in winter and spring to allow easy access for performing cultural practices when soils are wet. During dry summers when orchards are irrigated, mowing is used to control flushes of summer annual weeds which emerge after each subsequent irrigation. Just before harvest, summer annual weeds are mown or sprayed with glyphosate to aid almond harvest. Orchard floors must be weed-free at harvest to expedite almond drying and pickup after almonds are mechanically shaken to the ground. Recent research over four years (Connell et al., 2001) showed that an alternative weed management system using more herbicide saved time and labor and reduced the total number of mowings and preharvest glyphosate treatments needed. This system also prevented the buildup of common purslane (*Portulaca oleracea* L.) populations that were troublesome in conventional systems. The best alternative system consisted of spring-applied oryzalin at 1.1 kg a.i.$\cdot$ha^{-1} plus glyphosate at 1.12 kg a.i.$\cdot$ha^{-1}, followed by mowing in summer and preharvest spot treatment with glyphosate, as needed.

Mowing for Killing Cover Crops Before Planting or Transplanting No-Till Annual Crops

Fall-planted grass and legume cover crops suppress weeds, protect soil from runoff and erosion (Blevins, Thomas, and Cornelius, 1977), improve soil physical properties (Beale, Nutt, and Peele, 1955; McVay, Radcliffe, and Hargrove, 1989), increase organic matter (Blevins, Thomas, and Cornelius, 1977), reduce nitrate leaching to groundwater over winter (Stivers and Shennan, 1991), and decrease nitrogen requirements of later-sown crops, if the cover crops are nitrogen-fixing legumes (Hoyt and Hargrove, 1986).

Mowing can substitute for herbicides to kill cover crops before planting or transplanting no-till annual crops in spring. However, if cover crops are not completely killed, they deplete soil moisture or soil nitrogen needed for successful crop germination and emergence, or later compete with the crop. In Missouri, hairy vetch or rye cover crops which were rotary mown to a cutting height of about 8 cm before planting no-till corn controlled weeds [giant foxtail and common cocklebur (*Xanthium strumarium* L.)] better than mown soybean stubble, without herbicides (Johnson, Defelice, and Helsel, 1993). Although mowing could substitute for glyphosate at planting to control either a hairy vetch or rye cover crop, mowing did not substitute for preemergence herbicides for season-long weed control. Curran, Hoffman, and Werner (1994) verified that either tillage, mowing, or glyphosate plus 2,4-D before planting corn in May controlled a hairy vetch cover crop.

The type of mower influences cover crop kill. Flail mowing was less effective than either sicklebar mowing or undercutting (Creamer et al., 1995). Teasdale (1993) observed that if flail mowing within one week after planting corn did not completely kill a hairy vetch cover crop in spring, it regrew to compete with corn. Either flail mowing or rolling effectively killed cover crops before no-till transplanting broccoli in Maryland (Morse, 1999). Rolling lodged plants and crimped stems but kept shoots attached to roots. Rolling killed plants more slowly than mowing, delayed residue decomposition on the ground, and controlled weeds longer into the growing season (Lu et al., 2000). Weed control and yields of no-till broccoli were equal following either flail mowing or rolling treatments.

Mowing did not consistently kill cover crops in many studies. Inconsistent control between years and sites was partially attributed to differences in response between cover crop species when mown at different growth stages. Flail mowing at flowering killed hairy vetch better than at younger growth stages in Maryland (Teasdale, 1993). In Ohio, flail mowing hairy vetch at the early bud stage failed to control it before planting no-till corn (Hoffman, Regnier, and Cardina, 1993). Acceptable kills were achieved with later mowing in May and June, after flowering had finished. In other research, mowing (sicklebar or flail mowing) at the mid- to late-flowering stages killed cover crops best (rye, barley, hairy vetch, crimson clover, subterranean clover) (Creamer et al., 1995). Mowing winter wheat after heading with a sicklebar set at a 10 cm height killed wheat better than mowing before heading (Wilkins and Bellinder, 1996). Winter wheat was less susceptible than rye to mowing with a sicklebar mower at flowering. However, waiting to mow cover crops until they reach growth stages suitable for killing by mowing can delay crop planting past the best time for maximizing crop yields.

Mowing to Manage Perennial Forage-Living Mulch Intercropping Systems

Direct-seeding annual crops into living mulches of perennial forages which were suppressed by mowing has not been successful. Although mown living mulches reduced weed emergence and density, living mulches also directly reduced corn yields themselves, if they were inadequately suppressed (Martin, Greyson, and Gordon, 1999). In corn in Canada, a perennial forage-living mulch of white clover and mixed grasses was repeatedly mown to a height of 3 cm starting when it first became 12 cm tall and continuing until canopy closure. Even though mowing reduced competition between the living mulch and corn, the living mulch and weeds still reduced corn yields. The living mulch also delayed corn emergence and development, increasing the likelihood of later frost damage before harvest.

Mowing was used to suppress perennial forage-living mulches in transplanted hot pepper (*Capsicum annuum* L.) and okra [*Abelmoschus esculentus* (L.) Moench.] in Illinois (Biazzo and Masiunas, 2000). Four living mulches [white clover, red clover (*T. pratense* L.), perennial ryegrass, and canola (*Brassica napus* L.)] were mown with a rotary mower every other week to a 10 cm height beginning two weeks after transplanting. Although repeatedly mown living mulches of perennial ryegrass or white clover controlled weeds [velvetleaf (*Abutilon theophrasti* Medikus), common purslane, pigweed species (*Amaranthus* spp.), barnyardgrass, large crabgrass, and giant foxtail] and reduced late-emerging weed density best, the living mulches still decreased hot pepper and okra yields.

Mowing was used to suppress living mulches for producing white cabbage in Norway (Brandsaeter, Netland, and Meadow, 1998). But mowing living mulches (white clover, subterranean clover, or perennial ryegrass) once or twice per growing season did not increase cabbage yields compared to the unmown living mulch. Living mulches did not reduce early-season weed growth, but did reduce late-season weed growth [common lambsquarters, common hempnettle (*Galeopsis tetrahit* L.), and corn speedwell (*Veronica arvensis* L.)].

Goal 5: Mowing to Enhance Aesthetic Appeal of Landscaping

Type of mower, mowing height, and mowing frequency all influenced turf health, persistence, competitiveness with weeds, and aesthetic appeal (Polomski and Shaughnessy, 2001). Perennial turf grass species and varieties used for lawns, turf, or golf greens are usually mown uniformly close to soil surfaces and vary in mowing tolerance. Mowing height depends on turf species and region. Mowing height is chosen to encourage a dense competitive grass stand which resists weed invasions. Excessively low mowing can encourage weed encroachment into gaps. Lawn grasses with narrow leaves or a short horizontal growth habit can be mown lower than grasses with wide leaves or an erect, bunch-grass-type growth habit.

The rate of lawn regrowth following mowing determines the subsequent mowing frequency. If environmental stress, such as shading, drought, or temperature extremes, suppresses and delays lawn regrowth, then lawns should be mown less frequently and higher to favor lawn competition with weeds and minimize weed encroachment (Polomski and Shaughnessy, 2001). For example, mowing height and frequency interacted to influence weed control in golf greens. In North Carolina, yellow nutsedge (*Cyperus esculentus* L.) tuber production was completely inhibited in bermudagrass golf turf by mowing once a week to a 3.8 cm height or three times a week to a 1.3 cm height (Summerlin, Coble, and Yelverton, 2000). Yellow nutsedge shoot density, height, and horizontal spread were readily controlled by frequent, low mowing. But purple nutsedge (*Cyperus rotundus* L.), green kyllinga (*Kyllinga brevifolia* Rottb.), and false green kyllinga (*Kyllinga gracillima* Miq.) were resistant to these mowing regimes.

Following fertilization in spring, lawns require more frequent mowing than later in the growing season (Heckman et al., 2000; Polomski and Shaughnessy, 2001). Slow-release fertilizers reduce the number of mowings required. Soil nutrients limit turf growth and interact with mowing to influence turf regrowth and weed control (Heckman et al., 2000). For example, survival of Indian mock strawberry [*Duchesnea indica* (Andr.) Focke] infesting a tall fescue (*Festuca arundinacea* Schreb.) lawn was greatest when it was mown to a 6 cm height without NPK fertilization (Gray and Call, 1993). Survival was lowest when it was fertilized but not mown. In contrast, blue violet (*Viola* spp.) was eliminated by mowing but uninfluenced by NPK fertilization.

Vertical mowing is a management practice used to remove thatch buildup on turf (Callahan et al., 1997). A vertical mower having blades set 3.8 cm apart and operated at a depth to just groove soil surfaces was tested in Georgia on bermudagrass turf (Johnson, 1979). Thatch and loose debris were removed with a vacuum grass catcher after vertical mowing 0, 1, 2, 3, or 6 times per year starting in late February and continuing for three years. Vertical mowing encouraged summer annual weed growth more than winter annual weed growth. For example, vertical mowing alone favored large crabgrass encroachment over time because it thinned the turf and encouraged weed emergence. Vertical mowing plus DCPA herbicide reduced chickweed [*Stellaria media* (L.) Vill.] and henbit (*Lamium amplexicaule* L.) more than DCPA alone, but encouraged bromegrass (*Bromus* spp.) to increase. Vertical mowing and DCPA interacted to control weeds.

Goal 6: Direct Plant Succession to Restore Native Species Mixtures or Desirable Planted Species

In restoration ecology of wild land, infrequent mowing, often only once per growing season, is used to redirect plant succession when it is repeated over several years. It is used to prevent succession to woody shrubs, brush, trees, and woodlands, which invade grasslands after mowing is stopped (Piek, 1998). Mowing can also be used to prevent or reverse invasions by exotic herbaceous species. For example, mowing at the proper height, time, and frequency helped restore a short grass prairie in the Pacific Northwest by directing plant succession away from tall, exotic, invasive species to desirable native short grass species (Wilson and Clark, 2001). Mowing was more effective at 15 cm above soil surfaces than at 50 cm due to height differences between undesirable invasive species and native grass species. In these short grass prairies, mowing once per year for two to three years reduced tall oatgrass [*Arrhenatherum elatius* (L.) Presl] ground cover and encouraged perennial native grasses, such as California oatgrass (*Danthonia californica* Bol.) and Roemeri's fescue [*Festuca roemeri* (Pavlick) Alexeev]. Mowing once a year for three years when tall oatgrass flowered in late spring or early summer was effective for preventing seed production. Such success with mowing is specific to particular weeds and sites. In a different Oregon wetland prairie, either controlled burning or hand removal of woody species was more effective than mowing 8 to 12 cm above the ground for preventing invasions by woody species (Clark and Wilson, 2001). All three treatments were imposed in fall and repeated two years later. Mowing woody species followed by removal of mown residues did not reduce woody cover due to resprouting, and the practice encouraged invasion by exotic herbaceous species.

Because rangeland perennial grass species differ in regrowth rate and competitive ability following mowing, weed encroachment depends on both regrowth rate of competing species and the extent or intensity of defoliation by mowing (Sheley, Olson, and Larson, 1997). For example, diffuse knapweed (*Centaurea diffusa* Lam.) was better able to establish in mown than unmown bluebunch wheatgrass [*Pseudoroegneria spicata* (Pursh.) Schrb.] on rangeland in eastern Washington (Sheley, Olson, and Larson, 1997). But crested wheatgrass [*Agropyron cristatum* (L.) Gaertn.] regrew more readily than did bluebunch wheatgrass even when mowing completely defoliated plants. Diffuse knapweed established to low but equal levels in both mown and unmown crested wheatgrass.

Other goals for mowing are not well documented in the scientific literature. For example, California roadways are mown primarily to reduce the chance of fire and prevent it from spreading (Joe DiTomaso, 2002, personal communication). Mowing is timed to match initial grass senescence, which prevents grass recovery. Unfortunately, mowing at this time misses weeds below the mowing height, such as rosette-stage yellow starthistle (Benefield et al., 1999). Yellow starthistle takes advantage of increased light penetration to fill gaps in the mown grass canopy.

INTEGRATED WEED MANAGEMENT SYSTEMS INCLUDING MOWING IN COMBINATION WITH OTHER WEED MANAGEMENT METHODS

Mowing has been most often studied alone as a substitute for herbicides or tillage for controlling weeds. However, combinations or sequences of mowing, herbicides, and other nonchemical weed management practices improve weed control beyond that achieved with one method alone. The following are examples of new, experimental, integrated weed management systems which include mowing.

Combined Fluid Applicator-Mower

Recently, a commercial fluid applicator-mower (i.e., Burch Wet Blade) was introduced for mowing and simultaneously applying liquids (e.g., herbicides, growth regulators, fertilizer, biological control agents) in turf, pasture, roadsides, and rangeland (DeValerio et al., 2000). This fluid applicator-mower uses a closed liquid delivery system to apply herbicide to the lower surface of the cutting blade on a mower so that only wounded plant tissues are treated. The Burch Wet Blade uses ground speed sensing radar and a peristaltic pump-metering system to ensure consistent fluid flow to the mower blade.

Direct entry of herbicide into the vascular system of cut stems or leaves of mown plants will likely increase effectiveness of translocated herbicides more than contact herbicides (Wahlers, Burton, Maness, and Skroch, 1997a,b). Based on studies of radiolabeled herbicide uptake kinetics and whole plant autoradiography, Wahlers and colleagues (1997a) suggested that stem cutting caused xylem cavitation, which drew herbicide into the stem. Herbicides diffused laterally from the xylem into the phloem for later basipetal translocation.

This fluid applicator-mower has potential for reducing off-site movement of herbicides in drift. It may also have potential for reducing herbicide spray volume and dosage in agricultural systems that can be mown, but this requires further field testing and confirmation.

The Between-Row Mowing Weed Management System

The between-row mowing weed management system for annual row crops controls summer annual grass and broadleaf weeds with (1) band-applied, preemergence, soil-residual herbicides over crop rows; (2) two or more mowings between crop rows; and (3) shading by crop canopy closing (i.e., crop competition) in soybean (Donald, 2000a,b; Donald, Kitchen, and Sudduth, 2001) and corn (Donald, Kitchen, and Sudduth, 2001) grown in 76 cm rows in Missouri (see Figure 12.4). Mowing weeds close to the soil surface (< 3.8 cm) two or more times between crop rows killed or suppressed annual grass and broadleaf weeds, chiefly giant foxtail, common ragweed (*Ambrosia artemisiifolia* L.), and waterhemp species (*Amaranthus* spp.), if properly timed (Donald 2000a,b). Weeds growing within crop rows were controlled with herbicides, and any escapes were suppressed by early crop shading and competition. Weeds growing between crop rows were killed by close mowing, and any escapes were subsequently controlled by both a second mowing, as needed, and shading by crop canopy closure. Yields of soybeans and corn were as great with the between-row mowing weed management system as for weed-free checks. The between-row mowing weed management system is likely to be successful for competitive crops, such as corn, soybean, and grain sorghum [*Sorghum bicolor* (L.) Moench], which closed canopy and shaded weeds early in the growing season (Donald, 2000b). Early canopy closure followed by an extended period of shading before leaf drop helped suppress summer annual weeds. Extreme weather conditions, such as drought stress, which prevented canopy closure and shading, allowed weeds to escape (Donald 2000b; Donald, Kitchen, and Sudduth, 2001). Crop management that enhances crop competitiveness with weeds is critical to the success of this system (e.g., selection of tall or bushy crop cultivars which close canopy early, early planting, narrow row spacing, high seeding rate). Crops with greater capacity for yield component compensation, such as soybean or grain sorghum, are suitable for the between-row mowing weed management system. Crops that are unsuitable for this system include (1) crops grown in very narrow rows (i.e., "solid seeded" or drilled cereals, oilseeds, or forages); (2) crops that are noncompetitive (e.g., they vine, grow low, or lack a bushy growth habit) and fail to close canopy to shade weeds; (3) crops that require soil movement for their production or harvest; or (4) crops that are easily damaged by thrown soil. Between-row mowing is not adapted for most intercropping systems. Herbicide use is reduced 50 percent by banding because only 50 percent of the field area is sprayed. The between-row mowing weed management may have use in environmentally sensitive areas to help reduce soil erosion or surface water contamination by herbicides (Donald, 2000a). It also can be adapted for use in no-tillage farming (Donald, Kitchen, and Sudduth, 2001).

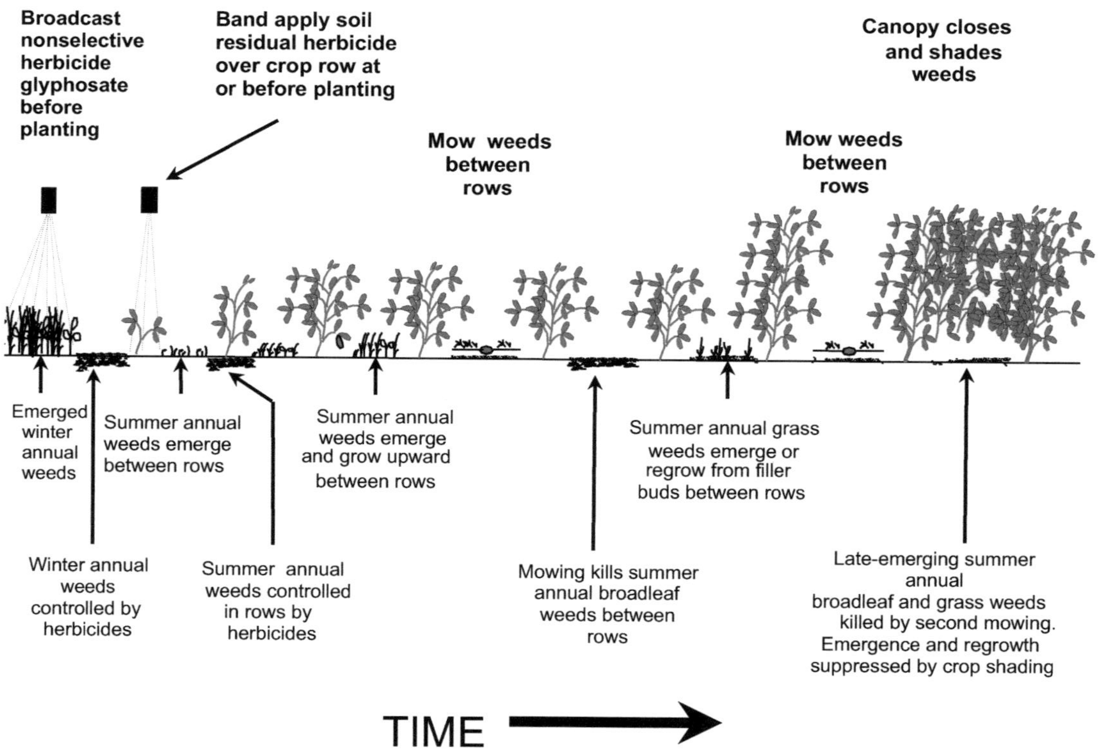

FIGURE 12.4. A diagram of the chief components of the between-row mowing weed management system for no-till row crops and how they contribute to weed management over time (Donald, Kitchen, and Sudduth, 2001).

Herbicides, Cultivation, and Mowing

Recently, mowing was used to supplement herbicides and cultivation to prevent weed seed production in peanut (*Arachis hypogaea* L.) (Wehtje et al., 1999). When weeds grew 20 cm above the peanut canopy, a tractor-pulled, rotary mower was used to cut seedheads off just above the canopy 8 to 13 weeks after planting. Mowing prevented peanut from being shaded by bristly starbur (*Acanthospermum hispidum* DC.), sicklepod [*Senna obtusifolia* (L.) Irwin & Barneby], and Florida beggarweed [*Desmodium tortuosum* (SW.) DC.]. Mowing also increased yield and net returns for one weed management system {i.e., paraquat plus 2,4-DB [4-(2,4-dichlorophenoxy) butanoic acid] applied four weeks after planting followed by cultivation at ten weeks after planting}. However, mowing did not improve yields in the highest-yielding systems which received the most weed management.

WHAT SCIENTIFIC KNOWLEDGE IS NEEDED TO BETTER USE MOWING TO ACHIEVE SEVERAL GOALS AT ONCE?

Mowing has not been systematically researched for weed management, and scientific principles for predictably optimizing mowing for weed control are largely lacking. Consequently, research on mowing has been largely empirical. This situation has to change if mowing is to be better used in integrated weed management.

Weed Classification Systems Based on Mowing Susceptibility, Tolerance, and Resistance

No databases summarize weed susceptibility, tolerance, or resistance to mowing at different growth stages, other than in turf (McCarty et al., 2001). Plant functional group analysis may help classify plant group response to mowing regimes in terms of susceptibility, tolerance, or resistance. In functional group analysis, ecologists classify groups of plants in terms of common responses to disturbance (i.e., mowing), rather than using taxonomic or phylogenetic relationships between species (Landsberg, Lavorel, and Stol, 1999). Plant functional groups or types are also termed plant guilds, structure-function groups, plant types, or adaptive syndromes (Wilson, 1999). Plants are classified based on similar functional traits used in responding to disturbance (Kleyer, 1999). Functional traits are identified and used in classification systems which allow plant populations to maximize their fitness when disturbed (Weiher et al., 1999). Rosenthal and Kotanen

(1994) classified disturbances as either abiotic or biotic. Abiotic disturbance includes environmental stress, such as drought, fire, mowing, herbicides, or soil disturbance, among others. Biotic disturbance includes interference, herbivory or grazing. Disturbance can also be characterized in terms of (1) distribution or spatial extent (Bellingham and Sparrow, 2000), (2) intensity or severity (i.e., extent of biomass removal), (3) frequency of disturbance, (4) depth below or above soil surfaces (Kleyer, 1999), (5) predictability (Bellingham and Sparrow, 2000), or (6) synergism with other factors (e. g., shading). In functional group analysis, plant response to disturbance often can be expressed on either a plant part, per plant, or per area basis, depending on scale (Harper, 1989). The modular metamer construction of plants (Porter, 1989) defines response to disturbance (Gautier, Varlet-Grancher, and Hazard, 1999). Plant functional group response to heavy grazing in grasslands can provide an initial model for mowing plant functional group characteristics. Grazing-tolerant plants share the following functional group traits:

1. small size,
2. prostrate growth habit,
3. meristems displayed low on plants,
4. small leaves,
5. hairy leaves,
6. high regrowth potential, including tillering,
7. plasticity in response to grazing, and
8. high fecundity (seed production) (Landsberg, Lavorel, and Stol, 1999).

Because plants differ in palatability and nutrient value, different herbivores can graze them to a different extent or not at all (McKinney and Fowler, 1991). There may not be a one-to-one correspondence among all grazing-tolerance-plant functional group traits because mowing is nonselective and uniform, whereas grazing is selective and nonuniform.

Models of Weed Phenology to Better Time Mowing for Different Goals

Mowing to prevent weed seed production is now entirely empirical and suffers from a lack of published summary databases. Models are needed to predict weed flowering phenology as a function of photoperiod and temperature (e.g., degree-day, heat sums, or thermal time), the chief environmental drivers controlling flowering. This information is needed to better time mowing (or combine harvest) for minimizing weed seed production and seed dispersal (i.e., maximize the proportion of the population that is in susceptible flowering stages when mown) in a range of agroecosystems. In Canada, seed shed of wild oat (*Avena fatua* L.) and spring wheat occur synchronously and can be modeled using thermal time (Shirtliffe, Entz, and Van Acker, 2000). These authors suggest that the proportion of weed seed removed from fields during combining could be manipulated by choice of crop, harvest method, and harvest timing relative to weed seed shed. Removing weed seed from fields by timing wheat harvest early before most weed seed had shed was suggested to minimize future weed problems on fields. Timing wheat harvest late after most weed seed had shed was suggested to limit weed seed dispersal across fields. There are obvious trade-offs between minimizing weed seed production and seed dispersal.

Mowing in Sequence or Combination with Herbicides for Integrated Weed Management

The sequential timing of mowing and herbicide application impacts weed control, especially for difficult-to-control perennial weeds in lawns, roadsides, pastures, and rangeland. Removing shoots of common dandelion too soon (i.e., two weeks) after applying 2,4-D at 1.1 kg a.i.$\cdot$ha^{-1} reduced later common dandelion control in lawns, presumably by limiting translocation of 2,4-D to roots (Mann, 1981). Some sequences of mowing followed by herbicides (2,4-D at 2.24 kg a.i.$\cdot$ha^{-1} or picloram at 0.56 kg a.i.$\cdot$ha^{-1}) partially controlled established Macartney rose infesting Texas pastures better than others, depending upon when herbicide treatment was timed after mowing (Haas, Lehman, and Morton, 1970). In this case, partial control was best when a year elapsed between mowing Macartney rose and subsequent 2,4-D application, compared to 2,4-D treatment of unmown checks.

JUSTIFICATIONS FOR USING MOWING IN INTEGRATED WEED MANAGEMENT

In all likelihood, mowing for weed control can be justified based on economics, energy use, and environmental quality. Data are needed to document advantages and justify implementation of new, experimental, integrated weed management systems that incorporate mowing. These justifications are needed to secure research funding for testing and implementing these approaches.

Economics and Energy Use of Mowing

Breakeven analysis for mowing requires knowledge of number of trips per growing season and costs for equipment, labor, fuel use, maintenance, and repairs. In general, the time required to treat equivalent areas and fuel use can be ranked as follows: herbicide application < mowing < secondary tillage (e.g., rotary hoeing or field cultivation).

Environmental Quality and Mowing

Vegetation can be mown where ground cover is required for soil erosion control and herbicide use would be restricted, impossible, or undesirable. Mowing also can be used where there are concerns about herbicide spray drift to nearby vegetation or overspraying and herbicide contamination of surface water or ground water wells. Mowing is advantageous near or in highly populated areas and suburban housing, where the public objects to "chemical trespass" from herbicide drift. Mowing also may have a place when weather conditions, such as wind speed and direction, air temperature and temperature inversions, dew, or frost, either cause drift or reduce herbicide efficacy for weed control.

Mowing can substitute for field cultivation, especially in perennial crops where cultivation might damage root systems or lead to soil erosion. Mowing has relatively low power requirements compared to cultivation and can be performed quickly, improving weed control timeliness (Anonymous, 1985). Fields can be mown before they can be cultivated after rainfall, because mowing does not require soils to be dry enough to pulverize in order to kill weeds. As noted earlier, mowing can control annual weeds that have become too large to kill with cultivation.

Mowing reduces wind or water erosion by keeping plant residues on soil surfaces. It is also compatible with many other soil conservation measures, such as no-tillage farming practices (Donald, Kitchen, and Sudduth, 2001), grassed waterways, or terraces. Compared to either herbicides or cultivation, mowing has few detrimental off-site environmental effects. However, it is noisy, and, like cultivation, mowing raises dust to contaminate air (Anonymous, 1985).

Unfortunately, wheel traffic from mowing can compact soil under permanent grass cover, especially on silty clay loam soils (Oliveira and Merwin, 2001). In an apple (*Malus domestica* Borkh.) orchard, biweekly mowing red fescue (*Festuca rubra* L.) induced soil compaction which lead to increased bulk density, lowered soil porosity, and reduced infiltration and volumetric soil water content, all of which reduced orchard root development, growth, and yield (Oliveira and Merwin, 2001). Other researchers did not observe compaction with repeated mowing on other soil types and locations (cited in Oliveira and Merwin, 2001). The impact of repeated mowing on soil structure and weed biology has not been studied.

REFERENCES

Abdul-Baki, A. A., R. D. Morse, T. E. Devine, and J. R. Teasdale (1997). Broccoli production in forage soybean and foxtail millet cover crop mulches. *HortScience* 32: 836-839.

Abdul-Baki, A. A. and J. R. Teasdale (1993). A no-tillage tomato production system using hairy vetch and subterranean clover mulches. *HortScience* 28: 106-108.

Aiken, G. E., S. E. Sladden, and D. I. Bransby (1995). Cutting height and frequency effects on composition, yield, and quality of a bermudagrass-crabgrass mixture. *Journal of Production Agriculture* 8: 79-83.

Aldrich, R. J. and R. J. Kremer (1997). *Principles in weed management,* Second edition. Ames: Iowa State University Press.

Amor, R. L. and R. V. Harris (1977). Control of *Cirsium arvense* (L.) Scop. by herbicides and mowing. *Weed Research* 17: 303-309.

Anderson, W. P. (1983). *Weed science: Principles,* Second edition. New York: West Publishing Company.

Anonymous (1985). *Principles of weed control in California.* California Weed Conference. Fresno, CA: Thomson Publications.

Ballare, C. L. and J. J. Casal (2000). Light signals perceived by crop and weed plants. *Field Crops Research* 67: 149-160.

Ballare, C. L., A. L. Scopel, C.M. Ghersa, and R. A. Sanchez (1987a). The demography of *Datura ferox* (L.) in soybean crops. *Weed Research* 27: 91-102.

Ballare, C. L., A. L. Scopel, C. M. Ghersa, and R. A. Sanchez (1987b). The population ecology of *Datura ferox* in soybean crops: A simulation approach incorporating seed dispersal. *Agriculture, Ecosystems, and Environment* 19: 177-188.

Bassett, I. J. and C. W. Crompton (1978). The biology of Canadian weeds. 32. *Chenopodium album* L. *Canadian Journal of Plant Science* 58: 1061-1072.

Beale, O. W., G. B. Nutt, and T. C. Peele (1955). The effects of mulch tillage on runoff, erosion, soil properties, and crop yields. *Soil Science Society of America Proceedings* 19: 244-247.

Beck, K. G. and J. R. Sebastian (2000). Combining mowing and fall-applied herbicides to control Canada thistle *(Cirsium arvense). Weed Technology* 14: 351-356.

Bellingham, P. J. and A. D. Sparrow (2000). Resprouting as a life history strategy in woody plant communities. *Oikos* 89: 409-416.

Benefield, C. B., J. M. Di Tomaso, G. B. Kyser, S. B. Orloff, K. R. Churches, D. B. Marcum, and G. A. Nader (1999). Success of mowing to control yellow starthistle depends on timing and plant's branching system. *California Agriculture* 53(2): 17-21.

Biazzo, J. and J. B. Masiunas (2000). The use of living mulches for weed management in hot pepper and okra. *Journal of Sustainable Agriculture* 16: 59-79.

Black, A. L. and F. H. Siddoway (1977). Winter wheat recropping on dryland as effected by stubble height and nitrogen fertilization. *Soil Science Society of America Journal* 41: 1186-1190.

Blevins, R. L., G. W. Thomas, and P. L. Cornelius (1977). Influence of no-tillage and nitrogen fertilization on certain soil properties after 5 years of continuous corn. *Agronomy Journal* 69: 383-386.

Bovey, R. W. and R. E. Meyer (1974). Mortality of honey mesquite and huisache seedlings from herbicides and top removal. *Weed Science* 22: 276-279.

Brabec, J. and P. Pysek (2000). Establishment and survival of three invasive taxa of the genus *Reynoutria* (Polygonaceae) in mown meadows: A field experimental study. *Folia Geobotanica* 35: 27-42.

Brandsaeter, L. O., J. Netland, and R. Meadow (1998). Yields, weeds, pests and soil nitrogen in a white cabbage-living mulch system. *Biological Agriculture and Horticulture* 16: 291-309.

Bryan, W. B. and T. A. Mills (1988). Effect of frequency and method of defoliation and plant size on the survival of multiflora rose. *Biological Agriculture and Horticulture* 5: 209-214.

Callahan, L. M., W. L. Sanders, J. M. Parham, C. A. Harper, L. D. Lester, and E. R. McDonald (1997). Comparative methods of measuring thatch on a creeping bentgrass green. *Crop Science* 37: 230-237.

Cavers, P. B. and M. Kane (1990). Responses of proso millet *(Panicum miliaceum)* seedlings to mechanical damage and/or drought treatments. *Weed Technology* 4: 425-432.

Clark, D. L. and M. V. Wilson (2001). Fire, mowing, and hand-removal of woody species in restoring a native wetland prairie in the Willamette Valley of Oregon. *Wetlands* 21: 135-144.

Connell, J. H., F. Colbert, W. Krueger, D. Cudney, R. Gast, T. Bettner, and S. Dallman (2001). Vegetation management options in almond orchards. *HortTechnology* 11: 254-257.

Coulson, S. J., J. M. Bullock, M. J. Stevenson, and R. F. Pywell (2001). Colonization of grassland by sown species: Dispersal versus microsite limitation in responses to management. *Journal of Applied Ecology* 38: 204-216.

Creamer, N. G., B. Plassman, M. A. Bennett, R. K. Wood, B. R. Stinner, and J. Cardina (1995). A method for mechanically killing cover crops to optimize weed suppression. *American Journal of Alternative Agriculture* 10: 157-162.

Curran, W. S., L. D. Hoffman, and E. L. Werner (1994). The influence of hairy vetch *(Vicia villosa)* cover crop on weed control and corn *(Zea mays)* growth and yield. *Weed Technology* 8: 777-784.

Currie, R. S. and T. F. Peeper (1988). Combine harvesting affects weed seed germination. *Weed Technology* 2: 499-504.

Dabney, S. M. (1998). Cover crop impacts on watershed hydrology. *Journal of Soil and Water Conservation* 53: 207-213.

Derscheid, L. A., R. L. Nash, and G. A. Wicks (1961). Thistle control with cultivation, cropping and chemicals. *Weeds* 9: 90-102.

DeValerio, J. T., R. Charudattan, J. J. Mullahey, W. H. Sherrod, and P. D. Roberts (2000). Biological control of *Solanum viarum* Dunal (tropical soda apple) by *Ralstonia solancearum* (E.F. Smith) Yabuuchi applied with the Burch Wet Blade mower system. *Abstracts Weed Science Society of America Meetings* 40: 28-29.

Donald, W. W. (1990). Management and control of Canada thistle *(Cirsium arvense). Reviews of Weed Science* 5: 193-250.

Donald, W. W. (2000a). Between-row mowing plus in-row band-applied herbicide for weed control in *Glycine max. Weed Science* 48: 487-500.

Donald, W. W. (2000b). Timing and frequency of between-row mowing and band-applied herbicide for annual weed control in soybean. *Agronomy Journal* 92: 1013-1019.

Donald, W. W., N. R. Kitchen, and K. A. Sudduth (2001). Between-row mowing plus banded herbicide to control annual weeds and reduce herbicide use in no-till soybean *(Glycine max)* and corn *(Zea mays). Weed Technology* 15: 576-584.

Evetts, L. L. and O. C. Burnside (1972). Germination and seedling development of common milkweed and other species. *Weed Science* 20: 371-378.

Feldman, S. R. and J. P. Lewis (1990). Output and dispersal of propagules of *Carduus acanthoides* L. *Weed Research* 30: 161-169.

Fleischer, S. J., M. J. Gaylor, R. Dickens, and D. L. Turner (1989). Roadside management of annual fleabane *(Erigeron annuus)* and wild carrot *(Daucus carota). Weed Technology* 3: 72-75.

Galloway, B. A. and L. A. Weston (1996). Influence of cover crop and herbicide treatment on weed control and yield in no-till sweet corn (*Zea mays* L.) and pumpkin (*Cucurbita maxima* Duch.). *Weed Technology* 10: 341-346.

Garay, A. H., C. Matthew, and J. Hodgson (1999). Tiller size/density compensation in perennial ryegrass miniature swards subject to differing defoliation heights and a proposed productivity index. *Grass and Forage Science* 54: 347-356.

Gautier, H., C. Varlet-Grancher, and L. Hazard (1999). Tillering response to the light environment and to defoliation in populations of perennial ryegrass (*Lolium perenne* L.) Selected for contrasting leaf length. *Annals of Botany* 83: 423-429.

Ghersa, C. M., R. L. Benech-Arnold, E. H. Satorre, and M. A. Martinez-Ghersa (2000). Advances in weed management strategies. *Field Crops Research* 67: 95-104.

Gonzalez-Andujar, J. L., J. N. Perry, and S. R. Moss (1999). Modeling effects of spatial patterns on the seed bank dynamics of *Alopercurus myosuroides. Weed Science* 47: 697-705.

Gray, E. and N. M. Call (1993). Fertilization and mowing on persistence of Indian mock strawberry *(Duchesnea indica)* and common blue violet *(Viola papilionacea)* in a tall fescue *(Festuca arundinacea)* lawn. *Weed Science* 41: 548-550.

Haas, R. H., S. K. Lehman, and H. L. Morton (1970). Influence of mowing and spraying dates on herbicidal control of Macartney rose. *Weed Science* 18: 33-37.

Hakansson, S. (1995a). Weeds in agricultural crops: 1. Life-forms and occurrence under Swedish conditions. *Swedish Journal of Agricultural Research* 25: 143-154.

Hakansson, S. (1995b). Weeds in agricultural crops: 2. Life-forms and occurrence in a European perspective. *Swedish Journal of Agricultural Research* 25: 155-161.

Hakansson, S. (1995c). Weeds in agricultural crops: 3. Life-forms, C3 and C4 photosynthesis and plant families in a global perspective. *Swedish Journal of Agricultural Research* 25: 163-171.

Harper, J. L. (1989). Canopies as populations. In G. Russell, B. Marshall, and P. G. Jarvis (Eds.), *Plant canopies: Their growth, form, and function* (pp. 105-128). New York: Cambridge University Press.

Heckman, J. R., H. Liu, W. Hill, M. DeMilia, and W. L. Anastasia (2000). Kentucky bluegrass responses to mowing practice and nitrogen fertility management. *Journal of Sustainable Agriculture* 15: 25-33.

Heisel, T., J. Schou, S. Christensen, and C. Andreasen (2001). Cutting weeds with a CO_2 laser. *Weed Research* 41: 19-29.

Hodgson, J. M. (1958). Canada thistle (*Cirsium arvense* Scop.) control with cultivation, cropping, and chemical sprays. *Weeds* 6: 1-11.

Hoffman, M. L., E. E. Regnier, and J. Cardina (1993). Weed and corn *(Zea mays)* response to a hairy vetch *(Vicia villosa)* cover crop. *Weed Technology* 7: 594-599.

Hoveland, C. S., R. G. Durham, and J. H. Bouton (1996). Weed encroachment in established alfalfa as effected by cutting frequency. *Journal of Production Agriculture* 9: 399-402.

Hoyt, G. D. and W. L. Hargrove (1986). Legume cover crops for improving crop and soil management in the Southern United States. *HortScience* 21: 397-402.

Hume, L. (1988). Long-term effects of 2,4-D application on plants: II. Herbicide avoidance by *Chenopodium album* and *Thlaspi arvense*. *Canadian Journal of Botany* 66: 230-235.

Hunt, D. (2001). *Farm power and machinery management,* Tenth edition. Ames: Iowa State University.

Jeffery, L. S. and L. R. Robison (1971). Growth characteristics of common milkweed. *Weed Science* 19: 193-196.

Jewiss, O. R. (1972). Tillering in grasses—Its significance and control. *Journal of the British Grassland Society* 27: 65-81.

Johnson, B. J. (1979). Vertical mowing and herbicide treatments on thatch in bermudagrass *(Cynodon dactylon)* turf. *Weed Science* 27: 14-17.

Johnson, G. A., M. S. Defelice, and Z. R. Helsel (1993). Cover crop management and weed control in corn *(Zea mays)*. *Weed Technology* 7: 425-530.

Kanafojski, C. and T. Karwowski (1972). *Agricultural machines, theory and construction,* Volume 2, *Crop-harvesting machines.* Warsaw, Poland: Foreign Publications Department of the National Center for Scientific, Technical, and Economic Information.

Katovich, E. J. S., R. L. Becker, and D. W. Ragsdale (1999). Effect of *Galerucella* spp. on survival of purple loosestrife *(Lythrum salicaria)* roots and crowns. *Weed Science* 47: 360-365.

Kleyer, M. (1999). Distribution of plant functional types along gradients of disturbance intensity and resource supply in an agricultural landscape. *Journal of Vegetation Science* 10: 697-708.

Landsberg, J., S. Lavorel, and J. Stol (1999). Grazing response groups among understory plants in arid rangelands. *Journal of Vegetation Science* 10: 683-696.

Lord, W. J. and E. Vlach (1973). Responses of peach trees to herbicides, mulch, mowing, and cultivation. *Weed Science* 21: 227-229.

Lu, Y.-C., K. B. Watkins, J. R. Teasdale, and A. A. Abdul-Baki (2000). Cover crops in sustainable food production. *Food Reviews International* 16: 121-157.

Magda, D. and M. Jarry (2000). Prediction of cutting effects on a population of *Chaerophyllum aureum*—A demographic approach. *Journal of Vegetation Science* 11: 485-492.

Mangan, F., R. DeGregorio, M. Schonbeck, S. Herbert, K. Guillard, R. Hazzard, E. Sideman, and G. Litchfield (1995). Cover cropping systems for Brassicas in the northeastern United States: 2. Weed, insect, and slug incidence. *Journal of Sustainable Agriculture* 5: 15-36.

Mann, H. (1981). Common dandelion *(Taraxacum officinale)* control with 2,4-D and mechanical treatments. *Weed Science* 29: 704-708.

Martin, R. C., P. R. Greyson, and R. Gordon (1999). Competition between corn and a living mulch. *Canadian Journal of Plant Science* 79: 579-586.

Maxwell, B. D. and C. Ghersa (1992). The influence of weed seed dispersion versus the effect of competition on crop yield. *Weed Technology* 6: 196-204.

McCanny, S. J. and P. B. Cavers (1988). Spread of proso millet (*Panicum miliaceum* L.) in Ontario, Canada: II. Dispersal by combines. *Weed Research* 28: 67-72.

McCarty, L. B., J. W. Everest, D. W. Hall, T. R. Murphy, and F. Yelverton (2001). *Color atlas of turfgrass weeds.* Chelsea, MI: Ann Arbor Press.

McCarty, M. K. and J. L. Hatting (1975). Effects of herbicides or mowing on musk thistle seed production. *Weed Research* 15: 363-367.

McIntyre, S., S. Diaz, S. Lavorel, and W. Cramer (1999). Plant functional types and disturbance dynamics—Introduction. *Journal of Vegetation Science* 10: 604-608.

McIntyre, S., S. Lavorel, J. Landsberg, and T. D. A. Forbes (1999). Disturbance response in vegetation toward a global perspective on functional traits. *Journal of Vegetation Science* 10: 621-630.

McKinney, K. K. and N. L. Fowler (1991). Genetic adaptation to grazing and mowing in the unpalatable grass *Cenchrus incertus. Oecologia* 88: 238-242.

McVay, K. A., D. E. Radcliffe, and W. L. Hargrove (1989). Winter legume effects on soil properties and nitrogen fertilizer requirements. *Soil Science Society of America Journal* 53: 1856-1862.

Mislevy, P., J. J. Mullahey, and F. G. Martin (1999). Preherbicide mowing and herbicide rate on tropical soda apple *(Solanum viarum)* control. *Weed Technology* 13: 172-175.

Monaco, T. J., S. C. Weller, and F. M. Ashton (2002). *Weed science: Principles and practices,* Fourth edition. New York: John Wiley and Sons.

Morse, R. (1999). Cultural weed management methods for high-residue/no-till production of transplanted broccoli (*Brassica oleracea* L. Gp. Italica). In A. Liptay, C.S. Vavrina, and G.E. Welbaum (Eds.), *Proceedings of the Sixth Symposium on Stand Establishment and ISHS Seed Symposium* (pp. 121-128). ISHS Acta

Horticulturae, Vol. 504. Roanoke, VA: International Society for Horticultural Science.

Morvan-Bertrand, A., J. Boucaud, and M. Prud'homme (1999a). Influence of initial levels of carbohydrates, fructans, nitrogen, and soluble proteins on regrowth of *Lolium perenne* L. cv. Bravo following defoliation. *Journal of Experimental Botany* 50: 1817-1826.

Morvan-Bertrand, A., J. Boucaud, and M. Prud'homme (1999b). Partitioning of reserve and newly assimilated carbon in roots and leaf tissues of Lolium perenne during regrowth after defoliation: Assessment by ^{13}C steady-state labeling and carbohydrate analysis. *Plant Cell and Environment* 22: 1097-1108.

Muenscher, W. C. (1980). *Weeds,* Second edition. Ithaca, NY: Cornell University Press.

Norris, R. F. and D. Ayres (1991). Cutting interval and irrigation timing in alfalfa: Yellow foxtail invasion and economic analysis. *Agronomy Journal* 83: 552-558.

O'Dogherty, M. J. (1982). A review of research on forage chopping. *Journal of Agricultural Engineering Research* 27: 267-289.

O'Dogherty, M. J. and G. E. Gale (1991a). Laboratory studies of the dynamic behaviour of grass, straw, and polystyrene tube during high speed cutting. *Journal of Agricultural Engineering Research* 49: 33-57.

O'Dogherty, M. J. and G. E. Gale (1991b). Laboratory studies of the effect of blade parameters and stem configuration on the dynamics of cutting grass. *Journal of Agricultural Engineering Research* 49: 99-111.

Oliveira, M. T. and I. A. Merwin (2001). Soil physical conditions in a New York orchard after eight years under different groundcover management systems. *Plant and Soil* 234: 233-237.

Paez, A., M. E. Gonzalez, X. Yrausquin, A. Salazar, and A. Casanova (1995). Water stress and clipping management effects on guineagrass: I. Growth and biomass allocation. *Agronomy Journal* 87: 698-706.

Persson, S. (1987). *Mechanics of cutting plant material.* ASAE Monograph Number 7. St. Joseph, MO: American Society of Agricultural Engineers.

Piek, H. (1998). The practical use of grazing in nature reserves in the Netherlands. In M. F. Wallis DeVries, J. P. Bakker, and S. E. Van Wieren (Eds.), *Grazing and conservation management* (pp. 253-272). Boston, MA: Kluwer Academic Publishers.

Pike, D. R. and J. F. Stritzke (1984). Alfalfa (*Medicago sativa*)-cheat (*Bromus secalinus*) competition. *Weed Science* 32: 751-756.

Polomski, B. and D. Shaughnessy (2001). Mowing lawns. Clemson University Extension Home and Garden Information Centre (HGIC) 1205. Available online at <http://hgic.clemson.edu/factsheets/HGIC1205.htm>.

Porter, J. R. (1989). Modules, models and meristems in plant architecture. In G. Russell, B. Marshall, and P. G. Jarvis (Eds.), *Plant canopies: Their growth, form and function* (pp. 143-159). New York: Cambridge University Press.

Raju, M. V. S. (1975). Experimental studies on leafy spurge (*Euphorbia esula* L.): I. Ontogeny and distribution of buds and shoots on the hypocotyl. *Botanical Gazette* 136: 254-261.

Raunkiaer, C. (1934). *The life forms of plants and statistical plant geography*. Oxford, United Kingdom: Clarendon Press.

Rew, L. J. and R. D. Cousens (2001). Spatial distribution of weeds in arable crops: Are current sampling and analytical methods appropriate? *Weed Research* 41: 1-18.

Rew, L. J., R. J. Froud-Williams, and N. D. Boatman (1996). Dispersal of *Bromus sterilis* and *Antrhiscus sylvestris* seed within arable field margins. *Agriculture, Ecosystems, and Environment* 59: 107-114.

Room, P. M., L. Maillette, and J. S. Hanan (1994). Module and metamer dynamics and virtual plants. *Advances in Ecological Research* 25: 105-157.

Rosenthal, J. P. and P. M. Kotanen (1994). Terrestrial plant tolerance to herbivory. *Trends in Ecology and Evolution* 9: 145-148.

Ross, M. A. and C. A. Lembi (1985). *Applied weed science,* First edition. Minneapolis, MN: Burgess Publishing Company.

Ross, M. A. and C. A. Lembi (1999). *Applied weed science,* Second edition. Upper Saddle River, NJ: Prentice Hall.

Roundy, B. A., R. B. Taylorson, and L. B. Sumrall (1992). Germination response of Lehmann lovegrass to light. *Journal of Range Management* 45: 81-84.

Sachs, T. and M. Hassidim (1996). Mutual support and selection between branches of damaged plants. *Vegetatio* 127: 25-30.

Schreiber, M. M. (1965). Effect of date of planting and stage of cutting on seed production of giant foxtail. *Weeds* 13: 60-62.

Schreiber, M. M. (1967). Effect of density and control of Canada thistle on production and utilization of alfalfa pasture. *Weeds* 15: 138-142.

Schwab, G. O., R. K. Frevert, T. W. Edminster, and K. K. Barnes (1981). *Soil and water conservation engineering,* Third edition. New York: John Wiley and Sons.

Scifres, C. J. and R. R. Hahn (1971). Response of honey mesquite seedlings to top removal. *Journal of Range Management* 24: 296-298.

Sequin, P., P. R. Peterson, C. C. Sheaffer, and D. L. Smith (2001). Physical sod suppression as an alternative to herbicide use in pasture renovation with clovers. *Canadian Journal of Plant Science* 81: 255-263.

Sheley, R. L., B. E. Olson, and L. L. Larson (1997). Effect of weed seed rate and grass defoliation level on diffuse knapweed. *Journal of Range Management* 50: 39-43.

Shirtliffe, S. J., M. H. Entz, and R. C. Van Acker (2000). *Avena fatua* development and seed shatter as related to thermal time. *Weed Science* 48: 555-560.

Smith, H. P. and L. H. Wilkes (1976). *Farm machinery and equipment,* Sixth edition. St. Louis, MO: McGraw-Hill Book Company.

Stivers, L. J. and C. Shennan (1989). Meeting the nitrogen needs for processing tomatoes through winter cover cropping. *Journal of Production Agriculture:* 4: 330-335.

Strykstra, R. J., R. M. Bekker, and G. L. Verweij (1996). Establishment of *Rhinanthus angustifolius* in a successional hayfield after seed dispersal by mowing machinery. *Acta Botanica Neerlandica* 45: 557-562.

Strykstra, R. J., G. L. Verweij, and J. P. Bakker (1997). Seed dispersal by mowing machinery in a Dutch brook valley system. *Acta Botanica Neerlandica* 46: 387-401.

Summerlin, J. R. Jr., H. D. Coble, and F. H. Yelverton (2000). Effect of mowing on perennial sedges. *Weed Science* 48: 501-507.

Sumrall, L. B., B. A. Roundy, and V. K. Winkel (1991). Influence of canopy removal by burning and clipping on emergence of *Eragrostis lehmanniana* seedlings. *Journal of Wildland Fire* 1: 34-40.

Teasdale, J. R. (1993). Reduced-herbicide weed management systems for no-tillage corn *(Zea mays)* in a hairy vetch *(Vicia villosa)* cover crop. *Weed Technology* 7: 879-883.

Thill, D. C., J. T. O'Donovan, and C. A. Mallory-Smith (1993). Integrated weed management strategies for delaying herbicide resistance wild oats. *Phytoprotection* 75(Suppl.): 61-70.

Tuck, C. R., M. J. O'Dogherty, D. E. Baker, and G. E. Gale (1991). Field experiments to study the performance of toothed disc mowing mechanisms. *Journal of Agricultural Engineering Research* 50: 93-106.

Turgeon, A. J. (Ed.) (1994). *Turf weeds and their control.* Madison, WI: American Society of Agronomy, Inc.

Vallentine, J. F. (1989). *Range development and improvements,* Third edition. New York: Academic Press, Inc.

Van Tienderen, P. H. and J. Van Der Toorn (1991a). Genetic differentiation between populations of *Plantago lanceolata:* I. Local adaptation in three contrasting habitats. *Journal of Ecology* 79: 27-42.

Van Tienderen, P. H. and J. Van Der Toorn (1991b). Genetic differentiation between populations of *Plantago lanceolata:* II. Phenotypic selection in a transplant experiment in three contrasting habitats. *Journal of Ecology* 79: 43-59.

Vengris, J., E. R. Hill, and D. L. Field (1966). Clipping and regrowth of barnyardgrass. *Crop Science* 6: 342-344.

Vila, M. and J. Terradas (1998). Neighbour effects on *Erica multifora* (Ericaceae) reproductive performance after clipping. *Acta Oecologica* 19: 139-145.

Wahlers, R. L., J. D. Burton, E. P. Maness, and W. A. Skroch (1997a). Physiological characteristics of a stem cut and blade delivery method of application. *Weed Science* 45: 746-749.

Wahlers, R. L., J. D. Burton, E. P. Maness, and W. A. Skroch (1997b). A stem cut and blade delivery method of herbicide application for weed control. *Weed Science* 45: 829-832.

Warwick, S. I. (1990). Genetic variation in weeds—With particular reference to Canadian agricultural weeds. In S. Kawano (Ed.), *Biological approaches and evolutionary trends in plants* (pp. 3-18). New York: Academic Press.

Wehtje, G., L. W. Wells, J. H. Choate, N. R. Martin Jr., and J. M. Curtis (1999). Mowing as a weed control supplement to herbicides and cultivation in peanut (*Arachis hypogaea* L.). *Weed Technology* 13: 139-144.

Weiher, E., A. van der Werf, K. Thompson, M. Roderick, E. Garnier, and O. Eriksson (1999). Challenging Theophrastus: A common core list of plant traits for functional ecology. *Journal of Vegetation Science* 10: 609-620.

Welton, F. A., V. H. Morris, and A. J. Hartzler (1929). Organic food reserves in relation to the eradication of Canada thistles. *Ohio Agricultural Experiment Station Bulletin* 441.

Wilkins, E. D. and R. R. Bellinder (1996). Mow-kill regulation of winter cereals for spring no-till crop production. *Weed Technology* 10: 247-252.

Willoughby, I. and H. G. McDonald (1999). Vegetation management in farm forestry: A comparison of alternative methods of inter-row management. *Forestry* 72: 109-121.

Wilson, B. J. (1970). Studies of the shedding of seed of *Avena fatua* in various cereal crops and the presence of this seed in the harvested material. In *Proceedings of the 10th British Weed Control Conference* (pp. 831-837). Worcester, UK: British Crop Protection Council.

Wilson, J. B. (1999). Guilds, functional types and ecological groups. *Oikos* 86: 507-522.

Wilson, M. V. and D. L. Clark (2001). Controlling invasive *Arrhenatherum elatius* and promoting native prairie grasses through mowing. *Applied Vegetation Science* 4: 129-138.

Zimdahl, R. L. (1993). *Fundamentals of weed science*. New York: Academic Press.

Chapter 13

Herbicide Fate Under Conservation Tillage, Cover Crop, and Edge-of-Field Management Practices

Martin A. Locke
Robert M. Zablotowicz
Mark A. Weaver

INTRODUCTION OF DIFFERENT RESIDUE MANAGEMENT SYSTEMS

Conservation management systems are more widespread due to environmental regulatory pressure and a desire to reduce input costs. Conservation management systems are defined as those that integrate practices to preserve or enhance soil and water resources. Some examples of conservation management practices include reduced tillage, cover crops, crop rotation, variable row spacing, and timing of crop planting. One objective of conservation management systems is to increase coverage of the soil surface by plant residues to a level of 30 percent or more. Many benefits accrue from plant residue accumulation on the soil surface, such as protection from erosion, conservation of soil moisture, and enhancement of soil tilth. Soil and plant microenvironments are created within these systems and can be manipulated to achieve desired benefits. For example, coverage of soil by plant residues in cover crop systems or minimization of soil disturbance by cultivation may inhibit germination and growth of weeds, in turn reducing the need for herbicides. However, Teasdale (1998) observed that inhibition of weeds by cover crops early in the growing season did not provide adequate control later in the season; thus, herbicides were required. The use of conservation management practices also may alter herbicide dissipation in soil

Mention of a trademark or proprietary product does not constitute a guarantee or warranty of the product by the USDA and does not imply their approval to the exclusion of other products that may also be suitable.

and plant residues, thereby influencing herbicide efficacy in controlling weeds. Sorption, degradation, movement in leachate or surface water, volatilization, and plant uptake are mechanisms of herbicide dissipation, and these mechanisms can be affected by management-influenced soil characteristics.

Although one goal of conservation management systems is to minimize input costs such as cultivation, we recognize that other inputs such as herbicides will likely remain as management tools within these systems. In this review, we will discuss herbicide dissipation in soil within the context of conservation management systems, evaluating how these practices modify herbicide fate. The discussion will focus on three areas of conservation management: reduced tillage, cover crops, and edge-of-field practices.

Tillage is considered to be reduced if at least one major cultivation practice is excluded or the intensity of the tillage operation is minimized. Other terms that describe forms of reduced tillage include no-tillage, zero tillage, ridge tillage, and strip tillage. Cover crops are used for a variety of purposes (e.g., weed inhibition, nutrient source, erosion control), and numerous species are utilized (Locke, Reddy, and Zablotowicz, 2002). Some of the more common cover crops include grass species such as rye, wheat, and ryegrass, and legumes such as clovers and vetches. Edge-of-field management practices are defined here as those practices at the perimeter of row crop fields that provide a buffer or barrier between the field and outlying areas. Some edge-of-field practices include vegetative buffer strips, vegetated furrows and ditches, and forested riparian areas.

EFFECTS OF CONSERVATION MANAGEMENT PRACTICES ON ORGANIC MATTER, MICROBIOLOGICAL CHARACTERISTICS, AND SOIL STRUCTURE

When left untilled, surface soil develops into a rich organic substrate to support microbial activity. Under a wide variety of cropping systems, total organic carbon in surface soils often increases as degree of tillage is reduced (Blevins et al., 1983; Doran, 1980; Franzluebbers, Hons, and Zuberer, 1994; Locke and Harper, 1991b; Locke, 1992; Locke, Gaston, and Zablotowicz, 1996), with only marginal changes in organic carbon at lower soil depths. The results of a study (Zablotowicz, Locke, et al., 2000) assessing the effects of long-term no-tillage on soil organic matter and microbial biomass in different soil depths are summarized in Table 13.1. After 11 years of continuous no-tillage practices, the organic carbon content of the surface 0 to 2 cm of no-tilled (NT) soil was 47 percent greater than that of the conven-

TABLE 13.1. Soil organic carbon, microbial biomass, fluorescein diacetate (FDA) hydrolysis, fluometuron sorption, and fluometuron degradation constants in long-term no-tillage and conventional tillage Dundee silt loam soil at four depths.

Soil depth (cm)	Organic carbon ($g \cdot kg^{-1}$)		Microbial biomass ($mg\ N \cdot kg^{-1}$)		FDA activity ($nmol \cdot g^{-1} \cdot h^{-1}$)		Freundlich sorption ($K_f \cdot L \cdot kg^{-1}$)		Fluometuron half-life (days)	
	NT	CT	NT	CT	NT	CT	NT	CT	NT	CT
0-2	30.1 a	20.3 b	175 a	85 b	1753 a	772 b	7.19	1.89	29.0	13.3
2-5	11.8 a	11.2 a	84 b	117 a	561 b	750 a	1.96	2.18	27.2	14.7
5-10	5.9 a	6.9 a	21 b	35 a	249 b	496 a	1.41	1.04	69.0	36.6
10-25	3.3 a	4.4 a	<10	<10	91 a	83 a	1.46	1.16	72.5	135.1

Source: Adapted from Zablotowicz, Locke, et al., 2000.
Note: Means followed by the same letter are not significantly different ($p = 0.05$).

tionally tilled (CT) soil, while no significant differences were observed at lower soil depths.

Higher soil moisture below the surface crust and labile organic carbon substrate in the soil surface provide an ideal environment for microbial activity, influencing both the size and diversity of microbial populations under conservation management. Indicators of microbial activity, such as microbial biomass (Franzluebbers, Hons, and Zuberer, 1994; Zablotowicz, Locke, et al., 2000), aerobic microbes, facultative anaerobes, and denitrifiers (Doran, 1980; Linn and Doran, 1984), are often higher in reduced-tillage surface soils. Cover crop residues or organic amendments can also enhance soil microflora (Wagner et al., 1995; Zablotowicz, Locke, and Smeda, 1998). Increased bacterial populations, especially gram-negative bacteria such as *Pseudomonas* spp., have been associated with residue managed soils compared to CT soils (Reddy, Zablotowicz, and Locke, 1995; Wagner et al., 1995; Zablotowicz, Locke, and Smeda, 1998). As organic substrate declines with soil depth, differences in microbial activity attributed to conservation tillage also diminish. For example, in the upper 2 cm of soil, microbial biomass was 106 percent greater than that of conventionally tilled soil (Zablotowicz, Locke, and Smeda, 1998). In lower 2 to 10 cm depths, however, microbial biomass was greater in CT compared to NT soils.

Associated with enhanced microbial populations, soil enzymatic activities from reduced-tillage systems are often higher relative to tilled soils (Dick, 1984; Doran, 1980; Reddy, Zablotowicz, and Locke, 1995; Zablotowicz, Hoagland, and Wagner, 1998). Understanding soil enzyme activity in soil is important in ascribing soil quality indexes (Dick, 1997; Masciandro and Ceccanti, 1999) and in predicting herbicide metabolism (Zablotowicz, Hoagland, and Wagner, 1998; Zablotowicz, Hoagland, et al., 2000). Some hydrolytic enzymes of importance to herbicide metabolism in soil include amidases, chlorohydrolases, esterases, and nitrilases (Hoagland and Zablotowicz, 2000). For example, measuring hydrolysis of the traditional esterase/lipase substrate, fluorescein diacetate, correlated with deesterification of the herbicide fenoxaprop-ethyl in soil (Zablotowicz, Hoagland, et al., 2000). Fluorescein acetate hydrolysis activity (FDA) was 127 percent greater than that of conventional tilled soil (Zablotowicz, Locke, et al., 2000). However, in lower 2 to 10 cm depths FDA hydrolytic activity was greater than in CT compared to NT soils. The enzyme aryl acylamidase was responsible for cleavage of the amide bond in acylanilide herbicides, and the activity of this enzyme was severalfold higher in NT versus CT soils, as was sulfatase and phosphatase (Zablotowicz, Hoagland, and Wagner, 1998). Oxidative enzymes such as peroxidases, phenol oxidases, and tyrosinases also may be important in oxidative coupling of aromatic pesticides and their metabolites to soil humus (Bollag, Myers, and Minard, 1992). The role of these oxidative enzymes and herbicide dissipation needs to be further assessed in conservation residue managed soils.

Soil profiles that are left undisturbed often develop large pores and channels as a result of roots decomposing in situ, leaving voids, microfauna activity (e.g., earthworms, insects), or the formation of cracks during drying in high shrink-swell soils. Large (> 0.25 mm) and more stable soil aggregates result from decomposing plant material, increasing porosity, and potential for water infiltration (Beare, Hendrix, and Coleman, 1994).

HERBICIDE SORPTION AND MOVEMENT IN CONSERVATION-MANAGED SYSTEMS

When herbicide is applied directly to soil or is washed off foliage, affinity for sorption by soil greatly influences its dissipation. Herbicide sorption in soil is strongly influenced by the organic content and composition in soil. Sorption to humic and other organic substances involves several binding mechanisms including ion exchange, hydrogen bonding, covalent bonding, charge transfer, hydrophobic partitioning, Van der Waals attraction, and ligand exchange. Because sorption is often positively correlated with bulk, electronic, and hydrophobic properties of the herbicide molecules (Reddy and Locke, 1994), these relationships can be extrapolated to conservation residue managed soils with elevated organic carbon levels. Herbicide sorption was consistently greater in surface soil from reduced tillage than from tilled areas (Locke, 1992; Brown et al., 1994; Locke, Zablotowicz, and Gaston, 1995; Reddy et al., 1995), primarily attributed to greater quantities of organic carbon rather than to a stronger affinity or energy of sorption by organic components in the soils. For example, fluometuron sorption was greatest in the surface 0 to 2 cm of NT soil and about fourfold higher than that of the same depth of CT soil (see Table 13.1). Collaborative studies are assessing the use of NMR techniques for understanding herbicide sorption mechanisms (Kingery et al., 2000). They determined that a higher number of sorption sites (functional groups) in humic material from NT soil may account for greater herbicide sorption rather than differences in the composition and character of sorption sites.

Sorption characteristics of several other herbicides under varying tillage and residue management systems have also been evaluated, e.g., acifluorfen, alachlor, bentazon, chlorimuron, cyanazine, metolachlor, and sulfentrazone (e.g., Locke, 1992; Locke, Gaston, and Zablotowicz, 1997; Reddy, Locke, and Gaston, 1997; Reddy, Locke, and Zablotowicz, 1997; Reddy and Locke, 1998). Comparative sorption studies of metabolites versus the parent compound were conducted for acifluorfen (Locke, Gaston, and Zablotowicz, 1997), bentazon (Gaston, Locke, and Zablotowicz, 1996), cyanazine (Reddy, Locke, and Zabltowicz, 1997), and fluometuron (Locke and Zablotowicz, 2000). The formation of certain metabolites and their role in increased sorption has been demonstrated. For example, sorption of the derivative amino-acifluorfen was found to be significantly higher than the herbicide acifluorfen (Locke, Gaston, and Zablotowicz, 1997).

Some research has shown that characteristics of reduced-tillage soil profiles enhance herbicide leaching (Hall, Murray, and Hartwig, 1989; Hall and Mumma, 1994; Isensee and Sadeghi, 1994). As discussed earlier, reduced-tillage conditions promote the development of macropore channels through soil profiles. Preferential flow through macropore channels, particularly during saturated conditions, results in some solute (herbicide) bypassing the bulk soil matrix. As solute flows rapidly through macropores, very limited interaction and sorption occurs between solute and organic or mineral soil constituents, thus enhancing solute movement in leachate to lower soil depths. As macropore channels age, organic residues may deposit along the inner lining. The depositions of organic residues increase the potential for sorption of solute in the leachate (Shipitalo et al., 1990; Stehouwer, Dick, and Traina, 1994).

Soil forms a dry crust at the surface, and the accumulation of plant residues helps reduce moisture evaporation from soil. If the soil surface is very dry at the time of herbicide application, sorption may be enhanced because of differential effects on physical and chemical properties of mineral and organic colloids, such as increased organic matter solubility, increased acidity, and exposure of hydrophobic components (Chiou, 1989; Sparks, 1990). Higher soil moisture conditions or precipitation soon after herbicide application increase the likelihood of herbicide movement under reduced tillage (Shipitalo et al., 1990; Isensee and Sadeghi, 1995).

Several factors contribute to reduced water and sediment runoff in conservation management systems. As plant residues on the soil surface slow water flow, the water has more time to infiltrate, especially through macropore channels if saturated conditions are present. In addition, the organic residues have a great capacity for absorbing and retaining water that would otherwise be lost in runoff. Sediments also have time to settle as water flow is reduced. Slowing water and sediment loss in conservation systems has an effect on herbicide loss. In some studies, total herbicide loss was reduced under conservation management (e.g., Hall and Mumma, 1994; Webster and Shaw, 1996). However, results have not always been consistent, resulting in mixed or lower herbicide losses occurring in conventional tillage (Gaynor, MacTavish, and Findlay, 1995; Myers, Wagger, and Leidy, 1995; Seta et al., 1993). These conflicting results in management impacts on herbicide runoff are attributed to variability associated with site-specific conditions, such as rainfall patterns, soil slope, or soil structural properties.

HERBICIDE DEGRADATION IN CONSERVATION-MANAGED SYSTEMS

Herbicide degradation in soil is controlled by the interaction of several factors, e.g., soil chemical and physical characteristics (texture, pH, moisture), climate (rainfall and temperature), the microbial community (magnitude of the population, presence of specific degraders, and their relative activity), and herbicide bioavailability. Thus, the effects of residue management practices on herbicide degradation depend upon these factors as well the chemical and physical characteristics of the particular herbicide, and the nature of the management practice. Three approaches are commonly used to assess management effects on herbicide degradation: (1) in vitro laboratory assays under controlled, usually static conditions; (2) laboratory column studies using intact soil cores; and (3) field dissipation studies. Understanding the fate of metabolites as well as the parent compound is very relevant in assessing environmental impacts of pesticides. Metabolite degradation is dependent on both the soil properties and chemical properties of the metabolite.

As reviewed elsewhere (Locke and Bryson, 1997), many well-designed laboratory studies have assessed comparative herbicide degradation in conventional and no-tillage soils. Overall, no clear conclusion on tillage effects and herbicide degradation can be generalized due to the wide range of herbicide chemistries and differences in cropping systems. Once initial degradation of herbicide occurs, however, differences have been observed in patterns of metabolite accumulation and degradation. Work with radiolabelled herbicide has determined that a large proportion of herbicide residues are not extractable from soil and that, in many cases, a greater proportion of applied herbicide or metabolites is trapped (nonextractable) in conservation tillage soils than conventional soils (Kells et al., 1980; Locke and Harper, 1991a,b; Reddy et al., 1995; Locke, Gaston, and Zablotowicz, 1996). Higher microbial activities and organic carbon levels found in conservation tillage soils may lead to greater polymerization and binding of herbicides or metabolites in humic material.

Initial studies by Kells and colleagues (1980) assessed the dissipation of ^{14}C-atrazine in CT and NT soils under field and laboratory conditions. These studies indicated that there was a more rapid incorporation of ^{14}C-atrazine into the nonextractable fraction of NT compared to CT soils. These effects were reversible by liming, and the authors concluded that chemical hydrolysis was responsible for enhanced degradation under NT conditions. Hydroxyatrazine, the product of chemical hydrolysis, is more prone to sorption by humic materials and not readily extracted by methanol (Lerch, Thurman, and Kruger, 1997). Levanon and colleagues (1994) observed delayed atrazine mineralization in CT compared to NT soil attributed to a lower initial population of atrazine-mineralizing microorganisms in the CT soil. However, a similar total mineralization was observed in both soils at the end of the incubation. Atrazine was more rapidly dissipated in the NT soil accompanied by a more rapid incorporation of ^{14}C into nonextractable residues in the NT soil compared to the CT soil.

The degradation of ^{14}C-ring labeled alachlor in surface soils of NT and CT Dundee silt loam was studied in a laboratory biometer flask study (Locke, Gaston, and Zablotowicz, 1996). Similar alachlor dissipation was observed in both soils. However, in the NT soil about 13 percent of the alachlor was mineralized during a 55-day incubation compared to 7 percent in the CT soils, with greater alachlor incorporated into nonextractable residues in the NT compared to CT the soil. In the CT soils, there was a greater accumulation of polar metabolites, e.g., the sulfonic acid derivative. Increased microbial activity in no-tillage soil was likely responsible for a shorter residence time for polar metabolites such as sulfonic acid found in NT soils.

Degradation of ^{14}C-chlorimuron was assessed in laboratory studies using three CT and NT surface soils (Reddy, Zablotowicz, and Locke, 1995). In a Miami silt loam and a Drummer silty clay loam with 72 and 140 percent greater organic matter content compared to their respective CT soils, greater mineralization was observed in the CT soils and a higher proportion of the ^{14}C was nonextractable in the NT soils.

The effects of tillage on bentazon degradation were studied in five soils with the duration of NT practices ranging from 3 to 18 years (Wagner et al., 1996); however, a significant effect of tillage was observed in only two of the soils, both of which had a long history of bentazon application. In a Dundee silt loam from Mississippi, more rapid bentazon dissipation and mineralization was found under CT soil compared to NT soil, while the inverse was observed in a Miami silt loam soil from Illinois. Gaston, Locke, and Zablotowicz (1996) and Gaston and colleagues (1996) also found that bentazon degradation was more rapid in Dundee CT soil than in NT soil, consistent with Wagner and colleagues (1996), and was leached more rapidly in the NT soil. There was higher organic carbon in the Dundee NT soil; thus for the same moisture content, water was held at a greater tension (Gaston, Locke, and Zablotowicz, 1996; Gaston et al., 1996). In addition, the NT soil had higher bulk density and higher volumetric water content;

thus it was more moist and likely less aerated. Initial degradation of bentazon in soil has been attributed to the ring hydroxylation by fungi (Huber and Otto, 1994), with greater degradation occurring under well-aerated soil conditions. More rapid bentazon degradation in the Dundee CT would have occurred under the more aerated conditions with relatively low organic carbon. As bentazon degrades, metabolites are less mobile and may become nonextractable, resulting in less bentazon leached in the CT soil.

The effect of a winter cereal rye cover crop and time of soil sampling on the potential for 2,4-D degradation was assessed in surface and subsurface of a Willamette silt loam soil (Bottomley et al., 1999). Populations of 2,4-D mineralizing bacteria were higher in the surface (0 to 20 cm) soil under cereal rye compared to winter fallow soil.

The effects of tillage and cover crop management on fluometuron degradation has been evaluated in several studies. In laboratory studies by Brown and colleagues (1994), fluometuron was more persistent in surface (0 to 4 cm) CT and NT soils managed under hairy vetch (half-life 71 and 78 days, respectively) compared to CT and NT soils with no cover crop (half-life 57 and 49 days, respectively). Other laboratory studies (Locke, Zablotowicz, and Gaston, 1995; Zablotowicz, Locke, and Hoagland, 1997) indicated fluometuron degraded more rapidly in surface (0 to 2 cm) NT and CT soils under a ryegrass cover crop compared to soils with no cover crop. Studies by Zablotowicz, Locke, and colleagues (2000) compared the degradation of fluometuron in NT and CT at four depths. In the upper 10 cm soil depths, fluometuron degraded more rapidly in CT compared to NT soil (see Table 13.1). Despite higher microbial biomass and activity in NT 0 to 2 cm soil, the increased sorptive capacity of the soil minimized the amount of fluometuron that was bioavailble for microbial degradation. In the lower 2 to 5 and 5 to 10 cm depths greater microbial activity and more labile organic matter in CT soils was responsible for greater fluometuron degradation compared to NT soils. Understanding interactions regulating bioavailability (sorption) and microbial activity is key in understanding herbicide transformations under NT practices.

Kinetics of metribuzin degradation in NT and CT soils were similar, but dynamics of metabolite accumulation and degradation differed (Locke and Harper, 1991a,b). Accumulations of extractable polar metabolites of metribuzin were greater in residue-managed soils than in CT and nonamended soils. Nonextractable herbicide residues were also greater in the NT.

The effect of conservation management on dissipation of herbicides under field conditions is difficult to assess (e.g., Ghadiri et al., 1984; Gaynor, Stone, and Vyn, 1987; Curran, Liebl, and Simmons, 1992; Monks and Banks, 1993) because many factors may counter one another's effects. Retention at the soil surface may increase herbicide residence time, since sorption to organic residue impedes further movement. Likewise, photolabile compounds remaining on the soil surface may be subject to photodegradation. However, if the herbicide moves through the dry crust and encounters more moist conditions and active microbial populations, degradation may be enhanced. More developed macropores and better aggregation in the conservation soils also may provide channels for increased herbicide mobility.

HERBICIDE RETENTION AND DEGRADATION IN PLANT RESIDUES

Postemergence herbicides are an important part of reduced-tillage and cover crop systems. Live plants as well as plant residues intercept much of the herbicide. The intercepted herbicide may be taken up by the plant or sorb to plant tissue (Reddy, Locke, and Howard, 1995; Reddy and Locke, 1996). Some studies indicate that crop and weed residues can have a higher capacity for herbicide sorption than soil (Dao, 1991; Reddy et al., 1995; Reddy, Locke, and Gaston, 1997), and as these residues age, sorption capacity may increase even further (Dao, 1991). Plant residues also intercept and retain preemergence herbicides. Banks and Robinson (1986) found that the efficacy of several chloroacetamide herbicides was reduced because they were intercepted by wheat straw residue.

Decomposing residues harbor enhanced microbial populations, and these organisms may cometabolize intercepted herbicides. For example, laboratory studies demonstrated enhanced fluometuron metabolism in association with plant residues (Zablotowicz, Locke, and Smeda, 1998). The rate of fluometuron degradation was affected by increased moisture content. However, 2,4-D was poorly degraded under the same cover crop residues compared to soil.

CONSERVATION EDGE-OF-FIELD PRACTICES AND HERBICIDE DISSIPATION

Although numerous practices within the field can influence the rate of herbicide degradation and sorption, waterway management and edge-of-field practices present a final opportunity to prevent pollution of surface water bodies. Edge-of-field zones such as vegetated buffer strips (VBS), constructed wetlands, riparian zones, and grass waterways are inherently transitory zones between the agroecosytems and the aquatic systems receiving the runoff water. Proper management of these transitional zones may provide an effective and practical means of environmental protection.

Edge-of-field buffer zones can be effective in minimizing herbicide movement to surface aquatic systems through a variety of mechanisms. In the simplest case, the buffer zones provide a spatial barrier to off-target herbicide drift. Even with proper nozzle selection and timing of application with respect to meteorological conditions, some drift is likely to occur (reviewed in Akesson and Yates, 1988; Bode, 1987). A vegetated field buffer will intercept much of the drift and thus protect potentially sensitive adjacent areas. For example, Robinson and colleagues (2000) recommended a 30 m buffer zone for the helicopter application of asulam herbicide to avoid biologically active levels, although the herbicide was detectable up to 200 m downwind. A 3 m unsprayed buffer reduced drift to a ditch by 95 percent in another study (Geert, de Snoo, and de Wit, 1998).

Many of the same principles that relate to conservation tillage practices are applicable to field borders. Just as reduced tillage and cover crop establishment promote herbicide retention within the field through increased affinity for organic matter, increased microbial activity, and increased infiltration through improved soil tilth, a growing body of research supports the view that the same processes are occurring at the field margins. For example, in a comparison between a VBS and an adjacent cultivated field, Staddon, Locke, and Zablotowicz (2001) found more than twice as much organic matter in the VBS. Soil from the VBS also had higher metolachlor sorption and higher rates of metolachlor degradation in vitro.

A primary means of herbicide movement out of fields is through movement of suspended soil with bound herbicide. Clearly, any edge-of-field practice that slowed the velocity of runoff water and trapped suspended sediment would likely reduce the loss of herbicide from the field (Dillaha et al., 1989; Deletic, 2001; Ghadiri, Rose, and Hogarth, 2001; Jin and Römkens, 2001; Smolikowski, Puig, and Roose, 2001). The same practices, such as grassed waterways and filter strips, that have long been employed to reduce soil erosion have been shown to effectively reduce runoff of herbicides. Even in the case of metolachlor and metribuzin, two fairly soluble herbicides, a 2 m fescue filter strip consistently reduced losses in all three years of study and three different tillage regimes (Webster and Shaw, 1996). Similarly, the effectiveness of switchgrass as a filter strip was demonstrated for metolachlor and atrazine (Mersie et al., 1999). Rankins, Shaw, and Boyette (2001) evaluated four grass species in a 30 cm filter strip, all of which reduced fluometuron and norflurazon concentrations in runoff water by at least 53 percent and 46 percent, respectively, relative to unfiltered controls. The effectiveness of filter strips is clearly related to their ability to remove sediment by filtration, reduce runoff velocity and increase infiltration rate, and, in one example with switchgrass, even create backwater (Dabney et al., 1993).

In addition to removal of herbicides sorbed on suspended soil, VBS influence herbicide fate through other mechanisms. The plants themselves will intercept herbicides by uptake of the compound or binding to the phyloplane and rhizosphere. Plants may enhance herbicide degradation by direct uptake and metabolism or indirectly via accelerated metabolism within the rhizosphere (Zablotowicz, Locke, and Hoagland, 1997; Zablotowicz and Hoagland, 1999). The rhizosphere effect may be due to the niche it provides for microflora. In rhizospheres, herbicides are subject to transformation by metabolic and cometabolic proccesses. Rhizosphere bacteria cometabolize herbicides by activity of enzymes such as aryl acylamidase (Hoagland, Zablotowicz, and Locke, 1994), glutathione *S*-transferase (Zablotowicz, Hoagland, and Locke, 1994; Zablotowicz et al., 1995), and aromatic nitroreductase (Zablotowicz, Locke, and Hoagland, 1997) among others. Through root death and decay, plants also provide aeration and channelization that may lead to improved infiltration of herbicides (Muñoz-Carpena, Parsons, and Gilliam, 1999). Others have suggested volatilization as a means of herbicide dissipation within VBS (Dillaha et al., 1989). Understanding the contribution of plants to enhanced herbicide degradation may aid in developing better management recommendations for edge-of-field practices.

Before runoff water and associated sediments, nutrients, and pesticides leave the agroecosystem, they often must travel though a network of ditches. These often-overlooked conduits provide a final opportunity to prevent the pollution of valued water bodies. "Ditch" is an imprecise term that encompasses a range of topographic, hydrologic, and vegetative features that may affect the fate of herbicides. Consequently, the utility of ditches is not easily generalized. One study demonstrated very rapid and thorough removal of atrazine from water in a Mississippi Delta agricultural drainage ditch (Moore et al., 2001). Just one hour after exposure, 61 percent of the atrazine was found in association with the vegetation, and the authors concluded that for their experimental conditions atrazine could be completely removed within 50 m of ditch length. Although their study is encouraging, it is noteworthy that the atrazine removal was largely plant mediated and their study was conducted in July in Mississippi. A cooler climate or a ditch that is not as heavily vegetated, as may be typical when atrazine is typically applied, may produce very different results. Another study with linuron in the Netherlands noted only 1 percent of the herbicide associating with the vegetation and 6 percent in the ditch sediment, with the remainder associated with the aqueous phase of the ditch (Crum, Aaldernik, and Brock, 1998).

Edge-of-field practices can provide numerous other benefits in addition to herbicide containment and dissipation. These uncropped, and often largely unmanaged, areas serve as remnant "natural" patches in the landscape with disproportionately high biodiversity (Moonen and Marshall, 2001; Marshall, 2002). These refugia are important in the ecology of small mammals (Chapman and Ribic, 2002), birds, including species with potential utility in biocontrol (Jobin, Choinière, and Bélanger, 2001), and bumblebees (Bäckman and Tiainen, 2002). The soils in the field margins are often enriched in organic matter and, in the cases of ditches, riparian zones, and grassed waterways, the saturated conditions can lead to anaerobic conditions which facilitate denitrification (Ettema, Lowrance, and Coleman, 1999; Watts and Seitzinger, 2000).

REFERENCES

Akesson, N.B. and W.E. Yates (1988). Effect of weather factors on the application of herbicides. In C.G. McWhorter and M.R. Gebhardt (Eds.), *Methods of applying herbicides* (pp. 335-344). Champaign, IL: Weed Science Society of America.

Bäckman, J.-P.C. and J. Tiainen (2002). Habitat quality of field margins in a Finnish farmland area for bumblebees (Hymenoptera: *Bombus* and *Psithyrus*). *Agriculture, Ecosystems and Environment* 89: 53-68.

Banks, P.A. and E.L. Robinson (1986). Soil reception and activity of acetochlor, alachlor, and metolachlor as affected by wheat *(Triticum aestivum)* straw and irrigation. *Weed Science* 34: 607-611.

Beare, M.H., P.F. Hendrix, and D.C. Coleman (1994). Water-stable aggregates and organic matter fractions in conventional and no-tillage soils. *Soil Science Society of America Journal* 58: 777-786.

Blevins, R.L., G.W. Thomas, M.S. Smith, W.W. Frye, and P. L. Cornelius (1983). Changes in soil properties after 10 years continuous non-tilled and conventionally tilled corn. *Soil and Tillage Research* 3: 135-146.

Bode, L.C. (1987). Spray application technology. In C.G. McWhorter and M.R. Gebhardt (Eds.), *Methods of applying herbicides* (pp. 85-110). Champaign, IL: Weed Science Society of America.

Bollag, J.-M., C.J. Myers, and R.D. Minard (1992). Biological and chemical interactions of pesticides with soil organic matter. *Science of the Total Environment* 123/124: 205-217.

Bottomley, P.J., T.J. Sawyer, L. Boersma, R.P. Dick, and D.D. Hemphill (1999). Winter cover crop enhances 2,4-D mineralization potential of surface and subsurface soil. *Soil Biology and Biochemistry* 31: 849-857.

Brown, B.A., R.M. Hayes, D.D. Tyler, and T.C. Mueller (1994). Effect of tillage and cover crop on fluometuron adsorbtion and degradation under controlled conditions. *Weed Science* 44: 171-175.

Chapman, E.W. and C.A. Ribic (2002). The impact of buffer strips and stream-side grazing on small mammals in southwestern Wisconsin. *Agriculture, Ecosystems and Environment* 88: 49-59.

Chiou, C.T. (1989). Theoretical considerations of the partition uptake of nonionic organic compounds by soil organic matter. In B.L. Sawhney and K. Brown (Eds.), *Reactions and movement of organic chemicals in soils* (pp. 1-29). Soil Science Society of America Special Publication No. 22. Madison, WI: Soil Science Society of America.

Crum, S.J.H., G.H. Aaldernik, and T.C.M. Brock (1998). Fate of the herbicide linuron in outdoor experimental ditches. *Chemosphere* 36: 2175-2190.

Curran, W.S., R.A. Liebl, and F.W. Simmons (1992). Effects of tillage and application method on clomazone, imazaquin, and imazethapyr persistence. *Weed Science* 40: 482-489.

Dabney, S.M., G.H. Dunn, K.T. Krishnegowda, and C.V. Alonso (1993). Stage-discharge relationships of narrow grass barriers. *Proceedings of the American Society of Agronomy* 85: 312.

Dao, T.H. (1991). Subsurface mobility of metribuzin as affected by crop residue placement and tillage method. *Journal of Environmental Quality* 24: 1193-1198.

Deletic, A. (2001). Modelling of water and sediment transport over grassed area. *Journal of Hydrology* 248: 168-182.

Dick, R.P. (1997). Soil enzyme assays as integrative indicators of soil quality. In C.E. Pankhurst, B.M. Doube, and V.V.S.R Gupta (Eds.), *Biological indicators of soil health* (pp. 121-156). Wallingford, UK: CAB International.

Dick, W.A. (1984). Influence of long-term tillage and crop rotation combinations on soil enzyme activities. *Soil Science Society of America Journal* 48: 569-574.

Dillaha, T.A., R.B. Reneau, S. Mostaghimi, and D. Lee (1989). Vegetative filter strips for agricultural non point source pollution control. *Transactions of the American Society of Agricultural Engineers* 32: 513-519.

Doran, J.W. (1980). Soil microbial and biochemical changes associated with reduced tillage. *Soil Science Society of America Journal* 44: 765-771.

Ettema, C.H., R. Lowrance, and D.C. Coleman (1999). Riparian soil response to surface nitrogen input: Temporal changes in denitrification, labile and microbial C and N pools, and bacterial and fungal respiration. *Soil Biology and Biochemistry* 31: 1609-1624.

Franzluebbers, A.J., F.M. Hons, and D.A. Zuberer (1994). Long-term changes in soil carbon and nitrogen pools in wheat management systems. *Soil Science Society of America Journal* 58: 1639-1645.

Gaston, L.A., M.A. Locke, S.C. Wagner, R.M. Zablotowicz, and K.N. Reddy (1996). Sorption of bentazon and degradation products in two Mississippi soils. *Weed Science* 44: 678-682.

Gaston, L.A., M.A. Locke, and R.M. Zablotowicz (1996). Sorption and degradation of bentazon in conventional- and no-till Dundee soil. *Journal of Environmental Quality* 25: 120-126.

Gaynor, J.D, D.C. MacTavish, and W.I. Findlay (1995). Atrazine and metolachlor loss in surface and subsurface runoff from three tillage treatments in corn. *Journal of Environmental Quality* 24: 246-256.

Gaynor, J.D., J.A. Stone, and T.J. Vyn (1987). Tillage systems and atrazine and alachlor residues on a poorly drained soil. *Canadian Journal of Soil Science* 67: 959-963.

Geert, R., G.R. de Snoo, and P.J. de Wit (1998). Buffer zones for reducing pesticide drift to ditches and risks to aquatic organisms. *Ecotoxicology and Environmental Safety* 41: 112-118.

Ghadiri, H., C.W. Rose, and W.L. Hogarth (2001). The influence of grass and porous barrier strips on runoff hydrology and sediment transport. *Transactions of the American Society of Agricultural Engineers* 44: 259-268.

Ghadiri, H., P.J. Shea, G.A. Wicks, and L.C. Haderlie (1984). Atrazine dissipation in conventional-till and no-till sorghum. *Journal of Environmental Quality* 13: 549-552.

Hall, J.K. and R.O. Mumma (1994). Dicamba mobility in conventionally tilled and non-tilled soil. *Soil and Tillage Research* 30: 3-17.

Hall, J.K., M.R. Murray, and N.L. Hartwig (1989). Herbicide leaching and distribution in tilled and untilled soil. *Journal of Environmental Quality* 18: 439-445.

Hoagland, R.E. and R.M. Zablotowicz (2000). Plant and microbial hydroytic enzymes related a pesticide metabolism. In J.C. Hall, R.E. Hoagland, and R.M. Zablotowicz (Eds.), *Pesticide metabolism in plants and microorganisms: Similarities and divergences* (pp. 58-88). ACS Symposium Series 777. Washington, DC: American Chemical Society.

Hoagland, R.E., R.M. Zablotowicz, and M.A. Locke (1994). Propanil metabolism in rice rhizospheres. In T.A. Anderson and J.R. Coats (Eds.), *Bioremediation through rhizosphere technology* (pp. 160-183). ACS Symposium Series 563. Washington, DC: American Chemical Society.

Huber, R. and S. Otto (1994). Environmental behaviour of bentazon. *Review of Environmental Contamination and Toxicology* 137: 111-141.

Isensee, A.R. and A.M. Sadeghi (1994). Effects of tillage and rainfall on atrazine levels in soil. *Journal of Environmental Quality* 42: 462-467.

Jin, C.-X. and M.J.M. Römkens (2001). Experimental studies of factors in determining sediment trapping in vegetative filter strips. *Transactions of the American Society of Agricultural Engineers* 44: 277-288.

Jobin, B., L. Choinière, and L. Bélanger (2001). Bird use of three types of field margins in relation to intensive agriculture in Québec, Canada. *Agriculture, Ecosystems and Environment* 84: 131-143.

Kells, J.J., C.E. Rieck, R.L. Blevins, and W.M. Muir (1980). Atrazine dissipation as affected by surface pH and tillage. *Weed Science* 28: 101-104.

Kingery, W.L., A.J. Simpson, M.H.B. Hayes, M.A. Locke, and R.P. Hicks (2000). The application of multi-dimensional NMR to the study of soil humic substances. *Soil Science* 165: 483-494.

Lerch, R.N., E.M. Thurman, and E.L. Kruger (1997). Mixed-mode sorption of hydroxylated atrazine degradation products in soil: A mechanism for bound residue. *Environmental Science and Technology* 31: 1539-1546.

Levanon, D., J.J. Meisinger, E.E. Codling, and J.L. Starr (1994). Impact of tillage on microbial activity and the fate of pesticides in the upper soil. *Water, Air and Soil Pollution* 72: 179-189.

Linn, D.M. and J. W. Doran (1984). Aerobic and anaerobic microbial populations in no-till and plowed soils. *Soil Science Society of America Journal* 48: 794-799.

Locke, M.A. (1992). Sorption-desorption kinetics of alachlor in surface soil from two soybean tillage systems. *Journal of Environmental Quality* 21: 558-566.

Locke, M.A. and C.T. Bryson (1997). Herbicide-soil interactions in reduced tillage and plant residue management systems. *Weed Science* 45: 307-320.

Locke, M.A., L.A. Gaston, and R.M. Zablotowicz (1996). Alachlor biotransformation and sorption in soil from two soybean tillage systems. *Journal of Agricultural and Food Chemistry* 44: 1128-1134.

Locke, M.A., L.A. Gaston, and R.M. Zablotowicz (1997). Acifluorfen sorption and sorption kinetics in soil. *Journal of Agricultural and Food Chemistry* 45: 286-293.

Locke, M.A. and S.S. Harper (1991a). Metribuzin degradation in soil: I—Effects of soybean residue, metribuzin level, and soil depth. *Pesticide Science* 31: 221-237.

Locke, M.A. and S.S. Harper (1991b). Metribuzin degradation in soil: II—Effects of tillage. *Pesticide Science* 31: 239-247.

Locke, M.A., K.N. Reddy, and R.M. Zablotowicz (2002). Weed management in conservation production systems. *Weed Biology and Management* 2: 123-132.

Locke, M.A. and R.M. Zablotowicz (2000). Comparative sorption of fluometuron and metabolites in conservation- and conventional-tillage soils. *Proceedings of the Weed Science Society of America* 40: 88.

Locke, M.A., R.M. Zablotowicz, and L.A. Gaston (1995). Fluometuron interactions in crop residue-managed soils. In W.L. Kingery and N. Buehring (Eds.), *Conservation farming: A focus on water quality* (pp. 55-58). Special Bull. 88-7. Starkville: Mississippi Agriculture and Forestry Experiment Station.

Locke, M.A., R.M. Zablotowicz, and L.A. Gaston (1996). Effect of tillage on alachlor sorption and biotransformation in soil. In H. Brown, G.W. Cussans, M.D. Devine, S.O. Duke, C. Fernandez-Quintanilla, A. Helweg, R.E. Labrada, M. Landes, P. Kudsk, and J.C. Streibig (Eds.), *Proceedings of the Second International Weed Control Congress,* June 25-28, Volume 1 (pp. 287-292). Copenhagen, Denmark: DJF, Flakkebjerg.

Marshall, E.J.P. (2002). Introducing field margin ecology in Europe. *Agriculture, Ecosystems and Environment* 89: 1-4.

Masciandro, G. and B. Ceccanti (1999). Assessing soil quality in different agroecosystems through biochemical and chemo-structural properties of humic substances. *Soil and Tillage Research* 51: 129-137.

Mersie, W., C.A. Seybold, C. McNamee, and J. Huang (1999). Effectiveness of switchgrass filter strips in removing dissolved atrazine and metolachlor from runoff. *Journal of Environmental Quality* 28: 816-821.

Monks, C.D. and P.A. Banks (1993). Effect of straw, ash and tillage on dissipation of imazaquin and imazethapyr. *Weed Science* 41: 133-137.

Moonen, A.C. and E.J. P. Marshall (2001). The influence of sown margin strips, management and boundary structure on herbaceous field margin vegetation in two neighbouring farms in southern England. *Agriculture, Ecosystems and Environment* 86: 187-202.

Moore, M.T., E.R. Bennett, C.M. Cooper, S. Smith Jr., F.D. Shields, C.D. Milam, and J.L. Farris (2001). Transport and fate of atrazine and lambda-cyhalothrin in an agricultural drainage ditch in the Mississippi Delta, USA. *Agriculture, Ecosystems and Environment* 87: 309-314.

Muñoz-Carpena, R., J.E. Parsons, and J.W. Gilliam (1999) Modeling hydrology and sediment transport in vegetative filter strips. *Journal of Hydrology* 214: 111-129.

Myers, J.L., M.G. Wagger, and R.B. Leidy (1995). Chemical movement in relation to tillage system and simulated rainfall intensity. *Journal of Environmental Quality* 24: 1183-1192.

Rankins, A. Jr., D.R. Shaw, and M. Boyette (2001). Perennial grass filter strips for reducing herbicide losses in runoff. *Weed Science* 49: 647-651.

Reddy, K.N. and M.A. Locke (1994). Prediction of soil sorption (Koc) of herbicides using semiempirical molecular properties *Weed Science* 42: 453-461.

Reddy, K.N. and M.A. Locke (1996). Imazaquin spray retention, foliar washoff, and runoff losses under simulated rainfall. *Pesticide Science* 48: 179-187.

Reddy, K.N. and M.A. Locke (1998). Sulfentrazone sorption, desorption, and mineralization in soils from two tillage systems. *Weed Science* 46: 494-500.

Reddy, K.N., M.A. Locke, and L.A. Gaston (1997). Tillage and cover crop effects on cyanazine adsorption and desorption kinetics. *Soil Science* 162: 501-509.

Reddy, K.N., M.A. Locke, and K.D. Howard (1995). Bentazon spray retention, activity, and foliar washoff in weed species. *Weed Technology* 9: 773-778.

Reddy, K.N., M.A. Locke, S.C. Wagner, R.M. Zablotowicz, L.A. Gaston, and R.J. Smeda (1995). Chlorimuron ethyl sorption and desorption kinetics in soils and herbicide-desiccated cover crop residues. *Journal of Agricultural and Food Chemistry* 43: 2752-2757.

Reddy, K.N., M.A. Locke, and R.M. Zablotowicz (1997) Soil type and tillage effects on sorption of cyanazine and degradation products. *Weed Science* 45: 727-732.

Reddy, K.N., R.M. Zablotowicz, and M.A. Locke (1995). Chlorimuron adsorption, and degradation in soils from conventional tillage and no-tillage systems. *Journal of Environmental Quality* 24: 760-767.

Robinson, R.C., R.G. Parsons, G. Barbe, P.T. Patel, and S. Murphy (2000). Drift control and buffer zones for helicopter spraying of bracken *(Pteridium aquilinum). Agriculture, Ecosystems and Environment* 79: 215-231.

Seta, A.K., R.L. Blevins, W.W. Frye, and B.J. Barfield (1993). Reducing soil erosion and agricultural losses with conservation tillage. *Journal of Environmental Quality* 22: 661-665.

Shipitalo, M.J., W.M. Edwards, W.A. Dick, and L.B. Owens (1990). Initial storm effects on macropore tranport of surface applied chemicals in no-till soil. *Soil Science Society of America Journal* 54: 1530-1536.

Smolikowski, B., H. Puig, and E. Roose (2001). Influence of soil protection techniques on runoff, erosion and plant production on semi-arid hillsides of Cabo Verde. *Agriculture, Ecosystems and Environment* 87: 67-80.

Sparks, D.L. (1990). *Soil physical chemistry,* Fourth edition, Volume 1. Boca Raton, FL: CRC Press, MS.

Staddon, W.J., M.A. Locke, and R.M. Zablotowicz (2001). Microbiological characteristics of a vegetative buffer strip soil and degradation and sorption of metolachlor. *Soil Science Society of America Journal* 65: 1136-1142.

Stehouwer, R.C., W.A. Dick, and S.J. Traina (1994). Sorption and retention of herbicides in vertically oriented earthworm and artificial burrows. *Journal of Environmental Quality* 23: 286-292.

Teasdale, J.R. (1998). Cover crops, smother plants, and weed management. In J.L. Hatfield., D.D. Buhler, and B.A. Stewart (Eds.), *Integrated weed and soil management* (pp. 247-270). Chelsea, MI: Ann Arbor Press.

Wagner, S.C., R.M. Zablotowicz, L.A. Gaston, M.A. Locke, and J. Kinsella (1996). Bentazon degradation in soil: Influence of tillage and history of bentazon application. *Journal of Agricultural and Food Chemistry* 44: 1593-1598.

Wagner, S.C., R.M. Zablotowicz, M.A. Locke, R.J. Smeda, and C.T. Bryson (1995). Influence of herbicide-desiccated cover crops on biological soil quality in the Mississippi Delta. In W.L. Kingery and N. Buehring (Eds.), *Conservation farming: A focus on water quality* (pp. 86-89). Special Bull. 88-7. Starkville: Mississippi Agriculture and Forestry Experiment Station.

Watts, S.H. and S.P. Seitzinger (2000). Denitrification rates in organic and mineral soils from riparian sites: A comparison of N_2 flux and acetylene inhibition methods. *Soil Biology and Biochemistry* 32: 1383-1392.

Webster, E.P. and D.R. Shaw (1996). Impact of vegetative filter strips on herbicide loss in runoff from soybean *(Glycine max). Weed Science* 44: 662-671.

Zablotowicz, R.M. and R.E. Hoagland (1999). Microbiological considerations in phytoremediation approaches for pesticide-contaminated soils and waters. In R.C. Rajak (Ed.), *Microbial biotechnology for sustainable development and productivity* (pp. 343-360). Jodhpur, India: Scientific Publishers.

Zablotowicz, R.M., R.E. Hoagland, and M.A. Locke (1994). Glutathione S-transferase activity in rhizosphere bacteria and the potential for herbicide detoxification. In T.A. Anderson, and J.R. Coats (Eds.), *Bioremediation through rhizosphere technology* (pp. 184-198). ACS Symposium Series 563. Washington, DC: American Chemical Society.

Zablotowicz, R.M., R.E. Hoagland, M.A. Locke, and W.J. Hickey (1995). Glutathione S-transferase activity and metabolism of glutathione conjugates by rhizosphere bacteria. *Applied and Environmental Microbiology* 61: 1054-1060.

Zablotowicz, R.M., R.E. Hoagland, W.J. Staddon, and M.A. Locke (2000). Effects of pH on chemical stability and de-esterification of fenoxaprop-ethyl by purified enzymes, bacterial extracts, and soils. *Journal of Agricultural and Food Chemistry* 48: 4711-4716.

Zablotowicz, R.M., R.E. Hoagland, and S.C. Wagner (1998). 2-Nitroacetanilide as substrate for determination of aryl acylamidase activity in soils. *Soil Biology and Biochemistry* 30: 679-686.

Zablotowicz, R.M., M.A. Locke, L.A. Gaston, and C.T. Bryson (2000). Interactions of tillage and soil depth on fluometuron degradation in a Dundee silt loam soil. *Soil and Tillage Research* 57: 61-68.

Zablotowicz, R.M., M.A. Locke, and R.E. Hoagland (1997). Aromatic nitroreduction of acifluorfen in soils, rhizospheres and pure cultures of rhizobacteria. In E.L. Kruger, T.A. Anderson, and J.R. Coats (Eds.), *Phytoremediation of soil and water contaminants* (pp. 38-53). ACS Symposium Series 664. Washington, DC: American Chemical Society.

Zablotowicz, R.M., M.A. Locke, and R.J. Smeda (1998). Degradation of 2,4-D and fluometuron in cover crop residues. *Chemosphere* 37: 87-101.

Chapter 14

Strategies for Developing Bioherbicides for Sustainable Weed Management

S. M. Boyetchko
E. N. Rosskopf

INTRODUCTION

Although the utilization of microbial plant pathogens for biological control of weeds was first conceptualized in the early 1900s, significantly greater attention and research in pursuit of specific fungal pathogens for control of target weed species did not begin until the 1970s. Several recent reviews have described the various types of biological control approaches and provide examples of pathogens that have been successfully commercialized or are at various stages of development (Charudattan, 1991; Watson, 1991; TeBeest, Yang, and Cisar, 1992; Boyetchko, 1999; Rosskopf, Charudattan, and Kadir, 1999; Charudattan, 2000; Boyetchko et al., 2002). Despite the accumulated wealth of information on biological weed control, examples of their application as components of integrated pest management (IPM) programs are often limited. This differs in comparison to the numerous studies on biocontrol of insects, but is most likely a result of the significantly greater and long-standing activity in biocontrol by entomologists over the past 100 years or so (Mason and Huber, 2002).

As crop production systems shift from traditional farming practices to more complex systems, weed communities shift in response to the various interactions among agronomic, environmental, and ecological selection pressures (Derksen, Blackshaw, and Boyetchko, 1996). Although biological control has often been described as an alternative weed control strategy, it can be argued that it serves a more important role as a component of an integrated pest management program, and not as a replacement for chemical pesticides and other weed management tactics. Integrated weed management (IWM) has been defined as the utilization and coordination of a variety of weed control strategies and encompasses cultural, chemical, mechanical, biological, and genetic measures (Shaw, 1982; Walker and Buchanan,

1982; Smith, 1991; Swanton and Weise, 1991; Derksen, Blackshaw, and Boyetchko, 1996; Buhler, 2002). Biological control therefore becomes one more weed control option in the toolbox for farmers to use. Unfortunately, few bioherbicide agents are at the stage of development where they can be integrated into weed control programs and thus represent an untapped and undervalued technology that needs to be fully exploited. This aspect underscores the need for continued and enhanced research to develop biological weed control agents for application in weed management programs. In addition, Buhler (2002) stressed that the complexity and diversity of weed populations and communities necessitate the integration of innovative and comprehensive weed management tactics.

Definition and Examples of Biological Weed Control

Biological control is generally defined as the intentional use of living organisms to suppress, reduce, or eradicate a pest population (Boyetchko, 1999). Three major strategies for microbial biological control have been used: the classical, augmentative, and inundative (or bioherbicide) approaches (Charudattan, 1991; Watson, 1991; Boyetchko, 1999; Rosskopf, Charudattan, and Kadir, 1999). Classical biological control involves a single introduction of a natural enemy, usually from the center of origin of the weed, into a geographical region where the natural enemies of the weed are absent. This approach is characterized by the release, establishment, and survival of the pathogen where weed control is expected to occur over a period of time, and it is generally considered more suitable for control of pasture and rangeland weeds. The bioherbicide approach, which will be the main focus of this review, is the artificial mass production and repeated application of natural enemies (pathogens) to target weeds that results in significant damage to the weed population while minimizing crop yield losses (Frantzen, Paul, and Müller-Schärer, 2001). Charudattan (1991) appropriately suggested that the distinguishing feature between classical and bioherbicide agents is not the fact that one is exotic (i.e., nonindigenous) while the another is native, but it is in the epidemiological characteristics between the two. Although pathogens developed as bioherbicides may or may not have coevolved within the center of origin of the weed, it has been suggested that new host-pathogen associations may yield more virulent and effective biocontrol agents. The augmentative approach is similar to the bioherbicide approach in that the inoculum is distributed through human intervention; however, the main difference is that the inoculum is not mass produced nor is it applied at the high inoculum rates using augmentative biological control as it is with bioherbicides.

A notable example of the bioherbicide approach is the commercialization of Collego, which consists of a foliar-applied fungal pathogen *Colletotrichum gloeosporioides* (Penz.) Penz. & Sacc. f.sp. *aeschynomene* for control of northern jointvetch [*Aeschynomene virginica* (L.) B.S.P.] in rice (*Oryza sativa* L.) and soybean [*Glycine max* (L.) Merr.] (TeBeest and Templeton, 1985). DeVine is a liquid formulation of the endemic soilborne fungus *Phytophthora palmivora* Butler for the control of stranglervine [*Morrenia odorata* (H. & A.) Lindl.] in the citrus-growing area of Florida (Burnett, Tucker, and Ridings, 1974; Charudattan, 1991; Boyetchko, 1999) and *Colletotrichum gloeosporioides* f.sp. *malvae* controls round-leaved mallow (*Malva pusilla* Sm.) and small-flowered mallow (*M. parviflora* L.). The aforementioned bioherbicides are applied in an analogous fashion to chemical herbicides and are specific to one or two target weed species. Other bioherbicides exhibiting similar features include Camperico, a foliar bacterial agent [*Xanthomonas campestris* (Pammel) Dowson pv. *poae*] to control *Poa annua* L. (annual bluegrass) on golf courses; Stumpout, a wood-decaying fungus [*Cylindrobasidium laeve* (Pers.) Chamuris] to control resprouting of wattle (*Acacia mearnsii* DeWild. and *A. pycnantha* Benth.) in South Africa; Dr. BioSedge, a host-specific rust fungus, *Puccinia canaliculata* (Schw.) Lagerh., to control specific biotypes of yellow nutsedge (*Cyperus esculentus* L.); and *Alternaria destruens* E.G. Simmons, a pathogen recently cleared by the U.S. Environmental Protection Agency (EPA) for registration as Smolder, to control dodder (*Cuscuta* L. spp.) (Phatak et al., 1983; Bewick et al., 1986; Beste et al., 1992; Imaizumi et al., 1997; Morris, Wood, and Den Breeÿen, 1998; Boyetchko, 1999; Rosskopf, Charudattan, and Kadir, 1999; Charudattan, 2000; EPA, 2005).

Although host specificity is considered one of the advantages of many bioherbicides in that it minimizes nontarget effects, it is also one of the major criticisms of them since most ecosystems are comprised of multispecies weed communities (Frantzen, Paul, and Müller-Schärer, 2001). The economics of controlling single-target weed species may therefore be cost prohibitive to farmers, particularly if the control of one target weed results in its replacement by other weed species. Industry and farmers generally view single-target weed management as difficult to justify in agroecosystems containing complex weed communities. Conversely, host specificity could also be viewed as an advantage in that it reduces the risk of the bioherbicide causing significant damage on nontarget organisms (i.e., crops). Despite the perception that all bioherbicides have a narrow spectrum of activity, some bioherbicides are currently marketed or in various stages of development exhibiting broad-spectrum activity against multiple weed targets. These include *Chondrostereum purpureum* (Pers. ex Fr.) Pouzar, *Pseudomonas syringae* van Hall pv. *tagetis* (PST), and *Sclerotinia sclerotiorum* (Lib.)

DeBary (De Jong, Scheepens, and Zadoks, 1990; Wall, 1994; Johnson, Wyse, and Jones, 1995, 1996; Shamoun et al., 1996; Dumas et al., 1997). *Chondrostereum purpureum* controls regrowth of hardwood tree species such as red alder (*Alnus rubra* Bong.), black cherry (*Prunus serotina* Ehrh.), poplar (*Populus* L. spp.), and white birch (*Betula papyrifera* Marsh.) when applied to cut stumps. PST has potential for control of Canada thistle [*Cirsium arvense* (L.) Scop.] and other Asteraceae weeds (Johnson, Wyse, and Jones, 1995, 1996; Bailey et al., 2000), while *S. sclerotiorum* is an omnivorous pathogen that controls a variety of weeds including Canada thistle, dandelion [*Taraxacum officinale* (L.) Web. ex Wigg.], and diffuse and spotted knapweeds (*Centaurea diffusa* Lam. and *C. maculosa* Lam.) (Brosten and Sands, 1986; Riddle, Burpee, and Boland, 1991; Mortensen and Hogue, 1995; Mortensen, 1998). Although nontarget effects are considered a high risk with many broad-spectrum bioherbicides, researchers have used molecular techniques, such as mutagenesis and genetic containment utilizing suicide genes, to bioengineer pathogens such as *S. sclerotiorum* with restricted nutritional requirements and nonsporulating traits (Sands, Ford, and Miller, 1990; Sands, Miller, and Ford, 1990; Sands and Miller, 1993).

Benefits of Biological Control for Sustainable Weed Management

Several factors that have precipitated the need to develop additional weed control strategies include the deregistration or phasing out of several older herbicides, rising costs associated with discovery and registration of new chemical herbicides, availability of chemicals for small markets (e.g., minor use registration), growing public awareness of environmental issues related to spray drift and pesticide residues in the soil, water, and food, and development of herbicide-resistant weed populations (Swanton, Harker, and Anderson, 1993; Bellinder, Gummesson, and Karlsson, 1994; Boyetchko, 1999; Beckie, Hall, and Tardif, 2001; Charudattan, 2001). Despite these new challenges facing farmers in modern agriculture, several opportunities have emerged. The shift from chemical herbicides as the predominant weed control option toward more integrated strategies that include biological control can provide producers with greater flexibility and help preserve the effectiveness of chemical herbicides. Biological weed control alleviates the heavy dependency on chemical herbicides while meeting the needs of organic and pesticide-free crop production (Scheepens, Müller-Schärer, and Kempenaar, 2001; Van Acker et al., 2001). Management of herbicide-resistant weed populations can be achieved through the application of bioherbicides that exhibit novel modes of action, thereby mitigating or delaying the development of new populations showing single or multiple resistance to specific chemical herbicides. Müller-Schärer and Frantzen (1996) proposed a "system management approach" that promotes the assessment of biological control based on improved crop yields and biological diversity. This perspective encourages a change in philosophy from a "clean crop" concept to management of weeds below economic thresholds at a population level.

Within the context of sustainable weed management, the aim of this review is to discuss the evolving role of biological weed control, more specifically bioherbicides, as one component of a weed management system. Although there are limited examples of bioherbicides that have been applied as part of an integrated weed management program, strategies for selection of bioherbicide candidates and their implementation into crop production systems will be explored based on relevant research. In some cases, where examples of bioherbicides are limited or not available, research using classical or augmentative agents will be presented in order to illustrate a point. The ultimate goal is to provide a framework that researchers and crop protection managers may consider when developing bioherbicides for the IWM system.

SELECTION OF BIOLOGICAL CONTROL STRATEGY

Upon the outset of developing a biological control strategy, careful consideration should be given to the target weed and the characteristics of the agroecosystem under which it grows (Charudattan and DeLoach, 1988). Due to cultural, climatic, edaphic, and biotic differences among various agroecosystems, it becomes very complex when selecting the appropriate pathogen for biological control, particularly as the choice of cropping practice influences weed communities and other biota. Charudattan (2001) further suggested that weeds with worldwide economic importance (e.g., *Chenopodium* L. spp., *Cyperus* L. spp., *Echinochloa* P. Beauv. spp., several grass weeds, and invasive tree species) be targets for biological control programs to ensure the interest of industry to commercialize bioherbicide products over the next ten years or so.

Criteria for Selection of Biological Control Candidates

Basic concepts in weed ecology to consider for weed management are the critical period of weed competition and time of emergence of the weed in relation to the crop (Weaver, 1984; O'Donovan et al., 1985; Aldrich, 1987; Swanton and Weise, 1991). Critical period considers (1) the length of time weed control is required to prevent crop yield losses and (2) the length of time crops can tolerate weeds before weed competition results in reductions in crop yield (Weaver, 1984; Swanton and Weise, 1991). The duration of the critical period will vary with the weed species and density, crop (and variety), and environmental conditions. Relative time of emergence of the weed compared to the crop can also have a dramatic impact on the severity of crop yield loss. For example, wheat (*Triticum aestivum* L.) and barley (*Hordeum vulgare* L.) yield changed by about 3 percent for each day that wild oat (*Avena fatua* L.) emerged before or after the crop (O'Donovan et al., 1985). This suggests that for every day that weed control is delayed, there is a significant impact on weed competition and resulting crop yield loss. These two ecological concepts may therefore predetermine which biological control approach may be more suitable: a preemergent or postemergent type of bioherbicide. This may be critical when considering that some foliar-applied biocontrol agents may take a longer period of time to suppress or control weeds due, in part, to environmental constraints such as duration of dew period or optimum temperature required for disease infection to occur. For example, *Colletotrichum truncatum* (Schw.) Andrus & Moore required 12 hours of dew to control hemp sesbania [*Sesbania exaltata* (Raf.) Rydh. ex. A.W. Hill] without a suitable formulation (Boyette, 1994).

Under natural conditions, *Colletotrichum gloeosporioides* f.sp. *malvae* causes disease on round-leaved mallow late in the growing season (Mortensen, 1988). Optimum weed control was achieved when a minimum of 20 hours of dew at 20 to 25°C was provided (Makowski, 1993). For control of velvetleaf (*Abutilon theophrasti* Medik.), 48 hours dew period was required to achieve the same level of weed control using a similar amount of pathogen inoculum. It may be argued that under a majority of environments, it is almost impossible to guarantee 20 or more hours of dew. In the case of *Alternaria cirsinoxia* Simmons & Mortensen for control of Canada thistle, at least eight hours of continuous leaf wetness was required to establish infection (Green and Bailey, 2000a,b). The pathogen infected plant tissues at temperatures ranging from 10 to 30°C under conditions of high relative humidity (>98 percent) or free water, indicating its infectivity under a broad range of temperature regimes, but long durations of intermittent leaf wetness was not conducive to high bioherbicidal activity. This pathogen may not have great potential in areas that have a semiarid climate, but may have greater applicability in geographical locations with greater levels of rainfall or higher humidity. In general, it is believed that more successful biocontrol agents are those that provide effective weed control with short dew period durations under a wide range of temperature regimes (TeBeest, 1991). Improvements in formulation may help to extend dew period or reduce dew period requirements, but unless the pathogen is effective under environmental conditions prevalent in the region of application, formulation and other technologies will not necessarily improve the efficacy of a moderately effective biocontrol agent.

Another factor to consider is the impact of the pathogen at the weed population level. Frantzen, Paul, and Müller-Schärer (2001) described how epidemics of the pathogen may be linked to intra- and interspecific plant competition. Competitiveness of a weed is partially determined by the rate of growth of the weed compared to the crop. A pathogen will therefore have greater impact on inter-specific competition if it infects and causes disease on the weed earlier and for a longer duration. Although foliar-applied fungal and bacterial pathogens are appropriate for postemergence application, weeds that emerge early in the growing season have a strong competitive advantage over the crop and may therefore be better targets for pathogens that act quickly. Soilborne pathogens (fungi and/or bacteria) used as preemergent bioherbicides may be more appropriate in this case due to better timing of application. For example, Kennedy and colleagues (1991) used a deleterious rhizobacterial strain (i.e., *Pseudomonas fluorescens* Migula D7 strain) to reduce the competition of downy brome (*Bromus tectorum* L.) in wheat. Although the weed was not eliminated from the field, the bacteria reduced the competitive nature of the weed in favor of the crop.

Plant architecture has also been indirectly used as a criterion for selection of biocontrol strategy. Hasan and Ayres (1990) suggested that foliar and stem pathogens are highly successful biocontrol agents because they can easily infect shoots, particularly at the growing point. Grass weeds represent a greater challenge because their meristems are covered by leaf sheaths that protect them from infection by foliar pathogens (Greaves and MacQueen, 1992; Auld and Morin, 1995). Although foliar-applied pathogens have generally been the bioherbicide of choice in the majority of cases, pathogens infecting at or below the soil surface represent another option for controlling Poaceae weeds and those weeds, mainly perennials, capable of regeneration through rhizomes and stolons. Many soilborne fusaria species have been investigated as either preemergent or postemergent bioherbicides (Jones and Hancock, 1990). This group of fungi has been viewed as highly suitable candidates because of their specialization on various hosts as *forme speciales*. Also, many soilborne fungi are not reliant on highly restrictive environmental conditions such as high humidity or free water. They are capable of causing root decay and lesions at or below the soil line, reducing weed emergence, and causing internal necrosis or wilting. Similar comparisons could be made for deleterious soil- or root-colonizing bacteria (Kremer and Kennedy, 1996; Boyetchko, 1999). The fact that soilborne pathogens are underexploited as potential bioherbicides may be due to the lack of understanding of their ecology and biology.

Assessing Population Structures

Agroecosystems consist of weed communities that are genetically heterogeneous in nature. The ability of individual strains or isolates of fungal and bacterial bioherbicide candidates to control these inherently diverse populations will be reflected in their field performance and their utility over a variety of geographical locations where weed control is required. Conversely, genetic characterization of the population structure of biocontrol agents will bring to light the potential genetic integration of virulence factors into a natural population and the impact it may have on nontarget plants. These are major considerations for implementation of bioherbicides in sustainable cropping systems.

Weed Population Structure

The process of choosing and developing a biological control agent for a particular weed or group of related weeds can be greatly enhanced by having a detailed understanding of the target weed population. Knowing the life cycle and growth habit of a particular weed can help a researcher select the optimal application regimes and target the plant at its weakest point both spatially and temporally. Another aspect that can significantly improve the chance of success with a biological control agent is understanding the population genetics of the weed, which is generally more variable than in crop plants. This factor often protects a given weed population from devastating epidemics that are commonly seen in genetically homogeneous crop monocultures. Determining the level of diversity in a weed population can provide crucial information regarding expectations of control over an entire population. As the development of molecular tools for determining genetic heterogeneity has progressed, its application to the field of weed biological control has become more commonplace. Information gathered with these tools has been used to identify pockets of genetically different weed populations that have been previously thought to have very little variability.

Although initially tested on single accessions of field bindweed (*Convolvulus arvensis* L.), Vogelgsang and colleagues (1999) evaluated the susceptibility of various accessions of the weed to *Phomopsis convolvulus* Ormeno-Nuñez, Reeleder, & A.K. Watson. The weed was reported to have intraspecific variations in morphological and physiological characteristics (Garcia-Baudin and Darmency, 1979), but the bioherbicide was capable of controlling the weed obtained from different geographical locations. Vogelgsang and colleagues (1999) pointed out that the study was conducted in a controlled environment and indicated the need to evaluate the efficacy of the bioherbicide under field conditions in order to further explore the susceptibility of morphologically variable weed accessions.

The system involving the registered rust bioherbicide *Puccinia canaliculata* for control of nutsedge is an example that further demonstrates the impact of genetic variation of weed populations on the effectiveness of biocontrol agents (Bruckart, Johnson, and Frank, 1988; Scheepens and Hoogerbrugge, 1991). The rust fungus collected from a particular geographical location infected only yellow nutsedge (*Cyperus esculentus* L.) from the same location and did not infect purple nutsedge (*C. rotundus* L.), although the latter was reported to be a host of the rust in other locations (Phatak, Callaway, and Vavrina, 1987). Previous studies (Horak and Holt, 1986; Horak, Holt, and Ellstand, 1987) using isozyme analysis on limited populations of purple and yellow nutsedges indicated that the genetic diversity within these species was extremely low. Similarly, Okoli and colleagues (1997), using the random amplified polymorphic DNA (RAPD) approach, found that purple nutsedge populations from the United States and the Caribbean region showed very little genetic variation, representing a very large clonal population. Conversely, the same RAPD technique revealed a high degree of interspecific variation between purple and yellow nutsedges and extensive variability within and between the yellow nutsedge populations sampled. This work underscores the need to reexamine the search for biological control agents for these weeds in which, as the authors suggest, different approaches should be sought for the two weed species. An obligate parasite, such as *P. canaliculata,* would be more successfully applied to the homogeneous populations found in purple nutsedge and facultative parasites for control of the diverse populations of yellow nutsedge.

It has become increasingly more common to use molecular techniques for characterizing weed populations at the onset of biological control projects. Information gained by assessing the genetic composition of a target weed population can assist in determining if it is advantageous to return to the center of origin of a target weed or if the target population is genetically distinct from the plants at the origin. The amplified fragment length polymorphism (AFLP) technique has been used to estimate genetic diversity of the highly aggressive weed *Rubus alcefolius* Poir., a target weed for classical biological control on La Reunion Island (Amsellem et al., 2000). The weed was believed to be introduced to the island directly from its native range in Asia or from Asia through Madagascar. The AFLP analysis revealed that the plants collected from the native range were grouped as one clade and had a high level of genetic diversity among individuals from the same location. The second clade was comprised of individuals from other areas of introduction, including Madagascar. The populations from Madagascar showed levels of diversity that were comparable to the populations found in Asia, while those from La Reunion had lower intralocality variation.

By investigating biological control agents for the pasture weed serrated tussock [*Nasella trichotoma* (Nees) Hack. ex Arechav.], Australian scientists used RAPD analysis and found populations of the weed from Argentina and Victoria and New South Wales in Australia to be genetically distinct (Hussaini, Lawrie, and McLaren, 2000). Morphological differences between Victoria populations and those found in New South Wales support the distinction of these two populations, and biological control agents are now being sought to control both populations.

Although the use of molecular techniques has elucidated many new relationships, classical approaches can also reveal extremely important information concerning interactions between pathogens with potential as weed biological control agents and their target weeds. Studies conducted to determine factors affecting infection of downy brome by the smut fungus *Ustilago bullata* Berk. resulted in the identification of resistance polymorphism both among and within four downy brome populations (Meyer, Nelson, and Clement, 2001). The weed populations were chosen from four contrasting habitats and bulk teliospore collections were conducted from each location. Highest infection levels resulted from inoculation with the locally collected *U. bullata* teliospores. One downy brome population was found to be completely resistant to bulk inoculum of the smut pathogen from any location other than the local population. In other populations, levels of susceptibility to bulk teliospore inoculations were variable when teliospores from another area were used, but were completely susceptible to smut inoculum collected locally. These inoculations established that the warm desert population was distinct and that it was composed of a single line. The authors reported that these findings had been confirmed through the use of molecular satellite markers. In one bulk inoculum source a minimum of two pathogen races were present. Although this pathosystem is not an example of a bioherbicide agent, it illustrates how this information will allow researchers to direct their efforts to control individual populations using a specific race of pathogen.

Population Structure of the Biocontrol Agents

Gathering information on the population structure of the biological control agent is often a crucial component in the development and release of an agent. Molecular tools that have been developed for use in the identification and phylogenetic analysis of fungi and bacteria are extremely helpful for characterizing pathogens to be used for biological control and for clarifying taxonomic relationships as well as providing information necessary for regulatory decisions, patenting procedures, and the development of particular strains as commercial products. Studies of the population structure can also be used to elucidate aspects of the ecology, biology, and pathogenicity of the organisms.

Sands and colleagues (1997), working with wilt of coca (*Erythroxylum coca* Lam.), isolated *Fusarium oxysporum* Sch.:Fr. and confirmed that it was the causal agent of the disease that was observed and previously reported in Hawaiian coca research plots. Through testing of the host range and vegetative compatibility of the Hawaiian isolate, it was determined that the fungus represented a new *forma specialis*. Using 10-base RAPD primers, the Hawaiian isolate was found to be closely related to the isolates from South America, suggesting that the fungus was introduced on the coca transplants that were used to establish the Hawaiian research plantation. Knowledge of the origin of the fungus, in this case, would establish that a novel isolate for biological control of coca was not being introduced into South America.

Cercospora rodmanii Conway and *C. piaropi* Tharp. have been studied as potential biological control agents for water hyacinth [*Eichhornia crassipes* (Mart.) Solms]. These fungi were originally described as two distinct species based on minor differences in conidial morphology and disease development. Tessmann and colleagues (2001) using partial DNA sequences from three protein-coding genes, elongation factor-1α, β-tubulin, and histone H3 found that separation of the species was not supported by the phylogenetic analysis. Based on combined datasets for 14 isolates from around the world, two clades were supported that provided differentiation based on geographical distribution rather than species separation. The description of *C. piaropi* was therefore amended to include *C. rodmanii* as a synonym. This significantly decreases the regulatory issues if this pathogen is to be developed for worldwide use. This example also brings to light an additional area for consideration. When a single isolate of a biological control agent is released over a large area, there is the possibility of gene flow from the inundatively applied isolate or strain to the existing, native population of the pathogen. The impact of this type of gene flow may have long-term effects on the population structure and result in modifications that possess the potential to threaten nontarget plant species. In addition, the "environmental fate" of the organism can be tracked if sufficient information regarding the genetic composition of the native population and the introduced population of the pathogen is known.

In the case of the fungus *Chondrostereum purpureum,* a biological control agent for hardwood tree species, an assessment of the population structure of the pathogen was required prior to application of the pathogen on a large scale. The first population analysis was performed using the large nontranscribed spacer region of the rDNA from 107 isolates from Europe, New Zealand, and North America. This region was amplified by polymerase chain reaction (PCR) and then analyzed using restriction fragment length polymorphisms (RFLPs). Through the type patterns generated, the North American isolates could be distinguished from the European and New Zealand isolates, but there was an equal distribution of types across central North America. Although the rate of gene flow could not be determined, the distribution of nuclear types led to the conclusion that gene flow had occurred throughout Canada, which was the proposed geographical area of release of the biological control agent (Ramsfield et al., 1996). Analysis of the mitochondrial DNA corroborated that gene flow had occurred across the continent (Ramsfield et al., 1999). Gosselin, Jobidon, and Bernier (1999) studied the genetic diversity among isolates of *C. purpureum* collected from four Canadian ecozones using RAPDs and concluded that more genetic diversity existed within an ecozone than between ecozones. This implies that the use of any native isolate would have little impact on the overall native population. The researchers went a further step in developing sequence-characterized amplified regions (SCAR) primer pairs. These primers could differentiate the biocontrol agent, isolate PFC2139, from other *C. purpureum* isolates. Prior to application, the SCAR primers were used to determine the frequency of the amplified region in the natural population. Eighteen months after the release of the biological control strain, the index of similarity of the prerelease samples and the postrelease samples to the biological control agent was compared. No significant change was found. In addition, recovery of the introduced *C. purpureum* strain was extremely low compared to isolation of wild types. The biological control agent was isolated from an area close to the release site and was found to be similar to a natural population of the fungus (Hintz, Becker, and Shamoun, 2001). An additional study of the same type in an area with a lower frequency of the SCAR amplified regions and a long-term evaluation of multiple releases would prove to be interesting.

Another possibility for investigating the dynamics of an introduced biocontrol agent is through gene tagging. A common method for tagging organisms is the generation of antibiotic-resistant mutants. The drawbacks to this approach are that the resistance can be transferred to other organisms existing in the environment and that the resistance may not be exclusive to the introduced organism. The insertion of a gene or gene fragment that can be tracked is one avenue of research that may provide answers to many questions related to the ecological significance of released pathogens. Although these methods lead to the production of a genetically modified organism, they do allow for precise monitoring of an organism. Several systems exist that can be applied to biological control organisms. These include the insertion of the *lux* gene for bioluminescence and the green florescent protein (GFP). These can be monitored by the detection of the unique DNA sequence, by amplification using PCR, or by detecting the light emitted by the organism (Kennedy, 1996; Poppenborg, Friehs, and Flaschel, 1997).

Another PCR-based method has been developed by Schaad and colleagues (2001) using a gene-tagging system for *Xanthomonas* spp. Through the insertion of the fatty acid desaturase (DES) gene fragment from *Pseudomonas syringae* pv. *phaseolicola,* a pigG::DES strain of *X. campestris* pv. *convolvuli* was generated. This organism is under development as a biological control agent for bindweed, *Convolvulus arvensis.* The genetically modified strain caused the same symptoms on bindweed that are characteristic of the wild type and showed normal growth on a variety of media. This system provided a highly sensitive detection method in which only DNA from viable cells was amplified using PCR amplification. In this system, the only organisms that would result in a false positive are *P. syringae* pv. *phaseolicola* and *P. syringae* pv. *actinidiae,* which are not likely to be found in the same locations where the biological control agent would be used. The authors point out that if an opportunity existed for these organisms to be present, they could be easily distinguished from the biological control agent using various selective media such as yeast-extract-dextrose carbonate calcium agar.

FORMULATIONS AND SPRAY APPLICATION

Two areas that have been recognized as important aspects for developing bioherbicide agents are formulation and application in the field. Foliar application of postemergent agents often encounters environmental constraints, particularly with respect to length of leaf wetness or dew. Formulations are needed to extend dew period requirements, increase moisture retaining properties, reduce the rate of evaporation of the spray droplet, and/or enhance the rate of infection of the microbial agent. Solid-based formulations are appropriate for stabilization of highly efficaceous propagules for extended periods prior to marketing and use by farmers and for delivery of preemergent bioherbicides. Delivery of these agents requires a pragmatic approach that does not compromise the practical aspects of application in production agriculture such as water volume, strategic placement of the formulated agent into the crop canopy, and retention on the target weed.

Formulation

Several recent reviews have focused on some of the advancements in formulation technology and provide many examples of typical foliar and soil formulations used for bioherbicides (Boyette et al., 1996; Greaves, Holloway, and Auld, 1998; Green et al., 1998; Boyetchko et al., 1999, 2002; Zidack and Quimby, 1999). In general, the development of bioherbicides has followed the chemical herbicide paradigm whereby they have been applied like a pesticide with or without a suitable formulation. Key criteria for a successful formulation include product stability and adequate shelf life, guaranteed propagule viability and efficacy, and ease of production and application. Two factors that may also predetermine the selection of appropriate formulation and application are the biology of the target weed and the bioherbicide agent (Zidack and Quimby, 1999). Post-emergent foliar application has been considered for many annual weed species, although they may also be targets for preemergent formulations, depending on their growth habit and competitive ability. Perennial weeds may require application of either systemic pathogens or soilborne pathogens (fungi or bacteria) encapsulated into granules. Some specific aspects of the biology of the microbial agent that may dictate the most appropriate choice of formulation include the ease of sporulation of fungal propagules in submerged liquid as opposed to those fungal agents that require solid-state fermentation, which may be more difficult and expensive. The duration of dew period required for germination and infection of the pathogen will necessitate the selection of appropriate formulation to overcome environmental limitations. Foliar-applied bacteria often require wounds or natural openings since they are not capable of directly penetrating leaf surfaces (Zidack, Backman, and Shaw, 1992). Many bacteria may require protection from UV radiation and desiccation because their propagules are not pigmented as are many fungal agents.

The main types of materials used in formulations include surfactants and adjuvants, emulsions and vegetable oils, hydrophilic polymers, and solid-based encapsulation materials (Green et al., 1998; Boyetchko et al., 1999; Zidack and Quimby, 1999). Surfactants aid in the dispersal and facilitate uniform distribution of the microbial propagule on the plant surface. They also alter leaf-wetting properties by modifying the surface tension of spray droplets at the plant-microbe interface and may have properties that serve as stickers, emulsifiers, and stabilizers. Several adjuvants have also been shown to increase bioherbicidal activity and survival of the microbial agent (Daigle and Cotty, 1991). Polyvinyl alcohol (PVA) and humectants such as psyllium (e.g., Metamucil) have been used as water-retaining compounds to reduce the rate of moisture loss (Greaves, Dutton, and Lawrie, 2000). Silwet L-77 and Silwet 408 are examples of organosilicone surfactants that facilitate penetration of bacterial and fungal bioherbicides such as *Ascochyta pteridis* Bres. and *Pseudomonas syringae* into plant stomata and hydathodes (Womack and Burge, 1993; Zidack and Backman, 1996; Bailey et al., 2000). Invert emulsions have been used to address the issue of moisture retention around the droplet containing the bioherbicide agent (Daigle and Connick, 1990; Womack and Burge, 1993; Greaves, Dutton, and Lawrie, 2000). These formulations are used to entrap water in order to reduce evaporation and retain sufficient moisture required for germination and infection of the microbial propagules. For example, a mixture of paraffin wax, mineral oil, soybean oil, and lecithin was used to enhance spore germination of *Alternaria cassiae* Jurair & Khan for control of sicklepod (*Cassia obtusifolia* L.) (Daigle et al., 1990). Application of the pathogen in the formulation also led to significantly higher virulence and infection using extremely low rates of inoculum (i.e., from 10 to 100 to a single spore per droplet) (Amsellem et al., 1990). The major disadvantages of these invert emulsions include phytotoxicity on nontarget plants and extremely high viscosity that makes application with conventional spray equipment difficult. Vegetable oils have also been explored with a variety of bioherbicide agents (Auld, 1993; Egley and Boyette, 1995; Klein, Auld, and Fang, 1995). They appear to have better leaf-wetting and water-retention properties, but their applicability using water volumes well below 500 to 1000 $L \cdot ha^{-1}$ needs to be fully evaluated. Several natural and synthetic hydrophilic polymers have also shown different abilities to retain moisture and enhance bioherbicidal activity and may provide some promise for reducing dew period requirements (Shabana et al., 1997).

Encapsulation of fungal bioherbicides into solid matrices are often used to apply bioherbicides at or below the soil surface. Many of these solid-based formulations have advantages by providing a buffer against harsh environmental conditions, containing carriers in the formulation that act as a nutrient source for the microbe, providing a stable shelf life that may be longer than that provided with liquid formulations, and having great versatility in the size and weight of granules for field application (Green et al., 1998; Boyetchko, 1999; Boyetchko et al., 2002). Sodium alginate has been used to entrap fungal propagules, along with other carriers and fillers (e.g., kaolin clay, soy flour, and cornmeal) for soil-applied bioherbicides (Walker and Connick, 1983; Weidemann, 1988; Bailey et al., 1997). "Pesta" is based on a wheat-gluten matrix consisting of semolina flour, kaolin, and fungal biomass that is produced into pasta-like sheets, dried and crumbled, and applied to the soil (Connick, Boyette, and McAlpine, 1991). By modifying moisture content and sugar concentration, the amount of fungal propagules was increased and stabilized and shelf life was extended to at least one year (Connick et al., 1996, 1998). More recently, the pesta-based technology was used to formulate a strain of *Pseudomonas fluorescens,* and preemergent application of this bioherbicide resulted in up to 90 percent reduction in emergence of green foxtail [*Setaria viridis* (L.) Beauv.] (Daigle, Connick, and Boyetchko, 2002). Zidack and Quimby (2002) reported that a Stabileze formulation (starch/oil/sucrose matrix) could also be used to stabilize pseudomonad bacteria. The addition of sucrose and/or oil to the formulation significantly increased bacterial survival, compared to the formulation without sucrose or oil. However, the bacteria formulated in the Stabileze were applied as postemergent bioherbicides, not to the soil, thereby necessitating the resuspension of the formulation in liquid. A hundredfold dilution of the Stabileze was required before the suspension was sprayable, resulting in a reduction of the titer of active ingredient (i.e., bacterial propagules) to a concentration that did not provide adequate inoculum required for weed control. The authors suggested that further modifications, such as reducing the amount of inert ingredients or the addition of a fermentation step prior to application, may remedy the situation.

Delivery and Application

Just as there are similarities between chemical pesticides and bioherbicides, there are inherent differences between the two which may have a significant impact on the choice of delivery system (Bateman, 1999; Bateman and Chapple, 2001). Similar to chemicals, biopesticide agents are expected to act immediately and provide relatively quick pest control, but living organisms have the potential to multiply in the environment through secondary cycling, thereby providing longer season control. More important, the specific mode of action will dictate the most appropriate delivery system of the biocontrol agent. However, it should be borne in mind that major deviations from existing application equipment for biopesticides may impede the adoption of the technology. Until such time that microbial-based technology is more widely used, biopesticides should initially be marketed in a form that is compatible with existing farm equipment in order to promote their use by farmers.

For the most part, the preferred method of application is with conventional sprayers for postemergent bioherbicides, usually with the addition of emulsifiers, stickers, spreaders, and other adjuvants to aid in the dispersal and adhesion onto the plant surface (Boyetchko et al., 1999). For agents (e.g., bacteria) that require wounds or natural openings to infect the plant, cutting or mowing was used to facilitate entry. Such is the case for *X. campestris* pv. *poae* (Camperico) where mowing of turfgrass was required to permit entry of the bacteria, thus ensuring adequate weed control (Imaizumi et al., 1997). For bioherbicides such as BioChon *(Chondrostereum purpureum)* and Stumpout *(Cylindrobasidium laeve)* wounds are created by cutting tree stumps and applying the formulated bioherbicide agent as a paste (De Jong, Scheepens, and Zadoks, 1990; Morris, Wood, and Den Breeÿen, 1998).

In the initial stages of identifying a promising foliar bioherbicide candidate, researchers often evaluate its efficacy by spraying until runoff and providing optimal conditions (e.g., dew periods greater than 12 hours) in order to assure the desired level of weed control (Lawrie et al., 2002; Lawrie, Greaves, Western, and Down, 1999; Greaves, Dutton, and Lawrie, 2000). Use of inappropriate application methods and simple formulations under controlled environmental conditions have led to less than spectacular results in the field. Many of the bioherbicides are applied at high water volumes (i.e., >500 $L \cdot ha^{-1}$) that are unrealistic and at rates not typical or even practical for use in agroecosystems (Boyetchko et al., 2002). For example, water volumes of 500 to 1000 $L \cdot ha^{-1}$ were required to obtain adequate spore germination of *Colletotrichum orbiculare* (Berk. & Mont.) Arx formulated in different vegetable and mineral oil emulsions (Klein, Auld, and Fang, 1995). Peng and colleagues (2001) also reported that application of a foliar mycoherbicide for control of green foxtail using water volumes of 100 to 800 $L \cdot ha^{-1}$ did not achieve the same level of control as that obtained by spraying with an airbrush until runoff. Moreover, a water volume of 1,960 $L \cdot ha^{-1}$ was required to achieve the same spray retention as that obtained with the airbrush. Carrier volume could be reduced by increasing spore concentrations and by selecting appropriate nozzles that generate finer spray droplets, thereby improving spray retention on the leaf surfaces. Boyette (1994) also demonstrated that the use of unrefined corn oil reduced water volume requirements from 500 to 5 $L \cdot ha^{-1}$ for *C. truncatum*. He speculated that the oil emulsion reduced the rate of evaporation of the droplets and improved the water retention properties.

One area in application technology that requires additional attention is effective spray deposition and retention (Lawrie et al., 1997). Most initial evaluations of bioherbicides are conducted with various aerosol-type sprayers (e.g., airbrush) which generate very small droplets, and many of these may not necessarily reach their target (Lawrie et al., 2002). Often, the lack of spray retention is compensated by the high water volumes applied, which often equate to the equivalent of 2000 $L \cdot ha^{-1}$ or more in the field. From a practical standpoint, application volumes tend to be in the range of 200 to 300 $L \cdot ha^{-1}$ or less, with low (3 to 150 $L \cdot ha^{-1}$) and ultra-low (0.5 to 3.1 $L \cdot ha^{-1}$) volumes being used for pesticides (Bateman, 1999; Boyetchko et al., 1999). Lawrie and colleagues (2002) indicated that at water volumes of 200 to 300 $L \cdot ha^{-1}$, fungal spores in the droplets generated would remain viable at first impact on the leaf surface. Bateman (1999) emphasized the importance of measuring and interpreting droplet size spectra for optimal dose transfer of biopesticides. It is extremely important to be aware of the required droplet size that carries the optimum inoculum load to the target weed, without compromising the practical aspects of application technology (i.e., water volume, application rate, travel speed, operator acceptability, etc.). For example, when comparing different application methods, Lawrie and colleagues (2002) found that a significantly greater number of fungal spores were retained on leaves of redroot pigweed (*Amaranthus retroflexus* L.) using an airbrush (786 $L \cdot ha^{-1}$) as opposed to those sprayed with a spinning disc (40 $L \cdot ha^{-1}$), hydraulic nozzle (105 $L \cdot ha^{-1}$), or twin-fluid nozzle (103 $L \cdot ha^{-1}$). Also, there were differences in spray retention depending on the type of plant tissue (e.g., cotyledons, leaves, stems). Spray droplets can be intercepted by crop canopy, resulting in reductions in spray deposition and retention on the target weed. Up to 84 percent reduction in spray retention of fungal spores onto *A. retroflexus* was observed when applied in a wheat canopy. When evaluating two potential mycoherbicides [*Stagonospora* (Sacc.) Sacc.] spp. and *Mycocentrospora acerina* (R. Hartig) Deighton}, 20 percent of the droplets that were less than 150 μm in diameter contained no propagules, while 60 percent of the larger droplets contained over 20 conidia (Lawrie et al., 1997). Only 14 percent of the smaller droplets contained only one or two conidia, and these droplets dried out very quickly, thus increasing the risk of reduced viability of the inoculum. One critical question to ask is what minimum and/or optimum number of spores per droplet are required to achieve infection and subsequent weed control? This will likely differ among bioherbicide agents and will be dependent on morphological and physical properties of the propagules. Spore size and shape may dictate optimum spray droplet size, and the hydrophilic and hydrophobic properties will influence the distribution of the propagules within the droplet.

Another aspect equally significant to consider when applying bioherbicides is the role of the propellant used to pressurize the sprayer (Rosskopf,

Charudattan, and Kadir, 1999; Charudattan, 2001). Most pressurized sprayers rely on CO_2 or compressed air as the propellent. Those sprayers using CO_2 have resulted in the acidification of the spray suspension due to production of carbonic acid under pressure (Rosskopf, Gaffney, and Charudattan, 1997). Significant reduction in spore germination was observed as a result of lower pH in the spray suspension when using a CO_2 sprayer with *Phomopsis amaranthicola* Rosskopf, Charudattan, & Shabana, a bioherbicide of pigweeds (*Amaranthus* spp.), and *Alternaria cassiae,* a bioherbicide of sicklepod; but these detrimental effects on bioherbicide performance were not observed with a compressed air sprayer. The impact on spore germination also caused reductions in disease severity and subsequent weed control. When another pathogen, *Microsphaeropsis amaranthi* (Ellis & Barthol.) Heiny & Mintz, was evaluated, there were no changes in spore germination or bioherbicidal activity as a result of CO_2 pressurization or pH. These findings suggest that application systems must be clearly evaluated to determine what impact they may or may not have on the bioherbicide.

Other novel application techniques that have been investigated include the use of the Burch Wet Blade mower (DeValerio et al., 2000; Charudattan, 2001). The bacterial agent *Ralstonia solanacearum* (Smith) Yabuuchi was applied with this mower for control of tropical soda apple (*Solanum viarum* Dunal). As vegetation is cut with the mower, a suspension of the bioherbicide is discharged through a fluid-delivery device onto the blade and delivered onto the cut surface. The PFC-ALDERWAK is another system for delivery of bioherbicides for vegetation management that was described by Dorworth (1995). This device is a hammerlike tool that is loaded with a pellet containing the biocontrol agent, in this case, *Nectria ditissima* Tul. & C. Tul., for control of red alder, and the pellet is discharged into the stem or trunk of the tree. The Burch Wet Blade and PFC-ALDERWAK could have other applications for vegetation management using bioherbicides such as BioChon, Chontrol, and Stumpout.

Application of bacterial and fungal bioherbicides to the soil may be achieved through soil drenches, seed treatments, or granules (Boyetchko, 1996; Boyetchko et al., 1999). The technique for delivery as granules may be similar to those applying fertilizers, but the choice of a specific application method will be dependent on the crop, weeds, and farming practices. Technology is available for seed treatment of crops, and coating the seed with various polymers and dormant microbes appears feasible. However, the flexibility and economics of seed treating a particular crop as compared to formulating into a granule for delivery into various crops must be evaluated. Weed seeds often reside throughout a soil profile at various depths. Therefore, selective targeting of the weeds through strategic placement of the bioherbicide inoculum, similar to fertilizer placement strategies, may be better achieved through the use of granules. Another possible delivery system is through the application of the bioherbicides onto crop residues in conservation tillage systems (Derksen, Blackshaw, and Boyetchko, 1996). For example, the Collego pathogen overwinters on unburied plant debris for at least seven months (TeBeest, 1982, 1991). It may be possible to achieve secondary spread of the pathogen, thereby promoting control of the weed with the bioherbicide under conservation tillage. Pseudomonad bacteria with weed-suppressive properties may also be applied onto crop residues. Stroo, Elliott, and Papendick (1988) reported greater populations of *Pseudomonas* spp. in zero-tillage plots compared to tilled plots. Many of the bacteria isolated from the crop residues produced deleterious metabolites that are inhibitory to weeds.

INTEGRATION IN WEED MANAGEMENT AND CROP PRODUCTION SYSTEMS

Weeds are an important component of agroecosystems, and the potential for bioherbicides will be realized as they become fully integrated into production agriculture. Cropping practices such as crop rotations, use of chemical pesticides and fertilizers, and other management practices including conventional and zero tillage will have an impact, whether negative or positive, on the performance of biological control agents. Whether through management of single weed species or multi-species weed communities, evaluation and enhancement of these weed management tools may be accomplished through synergizing with chemical herbicides and understanding their impact on weed and crop competition.

Synergizing with Chemical Herbicides

Implementation of biological control into crop production systems will be dependent on their ability to become integrated with other weed and pest management tools. This is particularly important with respect to the compatibility of these microbial agents with chemical pesticides. Enhanced efficacy of bioherbicides has been achieved through the utilization of various chemicals, including herbicides, and may prove to be an unexpected benefit with some bioherbicide candidates (Altman, Neate, and Rovira, 1990; Smith, 1991; TeBeest, 1996; Boyetchko, 1999; Rosskopf, Charudattan, and Kadir, 1999; Boyetchko et al., 2002).

Colletotrichum coccodes (Wallr.) Hughes causes disease and mortality in velvetleaf over a wide range of temperatures and dew period durations, with the optimum disease occurring at 24°C and a dew period duration of 24 hours (Wymore et al., 1988). Velvetleaf plants will continue to grow after defoliation by the pathogen if environmental conditions are suboptimal for disease development. If the bioherbicide is tank mixed with thidazuron, mortality of the weed is increased when compared to application of the fungus alone (Wymore, Watson, and Gotlieb, 1987; Hodgson et al., 1988). This enhanced activity has also been demonstrated with *Puccinia canaliculata* and paraquat on yellow nutsedge (Phatak et al., 1983; Phatak, 1984). Coapplication of the pathogen and herbicide resulted in 99 percent weed control compared to only 60 percent with rust alone and 10 percent with paraquat alone. Experiments with reduced rates of herbicides plus *Pseudomonas syringae* pv. *tagetis* (PST) clearly showed enhanced bioherbicidal activity for control of Canada thistle (Bailey et al., 2000). When PST was applied with glyphosate at 0.1 times the recommended rate, there were significant reductions in biomass of Canada thistle and greater symptoms of apical chlorosis, when compared to the PST alone and reduced rate of glyphosate alone.

A study examining compatibility of spores of *Chondrostereum purpureum* with pesticides and adjuvants showed that the majority of chemicals tested were fungitoxic to the bioherbicide at concentrations of 0.01 and 0.1 percent (v/v) (Prasad, 1994). Grant, Prusinkiewicz, Makowski, and colleagues (1990) reported that more than 50 percent of herbicides applied at 0.1 to 2 times the recommended rate did not affect germination and/or appressorial formation of conidia of *C. gloeosporioides* f.sp. *malvae*. At lower concentrations, many herbicides were less toxic to the fungal spores than at higher concentrations. Fungicides such as benomyl also reduced spore germination, but effects on the spores declined as concentrations of the fungicide decreased. Although generation of benomyl-resistant strains of bioherbicides such as *C. gloeosporioides* f.sp. *aeschynomene* have been developed, with the view of tank mixing for field application, the benomyl-resistant mutants were found to be less competitive than the wild-type strains and were not capable of causing similar levels of weed control (TeBeest, 1984; Luo and TeBeest, 1997). Greenhouse studies conducted by Holmström-Ruddick and Mortensen (1995) also reported that application of benomyl concurrently or 24 hours prior to application of a benomyl-resistant strain of *C. gloeosporioides* f.sp. *malvae* affected the ability of the pathogen to cause disease on round-leaved mallow, but postapplication of the fungicide did not reduce its bioherbicidal potential when compared to the wild-type strain. Experiments with *C. gloeosporioides* f.sp. *malvae* and split applications with several herbicides resulted in enhanced control of round-leaved mallow than with the bioherbicide alone (Grant, Prusinkiewicz, Mortensen, and Makowski, 1990). If the pathogen was applied 48 hours before the chemical herbicide, there was no detrimental effect on the bioherbicidal activity. However, tank-mixing of grass weed herbicides (i.e., diclofop and sethoxydim) with the pathogen reduced disease development on the weed, whereas split applications did not have a significant effect on the pathogen. When broadleaf herbicides (i.e., bromoxynil plus MCPA, 2,4-DB, imazethapyr, and metribuzin) were applied either as tank mixes or as split applications, there was no significant detrimental effect on the bioherbicidal activity.

Smith (1991) reported on the integrated application of Collego in greenhouse and field trials with several chemical pesticides. Tank mixing of the bioherbicide with propanil, benomyl, and fentin hydroxide resulted in reductions in infection of northern jointvetch, while postapplication of the chemicals a few days after the pathogen did not inhibit infection and disease development. In small- and large-plot field experiments, it was demonstrated that Collego could be tank mixed with acifluorfen and bentazon but not with 2,4-D without loss of bioherbicidal activity. Some fungicides used to control rice sheath blight could be applied one week after the biocontrol agent, but application of benomyl was restricted to at least three weeks after Collego. Application of various insecticides did not appear to have a detrimental effect on the efficacy of Collego.

These examples demonstrate the potential of integrating the bioherbicide with chemical pesticides and that timing of application may be critical when making recommendations on spray application. It is therefore important to evaluate the tolerance of each bioherbicide agent to a variety of chemical pesticides and the impact they may have on the performance of the pathogen in a field situation. These factors will be essential for the successful adoption of biocontrol technologies by producers into agricultural production systems.

Weed/Crop Competition

As discussed earlier, the measure of the success of a bioherbicide is its ability to control weeds at a population level in crop competition. For convenience, preliminary evaluation of the potential of most biocontrol agents is determined using a single-plant system based on evaluation of direct damage to weeds. Although this is a simpler process in the initial discovery phase of the pathogen, it clearly has its limitations for projecting the impact of the pathogen at a population level. Kadir and colleagues (1999) evaluated the efficacy of *Dactylaria higginsii* (Luttrell) M. B. Ellis on purple nutsedge in competition with tomato (*Lycopersicon esculentum* Mill.) under controlled environment conditions. They found that application of the pathogen at a concentration of 10^6 conidia·mL^{-1} reduced the interference of purple nutsedge in tomato. It was speculated that shading of the tomato plants contributed to high humidity conditions under the canopy, thus enhancing development of disease symptoms and secondary infection of the pathogen. In addition, diseased nutsedge plants were weaker and thinner and therefore less competitive and more susceptible to the pathogen. The efficacy assessment with the pressure of a competitive crop allowed a better understanding of the optimal dose of the bioherbicide, although the results need to be further validated under field conditions.

Kennedy and colleagues (1991) also clearly demonstrated the capability of deleterious soil bacteria to reduce the competitive nature of downy brome in a wheat field. Although the bacteria did not eradicate the weed, crop yield increases of 18 to 35 percent were observed. More important, biocontrol agents that may have been deemed not highly efficacious in early stages of evaluation were shown to be highly effective when applied to weeds grown in competition with a crop. Guntli and colleagues (1999) observed the inability of the pathogen *Stagonospora convolvuli* Dearn. & House to control hedge bindweed [*Calystegia sepium* (L.) R.Br.] when grown in competition with red clover (*Trifolium pratense* L.) under controlled environment conditions, despite severe necrosis caused on the weed. It was speculated that the absence of rain splash, an important factor for secondary spread of the pathogen, could have attributed to the lack of weed control. However, when hedge bindweed was inoculated with the pathogen in a maize field with red clover as a green cover, higher levels of disease and significant reductions in ground cover of the weed were observed, when compared to the untreated plots and plots with only the pathogen applied. Greater disease levels in the field compared to the greenhouse were also attributed to spread of secondary inoculum through rain splash.

The competitive ability of wild pansy (*Viola arvensis* Murray), a weed in spring and winter cereals, was significantly reduced when infected with *Mycocentrospora acerina* in a wheat field (Lawrie, Greaves, Down, and Lewis, 1999). Application of high inoculum rates of the pathogen caused high mortality of the weed and completely eliminated the impact of weed competition on crop yield. Suboptimal inoculum levels that caused apparent disease but no mortality of the weed resulted in a 28 percent reduction in the competitive ability of the weed when compared to the nontreated control. Postemergence application of the bioherbicide *Ascochyta caulina* (P. Karst.) Aa & Kesteren also reduced the competitiveness of common lambsquarters (*Chenopodium album* L.) in a corn cropping system (Kempenaar, Horsten, and Scheepens, 1996). Crop yield loss was reduced in corn but not in sugar beet (*Beta vulgaris* L.), indicating that greater speed and higher level of weed suppression may be required in sugar beet. Perhaps the time of emergence and critical period for sugar beet need to be considered more thoroughly when evaluating the potential of *A. caulina* as a bioherbicide.

Colletotrichum coccodes, a bioherbicide with limited capability of controlling velvetleaf, was found to have substantially higher bioherbicidal activity when the competition from the soybean crop was present (DiTommaso and Watson, 1995; DiTommaso, Watson, and Hallet, 1996). Interspecific competition resulted in greater reductions in seed yield of velvetleaf inoculated with *C. coccodes,* compared to velvetleaf grown in pure stands. These results were mostly attributed to stunting of the weed which permitted the soybean to outgrow it. Yield loss of soybean was also lower in plots where velvetleaf was inoculated with the bioherbicide compared to untreated plots. At high densities of the weed, inoculated velvetleaf plants exhibited significant reductions in plant growth, particularly due to lower nutrient acquisition as a result of greater intraspecific competition (DiTommaso and Watson, 1995). Paul and Ayres (1986, 1987) also reported that common groundsel (*Senecio vulgaris* L.) was not significantly affected by the rust *Puccinia lagenophorae* Cooke when grown in pure stands, but interspecific competition with lettuce reduced the competitive ability of the weed. DiTommaso, Watson, and Hallet (1996) strongly suggested the need to understand how the interaction between the foliar-applied fungal pathogens and the competing crop's morphology have a significant impact on the competitive nature of the target weed. These studies also demonstrate the importance of evaluating the potential of bioherbicide candidates in crop competition and that efficacy of bioherbicides can be significantly improved under these conditions.

Application in Cropping Systems

The ultimate test for the utility of a bioherbicide is its implementation into production agriculture. We have previously discussed some of the issues related to formulation and application and how weed competition and synergy with chemical pesticides may have an impact on the performance of a bioherbicide. Efforts to fully integrate biological control into crop production systems, which include application of pesticides, cultural practices, and crop rotations, need to be investigated to realize the potential of bioherbicides. For example, timing of chemical pesticides in relation to the bioherbicide had significant impact on the performance of the microbial agent (Grant, Prusinkiewicz, Mortensen, and Makowski, 1990; Smith, 1991; Holmström-Ruddick and Mortensen, 1995). Unfortunately, there are few examples of integrating bioherbicides in agroecosystems, at least from a systems approach.

Müller-Schärer and Frantzen (1996) described a "system management approach" for developing biological weed control agents for implementation into cropping systems. Although this system is associated with the augmentative approach to biological control and is viewed as being useful where a single weed species will be controlled in crops, it provides specific principles in weed ecology and disease epidemiology where the emphasis is on weed management by reducing the impact of the weed on crop yield and optimizing dispersal and spread of the biological control agent. Eradication of the weed was not the focus, but reducing its competitive advantage below economic thresholds was essential. Using the weed pathosystem *Puccinia lagenophorae-Senecio vulgaris* as a model, some of the key factors considered important to this approach were the life history of the weed as well as pathogen screening and epidemiology, and genetic structure of the weed and pathogen population. It was stressed that biological control agents should be viewed as stress factors of weeds and as components of the whole pest management strategy. The system management approach was extended to further include several biological control agents, including insects, and all interactions of the host (plant), environment, natural enemies (microbial and insect natural enemies), and humans (Frantzen and Hatcher, 1997). The approach evolved and considered intra- and interspecific competition and its link with pathogen epidemics (Frantzen, Paul, and Müller-Schärer, 2001). The combined effects of the pathogen on plant traits (i.e., biomass, reproductive capacity) and its impact on weed competitiveness were considered along with the dispersal and spread of the pathogen as an epidemic in the host (weed) population. Some of the environmental factors, such as lack of moisture (or dew period), and temperature as well as inoculum load may also be considered as part of the epidemiological factors of the pathogen. It is anticipated that the competitiveness of the weed population may therefore be predicted when considering the temporal and spatial distribution of the pathogen inoculum within the host population. Moreover, this model should provide the basis for evaluating the effectiveness of bioherbicide agents in sustainable agroecosystems.

Application of the bacterial agent PST for control of Canada thistle was evaluated as part of an integrated weed management program in conservation tillage (Hoeft et al., 2001). Two cultivars of soybean differing in competitiveness against weeds were tested for synergism with PST, and efficacy of the bioherbicide was compared to the herbicide bentazon. PST affected survival of Canada thistle, height, and reproductive ability, but weed suppression was more evident with the herbicide application. No effects between the soybean varieties were observed; therefore, any synergism between the competitive varieties and PST application was not evident. PST did, however, reduce the competitive ability of Canada thistle in soybean through reductions in the physical size of the surviving weed plants. It would have been interesting to determine whether any synergism existed between PST and bentazon, as has been observed with PST and glyphosate (Bailey et al., 2000). However, in the latter study, these results were observed only in the greenhouse and no synergism was observed in field trials (Boyetchko, unpublished data). Hoeft and colleagues (2001) indicated the potential of using PST in conservation-tillage systems and that further research is required to evaluate synergisms with other nonchemical weed management practices.

In another study, rhizobacteria (root-colonizing bacteria) and endorhizal bacteria were isolated from different cropping systems to determine which management system favored higher bacterial populations with weed-suppressive properties (Li and Kremer, 2000). Cropping systems ranged from a native prairie ecosystem to six cropping systems with different levels of input (i.e., based on level and intensity of tillage, fertilization, and chemical pesticide input) and diversity of crop rotation (e.g., monoculture to diverse annual and perennial crops). Bacteria which were randomly sampled from each cropping system were isolated, purified, and tested for phytotoxicity using a lettuce (*Lactuca sativa* L.) seedling bioassay. The highest populations of deleterious rhizobacteria (DRB) were isolated from a no-till system with *Zea mays-Glycine max-Triticum aestivum* cover crop rotation with the second highest populations of DRB isolated from an organic input system with continuous Virginia strawberry (*Fragaria virginiana* Duchesne). The cropping system with the lowest populations of DRB resided in the cool-season pasture containing grasses and legumes. This study clearly demonstrates the impact of a crop management system on the occurrence of weed-suppressive bacteria. It would be interesting to conduct follow-up studies to determine long-term effects of these cropping systems on weed communities as a result of DRB population dynamics. Furthermore, the impact of these cropping systems on the application of soil-applied bacterial bioherbicides (as preemergent agents) may provide valuable information on their integration in sustainable agroecosystems.

CONCLUSIONS

Bioherbicides have enormous potential as weed control strategies when a pragmatic approach to their integration in crop production systems is considered. Researchers in weed management need to clearly define the economic and ecological factors that contribute to crop yield loss and what additional weed control options may be used to minimize the impact of weed competitiveness. Although current agricultural practices incorporate the widespread use of chemical herbicides as a cost-effective method for reducing the impact of weeds in arable crops, the high performance of these chemicals may have resulted in the ultimate expectation that all crops must be weed-free. Based on this criterion, it becomes very difficult for other methods of weed control to gain acceptance by producers and industry (Hurle, 1997). The term *bioherbicide* itself may create the expectation that the end result will be complete weed eradication, similar to chemical herbicides, leaving doubts about the performance of the biocontrol agent when it is not achieved (Crump, Cother, and Ash, 1999). Nonetheless, the challenge will be to educate the agricultural community that a rational approach to weed management at a population level is to focus on reduced competition of weeds, not necessarily complete kill (Makowski, 1997; Auld and Morin, 1995). Over the past few decades, the scope and breadth of research in bioherbicides has expanded and there is an emerging awareness of the benefits of integrating them into sustainable crop production systems. Weed management is no longer viewed as a single-element approach (i.e., chemicals), and as crop production systems become more diverse, integrated weed management that uses a wide range of coordinated techniques (chemical, cultural, and biological approaches) becomes more feasible (Derksen, Blackshaw, and Boyetchko, 1996).

Perhaps a more critical evaluation of the most appropriate biocontrol strategy could be considered at the onset of a weed control program. It should be based on the most fundamental elements in weed ecology, time of emergence and critical period, instead of relying too heavily on the chance discovery of a "silver-bullet" pathogen. Selection of postemergent versus preemergent microbial agents may be based on the weed biology and ecology, thereby optimizing the chance of success in specific cropping systems. Unfortunately, the amount of information on the target weed population may be limited due to a decline in research support and expertise in weed biology and ecology. Detailed investigation of the weed population structure, through independent studies or extensive literature search prior to a bioherbicide project, will provide crucial information about the diversity of the entire weed population and the implications on the range of pathogen races or isolates required for controlling the target weed over a range of geographical locations. By the same token, the population structure of the bioherbicide agent will allow critical investigation of its ability to control heterogeneous weed populations and to determine whether indigenous or exotic pathogens must be collected and the potential risk of gene transfer among different populations of the same pathogen with inundative applications.

The environmental constraints that limit the field performance of the bioherbicide may be overcome through development of suitable formulations and application strategies. Current research in formulation has begun to target these areas. However, some issues that have yet to be clearly resolved, particularly for foliar-applied agents, are feasible technologies for reducing dew period duration or improving moisture retention of spray droplets as well as improving leaf-wetting ability and spray retention at practical application rates. Greaves, Dutton, and Lawrie (2000) clearly stated that in order for us to make advancements in formulation, we must progress from the use of simple formulations to more complex mixtures of ingredients to address these issues. Early evaluations of promising bioherbicide candidates should be made using realistic application systems that are similar to existing field application equipment. True use of integrated weed management will entail the combination of cultural practices (e.g., mechanical tillage, row spacing, and seeding rates.), addition of cover crops, use of competitive crop varieties, chemical pesticides, and biological controls. It is predicted that implementation of bioherbicides will increase as more agents are discovered and developed. Their full potential will be realized as teams of researchers with expertise in pest management and agronomy begin to integrate bioherbicides into crop production systems.

NOMENCLATURE

2,4-DB: 4-2,4-dichlorophenoxy)-butanoic acid
acifluorfen: 5-[2-chloro-4-(trifluoromethyl)phenoxyl]-2-nitrobenzoic acid
benomyl: methyl 1-(butylcarbamoyl)-2-benzimidazole-carbamate
bentazon: 3-(1-methylethyl)-(1*H*)-2,1,3-benzothiadiazin-4-(3*H*)-one 2,2- dioxide
bromoxynil: 3,5-dibromo-4-hydroxybenzonitrile ester of octanoic acid
diclofop: (±)-2-[4-(2,4-dichlorophenoxy)phenoxy]propanoic acid
fentin hydroxide: triphenyltin hydroxide
glyphosate: *N*-(phosphonomethyl)glycine
imazethapyr: (±)-2-[4,5-dihydro-4-methyl-4-(1-methylethyl)-5-oxo-1*H*-imidazol-2-yl]5-ethyl-3-pyridinecarboxylic acid
MCPA: (4-chloro-2-methylphenoxy)acetic acid
metribuzin: 4-amino-6-(1,1-dimethylethyl)-3-(methylthio)-1,2,4-triazin-5(4*H*)-one
paraquat: 1,1'dimethyl-4,4'-bipyridinium ion
propanil: *N*-(3,4-dichlorophenyl)propanamide
sethoxydim: 2-[1-ethoxyimino)butyl]-5-[2-(ethylthio)propyl]-3-hydroxy-2-cyclohexen-1-one
thidazuron: N-phenyl-N'-1,2,3-thiadiazol-5-yl-urea

REFERENCES

Aldrich, R.J. (1987). Predicting crop yield reductions from weeds. *Weed Technology* 1: 199-206.

Altman, J., S. Neate, and A.D. Rovira (1990). Herbicide-pathogen interactions and mycoherbicides as alternative strategies for weed control. In R.E. Hoagland (Ed.), *Microbes and microbial products as herbicides* (pp. 240-259). Washington, DC: American Chemical Society.

Amsellem, L., T. Le Bourgeois, J.L. Noyer, and M. Hossaert-McKey (2000). Comparison of genetic diversity of the weed *Rubus alcefolius* in its introduced and native areas. In N.R. Spencer (Ed.), *Proceedings of the Tenth International Symposium on Biological Control of Weeds* (pp. 253-260), July 4-14, 1999. Bozeman: Montana State University.

Amsellem, Z., A. Sharon, J. Gressel, and P.C. Quimby Jr. (1990). Complete abolition of high inoculum threshold of two mycoherbicides (*Alternaria cassiae* and and *A. crassa*) when applied in invert emulsion. *Phytopathology* 80: 925-929.

Auld, B.A. (1993). Vegetable oil suspension emulsions reduce dew dependence of a mycoherbicide. *Crop Protection* 12: 477-479.

Auld, B.A. and L. Morin (1995). Constraints in the development of bioherbicides. *Weed Technology* 9: 638-652.

Bailey, B.A., K.P. Hebbar, M. Strem, L.C. Darlington, and R.D. Lumsden (1997). An alginate prill formulation of *Fusarium oxysporum* Schlechtend: Fr. f.sp. *erythroxyli* for biocontrol of *Erythroxylum coca* var. *coca*. *Biocontrol Science and Technology* 7: 423-435.

Bailey, K.L., S.M. Boyetchko, J. Derby, W. Hall, K. Sawchyn, T. Nelson, and D.R. Johnson (2000). Evaluation of fungal and bacterial agents for biological control of Canada thistle. In N.R. Spencer (Ed.), *Proceedings of the Tenth International Symposium on Biological Control of Weeds* (pp. 203-208), July 4-14, 1999. Bozeman: Montana State University.

Bateman, R. (1999). Delivery systems and protocols for biopesticides. In R. Hall and J.J. Menn (Eds.), *Methods in biotechnology*, Volume 5, *Biopesticides: Use and delivery* (pp. 509-528). Totowa, NJ: Humana Press.

Bateman, R. and A. Chapple (2001). The spray application of mycopesticide formulations. In T.M. Butt, C. Jackson, and N. Magan (Eds.), *Fungi as biocontrol agents* (pp. 289-309). Wallingford, UK: CAB International.

Beckie, H.J., L.M. Hall, and F.J. Tardif (2001). Herbicide resistance in Canada—Where are we today? In R.E. Blackshaw and L.M. Hall (Eds.), *Integrated weed management: Explore the potential* (pp. 1-36). Saint-Anne-de-Bellevue, Québec: Expert Committee on Weeds.

Bellinder, R.R., G. Gummesson, and C. Karlsson (1994). Percentage-driven government mandates for pesticide reduction: The Swedish model. *Weed Technology* 8: 350-359.

Beste, C.E., J.R. Frank, W.L. Bruckart, D.R. Johnson, and W.E. Potts (1992). Yellow nutsedge *(Cyperus esculentus)* control in tomato with *Puccinia canaliculata* and pebulate. *Weed Technology* 6: 980-984.

Bewick, T.A., L.K. Binning, W.R. Stevenson, and J. Stewart (1986). Development of biological control for swamp dodder. *Proceedings of the North Central Weed Control Conference* 41: 24.

Boyetchko, S.M. (1996). Impact of soil microorganisms on weed biology and ecology. *Phytoprotection* 77: 41-56.

Boyetchko, S.M. (1999). Innovative applications of microbial agents for biological weed control. In K.G. Mukerji, B.P. Chamola, and K. Upadhyay (Eds.), *Biotechnological approaches in biocontrol of plant pathogens* (pp. 73-97). London: Kluwer Academic/Plenum Publishers.

Boyetchko, S., E. Pedersen, Z. Punja, and M. Reddy (1999). Formulations of biopesticides. In F.R. Hall and J.J. Menn (Eds.), *Methods in biotechnology,* Volume 5, *Biopesticides: Use and delivery* (pp. 487-508). Totowa, NJ: Humana Press.

Boyetchko, S.M., E.N. Rosskopf, A.J. Caesar, and R. Charudattan (2002) Biological weed control with pathogens: From search for candidates to applications. In G.G. Khachatourians and D.K. Arora (Eds.), *Applied mycology and biotechnology,* Volume II, *Applications of fungal biotechnology to food production* (pp. 239-274). New York: Elsevier Science.

Boyette, C.D. (1994). Unrefined corn oil improves the mycoherbicidal activity of *Colletotrichum truncatum* for hemp sesbania *(Sesbania exaltata)* control. *Weed Technology* 8: 526-529.

Boyette, C.D., P.C. Quimby Jr., A.J. Caesar, J.L. Birdsall, W.J. Connick Jr., D.J. Daigle, M.A. Jackson, G.H. Egley, and H.K. Abbas (1996). Adjuvants, formulations, and spraying systems for improvement of mycoherbicides. *Weed Technology* 10: 637-644.

Brosten, B.S. and D.C. Sands (1986). Field trials of *Sclerotinia sclerotiorum* to control Canada thistle *(Cirsium arvense). Weed Science* 34: 377-380.

Bruckart, W.L., D.L. Johnson, and J.R. Frank (1988). Bentazon reduces rust-induced disease in yellow nutsedge *(Cyperus esculentus). Weed Technology* 2: 229-303.

Buhler, D.D. (2002). Challenges and opportunities for integrated weed management. *Weed Science* 50: 273-280.

Burnett, H.C., D.P.H. Tucker, and W.H. Ridings (1974). *Phytophthora* root and stem rot of milkweed vine. *Plant Disease Reporter* 58: 355-357.

Charudattan, R. (1991). The mycoherbicide approach with plant pathogens. In D.O. TeBeest (Ed.), *Microbial control of weeds* (pp. 24-57). New York: Chapman and Hall, Inc.

Charudattan, R. (2000). Current status of biological control of weeds. In C.G. Kennedy and T.B. Sutton (Eds.), *Emerging technologies for integrated pest management: Concepts, research, and implementation* (pp. 269-288). St. Paul, MN: APS Press.

Charudattan, R. (2001). Biological control of weeds by means of plant pathogens: Significance for integrated weed management in modern agro-ecology. *BioControl* 46: 229-260.

Charudattan, R. and C.J. DeLoach Jr. (1988). Management of pathogens and insects for weed control in agroecosystems. In M.A. Altieri and M. Liebman (Eds.),

Weed management in agroecosystems: Ecological approaches (pp. 245-264). Boca Raton, FL: CRC Press.

Connick, W.J. Jr., C.D. Boyette, and J.R. McAlpine (1991). Formulation of mycoherbicides using a pasta-like process. *Biological Control* 1: 281-287.

Connick, W.J. Jr., D.J. Daigle, C.D. Boyette, and J.R. McAlpine (1996). Water activity and other factors that affect the viability of *Colletotrichum truncatum* conidia in wheat flour-kaolin granules ('Pesta'). *Biocontrol Science and Technology* 1: 277-284.

Connick, W.J. Jr., D.J. Daigle, A.B. Pepperman, K.P. Hebbar, R.D. Lumsden, T.W. Anderson, and D.C. Sands (1998). Preparation of stable, granular formulations containing *Fusarium oxysporum* pathogenic to narcotic plants. *Biological Control* 13: 79-84.

Crump, N.S., E.J. Cother, and G.J. Ash (1999). Clarifying the nomenclature in microbial weed control. *Biocontrol Science and Technology* 9: 89-97.

Daigle, D.J. and W.J. Connick Jr. (1990). Formulation and application technology for microbial weed control. In R.E. Hoagland (Ed.), *Microbes and microbial products as herbicides* (pp. 288-304). Washington, DC: American Chemical Society.

Daigle, D.J., W.J. Connick Jr., and S.M. Boyetchko (2002). Formulating a weed-suppressive bacterium in 'Pesta'. *Weed Technology* 16: 407-413.

Daigle, D.J., W.J. Connick Jr., P.C. Quimby Jr., J. Evans, B. Trask-Morrell, and F.E. Fulgham (1990). Invert emulsions: Carrier and water source for the mycoherbicide, *Alternaria cassiae*. *Weed Technology* 4: 327-331.

Daigle, D.J. and P.J. Cotty (1991). Factors that influence germination and mycoherbicidal activity of *Alternaria cassiae*. *Weed Technology* 5: 82-86.

De Jong, M.D., P.C. Scheepens, and J.C. Zadoks (1990). Risk analysis for biological control: A Dutch case study in biocontrol of *Prunus serotina* by the fungus *Chondrostereum purpureum*. *Plant Disease* 74: 189-194.

Derksen, D.A. R.E. Blackshaw, and S.M. Boyetchko (1996). Sustainability, conservation tillage and weeds in Canada. *Canadian Journal of Plant Science* 76: 651-659.

DeValerio, J.T., R. Charudattan, J.J. Mullahey, W.H. Sherrod, and P.D. Roberts (2000). Biological control of *Solanum viarum* Dunal (tropical soda apple) by *Ralstonia solanacearum* (E.F. Smith) Yabuuchi applied with the Burch Wet Blade mower system. *Weed Science Society of America Abstracts* 40: 28-29.

DiTommaso, A. and A.K. Watson (1995). Impact of a fungal pathogen, *Colletotrichum coccodes* on growth and competitive ability of *Abutilon theophrasti*. *New Phytologist* 131: 51-60.

DiTommaso, A., A.K. Watson, and S.G. Hallett (1996). Infection by the fungal pathogen *Colletotrichum coccodes* affects velvetleaf *(Abutilon theophrasti)*-soybean competition in the field. *Weed Science* 44: 924-933.

Dorworth, C.E. (1995). Biological control of red alder *(Alnus rubra)* with the fungus *Nectria ditissima*. *Weed Technology* 9: 243-248.

Dumas, M.T., J.E. Wood, E.G.. Mitchell, and N.W. Boyonoski (1997). Control of stump sprouting of *Populus tremuloides* and *P. grandidentata* by inoculation with *Chondrostereum purpureum*. *Biological Control* 10: 37-41.

Egley, G.H. and C.D. Boyette (1995). Water-corn oil emulsion enhances conidia germination and mycoherbicidal activity of *Colletotrichum truncatum*. *Weed Science* 43: 312-317.

Environmental Protection Agency (EPA) (2005). *Altemaria destruens* strain 059 (028301) fact sheet. Biopesticides active ingredient factsheet. Washington, DC: U.S. EPA. Available at <http://www.epa.gov/oppbppd1/biopesticides/ingredients/factsheets/factsheet_028816.htm>.

Frantzen, J., and P.E. Hatcher (1997). A fresh view on the control of the annual plant *Senecio vulgaris*. *Integrated Pest Management Reviews* 2: 77-85.

Frantzen, J., N.D. Paul, and H. Müller-Schärer (2001). The system management approach of biological weed control: Some theoretical considerations and aspects of application. *BioControl* 46: 139-155.

Garcia-Baudin, J.M., and H. Darmency (1979). Différences intraspécifiques chez *Convolvulus arvensis* L. *Weed Research* 19: 219-244.

Gosselin, L., R. Jobidon, and L. Bernier (1999). Genetic variability and structure of Canadian populations of *Chondrostereum purpureum*, a potential biophytocide. *Molecular Ecology* 8: 113-122.

Grant, N.T., E. Prusinkiewicz, R.M.D. Makowski, B. Holmström-Ruddick, and K. Mortensen (1990). Effect of selected pesticides on survival of *Colletotrichum gloeosporioides* f.sp. *malvae*, a bioherbicide for round-leaved mallow *(Malva pusilla)*. *Weed Technology* 4: 701-715.

Grant, N.T., E. Prusinkiewicz, K. Mortensen, and R.M.D. Makowski (1990). Herbicide interactions with *Colletotrichum gloeosporioides* f.sp. *malvae* a bioherbicide for round-leaved mallow *(Malva pusilla)* control. *Weed Technology* 4: 716-723.

Greaves, M.P., L. Dutton, and J. Lawrie (2000). Formulation of microbial herbicides. *Aspects of Applied Biology* 57: 171-178.

Greaves, M.P., P.J. Holloway, and B.A. Auld (1998). Formulation of microbial herbicides. In H.D. Burges (Ed.), *Formulation of microbial biopesticides: Beneficial microorganisms, nematodes, and seed treatments* (pp. 203-233). Dordrecht, the Netherlands: Kluwer Academic Publishers.

Greaves, M.P. and M.D. MacQueen (1992). Bioherbicides: Their role in tomorrow's agriculture. In I. Denholm, A.L. Devonshire, and D.W. Hollomon (Eds.), *Resistance '91: Achievement and developments in combating pesticide resistance* (pp. 295-306). London: Elsevier Applied Science.

Green, S. and K.L. Bailey (2000a). Effects of leaf maturity, infection site, and application rate of *Alternaria cirsinoxia* conidia on infection of Canada thistle *(Cirsium arvense)*. *Biological Control* 19: 167-174.

Green, S. and K.L. Bailey (2000b). Influence of moisture and temperature on infection of Canada thistle by *Alternaria cirsinoxia*. *Plant Disease* 84: 1126-1132.

Green, S., S.M. Stewart-Wade, G.J. Boland, M.P. Teshler, and S.H. Liu (1998). Formulating microorganisms for biological control of weeds. In G.J. Boland and L.D. Kuykendall (Eds.), *Plant-microbe interactions and biological control* (pp. 249-281). New York: Marcel Dekker, Inc.

Guntli, D., S. Burgos, I. Kump, M. Heeb, H.A. Pfirter, and G. Défago (1999). Biological control of hedge bindweed *(Calystegia sepium)* with *Stagonospora con-*

volvuli strain LA39 in combination with competition from red clover *(Trifolium pratense). Biological Control* 15: 252-258.

Hasan, S. and P.G. Ayres (1990). The control of weeds through fungi: Principles and prospects. *New Phytologist* 115: 201-222.

Hintz, W.E., E.M. Becker, and S.F. Shamoun (2001). Development of genetic markers for risk assessment of biological control agents. *Canadian Journal of Plant Pathology* 23: 13-18.

Hodgson, R.H., L.A. Wymore, A.K. Watson, R.H. Snyder, and A. Collette (1988). Efficacy of *Colletotrichum coccodes* and thidiazuron for velvetleaf *(Abutilon theophrasti)* control in soybean *(Glycine max). Weed Technology* 2: 473-480.

Hoeft, E.V., N. Jordan, J. Zhang, and D.L. Wyse (2001). Integrated cultural and biological control of Canada thistle in conservation tillage soybean. *Weed Science* 49: 642-646.

Holmström-Ruddick, B. and K. Mortensen (1995). Factors affecting the pathogenicity of a benomyl-resistant strain of *Colletotrichum gloeosporioides* f.sp. *malvae. Mycological Research* 99: 1108-1112.

Horak, M.J. and J.S. Holt (1986). Isozyme variability and breeding systems in populations of yellow nutsedge *(Cyperus esculentus). Weed Science* 35: 538-543.

Horak, M.J., J.S. Holt, and N.C. Ellstand (1987). Genetic variation in yellow nutsedge *(Cyperus esculentus). Weed Science* 35: 506-512.

Hurle, K. (1997). Concepts in weed control—How does biocontrol fit in? *Integrated Pest Management Reviews* 2: 87-89.

Hussaini, I.P., A.C. Lawrie and D.A. McLaren (2000). Pathogens on and variation in *Nasella trichotoma* in Australia. In N.R. Spencer (Ed.), *Proceedings of the Tenth International Symposium on Biological Control of Weeds* (pp. 269-280), July 4-14, 1999. Bozeman: Montana State University.

Imaizumi, S., T. Nishino, K. Miyabe, T. Fujimori, and M. Yamada (1997). Biological control of annual bluegrass (*Poa annua* L.) with a Japanese isolate of *Xanthomonas campestris* pv. *poae* (JT-P482). *Biological Control* 8: 7-14.

Johnson, D.R., D.L. Wyse, and K.J. Jones (1995). Efficacy of spring and fall applications of *Pseudomonas syringae* pv. *tagetis* for Canada thistle (*Cirsium arvense* L.) control in soybean. *Weed Science Society of America Abstracts* 35: 61.

Johnson, D.R., D.L. Wyse, and K.J. Jones (1996). Controlling weeds with phytopathogenic bacteria. *Weed Technology* 10: 621-624.

Jones, R.W. and J.G. Hancock (1990). Soilborne fungi for biological control of weeds. In R.E. Hoagland (Ed.), *Microbes and microbial products as herbicides* (pp. 276-286). Washington, DC: American Chemical Society.

Kadir, J.B., R. Charudattan, W.M. Stall, and T.A. Bewick (1999). Effect of *Dactylaria higginsii* on interference of *Cyperus rotundus* with *L. esculentum. Weed Science* 47: 682-686.

Kempenaar, C., P.J.F.M. Horsten, and P.C. Scheepens (1996). Growth and competitiveness of common lambsquarters *(Chenopodium album)* after foliar application of *Ascochyta caulina* as a mycoherbicide. *Weed Science* 44: 609-614.

Kennedy, A.C. (1996). Molecular biology of bacteria and fungi for biological control of weeds. In M. Gunasedaran and D.J. Weber (Eds.), *Molecular biology of the biological control of pests and diseases of plants* (pp. 155-172). Boca Raton, FL: CRC Press, Inc.

Kennedy, A.C., L.F. Elliott, F.L. Young, and C.L. Douglas (1991). Rhizobacteria suppressive to the weed downy brome. *Soil Science Society of America Journal* 55: 722-727.

Klein, T.A., B.A. Auld, and W. Fang (1995). Evaluation of oil suspension emulsions of *Colletotrichum orbiculare* as a mycoherbicide in field trials. *Crop Protection* 14: 193-197.

Kremer, R.J. and A.C. Kennedy (1996). Rhizobacteria as biocontrol agents of weeds. *Weed Technology* 10: 601-609.

Lawrie, J., M.P. Greaves, V.M. Down, and A. Chassot (1997). Some effects of spray droplet size on distribution, germination of and infection by mycoherbicide spores. *Aspects of Applied Biology* 48: 175-182.

Lawrie, J., M.P. Greaves, V.M. Down, and J.M. Lewis (1999). Effects of the plant-pathogenic fungus *Mycocentrospora acerina* (Hartig) Deighton on growth and competition of *Viola arvensis* (Murr.) in spring wheat. *Biocontrol Science and Technology* 9: 105-112.

Lawrie, J., M.P. Greaves, V.M. Down, and N.M. Western (2002). Studies of spray application of microbial herbicides in relation to conidial propagule content of spray droplets and retention on target. *Biocontrol Science and Technology* 12: 107-119.

Lawrie, J., M.P. Greaves, N.M. Western, and V. Down (1999). Application and formulation: Forgotten factors in the development of microbial herbicides. In *Proceedings of the 1999 Brighton Conference—Weeds* (pp. 289-296). Brighton, UK: British Crop Protection Council.

Li, J. and R.J. Kremer (2000). Rhizobacteria associated with weed seedlings in different cropping systems. *Weed Science* 48: 734-741.

Luo, Y. and D.O. TeBeest (1997). Infection components of wild-type and mutant strains of *Colletotrichum gloeosporioides* f.sp. *aeschynomene* on northern jointvetch. *Plant Disease* 81: 404-409.

Makowski, R.M.D. (1993). Effect of inoculum concentration, temperature, dew period, and plant growth stage on disease of round-leaved mallow and velvetleaf by *Colletotrichum gloeosporioides* f.sp. *malvae*. *Phytopathology* 83: 1229-1234.

Makowski, R.M.D. (1997). Foliar pathogens in weed biocontrol: Ecological and regulatory constraints. In D.A. Andow, D.W. Ragsdale, and R.F. Nyvall (Eds.), *Ecological interactions and biological control* (pp. 87-103). Boulder, CO: Westview Press.

Mason, P.G. and J.T. Huber (2002). *Biological control programmes in Canada, 1981-2000*. Wallingford, Oxon, UK: CABI Publishing.

Meyer, S.E., D.L. Nelson, and S. Clement (2001). Evidence for resistance polymorphism in the *Bromus tectorum-Ustilago bullata* pathosystem: Implications for biocontrol. *Canadian Journal of Plant Pathology* 23: 19-27.

Morris, M.J., A.R. Wood, and A. Den Breeÿen (1998). Development and registration of a fungal inoculant to prevent regrowth of cut wattle tree stump in South Africa, and a brief overview of other bioherbicide projects currently in progress.

In *Proccedings of the Fourth International Bioherbicide Workshop: Programme and Abstracts* (p. 15), August 6-7. Glasgow, Scotland: University of Strathclyde.

Mortensen, K. (1988). The potential of an endemic fungus, *Colletotrichum gloeosporioides*, for biological control of round-leaved mallow *(Malva pusilla)* and velvetleaf *(Abutilon theophrasti). Weed Science* 36: 473-478.

Mortensen, K. (1998). Biological control of weeds using microorganisms. In G.J. Boland and L.D. Kuykendall (Eds.), *Plant-microbe interactions and biological control* (pp. 223-248). New York: Marcel Dekker Inc.

Mortensen, K. and E.J. Hogue (1995). *Sclerotinia sclerotiorum* as a potential biological control agent for diffuse knapweed on dry rangeland in interior British Columbia. In E.S. Delfosse and R.R. Scott (Eds.), *Proceedings of Eighth International Symposium on Biological Control of Weeds* (pp. 397-406), February 2-7, 1992, Lincoln University, Cantebury, New Zealand. Melbourne, Australia: CSIRO Publishing.

Müller-Schärer, H. and J. Frantzen (1996). An emerging system management approach for biological weed control in crops: *Senecio vulgaris* as a research model. *Weed Research* 36: 483-491.

O'Donovan, J.T., E.A. de St. Remy, P.A. O'Sullivan, D.A. Dew, and A.K. Sharma (1985). Influence of the relative time of emergence of wild oat *(Avena fatua)* on yiled loss of barley *(Hordeum vlugare)* and wheat *(Triticum aestivum). Weed Science* 33: 498-503.

Okoli, C.A.N., D.G. Shilling, R.L. Smith, and T.A. Bewick (1997). Genetic diversity in purple nutsedge (*Cyperus rotundus* L.) and yellow nutsedge (*C. esculentus* L.) *Biological Control* 8: 111-118.

Paul, N.D. and P.G. Ayres (1986). Interference between healthy and rust-infected groundsel (*Senecio vulgaris* L.) within mixed populations of different densities and proportions. *New Phytologist* 104: 257-269.

Paul, N.D. and P.G. Ayres (1987). Effects of rust infection of *Senecio vulgaris* on competition with lettuce. *Weed Research* 27: 431-441.

Peng, G., T.M. Wolf, K.N. Byer, and B. Caldwell (2001). Spray retention on green foxtail (*Setaria viridis*) using airbrush and broadcast sprayers and its impact on the efficacy of a mycoherbicide agent. In H.W. Ni and G.Y. Zhen (Eds.), *The Proceedings of the 18th Asian-Pacific Weed Science Society Conference* (pp. 699-706), May 28-June 2. Beijing, China: Standards Press of China.

Phatak, S.C. (1984). Knock out nutsedge. *American Vegetable Grower* 32: 44-46.

Phatak, S.C., M.B. Callaway, and C.S. Vavrina (1987). Biological control and its integration in weed management systems for purple and yellow nutsedge (*Cyperus rotundus* and *C. esculentus*). *Weed Technology* 1: 84-91.

Phatak, S.C., H.D. Wells, D.R. Sumner, D.K. Bell, and N.C. Glaze (1983). Biological control of yellow nutsedge with the indigenous rust fungus *Puccinia canaliculata. Science* 219: 1446-1447.

Poppenborg, L., K. Friehs, and E. Flaschel (1997). The green fluorescent protein is a versatile reporter for bioprocess monitoring. *Journal of Biotechnology* 58: 79-88.

Prasad, R. (1994). Influence of several pesticides and adjuvants on *Chondrostereum purpureum*—A bioherbicide agent for control of forest weeds. *Weed Technology* 8: 445-449.

Ramsfield, T.D., E.M. Becker, S.M. Rathlef, Y. Tang, T.C. Vrain, S.F. Shamoun, and W.E. Hintz (1996). Geographic variation of *Chondrostereum purpureum* detected by polymorphisms in the ribosomal DNA. *Canadian Journal of Botany* 74: 1919-1929.

Ramsfield, T.D., S.F. Shamoun, Z.K. Punja, and W.E. Hintz (1999). Variation in the mitochondrial DNA of the potential biological control agent *Chondrostereum purpureum. Canadian Journal of Botany* 77: 1490-1498.

Riddle, G.E., L.L. Burpee, and G.J. Boland (1991). Virulence of *Sclerotinia sclerotiorum* and *S. minor* on dandelion. *Weed Science* 39: 109-118.

Rosskopf, E.N., R. Charudattan, and J.B. Kadir (1999). Use of plant pathogens in weed control. In T.S. Bellows and T.W. Fisher (Eds.), *Handbook of biological control* (pp. 891-918). San Diego, CA: Academic Press.

Rosskopf, E.N., J.F. Gaffney, and R. Charudattan (1997). The effect of spray propellant on the efficacy of bioherbicide candidates. *Weed Science Society of America Abstracts* 37: 62.

Sands, D.C., E.J. Ford, and R.V. Miller (1990). Genetic manipulation of broad host-range fungi for biological control of weeds. *Weed Technology* 4: 471-474.

Sands, D.C., E.J. Ford, R.V. Miller, B.K. Sally, M.K. McCarthy, T.W. Anderson, M.B. Weaver, C.T. Morgan, and A.L. Pilgeram (1997). Characterization of a vascular wilt of *Erythroxylum coca* caused by *Fusarium oxysporum* f.sp. *erythroxyli* forma specialis nova. *Plant Disease* 81: 501-504.

Sands, D.C. and R.V. Miller (1993). Altering the host range of mycoherbicides by genetic manipulation. In S.O. Duke, J.J. Menn, and J.R. Plimmer (Eds.), *Pest control with enhanced environmental safety* (pp. 101-109). Washington, DC: American Chemical Society.

Sands, D.C., R.V. Miller, and E.J. Ford (1990). Biotechnological approaches to control of weeds with pathogens. In R.E. Hoagland (Ed.), *Microbes and microbial products as herbicides* (pp. 184-190). Washington, DC: American Chemical Society.

Schaad, N.W., W. Song, S. Hutcheson, and F. Dane (2001). Gene tagging systems for polymerase chain reaction based monitoring of bacteria released for biological control of weeds. *Canadian Journal of Plant Pathology* 23: 36-41.

Scheepens, P.C. and A. Hoogerbrugge (1991). Host specificity of *Puccinia caniliculata,* a potential biological control agent for *Cyperus esculentus. Netherlands Journal of Plant Pathology* 97: 245-250.

Scheepens, P.C., H. Müller-Schärer, and C. Kempenaar (2001). Opportunities for biological weed control in Europe. *BioControl* 46: 127-138.

Shabana, Y.M., R. Charudattan, J.T. DeValerio, and M.A. Elwakil (1997). An evaluation of hydrophilic polymers for formulating the bioherbicide agents *Alternaria cassiae* and *A. eichhorniae. Weed Technology* 11: 212-220.

Shamoun, S.F., T.D. Ramsfield, G. Shrimpton, and W.E. Hintz (1996). Development of *Chondrostereum purpureum* as a mycoherbicide for red alder *(Alnus rubra)* in utility rights-of-way (Abstr.). In P. Comeau and G. Harper (Eds.), *Proceedings of the National Meeting of the Expert Committee on Weeds* (p. 199), December 9-12. Victoria, BC: Expert Committee on Weeds.

Shaw, W.C. (1982). Integrated weed management systems technology for pest management. *Weed Science* 30(Supplement 1): 2-12.

Smith, R.J. Jr. (1991). Integration of biological control agents with chemical pesticides. In D.O. TeBeest (Ed.), *Microbial control of weeds* (pp. 189-208). New York: Chapman and Hall.

Stroo, H.F., L.F. Elliott, and R.I. Papendick (1988). Growth, survival and toxin production of root-inhibitory pseudomonads on crop residues. *Soil Biology and Biochemistry* 20: 201-207.

Swanton, C.J., K.N. Harker, and R.L. Anderson (1993). Crop losses due to weeds in Canada. *Weed Technology* 7: 537-542.

Swanton, C.J. and S.F. Weise (1991). Integrated weed management: The rationale and approach. *Weed Technology* 5: 657-663.

TeBeest, D.O. (1982). Survival of *Colletotrichum gloeosporioides* f.sp. *aeschynomene* in rice irrigation water and soil. *Plant Disease* 66: 469-472.

TeBeest, D.O. (1984). Induction of tolerance to benomyl in *Colletotrichum gloeosporioides* f.sp. *aeschynomene* by ethyl methanesulfonate. *Phytopathology* 74: 864.

TeBeest, D.O. (1991). Ecology and epidemiology of fungal plant pathogens studied as biological control agents of weeds. In D.O. TeBeest (Ed.), *Microbial control of weeds* (pp. 97-114). New York: Chapman and Hall.

TeBeest, D.O. (1996). Biological control of weeds with plant pathogens and microbial pesticides. *Advances in Agronomy* 56: 115-137.

TeBeest, D.O. and G.E. Templeton (1985). Mycoherbicides: Progress in the biological control of weeds. *Plant Disease* 69: 6-10.

TeBeest, D.O., X.B. Yang, and C.R. Cisar (1992) The status of biological control of weeds with fungal pathogens. *Annual Review of Phytopathology* 30: 637-657.

Tessmann, D.J., R. Charudattan, H.C. Kistler, and E.N. Rosskopf (2001). A molecular characterization of *Cercospora* species pathogenic to water hyacinth and emendation of *C. piaropi*. *Mycologia* 99: 1108-1112.

Van Acker, R.C., D.A. Derksen, M.H. Entz, G. Martens, T. Andrews, and O. Nazarko (2001). Pesticide-free production: A reason to implement integrated weed management. In R.E. Blackshaw and L.M. Hall (Eds.), *Integrated weed management: Explore the potential* (pp. 61-73). Saint-Anne-de-Bellevue, Québec: Expert Committee on Weeds.

Vogelgsang, S., A.K. Watson, A. DiTommaso, and K. Hurle (1999). Susceptibility of various accessions of *Convolvulus arvensis* to *Phomopsis convolvulus*. *Biological Control* 15: 25-32.

Walker, H.L. and W.J. Connick Jr. (1983). Sodium alginate for production and formulation of mycoherbicides. *Weed Science* 31: 333-338.

Walker, R.H. and G.A. Buchanan (1982). Crop manipulation in integrated weed management systems. *Weed Science* 30(Supplement 1): 17-24.

Wall, R.E. (1994). Biological control of red alder using stem treatments with the fungus *Chondrostereum purpureum*. *Canadian Journal of Forest Research* 24: 1527-1530.

Watson, A.K. (1991). The classical approach with plant pathogens. In D.O. TeBeest (Ed.), *Microbial control of weeds* (pp. 3-23). New York: Chapman and Hall.

Weaver, S.E. (1984). Critical period of weed competition in three vegetable crops in relation to management practices. *Weed Research* 24: 317-325.

Weidemann, G.J. (1988). Effects of nutritional amendments on conidial production of *Fusarium solani* f.sp. *cucurbitae* on sodium alginate granules and on control of Texas gourd. *Plant Disease* 72: 757-759.

Womack, J.G. and M.N. Burge (1993). Mycoherbicide formulation and the potential for bracken control. *Pesticide Science* 37: 337-341.

Wymore, L.A., C. Poirier, A.K. Watson, and A.R. Gotlieb (1988). *Colletotrichum coccodes*, a potential bioherbicide for control of velvetleaf *(Abutilon theophrasti)*. *Plant Disease* 72: 534-538.

Wymore, L.A., A.K. Watson, and A.R. Gotlieb (1987). Interaction between *Colletotrichum coccodes* and thidiazuron for control of velvetleaf *(Abutilon theophrasti)*. *Weed Science* 35: 377-383.

Zidack, N.K. and P.A. Backman (1996). Biological control of kudzu *(Pueraria lobata)* with the plant pathogen *Pseudomonas syringae* pv. *phaseolicola*. *Weed Science* 44: 645-649.

Zidack, N.K., P.A. Backman, and J.J. Shaw (1992). Promotion of bacterial infection of leaves by an organosilicone surfactant: Implications for biological weed control. *Biological Control* 2: 111-117.

Zidack, N.K. and P.C. Quimby Jr. (1999). Formulation and application of plant pathogens for biological weed control. In F.R. Hall and J.J. Menn (Eds.), *Methods in biotechnology,* Volume 5, *Biopesticides: Use and delivery* (pp. 371-381). Totowa, NJ: Humana Press.

Zidack, N.K. and P.C. Quimby Jr. (2002). Formulation of bacteria for biological weed control using the stabileze method. *Biocontrol Science and Technology* 12: 67-74.

PHOTO 5.1. Sweet corn grown without cover crop (A); with hairy vetch (B) or rye + vetch (C) residues. Weed suppression by hairy vetch residue (D) is short-lived compared to rye + vetch (E).

PHOTO 5.2. Tomato and summer squash grown in rye-vetch mixed cover crop, Fayetteville, Arkansas. Yellow nutsedge infestation is apparent in plots without cover crop (A). Thick rye-vetch residue suppresses yellow nutsedge (B). Summer squash in a bare-ground plot (C) has less luxuriant vegetative growth than that in rye-vetch residue (D).

PHOTO 16.1. Weedy thimbleberry (*Rubus parviflorus* Nutt.) rapidly invading a young conifer regenerating site after harvesting and effectively competing with planted young conifer species and reducing the plantation productivity.

PHOTO 16.2. Severe foliar damage on *Rubus strigosus* Michx. = *R. idaeus* var. *strigosus* (Michx.) Focke, caused by the potential biocontrol agent *Fusarium avenaceum* (Fr.) Sacc.

PHOTO 16.3. Young conifer regenerating site in coastal British Columbia dominated by competing weedy salmonberry (*Rubus spectabilis* Pursh). Note that the young planted seedlings of Douglas fir [*Pseudotsuga menziesii* (Mirb.) Franco] were allowed to grow to "free-to-grow" stage after chemical and manual brushing treatments of *R. spectabilis*.

PHOTO 16.4. Weedy salal (*Gaultheria shallon* Pursh) rapidly invading a young conifer regenerating site after harvesting and effectively competing with planted young conifer species and reducing the plantation productivity.

PHOTO 16.5. Natural infection of salal by the potential biocontrol agent *Valdensinia heterodoxa* Peyronel, which results in severe foliar necroses and premature defoliation of the salal.

PHOTO 16.6. Inoculation of salal with *Valdensinia heterodoxa* (isolate PFC 2761) under controlled conditions (left) and untreated salal (right).

PHOTO 16.7. Basidiocarps "fruiting bodies" of *Chondrostereum purpureum* infecting red alder (*Alnus rubra* Bong.) stump.

PHOTO 16.8. Female plant of the western hemlock dwarf mistletoe [*Arecuthobium tsugense* (Rosendahl) G.N. Jones] infecting western hemlock [*Tsuga* heterophylla (Raf.) Sarg.] branch.

PHOTO 16.9. *Colletotrichum gloeosporiodes* (Penz.) Penz. & Sacc. infecting shoots and berries of *Arecuthobium tsugense.*

PHOTO 16.10. Perithecia of *Neonectria neomacrospora* C. Broth & Samuels, infecting the basal cup and swelling (endophytic system) of *Arecuthobium tsugense.*

PHOTO 16.11. Formulated paste product of *C. Purpureum* (isolate PFC 2139) as Chontrol packaged in a plastic squeeze bottle.

PHOTO 16.12. Application of *C. purpureum* by a topical treatment of cut stumps of big leaf maple (*Acer macrophyllum* Pursh).

PHOTO 17.1. *Chenopodium album* (A) and in a maize field (B).

PHOTO 17.2. Leaf puncture assay of *Ascochyta caulina* culture filtrate on *Chenopodium album*.

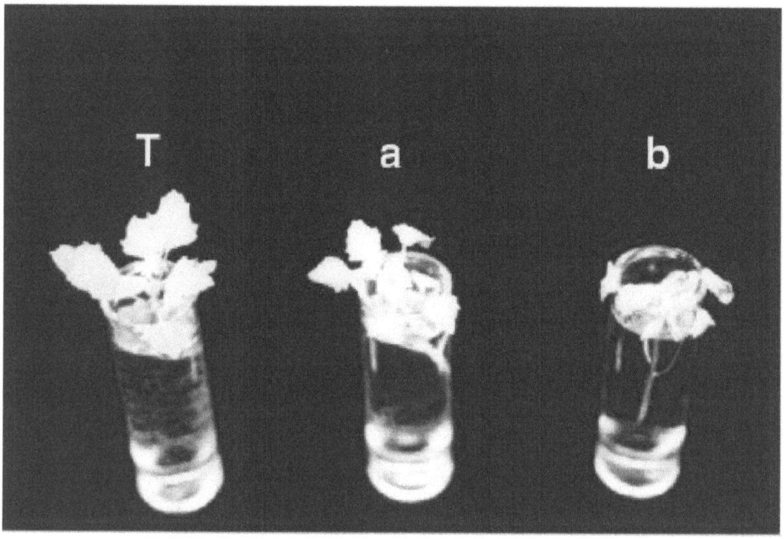

PHOTO 17.3. Effect of ascaulitoxin (a), *Ascochyta caulina* culture filtrate (b) on plant cuttings of *Chenopodium album* (T = test control).

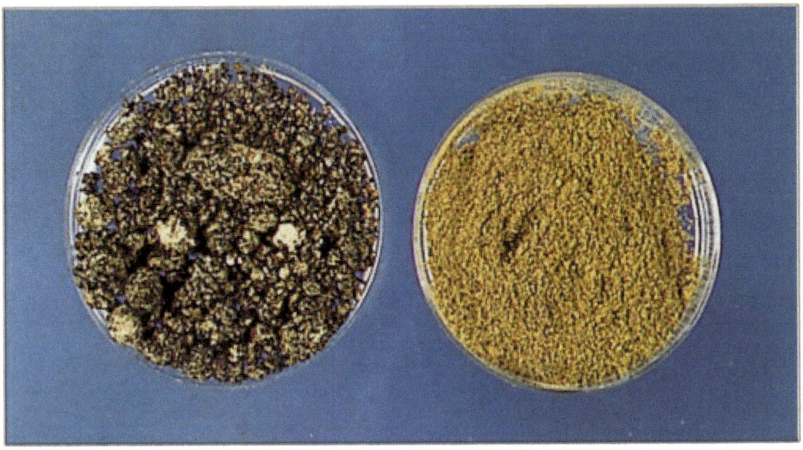

PHOTO 17.4. The *Ascochyta caulina* culture on bran (left) and the powder of the toxin mixture obtained from the purification of the liquid culture filtrates of the same fungus (right).

PHOTO 17.5. Effect of application of toxin mixture on *C. album* plants grown in greenhouse in pots: alone (left), together with *A. caulina* (right), *A. caulina* alone (center).

PHOTO 21.1. Different *Orobanche* spp. (*O. cumana, O. crenata, O. cernua, O. aegyptiaca, O. ramosa, O. minor, O. lutea, O. gracilis,* and *O. foetida,* from left to right and top to bottom). (*Source:* Adapted from Linke, Sauerborn, and Saxena, 1996. Used by permission of International Center for Agricultural Research in the Dry Areas.)

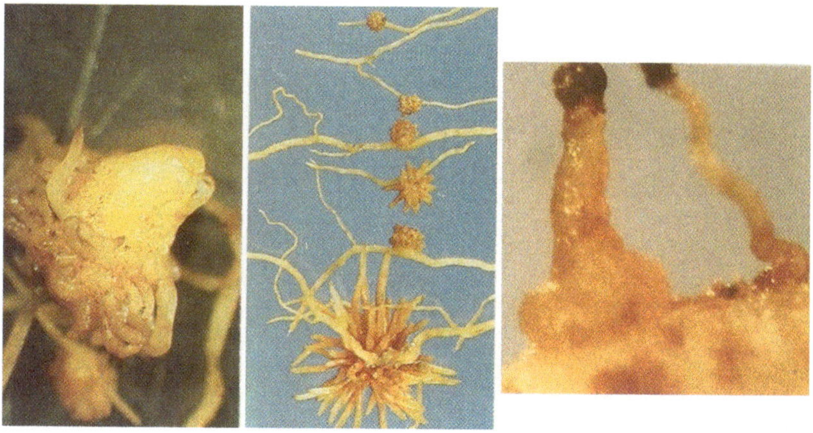

PHOTO 21.2. *Orobanche* haustoria at different developmental stages. (Used by permission of International Center for Agricultural Research in the Dry Areas, Aleppo, Syria.)

PHOTO 21.3. *Cuscuta campestris* spread on weed species along roadsides.

PHOTO 21.4. *Cuscuta monogyna* Vahl. (eastern dodder) on lemon [*Citrus limon* (L.) Burm.f.] in Jordan.

PHOTO 21.5. Oriental mistletoe (*Viscum cruciatum* Sieb.) on wild almond (*Amygdalus communis* L.).

PHOTO 21.6. *Loranthus acaciea* Zucc. parasitizing Christ's thorn [*Zizyphus spinachristi* (L.) Desf.].

PHOTO 21.7. *Loranthus acaciea* parasitizing Chinese date (*Zizyphus jujuba* Mill.).

PHOTO 21.8. *Osyris alba* parasitizing almond *(Amygdalus communis)*.

PHOTO 21.9. *Cistanche tubulosa.*

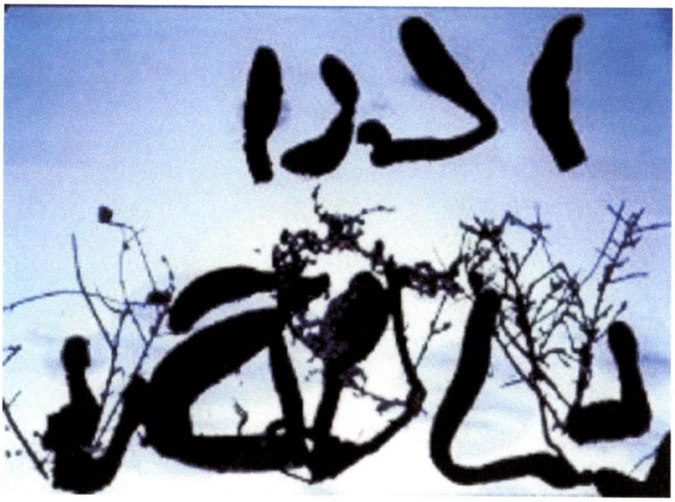

PHOTO 21.10. *Cynomorium coccineum* collected from *Salsola vermiculata* plants.

Chapter 15

Developing Microbial Weed Control Products: Commercial, Biological, and Technological Considerations

Karen L. Bailey
Edmund K. Mupondwa

INTRODUCTION

Biological control of weeds is a pest management practice that encompasses dynamic interactions among several independent, living systems. These systems consist of a target weed, a biological control agent, and the other nontarget plants (crops) and organisms in the area of application. An array of environmental conditions encountered by these living systems at any one location or in any given year may cause significant variation in the predicted responses (Kennedy and Kremer, 1996; Kremer, 1998). Despite this complexity, our expectations for biological control are to provide sufficient harm to the weed in a consistent manner to reduce the potential economic damage while not harming the other components in the system. Control is primarily measured by weed mortality, by the suppression of weed growth, or by the inhibition of life cycle development.

Inundative application methods, whereby a high propagule density of a microbial agent is released on a target weed to achieve rapid control, have thrust this type of biological control into the realm of product development and commercialization. It was assumed that inundative biological control could parallel the development and commercialization models of synthetic chemicals. The result has been successful commercialization of some microorganisms using the chemical pesticide paradigm, but many others have failed. By the early 1990s, over 100 organisms were described in the litera-

We sincerely thank Gloria Falk and Barbara Gradin of Agriculture & Agri-Food Canada, Saskatoon Research Centre, for their review of this manuscript. At the time of the writing of this paper, the authors were not recipients of industry funding.

ture as having potential for inundative biological control of weeds, but only two (DeVine and Collego) were successfully commercialized (Zorner, Evans, and Savage, 1993). By the twenty-first century, 8 pathogens were registered and/or commercialized for inundative biological weed control and another 54 have potential for further product development (Charudattan, 2001).

Several reviews have discussed the type of constraints encountered in the development of microbial weed control products during the 1980s and early 1990s (Charudattan, 1991; Greaves and MacQueen, 1992; Auld and Morin, 1995; Mortensen, 1998). These constraints have been described as follows:

1. *biological,* which includes factors such as host range limitations, host variability, resistance mechanisms in the host, interactions with other microorganisms, and compatibility with other pest management practices,
2. *environmental,* which are affected by moisture, temperature, and nutrient availability,
3. *technological,* which includes mass production methods, formulation, storage, and shelf life, and
4. *commercial,* which are affected by market size, protection of intellectual property, cost of production, and the regulatory process.

From an industry perspective, the most limiting constraints have been primarily technological and commercial, and to some degree biological. Inappropriate screening procedures that do not select a sufficient number of organisms adapted for the environment where it is applied and inconsistent performance of the microorganisms have been among the biological problems (Zorner, Evans, and Savage, 1993; Campbell, 1994; Marrone, 1999). For the development of microbial weed control products, the fermentation, stabilization, and delivery technologies have been the most limiting because the current technologies are not cost-effective, do not maintain agent viability and longevity, and cannot improve their ability to work consistently in a field environment (Zorner, Evans, and Savage, 1993). Furthermore, the regulatory process may lengthen the duration of the development process and ultimately associated costs. Commercialization is most often driven by the market rather than the product, so commercial success depends on asking the right questions, determining market potential and clients, and defining the limits for acceptable costs of the retail product (Cross and Polonenko, 1996).

With hindsight, both the problematic and the successful experiences encountered in the development of microbial weed control products, as discussed by Templeton, Smith, and TeBeest (1989), Zorner, Evans, and Savage (1993), and Cross and Polonenko (1996), provide a basis for assessing the approaches taken for evaluating the scientific and commercial potential. When industrial commercialization is the purpose, research and development programs conducted by basic researchers should also take into consideration those factors deemed important by the industry segment. This chapter examines the commercial, biological, and technological factors considered important by industry and discusses approaches that may help improve the chances of successful product commercialization.

COMMERCIAL CONSIDERATIONS

Marketplace Economics

Market Size: A Global Overview of the Pesticide Industry

To understand the market potential for microbial weed control products, it is useful to present an overview of the global pesticide industry. Available data for 1998 estimate the world market for pesticides at $38 billion (FAO various issues; Aspelin and Grube 1999; Mupondwa 2001). Figure 15.1 illustrates the size of the global pesticide market by region. The United States is the largest market and represents 33 percent of world sales, followed by Western Europe (25 percent), Asia (18 percent), and Latin Amer-

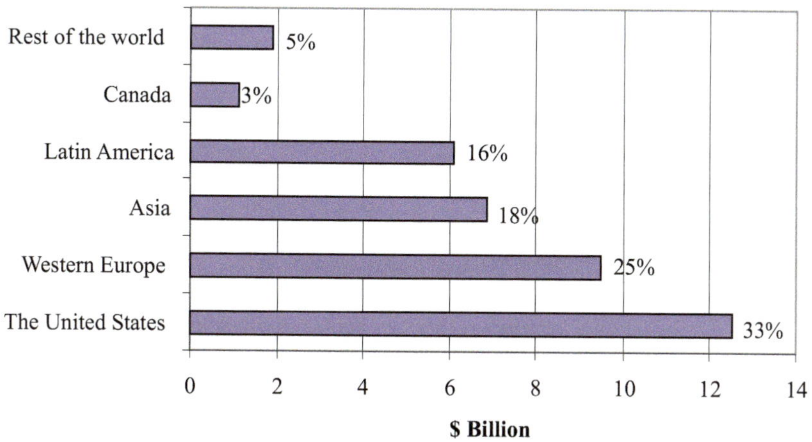

FIGURE 15.1. Global pesticide sales and percentage shares of global pesticide market.

ica (16 percent). The rest of the world accounted for the remaining 8 percent. According to the World Resources Institute (1998), Africa, Asia, and Latin America are emerging as major important markets for pesticides.

Herbicides account for the highest proportion of pesticide sales in the world (43 percent of total pesticides used), underscoring their importance in crop protection (see Figure 15.2). The United States is the largest market for herbicides and accounts for 39 percent of the global herbicide sales. This is followed by insecticides, which account for 33 percent. Insecticides are the second-largest sector in both sales volume and sales value (FAO various issues).

The pesticide industry is highly dependent on agriculture for its viability, as illustrated in Figure 15.3 using 1997 data from Canada and the United States. In Canada, agriculture accounts for over 90 percent of the value of all pesticides sold, with herbicides accounting for 80 percent. In the United States, agricultural needs accounted for over 70 percent of the value of all pesticides sold, with herbicides accounting for 82 percent (Crop Protection Institute of Canada, 2000; Environmental Protection Agency [EPA], 1998; U.S. Department of Agriculture, 1999). If one disaggregates herbicide use data in Canada and United States by crop, an interesting picture emerges. In Canada, wheat and canola (*Brassica napus* L.) account for 95 percent of herbicide applications. In the United States, corn is the largest crop, averaging 30 million hectares and exceeding any other crop in number of hectares treated with herbicides. This is followed by soybeans.

In terms of industry structure, it is well-known that the world pesticide industry is highly concentrated, with significant barriers to entry, and an extreme dependence on agriculture. Fewer than 100 firms supply a large num-

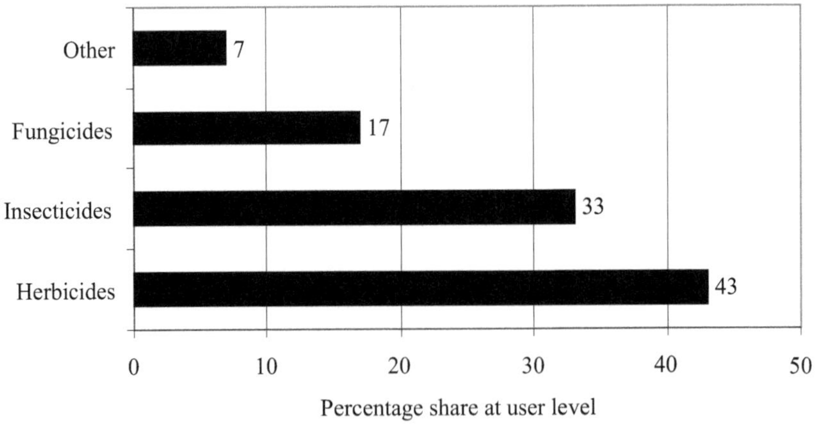

FIGURE 15.2. World pesticide sales and percent share at user level in 1997.

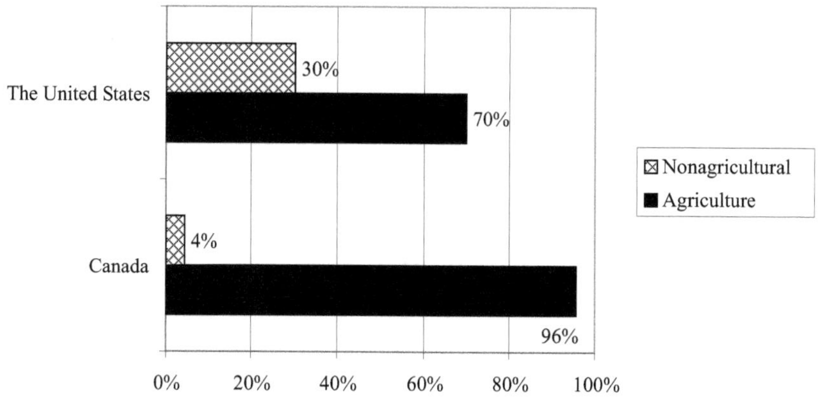

FIGURE 15.3. Annual pesticide usage by sector in Canada and the United States, 1999 data.

ber of active ingredients. These firms are characterized by a large degree of vertical integration, from production of formulations to basic materials. Industry Canada (1998) estimates that 15 firms in the United States supply 40 percent of global pesticide requirements. Table 15.1 presents the top ten companies that account for 80 percent of world sales.

Biopesticide Market Overview

Bioherbicides belong to a class of biological crop protection products called biopesticides. Based on use, biological crop protection plays a relatively minor role in crop protection. Currently, the global biopesticide market is valued at $400 million, or 1.5 percent of the total (Navon, 2000; FAO various issues; Gaugler, 1997). However, at present, over 90 percent of the world biopesticide market is dominated by a single product type, the bioinsecticide made from *Bacillus thuringiensis (Bt)* marketed for control of insect pests. Notwithstanding their small market share, biopesticides have a key distinguishing feature: they are projected to grow at 10 to 15 percent per annum (from $460 million in 2000 to approximately $2 to $4 billion) in the next 10 to 15 years compared to only 2 percent growth for chemical pesticides (Menn, 1996; Navon, 2000). The computed projected market shares based on these growth rates are highlighted in Figure 15.4. These growth rates for biopesticides are predicated on a number of factors that are discussed later in this section. Before discussing some of these factors, let us look at the commercial status of available bioherbicides.

TABLE 15.1. Top ten agrochemical companies.

Company	1998 sales $ million	Company percent share of total world sales
Novartis[a]	4,199	13.5
Monsanto	3,126	10.1
AstraZeneca[a] (Syngenta)	2,674	8.6
DuPont	2,518	8.1
AgrEvo/Aventis[b]	2,352	7.6
Bayer	2,254	7.3
Rhône-Poulenc[b]	2,202	7.1
American Cyanamid	2,119	6.8
Dow AgroSciences	2,050	6.6
BASF	1,855	6.0
Total top ten	25,349	81.7
World total	31,000	100.0

Source: Adapted from *Chemical Week,* May 12, 1999; Corporate and Security Exchange Information, various sources.
[a]Syngenta formed December 2000 via agribusiness merger of Novartis and AstraZeneca.
[b]Formed December 1999 via merger of Hoechst and Rhône-Poulenc life science businesses.

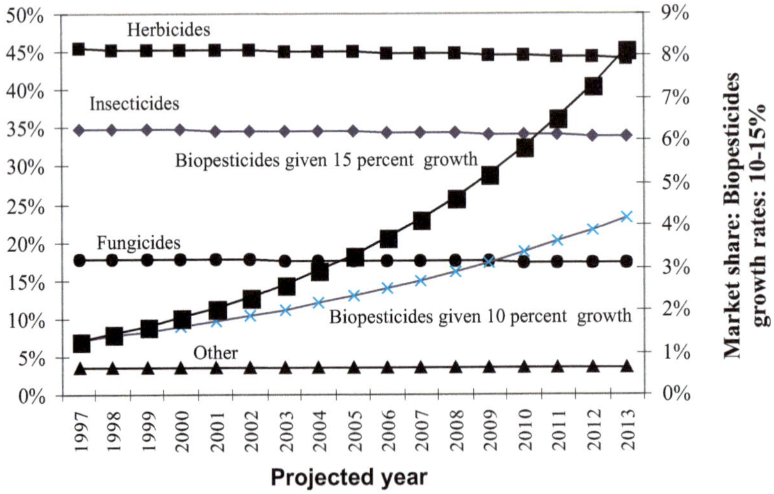

FIGURE 15.4. Projected growth in market shares: Synthetic pesticides versus biopesticides.

Commercial Status of Bioherbicides

Two fungal weed control products, DeVine and Collego, were first commercialized in 1982, following registration and labeling by the U.S. Environmental Protection Agency. The registration of these products generated tremendous interest in the bioherbicide concept. DeVine (from Abbott Laboratories, Chicago, Illinois), was registered in 1981 and was the first commercially available mycoherbicide (Kenney, 1986; Charudattan, 1991). It is a formulation of the endemic soilborne plant pathogen *Phytophthora palmivora* for control of strangler vine [*Morrenia odorata* (Hook & Arn.) Lindl.] in Florida citrus groves. The isolation was made from a dying strangler vine in 1973 (Burnett et al., 1973). Following isolation, Abbott Laboratories began cooperative research in 1977 with the Department of Plant Industries, Florida Department of Agriculture, with the goal of developing *P. palmivora* as a commercial bioherbicide. It is applied as a liquid containing chlamydospores and is capable of providing up to 96 percent weed control within ten weeks (Ridings, 1986). DeVine is considered an example of a successful mycoherbicide, and licensing of it has been recently acquired by Encore Technologies, Minnetonka, Minnesota. Collego was developed as a postemergence bioherbicide for control of northern jointvetch [*Aeschynomene virginica* (L.) B.S.P.], a leguminous weed in rice and soybeans, found especially in Arkansas (Bowers, 1986). It was developed from an isolate of *Colletotrichum gloeosporioides* f.sp. *aeschynomene* in a cooperative program between the University of Arkansas, the USDA, and The Upjohn Company. Currently, it is commercialized by Pharma and Upjohn Company, Kalamazoo, Michigan, and Encore Technologies of Minnetonka, Minnesota. Both fungi offered a number of positive attributes for weed control, including (1) high specificity for target weed; (2) lack of toxicity to crop plants or other nontarget organisms in the environment; and (3) relatively low cost of production.

The excellent efficacy of both DeVine and Collego led to very high expectations for other biological agents. In spite of this initial success, very few bioherbicides have been commercialized (Table 15.2). Between 1980 and 2000, only seven fungi and one bacterium were registered as bioherbicides in Canada (BioMal), the United States (DeVine, Collego, Dr. BioSedge), Japan (Camperico), the Netherlands (Biochon), and South Africa (Stumpout, Hakatak). In Canada, BioMal was the first bioherbicide to be registered. It was registered in 1985 by Philom Bios (Saskatoon, Saskatchewan) for the control of round-leaved mallow (*Malva pusilla* Sm.) in Canada and the United States. BioMal contains the fungus *C. gloeosporioides* f.sp. *malvae* and is based on a technology license from Agricul-

TABLE 15.2. List of registered and nonregistered bioherbicides in 2002.

Control agent	Trade name	Formulation	Target weed	Target crop	Company
Phytophthora palmivora	DeVine[a]	Liquid	Strangler vine [*Morrenia odorata* (H. & A.) Lindl.]	Citrus	Abbott Labs
Colletotrichum gloeosporioides f.sp *aeschynomene*	Collego[a]	Dry powder	Northern jointvetch [*Aeschynomene virginica* (L.) B.S.P.]	Rice, soybean	Encore Technologies
C. gloeosporioides f.sp *malvae*	BioMal[a] (proposed name Mallet)	Wettable powder	Round leaved mallow (*Malva pusilla* Sm.)	Various	Philom Bios (new licensee, Encore Technologies)
Puccinia canaliculata	Dr. BioSedge[a]	N//A	Yellow nutsedge (*Cyperus esculentus* L.)	Various	Not produced
C. gloeosporioides	LUBOA II[b]	Granular mixture	Dodder (*Cuscuta* spp.)	Various	Ningxia Region, China
C. gloeosporioides	Hakatak[a]	Granular	Silky hakea (*Hakea sericea* Schrad.)	Tree plantation	PPRI, South Africa
Cylindrobasidium laeve	Stumpout[a]		*Acacia* spp.	Tree plantation	PPRI, South Africa
C. coccodes	Velgo[b]	Water + sorbitol	Velvetleaf (*Abutilon theophrasti* Medik.)	Corn, soybean	Not produced
Alternaria cassiae	CASST[b]	Water + surfactant	Sicklepod (*Cassia obtusifolia* L.)	Soybean, cotton	Mycogen
Pseudomonas syringae pv. *tagetis*	NA[b]	Silwet L-77	Canada thistle [*Cirsium arvense* (L.) Scop.]	Various	Encore Technologies
Xathomonas campestris pv. *poaea*	Camperico[a]	N/A	Annual bluegrass (*Poa annua* L.)	Turfgrass	Japan Tobacco
Chondrostereum purpureum	Biochon[a], Myco-Tech[a]	Dry powder, paste	Broadleaf trees	Tree plantation	Koppert Biologicals, Myco-Forestis Corp.

[a]Registered herbicide.
[b]Nonregistered herbicide.

ture and Agri-Food Canada. However, in 1994, Philom Bios made a decision to suspend commercialization due to market limitations (Philom Bios, personal communication, 1994). Subsequently, the licence for its production was acquired by Encore Technologies (Minnetonka, Minnesota), but further development was suspended in 2002. The next bioherbicide registered was Dr. BioSedge by Tifton Innovation Corporation (Tifton, Georgia). It is derived from the fungus *Puccinia canaliculata* and was registered in the United States for control of yellow nutsedge (*Cyperus esculentus* L.). Dr. BioSedge is not commercially available. Camperico represents a bioherbicide targeted at the nonagricultural niche market. It is used in golf courses to control annual bluegrass (*Poa annua* L.). It is registered to Japan Tobacco Inc. of Yokohama. Camperico is based on an isolate of *Xanthomonas campestris,* a wilt-inducing bacterium. Biochon is based on the fungus *Chondrostereum purpureum* and is targeted at broadleaf trees in plantations for suppression of stump sprouting in blackberry (De Jong, de Scheepens, and Zadoks, 1990; Pilgeram and Sands, 1999). Under conditions of wounding, it also attacks the vascular (phloem) system of hardwood weeds, such as quaking aspen, alders, birch, pincherry, and maples. Biochon is marketed by Koppert Biological Systems of the Netherlands, and it has been used by the Canadian Forest Service to biologically control forest weeds such as wild cherry (*Prunus serotina* Ehrh.). Koppert Biological Systems has branches in Canada, France, Spain, the United States, Italy, the United Kingdom, New Zealand, and the Canary Islands for the marketing of Biochon, in addition to its other line of products, including the fungus *Trichoderma viride* which limits the spread of *C. purpureum* as a tree pathogen in orchards (www.koppertonline.ca/aboutus.asp). In 2002, Myco-Tech Paste (containing a naturally occuring strain of *Chondrostereum purpureum*) was approved for the inhibition of sprouting and regrowth of deciduous trees in rights-of-way and conifer release management programs by the Pest Management Regulatory Agency in Canada (www.pmra-arla.gc.ca/).

A number of microbial candidates that are known to be both safe and efficacious have not progressed beyond the R&D phase to be adopted by farmers and other users. In the next section, we address key economic factors that have impeded the adoption rate of bioherbicides.

Market Size

Market size has been recognized as a critical impediment to the success of bioherbicides in the marketplace ever since the first bioherbicides were registered and commercialized (Bowers, 1986; Charudattan, 1991, 2000a,b; Auld and Morin, 1995; Georgis, 1997; Mortensen, 1998; Menn and Hall,

1999; Rosskopf, Charudattan, and Kadir, 1999). Almost all the bioherbicides that have been registered and commercialized are exclusively targeted at relatively small niche markets in which the bioherbicide is capable of controlling only a very narrow spectrum of weeds (and more often than not, only one weed). Ironically, the high specificity of bioherbicides, which is one of their inherent advantages from an ecological perspective, represents a shortcoming from a commercial vantage point and a competitive disadvantage vis-à-vis chemical herbicides that generally have a broader spectrum of activity. This is because the number of uses for which a herbicide is registered represents a measure of potential revenue that can be generated from that herbicide. For instance, a herbicide that is registered for 50 crops can generate higher revenue than one registered for 5 crops, let alone one, assuming all crops have equal market size. Of course market size varies from crop to crop. A herbicide registered for large crops, such as maize, soybean, cotton, wheat, rice, and canola (rapeseed), will generate greater revenue than one registered for a smaller crop. This is a major reason why large agrochemical companies have tended to focus their R&D on herbicides that target major crops. By implication, these are chemical herbicides with a broad spectrum of weed control.

Large agrochemical companies are interested in large markets because of a fundamental barrier to entry in this market: enormous R&D costs and the stringent regulatory process. For the latter, see the discussion under regulatory framework. Over the past 30 years, research, development, and registration costs have increased dramatically. In 1976, it cost about $6 million and took four years to develop and register a pesticide. The cost of conventional pesticide development per successful candidate currently is estimated as high as $100 million and development time typically ranges between seven and ten years (Industry Canada, 1998). This escalation in cost has been cited as a primary reason why the number of new pesticides registered by the EPA decreased by 50 percent in 1987-1991 compared to 1972-1976. According to Ollinger and Fernandez-Cornejo (1995), each 10 percent increase in regulatory costs resulted in a 3 percent reduction in the number of new pesticides introduced.

The decrease in new pesticide products may also be due, in part, to the discovery of fewer novel modes of action. The rate of discovery of successful new pesticide molecules is 1 in every 20,000 (Lethbridge, 1989). In other words, 1 in 20,000 compounds screened becomes a candidate for commercialization, presenting additional challenges in commercializing even the most novel products. In most cases, even the most promising compounds have been rendered commercially unattractive, unless they have a corresponding high-value market. Cost considerations and the need for high return to shareholder value are partly responsible for concentration in the agrochemical industry, with a proliferation of consolidations, mergers, and acquisitions aimed at creating sufficient critical mass of products required to generate the target revenues and to support further R&D in pesticide science.

The United States provides a good example of this consolidation and structural change (Figure 15.5). Between 1980 and 1997, more than 70 start-up biopesticide companies were conducting biological pest control research and development (Kalaitzandonakes and Marks, 2000). Research and development expenditure by each company averaged between $20 and $60 million to develop a single technology. In addition, established companies such as Monsanto, American Cynamid, Ciba, and Rohm and Haas with a long tradition in synthetic pesticides joined the new biopesticide industry by making huge investments in biopesticide R&D programs. None of the start-up biopesticide companies were successful in taking any products to the market. The primary reasons were a narrow spectrum of activity and sometimes inconsistent efficacy, which limited the agricultural markets and was compounded by low profit margins. Most of these biotechnology companies had focused on only one very high risk technology. However, in due course, it became apparent that companies needed to have more than one product line to become established in the marketplace. As a result, significant reorganization in industry structure took place through bankruptcy, acquisition, or consolidation (Gaugler, 1997; see Table 15.3). An example is Biosys's consolidation effort that has led to the acquisition of three companies: AgriSense in 1992 (pheromone technology), CGI in 1995 (virus technology), and AgriDyne in 1996 (azadirachtin technology). Joint R&D ventures were another approach that some companies adopted to enhance their

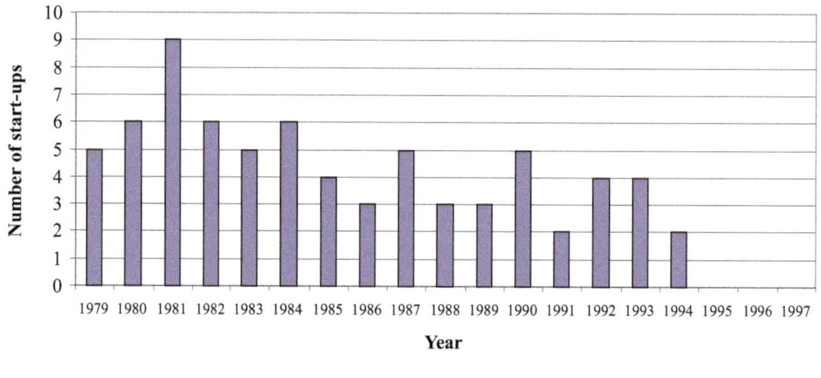

FIGURE 15.5. Number of new biopesticide companies between 1979 and 1997.

TABLE 15.3. Consolidation in the biopesticide industry.

IPO company	Product	Status of company
Biosys	Nematodes, viruses	Acquired by Thermo Trilogy
Entotech	Bacteria	Sold to Abbott Labs
Crop Genetic International	Viruses, bacteria	Acquired by another company
EcoScience	Fungi	Repeated downsizing
Ecogen	Bacteria, nematodes	Repeated downsizing
Mycogen	Fungi, initially bacteria	Acquired by Dow
Microbio	Nematodes, bacteria	
Mycotech	Fungi	

position. Earlier collaborations include those between Ecogen and Monsanto, and Mycogen and Dow Elanco.

Looking at the pesticide industry as a whole, the industry-wide escalation in R&D costs had a major impact on smaller firms, which invariably exited the pesticide industry altogether. The remaining firms diverted their R&D focus to large crop markets, away from minor crops such as horticulture. The latter has been the typical market at which most biopesticides have been targeted. However, as Georgis (1997) noted, markets for biopesticides in general continued to be too small, fragmented, and highly specialized, resulting in return on investment often being too small to be of interest to large companies. As a case in point, both DeVine and Collego occupied a very small niche with a limited area of application (Quimby, King, and Grey, 2002). In general, the niche markets to which these bioherbicides have been limited have required considerable price premiums to be profitable; otherwise, at commodity prices, the revenues would generally not justify the cost and risk of development compared to conventional herbicides.

The escalation in R&D costs of pesticides in itself also represents an opportunity for biopesticides. Large agrochemical companies require large markets in order to recover the high R&D expenditures incurred prior to full commercialization of conventional chemical pesticides. Biopesticides, on the other hand, require only a fraction of the R&D resources and time from development to commercialization. These comparatively lower R&D costs represent an opportunity for bioherbicides, especially for minor-use crops and other niche applications, as well as the decrease in new pesticide registrations that has ensued from stringent regulatory conditions and growing environmental and public concerns about pesticide use in urban areas.

The implications are that smaller companies will be the optimal vehicle for commercializing bioherbicides in niche markets. These small firms were the primary source of innovation in the development and commercialization of DeVine, Collego, and other novel bioherbicides. There is an opportunity to target high-value niche applications, especially nonagricultural uses such as golf courses, schools and colleges, recreational municipal parks, residential lawns, and vegetation management in rights-of-way. For instance, golf courses are intensively managed. The sheer expense of golf green construction means that they are designed to be used indefinitely. For this reason, they are intensively managed and receive high levels of crop protection chemicals. Heavy pesticide use on golf courses and in public amenity areas has been a target of regulatory scrutiny because of the health risks from prolonged exposure. Health issues associated with long-term pesticide exposure include eye, skin, pulmonary, neurologic, and gastrointestinal problems. Data on herbicide use in public amenities and golf courses are often aggregated by national statistical agencies. The Canadian turf and ornamental sector is valued at $6.86 million. The equivalent figure for the United States is aggregated under "Industrial, Commercial, and Government Uses." This category includes applications by owner/operators and custom/commercial applicators to industry, commercial, and government facilities, buildings, sites, and land, including lawns. Uses in this segment were valued at $748 million (EPA, 1998).

Although smaller firms have demonstrated the flexibility to respond rapidly to advances in bioherbicide science and technology, it is also widely accepted that large agrochemical companies are aware of the potential of bioherbicides as an important option available for farmers for weed control within an integrated pest management (IPM) setting. This is evident from some of the collaborative research and development agreements in the industry. Large firms possess enormous expertise in the commercialization of pesticides.

Of course, some challenges ahead for bioherbicides must continue to be addressed in terms of market penetration. A key argument for bioherbicides in niche applications is based on their perceived advantage of mammalian safety over chemicals. In the 1980s and early 1990s, bioherbicides in general were lucky to develop in a less competitive niche of the pesticide market. However, this advantage may be eroded by new developments in pesticide chemistry and may further challenge or restrict bioherbicides to fewer niche markets. As Georgis (1997) and Navon (2000) show, new chemistries and transgenic crops mean that biopesticides must now compete on a much larger playing field. Biopesticides must increasingly meet the key criteria of success, namely, product efficacy, price, ease of cultural adaptation, and customer service. A strategy for market positioning of bioherbicides is maintaining an attractive market share. Given their lower R&D and regulatory costs, bioherbicides can compete in high-value applications where the prevention of cosmetic injury to a crop is an important consideration.

In spite of these constraints to successful development and commercialization of bioherbicides, R&D continues seeking ways to overcome the constraints. Fungal bioherbicides often are so highly weed specific that the private sector is unable to risk venture capital for their development. Public-private sector partnerships are the key to helping overcome the technological limitations and small market size. In the United States, the commercial development of Collego and DeVine represented an example of a successful collaboration between the private and public sectors in terms of biological, technical, and commercial considerations. The model illustrates the strengths of the public sector in basic science, including the capacity to generate and disseminate fundamental scientific knowledge of biological interactions at the microorganism and ecosystem levels. It also demonstrates the private sector's capability to produce fungi on a large scale, protect the intellectual property, meet registration and regulatory requirements, and successfully market novel technologies.

Intellectual Property Protection

The economics of intellectual property rights and the patent system has been well reviewed in the economics literature (Arrow, 1962; Kitch, 1986; Landes and Posner, 1987, 1989; Gilbert and Shapiro, 1990; Griliches, 1990). The protection of intellectual property rights is fundamental to R&D investment and innovation in the development of bioherbicides. Because of the huge investment required to bring a new bioherbicide to the crop protection market, mechanisms are needed to protect the technology territory covered by the invention and allow inventors to recoup their costs. There are several ways in which an inventor can protect his or her invention: patents, trade secrets, copyrights, or trademarks. With respect to bioherbicides, patents or trade secrets would be the more logical method of protection and will be further discussed.

Patents

A patent is a legal grant by a government through its relevant patent office to an inventor which gives the inventor the legal right to exclude competitors from the technological space covered by the patent in that they cannot offer rival inventions without infringing the patent. Patent protection involving all forms of life—plants, animals, and microorganisms—has been an issue in international trade negotiations, especially with respect to the domestic enforcement of foreign intellectual property rights in countries that have poorly defined and inadequately enforced intellectual property laws. This deficiency in intellectual property protection currently is being addressed by several nations which have become signatories of two major international agreements in recent years: the 1992 Convention on Biological Diversity (CBD) (UNEP, 1992) and the 1993 Trade-Related Intellectual Property Rights (TRIPS) negotiated in the 1993 Uruguay Round of the Generalized Agreement on Tariffs and Trade (GATT) (UN, 1993). GATT was replaced on January 1, 1995, by the World Trade Organization (WTO). WTO rules are designed, inter alia, to reduce tariffs, prevent discrimination in trade, and improve market access.

There is no unanimity among all countries with respect to TRIPS (UNCTAD, 1996; Chin and Grossman, 1990; Deardorff, 1992; Helpman, 1993; Taylor, 1994; Yang, 1998; Perrin, 1999; Lesser et al., 2000). Most developing countries currently do not recognize any form of patent on biological resources and related technology. Most developing countries argue that the introduction of intellectual property rights into the WTO has nothing to do with trade and confers intellectual property rights on genetic resources and their derivatives to foreign inventors and results in adverse economic impacts on developing countries. Researchers need to be aware of this debate.

The 1993 WTO TRIPS agreement which came into effect in 1995 provides a 20-year life on a patent starting from the time the patent was filed. This term is already the standard in Europe. Special exemptions were allowed for some developing countries, most of whom are expected to fully implement the provisions of TRIPS by the year 2005.

Thus, an inventor has a 20-year monopoly (in a strict economic sense). In exchange for this monopoly, the patentee agrees to full disclosure of the invention. By conferring monopoly rights to the inventor that exclude rivalry, patents are considered to provide incentives for investment in costly and risky R&D. Because the patentee discloses all the novel information, the patent system facilitates the dissemination of new knowledge upon which competitors can build to develop improved versions of novel inventions. According to Arrow (1962) and Besen and Kirby (1989a,b), an efficient patent system aims to promote innovation and investment in R&D while minimizing losses due to market power and maximizing economic welfare. If potential innovators are limited in their ability to capture full value from their invention and patent law allowed competitors to easily "innovate around" an invention or produce complements to it, the incentive to create the innovation may be reduced significantly.

All patents must meet the statutory requirements of patentability. Under most patent systems, a patent application must satisfy a number of important criteria. The invention must be (1) novel; (2) inventive, i.e., not obvious to someone skilled in the art; (3) capable of industrial application (utility); and (4) sufficiently detailed so that someone who works in the same field could perform or produce the claimed invention. The latter only applies to the United States.

How do these patenting principles apply to fungal bioherbicides? In general, novelty in bioherbicide research involves microorganisms. Based on principles of patent law, the mere discovery of an organism is not patentable. However, most bioherbicides under development involve the isolation of a pathogen from the environment, such as *Phytophthora palmivora* (in the case of DeVine) or *Colletotrichum gloeosporioides* (in the case of Collego). The process used to isolate the pathogen from its source is patentable if the process of isolation is novel and involves an inventive step. If the pathogen produces a novel compound that can be characterized by structure or process, patentability may be extended to the discovery of the production of a new compound by a pathogen or microorganism. This fact explains why no known naturally occurring microorganisms have been patented on their own given the novelty argument.

In looking at patent literature on microbial cultures of naturally occurring organisms, most have been patented as biologically pure cultures. Of course, it is easier to obtain a patent for a genetically engineered organism because novelty can be demonstrated easily. The law pertaining to patenting living organisms is complex. Nevertheless, most patent offices do not prohibit the patenting of microbes that meet the patentability criteria. A 1980 landmark ruling by the U.S. Supreme Court opened the way for the patenting of living organisms, in the famous case of *Diamond v. Chakrabarty*. In this case, the court ruled that "anything under the sun that is made by man" should be patentable (provided the other requirements of patent law are met). Prior to this, the U.S. Patent Office rejected patent claims to living organisms on the grounds that they were not patentable subject matter.

Patent Costs

In discussing patents, it is important to consider one key factor: patent costs. Given the need for large markets for bioherbicides, patent strategy typically requires filing for extensive protection, such that patenting costs alone with sufficient geographic coverage can run from $50,000 to $500,000. Patenting costs of an invention of average complexity cost between $10,000 and $25,000 in Canada and the United States. Maintenance fees for the duration of the patent will be an additional $7,000 to $10,000 over the term of the patent. It is important to note that patent costs must be maintained. Nonpayment of patent maintenance fees will result in an abandoned patent. The cost of a European patent application covering 11 countries can cost between $30,000 and $50,000 (which includes filing, translation, and maintenance costs). A patent with extensive geographic coverage can cost from $250,000 to $500,000 over the life of the patent.

Feitelson, Payne, and Kim (1992) cite an example in which 39 patents were issued in the United States for the commercialization of *Bt* technology alone. However, protection of herbicides based on pathogens or other microorganisms could be risky, since the active ingredient can be isolated and identified. Most bioherbicide companies are small and their viability crucially depends on sufficient protection of their intellectual property. This leads to a related protection strategy: trade secrets.

Trade Secret

From a strategic point of view, a trade secret is another method of protecting an invention. The economic implications of trade secret law, especially as it relates to business information transmitted by persons, firms, and markets, have been well addressed (Grossman and Stiglitz, 1976; Kitch, 1980, 1986; Miller, 1989). Milgrim (1989) describes a trade secret as "any formula, pattern, device or compilation of information which is used in one's business, and which gives him an opportunity to obtain an advantage over competitors who do not know or use it." The U.S. Uniform Trade Secrets Act of 1979 describes it as

> information including a formula, pattern, compilation, program, device, method, technique, or process, that: i) derives independent economic value, actual or potential, from not being generally known to, and not being readily ascertainable by proper means by other persons who can obtain value from its disclosure or use, and ii) is the subject of efforts that are reasonable under the circumstances to maintain its secrecy. [§1(4)]

As a method of protecting intellectual property, inventors need to be aware of the fundamental difference between patent law and trade secret law. Trade secret law differs from patent law in three ways. First, the scope of protection under trade secret law varies among jurisdictions. According to Kitch (1980), some states in the United States base trade secret protection on a property theory, while others appeal to doctrines of tort, contract, constructive trust, or unjust enrichment. A number of states have criminalized the theft of a trade secret, making illegal both the use of "improper means" to discover the trade secret and the "misappropriation" of improperly discovered material. In general, a competitor who uses "improper means" of discovery, or an employee who breaches a confidentiality agreement with his or her employer under which a trade secret was disclosed, can be prosecuted. The available remedies are injunctive relief and damages.

Second, trade secret law lies outside the inducement/disclosure framework of patent law. While a patent holder is assigned rights in exchange for full public disclosure of the invention, trade secrets are, by definition, not disclosed. The law merely protects them against discovery by "improper means," implying that trade secrets are not protected against discovery by independent means or by reverse engineering.

Third, a trade secret has no expiration date, whereas a patent is assigned for a definite period. Trade secret protection rests exclusively on the commercial value of the matter to the claimant (Kitch, 1980). Trade secret law therefore includes a broader subject matter. Although there is no expiration date, the duration of trade secret protection is restricted only by the coincidence of independent discovery or by improper disclosure. For most companies that use trade secrets as a method of protection, the prospect of perpetual protection provides a real economic incentive to avoid the disclosure requirements of the patent regime. Most companies that use trade secrets have also set up internal mechanisms for ensuring nondisclosure.

Overall, trade secret law must be viewed as supplementing the patent system. In terms of economics, inventors may choose to protect their invention through trade secrets within the company if they believe that patent protection is too costly relative to the value of their invention, or that it will give them a reward substantially less than the benefit of their invention (Friedman, Landes, and Posner, 1991).

Patent Offices

Another important consideration for inventors relates to decisions about where to apply for patent protection, given that there is no such thing as a worldwide patent. In order to address the requirements of the global marketplace, most inventors will file patents in multiple jurisdictions. Patents can be filed in a number of jurisdictions: the Patent Cooperation Treaty (PCT), the European Patent Office (EPO), the Eurasia Patent Convention, or national patent offices such as the Canadian Intellectual Property Office (CIPO), the Indian Patent Office (IPO), the U.S. Patent and Trademark Office (USPTO).

The PCT, administered by the International Bureau of the World Intellectual Property Organization (WIPO), provides some of the broadest coverage, with 110 signatories in both developed and developing countries. It is one of several treaties that have been established in order to ensure a uniform process of granting patents across various national patent offices. The PCT has two important treaty provisions to which member countries can agree. These are Chapter I (International Application and International Search) and Chapter II (International Preliminary Examination).

Under Chapter I, patent applications in any of the contracting states may be filed as international applications under the treaty, thereby providing a convenient mechanism for filing applications for the same invention in many different countries at one time. For instance, submitting a PCT patent application at any national patent office, or with the International Bureau of WIPO, is the equivalent of applying to each of the patent offices in the countries that the applicant specifies. The applicant makes a single application which is subsequently converted into a number of national applications. The National Phase requires the applicant to pay local filing and translation costs. The main advantage for the applicant in filing a PCT application is that the costs of local filings and translations are deferred. As an example, a Canadian applicant needs to file only a single application in the Canadian Patent Office, with an opportunity to conduct at least a preliminary prosecution of the single patent application in one language (English or French), thus postponing the high cost of obtaining translation for up to 30 months from the priority date. A priority date is simply the date of the first filed application. The other distinct advantage is that such a PCT patent application has an international search provided by one of the international search bureaus. It is an extensive search that demonstrates the relevance of the cited references to the patent application. This international search is executed within three months of receipt of application, or nine months following the priority date, whichever is longer.

Chapter II provisions allow for preliminary examination of the patent application by contracting states bound by Chapter II. Thus, any applicant of a contracting state bound by Chapter II and whose international application has been filed with the relevant PCT office may make a demand for international preliminary examination. This treaty provision is in addition to the international search and essentially provides an opinion on the patentability of an invention. The provisions of Chapter II also allow a 30-month delay of filing, translation, and related fees from the priority date. Note that if a country is not a Chapter II signatory, the delay is only 20 months. As of April 1, 2002, all PCT applications will have the same deadline for national filing—30 months from the priority date. Chapter II will no longer be required to obtain an additional year to file national stage filings in the designated countries. However, Chapter II will remain an option for those who want to demand an international preliminary examination.

It may not be cost-effective to file a PCT application if the applicant is seeking protection in only a few countries. For a modest cost, however, a PCT application does preserve the applicant's option regarding foreign filing. Hence, it is cheaper to file one PCT application than to file individual applications in each country of interest. In addition, most PCT patent searches are more thorough given their international scope, and thus of higher quality, minimizing the probability of the application being rejected or challenged. This is because a PCT applicant also receives (within 16 months of the priority date) an International Search Report. This report contains no comments on the value of the invention but lists citations of prior art (patent and other literature) relevant to the claims of the PCT application and gives an indication of the relevance of the citations to questions of novelty and inventive step (nonobviousness). This information can be useful to an applicant in determining if the application should be continued in the designated national or regional patent offices. Note, however, the potential for infringement. Because the PCT application is not examined for at least 20 months, this may expose the invention to a potential infringer. This risk is particularly high in technology sectors with rapid technological innovations. Finally, it is prudent to be aware of jurisdictions that are not signatories to the PCT.

Patents can also be filed in the European Patent Office if an applicant is interested in filing only in European countries. A patent granted by the EPO is essentially a collection of national patents but with the advantage of needing only one application. The rights are designated in a number of member states of the European Patent Convention (EPC). Like a PCT, a European patent is required to be translated for most countries. It is important to point out that the EPO is not a European Union (EU) institution. It is independent of the EU and operated under the European Patent Convention. The cost of filing a patent application at the EPO depends on a number of factors, including the sheer size of the application, the number of claims, and whether the European application is based on a prior PCT application. It is normally estimated that if one requires protection in more than three or four designated countries, it is cheaper to file a single European patent application than it is to file separate national applications. It should be noted that a European patent can be filed as part of the PCT application.

Whichever route is used, the principles of patent law are very similar, although there may be fundamental differences in philosophy. For instance, the Canadian patent system is based on "first-to-file" while the U.S. patent system is based on "first-to-invent." The rest of the world also follows the "first-to-file" principle. Hence, within the PCT, the patent application is published 18 months following the date of first application, irrespective of the jurisdiction where the first application was filed. In Canada, patent applications are also published (or laid open to the public) 18 months after the priority date of filing. A first-to-invent system has major implications for patent enforcement and infringement cases. Consider two inventors both claiming that they were the first to invent a particular technology. The amount of data and information required to prove each inventor's claim to the invention can be onerous, as each one would have to present a detailed historical account of their research process as well as recourse to expensive expert witness testimony.

It is also important to note that patent protection involving forms of life—plants, animals, and microorganisms—represents a real challenge in terms of what agricultural economists refer to as "appropriability," i.e., the ability of a patent holder or licensee to capture profits from the innovation. Kalaitzandonakes and Hayenga (2000) and Teece (1987) have shown that the strength of intellectual property coverage and the ease of imitation influence the ability of agricultural biotechnology firms to appropriate profits from their invention. This is the case even in jurisdictions where the enforcement of intellectual property rights is well defined and enforced. These factors must be weighed in the choice of intellectual property protection strategy.

This weak appropriability is illustrated by many technologies involving life forms that have been commercialized in recent years. As noted by Kalaitzandonakes and Bjornson (1997) and Joly and deLooze (1996), these technologies have exhibited a high degree of technical imitation and low degree of appropriability. As expected, this was accompanied by a plethora of patent infringement suits among major agrochemical and life science companies (e.g., Monsanto, DeKalb, Mycogen, Novartis, Aventis, Pioneer, Rhone Poulenc) claiming patent rights to specific genes, processes, and organisms. Some of the cases are well-known and are worth summarizing (see Table 15.4).

In concluding this section, it is important to point out that given the risk and associated R&D expense of developing biopesticides, intellectual property protection is critical to the biopesticide industry. Inventors must be aware of the methods to protect their novel discoveries. They should understand the implications of public disclosure of their invention, including the risk of possible imitation or infringement in a given market. Specifically, how much of an innovation should be disclosed (with and without legal protection) and how much should be kept secret—given that enabling knowledge revealed through patent publication makes imitation feasible? Can the economic trade-off be quantified for various markets? The new economics of intellectual property which basically develops a new approach to answering these questions enables a better understanding of patent and commercialization strategies for these technologies.

TABLE 15.4. Selected patent infringement lawsuits in the life sciences sector.

Date	Plaintiff	Defendant	Details	Verdict
October 2001	DeKalb	DowAgroSciences	Dow AgroSciences' pipeline Bt corn products, including Herculex* I insect-resistant corn, which DowAgroSciences plans to introduce in 2002, violated multiple patents held by DeKalb which cover various technologies involving the use of Bt in corn.	Settlement agreement by two parties. Financial terms of the settlement were kept confidential.
February 1998	Mycogen	Monsanto, DeKalb, Delta, and Pine Land	Patent infringement of Bt technology. Sought $70 million in damages. Requested court to triple damages for Monsanto's alleged willful infringement.	Mycogen lost. Jury decided that Mycogen failed to prove that it was first to invent Bt technology. Patent invalidated.
June 1998	Monsanto	Mycogen and Novartis (formerly Ciba-Geigy)	Patent infringement of a Bt corn insect-resistance gene patent. Sought $6 million in damages from Novartis.	Jury found patent was literally infringed by defendants. But patent found to be unenforceable.[a] Defendants allowed continued use of Bt genes in competition with those licensed by plaintiff.
November 1998	Novartis	Monsanto and DeKalb	Patent infringement of genetically engineered corn.	Jury found patent was not infringed and that Novartis patent was invalid.
February 2004	Syngenta	Monsanto, Delta, and Pine Land	Patent infringement of methods transferring genes into dicot plants using Agrobacterium based vectors	Syngenta and Monsanto agreed to cross-license proprietary Agrobacterium mediated transformation technology

October 2004	Max Planck Institute, Bayer Crop Science, and Garching Innovation GmbH	Monsanto	Patent infringement involving Agrobacterium transformation methods in dicot plants	Settlement agreement to cross-license respective Agrobacterium mediated transformation
December 2004	Syngenta	Pioneer Hibred International, Monsanto, DeKalb, DowAgroSciences, and Mycogen	Patent infringement of synthetic Bt genes designed for increased expression in corn and methods for optimizing codons for more efficient expression of Bt insecticidal products, and products—YieldGard and Herculex	Syngenta lost. Judge held that defendants product use different codons and further Syngenta does not invent the invention.

[a] Novartis and Mycogen alleged that Monsanto's patent was invalid because the Bt gene technology was invented first by Agracetus, rather than Monsanto, and that Monsanto sought to conceal this by withdrawing Agracetus's patent application after acquiring Agracetus in 1996.

Registration and Regulations

Hitherto, we have presented business and economic issues related to the commercialization of both chemical pesticides and biopesticides. A fundamental step in the successful commercialization of new biological products is meeting regulatory requirements for their registration. There is often a misconception that all biopesticides are safer relative to synthetic ones, but the regulators are always concerned about due diligence in approval of all new products. Makowski (1997) identified possible production of mycotoxins by fungi used as bioherbicides. According to Beier (1990), mycotoxins are toxic compounds produced by fungi on grains and other foods that can pose health hazards to humans and animals. The potential bioherbicide *Phomopsis emicis* for use against the weed cornered-jack (*Emex australis* Steinh.) was found to produce sufficient amounts of a mycotoxin to be regarded as a threat to grazing livestock (Shivas et al., 1994). In the following section, we will discuss registration requirements for biopesticides, using the Canada Pest Management Regulatory Agency (PMRA) and the U.S. Environmental Protection Agency (EPA) as case studies.

PMRA, Health Canada, and the U.S. EPA

The Pest Management Regulatory Agency was established in April 1995 following the transfer of administration of the Pest Control Products Act from the Minister of Agriculture and Agri-Food to the Minister of Health. PMRA regulates all products designed to manage, destroy, attract, or repel pests that are used, sold, or imported into Canada. It also regulates other substances, such as formulants, adjuvants, and contaminants, that are contained in pest control products. The federal legislative authority for the regulation of pesticides in Canada is the Pest Control Products Act. The use of pesticides is also subject to regulation under provincial/territorial legislation.

The goal of the PMRA is to protect human health and the environment while supporting the competitiveness of agriculture, forestry, other resource sectors, and manufacturing. The agency is responsible for providing safe access to pest management tools while minimizing risks to human and environmental health. The agency is also dedicated to integrating the principles of sustainability into Canada's pest management regulatory regime. PMRA specific responsibilities can be summed up as:

1. to review applications for the registration of pest control products,
2. to conduct science-based health, environmental, and value (including efficacy) assessments of each pesticide before deciding if it should be approved for use in Canada,
3. to develop and implement policies and guidelines related to pest management,
4. to promote sustainable pest management,
5. to ensure efficiency in the processing of registration applications through such means as international harmonization and electronic submission and review of pesticide registration data,
6. to enforce compliance with the Pest Control Products Act, and
7. to disseminate information on pest management issues to the public.

In the United States, the EPA is responsible under the Federal Insecticide, Fungicide, and Rodenticide Act (FIFRA) for determining whether pesticides (chemical and biological) pose unreasonable risks to human health or the environment when used according to label directions. FIFRA also requires the EPA to balance the risks of pesticide use against the benefits to society and the economy when allowing a particular use.

Registration Requirements

Canadian data requirements are harmonized with the U.S. Environmental Protection Agency. Under the harmonization, Canada accepts all U.S. guidelines for conducting studies. Both countries require efficacy data to be generated. Five key facts must be established for the EPA and the PMRA to make proper appraisal of the benefits and potential hazards of a new pesticide. These are mammalian toxicity, classification of chemical origin or composition, mode of entry, mode of action, and formulation.

In recognition of the differences between conventional chemicals and organic-based products, both PMRA and EPA have created a special category called biochemical and microbial pesticides. Biochemical and microbial pesticides are generally distinguished from conventional chemical pesticides by their unique modes of action, low use volume, target species specificity, or natural occurrence. Therefore, biochemical and microbial pesticides are subject to a different set of data requirements than conventional pesticides. Novel microbial pesticides (i.e., genetically modified or non-indigenous microbial pesticides) are subject to additional data or information requirements than what is required for microbial pesticides that are based on indigenous or unaltered microorganisms.

The registration process usually begins with an Experimental Use Permit (EUP) to permit premarket field testing. Subsequently, the manufacturer submits an application for review. This application must include appropriate health and safety data. The regulatory agencies may require further testing to clarify specific health or environmental questions that may arise during registration review. This is a tedious process. For instance, the EPA annually reviews approximately 15,000 registration submissions. Most of these are for new formulations containing active ingredients which are already registered, new uses of existing products, or other label amendments to existing products. Only about 15 new chemicals (new active ingredients) are registered each year. Typically, 1 in 20,000 compounds screened will be labeled.

Table 15.5 summarizes the salient features of the tests and data required for product registration. These registration decisions are based on a review and evaluation of test data generated by the registrant in a variety of scientific areas. These general areas include toxicology and oncogenicity, ecological effects, environmental fate, and residue chemistry. Research programs are conducted according to Good Laboratory Practices (GLP) which are government-prescribed standards designed to ensure quality and integrity of the data supporting registration.

Pesticide Registration Costs

The PMRA and EPA registration process requires manufacturers to provide scientific data to substantiate that a proposed product is safe and poses no unreasonable adverse effects to human health or the environment. Tests pertaining to toxicology, reproduction disorders and abnormalities, and potential for tumors from exposure to the pesticide are required. Other required tests evaluate the effect of pesticides on aquatic systems and wildlife, farm worker health, and the environment. The agencies may require as many as 90 to 100 specific tests in all of these areas at the manufacturer's cost to substantiate the safety of the product. A manufacturer may spend five to seven years researching and developing a new active ingredient before a registration package is even submitted for consideration. Once the registration package is submitted, it may take up to three years or longer for the PMRA or EPA to review the data contained in the package and to approve or deny the registration.

All of these regulatory requirements affect the development time and cost of pesticide production. Ollinger and Fernandez-Cornejo (1995) estimated that the research and development of a new pesticide averages 11 years and can cost manufacturers between $50 and $70 million. They found

TABLE 15.5. Pesticide product registration: Tests and data requirements.

Data	Description
Wildlife and aquatic data	• Typical test animals: bobwhite quail, mallard duck, bluegill sunfish, rainbow trout, Daphnia • Includes acute, reproductive, and life cycle studies • Data required for both active ingredient and end-use product • Terrestrial and aquatic field testing may be required • Total of 27 different tests
Toxicology data	• Designed to gauge potential adverse effects on humans • Tests for acute, subchronic, and multigeneration effects through oral, dermal, and inhalation exposure • Includes testing for oncogenicity, teratogenicity, gene mutation, neurotoxicity, and reproductive effects • Total of 28 different tests
Environmental fate data	• How does a chemical break down and what are its breakdown products, for example: — Photo degradation in water and soil, hydrolysis, aerobic and anaerobic metabolism, leaching potential, volatility • Groundwater monitoring studies • Includes laboratory and field tests • Total of 25 different tests
Residual chemistry data	• Primarily concerned with residues on food and/or feed • Nature of residues in plants and animals • Represents multiple tests for each crop (each crop usually requires at least 8 sites) • Magnitude of residues in potable water, fish, irrigated crops, meat, milk, poultry (dermal treatment) • Analytical methods and standards • Total of 18 different tests
Other required tests	• Product chemistry — Description of manufacturing process, raw materials, analytical methods — 6 tests • Physical and chemical characteristics — Density, solubility, vapor pressure, pH, etc. — 21 tests • Plant protection data — Seed germination and vegetative vigor, aquatic plant growth — 8 tests

TABLE 15.5 *(continued)*

Data	Description
	• Re-entry data
	— Foliar and soil dissipation, dermal and inhalation dosimetry exposure
	— 4 tests
	• Nontarget insect data
	• Toxicity to honeybees, field testing for pollinators
	— 3 tests
	• Spray drift
	— Droplet size spectrum, drift field evaluation
	— 2 tests

Source: Adapted from the U.S. Environmental Protection Agency, 1998.

that regulation encourages the development of less-toxic pesticide materials but discourages new chemical registrations, encourages firms to abandon pesticide registrations for minor crops, and favors large firms over smaller ones. They also stated that regulation encourages firms to develop biopesticides as an alternative to chemical pesticides.

Regulatory Streamlining for Reduced-Risk Pesticides

The EPA has facilitated the development of biopesticides by establishing a tier approval system in which, under some circumstances, several tests are waived. These reduced regulation costs have helped lower the development costs of biopesticides (Ollinger and Fernandez-Cornejo, 1995). Biological pesticides include microbial pesticides, plant pesticides, and biochemical pesticides. The average time to register a biological pesticide was 11 months, compared to 38 months for a conventional pesticide.

Conventional pesticides can be classified as reduced risk by having low impact on human health, low toxicity to nontarget organisms, low potential for groundwater contamination, lower use rates, low pest resistance potential, and compatibility with integrated pest management practices. The average time to register a reduced-risk pesticide (other than biological pesticides) is 14 months. The EPA reports that it has registered between 13 and 19 reduced-risk pesticides per year from 1994 to 1998. The EPA has also facilitated the use of minimum-risk pesticides by establishing a process for exemption from costly FIFRA requirements. Thirty-one substances deemed to pose insignificant risks to human health and the environment have recently been deregulated. The EPA considered whether the substances were common foods, had nontoxic modes of action, were recognized by the FDA as safe, and had scientific evidence showing no significant adverse effects or persistence in the environment <http://www.epa.gov/pesticides/factsheets/alpha_fs.htm>.

Registration of a new formulation containing an active ingredient already registered or a new use for a currently registered active ingredient usually takes less time. Most pesticide registrations can cost significantly less than this, depending on their proposed uses; for example, a proposed use on cut flowers (enclosed area; nonfood use) will probably cost less than a proposed use on corn or soybeans (many hectares; food crops).

BIOLOGICAL CONSIDERATIONS

Discovery Process

The first step in the development of a bioherbicide begins with the discovery and screening process. After the commercialization of Collego, Templeton, Smith, and TeBeest (1989) said that the greatest constraint to developing mycoherbicides was inadequate field exploration for new suitable candidates. Since that time, extensive exploration for weed pathogens has been done (Charudattan, 1991; Kennedy and Kremer, 1996; Boyetchko, 1997; Charudattan, 2001). However, identifying important criteria to enable selection of organisms with the characteristics needed for commercialization has remained elusive. Screening procedures should go beyond the testing of Koch's postulates to confirm an organism's pathogenicity on a weed. Screening processes that integrate knowledge of the pathogen, host, and environment will greatly assist the selection process for lead candidates.

The high-throughput, in vitro screening models that evaluate microorganisms for their ability to produce natural products as used by pharmaceutical and agrichemical industries have been preferred by companies looking for biological control agents. These models are designed to look for specific chemical compounds and natural products (enzymes, antibiotics, phytotoxins) produced by the microorganisms and to evaluate between 2,000 and 20,000 isolates for every useful product found (Campbell, 1994). It requires a dedicated, sustained screening program to acquire the background knowledge and experience needed to collect and isolate microorganisms, determine fermentation and media requirements of many different species, and develop procedures to assist chemical dereplication of novel active compounds from known ones. The focus is on the selection of laboratory strains that can be fermented to produce significant quantities of the "active component."

Discovery of biological control agents by this approach alone may overlook the selection of suitable organisms because the model does not consider the prime objective of biological control, which is to deliver weed control under field conditions. Often laboratory tests and field tests show poor agreement; variability in activity may arise from using different media in the laboratory and from fluctuating environmental conditions outside the laboratory (Campbell, 1994). A combination of in vitro and in vivo testing will improve the chances of a suitable selection. Strategies for the detection of biocontrol organisms should include

1. defining the target weed(s) to work on,
2. selecting several thousand organisms from a broad range of environments,
3. conducting in vitro and complementary in vivo greenhouse screening using different combinations of hosts, soil types, environmental conditions, and microbial strains,
4. comparing activity to chemical standards, and
5. tabulating scores across all characteristics and then selecting lead candidates (Renwick and Powell, 1990).

Japan Tobacco, Inc. (Yokohama, Japan) is the only company to develop and commercialize a bioherbicide in Japan. It took Fujimori (1999) three years to find and screen 30,000 annual bluegrass plants to obtain 89 isolates of *Xanthomonas campestris* pv. *poae* of which one isolate was selected as being suitable for commercialization of the product Camperico for use in turf. From his company's perspective, the discovery of pathogenic isolates is not sufficient reason to pursue bioherbicide development unless other factors in the early screening process indicate that these isolates also have traits that will lead to successful commercialization. Their success was attained by developing a rapid, efficient screening process based on the production of specific enzymes/phytotoxins and evaluation of activity on small plants in tubes followed by greenhouse and field trials to substantiate the activity.

It has been debated that the number of organisms screened affects the success of selecting suitable organisms. Campbell (1994) found that many biocontrol programs screened only a few hundred organisms instead of a few thousand, thus diminishing the chances of finding the best ones. Marrone (1999) found that despite substantial university research on the genus *Bacillus* for biological control of plant diseases, relatively few strains were screened. To counter this approach, Agra Quest (Davis, California) designed a discovery and development pathway to improve the chances of selecting suitable organisms for development as microbial pesticides (Marrone, 1999). The first tier of their investigative pathway had activities flowing back and forth between the primary laboratory screen, natural product chemistry, and plant testing. The second tier of testing examined the relationships between plant testing, fermentation, formulation, and scale-up. Other important criteria for the system included setting parameters for closely defined targets from specific habitats and using specific taxonomic groups, selecting only hardy, high-producing strains with high levels of activity, optimizing the production process by activity not just cell yield, and formulating to increase activity by a minimum of 25 percent. Lead candidates were identified from comparisons to chemical pesticides on the basis of efficacy, cost, and ease of use. Using this in vivo screening methodology, Agra Quest was able to find and develop 1 to 3 biological products for every 700 to 5,000 organisms screened.

Screening for biological control organisms should utilize a series of general tests with defined criteria to determine whether the organism has acceptable characteristics for the defined purpose and area of use, instead of restricting the primary screen to only a few tests that limit the scope of determination. These tests should evaluate as many organisms as possible, from different environments, and with different modes of action (Campbell, 1994). Screening methods should be unique, setting different criteria and parameters based on the objectives in mind (Marrone, 1999), or too many organisms with mediocre ability or poor growth characteristics may be retained. At the Agriculture and Agri-Food Canada, Saskatoon Research Center, our primary-level screening for fungal pathogens of weeds consists of several steps:

1. confirming Koch' postulates with detached leaves of the host,
2. evaluating the pathogenic fungi for their growth characteristics in culture (propagule type, growth rate, sporulation rate) and mass production potential (i.e., liquid fermentation, solid state),
3. determining the preferred sites of infection on hosts for possible application and delivery methods (i.e., organisms are applied to roots, stems, leaves, flowers, seeds), and
4. determining efficacy and consistency of the response under a minimum of two dew period/temperature combinations, with and without a simple formulation consisting of a surfactant, nutritional source, and/or bonding agent in order to assess the pathogen's tolerance to environmental fluctuations.

Biocontrol agents are typically screened against only one target weed and important nontarget crops for safety (Boyetchko, 1996; Mortensen and Makowski, 1997). After efficacy on the target weed, a narrow host range is desirable from a registration and ecological point of view, although very narrow host ranges are not likely to be commercially viable. Early screens for host range testing should be expanded to include a broad spectrum of alternate weed hosts to explore other opportunities for control. Bourdôt and Harvey (1996) reported that *Sclerotinia sclerotiorum* had potential as a biological herbicide for controlling a few species of thistles in pasture grass and clover in Australia: 100 percent mortality of Canada or Californian thistle [*Cirsium arvense* (L.) Scop.] 12 days after infection, 75 percent for Scotch thistle [*Cirsium vulgare* (Savi.) Ten.], 58 percent for nodding thistle (*Carduus nutans* L.), and 8 percent for winged thistle (*Carduus tenuiflorus* Curt.). Later it was discovered that *S. sclerotiorum* could also control another economically important weed, bitou bush [*Chrysanthemoides monilifera* ssp. *rotundata* (L.) Nord. (DC.) Norl.], in coastal flora of eastern Australia (Cother, 2000).

Efficacy

After the organism has been selected, consideration should be given to maximizing efficacy and obtaining consistency of the response over a wide range of environments (ecozones). The disease triangle is the counterpoint for success in efficacy, namely the interactions among a virulent pathogen and a susceptible host in a conducive environment. Often there is a poor understanding of the underlying biology and ecology in the biocontrol system and of which components have the greatest influence on efficacy. Typically only simple models have been used to understand the system, with the true complexity often being ignored (Kremer, 1998).

To illustrate how a greater understanding of the biology can influence the selection of a bioherbicide candidate, the *Alternaria cirsinoxia-Cirsium arvense* pathogen-host system is used. In greenhouse studies *A. cirsinoxia* demonstrated three important traits: (1) high and rapid mortality to the weed, (2) safety to nontarget crop species, and (3) tolerance to fluctuating temperature extremes (Green, Mortensen, and Bailey, 2001). However, when the same level of efficacy could not be obtained under field conditions, a detailed study revealed several aspects of the pathogen's biology and the host-pathogen ecology that influenced efficacy. Moisture requirements of the pathogen were fairly rigid, needing a minimum of 8 hours continuous leaf wetness within 48 hours of the application, and these conditions did not frequently occur on the Canadian prairies (Green and Bailey,

2000b). Intermittent periods of leaf wetness were detrimental to the fungus. The pathogen could germinate on both young and old thistle leaves, but penetration occurred only on mature leaves and mostly at the leaf tip regardless of conidial density (Green and Bailey, 2000a). A study of the infection process supported the findings that moisture was a key factor in limiting efficacy, and it also demonstrated that the host responded to infection by rapidly depositing lignin and callose and mobilizing silicon at the site of penetration as host defense mechanisms to limit disease (Green, Bailey, and Tewari, 2001).

These studies suggest that preliminary screens for foliar-applied pathogens should apply fairly rigid tests for moisture requirements (such as setting a minimum leaf wetness period and using intermittent moisture conditions). It also becomes apparent that efficacy is not only influenced by the pathogen and environmental conditions, but also by the host. From weed biology and ecology studies on the host we learn about their life cycles, population dynamics, and effects of crop competition on the weed, but we have little understanding of the associated microbial relationships (Boyetchko, 1996; Kennedy and Kremer, 1996; Kremer, 1998). For biological control, these relationships are important. For example, distinct groups of bacteria have been found that are associated with specific weed species, suggesting host-specificity relationships (Kremer, 1987). In pathogen-crop systems, we recognize the relationship between pathogen virulence genotypes and variability in host susceptibility, but in biological control we largely ignore this concept of host genetic variability. Unless more resources are spent studying these basic processes (biology, genetics, ecology) in both the pathogen and the weed, our ability to manipulate the efficacy of the biological product will remain limited.

When host-specific phytotoxins are responsible for pathogenicity then efficacy may be enhanced by selecting isolates with greater virulence or by developing fermentation or growth processes that enhance chemical potency (Miller, 1991). The chemistry of many natural products developed for weed control (i.e., bialaphos, cinmethylin, and picloram) has shown that the half-life is shorter, lower concentrations are required, and the mode of action is different compared to synthetic chemicals currently used (Saxena and Pandey, 2001). The status of the use and efficacy of secondary microbial metabolites for weed control has been reviewed by Saxena and Pandey (2001); 11 of 12 natural herbicides undergoing various stages of commercial development provided control of at least two or more weeds. Non-host-specific phytotoxins (i.e., those with broad-spectrum use) have greater appeal for commercial development than host-specific ones since development is an expensive process.

Several other methods are being explored as ways to enhance efficacy, but none have reached the stage of commercial application. For example, the multiple-pathogen strategy would employ two or more organisms with complementary traits in an application cocktail. This approach may increase the number of target weeds that may be controlled with a single application or it could extend the range of environmental conditions under which maximum efficacy can be expressed (Campbell, 1994; Kremer, 1998; Charudattan, 2001). Chandramohan (1999) used a mixture of three fungal pathogens to control seven target weed species in citrus groves.

Enhanced efficacy has been observed from synergies between some biological organisms and chemical herbicides (Christy et al., 1993). Synergy is desirable in commercial applications since it may reduce chemical load on the environment and result in cost savings. The benefits of synergy may be derived by expanding host range, using reduced rates of application, or improving the efficacy of moderate pathogens. Wymore, Watson, and Gotlieb (1987) showed that the herbicide thidiazuron increased the penetration of *Colletotrichum coccodes* when applied together on velvetleaf (*Abutilon theophrasti* Medik.) resulting in greater weed control than either alone. Heiny (1994) found that the application rate of *Phoma proboscis* could be reduced from 10^8 to 10^7 conidia/mL when applied as a mixture with 2,4-D and MCPA to field bindweed (*Convolvulus arvensis* L.). The application of metribuzin with a fungal biocontrol agent reduced the fresh weight of scentless chamomile (*Matricaria perforata* Mérat) by 95 percent when compared to the agent alone and by 92 percent when compared to metribuzin alone (Peng, Byer, and Bailey, 2000). When metribuzin was applied as a pretreatment, the biocontrol agent caused 100 percent mortality on 10-leaf chamomile plants, whereas metribuzin alone or the biocontrol alone did not cause any mortality, thereby making a moderately effective pathogen into a highly effective one.

Genetic manipulations may be used to enhance efficacy of biocontrol microorganisms. Studies have shown that the enzyme cutinase has an important role in the host penetration process by many fungi (Kolattukudy, Podila, and Mohan, 1989; Podila et al., 1995). For some fungi, such as *Colletotrichum gloeosporioides,* pathogenicity is determined by cutinase, and mutants deficient of cutinase are nonpathogenic (Dickman and Patil, 1986). A study conducted in our lab at Agriculture and Agri-Food Canada, Saskatoon, showed that the biocontrol fungus *A. cirsinoxia* produced some nonspecific esterases to aid in the penetration of the cuticle of Canada thistle but did not produce cutinase. The fungus was genetically transformed with a cutinase gene, resulting in transformants that could break down cutin on plate bioassays whereas the wild-type isolates could not. The primary lesions on Canada thistle leaves when inoculated with the transformant were

46 percent longer than those lesions that developed from inoculation with the wild type. However, tests have not been conducted to determine whether penetration was faster in the presence of cutinase. Other possibilities for genetic manipulation include introducing gene sequences for phytotoxin production or overproduction of a phytotoxin into organisms that are adapted for a particular environment (Miller, 1991), using auxotrophy or nutritional dependency to limit spread of highly virulent pathotypes (Harvey et al., 1998), and transferring hypervirulence genes into the microorganisms (Gressel, 2001).

TECHNOLOGICAL CONSIDERATIONS

Technological problems are usually associated with aspects of production and formulation. These include

1. debating the use of liquid versus solid growth substrates, especially for fungi;
2. extending limited shelf life of spores, cells, and mycelial fragments;
3. refrigerating products for survival;
4. improving methods for down-stream processing and collection of viable biomass; and
5. minimizing costs associated with each of these stages to produce a high quality product.

These processes must be able to grow large quantities of the organism with high levels of efficacy and long shelf life at a low cost. Solutions are mostly attained by a trial and error process that is unique for each organism.

Typically industry has preferred the use of bacteria, yeasts, or other organisms that could be grown in high volume submerged liquid fermentation systems (as compared to solid growth substrates), harvesting the biomass by centrifugation or filtration, drying it by common methods (freeze, air, spray, or vacuum drying) for stability lasting about 18 months, and choosing its application as wettable powders, dusts, liquids, or granules using conventional pesticide application equipment (Stowell, 1991; Powell, 1992; Cross and Polonenko, 1996). However, many organisms, and fungi in particular, do not readily conform to this model. Camperico could not be stabilized as a freeze-dried product and had to be developed as a frozen product, giving it a shelf life of 1.5 years when stored at –35°C (Fujimori, 1999). Devine also had a short life, requiring made-to-order, refrigerated delivery of the product to the user (Kenney, 1986).

Difficulties in the commercial development of Collego were encountered by variable spore titer and viability, caking, and short shelf life, which required significant resources to overcome the production and formulation problems (Bowers, 1986). Despite these problems, these three products have been commercialized and successfully marketed. Solutions were attained by a trial and error process that was unique for each organism. In each case, industry maintained a strong commitment and took a more flexible approach in order to deliver this new technology.

Several excellent publications discuss the historical and current approaches being used for production, formulation, and delivery of biological control agents (Burges, 1998; Green et al., 1998; Boyetchko et al., 1999; Riesenberg and Guthke, 1999; Zidack and Quimby, 1999; Charudattan, 2001), yet relatively few people were working in these areas that convey new information publicly. A survey indicated that only 15 of 200 people working on biological weed control were working in the areas of production and formulation (Zorner, Evans, and Savage, 1993). Current consensus suggests that although significant advances have been made in formulation and application technologies, technological considerations remain the most limiting factor in the development of bioherbicides. Progress appears to be impeded by not being able to transfer the processes from one organism to another, possession of proprietary versus public knowledge, lack of systematic approaches based on characteristics of the pathogens, inadequate funding, and too few multidisciplinary collaborations (Green et al., 1998; Boyetchko et al., 1999; Zidack and Quimby, 1999).

CONCLUSIONS

The concept of commercial biological products for weed control is a desirable one that will be adopted by users if it can provide better control of weeds than other available products at a cost similar to or less than the other control methods, and if they perceive the biological products carry less hazard and risk to people and the environment (Cross and Polonenko, 1996). But before we can deliver these types of products to the users, a new paradigm must evolve that merges business and economics with science. This means publicly funded researchers and private-sector partners must understand both the marketplace and the product before commercialization will be successful. Because biological control involves living systems, the approaches to research, development, and commercialization must remain flexible, and may require the exploration and adoption of less-conventional processes (Kenney, 1986).

Powell and Jutsum (1993) concluded that successful biocontrol products will fill commercial niches where chemicals do not work or are not politically acceptable. Niche markets must support price premiums to be profitable, but searching for greater broad-spectrum activity or control of weeds that have large economic impact can make it more profitable to pursue commercialization. This is an example of biological considerations being closely intertwined with commercial considerations. Moreover, market penetration and maintenance of an attractive market share are necessary regardless of the niche. The ability to protect the invention (by patents and trade secrets) and incur lower development and regulatory costs for microbial products helps stimulate industry participation. When researchers/inventors recognize market-driven factors as forming important components of the research process then they will be able to place greater emphasis at project initiation on market economics to define specific objectives and criteria for use in the early screening and discovery process and carry the ideas through the research phase.

Learning about the biology, genetics, and ecology of the pathogen and weed system helps us to understand more about the product. Inconsistencies in product performance will result in a product that cannot maintain its market share. To understand these inconsistencies we need to learn why some pathogens work and others do not and then incorporate these new principles as criteria for early screening tests, production, and formulation. However, production and formulation advancements continue to be the most limiting factors to commercialization. Progress is impeded by too few researchers and a lack of research dollars, trade secret protection of successful processes, and often inflexible attitudes in accepting novel solutions. Those products which have been successfully commercialized have had unique, multidisciplinary collaborations between public and private sectors and both parties cooperated to find novel solutions that merged the demands of science and business.

REFERENCES

Arrow, K.J. (1962). *National Bureau of Economic Research: The rate and direction of inventive activity: Economic and social factors.* Princeton, NJ: Princeton University Press.

Aspelin, A.L. and A.H. Grube (1999). Pesticides industry sales and usage: 1996 and 1997 market estimates. United States Environmental Protection Agency, Office of Prevention Pesticides Protection and Toxic Substances 733-R-99-001. November.

Auld, B.A. and L. Morin (1995). Constraints in the development of bioherbicides. *Weed Technology* 9: 638-652.

Beier, R.C. (1990). Natural pesticides and bioactive components in foods. In G.W. Ware (Ed.), *Reviews of environmental contamination and toxicology,* Volume 113 (pp. 45-125). New York: Springer-Verlag Inc.

Besen, S.M. and S.N. Kirby (1989a). *Compensating creators of intellectual property: Collectives that collect.* RAND Corporation, R-3751-MF, November.

Besen, S.M. and S.N. Kirby (1989b). Private copying, appropriability, and optimal copying royalties. *Journal of Law and Economics* 32: 255-280.

Bourdôt, G.W. and I.C. Harvey (1996). The potential of the fungus *Sclerotinia sclerotiorum* as a biological herbicide for controlling thistles in pasture. *Plant Protection Quarterly* 11(Supplement 2): 259-262.

Bowers, R.C. (1986). Commercialization of Collego—An industrialist's view. *Weed Science* 34: 24-25.

Boyetchko, S.M. (1996). Impact of soil microorganisms on weed biology and ecology. *Phytoprotection* 77: 41-56.

Boyetchko, S.M. (1997). Principles of biological weed control with microorganisms. *HortScience* 32: 201-205.

Boyetchko, S.M., E. Pedersen, Z. Punja, and M. Reddy (1999). Formulations of biopesticides. In F.R. Hall and J.J. Menn (Eds.), *Methods in biotechnology,* Volume 5, *Biopesticides: Use and delivery* (pp. 487-508). Totowa, NJ: Humana Press.

Burges, H.D. (Ed.) (1998). *Formulations of microbial biopesticides.* Dordrecht, the Netherlands: Kluwer Academic Publishers.

Burnett, H.C., D.P.H. Tucker, M.E. Patterson, and W.H. Ridings (1973). Biological control of milkweed vine [*Morrenia odorata*] with a race of *Phytophthora citrophthora. Proceedings of Florida State Horticultural Society* 86: 111-115.

Campbell, R. (1994). Biological control of soil-borne diseases: Some present problems and different approaches. *Crop Protection* 13: 4-13.

Chandramohan, S. (1999). Multiple-pathogen strategy for bioherbicidal control of several weeds. Doctoral dissertation, University of Florida, Gainesville.

Charudattan, R. (1991). The mycoherbicide approach with plant pathogens. In D.O. TeBeest (Ed.), *Microbial control of weeds* (pp. 24-57). New York: Chapman and Hall, Inc.

Charudattan, R. (2000a). Biological control of weeds using plant pathogens: Accomplishments and limitations. *Crop Protection* 19: 691-695.

Charudattan, R. (2000b). Current status of biological control of weeds. In G. Kennedy and T.B. Sutton (Eds.), *Emerging technologies for integrated pest management: Concepts, research, and implementation* (pp. 269-288). St. Paul, MN: APS Press.

Charudattan, R. (2001). Biological control of weeds by means of plant pathogens: Significance for integrated weed management in modern agro-ecology. *BioControl* 46: 229-260.

Chin, J.C. and G.M. Grossman (1990). *Intellectual property rights and north-south trade.* Woodrow Wilson School Discussion Paper in Economics, No. 143. Princeton, NJ: Princeton University.

Christy, A.L., K.A. Herbst, S.J. Kostka, J.P. Mullen, and P.S. Carlson (1993). Synergizing weed biocontrol agents with chemical herbicides. In S.O. Duke, J.J.

Menn, and J.R. Plimmer (Eds.), *Pest control with enhanced environmental safety* (pp. 87-100). Washington, DC: American Chemical Society.

Cother, E.J. (2000). Pathogenicity of *Sclerotinia sclerotiorum* to *Chrysanthemoides monilifera* ssp. *rotunda* (Bitoubush) and selected species of the coastal flora in eastern Australia. *Biological Control* 18: 10-17.

Crop Protection Institute of Canada (2000). *Annual report.* Ottawa: Crop Life Canada Etobicoke.

Cross, J.V. and D.R. Polonenko (1996). An industry perspective on registration and commercialization of biocontrol agents in Canada. *Canadian Journal of Plant Pathology* 18: 446-454.

Deardorff, A.V. (1992). Welfare effects of global patent protection. *Economica* 59: 35-51.

De Jong, M.D., P.C. de Scheepens, and J.C. Zadoks (1990). Risk analysis for biological control: A Dutch case study in biocontrol of *Prunus serotina* by the fungus *Chondrostereum purpureum. Plant Disease* 74: 189-194.

Dickman, M.B. and S.S. Patil (1986). Cutinase deficient mutants of *Colletotrichum gloeosporioides* are nonpathogenic to papaya fruit. *Physiological and Molecular Plant Pathology* 28: 235-242.

Environmental Protection Agency (EPA) (1998). *Pesticide industry sales and usage.* Washington, DC: Office of Prevention, Pesticides and Toxic Substances.

Feitelson, J.S., J. Payne, and L. Kim (1992). *Bacillus thuringiensis:* Insects and beyond. *Biotechnology* 10: 271-275.

Friedman, D.D., W.M. Landes, and R.A. Posner (1991). Some economics of trade secret law. *Journal of Economic Perspectives* 5: 61-72.

Fujimori, T. (1999). Bioherbicides: The view from industry. In S. Hetherington and B.A. Auld (Eds.), *Proceedings of bioherbicides: Fresh perspectives, twelfth biennial conference of the Australasian Plant Pathology Society* (pp. 47-51). September 26. Canberra, New South Wales: Australasian Plant Pathology Society.

Gaugler, R. (1997). Alternative paradigms for commercializing biopesticides. *Phytoparasitica* 25: 3.

Georgis, R. (1997). Commercial prospects of microbial insecticides in agriculture. In H.F. Evans (Ed.), *Microbial insecticides: Novelty or necessity?; Proceedings of British Crop Protection Council No. 68* (pp. 243-252). Farnham, UK: British Crop Protection Council.

Gilbert, R. and C. Shapiro (1990). Optimal patent length and breadth. *RAND Journal of Economics* 21: 106-112.

Greaves, M.P. and M.D. MacQueen (1992). Bioherbicides: Their role in tomorrow's agriculture. In I. Denholm, A.L. Devonshire, and D.W. Hollomon (Eds.), *Resistance '91: Achievements and developments in combating pesticide resistance* (pp. 295-306). Essex, England: Elsevier Science Publishers, Inc.

Green, S. and K.L. Bailey (2000a). Effects of leaf maturity, infection site, and application rate of *Alternaria cirsinoxia* conidia on infection of Canada thistle *(Cirsium arvense). Biological Control* 19: 167-174.

Green, S. and K.L. Bailey (2000b). Influence of moisture and temperature on infection of Canada thistle by *Alternaria cirsinoxia. Plant Disease* 84: 1126-1132.

Green, S., K.L. Bailey, and J.P. Tewari (2001). The infection process of *Alternaria cirsinoxia* on Canada thistle *(Cirsium arvense)* and host structural defence responses. *Mycological Research* 105: 344-351.

Green, S., K. Mortensen, and K.L. Bailey (2001). Host range, temperature response, survival, and overwintering of *Alternaria cirsinoxia. Biological Control* 20: 57-64.

Green, S., S.M. Stewart-Wade, G.J. Boland, M.P. Teshler, and S.H. Liu (1998). Formulating microorganisms for biological control of weeds. In G.J. Boland and L.D. Kuykendall (Eds.), *Plant-microbe interactions and biological control* (pp. 249-281). New York: Marcel Dekker, Inc.

Gressel, J. (2001). Potential failsafe mechanisms against the spread and introgression of transgenic hypervirulent biocontrol fungi. *Trends in Biotechnology* 19: 149-154.

Griliches, Z. (1990). Patent statistics as economic indicators: A survey. *Journal of Economic Literature* 28: 1661-1707.

Grossman, S.J. and J.E. Stiglitz (1976). Information and competitive price systems. *American Economic Review, Papers and Proceedings* 66: 246-253.

Harvey, I.C., G.W. Bourdôt, D.J. Saville, and D.C. Sands (1998). A comparison of auxotrophic and wild strains of *Sclerotinia sclerotiorum* used as a mycoherbicide against Californian thistle *(Cirsium arvense). Biocontrol Science and Technology* 8: 73-81.

Heiny, D.K. (1994). Field survival of *Phoma proboscis* and synergism with herbicides for control of field bindweed. *Plant Disease* 78: 1156-1164.

Helpman, E. (1993). Innovation, imitation and intellectual property rights. *Econometrica* 61: 1247-1280.

Industry Canada (1998). *Pesticide industry profile.* Ottawa: Industry Canada.

Joly, P. and M. deLooze (1996). An analysis of innovation strategies and industrial differentiation through patent applications: The case of plant biotechnology. *Research Policy* 25: 1014-1027.

Kalaitzandonakes, N. and B. Bjornson (1997). Vertical and horizontal coordination in the agrobiotechnology industry: Evidence and implications. *Journal of Agricultural and Applied Economics* 29(1): 129-139.

Kalaitzandonakes, N. and M. Hayenga (2000). Structural change in the biotechnology and seed industrial complex: Theory and evidence. In W.H. Lesser (Ed.), *Proceedings of the International Consortium on Agricultural Biotechnology Research NE-165 Conference* (pp. 217-223). June 24-25, 1999, Washington, DC. Amherst, MA: Food Marketing Policy Centre, Dept. of Agricultural Economics, University of Connecticut and Dept. of Resource Economics, University of Massachusetts.

Kalaitzandonakes, N. and L. Marks (2000). Innovation dynamics and optimal licensing strategies in the agro-biotechnology industry. In W.H. Lesser (Ed.), *Proceedings of the International Consortium on Agricultural Biotechnology Research NE-165 Conference* (pp. 206-216). June 24-25, 1999, Washington, DC. Amherst, MA: Food Marketing Policy Centre, Dept. of Agricultural Economics, University of Connecticut and Dept. of Resource Economics, University of Massachusetts.

Kennedy, A.C. and R.J. Kremer (1996). Microorganisms in weed control strategies. *Journal of Production Agriculture* 9: 480-485.

Kenney, D.S. (1986). DeVine—The way it was developed—An industrialist's view. *Weed Science* 34: 15-16.

Kitch, E.W. (1980). The law and economics of rights in valuable information. *Journal of Legal Studies* 9: 683-724.

Kitch, E.W. (1986). Patents: Monopolies or property rights? *Research in Law and Economics* 8: 31-49.

Kolattukudy, P.E., G.K. Podila, and R. Mohan (1989). Molecular basis of the early events in plant-fungus interaction. *Genome* 31: 342-349.

Kremer, R.J. (1987). Identity and properties of bacteria inhabiting seeds of selected broadleaf weed species. *Microbial Ecology* 14: 29-37.

Kremer, R.J. (1998). Microbial interactions with weed seeds and seedlings and its potential for weed management. In J.L. Hatfield, D.D. Buhler, and B.A. Stewart (Eds.), *Integrated weed and soil management* (pp. 161- 179). Chelsea, MI: Ann Arbor Press.

Landes, W.M. and R.A. Posner (1987). Trademark law: An economic perspective. *Journal of Law and Economics* 30: 265-309.

Landes, W.M. and R.A. Posner (1989). An economic analysis of copyright law. *Journal of Legal Studies* 18: 325-366.

Lesser, W., G. Horstkotte-Wesseler, U. Lele, and D. Byerlee (2000). Intellectual property rights, agriculture, and the World Bank. In U. Lele, W. Lesser, and G. Horstkotte-Wesseler (Eds.), *Intellectual property rights in agriculture: The World Bank's role in assisting borrower and member countries* (pp. 1-21). Washington, DC: The World Bank.

Lethbridge, G. (1989). An industrial view of microbial inoculants for crop plants. *Special Publication of the Society of General Microbiology* 25: 11-28.

Makowski, R.M.D. (1997). Foliar pathogens in weed control: Ecological and regulatory constraints. In D.A. Andow, D.W. Ragsdale, and R.F. Nyvall (Eds.), *Ecological interactions and biological control* (pp. 87-103). Boulder, CO: Westview Press.

Marrone, P.G. (1999). Microbial pesticides and natural products as alternatives. *Outlook on Agriculture* 28: 149-154.

Menn, J.J. (1996). Biopesticides: Has their time come? *Journal of Environmental Science and Health B* 31(3): 383-389.

Menn, J.J. and F.R. Hall (1999). Biopesticides, present status and future prospects. In F.R. Hall and J.J. Menn (Eds.), *Methods in biotechnology,* Volume 5: *Biopesticides: Use and delivery* (pp. 1-10). Totawa, NJ: Humana Press.

Milgrim, R.M. (1989). *On trade secrets.* New York: Matthew Bender.

Miller, E. (1989). Antitrust restrictions on trade secret licensing: A legal review and economic analysis. *Law and Contemporary Problems* 52: 183-209.

Miller, J.D. (1991). Mycology, mycologists, and biotechnology. In D.L. Hawksworth (Ed.), *Frontiers in mycology* (pp. 225-240). Wallingford, UK: CAB International.

Mortensen, K. (1998). Biological control of weeds using microorganisms. In G.J. Boland and L.D. Kuykendall (Eds.), *Plant-microbe interactions and biological control* (pp. 223-248). New York: Marcel Dekker, Inc.

Mortensen, K. and R.M.D. Makowski (1997). Effects of *Colletotrichum gloeosporoides* f.sp. *malvae* on plant development and biomass of non-target field crops under controlled and field conditions. *Weed Research* 37: 351-360.

Mupondwa. E.K. (2001). Economic analysis of the biopesticide market: Business opportunities and potential for control of crop pests using biocontrol technologies. Special Economic Study, Saskatoon Research Centre, Agriculture and Agri-Food Canada.

Navon, A. (2000). *Bacillus thuringiensis* insecticides in crop protection—Reality and prospects. *Crop Protection* 19: 669-676.

Ollinger, M. and J. Fernandez-Cornejo (1995). *Regulation, innovation, and market structure in the U.S. pesticide industry.* Washington, DC: U.S. Department of Agriculture, Economic Research Service, AER-719.

Peng, G., K.N. Byer, and K.L. Bailey (2000). Potential control of *Matricaria perforata* using fungal pathogens plus herbicides. In *Proceedings of the Fifth International Bioherbicide Workshop,* June 5-6 (p. 12 abstract). Foz do IguaHu, Brazil: International Bioherbicide Group.

Perrin, R.K. (1999). Intellectual property rights and developing country agriculture. *Agricultural Economics* 21: 221-229.

Pilgeram, A.L. and D.C. Sands (1999). Mycoherbicides. In F.R. Hall and J.J. Menn (Eds.), *Methods in biotechnology,* Volume 5, *Biopesticides: Use and delivery* (pp. 359-370). Totara, NJ: Humana Press.

Podila, G.K., E. Rosen, M.J.D. San Francisco, and P.E. Kolattukudy (1995). Targeted secretion of cutinase in *Fusarium solani* f. sp. *pisi* and *Colletotrichum gloeosporioides. Phytopathology* 85: 238-242.

Powell, K.A. (1992). Biocontrol product fermentation, formulation, and marketing. In E.S. Tjamos (Ed.), *Biological control of plant diseases* (pp. 381-387). New York: Plenum Press.

Powell, K.A. and A.R. Jutsum (1993). Technical and commercial aspects of biocontrol products. *Pesticide Science* 37: 315-321.

Quimby, P.C., L.R. King, and W.E. Grey (2002) Biological control as a means of enhancing the sustainability of crop/land management systems. *Agriculture, Ecosystems and Environment* 88: 147-152.

Renwick, A. and K. Powell (1990). Biological control of soil-borne plant pathogens. *Brighton Crop Protection Conference 1990, Pests and Diseases* 1: 227-231.

Ridings, W.H. (1986). Biological control of stranglervine in citrus—A researcher's view. *Weed Science* 34(Supplement 1): 31-32.

Riesenberg, D. and R. Guthke (1999). High-cell-density cultivation of microorganisms. *Applied Microbiology and Biotechnology* 51: 422-430.

Rosskopf, E.N., R. Charudattan, and J.B. Kadir (1999). Use of plant pathogens in weed control. In T.W. Fisher, T.S. Bellows, L.E. Caltagirone, D.L. Dahlsten, C. Huffaker, and G. Gordh (Eds.), *Handbook of biological control* (pp. 891-918). San Diego, CA: Academic Press.

Saxena, S. and A.K. Pandey (2001). Microbial metabolites as eco-friendly agrochemicals for the next millennium. *Applied Microbiology and Biotechnology* 55: 395-403.

Shivas, R.G., J.G. Allen, J.A. Edgar, P.A. Cockrum, P.F. Gallagher, P.F. Ellis, and M. Harvey (1994). Production of phomopsin A by *Phomopsis emicis*. In S.M. Colegate and P.R. Dopling (Eds.), *Plant associated toxins: Agricultural, phytochemical, and ecological aspects* (pp. 161-166). Wallington, UK: CAB International.

Stowell, L. (1991). Submerged fermentation of biological herbicides. In D.O. TeBeest (Ed.), *Microbial control of weeds* (pp. 225-273). New York: Chapman and Hall.

Taylor, M.S. (1994). TRIPS, trade, and growth. *International Economic Review* 35: 361-381.

Teece, D. (1987). Profiting from technological innovation: Implications for integration, collaboration, licensing, and public policy. In D. Teece (Ed.), *The competitive challenge* (pp. 185-220). Cambridge, MA: Ballinger Publishing Co.

Templeton, G.E., R.J. Smith, and D.O. TeBeest (1989). Perspectives on mycoherbicides two decades after discovery of the Collego pathogen. In E.S. Delfosse (Ed.), *Proceedings of the Seventh International Symposium on Biological Control of Weeds* (pp. 553-558). March 6-11, Rome, Italy. Victoria, Australia: CSIRO Publications.

United Nations (UN) (1993). *Agreement establishing the World Trade Organization.* New York: United Nations.

United Nations Conference on Trade and Development (UNCTAD) (1996). *The TRIPS agreement and developing countries; UNCTAD/ITE/1. U.N. Sales No.96.II.D.10.* New York and Geneva: United Nations Publications.

United Nations Environment Program (UNEP) (1992). *Convention on biological diversity.* New York: United Nations.

U.S. Department of Agriculture (USDA) (1999). *Agricultural chemical use.* Washington, DC: National Agricultural Statistics Service and Economic Research Service.

World Resource Institute (WRI) (1998). World resources 1998-99: Environmental change and human health. Washington, DC: World Resource Institute, jointly with United Nations Environment Programme (UNEP), United Nations Development Programme (UNDP), and the World Bank (WB).

Wymore, L.A., A.K. Watson, and A.R. Gotlieb (1987). Interaction between *Colletotrichum coccodes* and thidiazuron for control of velvetleaf *(Abutilon theophrasti). Weed Science* 35: 377-383.

Yang, Y. (1998). Why do southern countries have little incentive to protect northern intellectual property rights? *Canadian Journal of Economics* 31: 800-816.

Zidack, N.K. and P.C. Quimby Jr. (1999). Formulation and application of plant pathogens for biological weed control. In F.R. Hall and J.J. Menn (Eds.), *Methods in biotechnology,* Volume 5, *Biopesticides: Use and delivery* (pp. 371-381). Totowa, NJ: Humana Press.

Zorner, P.S., S.L. Evans, and S.D. Savage (1993). Perspectives on providing a realistic technical foundation for the commercialization of bioherbicides. In S.O. Duke, J.J. Menn, and J.R. Plimmer (Eds.), *Pest control with enhanced environmental safety* (pp. 79-86). Washington, DC: American Chemical Society.

Chapter 16

Implementation of Weed Biocontrol in Forest Vegetation Management for Conifer Production

Simon F. Shamoun

INTRODUCTION

The demand for enhanced forest production and sustainability has increased the intensity of forest management (Wall, Prasad, and Shamoun, 1992) and the importance of vegetation competing with commercially valuable planted or naturally regenerating young conifer seedlings (Shamoun, 2000). The past few decades have seen significant changes in forest management, notably in site preparation, chemical herbicide use, and the development of new methods for forest practices and harvesting. The basic approaches to increase stand growth and fiber production are achieved through

1. rapidly establishing a new stand following harvest (planting practices);
2. maximizing the flow of site resources to the crop (controlling competing forest vegetation);
3. improving soil resources (fertilization);
4. selecting and breeding fast-growing genotypes; and
5. minimizing losses due to insects, diseases, fire, and forest weeds (stand protection).

All of these approaches have been actively researched over the past 35 years (Perry, 1998).

The need for management of forest vegetation in young conifer plantations has been established by numerous researchers and was comprehensively reviewed in several textbooks, international symposia proceedings, and hundreds of journal publications (Walstad and Kuch, 1987; Wagner, 1993; Wagner and

Thompson, 1998; Bovey, 2001; Wagner and Colombo, 2001). Competing forest vegetation ("forest weeds") suppresses the growth of commercial conifer species in young forest regeneration sites and reduces the quality and quantity of forest products (Shamoun, 2000). Control of the competing forest vegetation can take many forms, including mechanical or manual removal, and chemical herbicides. These methods have distinct disadvantages including environmental concerns, nontarget effects, and cost-effectiveness. This has necessitated a more intensive search for alternative management strategies for invasive forest weeds (Jobidon, 1991; Watson and Wall, 1995). One viable option is the use of naturally occurring plant pathogenic fungi as biological control agents or mycoherbicides (i.e., the use of formulated products of indigenous fungal plant pathogens that are applied directly to native target weeds), which if successful would be expected to yield an increase in early conifer growth rate and a shorter rotation age of naturally regenerated or planted commercially valuable conifer species (Wall and Hasan, 1996). The utilization of biological control method is particularly important in ecologically sensitive sites such as riparian zones and low-land forest areas, where chemical herbicide use is restricted (Shamoun, 2000; Harris and Shamoun, 2002).

Biological forest weed control strategy is the deliberate use of natural enemies (predators, parasites, and pathogens) to suppress the growth or reduce the population of a weed species to an acceptable density at which they no longer present an economic or environmental problem. Three major methods have been used for biological control of forest weeds: classical, inundative (bioherbicides/mycoherbicides), and silvicultural manipulation (augmentative). The classical biocontrol strategy has been used to control exotic forest weeds. The inundative approach is based on indigenous plant pathogens, whereby these pathogens are mass-produced and applied as formulated products for management of native weeds. The silvicultural method, which is based on a sound knowledge of the biology of plant pathogens, autecology of the target weeds, and the ecology of forest ecosystems, has been used very recently in a forestry setting (Markin and Gardner, 1993; Shamoun, 2000; Harris and Shamoun, 2002).

Biological control approaches are playing an increasingly important role in sustainable agriculture through integrated pest management (IPM) (Altieri, 1994). This method should be recognized as a tool in integrated forest vegetation management (IFVM) programs. Biological control strategy will be employed either independently or in combination with the traditional forest vegetation management tools, such as manual cutting, and reduced rates of the registered herbicides or plant growth regulators, if only to satisfy the wishes of the public in general (Wagner, 1993). From a practical, sociological, ecological, and economical viewpoint, a biological control strategy must be viewed as an essential component of IFVM to achieve silvicultural objectives using a variety of methods that are ecologically and environmentally sound, politically and socially desirable, and economically feasible (see Figure 16.1). Past and present examples of biological control research initiatives mainly by using fungal pathogens for management of forest weeds are reviewed in this chapter, detailing the concepts involved, techniques employed, and constraints to the development of commercial biological control products. Finally, prospects for the bioexploitation of fungal plant pathogens for management of forest weeds are critically discussed.

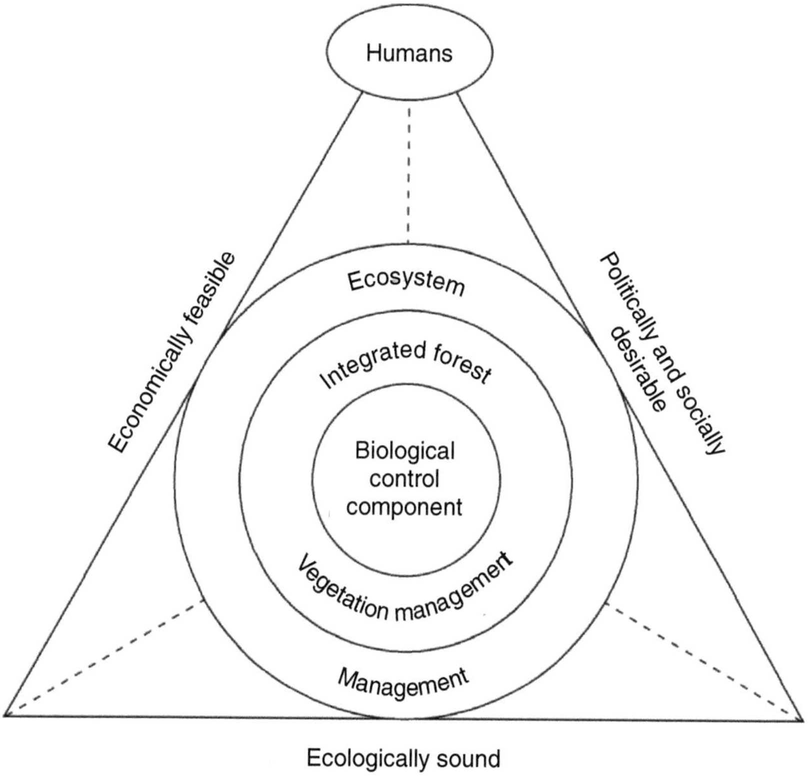

FIGURE 16.1. Biological control approach as an essential component of integrated forest vegetation management and its relation to forest ecosystem management.

NATURE OF COMPETING FOREST VEGETATION: "FOREST WEEDS"

Competing forest vegetation, known as "forest weeds," differ considerably from most agricultural weeds, especially in the northern hemisphere, in that they are mostly indigenous vigorous pioneer species that normally colonize forest land after disturbances such as fire, harvesting, wind, and flood. Disturbances in forest ecosystems are usually followed by a succession of pioneer species, beginning with herbaceous plants and followed by shrubs and fast-growing deciduous trees (Marks and Bormann, 1972; Wall and Hasan, 1999). In addition, forest weeds, as opposed to agricultural weeds, are generally perennials, and many of them arise from seeds that lie dormant in forest soils for many years, e.g., *Rubus* spp. and *Ribes* spp. (Wall, Prasad, and Shamoun, 1992; Oleskevich, Shamoun, and Punja, 1996). Others arise from wind-blown spores or seeds, e.g., bracken fern [*Pteridium aquilinum* (L.) Kuhn], red alder (*Alnus rubra* Bong.), and *Populus* spp. (Haeussler and Coates, 1986; Harrington, Zasada, and Allen, 1994). Pioneer species also become established vegetatively from roots and rhizomes, e.g., *Betula* spp. and *Rubus* spp. (Johansson, 1987; Oleskevich, Shamoun, and Punja, 1996). These pioneer species are generally short-lived and are eventually replaced by later successional species. Forest weeds can have beneficial effects such as protecting soil from erosion, providing thermal cover, enhancing soil chemical and physical properties (through symbiotic nitrogen fixers), and providing forage for wildlife and livestock. However, they can have detrimental effects on commercially valuable planted or naturally regenerated conifer seedlings, making their control necessary for maximum fiber productivity (Wagner and Zasada, 1991).

In addition to the pioneer species are shade-tolerant species, which can intensify in density and suppress the establishment of young conifer plantations after harvesting. These weeds, which include salal (*Gaultheria shallon* Pursh), kalmia (*Kalmia angustifolia* L.), beaked hazel (*Corylus cornuta* Marsh.), and salmon berry (*Rubus spectabilis* Pursh) (Wall 1977; Haeussler and Coates 1986; Oleskevich, Shamoun, and Punja, 1996) have special features, such as a thick glabrous cuticle on the foliage, that may impede the maximum uptake and entry by chemical herbicides and biological control agents. They can also thrive under light forest canopies occupying poor forest sites (Wall, Prasad, and Shamoun, 1992).

In addition to native competing forest vegetation, forest sites are increasingly invaded by exotic introduced weed species from other continents. Scotch broom [*Cytisus scoparius* (L.) Link] and gorse (*Ulex europaeus* L.) have colonized Australia, New Zealand, and the Pacific Northwest of North

America, and hinder growth of native plants and the establishment of conifer regeneration sites (Wall and Hasan, 1996; Peterson and Prasad, 1998; Clements, Peterson, and Prasad, 2001). Various blackberry species (*Rubus fruticosus* L. aggr.) have been introduced from Europe to Australia, Chile, and North America and continue to be a threat to native plants, ranges, and conifer plantations (Amor and Richardson, 1980). Similarly, paper bark tree [*Melaleuca quinquenervia* (Cav.) Blake] was introduced from Australia to south Florida, and has become an aggressive weed in Florida's Everglades ecosystem, displaced native vegetation, caused a deterioration of wildlife habitats, and created fire hazards (Bodle, Ferriter, and Thayer, 1994). Other tree weed species such as *Acacia* spp. and silky hakea (*Hakea sericea* Schrad.) were introduced from Australia to South Africa and are causing problems in water catchment areas (Morris, 1989).

Also competing with forest vegetation is a group of parasitic plants that have lost the ability to manufacture carbohydrates through photosynthesis; consequently, these plants adapted to a parasitic mode of existence which can adversely affect forests. Dwarf mistletoes (*Arceuthobium* spp.) are strongly host-specific, perennial, obligate parasites of conifers. Many of the economically important conifer species are attacked and damaged by dwarf mistletoes. Trees heavily infected by mistletoe show reduced quality of wood, height and diameter growth, and vigor, as well as predisposing the infected trees to attack by insects and decay fungi. A comprehensive review on aspects of biology, pathology, and systematics of dwarf mistletoes was published by Hawksworth and Wiens (1996). The proceedings of two symposia, Control Through Forest Management (Scharpf and Parmeter, 1978) and Dwarf Mistletoe Biology (Hawksworth and Scharpf, 1984), are also considered informative reviews. Most recently, comprehensive articles related to tissue culture of parasitic flowering plants and its application in agriculture and forestry, and publications on development of biological control strategies for management of dwarf mistletoes were reviewed (Shamoun, 1997; Ramsfield, Shamoun, and van der Kamp, 1999; Deeks, Shamoun, and Punja, 1999, 2001, 2002).

METHODOLOGY USED FOR DEVELOPMENT OF FUNGAL PLANT PATHOGENS AS MYCOHERBICIDES FOR FOREST WEEDS

The development of fungal plant pathogens as mycoherbicides for management of forest weeds will have to follow few precedents similar to the mycoherbicides for agricultural weeds. In general, it involves three phases (Templeton, 1982; Evans, Fröhlich, and Shamoun, 2001): discovery, development, and deployment. The discovery phase includes various activities such as collection of diseased forest weeds, isolation of fungal pathogens, demonstration of Koch's postulates, identification, and long-term storage of the purified fungal pathogens in liquid nitrogen. The development phase includes determination of conditions for optimum inoculum production and disease development, and determination of host range (including nontarget hosts). If these tests indicate that an effective product can be developed, steps may then be taken to register it for use. This process will usually involve patenting the technology. The deployment phase generally involves close interaction with the industrial sector for scale-up production, suitable formulation and application technology, testing for toxic or allergenic effects on animals, environmental and risk assessment studies after deployment, regulatory aspects, and marketing of the mycoherbicide product (Shamoun and Hintz, 1998b). Registration and commercialization are expensive and time-consuming processes, and the market for individual mycoherbicides may be small. For these reasons, many potentially good products may not reach the marketplace. Because this is a relatively new area of research in agriculture and forestry, constant modification of regulations governing registration will be necessary to balance potential risks against benefits.

HISTORY AND CURRENT STATUS OF RESEARCH AND DEVELOPMENT OF MYCOHERBICIDES FOR MANAGEMENT OF FOREST WEEDS

Numerous fungi have been tested for their potential as biological control agents of indigenous and exotic forest weeds over the past four decades. The following are selected examples of research initiatives for major biological control of forest weeds:

- The classic work by V. A. Wager in 1947 in South Africa suggested the potential use of the rust fungus *Kuehneola uredinis* (Link.) Arth. for control of bramble *(Rubus fruticosus bergii)* in Natal.
- Biological control of undesirable oaks (*Quercus* spp.) by the oak wilt fungus *Ceratocystis fagacearum* (Bretz) Hunt was tested as selective silvicide in pine plantations in Minnesota in the early 1950s and shown to be very effective (French and Schroeder, 1969).
- Control of the common persimmon (*Diospyros virginiana* L.), is a problem in pastures of the central and southeastern United States. A wilt fungus, *Cephalosporium diospyri* Crandall, has been applied to cut stumps, thereby preventing persimmon stump sprouting. The fun-

gus may also be applied to weed persimmons by "shooting" trees with impregnated buckshot (Wilson, 1965).

- To combat *Apiosporina morbosa* (Schw. ex Fr.) Arx., the causal agent of black knot disease of pin cherry (*Prunus pensylvanica* L. f.), healthy pin cherry trees on large clear-cut areas in eastern Canada were inoculated by placing diseased branches with canker during bud-break. This approach was effective in breaking the dominance by pin cherry, but culturing, inoculum production, and delivery systems were major constraints (Wall, 1986).
- A major forestry weed, bracken fern is widely distributed in North America and Europe. A formulation of *Ascochyta pteridis* (Bres.) Sacc. was effective in suppressing this weed (Webb and Lindow, 1987; Womack and Burge, 1993).
- The rust fungus *Phragmidium violaceum* (Schultz) Wint. was introduced from Europe to Australia and Chile and successfully used for control of blackberries of European origin (*Rubus fruticosus* L. aggr.) which are a serious problem to the habitat of native plants, pastures, forests, and national parks (Oehrens and Gonzalez, 1974; Bruzzese and Hasan, 1987).
- Regarding biological control of selected major weeds in Hawaiian forests, Trujillo, Latterell, and Rossi (1986) investigated the potential use of *Colletotrichum gloeosporioides* (Penz.) Penz. & Sacc. collected from Panama against the introduced forest weed Koster's curse [*Clidemia hirta* (L.) D. Don] in Hawaii. Under field conditions, inoculation of *Clidemia* with the fungus developed brown foliar lesions, which increased in size, causing severe premature defoliation and tip dieback. Gardner and Hodges (1983) showed that the rust fungus *Kuehneola uredinis* (Link) Arth. exerted some impacts on blackberry (*Rubus penetrans* Bailey = *R. argutus* Link.). Recently, Gardner and colleagues (1997) evaluated the rust fungus *Gymnoconia nitens* (Schw.) Kern & Thur. as a potential biological agent for alien *Rubus* spp. in Hawaiian forests. Experimental inoculation of alien *Rubus* spp. with the rust fungus led to early responses of chlorotic leaf flecking and puckering, leaf and stem contortion, and stem gall formation. However, the ability of this rust fungus to attack the two endemic species, namely akala/Hawai'i blackberry (*R. hawaiensis* Gray) and loganberry (*R. macraei* Gray), suggests that *G. nitens* would not be suitable for release in Hawaii as a biological control agent, at least on the islands with populations of native species.
- Regarding biological control of selected major forest weeds in Canadian forests, research has been concentrated at four Canadian Forest

Service establishments: Pacific Forestry Centre in Victoria, British Columbia; Northern Forestry Centre in Edmonton, Alberta; Great Lakes Forestry Centre in Sault Ste. Marie, Ontario; and Atlantic Forestry Centre in Fredericton, New Brunswick. In addition, active research is ongoing at the Quebec Ministry of Natural Resources (QMNR), as well as five Canadian universities: Laval University, McGill University, University of British Columbia, University of Victoria, and Simon Fraser University.

In the province of Quebec, research efforts on forest weed biocontrol programs were initiated in the late 1980s at McGill University–Macdonald Campus, Laval University, and the Quebec Ministry of Natural Resources with financial support from Quebec Hydro. From these research initiatives, a biological control agent, *Colletotrichum dematium* (Pers. ex Fr.) Grove, was found to be a potential mycoherbicide for management of fireweed [(*Chamerion*) *Epilobium angustifolium* (L.) Holub.] (Winder and Watson, 1994). Since 1988, research work at the QMNR has shown that the use of the secondary metabolites derived from allelopathic plants (e.g., barley, oat, and wheat straw mulches) have prevented the establishment of red raspberry (*R. idaeus* L.). The growth and vigor of planted black spruce [*Picea mariana* (Mill.) BSP] seedlings were not adversely affected by the treatments (Jobidon, Thibault, and Fortin, 1989a,b). Bialaphos is a peptide containing a phosphinothricin moiety and is a foliage active herbicide. This naturally occurring herbicidal compound, a fermentation product of *Streptomyces viridochromogenes,* has been found to have significant herbicidal effects on *R. idaeus* and *Kalmia angustifolia,* two important weeds in Canada and the northeastern United States (Jobidon, 1991). Bialaphos and similar microbially produced phytotoxins have potential new herbicides and hold a promising future as tools for management of forest weeds (Jobidon, 1991; Kenfield et al., 1989). In 1992, collaborative research work between QMNR, Laval University, and Quebec Hydro was established to investigate the potential use of *Colletotrichum purpureum* as a biological control agent of weedy hardwood species in rights-of-way sites (Gosselin and Jobidon, 1995; Jobidon, 1998).

A special research program on the development of biological control of forest weeds was established at the Canadian Forest Service Pacific Forestry Centre (CFS-PFC) in 1986 (Dorworth, 1990). Among target weeds considered were big-leaf maple (*Acer macrophyllum* Pursh), red alder (*Alnus rubra* Bong.), bluejoint reedgrass [*Calamagrostis canadensis* (Michx.) Beauv.], fireweed (*Epilobium angustifolium* L.), salal (*Gaultheria shallon* Pursh), quaking aspen (*Populus tremuloides* Michx.), *Rubus* spp., including wild red raspberry [*R. strigosus* Michx. = *R. idaeus* var. *strigosus*

(Michx.) Focke], thimbleberry (*R. parviflorus* Nutt.), salmonberry (*R. spectabilis* Pursh), and *Ribes* spp. (alternate host of white pine blister rust disease). Fungal plant pathogens from these weeds have been isolated, identified, and tested for their potential use as biological control agents (Dorworth, 1990; Wall and Shamoun, 1990; Wall, 1990; Oleskevich et al., 1998; Winder, 1999; Sutton, Shamoun, and Crous, 1996; Shamoun and Sieber, 2000; Shamoun et al., 2001).

Research activities at the Canadian Forest Service have been successful in obtaining six U.S. patents for different biological control pathosystems. These include biological control of *C. canadensis* with *Colletotrichum* and *Fusarium* species, as well as low-temperature basidiomycete *(Coprinus psychromorbidus)* (Winder, 1995; Mallett, 1999; Mallett, Schreiner, and Gaudet, 2000) control of *A. rubra* with *Nectria ditissima* Tul. (Dorworth, 1994); biological control of weed trees with *C. purpureum* (Wall, Prasad, and Sela, 1996); biological control of weedy *Rubus* spp. with *Fusarium avenaceum* (Shamoun and Oleskevich, 1999); and culture, formulation, and delivery technique of *Valdensinia heterodoxa* for use as a biological control agent of salal *(Gaultheria shallon)* (Vogelgsang and Shamoun, 2002). To date, CFS-PFC's efforts are underway to register these biological control agents and/or license these technologies through a "technology transfer agreement" with commercial partners or a consortium in order to make these agents commercially available.

Most recently, CFS-PFC established a collaborative research agreement with Weyerhauser Canada Ltd. to address the problem of invasive forest weeds in conifer regeneration sites. The targeted forest weeds are *Rubus* spp. (see Photo 16.1), including *R. idaeus* var. *strigosus, R. parviflorus,* and *R. spectabilis,* as well as salal (*G. shallon*). These forest weeds have been selected due to their capacity to rapidly invade reforestation sites and effectively reduce the growth and survival of many conifer species and the productivity of plantations. A promising biological control agent, *Fusarium avenaceum,* was identified, formulated on rice grain, and the inoculum combined with an organosilicone adjuvant. One or two applications cause substantial *Rubus* spp. foliar damage (see Photo 16.2). This allows the conifers to reach the "free to grow" stage in one to three years, and consequently increases conifer production (see Photo 16.3). In addition, the semi-shade-tolerant weed salal *(G. shallon),* which increases in abundance in cutovers on nutrient-poor soils, often forms a persistent barrier to young conifer regeneration (see Photo 16.4). *Valdensinia heterodoxa* is considered a promising potential candidate, as natural infection of this fungus results in severe foliar necroses and premature defoliation of salal in the field (see Photo 16.5). The agent combines the inoculation of the fungus *V. heterodoxa*-isolate (PFC 3027) of salal leaf and stem materials with the mycelium of the

PHOTO 16.1. Weedy thimbleberry (*Rubus parviflorus* Nutt.) rapidly invading a young conifer regenerating site after harvesting and effectively competing with planted young conifer species and reducing the plantation productivity. (See corresponding plate in central color section.)

PHOTO 16.2. Severe foliar damage on *Rubus strigosus* Michx. = *R. idaeus* var. *strigosus* (Michx.) Focke, caused by the potential biocontrol agent *Fusarium avenaceum* (Fr.) Sacc. (See corresponding plate in central color section.)

PHOTO 16.3. Young conifer regenerating site in coastal British Columbia dominated by competing weedy salmonberry (*Rubus spectabilis* Pursh). Note that the young planted seedlings of Douglas fir [*Pseudotsuga menziesii* (Mirb.) Franco] were allowed to grow to "free-to-grow" stage after chemical and manual brushing treatments of *R. spectabilis.* (See corresponding plate in central color section.)

fungus produced from cereal grain for control of salal (see Photo 16.6). Formulated products and delivery technologies of these fungi are patented, and further research is under way for their development as mycoherbicides for management of *Rubus* spp. and salal in conifer regeneration sites, respectively (Shamoun and Oleskevich, 1999; Vogelgsang and Shamoun, 2002).

Chondrostereum purpureum (Pers. ex Fr.) Pouzar (see Photo 16.7), a wood-inhabiting fungus that grows saprophytically in logs and stumps of numerous deciduous tree species, has been tested on hardwood weed species in conifer regeneration sites and utility rights-of-way (Wall, 1994; Shamoun and Hintz 1998a). A collaborative research agreement was established between CFS-PFC and its commercial partner MycoLogic Innovative Biologicals Inc., the University of Victoria, British Columbia, Canada,

PHOTO 16.4. Weedy salal (*Gaultheria shallon* Pursh) rapidly invading a young conifer regenerating site after harvesting and effectively competing with planted young conifer species and reducing the plantation productivity. (See corresponding plate in central color section.)

PHOTO 16.5. Natural infection of salal by the potential biocontrol agent *Valdensinia heterodoxa* Peyronel, which results in severe foliar necroses and premature defoliation of the salal. (See corresponding plate in central color section.)

PHOTO 16.6. Inoculation of salal with *Valdensinia heterodoxa* (isolate PFC 2761) under controlled conditions (left) and untreated salal (right). (See corresponding plate in central color section.)

PHOTO 16.7. Basidiocarps "fruiting bodies" of *Chondrostereum purpureum* infecting red alder (*Alnus rubra* Bong.) stump. (See corresponding plate in central color section.)

with financial support from BC Hydro, Natural Sciences and Engineering of Canada (NSERC), and the CFS. The aim of this research agreement is directed toward registration and commercialization of *C. purpureum* in North America as a mycoherbicide under the proposed trade name Chontrol (Shamoun and Hintz, 1998b). Similar efforts were successful by Dutch researchers and their commercial partner, Koppert Biological Systems, which produces *C. purpureum* as BioChon and markets it as a wood decay promoter and an effective biological control agent for American black cherry (*Prunus serotina* Ehrh.) (de Jong, Scheepens, and Zadocks, 1990; Ravensberg, 1998; de Jong, 2000). In addition, the use of endophytic fungi as potential mycoherbicides and the elucidation of their interaction with target weeds *A. rubra* and *Rubus* spp. has been proposed (Sieber, Sieber-Canavesi, and Dorworth, 1991; Shamoun and Sieber, 1993; Dorworth and Callan, 1996; Shamoun and Sieber, 2000). This area requires more attention in the biological control of forest weeds, as it holds great promise.

As described previously, another group that impedes the establishment of young conifer plantations is dwarf mistletoes (*Arceuthobium* spp.) (see Photo 16.8). Development of a biological control approach for management of dwarf mistletoes by using fungal pathogens has been under consideration for the past four decades in the Pacific Northwest of North America (Hawksworth and Wiens, 1996; Shamoun, 1997; Wicker, 1967; Muir, 1977). To date, research studies on biological control of dwarf mistletoes have been restricted almost entirely to collecting, identifying, and documenting the indigenous fungal parasites of all mistletoe species. No practical efforts have been made to utilize fungi as biological control agents.

At CFS-PFC, this research was reactivated in 1996 (Shamoun, 1997, 1999; Deeks, Shamoun, and Punja, 2001, 2002). Many fungal parasites are associated with dwarf mistletoes, which shows potential promise as biological control agents; these include the following:

- *Colletotrichum gloeosporioides* (Penz.) Penz. & Sacc. causes blight and dieback of shoots and berries, spreading quickly and destroying shoots and berries within dwarf mistletoe infection (see Photo 16.9) (Parmeter, Hood, and Scharpf, 1959; Muir, 1977; Kope, Shamoun, and Oleskevich, 1997; Shamoun, 1999; Ramsfield et al., 1999).
- *Neonectria neomacrospora* (C. Booth & Samuels) Mantiri & Samuuels (anamorph *Cylindrocarpon destructans* Wollenw.) (syn. *Nectria neomacrospora*) destroys the endophytic system and substantially reduces shoots and fruit production of western hemlock dwarf mistletoe in the Pacific Northwest of North America (Funk et al., 1973; Smith and Funk, 1980; Shamoun, 1997; Shamoun and DeWald, 2002) (see Photo 16.10).

- *Cylindrocarpon gillii* (D. E. Ellis) J.A. Muir (syn. *Septogloeum gillii*) causes early infection of dwarf mistletoes shoots characterized by small, yellowish-white lesions. These lesions enlarge, coalesce, erupt through the epidermis, and expose conspicuous masses of white spores; shoots tissues distal to lesions die.
- *Caliciopsis arceuthobii* (Peck) Barr [syn. *Wallrothiella arceuthobii* (Peck) Sacc.] fungus infects stigmas during anthesis, and later stigmas and apical portions of the fruit are replaced by black stroma. Normal fruit development is prevented and infected fruits fail to produce seeds (Hawksworth and Wiens, 1996).
- In South Africa, several fungal pathogens were evaluated for their potential as biological control for forest weeds. *Uromycladium tepperianum* (Sacc.) McAlp. is a gall-forming rust fungus that has damaging effects on orange wattle (*Acacia saligna* [Labill] Wendl.) introduced from Australia to South Africa. On some trees infected by the fungus, all the flowers were transformed into galls, causing tree mortality within a few months (Morris, 1997). An endemic strain of *Colletotrichum gloeosporioides* (Penz.) Penz. & Sacc. caused mortality in young seedlings of the woody shrub *Hakea sericea* Schrad. introduced from Australia to Cape Town, South Africa (Morris, 1989, 1991). Recently, research efforts were aimed at the production of a commercial mycoherbicide for control of several introduced Australian *Acacia* spp. The indigenous fungal pathogen *Cylindrobasidium laeve* (Pers.:Fr.) Chamuris [syn. *Corticium evolvens* (Fr.) Fr.], a white-rot fungus, has been tested as a treatment for stumps of black wattle (*A. mearnsii* De Wild.) and golden wattle (*A. pycnantha* Benth.), and methods of mass production of the commercial product Stumpout were recently registered (Morris, Wood, and Den Breeyen, 1998; Lennox, Morris, and Wood, 2000).
- For biological control of forest weeds in New Zealand, the main focus of the researchers is directed toward controlling the introduced scotch broom (*Cytisus scoparius* Link) and gorse (*Ulex europaeus* L.). Two potential candidate fungi, namely, *Fusarium tumidum* and *Pleiochaeta setosa,* were tested on these target weeds and have been found to be pathogenic. Studies are currently under way to determine their suitability as mycoherbicides (Johnston and Parkes, 1994; Frohlich, Morin, and Gianotti, 2000). Most recently, research has focused on use of native isolates of the white-rot fungus *Chondrostereum purpureum* as a potential mycoherbicide for scotch broom and gorse in New Zealand (G. Bourdot, AgResearch, New Zealand, personal communication, 2002).

PHOTO 16.8. Female plant of the western hemlock dwarf mistletoe [*Arceuthobium tsugense* (Rosendahl) G. N. Jones] infecting western hemlock [*Tsuga heterophylla* (Raf.) Sarg.] branch. (See corresponding plate in central color section.)

PHOTO 16.9. *Colletotrichum gloeosporiodes* (Penz.) Penz. & Sacc. infecting shoots and berries of *Arceuthobium tsugense.* (See corresponding plate in central color section.)

PHOTO 16.10. Perithecia of *Neonectria neomacrospora* C. Broth & Samuels, infecting the basal cup and swelling (endophytic system) of *Arecuthobium tsugense.* (See corresponding plate in central color section.)

DEVELOPMENT AND COMMERCIALIZATION OF MYCOHERBICIDE PRODUCTS FOR USE IN FORESTRY

To date, two mycoherbicide products are registered and available on the market for control of weedy tree species.

Stumpout

Stumpout is a formulated product of the white-rot fungus *Cylindrobasidium laeve* (Pers.:Fr.) Chamuris. Stumpout is registered in South Africa by the registrar of the Agricultural and Stock Remedies Act 36 of 1947 and currently is sold and used as a mycoherbicide to prevent regrowth of cut wattle stumps (Morris, Wood, and Den Breeyen, 1998). Several introduced Australian *Acacia* spp. have become important invasive weeds in South Africa. Unwanted wattle stands are often cleared by mechanical means. However, if left untreated, the stumps resprout, resulting in multistemmed trees. These are more difficult to fell and are certainly less useful than the original trees. Although chemical herbicides that kill the stumps are registered for use, a cheaper and environmentally safer management tool for controlling the regrowth was needed to reduce the high costs of clearing operations. The extensive use of herbicides along stream banks and watercourses is another sensitive issue that called for an alternative approach. *Cylindrobasidium laeve* is known from North America, where it colonizes fallen decaying branches and felled logs, causing a white rot (Ginns and Lefebvre, 1993). As the fungus is already present in South Africa, where it may have occurred for some time, it is not thought to pose a threat to commercial fruit tree orchards and conifer forestry plantations. Field trials on the efficacy of *C. laeve* as a potential biological control agent for cut wattle were conducted on large areas in South Africa. A formulated product of *C. laeve* was produced at the Plant Protection Research Institute (PPRI) research laboratory. Mass production of Stumpout and methods for the production of inoculum have been developed at PPRI (Morris, Wood, and Den Breeyen, 1998; Lennox, Morris, and Wood, 2000). Briefly, optimal basidiospores production was obtained by growing *C. laeve* on a modified potato-marmite-dextrose medium for three days at 25°C, and transferring small blocks of this agar to petri dishes containing small autoclaved discs of *A. mearnsii* wood (2 to 3 cm diameter, 2 mm thick), cut from young saplings, on a 1.5 percent water agar layer. A hole was cut in the lids of the petri dishes, over which was pasted autoclaved pieces of filter paper (Whatman No. 1) for aeration of the cultures. These plates were incubated at 19°C for 12 hours under fluorescent and near-UV light. The basidiospores are harvested by removing the colonized wood discs from the petri dishes, allowing them to dry for 30 minutes on a laminar flow bench, immersing them in mineral oil and shaking well (talc is added to the oil as a dispersant). The solution is then dispensed into small plastic sachets of approximately 10 ml, with a final minimum concentration of 2^{10} basidiospores/mL. During field trials the basidiospores are suspended in mineral oil or sunflower oil and applied to the cut stump surface of wattle trees with a small brush. Treated stump mortality was mostly around 80 to 100 percent, while untreated stump mortality was also high, probably because of natural background inoculum and colonization by other decay fungi. Treatments applied throughout the year appeared to be equally effective. This formulation of *C. laeve* was thus registered in South Africa as Stumpout for the treatment of wattle trees to prevent resprouting (Morris, Wood, and Den Breeyen, 1998; Lennox, Morris, and Wood, 2000).

BioChon/Chontrol/Myco-Tech

BioChon, Chontrol, and Myco-Tech are formulated products of the white-rot fungus *Chondrostereum purpureum*. BioChon was approved for use in the Netherlands to prevent the regrowth of cut stumps of American black cherry (*Prunus serotina* Ehrh.), while Chontrol and Myco-Tech are undergoing the registration process in North America. The fungus can infect most hardwood trees and shrubs, invading the cambium and sapwood and causing the silver leaf symptom (separation of leaf layers) in Rosaceae plants. The fungus is being effectively used in the Netherlands and other European countries by applying mycelial suspension to cut stumps (Scheepens and Hoogerbrugge, 1989; de Jong, Scheepens, and Zadocks, 1990). Currently Koppert Biological Systems markets BioChon in Europe as a wood decay promoter and as a potential biological control agent for woody weed trees and shrubs (Ravensberg, 1998; de Jong, 2000).

In Canada, a collaborative research project between Canadian Forest Service and MycoLogic Innovative Biologicals Inc. was established in an effort to register and commercialize the use of *C. purpureum* under the trade name Chontrol. The product is used for vegetation control in North American conifer regeneration areas and the utility right-of-way sector (Shamoun and Hintz, 1998b). A formulated product of *C. purpureum* was produced at CFS-PFC and a U.S. patent was obtained (Wall, Prasad, and Sela, 1996). Briefly, the procedure for the production of the formulated product is as follows: inoculum of *C. purpureum* (isolate CP-PFC 2139) was prepared from actively growing petri dish cultures growing on 1.25 percent malt extract agar. Inoculum plugs, 5 to 8 mm in diameter, were removed aseptically from the advancing zone of the culture, macerated, and suspended in 100 ml aliquots of 1 percent malt extract broth in Ehrlenmeyer flasks. Flasks were placed on 100 rpm rotary shakers at room temperature for two to five days and the contents then aseptically added to the dry formulation. The dry formulation consisted of 375 g talc, 100 g kaolin clay, 12.5 g corn starch, 5 g pectin, 5 g monosodium glutamate, 1 g monopotassium phosphate, and 1.5 g yeast extract. Ingredients were thoroughly mixed, placed in autoclaveable plastic bags, and autoclaved at 15 psi for 20 minutes. After cooling, the bags were inoculated with malt broth inoculum, with controls (formulation only) receiving sterile malt broth. After inoculation, the contents of each bag were agitated to incorporate the mycelial fragments and the bags were incubated at room temperature for three to eight weeks. During incubation, bags were agitated at weekly intervals to maintain uniform growth. Immediately before inoculation in the field, the contents of each bag were mixed with 600 mL of sterile 1 percent sucrose, 200 mL canola oil, 60 g finely powdered cellulose, and the yolks from two large eggs. The mixture was stirred vigorously to form a smooth paste and was applied to cut stumps in a plastic squeeze bottle (see Photo 16.11). The application of the Chontrol product is simply a topical treatment on the cut stumps of the target weedy hardwood species (see Photo 16.12).

The Canadian Forest Service (Atlantic, Great Lakes, and Pacific centers), British Columbia Ministry of Forests, Ontario Ministry of Natural Resources, and MycoLogic Innovative Biologicals Inc. collaborated and installed a coordinated set of national field trials aimed at providing field efficacy data critical to the registration and subsequent use of *C. purpureum.* This collaborative effort resulted in a common experimental protocol being implemented in three major Canadian provinces, namely New Brunswick, Ontario, and British Columbia. In each province, trembling aspen (*Populus tremuloides* Michx.) was studied as a species of national importance, while red maple (*Acer rubrum* L.), speckled alder [*Alnus rugosa* (Du Roi) Spreng.], and sitka alder [*Alnus viridis* ssp. *sinuata* (Regel) A. Love & Love] were studied as target weeds of regional importance, respectively (Harper et al., 1999; Pitt et al., 1999). Genetic characterization, epidemiology, and environmental fate studies were completed as essential components for the registration of *C. purpureum* (Ramsfield et al., 1996, 1999; Shamoun and Wall, 1996; de Jong et al., 1996; Becker et al., 1999; Becker, Ball, and Hintz, 1999; Hintz, Becker, and Shamoun, 2001). In separate studies conducted in the province of Quebec, *C. purpureum* was used as a biological control agent of unwanted tree species in the utility right-of-way sector (Gosselin and Jobidon, 1995). In addition, extensive genetic

PHOTO 16.11. Formulated paste product of *C. purpureum* (isolate PFC 2139) as Chontrol packaged in a plastic squeeze bottle. (See corresponding plate in central color section.)

PHOTO 16.12. Application of *C. purpureum* by a topical treatment of cut stumps of big leaf maple (*Acer macrophyllum* Pursh). (See corresponding plate in central color section.)

variability, population structure, and environmental fate investigations of *C. purpureum* were conducted in Quebec (Gosselin, Jobidon, and Bernier, 1996, 1999). The commercial partner of the CFS-PFC, MycoLogic Innovative Biologicals, Inc., has compiled all the relevant information and data related to registration requirements in the registration package submitted to the Pest Management Regulatory Agency (PMRA) in Canada and Environmental Protection Agency (EPA) in the United States for the registration of *C. purpureum* (isolate CP-PFC 2139) as Chontrol (Anonymous, 2002). In the province of Quebec, research activities are under way to register *C. purpureum* (isolate HQ1) under the trade name Myco-Tech (Anonymous 2002; Norbert, personal communication).

CONSTRAINTS IN THE DEVELOPMENT OF MYCOHERBICIDES FOR MANAGEMENT OF FOREST WEEDS

Several constraints in the development of commercial mycoherbicides need to be overcome. Mycoherbicides need to be predictable, easy to use, and provide to a certain extent a level of forest weed control comparable to the registered chemical herbicides before they will have general acceptance from industry and end users (Evans, Fröhlich, and Shamoun, 2001). The use of fungal plant pathogens as biological control agents is not risk free, which needs to be seriously and thoroughly assessed. Their safety, persistent survival, and potential for genetic change are major concerns to scientists, regulatory agencies, and the public at large (de Jong, Scheepens, and Zadocks, 1990; de Jong et al., 1996; Gosselin, Jobidon, and Bernier, 1999; Ramsfield et al., 1999; Hintz, Becker, and Shamoun, 2001). The following constraints in the development of mycoherbicides for forest weeds arise because of the desire to produce perfect analogs of chemical herbicides:

1. *Biological constraints,* including host specificity, host range, resistance mechanisms, and interaction with other organisms
2. *Environmental constraints,* involving aerial, soil, environments, as well as compatibility with other synergists including plant growth regulators, adjuvants, and pesticides
3. *Technological constraints,* involving production of active ingredients of organisms, formulation, shelf life, and application of the mycoherbicides
4. *Commercial constraints,* which include market size, patent protection and secrecy, cost of production, and strict regulations

CONCLUSIONS AND GENERAL PROSPECTS

In managed young conifer plantations, the establishment and growth of commercially valuable conifer species is often impeded by competing vegetation. Forest vegetation management often relies on manual cutting combined with chemical herbicide treatments to promote the growth of crop trees. The biological control strategy for management of competing vegetation forest weeds is poised to become an essential component of integrated forest vegetation management. Plant pathogenic fungi are presently considered the most promising biological control agents. The desire for an alternative to chemical herbicides that can be integrated with present vegetation management practices has promoted research into biological control strategies for forestry (Wagner, 1993). Traditionally, the classic biological control approach has been used against introduced forest weeds. Development of bioherbicides/mycoherbicides is a more promising strategy for management of indigenous forest weeds. Augmentative biological control has special relevance for forestry and therefore could be termed a "silvicultural manipulation" strategy that can be promoted by sound research programs based on ecology of forest ecosystems, biology of plant pathogens, and autecology of target weeds.

Recent advances in formulation technology, phytopathology, molecular biology, and silviculture have accelerated the commercialization and production of two biological control products, BioChon/Chontrol/Myco-Tech and Stumpout, for management of competing forest vegetation in the Netherlands, North America, and South Africa, respectively (Morris, Wood, and Den Breeyen, 1998; Ravensberg, 1998; Shamoun and Hintz, 1998a,b; Lennox, Morris, and Wood, 2000). Enhancement of the efficacy and safe use of biological control agents can be achieved by integrating them with manual brushing practices, such as application of BioChon/Chontrol/Myco-Tec or Stumpout on cut stump of hardwood tree weeds, or combining foliar pathogens such as *Fusarium* spp. with adjuvants/surfactants, growth regulators, or low doses of registered herbicides for application onto target weeds.

The use of fungal plant pathogens as biological control agents for management of forest weeds is not risk free, which needs to be seriously and thoroughly assessed. The main concerns to both regulatory agencies and the general public in using fungal plant pathogens for control of forest weeds are their safety, persistent survival, genetic instability, and threat to nontarget plants. Risk analysis based on epidemiological modeling systems of *C. purpureum* have indicated that risk is extremely low (de Jong, Scheepens, and Zadocks, 1990; de Jong et al., 1996). Recent advances in molecular and genetic technologies (i.e., PCR-DNA) have permitted the monitoring of biological control agent (e.g., *C. purpureum*) releases and provide a reliable tool for studying environmental fate and impact, and plant disease epidemiology (Ramsfield et al., 1996; Becker et al., 1999; Gosselin, Jobidon, and Bernier, 1999; Hintz, Becker, and Shamoun, 2001). These approaches can be used as a complementary tool to existing computer modeling systems. Current registration requirements for commercial biological control agents are based, in part, on presumed dangers based on theory. These requirements should be relaxed as we become more accustomed to using biological control agents. Most of the plant pathogens that have caused serious losses to forest trees were introduced accidentally in forest products and nursery stocks, such as chestnut blight, Dutch elm disease, and white pine blister rust (Manion, 1981). It was not done through using native plant pathogens via planned biological control programs (Cook et al., 1996). The potential use of biotechnological techniques to enhance the efficacy of biological control agents is a very promising strategy for development of mycoherbicides for forestry situations (Watson and Wall, 1995). From a practical, sociological, economical, and ecological viewpoint, mycoherbicide technology should be viewed as an essential component of integrated forest vegetation management that will be employed in combination with existing forest vegetation tools. Thus, it has been proven that biological control of forest weeds with fungal plant pathogens has good potential in the future; however, there are several constraints, including biological, environmental, technological, and commercial limitations, to the development of commercial mycoherbicides for forestry use. Current research activities on forest weed biocontrol should yield several improvements in forest vegetation management, including new commercial products and more widely acceptable approaches to forest management.

REFERENCES

Altieri, M.A. (1994). Sustainable agriculture. *Encyclopedia of Agricultural Science* 4: 239-247.

Amor, R.L. and R.G. Richardson (1980). The biology of Australian weeds: 2. *Rubus fruticosus* L. agg. *Journal of Australian Institute of Agricultural Science* 46: 87-97.

Anonymous (2002). *Proposal regulatory decision document PRDD2002-01,* Chondrostereum purpureum *(HQ1).* Ottawa, Ontario: Pest Management Regulatory Agency (PMRA), Health Canada. Available at <http://www.pmra-arla.gc.ca/English/pdf/prdd/prdd2002-ol-e.pdf>.

Becker, E.M., L.A. Ball, M.T. Dumas, D.G. Pitt, R.E. Wall, and W.E. Hintz (1999). *Chondrostereum purpureum* as a biological control agent in forest vegetation management: III. Infection survey of a national field trial. *Canadian Journal of Forest Research* 29: 859-865.

Becker, E.M., L.A. Ball, and W.E. Hintz (1999). PCR-based genetic markers for detection and infection frequency analysis of the biocontrol fungus *Chondrostereum purpureum* on sitka alder and trembling aspen. *Biological Control* 15: 71-80.

Bodle, M.J., A.P. Ferriter, and D.D. Thayer (1994). The biology, distribution, and ecological consequences of *Melaleuca quinquenervia* in the everglades. In S.M. Davis and J.C. Ogden (Eds.), *The Everglades, the ecosystem and its restoration* (pp. 341-355). Delray Beach, FL: St. Lucie Press.

Bovey, R.W. (2001). *Woody plants and woody plant management: Ecology, safety, and environment impact.* New York: Marcel and Dekker, Inc.

Bruzzese, E. and S. Hasan (1987). Infection of blackberry cultivars by the European blackberry rust, *Phragmidium violaceum. Journal of Horticultural Science* 64: 475-479.

Clements, D.R., D.J. Peterson, and R. Prasad (2001). The biology of Canadian weeds. 112. *Ulex europaeus L. Canadian Journal of Plant Science* 81: 325-337.

Cook, R.J., W. Bruckart, J. Coulson, M. Goettel, R. Humber, R. Lumsden, R. Maddox, J. McManus, M. Moose, L. Meyer, et al. (1996). Safety of microorganisms intended for pest and plant disease control: A framework for scientific evaluation. *Biological Control* 7: 333-351.

Deeks, S.J., S.F. Shamoun, and Z.K. Punja (1999). Tissue culture of parasitic flowering plants: Methods and applications in agriculture and forestry. *In Vitro Cellular Development and Biology—Plant* 35: 369-381.

Deeks, S.J., S.F. Shamoun, and Z.K. Punja (2001). In vitro germination and development of western hemlock dwarf mistletoe (*Arceuthobium tsugense* subsp. *tsugense*). *Plant Cell, Tissue and Organ Culture* 66: 97-105.

Deeks, S.J., S.F. Shamoun, and Z.K. Punja (2002). Histopathological of callus and germinating seeds of *Arceuthobium tsugense* subsp. *tsugense* infected by *Cylindrocarpon cylindroides* and *Colletotrichum gloeosporioides* in dual culture. *International Journal of Plant Sciences* 163: 765-773.

de Jong, M.D. (2000). The BioChon story: Deployment of *Chondrostereum purpureum* to suppress stump sprouting in hardwoods. *Mycologist* 14: 58-62.

de Jong, M.D., P.C. Scheepens, and J.C. Zadocks (1990). Risk analysis for biological control: A Dutch case study in biocontrol of *Prunus serotina* by the fungus *Chondrostereum purpureum. Plant Disease* 74: 189-194.

de Jong, M.D., S.T. Sela, S.F. Shamoun, and R.E. Wall (1996). Natural occurrence of *Chondrostereum purpureum* in relation to its use as a biological control agent in Canadian forests. *Biological Control* 6: 347-352.

Dorworth, C.E. (1990). Mycoherbicides for forest weed biocontrol: The PFC enhancement process. In C. Bassett, L.J. Whitehouse, and J.A. Zabkieweiz (Eds.), *Proceedings of the International Conference on Alternatives to Chemical Control of Weeds* (pp. 116-119). Forest Research Institute Bulletin no. 155. Rotorua, New Zealand: Forest Research Institute.

Dorworth, C.E. (1994). *Method for controlling red alder using* Nectria ditissima *ATCC 74260.* U.S. Patent No. 5,340,578.

Dorworth, C.E. and B.E. Callan (1996). Manipulation of endophytic fungi to promote their utility as vegetation biocontrol agents. In S.C. Redlin and L.M. Carris (Eds.), *Endophytic fungi in grasses and woody plants* (pp. 209-216). St. Paul, MN: American Phytopathological Society Press.

Evans, H.C., J. Fröhlich, and S.F. Shamoun (2001). Biological control: Weeds. In S.B. Pointing and K.D. Hyde (Eds.), *Bio-exploitation of filamentous fungi* (pp. 349-401). Fungal Diversity Research Series, Volume 6. Hong Kong, China: Fungal Diversity Press.

French, D.W. and P.B. Schroeder (1969). Oak wilt fungus, *Ceratocystis fagacearum*, as a selective silvicide. *Forest Science* 15: 198-203.

Frohlich, J., L. Morin, and A. Gianotti (2000). Exploring the host range of *Fusarium tumidum*, a candidate bioherbicide for gorse and broom in New Zealand. In N. R. Spencer (Ed.), *Proceedings of Tenth International Symposium on Biological Control of Weeds* July 4-14, 1999 (pp. 677-678). Bozeman: Montana State University.

Funk, A., R.B. Smith, and J.A. Baranyay (1973). Canker of dwarf mistletoe swellings on western hemlock caused by *Nectria fuckeliana* var. *macrospora*. *Canadian Journal of Forest Research* 3: 71-74.

Gardner, D.E. and C.S. Hodges (1983). Leaf rust caused by *Kuehneola uredinis* on native and non-native *Rubus* species in Hawaii. *Plant Disease* 67: 962-963.

Gardner, D.E., C.S. Hodges, E. Killgore, and R.C. Anderson (1997). An evaluation of the rust fungus *Gymnoconia nitens* as a potential biological control agent for alien *Rubus* species in Hawaii. *Biological Control* 10: 151-158.

Ginns, J. and M.N.L. Lefebvre (1993). *Lignicolous corticioid fungi (Basidiomycota) of North America, systematics, distribution, and ecology*. The Mycological Society of America—Mycologia Memoir No. 19. St. Paul, MN: American Phytopathological Society Press.

Gosselin, L. and R. Jobidon (1995). Biological control of deciduous tree species in rights-of-way by the fungus *Chondrostereum purpureum*. In R.E. Gaskin and J.A. Zabkiewicz (Eds.), *Popular summaries of Second International Conference on Forest Vegetation Management* (pp. 243-245). Forest Research Institute Bulltin no. 192. Rotura, New Zealand: Forest Research Institute.

Gosselin, L., R. Jobidon, and L. Bernier (1996). Assessment of genetic variation within *Chondrostereum purpureum* from Quebec by random amplified polymorphic DNA analysis. *Mycological Research* 100: 151-158.

Gosselin, L., R. Jobidon, and L. Bernier (1999). Genetic variability and structure of Canadian populations of *Chondrostereum purpureum*, a potential biophytocide. *Molecular Ecology* 8: 113-122.

Haeussler, S. and D. Coates (1986). *Autoecological characteristics of selected species that compete with conifers in British Columbia: A literature review*. Land management report no. 33. British Columbia, Canada: Ministry of Forests, Information Service Branch.

Harper, G.J., P.G. Comeau, W.E. Hintz, R.E. Wall, R. Prasad, and E.M. Becker (1999). *Chondrostereum purpureum* as a biological control agent in forest vegetation management—II. Efficacy on sitka alder and aspen in western Canada. *Canadian Journal of Forest Research* 29: 852-858.

Harrington, C.A., J.C. Zasada, and E.A. Allen (1994). Biology of red alder (*Alnus rubra* Bong.). In D.E. Hibbs, D.S. DeBell, and R.F. Tarrant (Eds.), *The biology and management of red alder* (pp. 3-22). Corvallis: Oregon State University Press.

Harris, P. and S.F. Shamoun (2002). Biological control of weeds in Canada: Results, opportunities and constraints. In R. Claudi, P. Nantel, and E. Muckle-Jeffs (Eds.), *Alien invaders in Canada's waters, wetlands, and forests* (pp. 291-302). Ottawa, Ontario: Natural Resource Canada, Canadian Forest Service.

Hawksworth, F.G. and R.F. Scharpf (Eds.) (1984). *Biology of dwarf mistletoes*. Proceedings of the symposium. General Technical Report RM-111. Fort Collins,

CO: U.S. Department of Agriculture Forest Service, Rocky Mountain Forest and Range Experiment Station.

Hawksworth, F.G. and D. Wiens (1996). *Dwarf mistletoes: Biology, pathology, and systematics*. Agriculture Handbook 709. Washington, DC: U.S. Department of Agriculture Forest Service.

Hintz, W.E., E.M. Becker, and S.F. Shamoun (2001). Development of genetic markers for risk assessment of biological control agents. *Canadian Journal of Plant Pathology* 23: 13-18.

Jobidon, R. (1991). Some future directions for biologically based vegetation control in forestry research. *The Forestry Chronicle* 67: 514-519.

Jobidon, R. (1998). Comparative efficacy of biological and chemical control of the vegetative reproduction in *Betula papyrifera* and *Prunus pensylvanica. Biological Control* 11: 22-28.

Jobidon, R., J.R. Thibault, and J.A. Fortin (1989a). Effect of straw residues on black spruce seedling growth and mineral nutrition, under greenhouse conditions. *Canadian Journal of Forest Research* 19: 1291-1293.

Jobidon, R., J.R. Thibault, and J.A. Fortin (1989b). Phytotoxic effect of barley, oat and wheat-straw mulches in eastern Quebec forest plantations: 2. Effects on nitrification and black spruce *(Picea mariana)* seedling growth. *Forest Ecology and Management* 29: 295-310.

Johansson, T. (1987). Development of stump suckers by *Betula pubescens* at different light intensities. *Scandinavian Journal of Forest Research* 2: 77-83.

Johnston, P.R. and S.L. Parkes (1994). Evaluation of the mycoherbicide potential of fungi found on broom and gorse in New Zealand. *Proceedings New Zealand Plant Protection Conference* 47: 121-124.

Kenfield, D.S., Y.F. Hallock, J. Clardy, and G.A. Strobel (1989). Curvulin, and *O*-methyl curvulinic acid: Phytotoxic metabolites of *Drechslera indica,* which causes necrosis on purslane and spiny amaranth. *Plant Science* 60: 123-127.

Kope, H.H., S.F. Shamoun, and C. Oleskevich (1997). First report of *Colletotrichum gloeosporioides* on *Arceuthobium tsugense* subsp. *tsugense* in Canada. *Plant Disease* 81: 1095 (Abstract).

Lennox, C.L., M.J. Morris, and A.R. Wood (2000). Stumpout—Commercial production of a fungal inoculant to prevent re-growth of cut wattle stumps in South Africa. In N.R. Spencer (Ed.), *Proceedings of the Tenth International Symposium on Biological Control of Weeds* July 4-14, 1999 (p. 140). Bozeman: Montana State University.

Mallett, K.I. (1999). Process for biological control of *Calamagrostis canadensis* using a low-temperature Basidiomycete. U.S. patent no. 5,993,802.

Mallett, K.I., K.A. Schreiner, and D.A. Gaudet (2000). Effect of cottony snow mold on mortality and biomass of *Calamagrostis canadensis* under controlled environmental conditions. *Biological Control* 18: 193-198.

Manion, P.D. (1981). *Tree disease concepts*. Englewood Cliffs, NJ: Prentice Hall, Inc.

Markin, G.P. and D.E. Gardner (1993). Status of biological control in vegetation management in forestry. *Canadian Journal of Forest Research* 23: 2023-2031.

Marks, P.L. and F.H. Bormann (1972). Re-vegetation following forest cutting: Mechanisms for return to steady-state nutrient cycling. *Science* 176: 914-915.

Morris, M.J. (1989). A method for controlling *Hakea sericea* Schrad. seedlings using the fungus *Colletotrichum gloeosporioides* (Penz.) Sacc. *Weed Research* 29: 449-454.

Morris, M.J. (1991). The use of plant pathogens for biological weed control in South Africa. *Agriculture, Ecosystem, and Environment* 37: 239-255.

Morris M.J. (1997). Impact of the gall-forming rust fungus *Uromycladium tepperianum* on the invasive tree *Acacia saligna* in South Africa. *Biological Control* 10: 75-82.

Morris, M.J., A.R. Wood, and A.D. Den Breeyen (1998). Development and registration of a fungal inoculant to prevent re-growth of cut wattle tree stumps in South Africa, and a brief overview of other bioherbicide projects currently in progress. In *Fourth International Bioherbicides Workshop—Programme and abstracts,* August 6-7 (p. 15). Glasgow, Scotland: University of Strathclyde. Abstract.

Muir, J.A. (1977). Effects of the fungal hyperparasite *Colletotrichum gloeosporioides* of dwarf mistletoe *(Arceuthobium americanum)* on young lodgepole pine. *Canadian Journal of Forest Research* 7: 579-583.

Oehrens, E.B. and S.M. Gonzalez (1974). Introduction de *Phragmidium violaceum* come factor de control biologico de zarzamora *(Rubus constrictus). Agro Sur* 2: 30-33.

Oleskevich, C., S.F. Shamoun, and Z.K. Punja (1996). The biology of Canadian weeds. No. 105. *Rubus strigosus* Michx., *R. parviflorus* Nutt., *R. spectabilis* Pursh. *Canadian Journal of Plant Science* 76: 187-201.

Oleskevich, C., S.F. Shamoun, R.F. Vesonder, and Z.K. Punja (1998). Evaluation of *Fusarium avenaceum* and other fungi for potential as biological control agents of invasive *Rubus* species in British Columbia. *Canadian Journal of Plant Pathology* 20: 12-18.

Parmeter, J.R., J.R. Hood, and R.F. Scharpf (1959). *Colletotrichum* blight of dwarf mistletoe. *Phytopathology* 49: 812-815.

Perry, D.A. (1998). Effect of vegetation management practices on forest ecosystems. In R.G. Wagner and D.G. Thompson (Comp.), *Popular summaries of the Third International Conference on Forest Vegetation Management* (pp. 7-9). Forest Research Information Paper No. 141. Ontario, Canada: Ontario Ministry of Natural Resources and Ontario Forest Research Institute.

Peterson, D.J. and R. Prasad (1998). The biology of Canadian weeds. 109. *Cystisus scoparius* (L.) Link. *Canadian Journal of Plant Science* 78: 497-504.

Pitt, D.G., M.T. Dumas, R.E. Wall, D.G. Thompson, L. Lanteigne, W.E. Hintz, G. Sampson, and R.G. Wagner (1999). *Chondrostereum purpureum* as a biological control agent in forest vegetation management: I. Efficacy on speckled alder, red maple, and aspen in eastern Canada. *Canadian Journal of Forest Research* 29: 841-851.

Ramsfield, T.D., E.M. Becker, S.M. Rathlef, Y. Tang, T.C. Vrain, S.F. Shamoun, and W.E. Hintz (1996). Geographic variation of *Chondrostereum purpureum* detected by polymorphisms in the ribosomal DNA. *Canadian Journal of Botany* 74: 1919-1929.

Ramsfield, T.D., S.F. Shamoun, Z.K. Punja, and W.E. Hintz (1999). Variation in the mitochondrial DNA of the potential biological control agent *Chondrostereum purpureum*. *Canadian Journal of Botany* 77: 1490-1498.

Ramsfield, T.D., S.F. Shamoun, and B.J. van der Kamp (1999). Fungal parasites of lodgepole pine dwarf mistletoe in British Columbia. *Canadian Journal of Plant Pathology* 21: 204 (Abstract).

Ravensberg, W.J. (1998). *BioChon, effective biological and environmentally friendly product*. Berkel en Rodenrijs, the Netherlands: Koppert Biological Systems.

Scharpf, R.F. and J.R. Parmeter (1978). *Proceedings, symposium on dwarf mistletoe control through forest management*. General technical report PSW-31. U.S. Department of Agriculture Forest Service, Pacific Southwest Forest and Range Experiment Station.

Scheepens, P.C. and A. Hoogerbrugge (1989). Control of *Prunus serotina* in forests with the endemic fungus *Chondrostereum purpureum*. In E.S. Delfosse (Ed.), *Proceedings of the Seventh International Symposium on Biological Control of Weeds* (pp. 545-551). Rome, Italy: Instituto Sperimentale per la Patologia Vegetale.

Shamoun, S.F. (1997). Development of biological control strategy for management of dwarf mistletoes. In R. Sturrock (Comp.), *Proceedings of the Forty-Fifth Western International Forest Disease Work Conference*, September 15-19, 1997 (pp. 36-42). Prince George, British Columbia, Canada: Pacific Forestry Centre, Candian Forest Service.

Shamoun, S.F. (1999). The occurrence and assessment of fungal parasites as potential biological control agents of western hemlock dwarf mistletoe. *Phytopathology* 89: S71.

Shamoun, S.F. (2000). Application of biological control to vegetation management in forestry. In N.R. Spencer (Ed.), *Proceedings of the Tenth International Symposium on Biological Control of Weeds* (pp. 87-96). Bozeman: Montana State University.

Shamoun, S.F. and L. DeWald (2002). Management strategies for dwarf mistletoes: Biological, chemical and genetic approaches. In B. Geils, J. Tovar, and B. Moody (Eds.), *Mistletoes of North American conifers* (pp. 75-82). General Technical Report RMRS-GTR-98. Ogden, UT: U.S. Department of Agriculture—U.S. Forest Service, Rocky Mountain Research Station.

Shamoun, S.F. and W.E. Hintz (1998a). Development and registration of *Chondrostereum purpureum* as a mycoherbicide for hardwood weeds in conifer reforestation sites and utility rights-of-way. In *IV International Bioherbicide Workshop—Programme and abstracts,* August 6-7 (p. 14). Glasgow, Scotland: University of Strathclyde.

Shamoun, S.F. and W.E. Hintz (1998b). Development of *Chondrostereum purpureum* as a biological control agent for red alder in utility rights-of-way. In R.G. Wagner and D.G. Thompson (Eds.), *Popular summaries of the Third International Conference on Forest Vegetation Management* (pp. 308-310). Forest

Research Information Paper No. 141. Ontario: Ontario Ministry of Natural Resources and Ontario Forest Research Institute.

Shamoun, S.F., D. Macey, R. Prasad, and R. Winder (2001). *Acer, Alnus, Betula, Populus* and *Prunus* spp., weedy hardwood trees (Aceraceae, Betulaceae, Salicaceae, Rosaceae). In P. Mason and J. Huber (Eds.), *Biological control programmes in Canada, 1981-2000* (pp. 283-290). Wallingford, UK: CABI Publishing.

Shamoun, S.F. and C. Oleskevich (1999). *Fusarium avenaceum* and its use as biological control agent for *Rubus* species. U.S. patent no. 5,985,648.

Shamoun, S.F. and T.N. Sieber (1993). Isozyme and protein patterns of endophytic and disease syndrome associated isolates of *Melanconium apiocarpum* and *M. marginale*. *Mycotaxon* 49: 151-166.

Shamoun, S.F. and T.N. Sieber (2000). Colonization of leaves and twigs of *Rubus parviflorus* and *Rubus spectabilis* by endophytic fungi in a reforestation site in British Columbia. *Mycological Research* 104: 841-845.

Shamoun, S.F. and R.E. Wall (1996). Characterization of Canadian isolates of *Chondrostereum purpureum* by protein content, API ZYM and isozyme analyses. *European Journal of Forest Pathology* 26: 333-342.

Sieber, T.N., F. Sieber-Canavesi, and C.E. Dorworth (1991). Endophytic fungi of red alder *(Alnus rubra)* leaves and twigs in British Columbia. *Canadian Journal of Botany* 69: 407-411.

Smith, R.B. and A. Funk (1980). Assessing *Nectria macrospora* as a biological control agent for hemlock dwarf mistletoe. *Bi-Monthly Progress Report* 36(3): 11-12.

Sutton, B.C., S.F. Shamoun, and P.W. Crous (1996). Two leaf pathogens of *Ribes* spp. in North America, *Quasiphloeospora saximontanesis* and *Phloeosporella ribis*. *Mycological Research* 100: 979-983.

Templeton, G.E. (1982). Biological herbicides: Discovery, development, deployment. *Weed Science* 30: 430-433.

Trujillo, E.E., F.M. Latterell, and A.E. Rossi (1986). *Colletotrichum gloeosporioides*, a possible biological control agent for *Clidemia hirta* in Hawaiian forests. *Plant Disease* 70: 974-976.

Vogelgsang, S. and S.F. Shamoun (2002). Growth, sporulation, and conidia discharge of *Valdensinia heterodoxa,* a foliar pathogen of salal, as influnced by temperature and photoperiod in vitro. *Mycological Research* 106: 480-490.

Wager, V.A. (1947). Can rust kill the bramble? *Farming in South Africa* 22: 831-832.

Wagner, R.G. (1993). Research directions to advance forest vegetation management in North America. *Canadian Journal of Forest Research* 23: 2317-2327.

Wagner, R.G. and S.J. Colombo (Ed.) (2001). *Regenerating the Canadian forest: Principles and practices for Ontario*. Markham, Ontario, Canada: Fitzhenry and Whiteside.

Wagner, R.G. and D.G. Thompson (Comps.) (1998). *Third International Conference on Forest Vegetation Management Popular summaries.* Forest Research Information Paper no. 141. Ontario, Canada: Ontario Forest Research Institute, Ontario Ministry of Natural Resources.

Wagner, R.G. and J.C. Zasada (1991). Integrating plant autecology and siuicultural activities to prevent forest vegetation management problems. *Forestry Chronicle* 67: 506-513.

Wall, R.E. (1977). *Ericaceous ground cover on cutover sites in southwestern Nova Scotia*. Canadian Forest Service Information Report NX-71. Fredericton, New Brunswick: Canadian Department of Fisheries and Environment, Canadian Forestry Service.

Wall, R.E. (1986). Effects of black knot disease on pin cherry. *Canadian Journal of Plant Pathology* 8: 71-77.

Wall, R.E. (1990). The fungus *Chondrostereum purpureum* as a silvicide to control stump sprouting in hardwoods. *Northern Journal of Applied Forestry* 7: 17-19.

Wall, R.E. (1994). Biological control of red alder using stem treatments with the fungus *Chondrostereum purpureum. Canadian Journal of Forest Research* 24: 1527-1530.

Wall, R.E. and S. Hasan (1996). Management of plant pathogens for vegetation management control in forestry. In S.P. Raychaudhuri and K. Maramorosch (Eds.), *Forest trees and palms—Diseases and control* (pp. 1-19). New Delhi, India: Oxford and IBH Publishing Co., PVT, Ltd.

Wall R.E. and S. Hasan (1999). Biological control of weeds in relation to the maintenance of stable and productive ecosystems. In S.P. Raychaudhuri and K. Maramorosch (Eds.), *Biotechnology and plant protection in forestry science* (pp. 109-130). New Delhi, India: Oxford and IBH Publishing Co. Pvt. Ltd.

Wall, R.E., R. Prasad, and S. Sela (1996). *Biological control for weed trees*. U.S. patent no. 5,587,158.

Wall, R.E., R. Prasad, and S.F. Shamoun (1992). The development and potential role of mycoherbicides for forestry. *Forestry Chronicle* 68: 736-741.

Wall, R.E. and S.F. Shamoun (1990). *Experiments on vegetation control with native pathogenic fungi in the southern interior of British Columbia*. Forest Resources Development Agreement Report 134. Victoria, BC: Canadian Forest Service and British Columbia Ministry of Forests.

Walstad, J.D. and P.J. Kuch (Eds.) (1987). *Forest vegetation management for conifer production*. New York: John Wiley and Sons, Inc.

Watson, A.K. and R.E. Wall (1995). Mycoherbicides: Their role in vegetation management in Canadian forests. In P.J. Charest and L.C. Duchesne (Eds.), *Recent progress in forest biotechnology in Canada* (pp. 74-82). Canadian Forest Service Information Report PI-X-120. Ontario: Canadian Forest Service.

Webb, R.R. and S.E. Lindow (1987). Influence of environment and variation in host susceptibility on a disease of bracken fern caused by *Aschocyta pteridis*. *Phytopathology* 77: 1144-1147.

Wicker, E.F. (1967). Appraisal of biological control of *Arceuthobium campylopodium* by *Colletotrichum gloeosporioides*. *Plant Disease Reports* 51: 311-313.

Wilson, C.L. (1965). Consideration of the use of persimmon wilt as a silvicide for weed persimmons. *Plant Disease Reports* 49: 789-791.

Winder, R.S. (1995). *Mycoherbicide and method for controlling* Calamagrostis canadensis. U.S. patent no. 5,472,690.

Winder, R.S. (1999). Evaluation of *Colletotrichum* sp. and *Fusarium* spp. as potential biological control agents for marsh reed grass *(Calamagrostis canadensis)*. *Canadian Journal of Plant Pathology* 21: 8-15.

Winder, R.S. and A.K. Watson (1994). A potential microbial control of fireweed (*Epilobium angustifolium*). *Phytoprotection* 75: 19-33.

Womack, J.G. and M.N. Burge (1993). Mycoherbicide formulation and the potential for bracken control. *Pesticide Science* 37: 337-341.

Chapter 17

Characterization of Phytotoxins from Phytopathogenic Fungi and Their Potential Use As Herbicides in Integrated Crop Management

Antonio Evidente
Mohamed A. Abouzeid

INTRODUCTION

The human-versus-weed battle began with the advent of agriculture when weed diffusion first caused great damage to agricultural fields, forests, and parks; important economical infrastructures; archaeological sites; and monumental areas. In fact, many plants of agrarian and forestal interest may die when weeds grow in the same field, causing heavy losses to crop production and reforestation. Weeds represent a serious problem to normal agrarian activities where their spread also reduces pasture areas, leading to reductions in animal food.

The control of weed spread has been achieved with agrochemicals, which must usually be applied annually or semiannually. These chemicals are frequently of low specificity and are weakly biodegradable, accumulating in plants and in drinkable water, producing heavy environmental pollution, or creating problems to human and animal health.

Biological control is compatible with the environment, with high specificity, and represents a long-term solution to weed control, especially for those weeds resistant to the chemical solution. The application of biological weed control offers significant opportunities not only for farmers, nature conservationists, and other vegetation managers but also for institutions and companies that wish to sell plant protection services and products, and for the general public that demands safe food and a visually attractive and diverse environment. New herbicides have become of great interest due to either the few natural-product-derived commercial herbicides already in use or the rapidly evolving resistance to current herbicides. This supports the need for more effort to be expended on natural-product-derived herbicides and makes attractive the prospect of evaluating a vast number of undiscovered or understudied natural compounds that are likely to have biological activity.

Numerous studies have been conducted on the use of natural enemies such as insects, nematodes, viruses, bacteria, and fungi for weed control, but particular interest has been directed to phytopathogenic fungi that could be applied with safety and simplicity. They have attracted attention due to the hazards they cause to the agricultural productivity of economic interest and the environmental damage for which they are responsible.

Plant pathogens are good sources of potent phytotoxins (Abbas and Duke, 1995), as they usually kill tissues before they consume them. In spite of this, those that kill weed species have received relatively little attention in natural-product-centered herbicide discovery efforts, with a notable exception. Maculosin, produced by *Alternaria alternata,* a pathogen responsible for diseased spotted knapweed (*Centaurea maculosa* Lam.), is the first phytotoxin with a high degree of host specificity. Moreover, other studies on the biological activity of this phytotoxin and its practical application as a knapweed control agent were described in previous work by Strobel and colleagues (1991).

Almost all fungal species produce phytotoxic metabolites. Phytotoxins are bioactive compounds with different chemical structures, mechanisms of action, host specificities, and biological and ecological impacts. Generally, these metabolites interfere with plant metabolism, and many are directly responsible for the process of pathogenesis while others are not considered toxins at all (Strobel, 1982; Graniti, Durbin, and Ballio, 1989; Ballio and Graniti, 1991; Evidente, 1997; Evidente and Motta, 2001).

Fungal phytotoxins may be used as a tool in weed control. Many of them are not specific to one weed species and may be considered potential natural herbicides in native forms or as derivatives and analogs. Either their direct application or their application with derivatives or/and analogs in combination with the pathogen may increase their herbicidal activity, so improving the efficacy in terms of pathogenesis, virulence, and herbicidal selectivity (Strobel, Sugawara, and Clardy, 1987; Ayer et al., 1989; Delfosse, 1990; Strobel, 1991; Strobel et al., 1991; Bottiglieri, Zonno, and Vurro, 2000; Duke et al., 2000).

Many weed studies conducted in the past few years have concentrated only on weed diffusion with respect to many economically important agrarian crops such as cereals (oats, barley, corn, maize, millet, rice, sorghum, and wheat), tomato, potato, tobacco, sugar beet, sugar cane, legumes, and forage. Intensive research is now directed toward the isolation of phytotoxins produced by some fungi pathogenic to weeds and belonging to important toxigenic genera, e.g., *Alternaria, Ascochyta, Drechslera, Fusarium, Phoma,* and *Pyrenephora.*

The structure and stereochemical determination of phytotoxins may help to plan and realize their total stereoselective synthesis in order to obtain amounts suitable to carry out experiments in greenhouses and fields for integrated management of weeds.

Some microbial toxins may be derivatized to modulate their biological activity in terms of phytotoxicity and selectivity. Regarding the analytical methods, phytotoxins may be used as biomarkers for in vitro selection of fungal species more useful in approaching new types of natural and safe herbicides.

This chapter gives a brief overview of the results obtained in the past few years by our research group, highlighting our efforts in developing weed biocontrol strategies using a combination of low-dose herbicides, fungi, and their phytotoxic secondary metabolites. The isolation and chemical and biological characterization of phytotoxic metabolites produced by some fungal pathogens of weeds of agrarian crops and pastures are illustrated. In particular, the work carried out to biologically control common lambsquarters *(Chenopodium album)* using *Ascochyta caulina,* the phytotoxins it produces, and low-dose herbicide are also described.

SOME PHYTOTOXINS FROM WEED PATHOGENIC FUNGI

Phytotoxins produced by plant pathogenic fungi are naturally occurring substances belonging to different groups of organic compounds with interesting biological activities. The fungal metabolites toxic for the infesting plants represent a useful tool to develop new methods for integrated weed management (IWM). Therefore, the isolation, the chemical and biological characterization, the synthesis, and the derivatization of phytotoxins allows to find new natural and safe herbicides (Ballio and Graniti, 1991; Evidente and Motta, 2001).

Some examples on the use of phytotoxic metabolites extracted and purified from different fungal species by our group and applied for weed control researches are illustrated in this section.

Putaminoxin (see Figure 17.1, part 1) is the main phytotoxin isolated from the organic extract of liquid culture filtrate of *Phoma putaminum,* a fungus isolated from infected leaves of annual fleabane [*Erigeron annuus* (L.) Pers.], which is an indigenous weed from North America widely found in fields and pastures all over Europe and commonly called annual fleabane.

This new 10-macrolide was characterized by spectroscopic methods, mainly 1D and 2D (one- and two-dimensional) ^1H- and ^{13}C- NMR (nuclear magnetic resonance) and HR-EIMS (high-resolution electron ionization mass spectrometry) as *(5S)*-5-hydroxy-9-propyl-6-nonen-9-olide (see Figure 17.1, part 1), and its structure was confirmed through converting the toxin into the corresponding 5-*O*-acetyl- and 6,7-dihydro-derivatives by standard acetylation and catalytic hydrogenation, respectively, and by the conversion into the 5-*O*-formylderivative, which was found as an artifact of the extraction procedure. The absolute stereochemistry of the secondary alcohol at C-5 was determined by applying GC (gas chromatography) Horeau's method (Evidente et al., 1995). At least other four structurally related metabolites, putamonixin B, C, D, and E, were isolated from the same fungal culture extract (Evidente et al., 1997; Evidente, Capasso, Andolfi, et al., 1998a). When assayed on annual fleabane by leaf-puncture assay, as well as on some weedy and cultivated nonhost plants, putaminoxin proved to be more toxic than the putaminoxin analogs and previously cited derivatives. The latter were all inactive due to a modification of the nonenolide ring and the alkyl side chain. On the basis of these results, the structural fea-

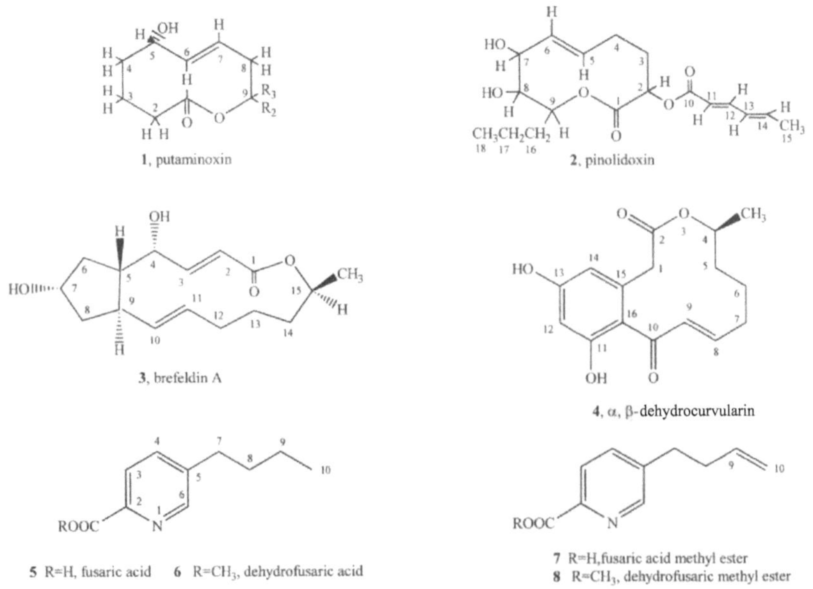

FIGURE 17.1. Phytotoxins produced by some fungi pathogenic to weeds: (1) putaminoxin *(Phoma putaminum);* (2) pinolidoxin *(Ascochyta pinodes);* (3) brefeldin A; (4) α, β-dehydrocurvularin *(Alternaria zinniae);* (5) fusaric; and (6) dehydrofusaric acids; and their methyl esters (7) and (8) *(Fusarium nygamai).*

tures that appeared to be of primary importance for the phytotoxic activity of the toxin were the presence of both the unchanged hydroxy group at C-5 and the alkyl side chain at C-9 (Evidente, Capasso, Andolfi, et al., 1998b).

Many of these structural features appear to be important for the activity of pinolidoxin (see Figure 17.1, part 2), a phytotoxic metabolite isolated from solid cultures of *Ascochyta pinodes,* a phytopathogenic fungus responsible for pea anthracnose (Evidente, Lanzétta, et al., 1993) in addition to three other minor correlated toxins that were characterized as 5,6-dihydro-, 7-epi-, and 5,6-epoxy-pinolidoxin (Evidente, Capasso, Abouzeid, et al., 1993). Pinolidoxin [2-(2,4-hexadienoyloxy)-7,8-dihydroxy-9-propyl-5-nonen-9-olide] is another nonspecific phytotoxic nonenolide which, being structurally related to Figure 17.1, part 1, may be proposed as a potential natural herbicide as well as putaminoxin. Therefore, a structure-activity relationships study was carried out using Figure 17.1, part 2, the three previously cited natural analogs, and three synthetic derivatives (the 7,8-*O,O*'-diacetyl-, the 7,8-*O,O*'-isopropylidene-, and 5,6,11,12,13,14-hexahydro-pinolidoxin). Pinolidoxin, compared to its analogs and derivatives, showed the highest phytotoxicity on both cultivated plants (pea [*Pisum sativum* L.]; sugar beet [*Beta vulgaris* (L.)] and weeds (showy crotolaria [*Crotalaria spectabilis* Roth]; fat hen [*Chenopodium album*]; buttercup oxalis [*Oxalis pes-caprae* (L.)], due to the presence of an unmodified diol system between C-7 and C-8 with the correct stereochemistry, functional groups, and the conformational freedom of the nonenolide ring (Evidente, Capasso, Andolfi, et al., 1998b). In view of their possible use as natural herbicides, the fungicide and zootoxic activities of both toxins (see Figure 17.1 parts 1 and 2) and their analogs and derivatives, as described, were assayed. The antifungal activity was assayed on *Geotrichum candidum,* a fungus very sensitive and easy to manipulate, largely used for bioassays of natural compounds. None of the compounds tested practically showed toxicity (Evidente, Capasso, Andolfi, et al., 1998b). Only the derivatives and analogs of pinolidoxin demonstrated zootoxic activity when assayed on larvae of brine shrimp (*Artemia salina* L.) where in particular the two derivatives (7,8-*O,O*'-diacetyl-, 7,8-*O,O*'-isoprophylidene-pinolidoxin) due to their higher lipophilicity, which allows them to penetrate the larvae membrane and then be activated by the hydrolysis of the protective group of the diol system between C-7 and C-8 (lethal metabolism). Furthermore, the reduced zootoxic activity of other derivatives and analogs of pinolidoxin would suggest that the toxicity on larvae of *A. salina* could be attributed to the integrity of the 7,8-diol system and the nonenolide ring of pinolidoxin, similar to that found for phytotoxicity. This means that the modification of the hexadienoyl side chain at C-2 does not affect activity (Evidente, Capasso, Andolfi, et al., 1998b).

The studies conducted on structure-activity relationships provide useful information on the variability of biological properties, with respect to the chemical structure, either the presence or absence of active group and/or chain. The availability of these metabolites in large amounts or the use of large-scale production systems could allow the testing of these toxins in greenhouses or field experiments to evaluate their potential practical application as new and original modified natural compounds (Evidente, Capasso, Andolfi, et al., 1998b).

Among the four species belonging to the genus *Xanthium,* an exotic annual weed largely spread in Australia, beach cocklebur (*X. occidentale* Bertol.), was the first to be reported and is becoming the most destructive and widespread species (Morin et al., 1994). The biological control of this noxious weed, mainly with the use of plant pathogens such as *Alternaria zinniae,* has been proposed. *Alternaria zinniae* has been considered a very good candidate to control Italian cocklebur (*X. italicum* Morretti) as well. Two main toxic metabolites were isolated from the highly phytotoxic culture filtrates of *A. zinniae*. They were chemically characterized by extensive application of spectroscopic methods (1D and 2D ^{1}H- and ^{13}C-NMR and HR-EIMS), while the structure and the configuration of the two toxins have been confirmed by X-ray analysis.

These data revealed the structure of the two toxins identical to brefeldin A (see Figure 17.1, part 3), the macrocyclic cytotoxic and antimicrobial metabolite previously isolated from other microscopic fungi (*Penicillium, Ascochyta, Alternaria,* and *Curvularia*) (Coombe, Jacobs, and Watson, 1968; Tietjen, Schaller, and Marten, 1983; Betina, 1992), and to the α,β-dehydrocurvularin (see Figure 17.1. part 4), the octaketide lactone also produced by a number of fungal species (*Alternaria, Aspergillus, Curvularia, Cercospora, Penicillium,* and *Stemphylium* spp.) (Caputo and Viola, 1977; Robeson and Strobel, 1985; Arai et al., 1989; Lai et al., 1989). Vurro and colleagues (1998) reported on the production of these two phytotoxins by a strain of *A. zinniae,* a good candidate for the biological control of *X. occidentale*.

A particular interest is directed to the activity of brefeldin A from the application point of view. Brefeldin A was particularly active in causing the fast appearance of large necrotic spots when assayed as a solution at 10^{-4} M by droplet application (15 μl) on previously punctured leaves and of *X. occidentale,* where leaves were kept in moist chambers for four days. Also on cotyledons, the toxin caused widespread necrosis. Brefeldin A was ineffective on punctured leaves of other nonhost plants (tree of heaven, *Ailanthus glandulosa* Desf.; common amaranth, *Amaranthus retroflexus* L.; pot marigold, *Calendula officinalis* L.; cucumber, *Cucumis sativus* L.; tomato, *Lycopersicon esculentum* Mill.; and black nightshade, *Solanum nigrum* L.).

The toxin solution assayed on the host plant cutting at the same concentration (10^{-4} M) caused the appearance of wide necrosis of leaf and cotyledons, and stems collapsed. In the spray assay, after one day the application of liquid culture filtrate of *A. zinniae* on host plants caused the appearance of necrosis, first at the apical edge and then on the whole leaf surface. The spray with brefeldin A solution resulted in the appearance of smaller necrosis on leaves, mainly affecting the leaf edges, and chlorosis (Vurro et al., 1998). α,β-dehydrocurvularin, in comparison to brefeldin A, showed a weak activity only in leaf puncture assay where it caused the appearance of small necrotic areas only when assayed at 10^{-3} M. The application of both brefeldin A and of α,β-dehydrocurvularin at up to 60 μg per disk caused no growth inhibition in *Geotrichum candidum*. Assayed on larvae of *A. salina,* brefeldin A caused 100 percent mortality when assayed at up to10^{-5} M, while no effect was observed for α,β-dehydrocurvularin even at concentrations ten times higher. The addition of 20 μl of culture filtrate/mL of sea water caused no effects on brine shrimp (Vurro et al., 1998).

The very good activity of both the culture filtrate *A. zinniae* and the toxin would suggest their possible direct application for controlling *Xanthium* plants. A suitable possibility could be the use of brefeldin A at very low concentration in a mycoherbicide suspension, together with *A. zinniae* conidia to block the defense reaction of *X. occidentale* and help the pathogen cause more severe disease, hence obtaining better control of this weed. Both compounds in brefeldin A and α,β-dehydrocurvularin could also indirectly improve the mycoherbicide production if used as biomarkers (i.e., there may be a positive correlation between toxin production and aggressiveness of the candidate mycoherbicide). More virulent strains of the pathogen could be selected simply by testing the in vitro production of toxic metabolites (Vurro et al., 1998).

Striga hermonthica (Del.) Benth., commonly called witchweed, is a parasitic weed that causes severe losses in many important cereal crops (corn, millet, rice, sorghum, and sugar cane); the loss in sorghum yield due to striga infestation may reach up to 70 percent. It is still very difficult to control this weed even using herbicides, cultural methods, and resistant crop varieties. Abbasher and Sauerborn (1992) suggested the use of pathogenic microrganisms, including *Fusarium nygamai* Burgess & Trimboli, proved to be particularly promising.

Two main phytotoxins were isolated from the acidic organic extract of culture filtrates of *F. nygamai*. They were identified as fusaric and 9,10-dehydrofusaric acids (see Figure 17.1, parts 5 and 7), using essentially spectroscopic methods (^{1}H- and ^{13}C-NMR and FAB [fast atom bombardment] and EIMS). Their corresponding methyl esters (see Figure 17.1, parts 6 and 8) were also isolated for the first time as naturally occurring compounds, but at very minute amounts (Capasso et al., 1996). The phytotoxicity of fusaric acids and their methyl esters were further investigated using biological assay on striga plants, seedlings, leaves, and seeds, in order to test their possible use as natural herbicides. The application of droplets of toxin solutions (8 μg per droplet) caused the appearance, after one day, of necrotic spots. There were no clear differences between the effects of different toxins. At 10^{-6} M, the toxins caused a dramatic reduction in seed germination, while at higher concentration (10^{-4} and 5×10^{-5} M) the total inhibition was observed. The use of these metabolites in combination with other cultural and biological methods could assist in controlling this weed (Zonno et al., 1996).

ASCOCHYTA CAULINA *PHYTOTOXINS*

Ascochyta caulina (P. Karst.) v.d. Aa & v. Kest. is a promising mycoherbicide for the control of *Chenopodium album* (see Photo 17.1) (Kempenaar, 1995; Kempenaar, Horsten, and Scheepens, 1996), which is also known as common lambsquarters or fat hen, a common weed of arable crops such as sugar beet and maize throughout the world (Holm et al., 1977).

The phytotoxins produced in vitro by a standard strain of *Ascochyta caulina* were hydrophilic, not being extractable by organic solvents with increasing polarity (*n*-hexane, $CHCl_3$, EtOAc, *n*-BuOH) and had a molecular weight lower than 1000 Da, as ascertained by dialysis experiments (Scheepens et al., 1997).

On this basis, the culture filtrates were lyophilized and an aliquot fractionated by a Bio-Gel P2 column eluted with ultra-pure water. Among the 14 homogeneous collected fractions, only 3 showed phytotoxic activity when assayed on host and nonhost plants. The fourth fraction proved to be a homogenous compound (235 mg/L) corresponding to the main toxin, which was named ascaulitoxin (see Figure17.2, part 9).

The structure of the ascaulitoxin, which had an amino acidic nature, was determined by spectroscopic methods (UV and FAB-MS), using essentially 1D and 2D ^{1}H and ^{13}C-NMR experiments, which allowed for assigning the chemical shifts to all protons and the corresponding carbons, and therefore to the toxin the structure of a new *N*-glucopyranoside of an unusual nonproteogenic *bis*-amino acid, namely the *N*2-β-D-glucopyranoside of the 2,4,7-triamino-5-hydroxyoctandioic acid (see Figure 17.2, part 9). This structure was confirmed by preparing the two partially penta- (see Figure 17.2, parts 10 and 11) and hexa-acetyl (see Figure 17.2, part 12) derivatives, whose spectroscopic data (HR-EI and FAB-MS and ^{1}H-NMR) were consistent to their structures and therefore assigned to the toxin (Evidente,

PHOTO 17.1. *Chenopodium album* (A) and in a maize field (B). (See corresponding plate in central color section.)

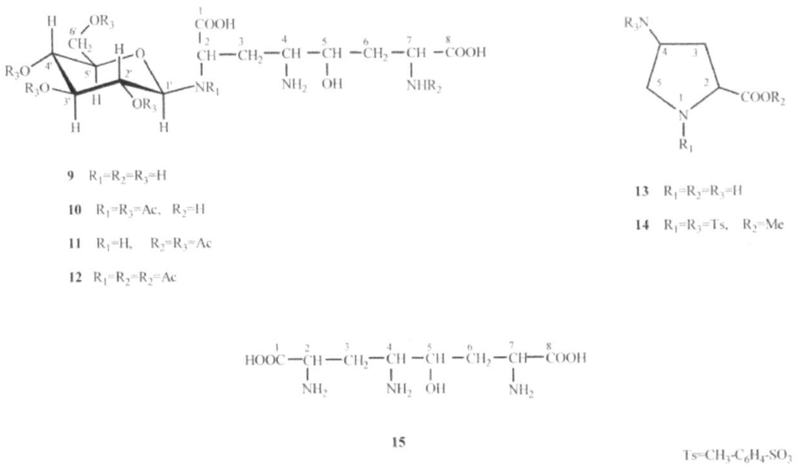

FIGURE 17.2. Structural formulae of ascaulitoxin (9); its derivatives (10, 11, 12, 14); aglycone (15); and 4-aminoproline (13).

Capasso, Cutignano, et al., 1998). The acetylation was carried out by pyridine and acetic anhydride at 90°C for 1 hectare and then left at room temperature for 4 days. The probable presence of a stable hydrogen bond between the C-4 NH_2 and the C-5 hydroxy group does not allow their derivatization.

The finding of an *N*-glucoside of an atypical *bis*-amino acid is not surprising because nonproteogenic amino acids frequently occur as free or peptide components of animals, higher plants, algae, and microorganisms including fungi (Jones, 1969; Bycroft, 1970, 1971; Bycroft and Higton, 1985).

Culture filtrate of *A. caulina* caused wide necrotic spots surrounded by chlorosis when assayed on the puncture leaves of *Chenopodium album* (see Photo 17.2). When assayed at 30 μg per droplet on punctured leaves of host and nonhost weed and cultivated plants, ascaulitoxin showed a different degree of phytotoxicity on the tested species (see Table 17.1). Ascaulitoxin caused the appearance of necrotic spots surrounded by chlorosis on the leaves of fat hen. Clear necrosis also appeared both on weeds (common sowthistle, *Sonchus oleraceous* L.; annual fleabane, *Erigeron annuus;* noogoora burr, *Xanthium occidentale;* tree of heaven *Ailanthus glandulosa*) and on cultivated plants (pea, *Pisum sativum* L.; cucumber, *Cucumis sativus*).

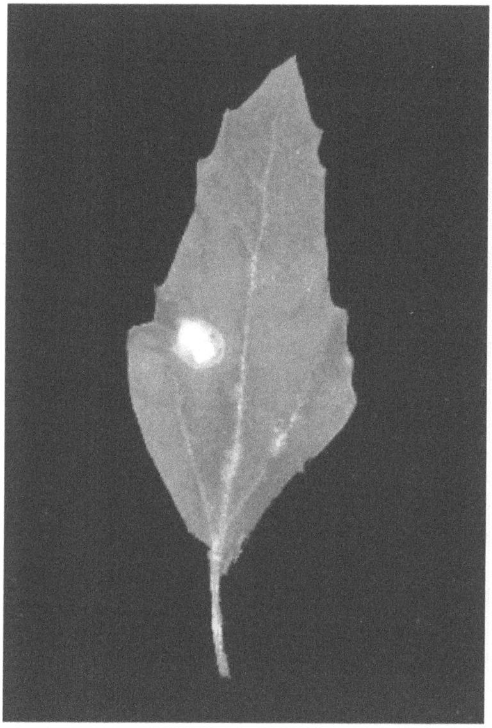

PHOTO 17.2. Leaf puncture assay of *Ascochyta caulina* culture filtrate on *Chenopodium album.* (See corresponding plate in central color section.)

Still clear, but in reduced size, were necrosis on tomato *(Lycopersicon esculentum)* and redroot pigweed *(Amaranthus retroflexus).*

When assayed, in comparison to *A. caulina* culture fitrates, on young fat hen cuttings (see Photo 17.3) at 8×10^{-4} M the toxin caused the necrosis of cotyledons, starting from their edges, while on young tomato cuttings ascaulitoxin caused a clear marginal necrosis of cotyledons and inhibited the production of adventitious roots, which were well developed in the control. At 10^{-4} M, the toxin caused 57 and 39 percent reduction in root elongation of fat hen and tomato seedlings, respectively. However, when assayed up to 100 μg per disc on fungi *(Geotrichum candidum)* as well as on bacteria (*Pseudomonas syringae* and *Escherichia coli*), ascaulitoxin showed no antimicrobial acitivity (Evidente, Capasso, Cutignano, et al., 1998).

Considering its interesting phytotoxicity on *Chenopodium album* and the lack of activity against fungi and bacteria, further studies are planned to

TABLE 17.1. Phytotoxic effect of ascaulitoxin by leaf puncture assay on host and nonhost plants.[a]

Latin name	Common name	Phytotoxicity[b]
Ailanthus glandulosa	Tree of heaven	++
Amaranthus retroflexus	Redroot pigweed	+
Chenopodium album	Fat hen	+++
Cucumis sativus	Cucumber	++
Erigeron annuus	Annual fleabane	++
Lycopersicon esculentum	Tomato	+
Pisum sativum	Pea	+
Sonchus oleraceus	Common sowthistle	++
Xanthium occidentale	Noogoora burr	++

[a]The toxin was assayed at concentration of 30 μg per droplet. The speed of symptom development varied between two and five days, depending on the species.
[b]Phytotoxicity index determined on the following scale: +, necrosis (0-1 mm); ++, necrosis (1-3 mm); +++, necrosis (3-5 mm).

evaluate the role of ascaulitoxin in plant disease development and on its mechanism of action. These aspects are important because the toxin could be used as a natural herbicide, either in combination with toxic culture filtrates of *A. caulina* or with the pathogen itself, as well as with other control methods of integrated weed management. However, any practical applications of this toxin appear to be seriously limited by the very low amounts of this metabolite present in the fungal culture filtrates. Therefore, effort should be first directed to devise a convenient and simple method of total synthesis. Because the naturally occurring toxin, as well as its aglycone, are only one of the possible sixteen stereoisomers, the determination of the relative configuration of the four chiral centers (C-2, C-4, C-5, and C-7) of ascaulitoxin (see Figure 17.2, part 9) appear to be the most pressing and relevant problem to establish its absolute stereochemistry and to realize its stereoselective synthesis.

The determination of the relative stereochemistry of the ascaulitoxin molecule was performed by NMR configuration analysis, based on the evaluation of the omo ($^3J_{HH}$)- and hetero ($^2J_{CH}$ and $^3J_{CH}$)-nuclear coupling constants, in combination with ROESY (rotating Overhauser effect spectroscopy) responses (Matsumori, Murata, and Tachibana, 1995; Matsumori et al., 1996, 1999; Wu et al., 2000; Bassarello, Bifulco, Zampella, et al., 2001). The method has proved particularly useful for the configuration analysis of polyoxygenated/polymethylated frameworks, typically found in polyketides of natural origin. We ascertained that the method was likely to

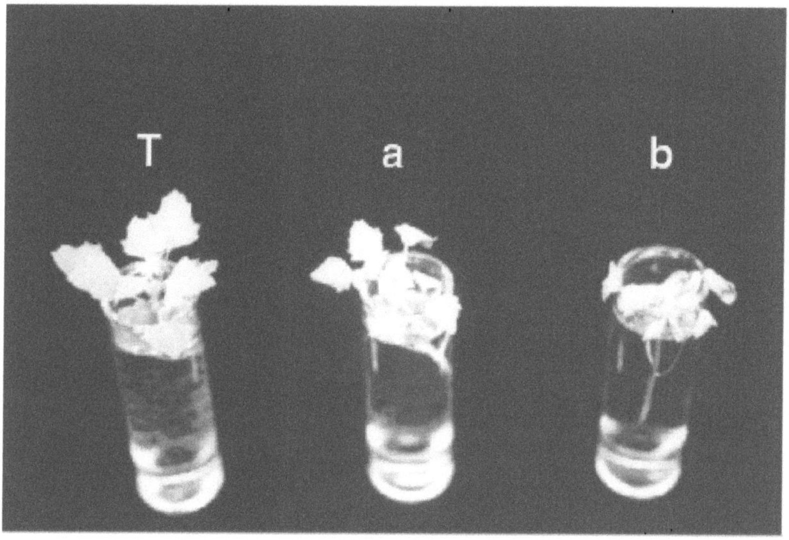

PHOTO 17.3. Effect of ascaulitoxin (a), *Ascochyta caulina* culture filtrate (b) on plant cuttings of *Chenopodium album* (T = test control). (See corresponding plate in central color section.)

display more breadth of application, since the need of oxygen (hydroxy, methoxy, etc.) substituents stemmed just on the need of suitable correlations between the values $^2J_{CH}$ couplings and the dihedral angles between the proton and the heteroatom attached to the carbon coupled to the proton. Consequently, efforts were made to test the applicability of this methodology to the nitrogen-substituted chains and the conformation, and the relative stereochemistry of the four chiral centers of ascaulitoxin was determined as reported in Figure 17.3, part 16 (Bassarello, Bifulco, Evidente, et al., 2001).

Considering that all phytotoxins produced by *A. caulina* have an amino acidic nature, the fractionation process to isolate the other toxins was modified. The culture filtrate was fractionated by cation-exchange chromatography, as described in detail later, which allowed the separation of the phytotoxins from the other eventual metabolites and the large amount of saccharose used as carbon source in the culture medium. The mixture of toxic metabolites was purified on a Bio-Gel P2 column, in the same conditions described previously, to collect five homogeneous fractions, of which the second, third, and fifth were phytotoxic. The second fraction corresponded to the ascaulitoxin, the third one to a further homogeneous metab-

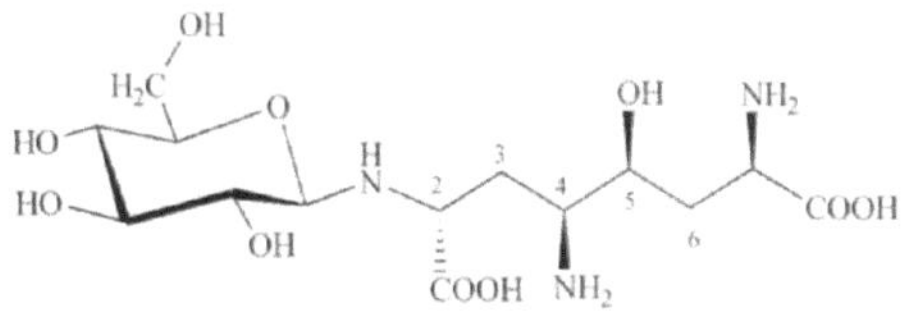

16

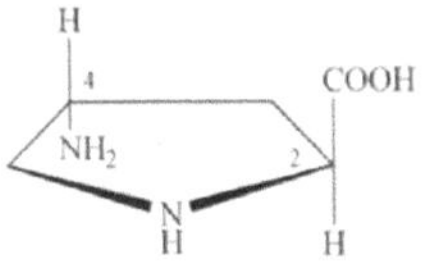

17 [*trans*-L-Enantiomer *(2S, 4R)*]

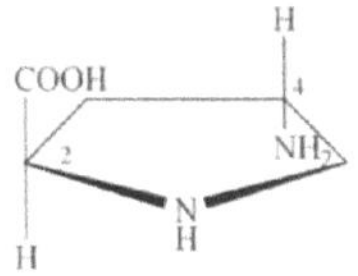

18 [*trans*-D-Enantiomer *(2R, 4S)*]

FIGURE 17.3. The relative and absolute stereochemistry of ascaulitoxin (16) and 4-aminoproline (17 and 18, respectively).

olite, which was characterized as 4-aminoproline (see Figure 17.2, part 13, 150 mg/L), and the latter consisted of a mixture of at least two metabolites.

The structure of this additional toxin (Figure 17.2, part 13) was determined using spectroscopic methods. In particular, its NMR data, compared to those of the *trans*-4-hydroxy-L-proline (see Figure 17.4, part 19), allowed assigning the structure of 4-aminoproline to the toxin. This structure was confirmed by preparing the corresponding N^1,N^4-ditosyl methyl ester derivative (see Figure 17.2, part 14). Considering that 4-aminoproline can exist as four stereoisomers, in particular the two couples of *cis*- and *trans*-enantiomers, and that biological activity of pyrrolidine derivatives depend on their substitution pattern, functionalization, and absolute configuration (Tanaka, Suzuki, and Sawanishi 1996; Thorbek, Hjeds, and Schaumburg, 1981), a study was undertaken to determine the stereochemistry of the toxin. Unfortunately, 4-aminoproline and the crystals of the corresponding N^1,N^4-ditosyl methyl ester derivative (see Figure 17.2, part 14) proved to be unsuitable for the X-ray analysis. Therefore the determination of the relative stereochemistry of the toxin was performed by a chemical method adopting a well-known synthetic procedure (Portoghese and Mikhail, 1966; Andreatta et al., 1967) as reported in Figure 17.4. In fact, its N^1,N^4-ditosyl methyl ester derivative (see Figure 17.2, part 14) showed different spectroscopic and chromatographic behavior when compared to the analog deriva-

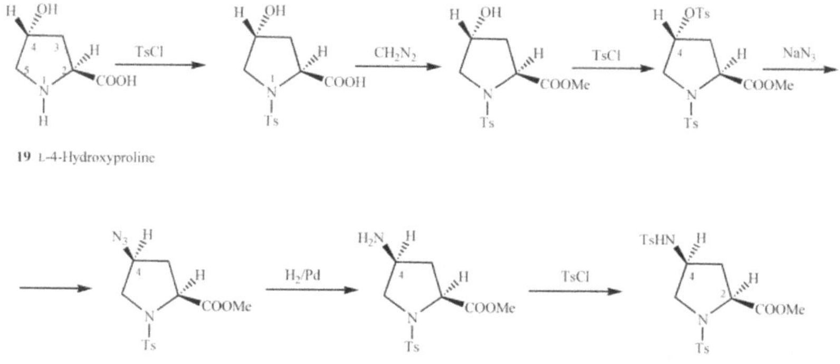

19 L-4-Hydroxyproline

20

FIGURE 17.4. The relative stereochemistry determination of 4-aminoproline.

tive of the *cis*-4-amino-L-proline (see Figure 17.4, part 20), synthesized by starting from *trans*-L-enantiomer according the reaction sequence reported in Figure 17.4. Therefore, the toxin is the *trans*-4-aminoproline, which, having an opposite optical rotation in respect to that reported for the *trans*-4-amino-L-proline (see Figure 17.3, part 17), proved to be the D-enantiomer (see Figure 17.3, part 18). The 4-aminoprolines are well-known as synthetic compounds (Mauger and Witkop, 1966; Andreatta et al., 1967), so this represents the first isolation of the *trans*-4-amino-D-proline as a naturally occurring compound and as a phytotoxin produced by *A. caulina* (Evidente et al., 2000).

Assayed at 1 mg/μL on punctured leaves (see Table 17.2), the toxin (*trans*-D-enantiomer) had a drastic effect on the host plant, causing the rapid appearance of large necrosis surrounding the puncture point. At a concentration five times lower, the toxin was still active but caused less necrosis. On other dicot leaves, the phytotoxicity varied from large necrotic areas (poppy, annual mercury, cucumber, wild cucumber), through medium ones (tree of heaven, tomato, common sowthistle), to small necrotic spots (black nightshade). An interesting aspect is the lack of toxicity when *trans*-D-enantiomer was assayed on several monocots, both cultivated (wheat, oat, barley) as well as wild (canarygrass, slender foxtail, wild oat). When tested at up to 10^{-5} M on cut young fat hen seedlings, the toxin caused wide necrosis and dryness of cotyledons, while no effects could be seen on stems (Evidente et al., 2000).

The toxin lacks antifungal and antibiotic activities when assayed at up to 50 μg per disk on *Geotrichum candidum* and on *Pseudomonas syringae* ssp. *syringae* and *E. coli,* as already described for the ascaulitoxin (Evidente,

TABLE 17.2. The phytotoxicity of the *trans*-4-amino-D-proline by leaf puncture assay on host and nonhost plants.[a]

Latin name	Common name	Phytotoxicity[b]
Ailanthus glandulosa Desf.	Tree of heaven	++
Alopecurus myosuroides Huds.	Slender foxtail	–
Avena fatua L.	Wild oat	–
Avena sativa L.	Oat	–
Chenopodium album L.	Fat hen	++++
Cucumis sativus L.	Cucumber	+++
Ecballium elaterium (L.) A. Rich	Wild cucumber	+++
Hordeum vulgare L.	Barley	–
Lycopersicon esculentum Mill.	Tomato	++
Mercurialis annua L.	Annual mercury	+++
Papaver rhoeas L.	Poppy	+++
Phalaris canariensis L.	Canarygrass	–
Solanum nigrum L.	Black nightshade	+
Sonchus oleraceus L.	Common sowthistle	++
Triticum durum Desf.	Durum wheat	–

[a]The toxin was assayed at concentration of 1 μg per μL. The speed of symptom development varied between two and five days, depending on the species.
[b]Phytotoxicity index determined on the following scale: –, no symptoms; +, necrosis (0-1 mm); ++, necrosis (1-3 mm); +++, necrosis (3-5 mm); ++++, necrosis (6-7 mm).

Capasso, Cutignano, et al. 1998), and has no zootoxicity when tested at up to 40 μg/mL of sea solution on brine shrimp larvae (*Artemia salina* L.) (Bottalico, Logrieco, and Visconti, 1989).

Considering the high phytotoxic activity of the *trans*-4-D-aminoproline (*trans*-D-enantiomer) on fat hen, the range of activity for other dicots and its inefficacy against monocots, in addition to the lack of antifungal, antibiotic, and zootoxic activity, investigations on the practical use of this toxin as natural herbicide, with or without the pathogen, promised to be very interesting. However, as already reported for ascaulitoxin, any practical application of *trans*-D-enantiomer appears to be seriously limited by the very low amounts of this metabolite purified from the fungal culture filtrates; therefore, any effort should be directed toward a simple synthesis similar to that of Figure 17.7 but starting from the appropriate, convenient, and commercially available stereoisomer of the 4-hydroxyproline.

Finally, the mixture of the most polar phytotoxic metabolites corresponding to the fifth group collected from gel-filtration chromatography (Bio-Gel P2 column) was successively purified by a reverse-phase preparative thin-layer chromatography (TLC) to obtain the main component, a third toxin, as a white homogeneous solid (150 mg/L), which in a preliminary spectroscopic investigation appeared to be similar to the 2,4,7-triamino-5-hydroxyoctandioic acid, the aglycone of the ascaulitoxin. In fact, the latter obtained by hydrogen fluoride (HF) hydrolysis of ascaulitoxin followed by purification with cation-exchange chromatography showed spectroscopic (^{1}H- and ^{13}C-NMR and ES [Electro Spray] MS) and chromatographic (TLC and ionic chromatography) properties identical to those of the third toxin. Thus, this latter can be formulated as the 2,4,7-triamino-5-hydroxyoctandioic acid (see Figure 17.2, part 15) (Evidente et al., 2001).

To purify the three phytotoxins produced by *A. caulina*, an alternative method based on ion exchange chromatography was developed to overcome difficulties and high cost, and to obtain each in a pure form (see Figure 17.5). In fact, after mild acidification (formic acid, 1M) the culture filtrates were applied to a cation-exchange chromatography (Dowex 50, H^+ form) column, which was first eluted with ultra-pure water to collect saccharose, which was used as carbon source in the culture medium, and other nonbasic compounds. The successive elution with ammonium hydroxide (1M) allowed collection of only a mixture of toxic metabolites (350 mg/L) (see Photo 17.4) (Vurro et al., 2001), which was used in greenhouse, field, and formulation experiments either alone or in combination with the

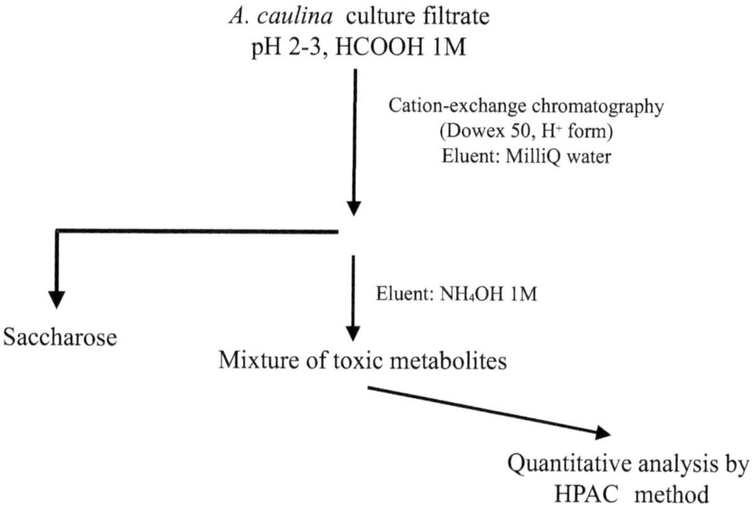

FIGURE 17.5. Fractionation of *Ascochyta caulina* culture filtrates by cation-exchange chromatography.

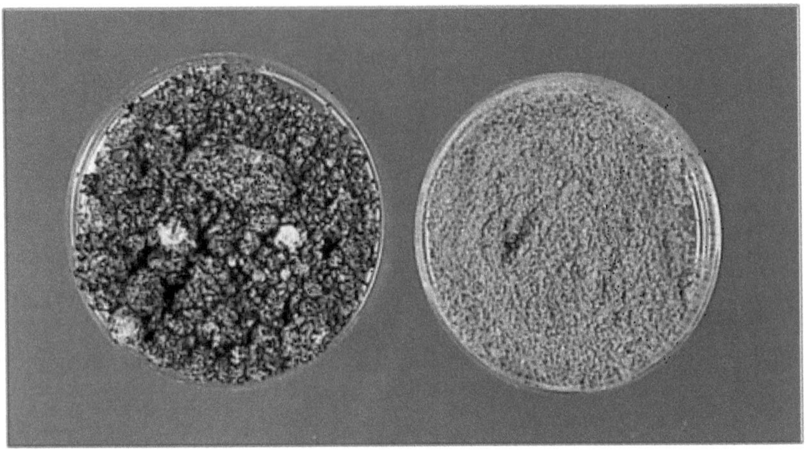

PHOTO 17.4. The *Ascochyta caulina* culture on bran (left) and the powder of the toxin mixture obtained from the purification of the liquid culture filtrates of the same fungus (right). (See corresponding plate in central color section.)

pathogen, its culture filtrates, or with a low-dose herbicide, to biocontrol host plants. The efficacy of the toxin mixture was compared with that of the culture filtrates alone or in combination with the fungus. In glasshouse experiments it showed the same toxicity as culture filtrates when applied at the same concentration as the latter (2 mg/mL). The phytotoxin mixture influenced the growth of *Chenopodium album* even at the lowest concentration. Greenhouse experiments also showed that the use of solutions from the toxin mixture (1 mg/mL) in conjunction with spores of *A. caulina* (at 10^6/mL) improved the biocontrol efficacy of this fungus by more than 30 percent (see Photo 17.5). Furthermore, the simultaneous application of toxins or fungal spores with low doses of herbicides at one-fifth of the labeled rate, such as metribuzin which acts as an inhibitor of photosynthesis at the level of photosystem II, and rimsulfuron which is an aceto-lactate synthase inhibitor, gave better results than single-agent treatments. The efficacy improvement of rimsulfuron, which is nearly ineffective against *C. album* when used at the labeled concentration, could have an interesting practical application in terms of management of herbicide resistance. Furthermore, exploration of toxin activity could expand the action spectrum of herbicides or biocontrol agents (Vurro et al., 2001).

Formulations containing different combinations of *A. caulina* conidia, its phytotoxins, and low-dose herbicides have been tested. A significant improvement in the efficacy of the fungus was achieved in glasshouse trials with an aqueous formulation containing PVA (polyvinyl alcohol, 0.1 per-

PHOTO 17.5. Effect of application of toxin mixture on *C. album* plants grown in greenhouse in pots: alone (left), together with *A. caulina* (right), *A. caulina* alone (center). (See corresponding plate in central color section.)

cent v/v), Psyllium (a plant-derived polysaccharide, 0.4 percent w/v), Sylgard 309 (a surfactant, 0.1 percent v/v), nutrients, and conidia (5×10^6/mL). Field trials have investigated the performance of *A. caulina* conidia applied at different development stages of *C. album* either as a single treatment or combined with sublethal doses of herbicides or with the fungal phytotoxins. With the available formulation, favorable weather conditions are needed to obtain infection in the field. The efficacy of the strain of *A. caulina* used so far has proved to be inadequate to justify its development as a bioherbicide. This is probably due to its low virulence (Netland et al., 2001).

Considering the amino acid nature of the toxins produced by the standard strain of *A. caulina* (ascaulitoxin, its aglycone, and *trans*-4-D-aminoproline), a specific method for the qualitative and quantitative analyses of these toxins by HPAC-PAD (high performance anionic chromatography-pulsed amperometric detector) on a dionex chromatograph was developed. This method allows a direct determination of amino acids avoiding the derivatization required for spectrophotometric and fluorescence detection methods (Martens and Frankenberger, 1992).

Ascaulitoxin, *trans*-4-D-aminoproline and 2,4,7-triamino-5-hydroxyoctandioic acid, were well separated on an AminoPac PA1 analytical column with an AminoPac PA1 guard column. The mobile phase consisted of 23 mM sodium hydroxide and 7 mM sodium tetraborate solution. These represented the best analysis conditions for the toxins (Figure 17.6). The calibration curves for metabolite determination were obtained using toxin samples purified from the culture filtrate of *A. caulina* standard strain according to the purification procedure just described. The developed method proved to be relatively simple, rapid, sensitive, and reproducible.

The application of this method to the culture filtrates of different strains of *A. caulina* for the qualitative and quantitative analyses of toxins produced in vitro required the preliminary separation of saccharose, which overlapped to the peak of ascaulitoxin. This preliminary purification was carried out by cation-exchange chromatograghy as described earlier, which furnished the mixture of the toxic metabolites. This was then analyzed by HPAC-PAD on a Dionex chromatograph, and in all cases the three toxins were well separated. The chromatographic profile obtained starting from the culture filtrates of the standard strain is given in Figure 17.7. A recovery rate close to 100 percent was observed for all toxins (Evidente et al., 2001). This method was applied to quantitatively estimate the toxin content in culture filtrates of ten different *A. caulina* strains, and the data obtained are illustrated in Figure 17.8. The strains synthesized all three toxins, and two strains (AC-2 and AC-14) were better producers than the standard one. Only one strain (AC-46) resulted in higher ascaulitoxin production, while

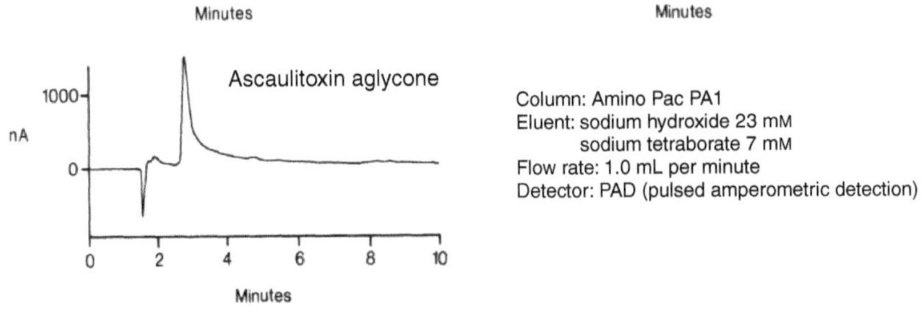

FIGURE 17.6. Chromatographic profile of ascaulitoxin, its aglycone, and *trans*-4-amino-D-proline by HPAC.

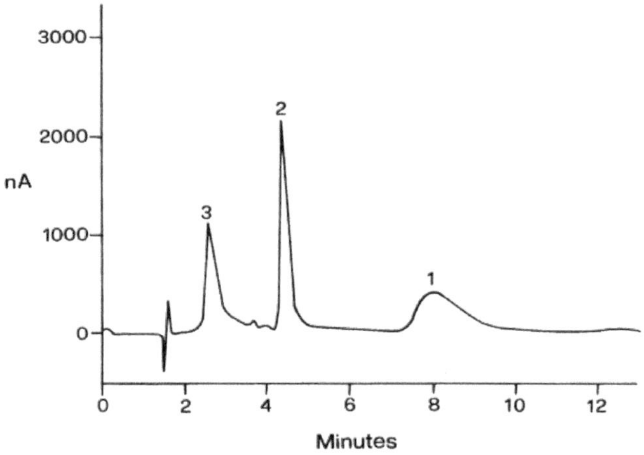

FIGURE 17.7. The HPAC-PAD chromatogram of the culture filtrate of *Ascochyta caulina* (strain AC-1) initially purified by cation-exchange chromatography. Key to peak identity: 1 = ascaulitoxin; 2 = *trans*-4-amino-D-proline; 3 = ascaulitoxin aglycone.

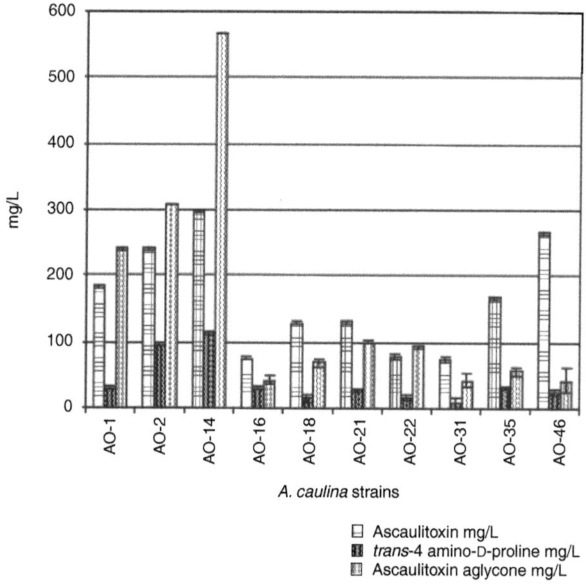

FIGURE 17.8. The toxins content in culture filtrates of different *Ascochyta caulina* strains.

the others caused lower production. It is also important to note the variation in toxin content reported between the tested culture filtrates of 74-298, 6-115, and 42-565 mg/L, for ascaulitoxin, *trans*-D-enantiomel, and aglycone, respectively, suggesting that there is no correlation in the ability of strains to produce different toxins. The developed HPAC-PAD analysis method proved to be efficient and sensitive for the identification and quantitative estimation of ascaulitoxin, its aglycone, and *trans*-4-amino-D-proline in crude samples (Evidente et al., 2001).

The main aim of this project funded by the European Commission and involving several research institutions, was either to isolate new promising strains of the pathogen or to enhance its efficacy which, although under standard conditions reached 70 percent weed control, was not sufficient for market interest. Unfortunately, a difficulty arises during weed pathogen management, which is the need to evaluate and compare the virulence of different strains for selecting the best isolates. Sometimes, non-host-specific toxins have been reported to be virulence factors (Proctor, Hohn and McCormick, 1995); this means, if a positive correlation is present between toxin accumulation and fungal virulence, selecting the most aggressive strains could be obtained through the evaluation of in vitro production of toxic metabolites. It is important to note that if the production of toxins is correlated to the strain virulence, then they could be used as biomarkers, and the quantitative evaluation of the toxins produced by the different strains could represent an important tool to identify and select the best strains to be used in biological and integrated management of *C. album.*

We conclude our discussion by highlighting some of our results obtained during the past years and their use in weed biocontrol strategies where further experiments are still in progress to overcome the important problems that arise during the practical application of phytotoxins in integrated crop management. Unfortunately, these are seriously limited by the very low amounts of bioactive compounds frequently produced by weed pathogenic fungi. Therefore, the streostructural determination of phytotoxins could assist in realizing their simple and convenient total synthesis, therefore furnishing amounts of phytotoxins sufficient to carry out experiments on their biological activities, mode of action, and toxicity. Furthermore, the synthetic phytotoxins could be used either to develop methods allowing the selection of fungal strains for mycoherbicide application or to prepare derivatives and analogs with modulated and/or increased biological activity and specificity, which could be used in greenhouse or field experiments in view of their potential practical application.

REFERENCES

Abbas, H.K. and S.O. Duke (1995). Phytotoxins from plant pathogens as potential herbicides. *Journal of Toxicology—Toxin Review* 14: 523-543.

Abbasher, A.A. and J. Sauerborn (1992). *Fusarium nygamai,* a potential bioherbicide for *Striga hermonthica* control in sorghum. *Journal of Biological Control* 2: 291-296.

Andreatta, R.H., V. Nair, A.V. Robertson, and W.R.J. Simpson (1967). Synthesis of *cis* and *trans* isomers of 4-chloro-L-proline, 4-bromo-L-proline, and 4-amino-L-proline. *Australian Journal of Chemistry* 20: 1493-1509.

Arai, K., B. J. Rawlings, Y. Yoshizawa, and C. Vederas (1989). Biosynthesis of antibiotic A26771B by *Penicillium turbatum* and dehydrocurvularin by *Alternaria cinerariae*: Comparison of stereochemistry of polyketide and fatty acid enoyl thiol ester reductase. *Journal of American Chemical Society* 111: 3391-3399.

Ayer, S.W., B.G. Isaac, D.M. Krupa, K.E. Crosby, L.J. Letendre, and R.J. Stonard (1989). Herbicidal compounds from microorganisms. *Pesticide Science* 27: 221-223.

Ballio, A. and A. Graniti (1991). Phytotoxins and their involvement in plant disease *Experientia* 47: 751-826.

Bassarello, C., G. Bifulco, A. Evidente, R. Riccio, and L. Gomez-Paloma (2001). Stereochemical studies on ascaulitoxin: Extension of the *J*-based NMR configurational analysis of a nitrogen substituted system. *Tetrahedron Letters* 42: 8611-8613.

Bassarello, C., G. Bifulco, A. Zampella, M.V. D'Auria, R. Riccio, and L. Gomez-Paloma (2001). Stereochemical sudies on sphinxolide: Advances in the *J*-based NMR determination of the relative configuration of flexible system. *European Journal of Organic Chemistry* 1: 39-44.

Betina, V. (1992). Biological effects of the antibiotic brefeldin A (decumbin, cyanein, ascotoxin, synergisidin): A retrospective. *Folia Microbiology* 37: 3-11.

Bottalico, A., A. Logrieco, and A. Visconti (1989). *Fusarium* species and their mycotoxins in infected cereals in the field and stored grains. In J. Chelkowsky (Ed.), *Fusarium: Mycotoxins, taxonomy and pathogenicity* (pp. 85-119). Amsterdam, the Netherlands: Elsevier.

Bottiglieri, A., M.C. Zonno, and M. Vurro (2000). I bioerbicidi contro le piante infestanti. *L'informatore agrario* 13: 69-73.

Bycroft, B.W. (1970). Amino-acids. In G.T. Young (Ed.), *Amino-acids, peptides, and proteins,* Volume 2 (pp. 1-5). London: The Chemical Society Burlington House.

Bycroft, B.W. (1971). Amino-acids. In G.T. Young (Ed.), *Amino-acids, peptides, and proteins,* Volume 3 (pp. 1-5). London: The Chemical Society Burlington House.

Bycroft, B.W. and A.A. Higton (1985). Amino acids, peptides and proteins. In R.H. Thomson (Ed.), *The chemistry of natural products* (pp. 238-297). Glasgow, UK: Blackie.

Capasso, R., A. Evidente, A. Cutignano, M. Vurro, M.C. Zonno, and A. Bottalico (1996). Fusaric and 9,10-dehydrofusaric acids and their methyl esters from *Fusarium nygamai*. *Phytochemistry* 41: 1035-1039.

Caputo, O. and F. Viola (1977). Isolation of α,β-dehydrocurvularin from *Aspergillus aureofulgens*. *Planta Medica* 31: 31-32.

Coombe, R.R., J.J. Jacobs, and T.R. Watson (1968). Constituents of some *Curvularia* species. *Australian Journal of Chemistry* 21: 783-788.

Delfosse, E.S. (Ed.) (1990). *Proceedings of the VII International Symposium on Biological Control of Weeds*. March 6-11. Rome, Italy: Ministero dell'Agricoltura e delle Foreste and Commonwealth Scientific and Industrial Research Organizations.

Duke, S.O., F.E. Dayan, J.G. Romagni, and A.M. Rimando (2000). Natural products as sources of herbicides: Current status and future trends. *Weed Research* 40: 99-111.

Evidente, A. (1997). Bioactive metabolites from phytopathogenic fungi and bacteria. In S.G. Pandalai (Ed.), *Recent research developments in phytochemistry* (pp. 255-292). Trivandrum, India: Research Signpost.

Evidente, A., A. Andolfi, M. Vurro, and M.C. Zonno (2001). Determination of *Ascochyta caulina* phytotoxins by high-performance anion exchange chromatography and pulsed amperometric detection. *Phytochemical Analysis* 12: 383-387.

Evidente A., A. Andolfi, M. Vurro, M.C. Zonno, and A. Motta (2000). *Trans*-4-aminoproline, a phytotoxic metabolite with herbicidal activity produced by *Ascochyta caulina*. *Phytochemistry* 53: 231-237.

Evidente, A., R. Capasso, M.A. Abouzeid, R. Lanzetta, M. Vurro, and A. Bottalico (1993). Three new toxic pinolidoxins from *Ascochyta pinodes*. *Journal of Natural Products* 56: 1937-1943.

Evidente, A., R. Capasso, A. Andolfi, M. Vurro, and M.C. Zonno (1998a). Putaminoxins D and E from *Phoma putaminum*. *Phytochemistry* 48: 941-945.

Evidente, A., R. Capasso, A. Andolfi, M. Vurro, and M.C. Zonno (1998b). Structure-activity relationship studies of putaminoxins and pinolidoxins, phytotoxic nonenolides produced by phytopathogenic *Phoma* and *Ascochyta* species. *Natural Toxins* 6: 183-188.

Evidente A., R. Capasso, A. Cutignano, O. Taglialatela-Scafati, M. Vurro, M.C. Zonno, and A. Motta (1998). Ascaulitoxin, a phytotoxic *bis*-amino acid *N*-glucoside from *Ascochyta caulina*. *Phytochemistry* 48: 1131-1137.

Evidente, A., R. Lanzetta, R. Capasso, A. Andolfi, A. Bottalico, M. Vurro, and M.C. Zonno (1995). Putaminoxin, a phytotoxic nonenolide from *Phoma putaminum*. *Phytochemistry* 40: 1637-1641.

Evidente, A., R. Lanzetta, R. Capasso, A. Andolfi, M. Vurro, and M.C. Zonno (1997). Putaminoxins B and C from *Phoma putaminum*. *Phytochemistry* 44: 1041-1045.

Evidente, A., R. Lanzetta, R. Capasso, M. Vurro, and A. Bottalico (1993). Pinolidoxin, a phytotoxic nonenolide from *Ascochyta pinodes*. *Phytochemistry* 34: 999-1003.

Evidente, A., and A. Motta (2001). Phytotoxins from fungi, pathogenic for agrarian, forestal and weedy plants. In C. Tringali (Ed.), *Bioactive compounds from natural sources* (pp. 473-525). London: Taylor and Francis.

Graniti, A., R.D. Durbin, and A. Ballio (Ed.) (1989). *Phytotoxins and plant pathogenesis*. NATO ASI Series, Series H, Volume 27. Berlin, Germany: Springer-Verlag.

Holm L.G., D.L. Pluckett, J.V. Pancho, and J.P. Herberger (1977). *The world's worst weeds (distribution and biology)* (pp. 84-91). Honoloulou: University Press of Hawaii.

Jones, J.H. (1969). Amino-acids. In G.T. Young (Ed.), *Amino-acids, peptides, and proteins* Volume 1 (pp. 1-3). London: The Chemical Society Burlington House.

Kempenaar, C. (1995). Studies on the biological control of *Chenopodium album* by *Ascochyta caulina*. Doctoral dissertation, University of Wageningen, the Netherlands.

Kempenaar, C., P.J.F.M. Horsten, and P.C. Scheepens (1996). Spore germination and disease development of *Chenopodium album* after application of *Ascochyta caulina* as a mycoherbicide. *European Journal of Plant Pathology* 102: 143-153.

Lai, S., Y. Shizuri, S. Yamamura, K. Kawai, Y. Terada, and H. Furukawa (1989). Novel curvularin-type metabolites of a hybrid strain me 0005 derived from *Penicillium citreo-viride* B. IFO 6200 and 4692. *Tetrahedron Letters* 30: 2241-2244.

Martens, D.A. Jr. and W.T. Frakenberger (1992). Pulsed amperometric detection of amino acids separated by anion exchange chromatography. *Journal of Liquid Chromatography* 15: 423-439.

Matsumori, N., D. Kaneno, M. Murata, H. Nakamura, and K. Tachibana (1999). Stereochemical determination of acyclic structures based on carbon-proton spin-coupling constants: A method of configuration analysis for natural products. *Journal of Organic Chemistry* 64: 866-876.

Matsumori, N., M. Murata, and K. Tachibana (1995). Conformational analysis of natural products using long-range carbon-proton coupling constants: Three-dimensional structure of okadaic acid solution. *Tetrahedron* 51: 12229-12238.

Matsumori, N., T. Nonomura, M. Sasaki, M. Murata, K. Tachibana, M. Satake, and T. Yasumoto (1996). Long-range carbon-proton coupling for stereochemical assignment of acyclic structures in natural products: Configuration of the C5-C9 portion of maitotoxin. *Tetrahedron Letters* 37: 1269-1272.

Mauger, A.B. and B. Witkop (1966). Analogs and homologs of proline and hydroxyproline. *Chemical Review* 66: 47-66.

Morin, L., B.A. Auld, J.F. Brown, and M.A. Cholili (1994). Pathogenic fungi occurring on the noogoora burr complex in Australia. *Proceedings Linnean Society NSW* 114: 133-148.

Netland, J., L.C. Dutton, M.P. Greaves, M. Baldwin, M. Vurro, A. Evidente, G. Einhorn, P.C. Scheepens, and L.W. French (2001). Biological control of *Chenopodium album* L. in Europe. *BioControl* 46: 175-196.

Portoghese, P.S. and A.A. Mikhail (1966). Bicyclic bases: Synthesis of 2,5-diazabicyclo[2.2.1]heptanes. *Journal of Organic Chemistry* 4: 1059-1062.

Proctor, R.H., T.M. Hohn, and S.P. McCormick (1995). Reduced virulence of *Gibberella zeae* caused by disruption of a trichothecene toxin biosynthetic gene. *Molecular Plant Microbe Interaction* 8: 593-601.

Robeson, D.J. and G.A. Strobel (1985). The identification of a major phytotoxic component from *Alternaria macrospora* as α,β-dehydrocurvularin. *Journal of Natural Products* 48: 139-141.

Scheepens, P.C., C. Kempenaar, C. Andreasen, T. Eggers, J. Netland, and M. Vurro (1997). Biological control of the annual weed *Chenopodium album,* with emphasis on the application of *Ascochyta caulina* as a microbial herbicide. *Integrated Pest Management Review* 2: 71-76.

Strobel, G.A. (1982). Phytotoxins. *Annual Review of Biochemistry* 51: 309-333.

Strobel, G.A. (1991). Il controllo biologico delle erbe infestanti. *Le Scienze* 277: 56-63. Italian edition of *Scientific American,* New York: Scientific American, Inc.

Strobel, G.A., D. Kenfield, G. Bunkers, F. Sugawara, and J. Clardy (1991). Phytotoxins as potential herbicides. *Experientia* 47: 819-826.

Strobel, G.A., F. Sugawara, and J. Clardy (1987). Phytotoxins from plant pathogens of weedy plants. In G.R. Waller (Ed.), *Allelochemicals: Role in agriculture and forestry* (pp. 516-523). ACS Symposium Series 330. Washington, DC: American Chemical Society.

Tanaka, K., H. Suzuki, and H. Sawanishi (1996). Asymmetric synthesis of (*2R*, *4S*)-4-amino-4-carboxy-2-methylpyrrolidine and (*2R*, *4S*)-4-amino-2-carboxy-2-ethylpirrolidine as novel 2-alkyl-substituted (-)-curcurbitine analogues. *Heterocycles* 43: 205-219.

Thorbek, P., H. Hjeds, and K. Schaumburg (1981). Syntheses and ^{1}H-NMR spectroscopic investigations of some pyrrolidine carboxylic acids designed as potential glial GABA uptake inhibitors. *Acta Chemica Scandinavica* B35: 473-479.

Tietjen, K.G., E. Schaller, and U. Matern (1983). Phytotoxins from *Alternaria cartami* Chowdhuri: Structural identification and physiological significance. *Physiological Plant Pathology* 23: 387-400.

Vurro, M., A. Evidente, A. Andolfi, M.C. Zonno, F. Giordano, and A. Motta (1998). Brefeldin A and α,β-dehydrocurvularin, two phytotoxins from *Alternaria zinniae*, a biocontrol agent of *Xanthium occidentale. Plant Science* 138: 67-79.

Vurro, M., M.C. Zonno, A. Evidente, A. Andolfi, and P. Montemurro (2001). Enhancement of efficacy of *Ascochyta caulina* to control *Chenopodium album* by use of phytotoxins and reduced rates of herbicides. *Biological Control* 21: 182-190.

Wu, M., T. Okino, L.M. Nogle, B.L. Marquez, R.T. Williamson, N. Sitachitta, F.W. Berman, T.F. Murray, K. McGough, R. Jacobs, et al. (2000). Structure, synthesis and biological properties of kalkitoxin, a novel neurotoxin from the marine cyanobacterium *Lyngbya majuscula. Journal of the American Chemical Society* 122: 12041-12042.

Zonno, M.C., M. Vurro, R. Capasso, A. Evidente, A. Cutignano, J. Suaerborn, and H. Thomas (1996). Phytotoxic metabolites produced by *Fusarium nygamai* from *Striga hermonthica.* In V.C. Moran and J.H. Hoffmann (Eds.), *Proceedings of the IX International Symposium on Biological Control of Weeds,* January 21-26 (pp. 223-226). Cape Town, South Africa: University of Cape Town.

Chapter 18

Applications of Soil and Rhizosphere Microorganisms in Sustainable Weed Management

Robert J. Kremer

INTRODUCTION

The goal of sustainable agricultural systems to manage weeds with no or very few chemicals relies partly on the use of biological controls to limit weed competition (Reganold, Papendick, and Parr, 1990). However, biological control as a single-tactic approach for controlling weeds in cultivated crops has not been as efficacious as those approaches involving herbicides and other cultural methods. The classic strategy of biological control in which host-weed-specific organisms are introduced into weed-infested sites and allowed to establish to population levels that regulate weed densities requires long periods (>1 year) for adequate control and is clearly unsuitable for use in annual crops. The inundative or mass-exposure strategy attempts to overwhelm a weed infestation with massive numbers of the biotic agent in order to attain weed control in the year of release and is the approach that has received the most attention in developing biological control for annual crops.

The inundative biological control strategy most applicable for use in annual crops is the *bioherbicide* approach, which involves application of weed pathogens in a manner similar to herbicide applications. Because many bioherbicides have been developed using selected plant pathogenic fungi, the term *mycoherbicide* is often used in reference to these fungal preparations. The best opportunities for the practical use of bioherbicides in sustainable agriculture are for those that effectively and consistently suppress economically important and difficult-to-control weeds.

Constraints to development of mycoherbicides include specific requirements of high humidity (dew period) and warm temperatures for maximum organism activity, which necessitates development of special formulations to assure effectiveness after delivery of the agent in the field. Bioherbicide formulations must maintain or enhance efficacy of the biocontrol agent yet be compatible with conventional field application systems (Boyetchko et al., 1998). Improved formulations for bioherbicides are under study to overcome humidity and temperature requirements. For example, a granular formulation of the fungal pathogen *Phomopsis convolvulus* applied preemergence to soil resulted in substantial reductions (90 to 100 percent) of field bindweed (*Convolvulus arvensis* L.) in field trials (Vogelgsang, Watson, and DiTomasso, 1998). Soilborne fungi have become important bioherbicide candidates because these fungi are applied directly to soils and reduce weed populations through decay of seeds prior to emergence or kill seedlings shortly after emergence (Jones and Hancock, 1990). A group of bacteria under intensive investigation for bioherbicidal potential are *deleterious rhizobacteria* (DRB), which differ from bacterial pathogens in that they are nonparasitic bacteria colonizing plant roots and suppress plant growth without invading the root tissues (Kremer and Kennedy, 1996).

The objectives of this chapter are to identify the role of soil and rhizosphere microorganisms in weed management, including their integration into current systems; to develop an understanding of factors affecting their successful use in both conventional and sustainable management systems; and to assess the prospects of developing strategies for exploiting native weed-suppressive microorganisms in biologically based weed management.

DELETERIOUS RHIZOBACTERIA

Deleterious rhizosphere bacteria are predominantly saprophytic organisms that live on or in plant roots, surviving on organic compounds released by plant root cells (Schippers, Bakker, and Bakker, 1987). Although these bacteria do not parasitize the plant as true pathogens, they may penetrate deep into the healthy root tissue, particularly in the intercellular spaces (Schippers, Bakker, and Bakker, 1987). They frequently produce phytotoxic metabolites that inhibit plant growth by disrupting various plant physiological processes (Alström and Burns, 1989; Bolton and Elliott, 1989). Electron microscopy of DRB colonizing the rhizoplane and endorhizal cells of leafy spurge (*Euphorbia esula* L.) revealed cell wall and membrane abnormalities apparently due to production of phytotoxins and/or enzymes by the bacteria, which consequently inhibited seedling growth (Souissi, Kremer, and White, 1997).

Many rhizobacteria produce a variety of secondary metabolites that influence physiological processes in plants and other microorganisms (Arshad and Frankenberger, 1998). Phytotoxins associated with the fluorescent pseudomonads, a diverse group of bacteria abundant in rhizospheres, have been most intensively studied (Arshad and Frankenberger, 1998; Dowling and O'Gara, 1994; Leisinger and Margraff, 1979). Phytotoxic compounds produced by rhizobacteria include volatile compounds (i.e., hydrogen cyanide), excessive quantities of plant growth hormones (i.e., auxins), siderophores, compounds with antimicrobial properties (i.e., phloroglucinol-type compounds), and other complex metabolites directly involved in phytotoxicity (Arshad and Frankenberger, 1998; Gurusiddaiah et al., 1994; Leisinger and Margraff, 1979). As more evidence accumulates that indicates metabolites produced by rhizobacteria play key roles in phytotoxicity, the ability to select isolates for effective biological control and to manipulate resident rhizobacteria for better activity in soils will improve.

Deleterious rhizobacteria have been largely overlooked as weed biocontrol agents; however, their greatest impact may be on management of weed problems specific to certain cropping systems. Isolates of DRB specific for management of downy brome (*Bromus tectorum* L.) in wheat (*Triticum aestivum* L.) in the Pacific Northwest are under development as bioherbicides (Kennedy et al., 1991). Several strains of rhizobacteria from over 2,000 accessions isolated from prairie soils in Canada have been selected for inhibition of downy brome and green foxtail [*Setaria viridis* (L.) Beauv.] seed germination and seedling growth (Boyetchko, 1999). Many of these isolates are undergoing development as biocontrol agents for spring and winter cereal cropping systems.

The rhizobacterial approach to weed management involves the application of inoculum in high enough numbers to the soil or on vegetative residues to achieve rapid initiation of growth-inhibitory activity (Kremer and Kennedy, 1996). The goal is not complete killing or eradication of the weed population but the reduction of the competitive pressure of the weeds growing with the crop. Decreased competitive ability includes inhibition of biomass accumulation, reduced density, and reduced seed production by target weeds.

Development of DRB As Biocontrol Agents

The major issues for the development of DRB as weed biocontrol agents are host specificity and efficacy (Kremer and Kennedy, 1996). Prior to commercial development, DRB isolates must be screened for conclusive specificity toward weed host(s) with no detrimental effects on nontarget plant species. *Pseudomonas fluorescens* strain D7 (Pf D7), the potential biological control DRB for downy brome, was tested for plant growth effects on 42 different grass and broadleaf species representing crops, weeds, and rangeland species common to the western or midwestern United States (Kennedy, Johnson, and Stubbs, 2001). Among all plant species, only downy brome root growth was significantly inhibited by Pf D7 in plant-soil bioassays. Because Pf D7 inhibits only downy brome, it may be an effective biocontrol agent that does not harm nontarget species.

In addition, it must be shown that the DRB agent can provide a satisfactory level of control at an acceptable application rate and that it can easily be integrated into standard pest control practices (Charudattan, 1989). From a commercial standpoint, extreme host specificity may be less desirable than the ability to suppress several weed species. However, the likelihood of increased potential for crop inhibition is very high, so only DRB with narrow host ranges have been field-tested to avoid possible detrimental effects on crops.

Several economically important weed species have been targeted in biocontrol projects in Canada and the United States involving DRB (see Table 18.1) (Kremer, 2002). DRB isolates showing potential for weed suppression have been identified for velvetleaf (*Abutilon theophrasti* Medik.), morning glory (*Ipomoea* spp.), cocklebur (*Xanthium strumarium* L.), pigweed (*Amaranthus* spp.), lambsquarters (*Chenopodium* spp.), smartweed (*Polygonum* spp.), and jimsonweed (*Datura stramonium* L.) (Kremer et al., 1990). Growth inhibition due to rhizobacteria has also been reported for grass weeds including downy brome (Kennedy et al., 1991; Boyetchko, 1999), Japanese brome (*Bromus japonicus* Thurb. ex Murr.) (Harris and Stahlman, 1996), jointed goatgrass (*Aegilops cylindrica* Host.) (Kennedy, Ogg, and Young, 1992), and foxtail species (*Setaria* spp.) (Boyetchko, 1999; Li and Kremer, 2000). Other target weeds for DRB suppression include leafy spurge and Russian knapweed [*Acroptilon repens* (L.) DC.] (Caesar, 1994). Most of the weed species under study infest cereal and row crops, but a few are perennial weeds of rangelands and forest ecosystems (Kremer, 2002). Selected DRB are intended for soil application but might also be applied directly to growing weeds as a postemergence control strategy. For example, cultures and cell-free supernatants of DRB strains sprayed on common chickweed [*Stellaria media* (L.) Vill.], common lambsquarters (*Chenopodium album* L.), and field pennycress (*Thlaspi arvense* L.) in greenhouse and field experiments reduced plant biomass and survival (Weissmann and Gerhardson, 2001). The preliminary results suggest DRB might be used for selective weed control in growing crops through a one-time foliar application or as a follow-up to initial soil incorporation of the same strains.

TABLE 18.1. Representative deleterious rhizobacteria selected for biological control of weeds.

Weed target	**Rhizobacteria**	**References**
Downy brome (*Bromus tectorum* L.)	*Pseudomonas, Xanthomonas*	Kennedy et. al., 1991; Harris and Stahlman, 1996; Boyetchko, 1999
Japanese brome (*Bromus japonicus* Thunb. ex Murr.)	*Pseudomonas, Xanthomonas, Enterobacter*	Harris and Stahlman, 1996; Kennedy, Ogg, and Young, 1992; Kennedy, Johnson, and Stubbs, 2001
Jointed goatgrass (*Aegilops cylindrica* Host.)	*Pseudomonas, Xanthomonas, Enterobacter*	Kennedy, Ogg, and Young, 1992; Harris and Stahlman, 1996
Giant foxtail (*Setaria faberi* Herrm.)	*Pseudomonas, Flavobacterium, Alcaligenes, Erwinia*	Kremer et al., 1990; Li and Kremer, 2000
Green foxtail [*Setaria viridis* (L.) Beauv.]	*Pseudomonas fluorescens*	Boyetchko, 1997, 1999
Wild oat (*Avena fatua* L.)	*Pseudomonas fluorescens*	Boyetchko, 1997
Field bindweed (*Convolvulus arvensis* L.)	*Pseudomonas, Alcaligenes*	Kremer et al., 1990
Leafy spurge (*Euphorbia esula* L.)	*Pseudomonas fluorescens, Flavobacterium balustinum, Agrobacterium*	Souissi and Kremer, 1994; Caesar, 1994
Morning glory (*Ipomoea* spp.)	*Flavobacterium, Alcaligenes, Pseudomonas*	Kremer et al., 1990; Li and Kremer, 2000
Redroot pigweed (*Amaranthus retroflexus* L.)	*Pseudomonas, Alcaligenes, Flavobacterium, Erwinia*	Kremer et al., 1990; Li and Kremer, 2000; Kennedy, Johnson, and Stubbs, 2001
Russian knapweed [*Acroptilon repens* (L.) DC.]	*Agrobacterium*	Caesar, 1994
Velvetleaf (*Abutilon theophrasti* Medik.)	*Pseudomonas, Flavobacterium*	Kremer et al., 1990; Li and Kremer, 2000

DRB IN INTEGRATED WEED MANAGEMENT

An expanded and long-term approach to weed control is integrated weed management in which all available strategies, including tillage, cultural practices, herbicides, allelopathy, and biological control, are used to reduce the weed seed bank in soil, prevent weed emergence, and minimize competition from weeds growing with crop plants (Aldrich and Kremer, 1997). Like chemical herbicides, bioherbicides may be most effective as a component in a multifaceted management program rather than as a single-tactic approach. Effective weed management offers several opportunities for integration of bioherbicides at the critical stages of weed development: as seeds in soil, as growing and competitive plants, and during seed production (Aldrich and Kremer, 1997). This may be the most promising situation for DRB to be considered as practical management options in cropping systems. Charudattan (2001) has suggested that more efforts are needed to identify and utilize rhizobacteria in systems integrating biocontrol agents with other control approaches.

To broaden the limited spectrum of activity of DRB, several tactics have been proposed for integration of these organisms with other weed management methods. Several cultural practices offer convenient application methods for integrating bioherbicides in cropping systems.

Integration with Chemicals

Weed growth suppression by DRB combined with herbicides applied at sublethal rates has met with some success (Greaves and Sargent, 1986). DRB inhibitory to downy brome and jointed goatgrass suppressed growth greater when combined with metribuzin and/or diclofop at less than label rates (Kremer and Kennedy, 1996). An understanding of the mechanisms of herbicide-DRB interactions will lead to strategies in which DRB selected for activity toward a weed can be paired with a specific chemical that increases the susceptibility of that weed to the DRB (Kremer, 1998). Use of DRB in this manner may develop into a systems-management approach that involves integration of bioherbicides and herbicides on a physiological basis to control economically important weeds in corn and other crops. This is currently under intensive evaluation as a potential integrated management system in Europe (Müller-Schärer, Scheepens, and Greaves, 2000). Successful development of these integrated strategies will increase efficacy of DRB agents, reduce herbicide inputs for weed control, and decrease potential environmental contamination.

Integration with Cultural Practices

Application of selected DRB during tillage may be effective in integrated weed management. Greater proportions of indigenous rhizobacteria inhibitory to downy brome and jointed goatgrass were detected under either conventional or reduced tillage compared to no-till (Kremer and Kennedy, 1996). Vegetative residues at or near the soil surface serve as substrates for production of weed-suppressive chemicals by DRB applied as bioherbicides directly to the residues. For example, numbers of wheat-inhibitory DRB greatly increased when applied to wheat straw, which promoted production of toxins inhibitory to wheat (Stroo, Elliott, and Papendick, 1988). Kremer (1998) suggested that application of DRB to surface vegetative residues to promote phytotoxin production might suppress weed growth prior to planting the crop, similar to a preemergence herbicide tactic.

Crop rotation is a practice that may also be manipulated to encourage development of specific inhibitory bacteria on weed roots. Previous work reporting a rotation effect in corn that was due partly to certain rhizobacteria specifically associated with corn roots illustrates the potential for using DRB to achieve suppression of weeds in crop rotation systems (Turco et al., 1990). Crop rotation may be manipulated to encourage development of specific inhibitory bacteria on weed roots. These observations suggest that application of selected DRB during tillage and in crop rotations enhance weed growth suppression and may be effective in integrated weed management (Kremer and Kennedy, 1996).

Weed seeds and seedlings may be attacked by selected DRB directly inoculated on crop seeds or by promoting colonization of crop roots by microbial agents applied at planting (Skipper, Ogg, and Kennedy, 1996). Crop roots not only may deliver DRB to adjacent roots of weeds but also maintain or even enhance DRB numbers for attack of seedlings emerging later in the season. Cover crops may also be used to integrate DRB into a weed management program by promoting DRB establishment in soils for control of weed seedlings prior to planting the main crop. Kremer (2000) showed that giant foxtail (*Setaria faberi* Herrm.) suppression was enhanced 20 to 30 percent when *Brassica* and yellow sweetclover [*Melilotus officinalis* (L.) Lam.] cover crops were combined with selected cultures of DRB at planting.

Management of Weed Seedbanks and Seedling Emergence

Prevention of seed germination and seedling emergence is fundamental for effective long-term weed management (Aldrich and Kremer, 1997). Enhancing weed seed mortality in soil through use of biotic agents or specific management systems has not been fully exploited (Cardina et al., 1999). Soil and rhizosphere microorganisms can play a significant role in reducing weed infestations by attacking seeds and seedlings before they become competitive with crop plants. Several approaches for managing the seed bank and seedling emergence have been described, including direct application of biotic agents to soil, crop residues, or to crop seeds to prevent emergence of weeds in the crop seed-germinating zone, and in combination with solarization for enhancing seed deterioration in soil (Kremer, 1993). Also, certain agrichemicals (ethephon, phthalimides, butylate, carbofuran) known to stimulate seed imbibition or germination can be incorporated into soil combined with selected seed-attacking microorganisms in biocides to kill germinating weed seeds (Kremer and Schulte, 1989).

SOIL AND RHIZOSPHERE BACTERIA IN SUSTAINABLE AGRICULTURE

Agricultural systems that avoid the use of chemical herbicides and promote restriction or banning of herbicides in areas deemed environmentally sensitive are most receptive to the adoption of bioherbicides. For simplicity, the term *sustainable agricultural systems* encompasses other designations including alternative, natural, organic, biological, ecological, and biodynamic farming systems, all of which involve a range of technological and management options to reduce costs, protect health and environmental quality, and enhance beneficial biological interactions and natural processes. In nearly all cases, few, if any, synthetic chemicals (including herbicides) are used. Sustainable agricultural systems offer the greatest opportunities to study and refine nonchemical weed management (Liebman and Gallandt, 1997), yielding valuable information useful in developing improved bioherbicides and advancing their use in broader biologically based weed management systems. Because the demand for bioherbicides for sustainable agriculture and ecosystem management is currently small, relative to chemical herbicides, such products may be provided most efficiently through small-scale, specialized industries or even on-farm production facilities focused on "niche markets" (Charudattan, 1990; Auld and Morin, 1995).

Biological Weed Management

Bioherbicides based on DRB and other soil microorganisms may be most effective in managing weeds as a component of a *biological weed management system* associated with sustainable agriculture. Biological weed management involves a diversity of biological approaches including allelopathy, bioherbicides, crop competition, and cultural practices to obtain equivalent reductions in weed densities as obtained with herbicides (Cardina, 1995). High inputs of organic amendments and green manure in sustainable agricultural systems promote the ability of crops to compete more vigorously with weeds, which intuitively suggests that the efficacy of bioherbicides would also be enhanced when used with these amendments (Gallandt et al., 1998). Indeed, the bioherbicidal fungus *Gliocladium virens* produces the phytotoxin viridiol when grown in organic substrates such as peat and composts, making this bioherbicide ideal for use in biological weed management (Jones and Hancock, 1990). Furthermore, this fungus also produces compounds effective against fungal plant pathogens, suggesting the possibility of developing biotic agents with efficacy toward multiple pests (Jones and Hancock, 1990).

Because sustainable agriculture emphasizes pesticide-free crop production, bioherbicides will likely be a major component in biological weed management. Many of the approaches are similar to those for integrated weed management except that herbicides are not involved. Therefore, bioherbicides that would not be used in conventional integrated weed management because of unacceptable efficacy or the long time needed for realizing an effect would be practical in biological weed management. For example, prevention of weed seed production and reduction of the seed bank could be attained using a bioherbicide consisting of seed pathogens applied to growing weeds (Medd and Campbell, 1996). However, the impact of the bioherbicide would not be evident for one to two years when noticeable decreases in weed seedling densities occur due to the reduced seed bank size.

Conservation Biological Control

Biocontrol of weeds is critical to sustainable agriculture systems. To accomplish this, a greater understanding of soil microbial ecology is needed before biocontrol can be reliably promoted as useful practice for sustainable agriculture. Specific microorganisms including DRB have been developed for agriculture, including biological control of plant pests to reduce chemical inputs. Investigations of DRB suggest that sustainable agricultural practices for many crops may be linked to management of the indigenous microbial communities for biological control of weeds, thus reducing dependence on selected DRB as inundative biocontrol agents. Development of agroecosystems with the capacity to suppress weeds using naturally occurring soil-weed interactions has received very little attention (Gallandt, Liebman, and Huggins, 1999). If successful, such management strategies would promote the development of beneficial soil microorganisms similar to a *conservation biological control* approach (Newman, Thompson, and Richman, 1998), which conceivably would address sustainable weed management with reduced or no herbicide inputs (Parr et al., 1992).

Agronomic practices that result in relatively high soil organic matter, such as organic farming and integrated cropping systems, support higher populations of weed DRB (Kremer and Li, 2003). Promotion of DRB growth in such systems could then be exploited in alternative weed management systems. For example, a study of crop management practices on claypan soils (Epiaqualfs) that involved reduced tillage, maintenance of high soil organic matter, and limited inputs of herbicides and insecticides found increased levels of DRB associated with weed seedlings that contributed to natural weed suppression (Li and Kremer, 2000). Growth of common lambsquarters in a potato (*Solanum tuberosum* L.) crop decreased by more than 75 percent in soils receiving annual inputs of green manure, composted plant residues, and beef manure compared to nonamended soils (Gallandt, Liebman, and Huggins, 1999). Involvement of antagonistic soil microbes was not indicated, although the associated increase in soil quality, including soil organic matter, was implicated in development of weed suppressiveness. Compost-amended soils planted to winter wheat exhibited 29 and 78 percent reductions in broadleaf and grassy weed densities, respectively, compared to field soils receiving inorganic fertilizer or no supplementary nutrient sources (Carpenter-Boggs, Reganold, and Kennedy, 2000). A red clover (*Trifolium pratense* L.) cover crop incorporated into soil with compost stimulated natural weed-suppressive microorganisms and, combined with allelochemicals released from the incorporated residues, enhanced control of wild mustard [*Brassica kaber* (DC.) L.C. Wheeler] (Conklin et al., 2002). Populations of DRB on seedling roots of several weed species were highest in organic farming and integrated cropping systems. This indicated that weed-suppressive bacteria were associated with soil management that promoted high levels of soil biological activity (Kremer and Li, 2003). The foregoing studies strengthen the possibility of using conservation biological control approaches in manipulating indigenous microbial communities for active weed-growth suppression in soils under sustainable management.

For native DRB to be effective, colonization and growth suppression of weed seedlings are required; however, suppression, although dependent on weed host, is also influenced by soil environmental factors. Soil and crop management practices greatly influence weed suppression by indigenous soil microorganisms (Gallandt, Liebman, and Huggins, 1999; Carpenter-Boggs, Reganold, and Kennedy, 2000). Cropping systems provide a practical means for managing beneficial microorganisms, including weed-suppressive DRB (see Figure 18.1). Rhizosphere microorganisms exist in a finely balanced equilibrium and are subject to a range of edaphic and plant-modifying influences. For example, addition of organic amendments to soil has consistently shifted the soil microbial equilibrium to enhance weed-suppressive activity (Gallandt, Liebman, and Huggins, 1999; Conklin et al., 2002; Kremer and Li, 2003). Further, activities of specific soil microorganisms that develop as a result of specific crop and soil management practices may be detected using assays for biological indicators, thereby linking improved soil quality to potential weed suppression (see Figure 18.1) (Kremer and Li, 2003).

CONCLUSION

DRB and other microorganisms applied to soil as biological control agents have not been a consistent and reliable strategy for weed manage-

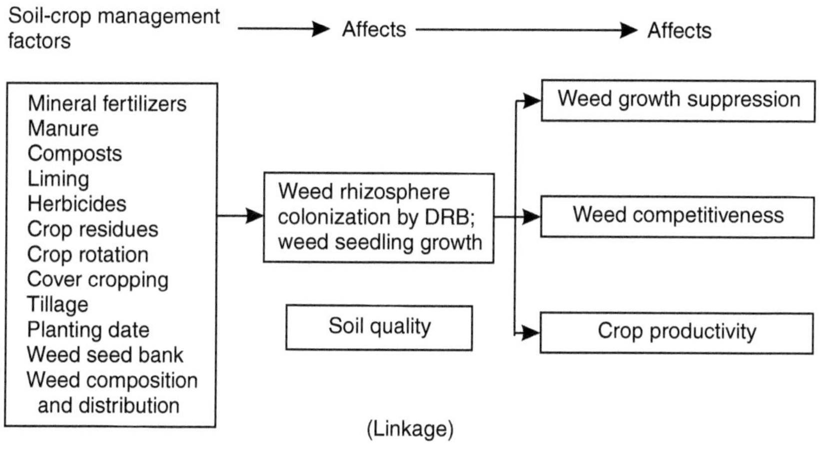

FIGURE 18.1. Factors in crop and soil management on temperate zone soils that may affect weed growth and competition and crop productivity indirectly through combined effects on the weed rhizosphere and soil quality.

ment, especially as a practical tool in cropping systems. Integration of biological control into current management systems with other weed control methods will best demonstrate the potential effectiveness of these agents. DRB as bioherbicides for use in sustainable agriculture will likely demonstrate the greatest effectiveness in biological weed management systems in the short term and generate impetus for continued discovery and development of bioherbicides for more widespread use (Kremer, 2002).

Alternatively, deliberate use of management practices that benefit natural weed-antagonistic microorganisms may lead to detrimental effects on weed population dynamics in production fields through seed and seedling mortality and growth suppression. Rather than relying on herbicides and inundative bioherbicides, exploiting inherent strengths of the cultural components within sustainable agricultural systems and by promoting naturally occurring weed-suppressive microorganisms will aid in implementing biologically based weed management that maximizes profits and sustains agroecosystems.

REFERENCES

Aldrich, R. J. and R. J. Kremer (1997). *Principles in weed management.* Ames: Iowa State University Press.

Alström, S. and R. G. Burns (1989). Cyanide production by rhizosphere bacteria as a possible mechanism of plant growth inhibition. *Biology and Fertility of Soils* 7: 232-238.

Arshad, M. and W. T. Frankenberger Jr. (1998). Plant growth-regulating substances in the rhizosphere: Microbial production and functions. *Advances in Agronomy* 62: 45-151.

Auld, B.A. and L. Morin (1995). Constraints in the development of bioherbicides. *Weed Technology* 9: 638-652.

Bolton, H. Jr. and L. F. Elliott (1989). Toxin production by a rhizobacterial *Pseudomonas* sp. that inhibits wheat root growth. *Plant and Soil* 114: 269-278.

Boyetchko, S. M. (1997). Efficacy of rhizobacteria as biological control agents of grassy weeds. In *Proceedings of the Soils and Crops Conference* (pp. 460-462). Saskatoon, Saskatchewan, Canada: Extension Division University of Saskatchewan.

Boyetchko, S. M. (1999). Innovative applications of microbial agents for biological weed control. In K. G. Mukerji, B. P. Chamola, and R. K Upadhyay (Eds.), *Biotechnological approaches in biocontrol of plant pathogens* (pp. 73-97). London: Plenum Press.

Boyetchko, S. M., E. Pedersen, Z. Punja, and M. Reddy (1998). Formulations of biopesticides. In F. R. Hall and J. J. Menn (Eds.), *Methods in biotechnology,* Volume 5: *Biopesticides: Use and delivery* (pp. 487-508). Totowa, NJ: Humana Press.

Caesar, A. J. (1994). Pathogenicity of *Agrobacterium* species from the noxious rangeland weeds *Euphorbia esula* and *Centaurea repens*. *Plant Disease* 78: 796-800.

Cardina, J. (1995). Biological weed management. In A.E. Smith (Ed.), *Handbook of weed management systems* (pp. 279-341). New York: Marcel Dekker, Inc.

Cardina, J., T. M. Webster, C. P. Herms, and E. E. Regnier (1999). Development of weed IPM: Levels of integration for weed management. *Journal of Crop Production* 2: 239-267.

Carpenter-Boggs, L., J. P. Reganold, and A. C. Kennedy (2000). Biodynamic preparations: Short-term effects on crops, soils, and weed populations. *American Journal of Alternative Agriculture* 15: 110-118.

Charudattan, R. (1989). The mycoherbicide approach with plant pathogens. In D. O. TeBeest (Ed.), *Microbial control of weeds* (pp. 24-57). New York: Chapman and Hall.

Charudattan, R. (1990). Pathogens with potential for weed control. In R. E. Hoagland (Ed.), *Microbes and microbial products as microbial herbicides* (pp. 133-154). Washington, DC: American Chemical Society.

Charudattan, R. (2001). Biological control of weeds by means of plant pathogens: Significance for integrated weed management in modern agro-ecology. *BioControl* 46: 229-260.

Conklin, A. E., M. S. Erich, M. Liebman, D. Lambert, E. R. Gallandt, and W. A. Halteman (2002). Effects of red clover *(Trifolium pratense)* green manure and compost soil amendments on wild mustard *(Brassica kaber)* growth and incidence of disease. *Plant and Soil* 238: 245-256.

Dowling, D. N. and F. O'Gara (1994). Metabolites of *Pseudomonas* involved in the biocontrol of plant disease. *Trends in Biotechnology* 12: 133-141.

Gallandt, E. R., M. Liebman, S. Corson, G. A. Porter, and S. D. Ullrich (1998). Effects of pest and soil management systems on weed dynamics in potato. *Weed Science* 46: 238-248.

Gallandt, E. R., M. Liebman, and D. R. Huggins (1999). Improving soil quality: Implications for weed management. *Journal of Crop Production* 2: 95-121.

Greaves, M. P. and J. A. Sargent (1986). Herbicide-induced microbial invasion of plant roots. *Weed Science* 34(Suppl. 1): 50-53.

Gurusiddaiah, S., D. R. Gealy, A. C. Kennedy, and A. G. Ogg Jr. (1994). Isolation and characterization of metabolites from *Pseudomonas fluorescens* D7 for control of downy brome *(Bromus tectorum)*. *Weed Science* 42: 492-501.

Harris, P. A. and P. W. Stahlman (1996). Soil bacteria as selective biological control agents of winter annual grass weeds in winter wheat. *Applied Soil Ecology* 3: 275-281.

Jones, R. W. and J. G. Hancock (1990). Soilborne fungi for biological control of weeds. In R. E. Hoagland (Ed.), *Microbes and microbial products as microbial herbicides* (pp. 276-286). Washington, DC: American Chemical Society.

Kennedy, A. C., L. F. Elliott, F. L. Young, and C. L. Douglas (1991). Rhizobacteria suppressive to the weed downy brome. *Soil Science Society of America Journal* 55: 722-727.

Kennedy, A. C., B. N. Johnson, and T. L. Stubbs (2001). Host range of a deleterious rhizobacterium for biological control of downy brome. *Weed Science* 49: 792-797.

Kennedy, A. C., A. G. Ogg Jr., and F. L. Young (1992). Biocontrol of jointed goatgrass. U.S. patent number 5,163,991. November 17.

Kremer, R. J. (1993). Management of weed seed banks with microorganisms. *Ecological Applications* 3: 42-52.

Kremer, R. J. (1998). Microbial interactions with weed seeds and seedlings and its potential for weed management. In J. L. Hatfield, D. D. Buhler, and B. A. Stewart (Eds.), *Integrated weed and soil management* (pp. 161-179). Chelsea, MI: Ann Arbor Press.

Kremer, R. J. (2000). Growth suppression of annual weeds by deleterious rhizobacteria integrated with cover crops. In N. R. Spencer (Ed.), *Proceedings of the Tenth International Symposium on Biological Control of Weeds* (pp. 931-940). Bozeman: USDA-ARS and Montana State University.

Kremer, R. J. (2002). Bioherbicides: Potential successful strategies for weed control. In O. Koul and G. Dhaliwal (Eds.), *Microbial biopesticides* (pp. 307-323). London: Taylor and Francis.

Kremer, R. J., M. F. T. Begonia, L. Stanley, and E. T. Lanham (1990). Characterization of rhizobacteria associated with weed seedlings. *Applied and Environmental Microbiology* 56: 1649-1655.

Kremer, R. J. and A. C. Kennedy (1996). Rhizobacteria as biological control agents of weeds. *Weed Technology* 10: 601-609.

Kremer, R. J. and J. Li (2003). Developing weed suppressive soils through improved soil quality management. *Soil and Tillage Research* 72: 193-202.

Kremer, R. J. and L. K. Schulte (1989). Influence of chemical treatment and *Fusarium oxysporum* on velvetleaf *(Abutilon theophrasti)* seed viability. *Weed Technology* 3: 369-374.

Leisinger, T. and R. Margraff (1979). Secondary metabolites of the fluorescent pseudomonads. *Microbiological Reviews* 43: 422-442.

Li, J. and R. J. Kremer (2000). Rhizobacteria associated with weed seedlings in different cropping systems. *Weed Science* 48: 734-741.

Liebman, M. and E. R. Gallandt (1997). Many little hammers: Ecological management of crop-weed interactions. In L. E. Jackson (Ed.), *Ecology in agriculture* (pp. 291-343). San Diego, CA: Academic Press.

Medd, R. W. and M. A. Campbell (1996). A rationale for the use of a non-specific fungal seed pathogen to control annual grass-weeds of arable lands. In V. C. Moran and J. H. Hoffmann (Eds.), *Proceedings of the Ninth International Symposium on Biological Control of Weeds* (pp. 193-197). Stellenbosch, South Africa: University of Capetown.

Müller-Schärer, H., P. C. Scheepens, and M. P. Greaves (2000). Biological control of weeds in European crops: Recent achievements and future work. *Weed Research* 40: 83-98.

Newman, R. M., D. C. Thompson, and D. B. Richman (1998). Conservation strategies for the biological control of weeds. In P. Barbosa (Ed.), *Conservation biological control* (pp. 371-396). San Diego, CA: Academic Press.

Parr, J. F., R. I. Papendick, S. B. Hornick, and R. B. Meyer (1992). Soil quality: Attributes and relationship to alternative and sustainable agriculture. *American Journal of Alternative Agriculture* 7: 5-11.

Reganold, J. P., R. I. Papendick, and J. F. Parr (1990). Sustainable agriculture. *Scientific American* 262(6): 112-120.

Schippers, B., A. W. Bakker, and P. A. Bakker (1987). Interactions of deleterious and beneficial rhizosphere microorganisms and the effect of cropping practices. *Annual Review of Phytopathology* 25: 339-358.

Skipper, H. D., A. G. Ogg Jr., and A. C. Kennedy (1996). Root biology of grasses and ecology of rhizobacteria for biological control. *Weed Technology* 10: 610-620.

Souissi, T. and R. J. Kremer (1994). Leafy spurge *(Euphorbia esula)* cell cultures for screening deleterious rhizobacteria. *Weed Science* 42: 310-315.

Souissi, T., R. J. Kremer, and J. A. White (1997). Scanning and transmission electron microscopy of root colonization of leafy spurge (*Euphorbia esula* L.) seedlings by rhizobacteria. *Phytomorphology* 47: 177-193.

Stroo, H. F., L. F. Elliott, and R. I. Papendick (1988). Growth, survival and toxin production of root-inhibitory pseudomonads on crop residues. *Soil Biology and Biochemistry* 20: 201-207.

Turco, R. F., M. Bischoff, D. P. Breakwell, and D. R. Griffith (1990). Contribution of soilborne bacteria to the rotation effect in corn. *Plant and Soil* 122: 115-120.

Vogelgsang, S., A. K. Watson, and A. DiTommaso (1998). Effect of soil incorporation and dose on control of field bindweed *(Convolvulus arvensis)* with the preemergence bioherbicide *Phomopsis convolvulus*. *Weed Science* 46: 690-697.

Weissmann, R. and B. Gerhardson (2001). Selective plant growth suppression by shoot application of soil bacteria. *Plant and Soil* 234: 159-170.

Chapter 19

Herbicide-Resistant Crops and Weed Management

Krishna N. Reddy
Clifford H. Koger

INTRODUCTION

Ever since humans began to settle in one place and cultivate food crops, they have continually developed and improved methods to control weeds to protect crops. Weed control methods have evolved over time, beginning with hand weeding (10,000 B.C.), primitive hand tools (6,000 B.C.), animal-powered implements (1,000 B.C.), mechanically powered implements (1920), biological control (1930), and chemical weed control (1947) (Hay, 1974). The era of chemical weed control began in 1947 with the introduction of 2,4-D and MCPA [(4-chloro-2-methylphenoxy)acetic acid] herbicides (Hay, 1974). The development of safe, effective, and relatively inexpensive herbicides, coupled with advances in application technology during the past six decades, has provided many successful weed management options in crop production. Herbicides provide cost-effective and efficient control of weeds and have dominated other weed control methods in recent years.

Crops made resistant to herbicides using biotechnology expand the utility of proven postemergence (POST), previously nonselective, broad-spectrum herbicides. In the mid-1990s, several herbicide-resistant crops (HRCs) were developed and commercialized. The HRCs contain genes that impart resistance to the herbicides. HRCs offer farmers simplicity and flexibility to manage weeds and are readily accepted in North America and are slowly making inroads into parts of Asia, Africa, Australia, Europe, and South America. HRCs have sparked heated debate all over the world, and

---

Mention of a trademark or proprietary product does not constitute a guarantee or warranty of the product by the U.S. Department of Agriculture and does not imply their approval to the exclusion of other products that may also be suitable.

the literature on HRCs is flooded with different points of view. Various aspects of HRCs have been the topic of several recent books and reviews (e.g., Dyer et al., 1993; Dekker and Duke, 1995; McLean and Evans, 1995; Duke, 1996, 2002; Darmency, 1997; Donn, 1997; Powles et al., 1997; Dunwell, 1999; Warwick, Beckie, and Small, 1999; Gressel and Rotteveel, 2000; Reddy 2001b; Duke et al., 2002; Gressel, 2002). This chapter will not summarize in detail the development of HRCs, nor will it discuss consumer reactions to genetically modified crops. Rather, it presents a brief summary of HRCs on the market, the trends in their adoption, and their impact on weed management in field crops. It will use information from the usual scientific publications as well as information obtained from various Web sites. Information from Web sites is of great value because it provides a current view of commercial intentions not often available in the published literature.

HERBICIDE-RESISTANT CROPS AND THEIR ADOPTION

Herbicide-resistant crops as a weed management tool is an extension of chemical weed control with one very important difference: HRCs survive the chemical treatment that previously would have killed the crop along with targeted weeds. Most transgenic HRCs have been created through the stable integration of a foreign gene using molecular biological techniques and plant transformation (Dyer, 1996; Donn, 1997). However, several nontransgenic HRCs also have been developed using traditional plant breeding techniques as well as hybridization with conspecific herbicide-resistant relatives (Dyer, 1996). Transgenic and nontransgenic HRCs on the market and in development have been listed in Table 19.1. Most common transgenic HRCs are resistant to glyphosate. These crops include 'Roundup Ready' canola (*Brassica napus* L.), corn (*Zea mays* L.), cotton (*Gossypium hirsutum* L.), and soybean [*Glycine max* (L.) Merr.]. Other transgenic HRCs include 'Liberty Link' canola and corn resistant to glufosinate, and BXN cotton resistant to bromoxynil. Traditionally bred HRCs include IMI corn and 'Clearfield' rice (*Oryza sativa* L.) resistant to imidazolinones, SR corn resistant to sethoxydim, and STS soybean resistant to sulfonylureas. Although several HRCs are available, only glyphosate- and glufosinate-resistant crops appear to have the greatest potential for adoption. These two herbicides are nonselective so that simpler weed management programs can be designed based on one herbicide compared to three to five different herbicides commonly used in nontransgenic crops.

The creation of HRCs is only a part of the success. For HRCs to succeed they must be adopted by farmers. U.S. farmers have adopted HRCs rapidly

TABLE 19.1. Transgenic and nontransgenic herbicide-resistant crops commercialized and in development.

Herbicide/chemical family	Crop	Trade name/trait	Year available/status
Transgenic			
Bromoxynil	Canola	BXN canola	1999
	Cotton	BXN cotton	1995
	Cotton	Stacked-trait (Bt/BXN)	In development
Glyphosate	Canola	Roundup Ready canola	1996
	Corn	Roundup Ready corn	1998
	Corn	Roundup Ready/YieldGard (stacked-trait)	2000
	Cotton	Roundup Ready cotton	1997
	Cotton	Bollgard/Roundup Ready (stacked-trait)	2000
	Soybean	Roundup Ready soybean	1996
	Alfalfa	Roundup Ready alfalfa	In development
	Lettuce	Roundup Ready lettuce	In development
	Potato	Roundup Ready potato	In development
	Sugar beet	Roundup Ready sugar beet	In development
	Wheat	Roundup Ready wheat	In development
Glufosinate	Canola	Liberty Link canola	1995
	Corn	Liberty Link corn	1997
	Corn	Stacked-trait (Bt/Liberty link)	In development

TABLE 19.1 *(continued)*

Herbicide/chemical family	Crop	Trade name/trait	Year available/status
	Cotton	Liberty Link cotton	2003
	Soybean	Liberty Link soybean	1996
	Sugar beet	Liberty Link sugar beet	In development
	Rice	Liberty Link rice	In development
Sulfonylureas	Cotton	–	In development
	Flax	–	In development
Nontransgenic			
Sethoxydim	Corn	SR corn	1996
Imidazolinones	Corn	Clearfield corn	1993
	Canola	–	1997
	Rice	Clearfield rice	2001
	Sugar beet	Clearfield sugar beet	In development
	Wheat	Clearfield wheat	In development
Sulfonylureas	Soybean	ST soybean	1994
Triazines	Canola	–	1984

Source: BIO, 2004; Duke et al., 2002; USFDA, 2005.

since their introduction in the mid-1990s. The majority of HRCs planted worldwide are grown in the United States where the area planted to HRCs is increasing every year (see Figure 19.1). Although adoption of HRCs was rather slow in the beginning, the transgenic crop area has been on the rise in recent years as more countries have approved HRCs for commercialization.

United States of America

In the United States, transgenic soybean resistant to glyphosate was commercialized for planting in 1996. As is evident in Figure 19.1, glyphosate-resistant soybean has been widely adopted by American farmers since its introduction and the area planted with it increased from 2 percent in 1996 to 75 percent in 2002 and to 85 percent of the total planted area in 2004 (BIO, 2004; USDA, 2005).

Three types of genetically modified cotton developed by Monsanto are currently on the market: (1) 'Bollgard' single-trait insect-protected cotton (Bt cotton), which provides season-long protection against cotton bollworm, pink bollworm, and tobacco budworm; (2) 'Roundup Ready' single-trait cotton, which enables farmers to apply glyphosate on cotton; and (3) 'Bollgard/Roundup Ready' stacked-trait cotton, which provides both insect

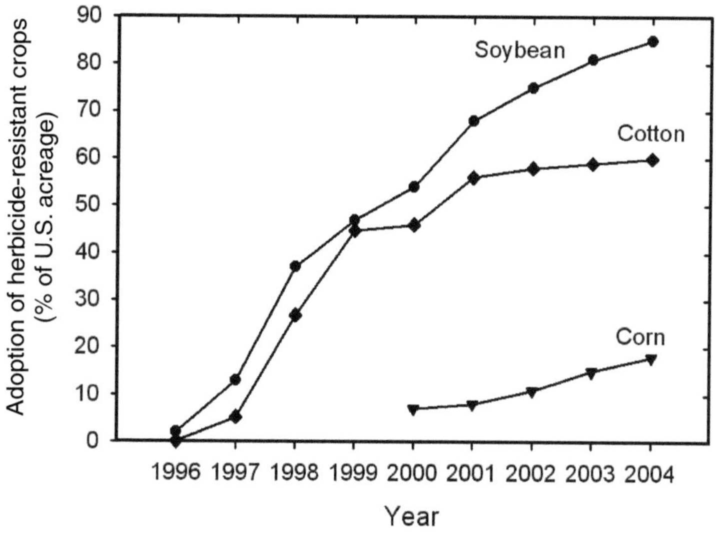

FIGURE 19.1. Adoption of herbicide-resistant corn (▼), herbicide-resistant cotton (◆), and herbicide-resistant soybean (●) in the United States, 1996-2002. (*Source:* BIO, 2004; Gianessi et al., 2002; ERS, 2004; USDA, 2005.)

protection and glyphosate tolerance in a single plant (Monsanto, 2004). Bromoxynil-resistant cotton (BXN cotton) from Rhône-Poulenc Ag Company (now Bayer Corporation) was commercialized in 1995 and discontinued in 2004. The U.S. cotton area planted to herbicide-resistant cotton has increased from 0.1 percent in 1995 to 60 percent in 2004 (Gianessi et al., 2002; USDA, 2005).

Monsanto introduced 'YieldGard' corn borer insect-protected corn (Bt corn) in 1997, glyphosate-resistant ('Roundup Ready') corn in 1998, and 'Roundup Ready/YieldGard' corn borer stacked-trait corn in 2000 (Monsanto, 2004). Glufosinate-resistant ('Liberty Link') corn was introduced in 1997 by AgrEvo USA Company (now Bayer Corporation). Other non-transgenic corn such as IMI corn tolerant to imidazolinones and SR corn tolerant to sethoxydim were also grown on a limited area. The U.S. corn area planted to herbicide-resistant corn, including stacked-trait corn, increased from 7 percent in 2000 to 45 percent in 2004 (ERS, 2004).

World

In 2004, six countries grew 98 percent of the 81 million hectares (ha) of global transgenic crop area (James, 2004). Most of the transgenic crops were grown in the United States. In 2004, the United States had about 59 percent of the global transgenic crop area, followed by Argentina (20 percent), Canada (6 percent), China (5 percent), Brazil (6 percent), and Paraguay (2 percent). The remaining 1 percent was grown in Australia, Bulgaria, Colombia, Germany, Indonesia, Mexico, Romania, Spain, South Africa, Philippines, India, Honduras, and Uruguay (James, 2003). Transgenic soybean was the most dominant among all transgenic crops grown globally. In 2004, transgenic soybean was planted to 48.4 million ha (60 percent of global transgenic area), followed by transgenic corn (23 percent), transgenic cotton (11 percent), and transgenic canola (6 percent). Herbicide resistance was the dominant trait, with insect resistance being the second. In 2004, herbicide-resistant crops (soybean, corn, canola, and cotton) occupied 58.6 million ha (72 percent of the global transgenic area), followed by Bt crops (19 percent), and stacked genes for herbicide and insect resistance traits (8.4 percent). The global transgenic crop area will likely increase as increasing numbers of farmers plant genetically modified crops and more countries approve commercialization. For example, India approved commercialization of its first genetically modified crop, Bt cotton, in 2002. Colombia (Bt cotton), Honduras (Bt corn), and Germany (Bt corn) have recently introduced transgenic crops (James, 2004). In 2004, the European Union Commission approved 17 corn varieties with insect resistance, making it the first biotech crop to be approved for planting in all 25 EU countries (James, 2004).

IMPACT OF HRCs ON WEED MANAGEMENT

Several HRCs are on the market or under development in North America. Currently, herbicide-resistant varieties of five major agronomic crops have been commercialized. This chapter is focused on the impact of herbicide-resistant varieties on weed management in corn, soybean, cotton, rice, and canola.

Corn

Corn producers deal with a multitude of weed species that must be managed to achieve profitable production. Prior to commercialization of HRCs, the standard weed control program was centered around chloroacetamide and/or triazine herbicides applied PRE to control grass and broadleaf weeds in corn. However, POST weed control is often warranted even when herbicides of these two classes are applied PRE. Nicosulfuron and primisulfuron applied POST control several annual and perennial grasses as well as a multitude of broadleaf weed species (Ngouajio and Hagood, 1993); while dicamba, bentazon, and atrazine have been the predominant POST broadleaf herbicides used in corn. Nicosulfuron interacts with organophosphate insecticides, commonly used for rootworm (*Diabrotica* spp.) control, which often results in corn injury (Morton et al., 1994). In addition, acetolactate synthase (ALS)-resistant weeds such as shattercane [*Sorghum bicolor* (L.) Moench], common cocklebur (*Xanthium strumarium* L.), common sunflower (*Helianthus annuus* L.), and common waterhemp (*Amaranthus rudis* Sauer) (Heap, 2005) have limited the effectiveness of ALS inhibitors, nicosulfuron, and primisulfuron. Several annual grasses such as broadleaf signalgrass [*Brachiaria platyphylla* (Griseb.) Nash], shattercane, and Texas panicum (*Panicum texanum* Buckl.) are extremely difficult to manage with chloroacetamide and triazine herbicides. In addition, concerns about the residues of alachlor, metolachlor, and atrazine in water sources have increased in recent years (Buehler et al., 1993; Blanchard and Donald, 1997; Ghidey, Alberts, and Lerch, 1997; Senseman et al., 1997). Concerns of drift and volatility and subsequent injury to sensitive crops such as soybean, cotton, tobacco *(Nicotiana tabacum),* and vegetable crops limits the use of dicamba.

Nevertheless, most weeds can be adequately managed with careful selection of herbicide programs in nontransgenic corn. However, HRCs that are commercially available provide more management options for certain problematic weeds and also provide weed management alternatives for chloroacetamide and/or triazine herbicides in areas where risk of ground and surface water contamination is high or where these herbicides' use is restricted. Currently, four types of herbicide-resistant corn varieties are commercially available.

Imidazolinone-Resistant Corn

Imidazolinone herbicides inhibit ALS, which is the first enzyme common to the biosynthesis of the branched-chain amino acids isoleucine, valine, and leucine (LaRossa and Schloss, 1984). ALS is the target enzyme for five commercial herbicide chemistry classes: imidazolinones (IMI), pyrimidinylthiobenzoates, sulfonylaminocarbonyltriazolinones, sulfonylureas, and triazolopyrimidine sulfonanilides (Vencill, 2002). ALS-inhibiting herbicides are highly active inhibitors of a single target enzyme, exhibit low mammalian toxicity, and can be used at very low rates for effective control of a broad spectrum of grass and broadleaf weeds (Beyer et al., 1988; Saari and Mauvais, 1996; Shaner, Bascomb, and Smith, 1996).

Commercial imidazolinone corn varieties have been developed through different selection mechanisms. Resistance is conferred by one of three genes that code for an insensitive ALS enzyme (Currie, Kwon, and Penner, 1995). Pioneer Hi-Bred International Inc. introduced imidazolinone-resistant hybrids with tolerance gene *XA17* (Currie, Kwon, and Penner, 1995). Other seed companies have since developed several imidazolinone-resistant corn hybrids using two unnamed resistance genes.

IMIs as well as other ALS-inhibiting herbicides have POST activity on many troublesome weed species in corn production. ALS-inhibiting herbicides have activity on numerous annual grass weed species, johnsongrass [*Sorghum halepense* (L.) Pers.], morning glories (*Ipomoea* spp.), velvetleaf (*Abutilon theophrasti* Medikus), pigweeds (*Amaranthus* spp.), and nutsedges (*Cyperus* spp.) (Hart, Lignowski, and Taylor, 1991; Ngouajio and Hagood, 1993; Richburg, Wilcut, and Wiley, 1995; Shaner, Bascomb, and Smith, 1996; Wilcut et al., 1996; Wilcut, Richburg, and Walls, 1999). Some ALS-inhibiting herbicides such as IMIs have no activity on many leguminous weed species such as sicklepod [*Senna obtusifolia* (L.) Irwin & Barnaby], hemp sesbania [*Sesbania exaltata* (Raf.) Rydb. ex A.W. Hill], and Florida beggarweed [*Desmodium tortuosum* (Sw.) DC.] (Wilcut et al., 1996). However, other ALS-inhibiting herbicides such as imazapic and nicosulfuron have activity on these troublesome species. Potential benefits of imidazolinone-resistant corn include more ALS herbicide and application (PRE and POST) options to control problem weeds, a new approach to the management of herbicide carryover, and more ALS herbicides to control or delay the development of resistant weeds to other herbicides.

Glufosinate-Resistant Corn

Glutamine synthetase is an enzyme that catalyzes the conversion of glutamate plus ammonium to glutamime as a part of nitrogen metabolism in plants (Vasil, 1996). Glufosinate, a contact, nonselective herbicide, inhibits glutamine synthatase. Ammonia levels in sensitive plants increase rapidly after glufosinate application, resulting in membrane disruption, inhibition of photorespiration and photosynthesis, chloroplast disruption, and eventually plant death (Vasil, 1996).

Advances in genetic transformation technology, specifically ballistic propulsion, have allowed scientists to insert a gene that imparts glufosinate resistance in plants. Phospinothricin, an active ingredient of glufosinate, inhibits glutamine synthetase, preventing the incorporation of ammonia into amino acids. Glufosinate resistance in plants was conferred by using the bialophos resistance (BAR) gene or the phosphinothricin acetyltransferase (PAT) gene that codes for the enzyme phospinothricin acetyltransferase (Droge, Broer, and Pulher, 1992; Vasil, 1996). This enzyme acetylates the free amino group of glufosinate, thus rendering glufosinate inactive in glufosinate-resistant plants (Droge, Broer, and Pulher, 1992). A gene from *Streptomyces viridochromogenes* that codes for the phosphinothricin acetlytransferase enzyme was inserted into several corn hybrids. Corn containing this gene is resistant to glufosinate at normal use rates.

Prior to the development of glufosinate-resistant corn hybrids, glufosinate was applied as a nonselective broad-spectrum herbicide in noncrop areas or as a "burndown" herbicide to control vegetation at planting in no-till or stale seedbeds (Lanie et al., 1994; Hydrick and Shaw, 1995). Glufosinate provides POST control of most troublesome annual and perennial grass and broadleaf weeds in glufosinate-resistant corn (Culpepper and York, 1999b; Tharp and Kells, 2001; Shrestha et al., 2001). Lack of residual activity often warrants sequential POST applications of glufosinate to achieve adequate season-long weed control (Krausz et al., 1999; Hamill et al., 2000; Ritter and Menbere, 2001). In addition, when weed pressure is high, a residual PRE herbicide such as alachlor, atrazine, or metolachlor followed by glufosinate (Culpepper and York, 1999b; Krausz et al., 1999) or POST application of glufosinate plus atrazine or metolachlor (Culpepper and York, 1999b; Hamill et al., 2000) is necessary. A system utilizing glufosinate would be an effective alternative to standard herbicide programs where environmental regulations limit the use of chloroacetamides or triazine herbicides. Growers tend to prefer a total POST weed management program in crop rotation systems where carryover of herbicides from the previous crop can injure sensitive rotational crops.

Glyphosate-Resistant Corn

Glyphosate inhibits biosynthesis of aromatic amino acids (phenylalanine, tyrosine, and tryptophan), which leads to arrest of protein production and prevention of formation of secondary products (Franz, Mao, and Sikorski, 1997). Glyphosate inhibits the enzyme 5-enolpyruvylshikimate-3-phosphate synthase (EPSPS) in the shikimic acid pathway (Amrhein, Schab, and Steinrücken, 1980). EPSPS catalyzes the reaction of shikimate-3-phosphate and phosphoenolpyruvate to form 5-enolpyruvylshikimate-3-phosphate and phosphate. EPSPS is present in all plants, bacteria, and fungi, but not in animals (Padgette et al., 1995). Glyphosate is toxicologically and environmentally benign (low toxicity to nontarget organisms, low or no groundwater movement, and limited persistence).

Glyphosate-resistant corn varieties were developed using ballistic propulsion technology to insert a resistant EPSPS gene. The resistant EPSPS gene consists of 445 amino acids, which differs from the wild-type EPSPS found in sensitive plants by two amino acids (Monsanto, 2004). This results in an EPSPS protein with greater than 99.3 percent sequence identity to that produced naturally in corn (Monsanto, 2004). The resistant EPSPS gene is expressed throughout a resistant corn plant and has a low affinity for glyphosate compared to the wild-type EPSPS enzyme. Several glyphosate-resistant corn varieties are commercially available to producers from multiple seed companies (e.g., Asgrow Seed Company, Delta King Seed Company, Dekalb Genetics Corporation, and Pioneer Hi-Bred International).

Prior to introduction of glyphosate-resistant crops, glyphosate was used as a nonselective POST herbicide to manage annual and perennial vegetation in either noncrop situations or in cropping systems to control existing vegetation at crop planting (Moomaw and Martin, 1985; Lanie et al., 1994; Krausz, Kapusta, and Matthews, 1996). Glyphosate-resistant corn provides producers the means to control most annual and perennial weeds in corn (Johnson et al., 2000). Glyphosate is an effective tool for managing problem weeds in corn such as morning glories, velvetleaf, common lambsquarters (*Chenopodium album* L.), large crabgrass [*Digitaria sanguinalis* (L.) Scop.], and common cocklebur (Tharp and Kells, 1999; Johnson et al., 2000). In many cases, a total POST glyphosate system provided weed control equal to that of a standard herbicide program consisting of a residual chloroacetamide or triazine herbicide followed by standard POST herbicides such as nicosulfuron, halosulfuron, bentazon, or dicamba (Johnson et al., 2000). However, in cases of extreme weed pressure and/or irrigated corn, PRE herbicides followed by glyphosate POST are often needed in order to achieve season-long weed control. This is especially true when applying only one glyphosate application or where wide rows are utilized, as glyphosate cannot be applied over the top of corn after reaching 76 cm in height (Johnson et al., 2000; Monsanto, 2004). Glyphosate-resistant corn with or without PRE herbicides can provide producers with another weed management tool.

Sethoxydim-Resistant Corn

Sethoxydim is a member of the cyclohexanedione herbicide family. This family, along with herbicides of the aryloxyphenoxyproprionate family, inhibit acetyl-coenzyme A carboxylase (ACCase) in susceptible monocotyledonous species. ACCase catalyzes the adenosine triphosphate (ATP)-dependent carboxylation of acetyl-CoA to form malonyl-CoA in the first step of fatty acid synthesis (Incledon and Hall, 1997). In general, monocotyledonous plants are susceptible to ACCase-inhibiting herbicides. ACCase is not inhibited by these herbicides in dicots and also in some naturally herbicide-resistant monocots. Herbicide resistance due to insensitivity of ACCase to herbicide inhibition is the most common mechanism of resistance (Burton et al., 1991; Somers, 1996).

Most of the ACCase-inhibiting herbicides are characterized by POST activity, low use rates, and low toxicity to nontarget organisms. These characteristics, in part, provided the rationale for developing corn varieties that are resistant to ACCase-inhibiting herbicides (Somers, 1996). Sethoxydim-resistant corn varieties have been regenerated from tissue cultures selected for callus growth in the presence of lethal doses of sethoxydim and herbicide-resistance trait conferred by altered ACCase (Somers, 1996). Nontransgenic sethoxydim-resistant corn varieties were commercialized in 1996.

Prior to sethoxydim-resistant corn, ACCase-inhibiting herbicides were primarily used POST to control susceptible grass species in dicotyledonous crops (Vencill, 2002). Sethoxydim is lethal to most annual and perennial grass weed species that have not developed resistance to ACCase herbicides (Byrd and York, 1987; Jordan, and Frans, and McClelland, 1993; Vidrine, Reynolds, and Blouin, 1995; Jordan, Griffin, et al., 1997). Sethoxydim-resistant corn provides producers with an effective tool to control troublesome grass weed species (e.g., johnsongrass) in corn. Repeated use of

ACCase-inhibiting herbicides upon introduction in the early 1980s has resulted in high selection pressure favoring at least 34 grass weed species in 27 countries that are resistant to one or more of this type of herbicides (Heap, 2005). In the majority of these cases, the resistant weed biotypes have developed an insensitive form of the ACCase enzyme (Devine, 1997). Due to the lack of broadleaf activity, broadleaf herbicides are needed in sethoxydim-resistant corn. However, many broadleaf herbicides applied in mixtures antagonize the activity of ACCase-inhibiting herbicides (Hatzios and Penner, 1985; Holshouser and Coble, 1990). This antagonistic effect is often reduced or avoided by applying the ACCase-inhibiting herbicide and broadleaf herbicide sequentially, increasing the rate of the ACCase-inhibiting herbicide, or by using adjuvants (Jordan and York, 1989; Corken et al., 1999).

Soybean

Prior to introduction of herbicide-resistant soybean, the standard herbicide program in soybean consisted of residual PRE herbicides followed by POST herbicides. PRE herbicides were typically dinitroaniline and/or imidazolinone herbicides, while the POST herbicides such as imidazolinones, sulfonylureas, diphenylethers, aryloxyphenoxypropionates, and cyclohexanediones were applied alone or in combination. Currently, at least 80 herbicides or prepackaged herbicide mixtures are registered for use in soybean (Anonymous, 2000), and it is not uncommon for a farmer to apply three to five different herbicides to manage weeds in soybean.

Weed resistance in soybean production remains a concern. This is especially true for the ALS- and ACCase-inhibiting herbicides, which are commonly used worldwide to manage broadleaf and grass weeds in soybean. Worldwide, 90 and 34 known weed species are resistant to ALS- and ACCase-inhibiting herbicides, respectively (Heap, 2005). Members of these two herbicide families are used extensively in other crops such as corn, cotton, and rice, so in some cases simply rotating crops without changing herbicide mode of action is not a solution for managing weed resistance. Even though most weeds in nontransgenic soybean can be adequately managed with existing herbicides, HRCs can provide producers with additional tools in managing problematic perennial weed species as well as weeds resistant to other herbicides. Currently two types of herbicide-resistant soybean are commercially available to producers.

Glyphosate-Resistant Soybean

Glyphosate resistant soybean varieties were developed using ballistic propulsion technology to insert the glyphosate resistant *cp4* EPSPS gene into soybean (Padgette et al., 1995). The *cp4* EPSPS gene was derived from the common soil bacterium *Agrobacterium* sp. strain and is a member of the class of EPSPS proteins found ubiquitously in plants and microorganisms (Monsanto, 2004). A multitude of glyphosate-resistant soybean varieties are available commercially from several seed companies. In most cases, one to as many as all of new soybean varieties developed by seed companies are resistant to glyphosate. U.S. seed companies currently marketing glyphosate-resistant soybean varieties include Asgrow Seed Company, Delta King Seed Company, Delta and Pine Land Company, Hartz seed company, and Pioneer Hi-Bred International.

Glyphosate-resistant soybean as a weed management tool was the topic of a recent review by Reddy (2001b). Glyphosate-resistant soybean provides producers unprecedented simplicity and flexibility to control weeds because glyphosate is a nonselective herbicide (Wilson and Worsham, 1988; Krausz, Kapusta, and Matthews, 1996). Glyphosate can be used to manage a broad spectrum of annual and perennial weeds with minimal concern for transgenic crop damage. Glyphosate has exceptional activity on most troublesome weeds in soybean such as broadleaf signalgrass, barnyardgrass [*Echinochloa crus-galli* (L.) Beauv.], johnsongrass, large crabgrass, common lambsquarters, common cocklebur, prickly sida (*Sida spinosa* L.), sicklepod, hemp sesbania, velvetleaf, pigweeds, and morning glories (Ateh and Harvey, 1999; Gonzini, Hart, and Wax, 1999; Culpepper et al., 2000, 2001; Reddy and Whiting, 2000; Norsworthy and Oliver, 2001; Reddy, 2001c; Young et al., 2001). Glyphosate also has activity on perennial weeds such as redvine [*Brunnichia ovata* (Walter) Shinness] (Reddy, 2000), trumpetcreeper [*Campsis radicans* (L.) Seem. ex Bureau] (Chachalis, Reddy, and Elmore, 2001), yellow nutsedge (*Cyperus esculentus* L.), and purple nutsedge (*Cyperus rotundus* L.) (Akin and Shaw, 2001). Use of glyphosate-resistant soybean offers an option of POST management of grass and broadleaf weeds with a single timely application of glyphosate (Ateh and Harvey, 1999; Nelson and Renner, 1999; Webster, Bryant, and Earnest, 1999; Reddy and Whiting, 2000; Reddy 2001c). Prior to glyphosate-resistant soybean, total POST weed control programs were more dependent on weed species, weed size, tank-mix partners, and adjuvant (DeFelice et al., 1989; Monks, Wilcut, and Richburg, 1993; Jordan, York, et al., 1997; Nelson, Renner, and Penner, 1998; Webster, Bryant, and Earnest, 1999). Use of a single herbicide such as glyphosate eliminates the possibility of antagonism that often occurs with the tank mixing of grass and broadleaf herbicides (Vidrine, Reynolds, and Blouin, 1995).

Glyphosate can be applied POST as needed, regardless of the crop growth stage with no risk of crop injury (Corrigan and Harvey, 2000). Glyphosate-resistant soybean fits easily in no-till and narrow-row soybean systems, thereby allowing producers to eliminate prophylactic application of PRE herbicides and reducing the number of POST herbicide applications due to quicker canopy closure (Wiesbrook et al., 2001). Unrestrictive POST glyphosate applications to glyphosate-resistant soybean provide early-season weed management at a time when weeds are most detrimental to the crop. Other benefits of glyphosate weed management systems include minimal environmental concern, no crop rotational restrictions, less dependency on soil-applied herbicides, and an alternative tool to manage weeds resistant to other herbicides (Burnside, 1992; Wyse, 1992; Wilcut et al., 1996; Reddy, 2001b).

The disadvantage of glyphosate-resistant soybean is the lack of residual soil activity of glyphosate. Multiple applications of glyphosate are often needed to manage multiple weed flushes, especially if a residual soil-applied herbicide was not used (Corrigan and Harvey, 2000). Producers are concerned about their ability to apply glyphosate in a timely manner in the event of rainy or wet field conditions. Research has also shown glyphosate efficacy on most weeds to be strongly correlated with weed size (Jordan, York, et al., 1997; Young et al., 2001). This is especially true for some troublesome broadleaf weeds such as pitted morning glory (*Ipomoea lacunosa* L.) and hemp sesbania (Lanie et al., 1994; Jordan, York, et al., 1997; Norsworthy and Oliver, 2002a,b). Adding residual PRE herbicides at soybean planting under certain conditions can provide an extended window for glyphosate POST applications, but PRE herbicides tend to increase weed control costs compared to glyphosate POST-only programs. In addition, overdependance on glyphosate will more likely increase the selection pressure on certain weed populations to develop resistance.

Sulfonylurea-Resistant Soybean

Sulfonylureas, similar to other ALS-inhibiting herbicide, are widely used because they effectively control a broad spectrum of grass and broadleaf weeds, are highly active inhibitors of a single target enzyme, exhibit low mammalian toxicity, and are used at very low rates (Beyer et al., 1988; Saari and Mauvais, 1996). The high degree of crop selectivity of sulfonylureas in general has diminished the need for developing sulfonylurea-resistant crops. However, scientific interest coupled with concerns about selectivity in market niches and residual activity in crop rotations led to development of several sulfonylurea-resistant crops (Saari and Mauvais, 1996). Currently, several soybean varieties are resistant to sulfonylurea herbicides. The varieties were developed from a Dupont soybean variety that was the first variety resistant to sulfonylurea herbicides (Simpson and Stoller, 1996). The first sulfonylurea-resistant soybean variety was developed through seed mutagenesis followed by selection in the M1 population for resistance to chlorosulfuron (Sebastian et al., 1989). The basis for resistance is an alteration in the gene-coding sequence for ALS, resulting in an enzyme that is resistant to certain sulfonylurea herbicides.

Prior to development of sulfonylurea-resistant soybean, many problem weeds such as morning glories and pigweeds could be managed with sulfonylurea herbicides, but at the expense of marked soybean injury. Problem weeds such as morning glories, pigweeds, browntop millet [*Brachiaria ramosa* (L.) Stapf.], common cocklebur, velvetleaf, common lambsquarters, sicklepod, hemp sesbania, hyssop spurge (*Euphorbia hyssopifolia* L.), prickly sida, and yellow nutsedge can be managed adequately without injuring soybean (Culpepper et al., 1997; Reddy and Whiting, 2000). Overall, applying sulfonylurea herbicides alone or in combination with other herbicides provides more broad-spectrum POST weed management options and better crop safety.

A limitation of sulfonylurea-resistant soybean systems is lack of or limited activity on grass species. In order to control grasses, an ACCase-inhibiting herbicide is applied with the sulfonylurea herbicide. Mixing these two types of herbicides often results in reduced activity of the grass herbicide. Thus, when both types of herbicides are required to manage existing weeds, the herbicides are applied sequentially or the ACCase-herbicide rate is increased. Even though sulfonylurea-resistant soybean is an effective weed management tool, repeated applications of sulfonylureas and other ALS-inhibiting herbicides can lead to selection pressure for weed resistance. However, sulfonylurea-resistant soybean can be a valuable asset in a well-planned crop and/or herbicide mode of action rotation program.

Cotton

Prior to introduction of herbicide-resistant cotton varieties, a typical herbicide program in nontransgenic cotton consisted of soil-applied preplant incorporated and/or PRE herbicides followed by pyrithiobac and/or MSMA (monosodium methanearsonate) applied POST over the top of cotton and concluded with MSMA plus a soil-applied herbicide such as fluometuron, prometryn, or diuron applied POST directed toward the base of cotton plants (Snipes et al., 1984; Byrd and York, 1987; Culpepper and York, 1998, 1999a; Askew and Wilcut, 1999; Faircloth et al., 2001). Fluometuron is the most common PRE herbicide used to control annual weeds in cotton. Several ACCase inhibitors are applied POST over the top of cotton to control grass weeds.

Prior to a pyrithiobac label in 1996, fluometuron and MSMA were the only POST herbicides labeled for over-the-top management of broadleaf weeds. However, these herbicides often resulted in cotton injury, delayed maturity, and reduced yield (Guthrie and York, 1989). Fluometuron plus MSMA applied POST directed has been the standard treatment for cotton less than 15 cm tall (Culpepper and York, 1999a). However, extra time, specialized equipment, and time-consuming techniques are often necessary to POST direct these herbicides to prevent herbicide contact to cotton foliage (Askew and Wilcut, 1999). The limited POST over-the-top herbicide options to control broadleaf weeds are exacerbated by reduced early-season cotton vigor, wide row spacings (76 to 102 cm), and delayed shading of row middles, which typically does not occur until at least 75 to 90 days after planting (Pline et al., 2001). Furthermore, the long, humid, and warm growing season, common to cotton production in the United States, is conducive to a broad spectrum of rapidly growing broadleaf weeds with extended emergence patterns. Cotton is more susceptible to yield reductions from weed interference than are corn or soybean (Coble and Byrd, 1992). Pyrithiobac POST over-the-top controls certain broadleaf weeds, but it has no activity on grass weeds and little activity on certain broadleaf weeds such as sicklepod (Paulsgrove and Wilcut, 2001; Reddy, 2001a). These factors, coupled with limited availability of POST herbicide options and the lack of consistent performance from prophylactic soil-applied herbicides, make weed management in cotton more difficult than in most other major agricultural crops.

Herbicide-resistant cotton varieties can be integrated into existing weed management systems. Two types of herbicide-resistant cotton varieties are currently available to producers. Both consist of herbicides with no residual soil activity but provide additional tools for POST weed control in cotton. However, PRE herbicides with residual activity such as dinitroanilines or fluometuron herbicides are often applied in a herbicide-resistant cotton weed control program (Askew and Wilcut, 1999; Reddy, 2001a). Reports are conflicting as to whether the herbicide-resistant cotton program can negate the necessity of a residual POST directed herbicide (Faircloth et al., 2001).

Glyphosate-Resistant Cotton

Glyphosate-resistant cotton varieties were developed using techniques similar to those of glyphosate-resistant soybean (Monsanto, 2004). Glyphosate applied POST over the top of glyphosate-resistant cotton provides effective control of a broad spectrum of annual and perennial grass and broadleaf weeds (Wilcut et al., 1996; Askew and Wilcut, 1999; Faircloth et al., 2001). Glyphosate can be applied POST over the top from emergence through the four-leaf stage of cotton (Monsanto, 2004). After the four-leaf growth stage, glyphosate can be applied as POST directed spray to avoid potential fruit abortion (Pline et al., 2001; Monsanto, 2004). Sequential applications must be at least ten days apart, and cotton must have at least two nodes of incremental growth between applications. Sequential applications of glyphosate are often required, because glyphosate has no residual activity. Some research has shown that a total POST glyphosate herbicide program is sufficient for season-long weed management (Askew and Wilcut, 1999).

Glyphosate alone can control such weeds as purple nutsedge and sicklepod that before glyphosate-resistant cotton were very difficult to manage. In some cases, the use of soil-applied or POST directed residual herbicides in conjunction with glyphosate in glyphosate-resistant cotton can be beneficial (Askew and Wilcut, 1999). This is especially true when cotton is planted in wide rows and late-season weed flushes occur. Overall, glyphosate can potentially reduce or eliminate soil-applied herbicides and reduce total herbicide input (Culpepper and York, 1998), provide greater flexibility in crop rotation programs, and offer the opportunity to control weeds that prior to glyphosate-resistant cotton were difficult to manage with POST herbicides.

Bromoxynil-Resistant Cotton

Bromoxynil is a member of the benzonitrile or nitrile herbicide family and is responsible for blocking electron transport and in turn ATP formation in photosystem II of susceptible plants. Bromoxynil is active on certain broadleaf weeds only and labeled for weed management in most monocotyledonous crops. Bromoxynil is a contact herbicide with little or no translocation within plants upon foliar application (Stalker et al., 1996). Upon application to tolerant plants, bromoxynil is rapidly metabolized into 3,5-dibromo 4-hydroxybenzamide, which is further degraded by enzymes that cleave the bromine and benzoic acid ring structure. These metabolites are at least 100-fold less toxic to plants than bromoxynil. A chimeric gene that codes for the nitrilase enzyme was derived from the soil bacterium *Klebsiella ozaenae*. Nitrilase enzyme rapidly detoxifies bromoxynil. This gene was introduced into cotton varieties via *Agrobacterium* transformation to produce bromoxynil-resistant cotton (Stalker et al., 1996). The nitrilase enzyme has a strict specificity for bromoxynil and not other benzonitrile compounds (Stalker et al., 1996). A standard field use rate is 0.56 kg a.i.·ha^{-1} per application with a maximum of three applications per season. Research with radiolabeled bromoxynil has determined that bromoxynil is rapidly metabolized to inactive metabolites in plants of bromoxynil-resistant cotton in the same fashion as in naturally tolerant plants (Stalker et al., 1996). Bromoxynil-resistant cotton is marketed by Stoneville Pedigree seed company. Bromoxynil-resistant cotton was discontinued in 2004.

Bromoxynil-resistant cotton provides producers a means to control a broad spectrum of broadleaf weeds, which are difficult to manage with limited POST herbicide options in nontransgenic cotton. Bromoxynil has excellent activity on troublesome broadleaf weeds such as morning glories, prickly sida, common cocklebur, hemp sesbania, common lambsquarters, jimsonweed (*Datura stramonium* L.), and velvetleaf (Culpepper and York, 1999a; Paulsgrove and Wilcut, 2001; Reddy, 2001a). However, bromoxynil has limited to no activity on common purslane (*Portulaca oleracea* L.), pigweeds, and sicklepod (Reddy, 2001a). An ACCase-inhibiting herbicide is invariably required in bromoxynil-resistant cotton, because bromoxynil has no activity on grass weeds. Bromoxynil and ACCase inhibitors may have to be applied separately, as mixing these often results in antagonism on grass weeds (Culpepper and York, 1997). Bromoxynil can replace most standard POST over the top and POST-directed herbicides normally applied in nontransgenic cotton depending upon weed species. Soil-applied residual PRE herbicides are necessary in bromoxynil-resistant cotton to reduce early-season weed interference and increase the utility of bromoxynil-resistant cotton system (Culpepper and York, 1997; Reddy, 2001a).

Rice

Rice producers are confronted with a barrage of grass and broadleaf weeds. Some of the most troublesome weeds in rice production systems include red rice (*Oryza sativa* L.), barnyardgrass, junglerice [*Echinochloa colona* (L.) Link], sprangletop species (*Leptochloa* spp.), large crabgrass, broadleaf signalgrass, yellow nutsedge, morning glory species, hemp sesbania, white eclipta (*Eclipta prostrata* L.), jointvetch species (*Aeschynomene* spp.), and a multitude of aquatic weeds such as ducksalad [*Heteranthera limosa* (Sw.) Willd.], common duckweed (*Lemna minor* L.), and alligator weed [*Alternanthera philoxeroides* (Mart.) Griseb.]. Today, rice producers have more herbicides at their disposal for weed management than ever before. A standard herbicide program typically consists of clomazone, pendimethalin, thiobencarb, or quinclorac applied PRE or delayed-PRE followed by single or combined POST application(s) of propanil, molinate, thiobencarb, and/or fenoxaprop followed by flooding for POST weed management. In addition, herbicides such as triclopyr, acifluorfen, and bentazon are applied for management of select broadleaf weeds. Most of the weed species previously mentioned can be managed adequately with available herbicides and proper water management (Crawford and Jordan, 1995; Ntanos, Koutroubas, and Mavrotas, 2000). One exception to this, however, is red rice. Red rice is a conspecific weed very similar genetically to cultivated rice (Vaughan et al., 2001). It has been very difficult to identify a herbicide that can be used to selectively kill red rice without causing injury to cultivated rice. This was one of the reasons for the development of cultivated rice resistant to herbicide(s). Prior to the development of herbicide-resistant rice, water management and crop rotation were the primary means of managing red rice. These cultural practices have had limited success for management of red rice. Currently one type of herbicide-resistant rice is available to rice producers.

Imidazolinone-Resistant Rice

The metabolic pathway of imidazolinone herbicides in rice is similar to that of corn (see section on imidazolinone-resistant corn). Imidazolinone-resistant rice was developed in 1993 by identifying and isolating a rice plant via chemical mutagenesis that was tolerant to imazethapyr herbicide (Croughan et al., 1995). Imidazolinone-resistant rice, commercially known as 'Clearfield Rice', survive imazethapyr treatment. Imazethapyr is a broad-spectrum herbicide primarily used in soybean and peanut prior to development of imidazolinone-resistant rice. It can be applied PRE and POST to kill grass and broadleaf weeds (Richburg, Wilcut, and Wehtje, 1993; Grichar, 1994; Hart and Wax, 1996). Imazethapyr applied POST kills susceptible weeds and provides residual activity on later-germinating weeds (Masson and Webster, 2001).

Imazethapyr is readily degraded by soil microbes under aerobic conditions and does not leach under field conditions (Mangels, 1991). The typical imidazolinone-resistant rice system consists of applying 140 $g{\cdot}ha^{-1}$ imazethapyr at different application timings such as PRE only (140 $g{\cdot}ha^{-1}$) or PRE (70 $g{\cdot}ha^{-1}$) followed by a POST (70 $g{\cdot}ha^{-1}$) application (Masson and Webster, 2001). These systems have resulted in 90 to 100 percent control of red rice, yellow nutsedge, barnyardgrass, morning glories, and *Leptochloa* spp. (Masson and Webster, 2001). Imidazolinone-resistant rice provides producers with a tool to manage red rice, and PRE and POST management options for other problematic rice weeds.

The major concern with imidazolinone-resistant rice is the potential for outcrossing of imidazolinone-resistant rice to native red rice. This is possible if the imidazolinone-resistant rice system is not 100 percent successful in killing all existing red rice plants. The chance of red rice and imidazolinone-resistant rice hybridization exists because the two are so genetically similar (Vaughan et al., 2001). Another concern is red rice and other weed species developing ALS resistance. Proper rotation of crops and herbicides with different mode of action must be implemented in order to reduce the risk of developing ALS-resistant weeds in an imidazolinone-resistant rice system.

Canola

Triazine-resistant canola was one of the first commercially developed HRCs. The first triazine-resistant canola variety was released on a commercial basis in Canada in 1984; however, no triazine-resistant canola variety with significantly improved photosynthetic efficiency compared to triazine-susceptible canola was ever developed (Hall, Donnelly-Vanderloo, and Hume, 1996). Although triazine-resistant canola can be considered technically successful, this crop had short-lived success because of rapid development of new herbicides, concern about triazines in the environment, and lower crop yield and quality due to the physiological changes brought about by the resistance trait (Hall, Donnelly-Vanderloo, and Hume, 1996). Currently, four types of herbicide-resistant canola varieties are available to producers. Herbicide-resistant canola was obtained through the insertion of novel genes (bromoxynil, glufosinate, and glyphosate resistant) or by mutagenesis (imidazolinone resistance). Most of the herbicide-resistant canola grown is canola (*Brassica napus* L.), but glufosinate- and glyphosate-resistant birdsrape mustard (*B. rapa* L.) varieties are also grown commercially in Canada (Simard et al., 2002).

Prior to introduction of herbicide-resistant canola, the typical herbicide program consisted of PRE application of a dinitroaniline herbicide such as trifluralin followed by ethametsulfuron and/or an ALS-inhibiting herbicide applied POST (Zollinger, 2002). Ethametsulfuron was the only herbicide that provided activity on broadleaf weeds. It provides adequate activity on wild mustard [*Brassica kaber* (DC.) L.C.Wheeler], flixweed [*Descurainia sophia* (L.) Webb. ex Prantl], smartweed (*Polygonum* spp.), and suppression of redroot pigweed (*Amaranthus retroflexus* L.) and field pennycress (*Thlaspi arvense* L.).

Glufosinate- and glyphosate-resistant canola can provide nonresidual broad-spectrum POST management of annual and perennial weeds. These two systems have excellent activity on problem weeds such as common lambsquarters, common purslane, dandelion [*Taraxacum officinale* Weber ex Wiggers], field pennycress, green foxtail [*Setaria viridis* (L.) Beauv.], quackgrass [*Elytrigia repens* (L.) Nevski], and wild mustard (Martin, Van Acker, and Friesen, 2001). Imazamox POST applications in imidazolinone-resistant canola can manage existing grass and broadleaf weeds and also provide residual activity (Zollinger, 2002). Herbicide-resistant canola varieties have provided producers with much-needed weed management tools that before the introduction of herbicide-resistant canola were not available.

BENEFITS AND RISKS OF HRCs AS WEED MANAGEMENT TOOLS

The farmers who have many options with herbicides must have seen some economic and weed control benefits from the use of HRCs; otherwise, the rapid increase in area planted to HRCs in recent years would not have occurred. Simple and flexible weed control programs can be designed based on one (e.g., glyphosate, glufosinate) or a few (e.g., bromoxynil, imidazolinones, sulfonylureas which need PRE herbicide partners for total weed control) herbicides. A broad spectrum of weeds can be controlled with glyphosate in glyphosate-resistant crops and glufosinate in glufosinate-resistant crops. HRCs provide greater flexibility in application timing. Glyphosate can be applied from emergence to flowering in glyphosate-resistant soybean (Reddy 2001b; Monsanto, 2004) and up to the four-leaf stage over the top and post directed beyond the four-leaf stage in glyphosate-resistant cotton (Monsanto, 2004). Extensive research in the United States has shown that two timely applications of glyphosate can provide effective control of weeds in glyphosate-resistant soybean. Farmers can eliminate prophylactic applications of PRE herbicides in glyphosate- and glufosinate-resistant crops depending upon weed species and their density. Carryover or soil persistence of PRE herbicides is a major concern in crop rotation. Well-designed herbicide programs utilizing HRCs can eliminate crop rotation restrictions. Glyphosate or glufosinate POST-only programs provide farmers with the freedom to choose a rotational crop for the following year and also freedom to plant a second crop within a season if the first crop fails to establish due to adverse weather conditions.

HRCs provide opportunity to manage problem weeds. For example, morning glory species with bromoxynil; purple nutsedge and sicklepod with glyphosate; and red rice with imidazolinones can be managed using HRCs. HRCs offer a tremendous advantage in the management of weeds resistant to other herbicides. In 59 countries, 178 weed species (107 dicots and 71 monocots) have evolved resistance to one or more herbicides (Heap, 2005). One hundred seven resistant weed biotypes are found in the United States. Examples are common cocklebur resistant to ALS inhibitors and organoarsenicals; goosegrass [*Eleusine indica* (L.) Gaertn.] resistant to ACCase inhibitors, ALS inhibitors, bipyridiliums, dinitroanilines, and glyphosate; horseweed [*Conyza canadensis* (L.) Cronq.] resistant to ALS inhibitors, bipyridiliums, and glyphosate; johnsongrass resistant to ACCase inhibitors, ALS inhibitors, and dinitroanilines; and pigweeds resistant to ALS inhibitors (Heap, 2005).

The most notable impact of adopting HRCs is reduction in weed control cost even though cost of seed is higher due to a technology fee. Glyphosate-based programs are price competitive with existing herbicide programs in nontransgenic crops. Introduction of competitively priced glyphosate-based programs resulted in manufacturers of other herbicides dropping their prices, in some cases by 40 percent. This resulted in an estimated savings of $216 million in herbicide expenditures for U.S. soybean farmers in 1999 compared to 1995, the year before glyphosate-resistant soybean was introduced (Carpenter and Gianessi, 2001). The estimated savings include the technology fee paid for glyphosate-resistant soybean. Herbicide use patterns have changed dramatically with the introduction of HRCs. As adoption of glyphosate-resistant soybean has increased, the average annual rate of glyphosate application has increased from 0.19 $kg \cdot ha^{-1}$ in 1996 to 0.48 $kg \cdot ha^{-1}$ in 1998, while all other herbicides combined dropped from about 1.12 $kg \cdot ha^{-1}$ to 0.64 $kg \cdot ha^{-1}$ (Heimlich et al., 2000). This translates to a decline of nearly 10 percent in the overall rate of herbicide use in soybean during the period. Based on regression analysis in soybean, an estimated 2.5 million kg of glyphosate is substituted for 3.3 million kg of other herbicides, such as imazethapyr, pendimethalin, and trifluralin. In general, herbicides that glyphosate replaced are at least three times as toxic and persist in the environment nearly twice as long as glyphosate (Heimlich et al., 2000).

Despite the benefits, some perceived risks are associated with the use of HRCs. Volunteer plants from a previous crop, arising from seed shattering or seeds being dropped during harvest, can become a weed in a succeeding crop, e.g., volunteer corn plants in soybean or cotton fields. In a corn-soybean or corn-cotton rotation system, farmers planting glyphosate-resistant soybean or glyphosate-resistant cotton after glyphosate-resistant corn will not be able to control volunteer glyphosate-resistant corn with glyphosate.

This will require an alternative control strategy such as use of either a PRE herbicide or another POST herbicide. Certain weed species are naturally tolerant to the herbicide. The weed species most likely to increase in frequency over time are those that either have a natural tolerance to the herbicide or are only partially controlled. Within three years of continuous BXN cotton production, a weed shift toward common purslane, sicklepod, and yellow nutsedge occurred from repeated use of bromoxynil (Reddy, 2004). Weed species shifts will be a greater problem if crops with the same herbicide-resistant trait are rotated (e.g., rotation among glyphosate-resistant corn, glyphosate-resistant soybean, and glyphosate-resistant cotton).

Continuous use of a single herbicide or herbicides with the same mode of action to control weeds can increase selection pressure to develop resistant weed biotypes. This has been documented with the selection of triazine-resistant, ACCase-inhibitor-resistant, and ALS-inhibitor-resistant weeds (Heap, 2005; Heap and LeBaron, 2001). Resistance to glyphosate in rigid ryegrass (*Lolium rigidum* Guad.) in Australia, South Africa, and the United States and goosegrass in Malaysia has been reported (Heap, 2005). Recently, VanGessel (2001) reported that within three years of using only glyphosate for weed management in glyphosate-resistant soybean, populations of horseweed had developed an 8- to 13-fold glyphosate resistance compared to susceptible populations.

The likelihood of development of herbicide-resistant weeds through gene flow from HRCs to their wild relatives is a strong possibility but varies greatly among different crops (Keeler, Turner, and Bolick, 1996; Darmency, 1997; Gressel and Rotteveel, 2000). For some crops, the probability of gene transfer is very high, for example, cultivated rice to red rice and canola to wild *Brassica* spp. For others such as corn, cotton, and soybean, the probability of lateral gene transfer is low. From a weed management perspective, the mere development of herbicide-resistant weeds is not necessarily a problem in and of itself. The herbicide-resistant weeds become a problem only when the herbicide to which it is resistant is applied. Nevertheless, the risk of development of a herbicide-resistant weed through either gene flow or selection pressure in HRCs is similar to selection of herbicide-resistant weeds due to repeated use of a selective herbicide in naturally tolerant crop. Transgenic crops represent a revolutionary breakthrough in weed control technology (Baldwin, 2000), and herbicides are an integral part of modern agriculture. Therefore, the farmers should not completely rely on HRCs to manage weeds and must employ simple measures, such as rotation of crops and/or herbicides from different modes of action.

The HRCs have impacted herbicide use as is evident from increased use of glyphosate with a concomitant decrease in the use of other herbicides. In response to the lower cost of glyphosate, other agrochemical industries have decreased the price of their herbicides to remain competitive. Both the herbicide market and profit margin are shrinking. Because the discovery and development of a new compound is expensive and the new product must be cost-effective to survive in the existing herbicide market, agrochemical industries are reluctant to invest in a shrinking market (Shaner, 2000). Currently, fewer new herbicides are being developed. The increased acquisitions and mergers among agrochemical industries may reduce competition for discovery of new herbicides.

CONCLUSIONS

In summary, HRCs are not evil, as opponents of genetically modified organisms have portrayed. HRCs provide farmers with another tool to manage weeds. Benefits and risks are associated with the use of HRCs, and the benefits outweigh their risks as a weed management tool. Simple and flexible weed control programs can be designed based on one or a few herbicides. HRCs should not be relied on to the exclusion of other weed control methods and should be used within integrated weed management systems. Prudent use of a HRCs in combination with other weed management tactics will most likely prolong their utility as a weed management tool.

REFERENCES

Akin, D.S. and D.R. Shaw (2001). Purple nutsedge *(Cyperus rotundus)* and yellow nutsedge (*Cyperus esculentus*) control in glyphosate-tolerant soybean *(Glycine max). Weed Technology* 15: 564-570.

Amrhein, N., J. Schab, and H.C. Steinrücken (1980). The mode of action of the herbicide glyphosate. *Naturwissenschaften* 67: 356-357.

Anonymous (2000). *Crop protection reference,* Sixteenth edition. New York: C and P Press.

Askew, S.D. and J.W. Wilcut (1999). Cost and weed management with herbicide programs in glyphosate-resistant cotton *(Gossypium hirsutum). Weed Technology* 13: 308-313.

Ateh, C.M. and R.G. Harvey (1999). Annual weed control by glyphosate in glyphosate-resistant soybean *(Glycine max). Weed Technology* 13: 394-398.

Baldwin, F.L. (2000). Transgenic crops: A view from the US extension service. *Pest Management Science* 56: 584-585.

Beyer, E.M. Jr., M.J. Duffy, J.V. Hay, and D.D. Schlueter (1988). Sulfonylureas. In P.C. Kearney and D.D. Kaufman (Eds.), *Herbicides: Chemistry, degradation, and mode of action* (pp. 117-189). New York: Marcel Dekker, Inc.

Biotechnology Industry Organization (BIO) (2004). Biotechnology information, advocacy, and business support. Available online at <www.bio.org>.

Blanchard, P.E. and W.W. Donald (1997). Herbicide contamination of groundwater beneath claypan soils in north-central Missouri. *Journal of Environmental Quality* 26: 1612-1621.

Buehler, D.D., G.W. Randall, W.C. Koskinen, and D.L. Wyse (1993). Atrazine and alachlor losses from subsurface tile drainage of clay loam soil. *Journal of Environmental Quality* 22: 583-588.

Burnside, O.C. (1992). Rationale for developing herbicide-resistant crops. *Weed Technology* 6: 621-625.

Burton, J.D., J.W. Gronwald, R.A. Keith, D.A. Somers, B.G. Greenenbach, and D.L. Wyse (1991). Kinetics of inhibition of acetyl-coenzyme A carboxylase by sethoxydim and halyoxyfop. *Pesticide Biochemistry and Physiology* 39: 100-109.

Byrd, J.D. Jr. and A.C. York (1987). Annual grass control in cotton *(Gossypium hirsutum)* with fluazifop, sethoxydim, and selected dinitroaniline herbicides. *Weed Science* 35: 388-394.

Carpenter, J.E. and L.P. Gianessi (2001). Agricultural biotechnology: Updated benefit estimates. Washington, DC: National Center for Food and Agricultural Policy. Available online at <www.ncfap.org/reports/biotech/updatedbenefits.pdf>.

Chachalis, D., K.N. Reddy, and C.D. Elmore (2001). Characterization of leaf surface, wax composition, and control of redvine and trumpetcreeper with glyphosate. *Weed Science* 49: 156-163.

Coble, H.D. and J.D. Byrd (1992). Interference of weeds with cotton. In C.G. McWhorter and J.R. Abernathy (Eds.), *Weeds of cotton: Characterization and control* (pp. 73-84). Memphis, TN: The Cotton Foundation.

Corken, C.B., D.L. Jordan, J.L. Griffin, P.R. Vidrine, B.J. Williams, and D.B. Reynolds (1999). Influence of adjuvants on interactions of sethoxydim with selected broadleaf herbicides used in corn *(Zea mays). Weed Technology* 13: 821-824.

Corrigan, K.A. and R.G. Harvey (2000). Glyphosate with and without residual herbicides in no-till glyphosate-resistant soybean *(Glycine max). Weed Technology* 14: 569-577.

Crawford, S.H. and D.L. Jordan (1995). Comparison of single and multiple applications of propanil and residual herbicides in dry-seeded rice *(Oryza sativa). Weed Technology* 9: 153-157.

Croughan, T.P., H.S. Utomo, D. Sanders, and M. Braverman (1995). Assessment of imidazolinone-resistant rice. *Rice Experiment Station Annual Progress Report* [Baton Rouge, Louisiana State University] 87: 491-525.

Culpepper, A.S., A.E. Giminez, A.C. York, R.B. Batts, and J.W. Wilcut (2001). Morningglory (*Ipomoea* spp.) and large crabgrass *(Digitaria sanguinalis)* control with glyphosate and 2,4-DB mixtures in glyphosate-resistant soybean *(Glycine max). Weed Technology* 15: 56-61.

Culpepper, A.S. and A.C. York (1997). Weed management in no-tillage bromoxynil-tolerant cotton *(Gossypium hirsutum). Weed Technology* 11: 335-345.

Culpepper, A.S. and A.C. York (1998). Weed management in glyphosate-tolerant cotton. *Journal of Cotton Science* 2: 174-185.

Culpepper, A.S. and A.C. York (1999a). Weed management and net returns with transgenic, herbicide-resistant, and nontransgenic cotton *(Gossypium hirsutum). Weed Technology* 13: 411-420.

Culpepper, A.S. and A.C. York (1999b). Weed management in glufosinate resistant corn *(Zea mays). Weed Technology* 13: 324-333.

Culpepper, A.S., A.C. York, R.B. Batts, and K.M. Jennings (1997). Sicklepod *(Senna obtusifolia)* management in an ALS-modified soybean *(Glycine max). Weed Technology* 11: 164-170.

Culpepper, A.S., A.C. York, R.B. Batts, and K.M. Jennings (2000). Weed management in glufosinate- and glyphosate-resistant soybean *(Glycine max). Weed Technology* 14: 77-88.

Currie, R.S., C.S. Kwon, and D. Penner (1995). Magnitude of imazethapyr resistance of corn *(Zea mays)* hybrids with altered acetolactate synthase. *Weed Science* 43: 578-583.

Darmency, H. (1997). Gene flow between crops and weeds: Risk for new herbicide resistant weeds? In R. De Parado, J. Jorrín, and L. García-Torres (Eds.), *Weed and crop resistance to herbicides* (pp. 239-248). Dordrecht, the Netherlands: Kluwer Academic Publications.

DeFelice, M.S., W.B. Brown, R.J. Aldrich, B.D. Sims, D.T. Judy, and D.R. Guethle (1989). Weed control in soybeans *(Glycine max)* with reduced rates of postemergence herbicides. *Weed Science* 37: 365-374.

Dekker, J. and S.O. Duke (1995). Herbicide-resistant field crops. *Advances in Agronomy* 54: 69-116.

Devine, M.D. (1997). Target-site based resistance to ACCase inhibitors. In R. De Prado, J. Jorrin, and L. Garcia-Torres (Eds.), *Weed and crop resistance to weeds* (pp. 61-69). Dordrecht, the Netherlands: Kluwer Academic Publishers.

Donn, G. (1997). Herbicide resistant crops generated by biotechnology. In R. De Prado, J. Jorrin, and L. Garcia-Torres (Eds.), *Weed and crop resistance to herbicides* (pp. 217-227). Dordrecht, the Netherlands: Kluwer Academic Publications.

Droge, W., I. Broer, and A. Pulher (1992). Transgenic plants containing the phosphinothricin-*N*-acetyltransferase gene metabolize the herbicide L-phosphinothricin (glufosinate) differentially from untransformed plants. *Planta* 187: 142-151.

Duke, S.O. (Ed.) (1996). *Herbicide-resistant crops: Agricultural, environmental, economic, regulatory, and technical aspects.* Boca Raton, FL: CRC Press, Inc. and Lewis Publishers.

Duke, S.O. (2002). Herbicide-resistant crops. In D. Pimentel (Ed.), *Encyclopedia of pest management* (pp. 358-360). New York: Marcel Dekker.

Duke, S.O., B.E. Scheffler, F.E. Dayan, and W.E. Dyer (2002). Genetic engineering crops for improved weed management traits. In K. Rajasekaran, T. J. Jacks, and J.W. Finley (Eds.), *Crop biotechnology: American Chemical Society Symposium Series* 829: 52-56.

Dunwell, J.M. (1999). Transgenic crops: The next generation, or an example of 2020 vision. *Annals of Botany* 84: 269-277.

Dyer, W.E. (1996). Techniques for producing herbicide-resistant crops. In S.O. Duke (Ed.), *Herbicide-resistant crops: Agricultural, environmental, economic, regulatory, and technical aspects* (pp. 37-51). Boca Raton, FL: CRC Press, Inc. and Lewis Publishers.

Dyer, W.E., F.D. Hess, J.S. Holt, and S.O. Duke (1993). Potential benefits and risks of herbicide-resistant crops produced by biotechnology. *Horticultural Reviews* 15: 367-407.

Economic Research Service (ERS) (2004). Adoption of genetically engineered crops in US. Washington, DC: Economic Research Service, U.S. Department of Agriculture. Available at <http://www.ers.usda.gov/data/biotechcrops/adoption.htm>.

Faircloth, W.H., M.G. Patterson, C.D. Monks, and W.R. Goodman (2001). Weed management programs for glyphosate-tolerant cotton *(Gossypium hirsutum). Weed Technology* 15: 544-551.

Franz, J.E., M.K. Mao, and J.A. Sikorski (1997). *Glyphosate: A unique global herbicide.* American Chemical Society Monograph 189. Washington, DC: American Chemical Society.

Ghidey, F., E.E. Alberts, and R.N. Lerch (1997). Spatial and temporal variability of herbicides in a claypan soil watershed. *Journal of Environmental Quality* 26: 1555-1563.

Gianessi, L.P., C.S. Silvers, S. Sankula, and J.E. Carpenter (2002). Plant biotechnology: Current and potential impact for improving pest management in U.S. agriculture, an analysis of 40 case studies. Washington, DC: National Center for Food and Agricultural Policy. Available online at <www.ncfap.org/40case studies.htm>.

Gonzini, L.C., S.E. Hart, and L.M. Wax (1999). Herbicide combinations for weed management in glyphosate-resistant soybean *(Glycine max). Weed Technology* 13: 354-360.

Gressel, J. (2002). *Molecular biology of weed control.* New York: Taylor and Francis Inc.

Gressel, J. and T. Rotteveel (2000). Genetic and ecological risks from biotechnologically-derived herbicide-resistant crops: Decision trees for risk assessment. *Plant Breeding Reviews* 18: 251-303.

Grichar, W.J. (1994). Spiny amaranth (*Amaranthus spinosus* L.) control in peanut (*Arachis hypogaea* L.). *Weed Technology* 8: 199-202.

Guthrie, D.S. and A.C. York (1989). Cotton *(Gossypium hirsutum)* development and yield following fluometuron postemergence applied. *Weed Technology* 3: 501-504.

Hall, J.C., M.J. Donnelly-Vanderloo, and D.J. Hume (1996). Triazine-resistant crops: The agronomic impact and physiological consequences of chloroplast mutation. In S.O. Duke (Ed.), *Herbicide-resistant crops: Agricultural, environmental, economic, regulatory, and technical aspects* (pp. 107-126). Boca Raton, FL: CRC Press, Inc. and Lewis Publishers.

Hamill, A.S., S.Z. Knezevic, K. Chandler, P.H. Sikkema, F.J. Tardif, A. Shrestha, and C.J. Swanton (2000). Weed control in glufosinate-resistant corn *(Zea mays). Weed Technology* 14: 578-585.

Hart, R.G., E.M. Lignowski, and F.R. Taylor (1991). Imazethapyr herbicide. In D.L. Shaner and S.L. O'Connor (Eds.), *The imidazolinone herbicides* (pp. 247-256). Boca Raton, FL: CRC Press.

Hart, S.E. and L.M. Wax (1996). Dicamba antagonizes grass weed control with imazethapyr by reducing foliar absorption. *Weed Technology* 10: 828-834.

Hatzios, K.K. and D. Penner (1985). Interactions of herbicides with other agrichemicals in higher plants. *Reviews of Weed Science* 1: 16-63.

Hay, J.R. (1974). Gains to the grower from weed science. *Weed Science* 22: 439-442.

Heap, I. (2005). The international survey of herbicide resistant weeds. Available online at <www.weedscience.org/in.asp>.

Heap, I.M. and H. LeBaron (2001). Introduction and overview of resistance. In S.B. Powles and D.L. Shaner (Eds.), *Herbicide resistance and world grains* (pp. 1-22). Boca Raton, FL: CRC Press.

Heimlich, R.E., J. Fernandez-Cornejo, W. McBride, C. Klotz-Ingram, S. Jans, and N. Brooks (2000). Genetically engineered crops: Has adoption reduced pesticide use? *Agricultural Outlook* (August): 13-17. Available online at <www.ers.usda.gov/publications/agoutlook/aug2000/ao273f.pdf>.

Holshouser, D.L. and H.D. Coble (1990). Compatibility of sethoxydim with five postemergence broadleaf herbicides. *Weed Technology* 4: 128-133.

Hydrick, D.E. and D.R. Shaw (1995). Non-selective and selective herbicide combinations in stale seedbed soybean *(Glycine max). Weed Technology* 9: 158-165.

Incledon, B.J. and J.C. Hall (1997). Acetyl-coenzyme A carboxylase: Quaternary structure and inhibition by graminicidal herbicides. *Pesticide Biochemistry and Physiology* 57: 255-271.

James, C. (2003). Global review of commercialized transgenic crops: 2003. ISAAA Briefs No. 30: Preview. Ithaca, NY: ISAAA. Available online at <www.isaaa.org/publications/pdfs/isaaabriefs/Briefs%2024.pdf>.

James, C. (2004). Preview: Global status of commercialized biotech/GM crops: 2004. ISAAA briefs no. 32. Ithaca, NY: ISAAA. Available at <http://www.isaaa.org/kc/CBTnews/press_release/briefs32/ESummary/Executive%Summary%20 (English).pdf>.

Johnson, W.G., P.R. Bradley, S.E. Hart, M.L. Buesinger, and R.E. Massey (2000). Efficacy and economics of weed management in glyphosate-resistant corn *(Zea mays). Weed Technology* 14: 57-65.

Jordan, D.L., R.E. Frans, and M.R. McClelland (1993). Total postemergence herbicide programs in cotton *(Gossypium hirsutum)* with sethoxydim and DPX-PE350. *Weed Technology* 7: 196-201.

Jordan, D.L., J.L. Griffin, P.R. Vidrine, D.R. Shaw, and D.B. Reynolds (1997). Comparison of graminicides applied at equivalent costs in soybean *(Glycine max). Weed Technology* 11: 804-809.

Jordan, D.L. and A.C. York (1989). Effects of ammonium fertilizers and BCH 81508 S on antagonism with sethoxydim plus bentazon mixtures. *Weed Technology* 3: 450-454.

Jordan, D.L., A.C. York, J.L. Griffin, P.A. Clay, P.R. Vidrine, and D.B. Reynolds (1997). Influence of application variables on efficacy of glyphosate. *Weed Technology* 11: 354-362.

Keeler, K.H., C.E. Turner, and M.R. Bolick (1996). Movement of crop transgenes into wild plants. In S.O. Duke (Ed.), *Herbicide-resistant crops: Agricultural, environmental, economic, regulatory, and technical aspects* (pp. 303-330). Boca Raton, FL: CRC Press, Inc. and Lewis Publishers.

Krausz, R.F., G. Kaputsa, and J.L. Matthews (1996). Control of annual weeds with glyphosate. *Weed Technology* 10: 957-962.

Krausz, R.F., G. Kaputsa, J.L. Matthews, J.L. Baldwin, and J. Maschoff (1999). Evaluation of glufosinate-resistant corn *(Zea mays)* and glufosinate: Efficacy on annual weeds. *Weed Technology* 13: 691-696.

Lanie, A.J., J.L. Griffin, P.R. Vidrine, and D.B. Reynolds (1994). Weed control with non-selective herbicides in soybean *(Glycine max)* stale seedbed culture. *Weed Technology* 8: 159-164.

LaRossa, R.A. and J.V. Schloss (1984). The sulfonylurea herbicide sulfometuron methyl is an extremely potent and selective inhibitor of acetolactate synthase in *Salmonella typhimurium* (pesticide mode of action). *Journal of Biological Chemistry* 259: 8753-8757.

Mangels, G. (1991). Behavior of the imidazolinone herbicides in soil: A review of the literature. In D.L. Shaner and S.L. O'Conner (Eds), *The imidazolinone herbicides* (pp. 191-210). Boca Raton, FL: CRC Press.

Martin, S.G., R.C. Van Acker, and L. F. Friesen (2001). Critical period of weed control in spring canola. *Weed Science* 49: 326-333.

Masson, J.A. and E.P. Webster (2001). Use of imazethapyr in water-seeded imidazolinone-tolerant rice *(Oryza sativa)*. *Weed Technology* 15: 103-106.

McLean, G.D. and G. Evans (Eds.) (1995). *Herbicide-resistant crops and pastures in Australian farming systems.* Canberra, Australia: Bureau of Resources Sciences.

Monks, C.D., J.W. Wilcut, and J.S. Richburg III (1993). Broadleaf weed control in soybean *(Glycine max)* with chlorimuron plus acifluorfen or thifensulfuron mixtures. *Weed Technology* 7: 317-321.

Monsanto (2004). U.S. agriculture. Available online at <www.Monsanto.com/Monsanto/us_ag/layout/default.asp>.

Moomaw, R.S. and A.R. Martin (1985). Herbicide evaluations for no-till soybean *(Glycine max)* production in corn *(Zea mays)* residue. *Weed Science* 33: 679-685.

Morton, C.A., R.G. Harvey, J.L. Wedberg, J.J. Kells, D.A. Landis, and W.E. Lueschen (1994). Influence of corn rootworm insecticides on the response of field corn *(Zea mays)* to nicosulfuron. *Weed Technology* 8: 289-295.

Nelson, K.A. and K.A. Renner (1999). Weed management in wide- and narrow-row glyphosate-resistant soybean. *Journal of Production Agriculture* 12: 460-465.

Nelson, K.A., K.A. Renner, and D. Penner (1998). Weed control in soybean *(Glycine max)* with imazamox and imazethapyr. *Weed Science* 46: 587-594.

Ngouajio, M. and E.S. Hagood Jr. (1993). Weed control in corn *(Zea mays)* with primisulfuron as influenced by rate, timing, and herbicide combinations. *Weed Technology* 7: 65-69.

Norsworthy, J.K. and L.R. Oliver (2001). Effect of seeding rate of drilled glyphosate-resistant soybean *(Glycine max)* on seed yield and gross profit margin. *Weed Technology* 15: 284-292.

Norsworthy, J.K. and L.R. Oliver (2002a). Hemp sesbania interference in drill-seeded glyphosate-resistant soybean. *Weed Science* 50: 34-41.

Norsworthy, J.K. and L.R. Oliver (2002b). Pitted morningglory interference in drill-seeded glyphosate-resistant soybean. *Weed Science* 50: 26-33.

Ntanos, D.A., S.D. Koutroubas, and C. Mavrotas (2000). Barnyardgrass *(Echinochloa crus-galli)* control in water-seeded rice *(Oryza sativa)* with cyhalofop-butyl. *Weed Technology* 14: 383-388.

Padgette, S.R., K.H. Kolacz, X. Re Delannay, B.J. LaVallee, C.N. Tinius, W.K. Rhodes, Y.I. Otero, G.F. Barry, D.A. Eichholtz, V.M. Peschke, et al. (1995). Development, identification, and characterization of a glyphosate-tolerant soybean line. *Crop Science* 35: 1451-1461.

Paulsgrove, M.D and J.W. Wilcut (2001). Weed management with pyrithiobac preemergence in bromoxynil-resistant cotton. *Weed Science* 49: 567-570.

Pline, W.A., A.J. Price, J.W. Wilcut, K.L. Edmisten, and R. Wells (2001). Absorption and translocation of glyphosate in glyphosate-resistant cotton as influenced by application method and growth stage. *Weed Science* 49: 460-467.

Powles, S.B., C. Preston, I.B. Bryan, and A.R. Jutsum (1997). Herbicide resistance: Impact and management. *Advances in Agronomy* 58: 57-93.

Reddy, K.N. (2000). Factors affecting toxicity, absorption, and translocation of glyphosate in redvine *(Brunnichia ovata)*. *Weed Technology* 14: 457-462.

Reddy, K.N. (2001a). Broadleaf weed control in ultra narrow row bromoxynil-resistant cotton *(Gossypium hirsutum)*. *Weed Technology* 15: 497-504.

Reddy, K.N. (2001b). Glyphosate-resistant soybean as a weed management tool: Opportunities and challenges. *Weed Biology and Management* 1: 193-202.

Reddy, K.N. (2001c). Weed management in transgenic soybean resistant to glyphosate under conventional tillage and no-tillage systems. *Journal of New Seeds* 3(1): 27-40.

Reddy, K.N. (2004). Weed control and species shift in bromoxynil- and glyphosate-resistant cotton *(Gossypium hirsutum)* rotation systems. *Weed Technology* 18: 131-139.

Reddy, K.N. and K. Whiting (2000). Weed control and economic comparisons of glyphosate-resistant, sulfonylurea-tolerant, and conventional soybean *(Glycine max)* systems. *Weed Technology* 14: 204-211.

Richburg, J.S. III., J.W. Wilcut, and G.R. Wehtje (1993). Toxicity of imazethapyr to purple *(Cyperus rotundus)* and yellow nutsedges *(Cyperus esculentus)*. *Weed Technology* 7: 900-905.

Richburg, J.S. III, J.W. Wilcut, and G.L. Wiley (1995). AC 263,222 and imazethapyr rates and mixtures for weed management in peanut *(Arachis hypogaea)*. *Weed Technology* 9: 801-806.

Ritter, R.L. and H. Menbere (2001). Weed management systems utilizing glufosinate-resistant corn *(Zea mays)* and soybean *(Glycine max). Weed Technology* 15: 89-94.

Saari, L.L. and C.J. Mauvais (1996). Sulfonylurea herbicide-resistant crops. In S.O. Duke (Ed.), *Herbicide-resistant crops: Agricultural, environmental, economic, regulatory, and technical aspects* (pp. 127-142). Boca Raton, FL: CRC Press, Inc. and Lewis Publishers.

Sebastian, S.A., G.M. Fader, J.F. Ulrich, D.R. Forney, and R.S. Chaleff (1989). Semidominant soybean mutation for resistance to sulfonylurea herbicides. *Crop Science* 29: 1403-1408.

Senseman, S.A., T.L. Lavy, J.D. Mattice, E.E. Gbur, and B.W. Skulman (1997). Trace level pesticide detections in Arkansas surface waters. *Environmental Science and Technology* 31: 395-401.

Shaner, D.L. (2000). The impact of glyphosate-tolerant crops on the use of other herbicides and on resistance management. *Pest Management Science* 56: 320-326.

Shaner, D.L., N.F. Bascomb, and W. Smith (1996). Imidazolinone-resistant crops: Selection, characterization, and management. In S.O. Duke (Ed.), *Herbicide-resistant crops: Agricultural, environmental, economic, regulatory, and technical aspects* (pp. 143-157). Boca Raton, FL: CRC Press, Inc. and Lewis Publishers.

Shrestha, A., I. Rajcan, K. Chandler, and C.J. Swanton (2001). An integrated weed management strategy for glufosinate-resistant corn *(Zea mays). Weed Technology* 15: 517-522.

Simard, M.-J., A. Légère, D. Pageau, J. Lajeunesse, and S. Warwick (2002). The frequency and persistence of volunteer canola *(Brassica napus)* in Québec cropping systems. *Weed Technology* 16: 433-439.

Simpson, D.M. and E.W. Stoller (1996). Physiological mechanisms in the synergism between thifensulfuron and imazethapyr in sulfonylurea-tolerant soybean *(Glycine max). Weed Science* 44: 209-214.

Snipes, C.E., R.H. Walker, T. Whitwell, G.A. Buchanan, J.A. McGuire, and N.R. Martin (1984). Efficacy and economics of weed control methods in cotton *(Gossypium hirsutum). Weed Science* 32: 95-100.

Somers, D.A. (1996). Aryloxyphenoxypropionate- and cyclohexanedione-resistant crops. In S.O. Duke (Ed.), *Herbicide-resistant crops: Agricultural, environmental, economic, regulatory, and technical aspects* (pp. 175-188). Boca Raton, FL: CRC Press, Inc. and Lewis Publishers.

Stalker, D.M., J.A. Kiser, G. Baldwin, B. Coulombe, and C.M. Houck (1996). Cotton weed control using the BXN system. In S.O. Duke (Ed.), *Herbicide-resistant crops: Agricultural, environmental, economic, regulatory, and technical aspects* (pp. 93-105). Boca Raton, FL: CRC Press, Inc. and Lewis Publishers.

Tharp, B.E. and J.J. Kells (1999). Influence of herbicide application rate, timing, and interrow cultivation on weed control and corn *(Zea mays)* yield in glufosinate-resistant and glyphosate-resistant corn. *Weed Technology* 13: 807-813.

Tharp, B.E. and J.J. Kells (2001). Effect of glufosinate-resistant corn *(Zea mays)* population and row spacing on light interception, corn yield, and common lambsquarters *(Chenopodium album)* growth. *Weed Technology* 15: 413-418.

United States Department of Agriculture (USDA) (2005). National Agricultural Statistics Service. Available online at <usda.mannlib.cornell.edu/reports/nass/field/pcp-baa>.

United States Food and Drug Administration (USFDA) (2005). List of completed consultations on bioengineered foods. Available online at <www.cfsan.fda.gov/~lrd/biocon.htm>.

VanGessel, M.J. (2001). Glyphosate-resistant horseweed from Delaware. *Weed Science* 49: 703-705.

Vasil, I.K. (1996). Phosphinothricin-resistant crops. In S.O. Duke (Ed.), *Herbicide-resistant crops: Agricultural, environmental, economic, regulatory, and technical aspects* (pp. 85-91). Boca Raton, FL: CRC Press, Inc. and Lewis Publishers.

Vaughan, L.K., B.V. Ottis, A.M. Prazak-Havey, C.A. Bormans, C. Sneller, J.M Chandler, and W.D. Park (2001). Is all red rice found in commercial rice really *Oryza sativa? Weed Science* 49: 468-476.

Vencill, W.K. (Ed.) (2002). *Herbicide handbook.* Lawrence, KS: Weed Science Society of America.

Vidrine, P.R., D.B. Reynolds, and D.C. Blouin (1995). Grass control in soybean (*Glycine max*) with graminicides applied alone and in mixtures. *Weed Technology* 9: 68-72.

Warwick, S.I., H.J. Beckie, and E. Small (1999). Transgenic crops: New weed problems for Canada. *Phytoprotection* 180: 71-84.

Webster, E.P., K.J. Bryant, and L.D. Earnest (1999). Weed control and economics in nontransgenic and glyphosate-resistant soybean *(Glycine max). Weed Technology* 13: 586-593.

Wiesbrook, M.L., W.G. Johnson, S.E. Hart, P.R. Bradley, and L.M. Wax (2001). Comparison of weed management systems in narrow-row, glyphosate- and glufosinate-resistant soybean *(Glycine max). Weed Technology* 15: 122-128.

Wilcut, J.W., H.D. Coble, A.C. York, and D.W. Monks (1996). The niche for herbicide-resistant crops in U.S. agriculture. In S.O. Duke (Ed.), *Herbicide-resistant crops: Agricultural, environmental, economic, regulatory, and technical aspects* (pp. 213-230). Boca Raton, FL: CRC Press, Inc. and Lewis Publishers.

Wilcut, J.W., J.S. Richburg III, and F.R. Walls Jr. (1999). Response of johnsongrass *(Sorghum halepense)* and imidazolinone-resistant corn *(Zea mays)* to AC 263,222. *Weed Technology* 13: 484-488.

Wilson, J.S. and A.D. Worsham (1988). Combinations of nonselective herbicides for difficult to control weeds in no-till corn, *Zea mays,* and soybean, *Glycine max. Weed Science* 36: 648-652.

Wyse, D.L. (1992). Future impact of crops with modified herbicide resistance. *Weed Technology* 6: 665-668.

Young, B.G., J.M. Young, L.C. Gonzini, S.E. Hart, L.M. Wax, and G. Kaputsa (2001). Weed management in narrow-and wide-row glyphosate-resistant soybean *(Glycine max). Weed Technology* 15: 112-121.

Zollinger, R.K. (2002). *North Dakota weed control guide.* Fargo: North Dakota State University Extension Service.

Chapter 20

Strategies for Managing Herbicide-Resistant Weeds

Hugh J. Beckie
Gurjeet S. Gill

INTRODUCTION

The main factors responsible for major herbicide-resistant (HR) weed problems worldwide are (1) multiple applications of highly efficacious herbicides with the same mode of action; (2) annual weed species that occur at high densities, are widely distributed, are genetically variable, are prolific seed producers, and have efficient gene (seed or pollen) distribution systems; and (3) simple cropping systems that favor a few dominant weed species (Owen, 2001a; Thill and Lemerle, 2001). The global occurrence of HR weed biotypes is continually updated (Heap, 2005). Resistance-prone species include annual ryegrass (*Lolium rigidum* Gaudin), blackgrass (*Alopecurus myosuroides* Huds.), wild oat (*Avena* spp.), common lambsquarters (*Chenopodium album* L.), kochia [*Kochia scoparia* (L.) Schrad.], and *Amaranthus* spp. The majority of cases of field-selected herbicide resistance are conferred by a single gene with a high degree of dominance (Powles et al., 1997), although there are numerous exceptions.

In different agroecosystems of the world, farmers have been reluctant to proactively manage weeds to delay the selection for herbicide resistance. If the cost and effort of preventing/delaying resistance are perceived or estimated to be the same as managing HR weeds, farmers are reluctant to change their weed management program until after resistance has occurred. Farmers usually adopt integrated weed management (IWM) strategies only after the development of resistance. Multiple-group resistance in grass weed species has been the chief impetus for the adoption of IWM (Powles et al., 1997, 2000). The lack of proactive management of the evolution of HR weed populations may be due to the inability of farmers to assess the economic risk associated with HR weeds (Rotteveel, de Goeij, and van Gemerden, 1997) and, in some cases, the unavailability of alternative herbicide groups. There is also the uncertainty of how much of a delay in resistance can be achieved by implementing an IWM program. Socioeconomic factors in developed countries, such as increasing size of farms with concomitant limited labor and time availability and preference for cropping systems based on lifestyle choice and cash flow of farmers (e.g., continuous cropping of annual crops), have reinforced a heavy reliance on herbicides as the dominant weed control method (Friesen et al., 2000).

Prevention can cost significantly less than dealing with resistance once it fully develops, where multiple-group herbicide resistance occurs, or where few alternative herbicides are available (Orson, 1999). The greatest direct cost of resistance to the farmer can occur during the first year of poor weed control (Peterson, 1999). Thereafter, the cost of resistance is minimal if cost-effective alternative herbicides are available, such as those for control of biotypes resistant to auxinic and photosystem I disrupting herbicides. In contrast, the number of herbicide options for control of some multiple-group HR biotypes may be limited, and those that are available usually increase costs. For example, most or all alternative herbicides to control multiple-group HR biotypes of wild oat (*Avena fatua* L.) or green foxtail [*Setaria viridis* (L.) Beauv.] in western Canada increase costs to farmers (Beckie, Thomas, and Légère, 1999; Beckie et al., 1999).

To manage resistance, farmers first use alternative herbicides (i.e., addition of a tank-mix partner or switching to herbicides with a different mode of action). With the cost of discovering, developing, and marketing a novel herbicide at approximately US$100 million in 2000 (Dunan and Westra, 2000), farmers should not expect many compounds with novel modes of action to be commercialized in the next few years. In some situations, herbicides that selected for resistance may continue to be used because of their cost-effective control of non-HR weed species (e.g., triazines applied to land with triazine-HR biotypes); the addition of a herbicide to control the HR weed species, however, will increase costs to the farmer (Peterson, 1999).

One of the underlying principles of any management strategy is to reduce the selection pressure for resistance evolution by any one selecting agent while maintaining adequate weed control. Selection pressure has the greatest impact on resistance evolution and is a factor that farmers can control. Selection pressure imposed by the application of a herbicide is the product of selection intensity (efficacy) and selection duration (persistence in soil) (Putwain, 1982). Selection pressure against a weed population over time, resulting in increasing frequency of HR individuals that collectively possess one or more resistance mechanisms, is a function of frequency of application. Mathematically, the selection pressure of a herbicide on a target weed species in a population has been defined as the proportion of HR plants divided by the proportion of herbicide-susceptible (HS) plants that remain after exposure to the herbicide (Gressel and Segel, 1982). These proportions are equal to one minus the effective kill, which is based on seed production (Gressel and Segel, 1982; Beckie and Morrison, 1993). By definition, the selection pressure can be reduced only by lowering the effective kill of HS plants or increasing the effective kill of HR plants. Any weed control practice that controls HS and HR genotypes equally eliminates selection pressure. Diversification of selection pressures on weed populations, such as varying the type and timing of herbicide application (e.g., preseeding, in-crop, pre- and postharvest), integrating cultural/mechanical weed management strategies with reduced herbicide use, and diversifying the cropping system as a whole, is required to reduce the selection pressure of any one selecting agent (Boerboom, 1999).

In this chapter, we summarize strategies to delay or manage HR weeds in input-intensive cropping systems. We emphasize herbicide-based strategies because herbicides are expected to continue to be the dominant weed control tool in these cropping systems in the future. As nonherbicide weed management practices are discussed extensively elsewhere in this book, we restrict our discussion to nonherbicide strategies that have proven effective in managing HR weeds. Case studies are described from different parts of the world, illustrating the integration of herbicide and nonherbicide strategies in successfully managing HR weeds.

HERBICIDE STRATEGIES

Herbicide Rotations and Sequences

We define *herbicide sequence* as two or more applications of herbicides within one crop, whereas *herbicide rotation* is the application of herbicides to multiple crops over several growing seasons.

Performance and cost of herbicides usually rank higher than mode of action when farmers select a herbicide. The lack of suitable herbicide options associated with crop rotation can be an impediment to herbicide group rotation (Bourgeois, Morrison, and Kelner, 1997; Légère et al., 2000). The level of adoption of herbicide group rotation for resistance management varies among farmers from different regions of the world and may be related to the perceived or de facto seriousness of the resistance problem. In western Canada in 1998, less than one-half of farmers practiced herbicide group rotation, even though awareness was high (Beckie, Chang, and Stevenson, 1999). In Australia that year, the adoption rate was 85 percent; mode-of-action labeling on herbicide containers in Australia since 1997 has contributed to the adoption of herbicide group rotation for resistance management (Shaner, Howard, and Chalmers, 1999). It was the most common resistance management practice cited by farmers in survey questionnaires conducted in some developed countries (Shaner, Howard, and Chalmers, 1999; H. J. Beckie, unpublished data).

Evolution of target-site resistance in weed biotypes is attributed to frequent use of herbicides of the same mode of action and the ease of selection by these modes of action (LeBaron and McFarland, 1990; Beckie, Hall, and Tardif, 2001). Knowledge of resistance risk could be an incentive for farmers to practice herbicide rotations or sequences to delay the rate of evolution of resistance.

The ease of selection by a herbicide mode of action is governed by several factors. The selection pressure (efficacy, persistence) imposed on target weed species by a herbicide mode of action is the most important factor affecting the rate of evolution of resistance. The slow evolution of resistance in weed biotypes to auxinic herbicides, first introduced in 1946, has been partially attributed to relatively low selection pressure (Coupland, 1994). Nonpersistent herbicides generally exert less selection pressure than those that control successive flushes of germinating weeds throughout the growing season. The contribution of persistence to selection pressure, however, depends on timing of herbicide application and the germination characteristics of the species in a geographic region. The soil residual activity of herbicides did not strongly influence selection pressure on wild oat, estimated by the reduction in seed yield during the growing season (i.e., effective kill), in a competitive crop (canola, *Brassica napus* L.) in western Canada (Beckie and Holm, 2002) (see Figure 20.1). The effective kill of wild oat by residual herbicides was the same as or lower than that of nonresidual herbicides. In the relatively short growing season in the Northern Great Plains, few wild oat plants may emerge after postemergence application of a nonresidual herbicide and produce viable seed in a competitive crop.

Whereas a single mutation can confer resistance to single-site-of-action herbicides, multiple mutations within a plant are often needed to confer resistance to herbicides with more than one site of action, such as chloracetamide and auxinic herbicides (Coupland, 1994; Foes et al., 1998). Individuals in an unselected population with multiple mutations for resistance generally would be rare (Preston and Mallory-Smith, 2001). The frequency of HR alleles in unselected populations defines the starting point for resistance evolution and thus dictates the length of time for resistance to evolve to noticeable levels. An unusually low rate of mutation of the locus conferring resistance or, alternatively, few fit mutations are speculated to contribute to the slow evolution of resistance to synthetic auxins and other herbicides such as glyphosate (Jasieniuk, Morrison, and Brulé-Babel, 1995;

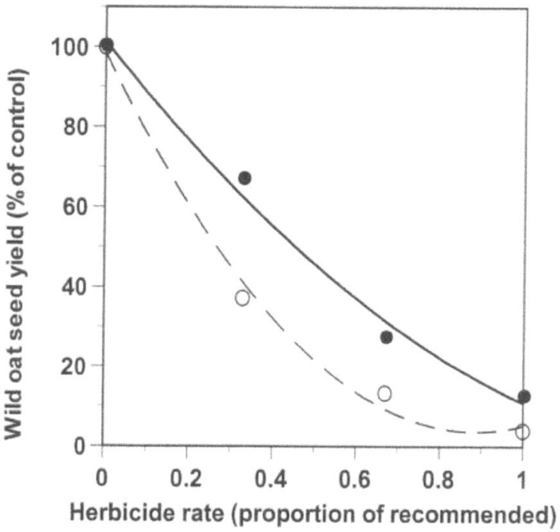

FIGURE 20.1. Response of wild oat (*Avena fatua* L.) seed yield to increasing rates of residual (R) and nonresidual (NR) herbicides in canola (*Brassica napus* L.) in western Canada: ethalfluralin (R), triallate (R), and glufosinate (NR) (represented by a solid line; response curves for these three herbicides did not differ significantly); and sethoxydim (NR), imazamox/imazethapyr (R), and glyphosate (NR) (represented by a dashed line; response curves for these three herbicides did not differ significantly). *Source:* Adapted from Beckie and Holm, 2002.

Gressel, 1999). Fit mutations are more probable for noncompetitive inhibitors of target-site enzymes, such as acetyl-coenzyme A carboxylase (ACCase) and acetolactate synthase (ALS) inhibitors, where the herbicide binding site is different from the active site. Multiple mutations of the ACCase and ALS target sites increases the frequency of occurrence of HR alleles in unselected populations (Murray, Brulé-Babel, and Morrison, 1996). In contrast, most mutations conferring resistance to glyphosate, a competitive inhibitor of EPSPS (enolpyruvylshikimate-3-phosphate synthase), and glufosinate, a competitive inhibitor of glutamine synthetase (Rasche et al., 1995), are believed to be lethal.

The higher the risk of a herbicide mode of action of selecting for resistance, as defined by the number of applications before resistance becomes detectable, the less often herbicides from that group should be applied in sequences or rotations (see Figure 20.2). This approach assumes that a particular herbicide mode of action is effective for a set number of applications prior to the onset of target-site resistance. The use of herbicide "shots" is ap-

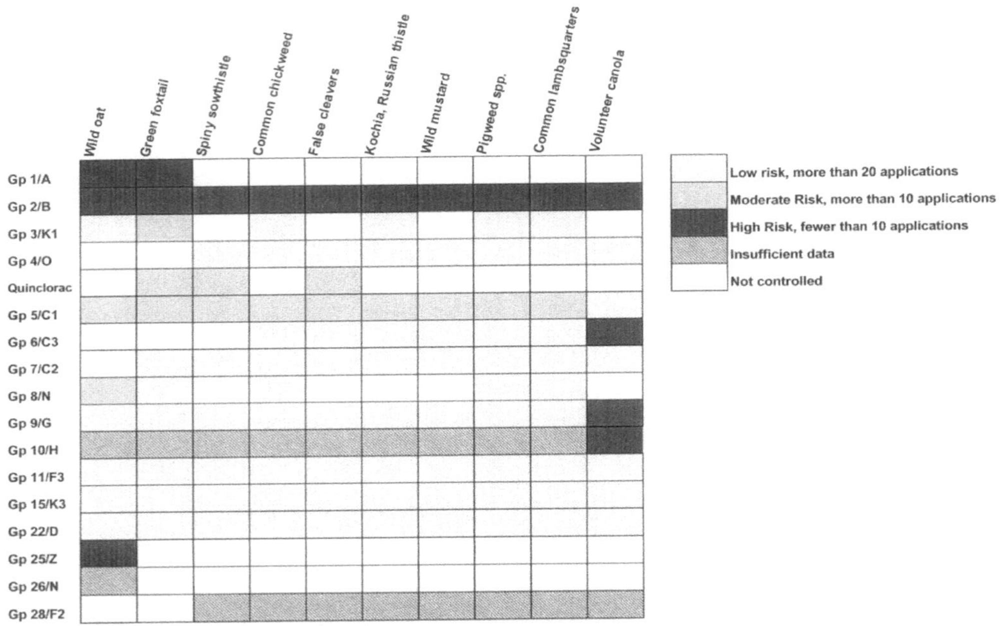

FIGURE 20.2. Classification of herbicide mode of action by risk (high, moderate, and low) of selection for target-site resistance in specific weed species in Canada. Scientific names of weed species (left to right): *Avena fatua* L.; *Setaria viridis* (L.) Beauv.; *Sonchus asper* (L.) Hill; *Stellaria media* (L.) Vill.; *Galium spurium* L.; *Kochia scoparia* (L.) Schrad., *Salsola pestifer* Nels.; *Sinapis arvensis* L.; *Amaranthus* spp.; *Chenopodium album* L.; *Brassica napus* L., *B. rapa* L. *Source:* Adapted from Beckie, Hall, and Tardif, 2001; numerical and alphabetical herbicide groups are described in Retzinger and Mallory-Smith, 1997, and Schmidt, 1997, respectively.

propriate in economic models and farm management decision aids (Diggle and Neve, 2001). It is widely agreed that Group 1/A and 2/B herbicides pose a high risk for selecting HR biotypes relative to herbicides from other groups (LeBaron and McFarland, 1990; Dellow et al., 1997; Gressel, 1997; Heap, 1999). In sequences, lower-risk, nonselective herbicides, such as Group 22/D (paraquat/diquat) or Group 9/G (glyphosate) should be used preseeding to reduce the number of weeds selected with in-crop herbicides that pose a higher risk.

Anecdotal information, namely field histories of herbicide use, usually form the basis for assessing risk of selecting for resistance based on a herbicide's mode of action. Only one long-term experiment has examined the effect of frequency of herbicide use on the evolution of resistance. In an experiment conducted from 1979 to 1998, resistance in wild oat to triallate occurred after 18 years where the herbicide was applied annually in continuous spring wheat (*Triticum aestivum* L.), but not where it was applied 10 times in a wheat-fallow rotation over the same period (see Table 20.1).

Trends discerned in the cross-resistance patterns of weeds resistant to herbicides of the same mode of action may be used as a guide for strategic herbicide use. For both wild oat and green foxtail, incidence of aryloxyphenoxypropionate (APP) resistance in HR biotypes is markedly higher than cyclohexanedione (CHD) resistance (Mansooji et al., 1992; Seefeldt et al., 1994; Beckie, Thomas, and Légère, 1999; Beckie et al., 1999; Cocker et al., 2000; Beckie, Thomas, and Stevenson, 2002). Thus as a short-term strategy to manage ACCase target-site resistance in populations of these species, CHDs have a higher probability of success. Documentation of patches with different cross-resistance patterns within a field (Bourgeois, Kenkel, and Morrison, 1997; Andrews, Morrison, and Penner, 1998), however, illustrates the short-term success of this approach. In many other weed

TABLE 20.1. Level of seeding triallate resistance in percentage in wild oat (*Avena fatua* L.) from two long-term crop rotations (1979-1998) at Saskatoon, Saskatchewan, Canada.

	Year that wild oat seeds were collected		
Plot[a]	**1996**	**1997**	**1998**
W, no herbicide	0	0	0
W-F, no herbicide	0	–	0
W, herbicide	0	3	14
W-F, herbicide	0	–	0

Source: Adapted from Beckie and Jana, 2000.
[a]W: continuous spring wheat (*Triticum aestivum* L.); W-F: spring wheat-fallow rotation (F phase present in 1997).

species, clethodim has controlled ACCase inhibitor-HR biotypes (Bradley and Hagood, 2001).

There is limited direct epidemiological evidence for the utility of herbicide group rotations in delaying the evolution of target-site resistance. Resistance is often attributed to lack of herbicide group rotation, i.e., frequent or repeated use of herbicides of the same mode of action. Examples where herbicide group rotation has been credited with preventing or delaying resistance in weeds include triazine-HR weeds in North America (Stephenson et al., 1990), isoproturon-HR littleseed canarygrass (*Phalaris minor* Retz.) in India (Singh, Kirkwood, and Marshall, 1999), ACCase inhibitor-HR wild oat in North America (Légère et al., 2000), ALS inhibitor resistance in wild radish (*Raphanus raphanistrum* L.) in Australia (Hashem, Bowran, et al., 2001), paddy weed (*Lindernia micrantha* D. Don) in Japan (Itoh, Wang, and Ohba, 1999), and weeds in field crops in Europe (Hartmann et al., 2000; Heap, 2005).

Multiple-group herbicide resistance can be conferred by non-target-site mechanisms, such as enhanced metabolism. Metabolic resistance has been reported more frequently in monocot weeds and crops than in dicot plants (Werck-Reichhart, Hehn, and Didierjean, 2000). Metabolism-based resistance to herbicides of different modes of action will clearly limit the effectiveness of herbicide group rotation as a tool to delay resistance. Testing suspected populations to determine resistance patterns is even more important where multiple-group resistance is suspected and will identify remaining herbicide options for farmers (Beckie et al., 2000). Herbicides that are not readily metabolized in the plant are unlikely to select for metabolism-based resistance. For example, the low incidence of dinitroaniline resistance may be due to the paucity of detoxification mechanisms in target plants (Holt, Powles, and Holtum, 1993). Sulfometuron is slowly metabolized in plants and has been used to discriminate between target-site and metabolic resistance in annual ryegrass (Boutsalis and Powles, 1995). Herbicides that are detoxified via pathways different than that mediated by cytochrome P450 oxygenases, that are metabolized slowly, or that are not metabolized (e.g., glyphosate, glufosinate, paraquat) will lessen the chance of selecting for multiple-group (metabolism-based) HR weed populations. Thus, nonselective herbicides used in HR crops will be an important tool for managing metabolic resistance in grass weed biotypes in the future.

Herbicide-Resistant Crops

Expansion of crops with novel traits conferring resistance to existing herbicides (e.g., glyphosate) will increasingly influence future herbicide-use patterns (see Table 20.2). Herbicide resistance was the dominant trait (73 percent) in transgenic crops grown in 2003. The HR crop constituting the largest area worldwide that year was soybean [*Glycine max* (L.) Merr.] (41.4 million ha or 61 percent of the area planted to transgenic crops, including insect resistance). Other important HR field crops include canola

TABLE 20.2. Transgenic crops[a] grown in 1999 and 2000, listed by country, trait, and crop.

	2000		2003	
	ha x 10^6	**%**	**ha x 10^6**	**%**
Country				
United States	30.3	68	48.2	63
Argentina	10.0	23	13.9	21
Canada	3.0	7	4.4	6
Brazil	–	–	3.0	4
China	0.5	1	2.8	4
South Africa	0.2	<1	0.4	1
Other	0.4	1	0.4	<1
Total countries	13.0		18.0	
Total area	44.2		67.7	
By trait				
Herbicide resistance (HR)	32.7	74	49.7	73
Bt[b]	8.3	19	12.2	18
HR + Bt	3.2	7	5.8	9
By trait and crop				
HR soybean *(Glycine max)*	25.8	59	41.4	61
Bt maize *(Zea mays)*	6.8	15	9.3	13
HR canola *(Brassica napus)*	2.8	6	3.6	5
HR maize	2.1	5	3.2	5
HR cotton *(Gossypium hirsutum)*	2.1	5	1.5	2
Bt + HT cotton	1.7	4	2.6	4
Bt cotton	1.5	3	3.1	5
Bt + HR maize	1.4	3	3.2	5

Source: Adapted from James, 2000, 2003.
[a]Excludes HR crops developed using nontransgenic techniques (e.g., imidazolinone-HR).
[b]Bt: *Bacillus thuringiensis.*

(B. napus) (5 percent by area), maize (*Zea mays* L.) (5 percent), and cotton (*Gossypium hirsutum* L.) (2 percent). Crops with stacked resistance traits (herbicide plus insect resistance) are also gaining importance, and their area increased by 2.7 million ha between 2000 and 2003.

Herbicide-resistant crops can slow the selection of HR weeds by increasing crop and herbicide rotation options. Herbicide-resistant crops could increase profitability and the use of environmentally benign herbicides (Burnside, 1992; Devine and Buth, 2001). Potentially frequent use of herbicides in HR crops, however, may select for new HR weed biotypes. Evolved weed resistance through selection pressure in HR crops is a greater risk than increased invasiveness of HR crops or introgression of HR traits into genomes of related weed species (Warwick, Beckie, and Small, 1999; Beckie, Hall, and Tardif, 2001). Potential impact of HR crops is largely dependent on the herbicide group and cropping area. Weeds resistant to Group 6/C3 (benzonitriles), Group 9/G (glyphosate), and Group 10/H (glufosinate) herbicides are very rare (Heap, 2001). Increasing the use of these products will slow the selection of weed resistance to herbicides of other modes of action, including Group 1/A (ACCase inhibitors), Group 2/B (ALS inhibitors), and Group 5/C1 (triazines). Frequent use of herbicides from Group 6/C3, 9/G, or 10/H, however, would increase the selection for very rare resistance genes. Since the introduction of glyphosate-HR crops, biotypes of four species resistant to glyphosate have been reported (Heap, 2001), suggesting that other weeds may also be selected for resistance. Given the importance of glyphosate in reduced-tillage cropping, repeated glyphosate use should be dissuaded. The use of Group 2/B herbicides in imidazolinone-HR crops will increase the selection for Group 2/B inhibitor-HR broadleaf and grass weeds.

Volunteer crops are common weeds, and their weediness depends on species, management practices, seed shatter prior to harvest, and disbursement of seed at harvest (Beckie, Hall, and Warwick, 2001). Chemical control options are generally more limited if volunteers are HR. Soybean is generally regarded as having no weedy attributes. Soybean seed rarely exhibits dormancy, and the species usually does not occur as a volunteer in the year following crop cultivation (Frick and Thomas, 1992; Anonymous, 1996). If they do occur, they are poor competitors and are easily controlled chemically or mechanically. Some situations, however, may favor occurrence of volunteer soybean. In soybean/cotton rotations in the southeastern United States, volunteer glyphosate-HR soybean can often be found in glyphosate-HR cotton, but densities are typically low and not of economic concern (Culpepper, York, and Stewart, 2002). Alternatively, dry conditions after soybean harvest can prevent autumn germination and subsequent mortality of plants by freezing temperatures during winter and increase volunteer soybean abundance in the following crop. A similar effect of soil moisture was observed for canola emergence patterns (Légère et al., 2001). Volunteer cotton usually does not occur in succeeding crops (J. Wilcut, personal communication); HR volunteers that do persist the following year can be controlled by alternative herbicides. Seeds of maize exhibit little dormancy, but the species can occur as volunteers in the year following cultivation of the crop (Anonymous, 1994). HR maize volunteers (excluding ACCase inhibitor-HR) can be controlled in soybean by several ACCase inhibitors, including diclofop, sethoxydim, and quizalofop. Volunteer spring wheat can persist for multiple years in the seed bank (see Figure 20.3). Control of imidazolinone-HR spring wheat is unlikely to be more difficult than non-HR cultivars. Control of volunteers of glyphosate-HR wheat would be limited both preseeding and in-crop. ACCase inhibitor use might increase for control of these volunteers.

Glyphosate-HR canola *(B. napus)* remains uncontrolled where glyphosate is used alone for preseeding or chemical fallow weed control. Similarly, imidazolinone-HR canola is uncontrolled in crops that receive only ALS inhibitors. All volunteer canola, however, can be controlled by alternative herbicides, including several inexpensive auxinic herbicides, such as 2,4-D and MCPA, or by using herbicide mixtures or nonchemical weed

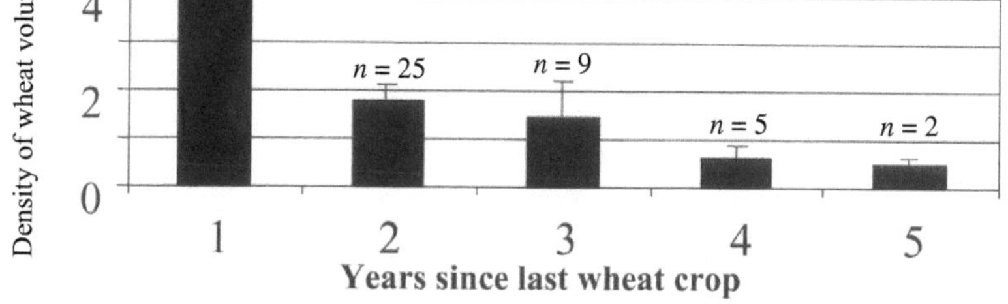

FIGURE 20.3. The persistence of volunteer spring wheat (*Triticum aestivum* L.) in cropping systems in western Canada (*n* = number of fields; bars = standard error). Frequency of fields where present, by year: 1: 31 percent; 2: 15 percent; 3: 10 percent; 4: 10 percent; 5: 9 percent). *Source:* Thomas and Leeson, 1999; adapted from Beckie, Hall, and Warwick, 2001.

control strategies. Although numerous herbicides control volunteers in cereal crops, few products are available for use in some oilseed or annual legume crops. If these crops are grown, careful crop rotation planning is required.

If crops are partial or obligate outcrossers and grown in close proximity to non-HR crops, HR genes may be transferred to non-HR parental plants and hybrid seed formed (Champolivier et al., 1999; Simpson et al., 1999). Farmers can experience unanticipated HR volunteers. Species with outcrossing potential include canola *(B. rapa, B. napus)* and maize. If more than one HR trait has been developed for a species and the species is partially or completely outcrossing, pollen flow could create multiple-HR volunteers. Multiple-HR canola *(B. napus)* volunteers in farmers' fields in Canada have been reported (Hall et al., 2000; Beckie, Hall, and Warwick, 2001).

Herbicide Mixtures

Acceptance of herbicide mixtures by farmers has been aided by cost-incentive programs from industry, labeling requirements, and the rapid development of resistance in specific cases. The admixed herbicides may be applied at lower rates, especially when interacting synergistically (Gressel, 1990).

If mixing partners of different modes of action do not meet the criteria of similar persistence and efficacy, and different propensity for selecting for resistance in target species, the effectiveness of mixtures for delaying target-site resistance will be reduced. For example, a mixture of an ALS inhibitor, chlorimuron, and metribuzin for resistance management of common waterhemp (*Amaranthus rudis* Sauer) is not effective because chlorimuron is more persistent than metribuzin and common waterhemp has uneven and season-long emergence (Sprague, Stoller, and Wax, 1997). Mixtures can inadvertently accelerate the evolution of multiple resistance (two or more resistance mechanisms conferred by multiple genes) if they fail to meet basic criteria for resistance management and are applied repeatedly (Rubin, 1991). To effectively delay metabolic resistance, the mixing partners must be degraded via different biochemical pathways (Wrubel and Gressel, 1994). Unfortunately, information on the mode of degradation of herbicides in plants is not available to most farmers.

Challenges to farmer adoption of mixtures for resistance management include increased cost, particularly for grass weed control, and availability of suitable mixing partners that meet the criteria outlined previously and are equally effective against the weed species targeted by farmers (see Table

TABLE 20.3. Tank mixture of an ALS inhibitor herbicide, thifensulfuron + tribenuron (formulated mixture), with an auxinic herbicide, 2,4-D, in cereal crops for delaying weed resistance.

	Weed species controlled	
Crop	**Thifensulfuron/ tribenuron**	**2,4-D**
Amaranthus retroflexus L. (redroot pigweed)	✔	✔
Ambrosia artemisiifolia L. (common ragweed)		✔
Capsella bursa-pastoris (L.) Medikus (shepherd's-purse)	✔	✔
Chenopodium album L. (common lambsquarters)	✔	✔
Crepis tectorum L. (narrowleaf hawksbeard)	✔	✔
Descurainia sophia (L.) Webb ex Prantl (flixweed)	✔	✔
Fagopyrum tataricum (L.) J. Gaertn. (tartary buckwheat)	✔	
Galeopsis tetrahit L. (common hempnettle)	✔	
Kochia scoparia (L.) Schrad. (kochia)	✔	✔
Lactuca serriola L. (prickly lettuce)		✔
Lappula echinata Gilib. (blue bur/European sticktight)		✔
Neslia paniculata (L.) Desv. (ball mustard)	✔	✔
Polygonum convolvulus L. (wild buckwheat/ black bindweed)	✔	
Polygonum persicaria L. (ladysthumb)	✔	
Polygonum scabrum Moench (green smartweed)	✔	
Salsola iberica Sennen & Pau (Russian thistle)	✔	✔
Saponaria vaccaria L. (cow cockle)	✔	
Senecio vulgaris L. (common groundsel)	✔	
Sinapis arvensis L. (wild mustard)	✔	✔
Spergula arvensis L. (corn spurry)	✔	
Stellaria media (L.) Vill. (common chickweed)	✔	
Thlaspi arvense L. (field pennycress)	✔	✔
Xanthium strumarium L. (common cocklebur)		✔
Brassica napus L./*B. rapa* L. (volunteer canola[a])	✔	✔
Helianthus annuus L. (volunteer sunflower)	✔	

Source: Adapted from Anonymous, 2002.
[a]Except imidazolinone-resistant varieties.

20.3). The ALS inhibitor herbicide thifensulfuron plus tribenuron (formulated mixture) is popular among farmers for controlling broadleaf weeds in cereal crops in Canada; the auxinic herbicide 2,4-D is registered as a tank mix with this ALS inhibitor (Anonymous, 2002). Eleven weed species are controlled by both mixing partners. Of these, ball mustard [*Neslia paniculata* (L.) Desv.], kochia, redroot pigweed (*Amaranthus retroflexus* L.), Russian thistle (*Salsola iberica* Sennen & Pau), field pennycress (*Thlaspi arvense* L.), and wild mustard (*Sinapis arvensis* L.) have ALS inhibitor-HR biotypes in Canada. A mixture of these two herbicides should markedly delay ALS inhibitor resistance in these species, particularly those that are highly self-pollinated. The benefits of this mixture for managing target-site resistance in species that are controlled by both partners are (1) 2,4-D poses a low risk for selecting for resistance; (2) the mixing partners have similar (short) soil residual activity; and (3) 2,4-D is inexpensive. The rate of 2,4-D used in the mixture, however, may result in reduced efficacy on some species, such as redroot pigweed and Russian thistle, compared to that of the ALS inhibitor herbicide. Common chickweed [*Stellaria media* (L.) Vill.] and common hempnettle (*Galeopsis tetrahit* L.), which have numerous ALS inhibitor-HR populations, are controlled only by the sulfonylurea herbicide. A total of 13 species (excluding volunteer sunflower, *Helianthus annuus* L.) are controlled by only one mixing partner. This example shows the inherent limitations of mixtures in delaying target-site resistance. In addition to meeting the criteria for resistance management outlined earlier, adoption by farmers as a strategy to delay resistance will depend largely on their assessment of the economic threat of resistance in specific weeds that are abundant in their fields and are known to have HR biotypes in their region.

Anecdotal evidence supports the usefulness of mixtures in resistance management. Mixtures with ALS inhibitors have successfully delayed resistance in weeds in rice (*Oryza sativa* L.) in Japan and in field crops in Europe (Gressel, 1997; Itoh, Wang, and Ohba, 1999; Heap, 2001). Farmers who included mixtures of herbicides with different modes of action coupled with various cultural practices were less likely to select ALS inhibitor-HR weed populations (Shaner, Feist, and Retzinger, 1997). *Chenopodium* and *Amaranthus* spp., which have often evolved triazine resistance when triazines were used alone, rarely have been reported to evolve resistance where atrazine/chloracetamide mixtures were used for over 20 years in monoculture maize in North America (Wrubel and Gressel, 1994). Atrazine is applied at a lower rate in this mixture, thus reducing selection pressure. Pendimethalin is an effective partner for propanil to delay or manage resistance in jungle rice [*Echinochloa colona* (L.) Link] in rice in Central America (Valverde, 1996; Riches et al., 1997).

Some carbamates and organophosphates competitively inhibit aryl acylamidase, the enzyme responsible for propanil metabolism in propanil-HR junglerice. This inhibition can result in synergistic effects. A formulation of propanil and piperophos, a phosphoric herbicide, was first marketed in 1995 in Costa Rica and cost-effectively controls propanil-HR junglerice while achieving selectivity in rice (Valverde, 1996; Valverde et al., 1999). Similarly, synergists such as anilofos, piperophos, or carbaryl gave excellent control of propanil-HR barnyardgrass [*Echinochloa crus-galli* (L.) Beauv.] in rice in Arkansas (Daou and Talbert, 1999; Norsworthy et al., 1999; Norsworthy, Talbert, and Hoagland, 1999; Talbert et al., 2000).

Herbicide rotations, sequences, or mixtures generally have the greatest effect in delaying resistance when the mechanism conferring resistance is target-site based, target weed species are highly self-pollinated, and seed spread is restricted. Multiple resistance can arise within a weed population through a change in selection history (usually sequential selection), through selection of multiple mechanisms by a single herbicide, or through crossing of individuals containing different resistance mechanisms (Hall, Holtum, and Powles, 1994; Preston and Mallory-Smith, 2001). Based on a compounded-resistance frequency model, the probability of HR mutants with multiple mechanisms of resistance (target-site based) in an unselected population is the product of the probabilities of resistance to each affected herbicide group and thus is rare (Wrubel and Gressel, 1994). Frequent use of herbicides in a field over time can enrich HR populations with different resistance mechanisms; outcrossing between populations in close proximity that possess different HR mechanisms can result in plants with multiple resistance. Spread of HR seed within and among fields can also aid this process.

Exploiting Reduced Fitness or Negative Cross-Resistance in Herbicide-Resistant Biotypes

Reduced fitness of triazine-HR plants compared with HS plants, documented frequently in the 1970s, resulted in optimistic predictions that this "cost of resistance" would also occur in biotypes resistant to herbicides of other modes of action. The target-site mutation conferring triazine resistance reduces photosynthetic efficiency, which is often manifested by decreased plant productivity and competitiveness (i.e., reduced fitness). Upon discontinuation of triazine herbicides, reduced fitness of HR compared with HS biotypes was predicted to reverse the evolution of resistance at a rate dependent on the fitness differential between biotypes. Unfortunately, resistance in weeds to other modes of action is generally not accompanied by reduced fitness (Holt and Thill, 1994). Lack of reduced fitness in HR biotypes has been inferred from little decline observed in the proportion of HR:HS individuals measured in fields over time after use of the selecting herbicide was discontinued (Andrews and Morrison, 1997). For noncompetitive inhibitors of target-site enzymes, such as ACCase or ALS inhibitors, the various sites of mutations for resistance are not near the active site of the enzyme and thus no fitness loss occurs due to lower affinity for the normal substrates (Wrubel and Gressel, 1994; Gressel, 1999).

Negative cross-resistance, i.e., HR plants are more sensitive to a herbicide than HS plants, has been documented in several triazine-HR weed species (see Table 20.4). Some herbicides that inhibit photosystem II bind more efficiently to the mutant triazine binding domain than to the wild (HS) type. Triazine-HR weeds frequently show negative cross-resistance to other photosystem II inhibitors, such as bentazon and pyridate. Triazine-HR weeds can also exhibit negative cross-resistance to herbicides that do not affect photosystem II (Table 20.4). Explanations for this phenomenon depend

TABLE 20.4. Examples of negative cross-resistance in triazine-resistant weed species.

Weed species	Herbicide	Group	HR/HS[a]	Source[b]
Amaranthus hybridus L.	Bentazon	6/C3	0.40	Jordon et al. (1999)
Chenopodium album L.	Bentazon	6/C3	0.21	Parks et al. (1996)
	Bromoxynil	6/C3	0.64	
	Dicamba	4/O	0.62	
	Pyridate	6/C3	0.61	
Echinochloa crus-galli (L.) Beauv.	Fluazifop	1/A	0.03	Gadamski et al. (2000)
	Sethoxydim	1/A	0.47	
	Metolachlor	15/K3	0.49	
	Chlorsulfuron	2/B	0.67	
	Glufosinate	10/H	0.63	
Conyza canadensis (L.) Cronq.	Bentazon	6/C3	0.53	
	Pyridate	6/C3	0.14	
	MCPA	4/O	0.12	
	Fluroxypyr	4/O	0.52	
	Glyphosate	9/G	0.38	
	Diquat	22/D	0.54	

[a]Based on the ratio of doses required to reduce biomass or mortality of HR and HS biotypes by 50 percent relative to the untreated controls.
[b]Information adapted from these sources.

on the specific herbicide but are largely speculative. The potential combined value of negative cross-resistance and general lack of fitness of triazine-HR biotypes in managing triazine resistance in weeds has yet to be realized (Gadamski et al., 2000). Negative cross-resistance has also been observed in nontriazine-HR biotypes. For example, an imidazolinone-HR smooth pigweed (*Amaranthus hybridus* L.) biotype was tenfold more sensitive to cloransulam-methyl, an ALS inhibitor, compared with an HS biotype (Poston et al., 2001).

Reduced Herbicide Rates Lowering Efficacy

Model simulations have concluded that it is not profitable to reduce herbicide rates to reduce selection pressure (efficacy or persistence) for resistance, unless accompanied by a compensating increase in other types of weed control (Diggle and Neve, 2001). The resulting increase in the density of HS weeds would reduce crop yield and quality and increase weed seed return to the seed bank (Gorddard, Pannell, and Hertzler, 1996; Morrison and Friesen, 1996) (see Figure 20.4). These model simulations have used input parameters based on field data, but the outputs have not been compared (validated) with field data.

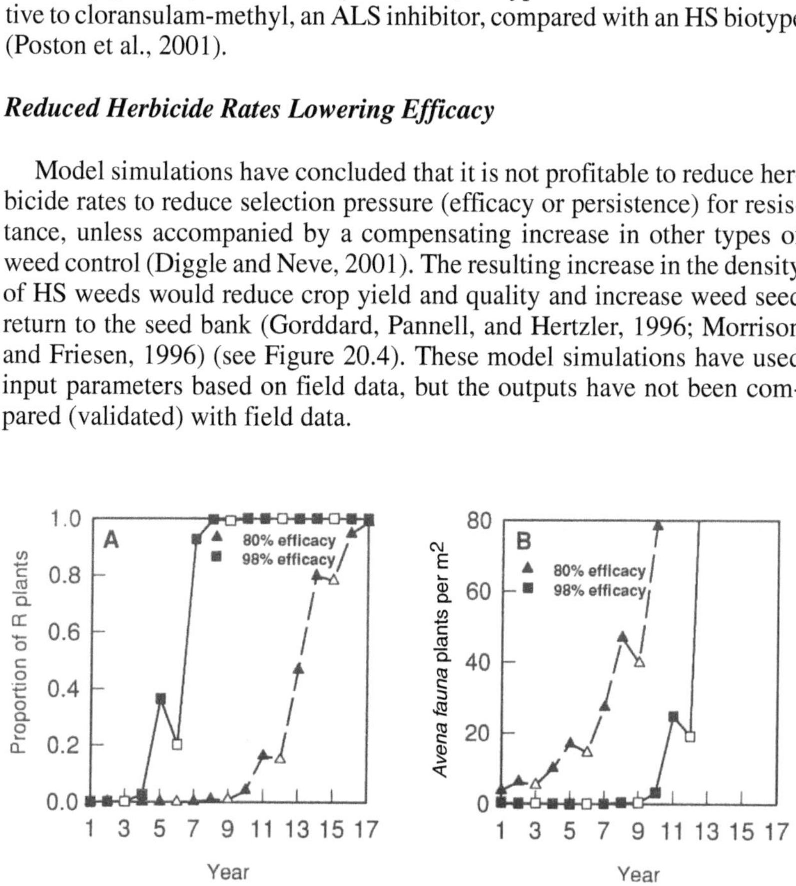

FIGURE 20.4. Model simulation of the effects of reducing herbicide efficacy in delaying the evolution of resistance in a wild oat (*Avena fatua* L.) population (A) and on wild oat densities (B). The herbicide is applied twice in a three-year period (closed symbols); the open symbols represent an "off" year. In each off year it was assumed that alternative control measures provided a 90 percent reduction in both HS and HR seed return. *Source:* Reproduced with permission from Morrison and Friesen, 1996.

Reduced herbicide rates may select for polygenic or quantitative (metabolic) resistance, resulting in small, incremental accumulation of HR alleles in a population (Cavan, Cussans, and Moss, 1999). Quantitative resistance, involving multiple nuclear genes with additive action, has been documented or postulated in HR weed populations, such as chlortoluron-HR blackgrass in the United Kingdom (Chauvel and Gasquez, 1994; Hall, Holtum, and Powles, 1994; Willis et al., 1997; Cavan, Cussans, and Moss, 1999), dicamba-HR kochia in North America (Dyer, Jasieniuk, and Maxwell, 2000; Westra, Nissen, and Karam, 2000; Cranston et al., 2001), and isoproturon-HR littleseed canarygrass in India (Malik and Singh, 1995; Singh, Kirkwood, and Marshall, 1998, 1999; Kulshrestha, Singh, and Yaduraju, 1999). Lower than recommended rates have been implicated or speculated as the causal factor in resistance in these biotypes. It has been suggested that soil-applied herbicides are more likely to select for quantitative resistance because late-emerging weeds are exposed to lower doses that allow accumulation of HR alleles (Maxwell and Mortimer, 1994; Gressel, 1997). Based on unvalidated model simulations, Gressel (1995) and Gardner, Gressel, and Mangel (1998) advocated a strategy of revolving doses to delay the evolution of major monogene (target-site) resistance and quantitative resistance.

The consensus among weed scientists involved in weed resistance is that reduced-rate application that lowers efficacy is not a good weed or weed-resistance management strategy unless accompanied by enhanced weed suppression by nonherbicide methods. In a recent study in western Canada, seeding barley (*Hordeum vulgare* L.) at relatively high rates resulted in optimum grain yields, undiminished economic returns, and effective wild oat management when tralkoxydim was used at lower than recommended rates (O'Donovan et al., 2001). Are there opportunities, however, to reduce rates without significantly lowering herbicide efficacy? In the past, label rates were frequently based on the amount needed to control the least sensitive weed, while other weeds on the product label may be sensitive at much lower rates. Thus, the selection pressure on these very sensitive species can be extremely high. Gradually, more sophisticated information is being, and should be, provided to farmers to adjust label rates according to prevailing environment conditions and herbicide sensitivity of the species present in their fields. For example, reduced but effective ALS inhibitor herbicide rates used to control common chickweed in Europe compared with those used in western Canada doubled the time for resistance evolution by reducing the time those herbicides remain active in the soil (Kudsk, Mathiassen, and Cotterman, 1995; Beckie, Hall, and Tardif, 2001). However, potential benefits from such a strategy would depend on seedling emergence pattern of species present, which determines population exposure to the herbicide.

Site-Specific Herbicide Application

Site-specific management using a global positioning system (GPS) may be a potentially useful tool for monitoring and managing HR weed patches at early stages of development in a field over time. Preliminary research indicates that prevention of seed production in HR wild oat patches at early stages of development by alternative herbicides or nonchemical methods is effective in restricting their spread (H. J. Beckie, unpublished data). Unfortunately, most farmers in North America fail to detect small HR patches (Beckie et al., 1999); field scouting after in-crop herbicide application to locate uncontrolled weeds is not generally performed because of either lack of awareness of the benefit of this practice or inconvenience due to large acreages of farms.

Site-specific herbicide application, utilizing economic thresholds to dictate application areas of a field, would allow some reduction in the overall selection pressure. Its effect on the rate of evolution of resistance would depend on the frequency of herbicide application to specific areas of a field over time and the proportion of the field treated each year. If application frequency of herbicides to specific areas of a field (e.g., lower slope areas) is similar to conventional herbicide application, gene (seed, pollen) flow from these field areas to those treated less frequently may negate any potential benefits of the technology. Furthermore, if these patches contain the majority of the weed population present in the field, then treating the weed patches will still exert a selection pressure similar to that of a blanket application. Leaving refugia for HS individuals to dilute the proportion of HR plants in a population will not be effective because the recessive control of resistance in outcrossing weed species is rare (Jasieniuk, Brulé-Babel, and Morrison, 1996). The only documented case of recessive inheritance of resistance in an outcrossing species was that of a picloram-resistant yellow starthistle (*Centaurea solstitialis* L.) biotype found in Washington State (Sterling et al., 2002).

NONHERBICIDE STRATEGIES

Agronomic practices affect weed densities and seed production but will reduce rates of resistance evolution only where they increase the effective kill of HR plants, decrease the effective kill of HS plants (indirectly by permitting reduced herbicide use), or eliminate selection pressure (control HS and HR plants equally) where herbicides are not used. Nonherbicide practices may increase the effective kill of HR plants in situations where differences exist in the population dynamics of HR and HS biotypes. Seeds of triallate-HR wild oat in North America are often less dormant than those of HS populations (O'Donovan et al., 1999). Greater and more rapid emergence of HR individuals compared with HS individuals, analogous to that of ALS inhibitor-HR kochia biotypes (Dyer, Chee, and Fay, 1993), may be potentially exploited for selective HR biotype control by tillage or nonselective herbicides prior to seeding. Agronomic practices that allow reduced herbicide use in cropping systems, while maintaining economically acceptable weed population densities, will delay resistance by reducing the number of HR alleles in a population (starting point for resistance) and the rate of enrichment of resistance. Where high levels of HR alleles are believed to be present in unselected populations, such as ALS inhibitor resistance in common waterhemp and Palmer amaranth [*Amaranthus palmeri* (S.) Wats.] in North America (Peterson, 1999) or *Lolium* spp. in Europe and Australia (Matthews and Powles, 1992; Maxwell and Mortimer, 1994; Dinelli, Bonetti, and Catizone, 2000), it is important to maintain low population densities via nonchemical methods or by using herbicides with low propensity to select these HR alleles.

The issues of economic risk, labor availability, and time management will infuence the adoption of cultural/mechanical practices for resistance weed management. Evolution of resistance in weed populations has not led to decreased herbicide use or a marked increase in nonchemical control methods, except in cases of multiple-group resistance in grass weeds (Powles et al., 1997; Preston and Mallory-Smith, 2001). We briefly summarize some nonherbicide practices that have proven effective in managing HR weeds. Used singly, none of these nonherbicide practices consistently provides an adequate level of weed control. When used in combination, however, effective weed management can be achieved (Matthews, 1994; Gill and Holmes, 1997).

Crop Rotation

Crop rotations are dictated primarily by profit potential and not management of HR weeds. Crop rotation, however, is frequently cited as one of the most influential factors in delaying or managing HR weeds (Stephenson et al., 1990; Carey, Hoagland, and Talbert, 1995; Bourgeois, Morrison, and Kelner, 1997; Gill and Holmes, 1997; Powles et al., 1997; Ritter and Menbere, 1997; Shaner, Howard, and Chalmers, 1999; Singh, Kirkwood, and Marshall, 1999; Hartmann et al., 2000). Diversity in crop sequences may reduce weed populations directly, or indirectly by facilitating herbicide rotation. Inclusion of competitive crops and varieties in crop rotations is viewed by farmers as important in resistance management (Bourgeois,

Morrison, and Kelner, 1997; Shaner, Howard, and Chalmers, 1999). The potential value of crop sequencing to delay or manage HR weeds will not be realized, however, unless accompanied by diversification in herbicide use. Repeated use of herbicides with the same mode of action will negate weed-suppression benefits associated with crop sequencing. Crop rotations had little influence on occurrence of Group 1/A inhibitor-HR wild oat in western Canada because farmers frequently applied these herbicides to cereal, oilseed, and annual legume crops that dominate cropping systems (Légère et al., 2000). Occurrence of resistance in wild oat was the lowest in rotations where frequency of fallow was the highest due to reduced frequency of herbicide use. Similarly, despite diversity in crop rotations used in a field in Western Australia, repeated triazine use in the different crops selected for triazine resistance in wild radish (Hashem, Dhammu, et al., 2001).

Seeding Date and Rate

Delayed seeding is often promoted for the control of some HR grass species in Europe, such as winter wild oat [*Avena sterilis* ssp. *ludoviciana* (Durieu) Nyman], hood canarygrass (*Phalaris paradoxa* L.), and blackgrass (Orson, 1999; Sattin, Gasparetto, and Campagna, 2001). In North America, delayed seeding is recommended to manage triallate-HR wild oat populations because they tend to emerge earlier in the growing season than the wild type. Delayed seeding of rice in Central America is commonly used to reduce HR junglerice population densities (Valverde, Riches, and Caseley, 2000). In Australia, delayed seeding has been integrated with other control tactics to manage HR annual ryegrass (Powles and Matthews, 1996; Gill and Holmes, 1997). With the widespread appearance of HR annual ryegrass, many Australian farmers are routinely increasing crop seeding rates by 20 to 40 percent, resulting in higher densities of crop plants to aid crop competitiveness (Gill and Holmes, 1997; Powles, 1997).

Tillage

Owen (2001b) reviewed the role of tillage and mechanical tactics in managing HR weed populations. The judicious use of tillage has been cited by farmers (Bourgeois, Morrison, and Kelner, 1997) and researchers (Orson and Livingston, 1987; Stephenson et al., 1990; Peterson, 1999) as an important practice to delay or manage resistance. Tillage may substitute for herbicide use or influence seed bank dynamics. For example, plowing to bury weed seeds to reduce germination is recommended for management of HR blackgrass in Europe (Orson and Livingston, 1987; Moss, 1997). Timely tillage can also stimulate weed germination prior to crop planting (Gill and Holmes, 1997). Tillage in some situations, however, can increase the persistence of seeds in soil. In Europe, tillage soon after seed shedding can instill a light requirement in canola seed, inducing dormancy and thus increased persistence of volunteers (Pekrun, Hewitt, and Lutman, 1998). In zero-tillage systems in North America, large quantities of crop residues may create conditions that similarly result in induction of secondary dormancy in canola seed, resulting in greater abundance of volunteers (Légère et al., 2001)

Reducing Gene Dispersal

Herbicide-resistant weed seed spread within and among fields has been documented (Stephenson et al., 1990; Ritter and Menbere, 1997; Andrews, Morrison, and Penner, 1998; Li et al., 2000). Fields within farms are more likely to have HR weeds than randomly picked fields, indicating movement of HR seed between fields via equipment (Anderson, Roeth, and Martin, 1996) or similar selection pressure among fields within a farm. Sharing of equipment among farmers has also been implicated in resistance (Debreuil, Friesen, and Morrison, 1996). Spread by farm operations can be significant via machinery, noncomposted manure, silage, or contaminated commercial grain stocks or feed (Stephenson et al., 1990; Ritter and Menbere, 1997), and is generally greater than natural gene dispersal. For example, wild oat can be spread more than 150 m by a combine harvester (Shirtliffe, Entz, and Maxwell, 1998). Spread of resistance among wild oat (*Avena* spp.) patches within 350 m of one another was documented in the United Kingdom (Cavan, Biss, and Moss, 1998). Wind dispersal of light-seeded species can also spread resistance rapidly, such as occurs in prickly lettuce (*Lactuca serriola* L.) (Rieger, Stone, and Preston, 2001) and horseweed [*Conzya canadensis* (L.) Cronq.] (VanGessel, 2001). Wind can efficiently transport kochia and Russian thistle tumbleweeds for long distances (Mallory-Smith, Thill, and Stallings, 1993). As the frequency of occurrence of resistance increases in a region, there are more opportunities for pollen and seed movement to influence occurrence, in addition to selection. Seed movement is probably responsible for the majority of gene flow in weed populations (Diggle and Neve, 2001). It is difficult to control the spread of resistance via pollen flow, especially when resistance is endowed by a single dominant or semidominant gene (Smeda, Currie, and Rippee, 2000). Transmission of the resistance trait via pollen may depend on the type of resistance mechanism. For example, major target-site resistance (nuclear) genes (e.g., ALS, ACCase inhibitors) were expressed in pollen of annual ryegrass and blackgrass, but non-target-site resistance mechanisms were not (Richter and Powles, 1993; Letouzé and Gasquez, 1999).

Management practices that limit the spread of HR seed can slow the occurrence of resistance. In western Canada, farmers who reported practicing weed sanitation (e.g., cleaning harvesting and tillage equipment when moving between fields, covering the box of grain trucks, mowing or spraying ditches or uncontrolled weed patches, etc.) were less likely to have HR wild oat than those who were less careful (Légère et al., 2000). Cleaning equipment when moving among fields, and mowing weed patches, ditches, and headlands ranked fourth and fifth, respectively, in importance among resistance management practices cited by farmers in western Canada (Bourgeois, Morrison, and Kelner, 1997). If the HR population covers a wide area across the field, management should focus on reducing seed return and spread by using low-risk herbicides in conjunction with cultural practices, such as cutting the crop (hay, silage, or green manure) before or soon after flowering of the HR weed species, growing competitive annual crops or perennial crops, or collecting weed seeds at harvest. Although weed seed catching/removal at harvest is effective in managing HR weeds (Matthews, 1994; Gill, 1997; Gill and Holmes, 1997), farmer adoption is low (Thill, O'Donovan, and Mallory-Smith, 1994; Powles, 1997; Shaner, Howard, and Chalmers, 1999). In Australia, there has been wide adoption of seed-set control techniques to manage resistance in annual ryegrass in pastures and annual legume crops. This strategy is described in the following section.

INTEGRATING HERBICIDE AND NONHERBICIDE STRATEGIES: CASE STUDIES

In this section, case studies describe the successful integration of herbicide and nonherbicide strategies for sustainable management of multiple-group HR grass weed species. These case studies illustrate their management at the field/paddock level (annual ryegrass and wild oat) or at a regional level (littleseed canarygrass). Populations of such HR biotypes threaten future cereal grain production in different areas of the world unless agronomic practices and cropping systems that favored their selection are altered to achieve reduced dependence on herbicide use while maintaining the required level of weed control.

Annual Ryegrass [Lolium rigidum *(L.) Gaudin*] *in Australia*

After the first report of herbicide resistance in annual ryegrass (Heap and Knight, 1982), resistance in this weed species has continued to increase rapidly. The recent confirmation of glyphosate resistance in annual ryegrass has further aggravated anxiety among farmers in Southern Australia, principally due to the vital role this herbicide plays in local farming systems (Powles et al., 1998). The following is a brief description of the techniques that farmers in Southern Australia have been using to manage HR populations of this troublesome weed. These techniques are followed by two case studies.

Techniques to Manage Resistance

Pastures and resistant weed management. Traditionally, Australian agriculture was based on the rotation of crop-pasture systems. From a HR weed management viewpoint, pastures provide an excellent opportunity for weed kill through grazing and without the use of selective herbicides. Research undertaken in pastures has repeatedly shown a high level of effectiveness of spring grazing in reducing annual ryegrass infestation in the subsequent wheat crop. For example, Pearce and Holmes (1976) showed large reductions (88 to 96 percent) in subsequent annual ryegrass density where pastures were heavily grazed in the spring during the flowering and seed-production stages of the weed. The combination of grazing animals with nonselective herbicides can be used to obtain large and rapid declines in weed populations (Gill and Holmes, 1997). However, with the decline in wool prices over the past 10 to 15 years, economic imperatives have weighed against the widespread adoption of this strategy for HR weed management.

Alternative herbicides for in-crop management of resistant weed populations. Even with widespread resistance to Group 1/A and 2/B herbicides, use of alternative herbicides is central to designing effective resistance management programs. In cereal crops in Southern Australia, use of trifluralin has continued to increase and has enabled farmers to crop profitably. Estimates indicate that trifluralin use in Australia has increased by 80 percent over the past ten years (Nufarm Australia, personal communication, 2002). It is intriguing as to why resistance to this herbicide in annual ryegrass is still relatively uncommon. Perhaps the answer may lie in its relatively low kill rate (70 to 80 percent); in conjunction with good agronomy, however, this level of weed control has proved adequate for profitable crop production. Some of the attractiveness of HR crops is also related to their usefulness for the management of HR weeds. Triazine-HR canola is a good example. Despite its yield penalty that is associated with triazine resistance, the crop is grown over large areas of Southern Australia because it enables the use of atrazine and simazine for the management of hard-to-control weeds. It is anticipated that weed control considerations will result in rapid adoption of glyphosate-HR crops, whenever they become available in Australia.

Seed-set prevention. Preventing seed production in weeds is central to all HR weed management programs. Part of the attraction of pastures for resistance management is based on the option of being able to use nonselective herbicides for seed-set control. Reluctance on the part of farmers to switch to pastures due to economic concerns has resulted in the development of this technique for use in annual legume crops (pulses) and is commonly referred to as "crop topping." Seed-set prevention in cropping systems can also be achieved by growing crops (e.g., oat, *Avena sativa* L.) that can be cut for hay before weeds have produced viable seeds. Both hay cutting and crop topping have been shown to provide large reductions in weed seed production, particularly that of annual ryegrass (Gill and Holmes, 1997). Adoption of such practices has enabled farmers to practice profitable continuous cropping without sacrificing weed control.

Nonchemical methods. Several control tactics can be used to deplete the size of the seed bank of annual ryegrass prior to sowing the crop. These include stubble burning, autumn cultivation to stimulate weed germination, and delayed sowing to maximize presowing weed kill (Gill and Holmes, 1997). Effectiveness of these techniques tends to be sensitive to prevailing environmental conditions, and they can also have a negative impact on soil health and sustainability. Faced with rampant HR weed populations, some farmers have felt compelled to revive some of these old techniques.

Capture of weed seeds during the harvest operation is a technique that may have never been used by Australian farmers if resistance had not threatened productivity and farm profitability. Seeds of some weed species are held firmly by inflorescence structures well after maturity and until harvest. Such weed species, which includes annual ryegrass, offer the opportunity to collect weed seed output during harvest and either remove it from the field or destroy it through milling or some other appropriate process. In South Australia, Matthews and Powles (1996) reported a 52 percent reduction in annual ryegrass seed bank due to capture of weed seeds at harvest. In Western Australia, Gill (1996) recorded a 60 to 80 percent removal of annual ryegrass seed production, which reduced weed infestation in the subsequent crop by 73 percent.

Both increased crop density and selection of competitive crops and genotypes can play an important role in reducing the impact of annual ryegrass on the crop as well as suppressing fecundity of the weed. Our research in South Australia has clearly indicated that wheat is currently one of the weakest links in the management of weed populations. Research by Medd and colleagues (1987) and later unpublished research by others has shown that increasing crop density can significantly enhance weed-competitive effects of wheat. The combination of increased crop density with herbicide use is particularly useful when using products with relatively low efficacy against annual ryegrass, such as trifluralin. Even within bread wheats, cultivars differ in their ability to compete with weeds (Lemerle et al., 1996). A combination of weed-competitive genotypes with high plant density can be a powerful tool in IWM programs. Coleman, Gill, and Rebetzke (2001) have recently shown strong genetic correlations between wheat plant traits and suppression of annual ryegrass growth. Quantitative trait loci (QTL) for traits in wheat associated with weed competitiveness have been identified. As confidence in these QTL improves with further research in different genetic backgrounds, wheat breeders will be able to use these markers to select for weed-competitive crop genotypes (Coleman, Gill, and Rebetzke, 2001).

Populations of weeds with high fecundity potential such as annual ryegrass can increase rapidly after control failures caused by resistance. Therefore, resistance management programs need to focus not only on seedling kill to avoid yield reduction but also on reducing reproductive output from surviving plants in order to manage populations in the long term. Annual ryegrass has a relatively short-lived seed bank, which makes it possible to achieve a large decline in weed populations within one to two growing seasons.

Case Study #1

This case study highlights several important issues related to the management of HR populations of this weed (see Table 20.5). Selection of a poorly competitive crop, in this case durum wheat (*Triticum aestivum* L.), and use of diclofop to which this population had already evolved resistance, resulted in a 40-fold increase in the annual ryegrass seed bank. Such large infestations can be extremely difficult to manage within a grain crop rotation. Consequently, the farmer decided to implement a three-year pasture phase, which has been shown by earlier research to be a reliable and low-risk option (Pearce and Holmes, 1976; Gill and Holmes, 1997). A combination of heavy grazing, mechanical topping, and spray topping in the final

TABLE 20.5. Case studies: Changes in the seed bank of two herbicide-resistant populations of annual ryegrass [*Lolium rigidum* (L.) Gaudin] in South Australia: Role of pasture phase (a) and management in continuous cropping (b).

	Case study #1		Case study #2	
Year	Management	Postcrop density (m^{-2})	Management	Postcrop density (m^{-2})
1993	Faba bean (*Vicia faba* L.)	430	Chickpea (*Cicer arietinum* L.)	170
1994	Durum wheat (*Triticum aestivum* L.) 1. 2 cultivations 2. Trifluralin 3. Diclofop	16,900 (39)[a]	Triticale (*X. Triticosecale* Wittmack) 1. 1 cultivation 2. Glyphosate 3. Trifluralin 4. Diclofop	2,670 (16)
1995	Pasture 1. Mechanical topping	10,600 (0.63)	Pea (*Pisum sativum* L.) 1. 2 cultivations 2. Glyphosate x 3 3. Trifluralin 4. Simazine 5. Sethoxydim 6. Crop topping (paraquat)	4,000 (1.50)
1996	Pasture 1. Mechanical topping	8,000 (0.76)	Wheat 1. 1 cultivation 2. Glyphosate x 2 3. Triasulfuron 4. Tralkoxydim	1,000 (0.25)
1997	Pasture 1. Mechanical topping 2. Spray topping with glyphosate	2,290 (0.28)	Oat (*Avena sativa* L.) hay 1. 1 cultivation 2. Glyphosate 3. Diuron + metolachlor 4. Cut for hay 5. Spray topping regrowth	420 (0.42)
1998	Canola (*Brassica napus* L.) 1. Trifluralin 2. Clethodim + haloxyfop	570 (0.25)	Pea 1. cultivation 2. Diquat + paraquat 3. Glyphosate 4. Trifluralin	170 (0.40)

TABLE 20.5 *(continued)*

	Case study #1		Case study #2	
Year	Management	Postcrop density (m^{-2})	Management	Postcrop density (m^{-2})
			5. Diuron + simazine	
			6. Fluazifop + butroxydim	
			7. Crop topping (paraquat)	
1999	Wheat	5	Durum wheat	330
	1. Windrows burnt	(0.01)	1. 1 cultivation	(1.99)
	2. Glyphosate		2. Glyphosate	
			3. Trifluralin	

Source: Unpublished data provided by T. Craddock.
[a]Figures in parentheses are lambda values, where lambda = year-two seed bank divided by year-one seed bank; values > 1 indicate a population increase.

year of pasture resulted in a 78 percent decline in the weed seed bank, making it possible for the farmer to resume cropping. Use of appropriate herbicides and nonchemical methods in 1998 and 1999 enabled this farmer to maintain the decline in the weed seed bank. A seed bank density of five seeds per m^2 was measured after the final crop (wheat) in the rotation.

Case Study #2

Failure of diclofop to control annual ryegrass in a triticale (*X. Triticosecale* Wittmack) crop in 1994 because of resistance resulted in a 16-fold increase in the weed seed bank (see Table 20.5). The weed population increased further after a pea (*Pisum sativum* L.) crop, even though selective herbicides from three different mode-of-action groups were used and followed by paraquat for seed-set control. Use of triasulfuron (Group 2/B) in 1996, to which this population was still susceptible, and tralkoxydim resulted in a 75 percent decline in the seed bank of annual ryegrass. The oat hay crop in 1997 was sprayed with a mixture of diuron and metolachlor and cut early to prevent any viable seed-set. There was a further decline in the annual ryegrass seed bank after the 1998 pea crop, which was treated with several pre- and postemergence herbicides as well as paraquat to prevent seed-set (crop topping). This case study illustrates successful management of Group 1/A inhibitor-HR annual ryegrass within the cropping rotation, provided tactics that provide reliable control of weed seed-set such as oat hay and crop topping are effectively utilized.

*Wild Oat (*Avena fatua* L.) in North America*

Herbicide-resistant wild oat in North America was first discovered in northwestern Manitoba in 1990 (Heap et al., 1993). Wild oat is well adapted to the cool, moist conditions that prevail in this region. Frequency of occurrence in fields and population densities of the weed in Manitoba are higher in this region than in other regions of the province. Consequently, herbicide use to control wild oat is also the highest in this region (Heap et al., 1993). In 1994, two populations from northwestern Manitoba were found to be resistant to herbicides from three groups: ACCase inhibitors (Group 1/A), imazamethabenz (Group 2/B), and flamprop (Group 25/Z) (Morrison et al., 1995). The case study outlined next is based on one farmer's experience in managing multiple-group HR wild oat on his farm, located in northwestern Manitoba.

The cropping history and herbicide use associated with the occurrence of multiple-group HR wild oat in one of the fields on the farm is typical of fields in western Canada with evolved wild oat resistance (see Table 20.6). ACCase inhibitors (diclofop, sethoxydim) had been applied seven times

TABLE 20.6. Cropping history and grass herbicide use associated with the occurrence of multiple-group herbicide-resistant wild oat (*Avena fatua* L.) in a 64 ha field on a farm in northwestern Manitoba, Canada.

Year	Crop	Herbicide	Group
1998	canola (*Brassica napus* L.)	glyphosate	9/G
1997	spring wheat (*Triticum aestivum* L.)	imazamethabenz[a]	2/B
		fenoxaprop	1/A
1996	canola	sethoxydim	1/A
1995	spring wheat	difenzoquat	8/Z
1994	spring wheat	imazamethabenz	2/B
1993	canola	sethoxydim	1/A
1992	spring wheat	triallate + trifluralin	8/N, 3/K1
1991	spring wheat	diclofop	1/A
1990	spring wheat	diclofop	1/A
1989	spring wheat	diclofop	1/A
1988	spring wheat	difenzoquat	8/Z
1987	flax (*Linum usitatissimum* L.)[b]	sethoxydim	1/A
	barley (*Hordeum vulgare* L.)	difenzoquat	8/Z
1986	spring wheat	diclofop	1/A

[a]Fenoxaprop was applied first and controlled wild oat poorly; imazamethabenz was subsequently applied.
[b]Flax and barley were grown on 32 ha each.

over an 11-year period between 1986 and 1996. Diclofop and sethoxydim, introduced in the late 1970s and early 1980s, had performed very well in controlling wild oat populations on the zero-tillage farm. Group 1/A resistance was first suspected in 1997 when fenoxaprop failed to control wild oat in spring wheat. Multiple wild oat patches in the field remained uncontrolled. Imazamethabenz was subsequently applied in the same growing season but controlled the weed poorly. The herbicide had been used only twice. Resistance test results indicated that the population was resistant to fenoxaprop (Group 1/A), imazamethabenz (Group 2/B), triallate/ difenzoquat (Group 8/N, Z), and flamprop (Group 25/Z). Triallate and difenzoquat had been used periodically in the field (four applications between 1986 and 1995). Flamprop, however, had not been applied between 1986 and 1998. ACCase inhibitor resistance was not due to target-site mutation (Beckie et al., 1999). Metabolic resistance is the likely mechanism responsible for resistance (H. J. Beckie, unpublished data). Frequent use of ACCase inhibitors may have selected for individuals possessing the ability to rapidly metabolize these herbicides, as well as those with different modes of action. By the spring of 1998 prior to herbicide application, HR wild oat covered a large proportion of the 64 ha field and also occurred in other fields on the farm. Glyphosate-HR canola (*B. napus*) was grown in 1998 and controlled HR wild oat well.

The farmer realized that a drastic change in his cropping system was required to sustain the economic viability of his farm. To manage resistance, the farmer altered his zero-tillage cropping system from one based on near monoculture of cereals (spring wheat) to a six-year sequence of diverse crops (see Figure 20.5). A short-term (three-year) alfalfa (*Medicago sativa*

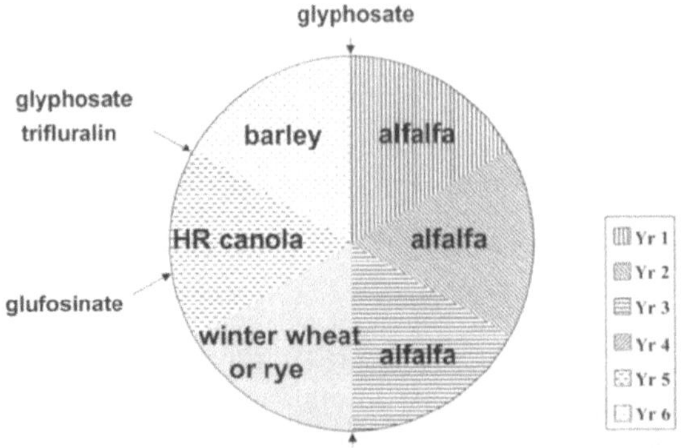

FIGURE 20.5. Cropping system used by a farmer in Manitoba, Canada, in response to quadruple-group herbicide-resistant wild oat (*Avena fatua* L.).

L.) stand has markedly reduced wild oat recruitment from the seed bank. A survey in southern Manitoba in 1993 and 1994 had documented the ability of three- to six-year alfalfa stands to reduce wild oat populations in cropping systems through crop competition and cutting regime of the crop for hay (Ominski, Entz, and Kenkel, 1999). The survey found that wild oat densities were reduced by 96 percent in cereal fields following alfalfa as compared with continuous cereal-cropped fields; the corresponding reduction in frequency of fields with wild oat and the mean proportion of a field infested with wild oat were 50 and 84 percent, respectively. Alfalfa is followed by fall-sown cereals (winter wheat or winter rye, *Secale cereale* L.), which have replaced spring wheat in the rotation. Winter cereals have eliminated the need for a wild oat herbicide application in the spring because of early-season vigorous growth and competitiveness. Herbicide-resistant canola (glufosinate-HR hybrid variety) follows the cereal crop in the rotation and is seeded as early as possible in spring to optimize yield potential and eliminate the need for a preseeding glyphosate application. Zand and Beckie (2002) found that hybrid canola cultivars were twice as competitive against wild oat than open-pollinated cultivars when weed interference was relatively high. Spring-sown barley succeeds canola in the rotation, using trifluralin (Group 3/K1) applied preplant (fall-applied granular formulation) without incorporation in soil to control wild oat. Following barley harvest, alfalfa is seeded in late summer. Seeding rates of cereals are typically 25 to 50 percent above recommended rates to increase stand density and aid crop competition against wild oat. This cropping system has allowed the farmer to maintain or increase net returns and reduce economic risk compared with the previous system. By diversifying the crop rotation to include perennial forages and weed-competitive annual crops that differ in phenology (winter and spring) and type (monocots and dicots), and including agronomic practices such as increased crop seeding rates, precision fertilizer placement (i.e., banding fertilizer), and using clean, vigorous crop seed, the farmer has markedly reduced wild oat population densities. This change in farming system, in turn, has allowed a reduction in herbicide use. Herbicides used in the cropping system are rotated by mode of action and pose a moderate (trifluralin) to low risk (glyphosate, glufosinate) of selecting for target-site-based HR weed biotypes. Because these herbicides are not readily metabolized in plants, they pose a low risk for selecting for metabolic-based HR biotypes. Typical of zero-tillage systems, the farmer relies heavily on glyphosate as a burnoff treatment prior to seeding alfalfa and cereals, and for terminating the alfalfa stand; glyphosate is usually applied four times during the six-year crop rotation.

Littleseed Canarygrass (**Phalaris minor** *Retz.*) *in India*

After the success of the Green Revolution, the farming systems in India thrived on the strength of selective herbicides with little attention to crop diversification and alternative weed management practices. Isoproturon (IPU) was used as the sole selective herbicide for littleseed canarygrass control for more than 15 years before it succumbed to resistance (Malik and Singh, 1995). In hindsight, this was a fairly predictable development, but it caught local researchers, policymakers, and farmers by surprise and resulted in major losses in wheat productivity. Official estimates in 1997 showed that more than 450,000 ha of farming land in Haryana and Punjab was affected by HR populations. Satisfactory control of the weed with IPU declined from a high of 78 percent in 1990 to less than 22 percent in 1993 (Malik, Gill, and Hobbs, 1998). Yield losses in severely affected districts ranged from 30 percent to complete weed dominance, and the crop had to be hand harvested for animal feed. It was in this desperate background that the Australian-Indian resistance management program was conceived and undertaken. The project has been a real success story, resulting in large gains in wheat productivity and farm incomes. Key aspects of the resistance management program are summarized here.

Alternative Herbicides

Unlike Australia where no new selective herbicide chemistry has been released for nearly 20 years, herbicide options in India have expanded dramatically in the past five years. The timing of the release of alternative herbicides in the Indian market was rather fortuitous and occurred at the peak of IPU resistance. Project researchers were obviously mindful of the urgent need of these herbicides and were able to facilitate their fast tracking through the regulatory, research, and advisory systems. Clodinafop, fenoxaprop, and sulfosulfuron have now been available to local farmers for the past three to four years and have had an enormous impact on weed control in wheat in the region. However, there are some worrisome warning signs. Some populations are showing a rise in GR_{50} (dose reducing weed growth by 50 percent relative to untreated plants) for fenoxaprop. Monitoring of targeted populations is now undertaken routinely and is expected to detect emerging problems before they affect large areas of farmland.

Crop Rotations

Littleseed canarygrass has a relatively short-lived seed bank, and forage crops such as berseem (*Trifolium alexandrinum* L.) can reliably achieve a large reduction in weed populations (Malik, Gill, and Hobbs, 1998). For most farmers, however, berseem is a nontradeable commodity and can be grown only for consumption by the landowner's animal flock. Consequently not more than 5 to 10 percent of the farm area can be planted to forage crops and the rest of the land needs to be sown to cash crops, such as wheat. Other rotational crops that have been shown to be effective in weed population management are sunflower and sugarcane (*Saccharum officinarum* L.). However, marketing of these crops is considered risky and the area cultivated to these crops tends to fluctuate widely. Due to the dietary preferences of the populace, wheat and rice are likely to remain key crops in the region, and resistance management programs can have local relevance only if they develop strategies that are effective within the rice-wheat system.

Adoption of Zero-Tillage Seeding System

Farmers in northwest India have enthusiastically embraced the zero-tillage system of planting wheat. The area under zero-tillage sown wheat has increased from two ha in 1997 to more than 40,000 ha in 2000 (R. K. Malik, personal communication, 2001). Zero tillage by itself is not the solution to the management of HR littleseed canarygrass, but it has been shown to create an environment that is conducive to the management of this problem. Some of the benefits of zero tillage arise from earlier sowing of crops under this system. Farmers can sow their wheat crops nearly a month earlier than would be feasible under the traditional system that often receives up to seven preseeding tillage operations. The ambient temperatures at seeding for these early-sown crops are higher and are not conducive for littleseed canarygrass germination. This microclimate often gives wheat a two- to three-week growth advantage over the weed and tilts the competitive balance in favor of the crop. However, the impact of zero tillage that is related to economics of cropping has had a much bigger effect on HR littleseed canarygrass management. Case studies undertaken have shown that net savings in fuel, machinery, and labor costs under zero tillage can exceed Rs 1,700 per ha (US$35 per ha) (R. K. Malik, personal communication, 2001). Resource-poor farmers have been able to use these savings to purchase new herbicide chemistry that has provided effective and reliable control of IPU-HR littleseed canarygrass. It is very clear that without these savings in crop establishment costs, local farmers would not have widely adopted alternative herbicides, which cost nearly five times more than IPU.

CONCLUSION

Proactive or reactive management for herbicide resistance in weeds must consider (1) the relative risks of herbicides of different modes of action to select for target-site resistance and the differing propensity of herbicides to be metabolized in HR biotypes when rotating among herbicides; (2) must meet basic criteria for effective herbicide mixtures; and (3) should incorporate agronomic practices in cropping systems that help reduce weed seed production and spread. Crops resistant to nonselective herbicides can be useful in achieving these objectives. Overreliance on HR crops, however, may maintain conditions that lead to resistance, namely simplified cropping systems favoring a few dominant weed species and frequent use of single site-of-action herbicides.

The extent to which farmers alter their current farming system to manage resistance depends on the nature and degree of infestation of an HR biotype. In many cases, simply switching to an alternative herbicide will cost-effectively control the HR population. For serious resistance problems, i.e., heavy infestations of multiple-group HR grass weed species, a long-term, cropping systems approach will be required. The case studies each illustrated a different approach to IWM, based on regional agroecological conditions, biology and ecology of the weed species with evolved resistance, and agronomic and socioeconomic considerations by farmers. In all these situations, herbicides remain as the dominant weed control method. Changes in cropping systems and practices in each situation, however, have resulted in less herbicide used and thus a reduction in selection pressure for resistance. Although different in approach, each case study showed that even serious weed resistance problems can be managed successfully if farmers are receptive to change and are committed to the long-term economic viability of their farms. To make this transition, they will need support. Much greater research and extension efforts by government and industry worldwide will be necessary to assist farmers in adapting their farming systems to effectively manage increasing incidence and complexity of herbicide resistance in weeds.

REFERENCES

Anderson, D.D., F.W. Roeth, and A.R. Martin (1996). Occurrence and control of triazine-resistant common waterhemp *(Amaranthus rudis)* in field corn *(Zea mays)*. *Weed Technology* 10: 570-575.

Andrews, T.S. and I.N. Morrison (1997). The persistence of trifluralin resistance in green foxtail *(Setaria viridis)* populations. *Weed Technology* 11: 369-372.

Andrews, T.S., I.N. Morrison, and G.A. Penner (1998). Monitoring the spread of ACCase inhibitor resistance among wild oat *(Avena fatua)* patches using AFLP analysis. *Weed Science* 46: 196-199.

Anonymous (1994). Regulatory directive dir94-11: The biology of *Zea mays* L. (corn/maize). A companion document to the assessment criteria for determining environmental safety of plants with novel traits. Ottawa, Ontario: Canadian Food Inspection Agency (CFIA), Plant Health and Production Division, Plant Biosafety Office. Available online at <http://www.inspection.gc.ca/english/plaveg/bio/dir/dir9411e.shtml>.

Anonymous (1996). Regulatory directive T-1-10-96: The biology of *Glycine max* (L.) Merr. (soybean). A companion document to the assessment criteria for determining environmental safety of plants with novel traits. Ottawa, Ontario: Canadian Food Inspection Agency (CFIA), Plant Health and Production Division, Plant Biosafety Office. Available online at <http://www.inspection.gc.ca/english/plaveg/bio/dir/t11096e.shtml>.

Anonymous (2002). *Guide to crop protection 2002: Weeds, plant diseases, insects.* Bi-provincial publication. Regina: Saskatchewan Agriculture and Food; Winnipeg: Manitoba Agriculture and Food.

Beckie, H.J., F. Chang, and F.C. Stevenson (1999). The effect of labeling herbicides with their site of action: A Canadian perspective. *Weed Technology* 13: 655-661.

Beckie H.J., L.M. Hall, and F.J. Tardif (2001). Herbicide resistance in Canada—Where are we today? In R.E. Blackshaw and L.M. Hall (Eds.), *Integrated weed management: Explore the potential* (pp. 1-36). Sainte-Anne-de-Bellevue, QC: Expert Committee on Weeds.

Beckie, H.J., L.M. Hall, and S.I. Warwick (2001). Impact of herbicide-resistant crops as weeds in Canada. In *Proceedings, Brighton Crop Protection Conference—Weeds* (pp. 135-142). Farnham, UK: British Crop Protection Council.

Beckie, H.J., I.M. Heap, R.J. Smeda, and L.M. Hall (2000). Screening for herbicide resistance in weeds. *Weed Technology* 14: 428-445.

Beckie, H.J. and F.A. Holm (2002). Response of wild oat *(Avena fatua)* to residual and non-residual herbicides in canola *(Brassica napus)* in western Canada. *Canadian Journal of Plant Science* 82: 797-802.

Beckie, H.J. and S. Jana (2000). Selecting for triallate resistance in wild oat. *Canadian Journal of Plant Science* 80: 665-667.

Beckie, H.J. and I.N. Morrison (1993). Effective kill of trifluralin-susceptible and -resistant green foxtail *(Setaria viridis). Weed Technology* 7: 15-22.

Beckie, H.J., A.G. Thomas, and A. Légère (1999). Nature, occurrence, and cost of herbicide-resistant green foxtail *(Setaria viridis)* across Saskatchewan ecoregions. *Weed Technology* 13: 626-631.

Beckie, H.J., A.G. Thomas, A. Légère, D.J. Kelner, R.C. Van Acker, and S. Meers (1999). Nature, occurrence, and cost of herbicide-resistant wild oat *(Avena fatua)* in small-grain production areas. *Weed Technology* 13: 612-625.

Beckie, H.J., A.G. Thomas, and F.C. Stevenson (2002). Survey of herbicide-resistant wild oat *(Avena fatua)* in two townships in Saskatchewan. *Canadian Journal of Plant Science* 82: 463-471.

Boerboom, C.M. (1999). Nonchemical options for delaying weed resistance to herbicides in Midwest cropping systems. *Weed Technology* 13: 636-642.

Bourgeois, L., N.C. Kenkel, and I.N. Morrison (1997). Characterization of cross-resistance patterns in acetyl-CoA carboxylase inhibitor resistant wild oat *(Avena fatua)*. *Weed Science* 45: 750-755.

Bourgeois, L., I.N. Morrison, and D. Kelner (1997). Field and grower survey of ACCase resistant wild oat in Manitoba. *Canadian Journal of Plant Science* 77: 709-715.

Boutsalis, P. and S.B. Powles (1995). Resistance of dicot weeds to acetolactate synthase (ALS)-inhibiting herbicides in Australia. *Weed Research* 35: 149-155.

Bradley, K.W. and E.S. Hagood Jr. (2001). Identification of a johnsongrass *(Sorghum halepense)* biotype resistant to aryloxyphenoxypropionate and cyclohexanedione herbicides in Virginia. *Weed Technology* 15: 623-627.

Burnside, O.C. (1992). Rationale for developing herbicide-resistant crops. *Weed Technology* 6: 621-625.

Carey, V.F. III, R.E. Hoagland, and R.E. Talbert (1995). Verification and distribution of propanil-resistant barnyardgrass *(Echinochloa crus-galli)* in Arkansas. *Weed Technology* 9: 366-372.

Cavan, G., P. Biss, and S.R. Moss (1998). Herbicide resistance and gene flow in wild-oats (*Avena fatua* and *Avena sterilis* ssp. *ludoviciana*). *Annals of Applied Biology* 133: 207-217.

Cavan, G., J. Cussans, and S.R. Moss (1999). Modelling strategies to prevent resistance in black-grass *(Alopecurus myosuroides)*. In *Proceedings, Brighton Crop Protection Conference—Weeds* (pp. 777-782). Farnham, UK: British Crop Protection Council.

Champolivier, J., J. Gasquez, A. Messéan, and M. Richard-Molard (1999). Management of transgenic crops within the cropping system. In P.J.W. Lutman (Ed.), *Gene flow and agriculture: Relevance for transgenic crops* (pp. 233-240). Farnham, Surrey, UK: British Crop Protection Council.

Chauvel, B. and J. Gasquez (1994). Relationships between genetic polymorphism and herbicide resistance within *Alopecurus myosuroides* Huds. *Heredity* 72: 336-344.

Cocker, K.M., J.O.D. Coleman, A.M. Blair, J.H. Clarke, and S.R. Moss (2000). Biochemical mechanisms of cross-resistance to aryloxyphenoxypropionate and cyclohexanedione herbicides in populations of *Avena* spp. *Weed Research* 40: 323-334.

Coleman, R.K., G.S. Gill, and G.J. Rebetzke (2001). Identification of quantitative trait loci for traits conferring weed competitiveness in wheat (*Triticum aestivum* L.). *Australian Journal of Agricultural Research* 52: 1235-1246.

Coupland, D. (1994). Resistance to the auxin analog herbicides. In S.B. Powles and J.A.M. Holtum (Eds.), *Herbicide resistance in plants—Biology and biochemistry* (pp. 171-214). Boca Raton, FL: CRC Press.

Cranston, H.J., A.J. Kern, J.L. Hackett, E.K. Miller, B.D. Maxwell, and W.E. Dyer (2001). Dicamba resistance in kochia. *Weed Science* 49: 164-170.

Culpepper, A.S., A.C. York, and A.M. Stewart (2002). Control of glyphosate-resistant soybean *(Glycine max)* with MSMA, pyrithiobac, and CGA-362622. *Weed Science Society of America Abstracts* 42: 4.

Daou, H. and R.E. Talbert (1999). Control of propanil-resistant barnyardgrass *(Echinochloa crus-galli)* in rice *(Oryza sativa)* with carbaryl/propanil mixtures. *Weed Technology* 13: 65-70.

Debreuil, D.J., L.F. Friesen, and I.N. Morrison (1996). Growth and seed return of auxin-type herbicide resistant wild mustard *(Brassica kaber)* in wheat. *Weed Science* 44: 871-878.

Dellow, J.J., M. Incerti, R. Britton, and A. Bishop (1997). Herbicide resistance extension strategy for the south eastern wheat belt of New South Wales, Australia. In H. Brown, G.W. Cussans, M.D. Devine, S.O. Duke, C. Fernandez-Quintanilla, A. Helweg, R.E. Labrada, M. Landes, P. Kudsk, and J.C. Streibig (Eds.), *Proceedings,* Volume 2, *Second International Weed Control Congress, Copenhagen, Denmark* (pp. 487-492). Flakkebjerg, Slagelse, Denmark: Dept. Weed Control and Pesticide Ecology.

Devine, M.D. and J.L. Buth (2001). Advantages of genetically modified canola: A Canadian perspective. In *Proceedings, Brighton Crop Protection Conference—Weeds* (pp. 367-372). Farnham, UK: British Crop Protection Council.

Diggle, A.J. and P. Neve (2001). The population dynamics and genetics of herbicide resistance—A modeling approach. In S.B. Powles and D.L. Shaner (Eds.), *Herbicide resistance and world grains* (pp. 61-99). New York: CRC Press.

Dinelli, G., A. Bonetti, and P. Catizone (2000). Investigation on biodiversity of Italian *Lolium* spp. populations susceptible and resistant to herbicides. In *Abstracts of the Third International Weed Science Congress, Foz do Iguassu, Brazil,* June 6-11 (p. 146). [CD-ROM.] Corvallis, OR: International Weed Science Society.

Dunan, C. and P. Westra (2000). What risks do herbicide resistant weeds present to chemical companies? In *Abstracts of the Third International Weed Science Congress, Foz do Iguassu, Brazil,* June 6-11 (pp. 155-156). [CD-ROM.] Corvallis, OR: International Weed Science Society.

Dyer, W.E., P.W. Chee, and P.K. Fay (1993). Rapid germination of sulfonylurea-resistant *Kochia scoparia* L. accessions is associated with elevated seed levels of branched chain amino acids. *Weed Science* 41: 18-22.

Dyer, W.E., M.A. Jasieniuk, and B.D. Maxwell (2000). Stress tolerance in *Kochia scoparia* L.: Phenotypic plasticity or genetic adaptation? In *Abstracts of the Third International Weed Science Congress, Foz do Iguassu, Brazil,* June 6-11 (pp. 147-148P). [CD-ROM.] Corvallis, OR: International Weed Science Society.

Foes, M.J., L. Liu, P.J. Tranel, L.M. Wax, and E.W. Stoller (1998). A biotype of common waterhemp *(Amaranthus rudis)* resistant to triazine and ALS herbicides. *Weed Science* 46: 514-520.

Frick B. and A.G. Thomas (1992). Weed surveys in different tillage systems in southwestern Ontario field crops. *Canadian Journal of Plant Science* 72: 1337-1347.

Friesen, L.F., T.L. Jones, R.C. Van Acker, and I.N Morrison (2000). Identification of *Avena fatua* populations resistant to imazamethabenz, flamprop, and fenoxaprop-P. *Weed Science* 48: 532-540.

Gadamski, G., D. Ciarka, J. Gressel, and S.W. Gawronski (2000). Negative cross-resistance in triazine-resistant biotypes of *Echinochloa crus-galli* and *Conzya canadensis*. *Weed Science* 48: 176-180.

Gardner, S.N., J. Gressel, and M. Mangel (1998). A revolving dose strategy to delay the evolution of both quantitative vs major monogene resistances to pesticides and drugs. *International Journal of Pest Management* 44: 161-180.

Gill, G.S. (1996). Management of herbicide resistant ryegrass in Western Australia—Research and its adoption. In R.C.H. Shepherd (Ed.), *Proceedings of the Eleventh Australian Weeds Conference* (pp. 542-545). Melbourne, Australia: Weed Science Society of Victoria.

Gill, G.S. (1997). Prevention and control of herbicide resistant weeds in Australia. In R. De Prado, J. Jorrín, and L. García-Torres (Eds.), *Weed and crop resistance to herbicides* (pp. 305-313). London: Kluwer Academic Publishers.

Gill, G.S. and J.E. Holmes (1997). Efficacy of cultural control methods for combating herbicide-resistant *Lolium rigidum*. *Pesticide Science* 51: 352-358.

Gorddard, R.J., D.J. Pannell, and G. Hertzler (1996). Economic evaluation of strategies for management of herbicide resistance. *Agricultural Systems* 51: 281-298.

Gressel, J. (1990). Synergizing herbicides. *Reviews of Weed Science* 5: 49-82.

Gressel, J. (1995). Creeping resistances: The outcome of using marginally-effective or reduced rates of herbicides. In *Proceedings, Brighton Crop Protection Conference—Weeds* (pp. 587-590). Farnham, UK: British Crop Protection Council.

Gressel, J. (1997). Burgeoning resistance requires new strategies. In R. De Prado, J. Jorrín, and L. García-Torres (Eds.), *Weed and crop resistance to herbicides* (pp. 3-14). London: Kluwer Academic Publishers.

Gressel, J. (1999). Modern herbicide design can lead to obsolescence via resistance. *Weed Science Society of America Abstracts* 39: 76-77.

Gressel, J. and L.A. Segel (1982). Interrelating factors controlling the rate of appearance of resistance: The outlook for the future. In H.M. LeBaron and J. Gressel (Eds.), *Herbicide resistance in plants* (pp. 325-347). New York: John Wiley and Sons.

Hall, L.M., J.A.M. Holtum, and S.B. Powles (1994). Mechanisms responsible for cross resistance and multiple resistance. In S.B. Powles and J.A.M. Holtum (Eds.), *Herbicide resistance in plants: Biology and biochemistry* (pp. 243-261). Boca Raton, FL: CRC Press.

Hall, L., K. Topinka, J. Huffman, L. Davis, and A. Good (2000). Pollen flow between herbicide-resistant *Brassica napus* (Argentine canola) is the cause of multiple resistant *B. napus* volunteers. *Weed Science* 48: 688-694.

Hartmann, F., I. Lánszki, L. Szentey, and A. Tóth (2000). Resistant weed biotypes in Hungary. In *Abstracts of the Third International Weed Science Congress, Foz do Iguassu, Brazil,* June 6-11 (p. 138). [CD-ROM.] Corvallis, OR: International Weed Science Society.

Hashem, A., D. Bowran, T. Piper, and H. Dhammu (2001). Resistance of wild radish *(Raphanus raphanistrum)* to acetolactate synthase-inhibiting herbicides in the Western Australia wheat belt. *Weed Technology* 15: 68-74.

Hashem, A., H.S. Dhammu, S.B. Powles, D.G. Bowran, T.J. Piper, and A.H. Cheam (2001). Triazine resistance in a biotype of wild radish *(Raphanus raphanistrum)* in Australia. *Weed Technology* 15: 636-641.

Heap, I.M. (1999). International survey of herbicide-resistant weeds: Lessons and limitations. In *Proceedings, Brighton Crop Protection Conference—Weeds* (pp. 769-776). Farnham, UK: British Crop Protection Council.

Heap, I.M. (2005). International survey of herbicide resistant weeds. Available online at <http://www.weedscience.org/in.asp>.

Heap, I.M., B.G. Murray, H.A. Loeppky, and I.N. Morrison (1993). Resistance to aryloxyphenoxypropionate and cyclohexanedione herbicides in wild oat *(Avena fatua). Weed Science* 41: 232-238.

Heap, J. and R. Knight (1982). A population of ryegrass tolerant to the herbicide diclofop-methyl. *Journal of Australian Institute of Agricultural Science* 48: 156-157.

Holt, J.S., S.B. Powles, and J.A.M. Holtum (1993). Mechanisms and agronomic aspects of herbicide resistance. *Annual Reviews of Plant Physiology and Plant Molecular Biology* 44: 203-229.

Holt, J.S. and D.C. Thill (1994). Growth and productivity of resistant plants. In S.B. Powles and J.A.M. Holtum (Eds.), *Herbicide resistance in plants: Biology and biochemistry* (pp. 299-316). Boca Raton, FL: CRC Press.

Itoh, K., G.X. Wang, and S. Ohba (1999). Sulfonylurea resistance in *Lindernia micrantha*, an annual paddy weed in Japan. *Weed Research* 39: 413-423.

James, C. (2000). *Global status of commercialized transgenic crops: 2000.* ISAAA Briefs No. 23: Preview. Ithaca, NY: ISAAA. Available online at <http://www.isaaa.org/kc/Publications/pdfs/isaaabriefs/Briefs%2023.pdf>.

James, C. (2003). *Global status of commercialized transgenic crops: 2003.* ISAAA Briefs No. 30: Preview. Ithaca, NY: ISAAA. Available online at <http://www.isaaa.org/kc/Publications/pdfs/isaaabriefs/Briefs%2030.pdf>.

Jasieniuk, M., A.L. Brulé-Babel, and I.N. Morrison (1996). The evolution and genetics of herbicide resistance in weeds. *Weed Science* 44: 176-193.

Jasieniuk, M., I.N. Morrison, and A.L. Brulé-Babel (1995). Inheritance of dicamba resistance in wild mustard *(Brassica kaber). Weed Science* 43: 192-195.

Jordon, N., M. Kelrick, J. Brooks, and W. Kinerk (1999). Biorational management tactics to select against triazine-resistant *Amaranthus hybridus:* A field trial. *Journal of Applied Ecology* 36: 123-132.

Kudsk, P., S.K. Mathiassen, and J.C. Cotterman (1995). Sulfonylurea resistance in *Stellaria media* [L.] Vill. *Weed Research* 35: 19-24.

Kulshrestha, G., S.B. Singh, and N.T. Yaduraju (1999). Mechanism of isoproturon resistance: The metabolism of isoproturon in susceptible and resistant biotypes of *Phalaris minor.* In *Proceedings, Brighton Crop Protection Conference—Weeds* (pp. 167-172). Farnham, UK: British Crop Protection Council.

LeBaron, H.M. and J. McFarland (1990). Herbicide resistance in weeds and crops. In M.B. Green, H.M. LeBaron, and W.K. Moberg (Eds.), *Managing resistance to agrochemicals: From fundamental research to practical strategies* (pp. 336-352). Washington, DC: American Chemical Society.

Légère, A., H.J. Beckie, F.C. Stevenson, and A.G. Thomas (2000). Survey of management practices affecting the occurrence of wild oat *(Avena fatua)* resistance to acetyl-CoA carboxylase inhibitors. *Weed Technology* 14: 366-376.

Légère, A., M.J. Simard, A.G. Thomas, D. Pageau, J. Lajeunesse, S.I. Warwick, and D.A. Derksen (2001). Presence and persistence of volunteer canola in Canadian cropping systems. In *Proceedings, Brighton Crop Protection Conference—Weeds* (pp. 143-148). Farnham, UK: British Crop Protection Council.

Lemerle, D., B. Verbeek, R.D. Cousens, and N.E. Coombes (1996). The potential for selecting wheat varieties strongly competitive against weeds. *Weed Research* 36: 503-513.

Letouzé, A. and J. Gasquez (1999). A pollen test to detect ACCase target-site resistance within *Alopecurus myosuroides* populations. *Weed Research* 40: 151-162.

Li, C., F.C. Yeh, I.N. Morrison, and T.S. Andrews (2000). Tracing the movement of herbicide resistance in fields infested with *Setaria viridis* L. using amplified fragment length polymorphisms. In *Abstracts of the Third International Weed Science Congress, Foz do Iguassu, Brazil,* June 6-11 (p. 144). [CD-ROM.] Corvallis, OR: International Weed Science Society.

Malik, R.K., G. Gill, and P.R. Hobbs (1998). Herbicide resistance—A major issue for sustainable wheat productivity in rice-wheat cropping systems in the Indo-Gangetic plains. Rice-Wheat Consortium Paper Series 3. New Delhi, India: Rice-Wheat Consortium for the Indo-Gangetic Plains.

Malik, R.K. and S. Singh (1995). Littleseed canarygrass *(halaris minor)* resistance to isoproturon in India. *Weed Technology* 9: 419-425.

Mallory-Smith, C.A., D.C. Thill, and G.P. Stallings (1993). Survey and gene flow in acetolactate synthase resistant kochia and Russian thistle. In *Proceedings, Brighton Crop Protection Conference—Weeds* (pp. 555-558). Farnham, UK: British Crop Protection Council.

Mansooji, A.M., J.A. Holtum, P. Boutsalis, J.M. Matthews, and S.B. Powles (1992). Resistance to aryloxyphenoxypropionate herbicides in two wild oat species (*Avena fatua* and *Avena sterilis* ssp. *ludoviciana*). *Weed Science* 40: 599-605.

Matthews, J.M. (1994). Management of herbicide resistant weed populations. In S.B. Powles and J.A.M. Holtum (Eds.), *Herbicide resistance in plants: Biology and biochemistry* (pp. 317-335). Boca Raton, FL: CRC Press.

Matthews, J.M. and S.B. Powles (1992). Aspects of the population dynamics of selection for herbicide resistance in *Lolium rigidum* (Gaud). In J.H. Combellack, K.J. Levick, J. Parsons, and R.G. Richardson (Eds.), *Proceedings of the First International Weed Control Congress* (pp. 318-320). Melbourne, Australia. Corvallis, OR: International Weed Science Society.

Matthews, J.M. and S.B. Powles (1996). Managing herbicide resistant annual ryegrass, southern Australian research. In R.C.H. Shepherd (Ed.), *Proceedings of the Eleventh Australian Weeds Conference* (pp. 537-541). Melbourne, Australia: Weed Science Society of Victoria.

Maxwell, B.D. and A.M. Mortimer (1994). Selection for herbicide resistance. In S.B. Powles and J.A.M. Holtum (Eds.), *Herbicide resistance in plants: Biology and biochemistry* (pp. 1-25). Boca Raton, FL: CRC Press.

Medd, R.W., B.A. Auld, D.R. Kemp, and R.D. Murison (1987). The influence of wheat density and spatial arrangement on annual ryegrass, *Lolium rigidum* Gaudin, competition. *Australian Journal of Agricultural Research* 36: 361-371.

Morrison, I.N., L. Bourgeois, L. Friesen, and D. Kelner (1995). Betting against the odds: The problem of herbicide resistance. In T.L. Roberts (Ed.), *Proceedings of the 1995 Western Canada Agronomy Workshop* (pp. 159-164). Red Deer, Alberta: Potash and Phosphate Institute of Canada.

Morrison, I.N. and L.F. Friesen (1996). Herbicide resistant weeds: Mutation, selection, misconception. In H. Brown, G.W. Cussans, M.D. Devine, S.O. Duke, C. Fernandez-Quintanilla, A. Helweg, R.E. Labrada, M. Landes, P. Kudsk, and J.C. Streibig. (Eds.), *Proceedings, Second International Weed Control Congress, Copenhagen, Denmark* (pp. 1-9). Flakkebjerg, Slagelse, Denmark: Dept. Weed Control and Pesticide Ecology.

Moss, S.R. (1997). Strategies for the prevention and control of herbicide resistance in annual grass weeds. In R. De Prado, J. Jorrín, and L. García-Torres (Eds.), *Weed and crop resistance to herbicides* (pp. 283-290). London: Kluwer Academic Publishers.

Murray, B.G., A.L. Brulé-Babel, and I.N. Morrison (1996). Two distinct alleles encode for acetyl-CoA carboxylase inhibitor resistance in wild oat *(Avena fatua). Weed Science* 44: 476-481.

Norsworthy, J.K., J.S. Rutledge, R.E. Talbert, and R.E. Hoagland (1999). Agrichemical interactions with propanil on propanil-resistant barnyardgrass *(Echinochloa crus-galli). Weed Technology* 13: 296-302.

Norsworthy, J.K., R.E. Talbert, and R.E. Hoagland (1999). Chlorophyll fluorescence evaluation of agrochemical interactions with propanil on propanil-resistant barnyardgrass *(Echinochloa crus-galli). Weed Science* 47: 13-19.

O'Donovan, J.T., K.N. Harker, G.W. Clayton, J.C. Newman, D. Robinson, and L.M. Hall (2001). Barley seeding rate influences the effects of variable herbicide rates on wild oat. *Weed Science* 49: 746-754.

O'Donovan, J.T., J.C. Newman, R.E. Blackshaw, K.N. Harker, D.A. Derksen, and A.G. Thomas (1999). Growth, competitiveness, and seed germination of triallate/difenzoquat-susceptible and -resistant wild oat populations. *Canadian Journal of Plant Science* 79: 303-312.

Ominski, P.D., M.H. Entz, and N. Kenkel (1999). Weed suppression by *Medicago sativa* in subsequent cereal crops: A comparative survey. *Weed Science* 47: 282-290.

Orson, J.H. (1999). The cost to the farmer of herbicide resistance. *Weed Technology* 13: 607-611.

Orson, J.H. and D.B.F. Livingston (1987). Field trials on the efficacy of herbicides on resistant black-grass *(Alopecurus myosuroides)* in different cultivation regimes. In *Proceedings, Brighton Crop Protection Conference—Weeds* (pp. 887-894). Farnham: British Crop Protection Council.

Owen, M.D.K. (2001a). Importance of weed population shifts and herbicide resistance in the Midwest USA corn belt. In *Proceedings, Brighton Crop Protection Conference—Weeds* (pp. 407-412). Farnham: British Crop Protection Council.

Owen, M.D.K. (2001b). World maize/soybean and herbicide resistance. In S.B. Powles and D.L. Shaner (Eds.), *Herbicide resistance and world grains* (pp. 101-163). New York: CRC Press.

Parks, R.J., W.S. Curran, G.W. Roth, N.L. Hartwig, and D.D. Calvin (1996). Herbicide susceptibility and biological fitness of triazine-resistant and susceptible common lambsquarters *(Chenopodium album). Weed Science* 44: 517-522.

Pearce, G.A. and J.E. Holmes (1976). The control of annual ryegrass. *Journal of Agriculture Western Australia* 17: 77-81.

Pekrun, C., J.D.J. Hewitt, and P.J.W. Lutman (1998). Cultural control of volunteer oilseed rape *(Brassica napus). Journal of Agricultural Science* 130: 155-163.

Peterson, D.E. (1999). The impact of herbicide-resistant weeds on Kansas agriculture. *Weed Technology* 13: 632-635.

Poston, D.H., J. Wu, K.K. Hatzios, and H.P. Wilson (2001). Enhanced sensitivity to cloransulam-methyl in imidazolinone-resistant smooth pigweed. *Weed Science* 49: 711-716.

Powles, S.B. (1997). Success from adversity: Herbicide resistance can drive changes to sustainable weed management systems. In *Proceedings, Brighton Crop Protection Conference—Weeds* (pp. 1119-1126). Farnham, UK: British Crop Protection Council.

Powles, S.B., D.F. Loraine-Colwill, J.J. Dellow, and C. Preston (1998). Evolved resistance to glyphosate in rigid ryegrass *(Lolium rigidum)* in Australia. *Weed Science* 46: 604-607.

Powles, S.B. and J.M. Matthews (1996). Integrated weed management for the control of herbicide-resistant annual ryegrass *(Lolium rigidum)*. In H. Brown, G.W. Cussans, M.D. Devine, S.O. Duke, C. Fernandez-Quintanilla, A. Helweg, R.E. Labrada, M. Landes, P. Kudsk, and J.C. Streibig (Eds.), In *Proceedings,* Volume 2, *Second International Weed Control Congress, Copenhagen, Denmark* (pp. 407-414). Flakkebjerg, Slagelse, Denmark: Dept. Weed Control and Pesticide Ecology.

Powles, S.B., M. Monjardino, R. Llewellyn, and D. Pannell (2000). Proactive versus reactive herbicide resistance management: Understanding the economic sense of herbicide conservation versus exploitation. In *Abstracts of the Third International Weed Science Congress, Foz do Iguassu, Brazil,* June 6-11 (pp. 148-149). [CD-ROM.] Corvallis, OR: International Weed Science Society.

Powles, S.B., C. Preston, I.B. Bryan, and A.R. Jutsum (1997). Herbicide resistance: Impact and management. *Advances in Agronomy* 58: 57-93.

Preston, C. and C.A. Mallory-Smith (2001). Biochemical mechanisms, inheritance, and molecular genetics of herbicide resistance in weeds. In S.B. Powles and D.L. Shaner (Eds.), *Herbicide resistance and world grains* (pp. 23-60). New York: CRC Press.

Putwain, P.D. (1982). Herbicide resistance in weeds—An inevitable consequence of herbicide use? In *Proceedings, Brighton Crop Protection Conference—Weeds* (pp. 719-728). Farnham, UK: British Crop Protection Council.

Rasche, E., J. Cremer, G. Donn, and J. Zink (1995). The development of glufosinate ammonium tolerant crops into the market. In *Proceedings, Brighton Crop Protection Conference—Weeds* (pp. 791-800). Farnham, UK: British Crop Protection Council.

Retzinger, E.J. and C. Mallory-Smith (1997). Classification of herbicides by site of action for weed resistance management strategies. *Weed Technology* 11: 384-393.

Riches, C.R., J.S. Knights, L. Chaves, J.C. Caseley, and B.E. Valverde (1997). The role of pendimethalin in the integrated management of propanil-resistant *Echinochloa colona* in Central America. *Pesticide Science* 51: 341-346.

Richter, J. and S.B. Powles (1993). Pollen expression of herbicide target site resistance genes in annual ryegrass *(Lolium rigidum). Plant Physiology* 102: 1037-1041.

Rieger, M.A., L. Stone, and C. Preston (2001). Herbicide resistant gene movement on a landscape scale with *Lactuca serriola. Weed Science Society of America Abstracts* 41: 125.

Ritter, R.L. and H. Menbere (1997). Distribution and management of triazine-resistant weeds in the Mid-Atlantic region of the U.S.A. In *Proceedings, Brighton Crop Protection Conference—Weeds* (pp. 1147-1152). Farnham, UK: British Crop Protection Council.

Rotteveel, T.J.W., J.W.F.M. de Goeij, and A.F. van Gemerden (1997). Toward the construction of a resistance risk evaluation scheme. *Pesticide Science* 51: 407-411.

Rubin, B. (1991). Herbicide resistance in weeds and crops, progress and prospects. In J.C. Caseley, G.W. Cussans, and R.K. Atkin (Eds.), *Herbicide resistance in weeds and crops* (pp. 387-414). Oxford, UK: Butterworth-Heinemann.

Sattin, M., M.A. Gasparetto, and C. Campagna (2001). Situation and management of *Avena sterilis* ssp *ludoviciana* and *Phalaris paradoxa* resistant to ACCase inhibitors in Italy. In *Proceedings, Brighton Crop Protection Conference—Weeds* (pp. 755-762). Farnham, UK: British Crop Protection Council.

Schmidt, R.R. (1997). HRAC classification of herbicides according to mode of action. In *Proceedings, Brighton Crop Protection Conference—Weeds* (pp. 1133-1140). Farnham, UK: British Crop Protection Council.

Seefeldt, S.S., D.R. Gealy, B.D. Brewster, and E.P. Fuerst (1994). Cross-resistance of several diclofop-resistant wild oat (*Avena fatua*) biotypes from the Willamette Valley of Oregon. *Weed Science* 42: 430-437.

Shaner, D.L., D.A. Feist, and E.J. Retzinger (1997). SAMOA: One company's approach to herbicide-resistant weed management. *Pesticide Science* 51: 367-370.

Shaner, D.L., S. Howard, and I. Chalmers (1999). Effectiveness of mode of action labelling for resistance management: A survey of Australian farmers. In *Proceedings, Brighton Crop Protection Conference—Weeds* (pp. 797-802). Farnham, UK: British Crop Protection Council.

Shirtliffe, S.J., M.H. Entz, and B.D. Maxwell (1998). The effect of chaff collection and harvest timing on the seed spread of wild oat *(Avena fatua). Weed Science Society of America Abstracts* 38: 39.

Simpson, E.C., C.E. Norris, J.R. Law, J.E. Thomas, and J.B. Sweet (1999). Gene flow in genetically modified herbicide tolerant oilseed rape *(Brassica napus)* in the UK. In P.J.W. Lutman (Ed.), *Gene flow and agriculture: Relevance for transgenic crops* (pp. 75-81). Farnham, Surrey, UK: British Crop Protection Council.

Singh, S., R.C. Kirkwood, and G. Marshall (1998). Effect of the monooxygenase inhibitor piperonyl butoxide on the herbicidal activity and metabolism of isoproturon in herbicide resistant and suceptible biotypes of *Phalaris minor* and wheat. *Pesticide Biochemistry and Physiology* 59: 143-153.

Singh, S., R.C. Kirkwood, and G. Marshall (1999). Biology and control of *Phalaris minor* Retz. (littleseed canarygrass) in wheat. *Crop Protection* 18: 1-16.

Smeda, R.J., R.S. Currie, and J.H. Rippee (2000). Fluazifop-P resistance expressed as a dominant trait in sorghum *(Sorghum bicolor). Weed Technology* 14: 397-401.

Sprague, C.L., E.W. Stoller, and L.M. Wax (1997). Response of an acetolactate synthase (ALS)-resistant biotype of *Amaranthus rudis* to selected ALS-inhibiting and alternative herbicides. *Weed Research* 37: 93-101.

Stephenson, G.R., M.D. Dykstra, R.D. McLaren, and A.S. Hamill (1990). Agronomic practices influencing triazine-resistant weed distribution in Ontario. *Weed Technology* 4: 199-207.

Sterling, T.M., I. Ray, A.D. Vallotton, and R.P. Sabba (2002). Recessive inheritance of picloram resistance in yellow starthistle (*Centaurea solstitialis* L.). *Weed Science Society of America Abstracts* 42: 11.

Talbert, R.E., L.A. Schmidt, J.S. Rutledge, E.E. Scherder, M.L. Lovelace, N.W. Buehring, and F.L. Baldwin (2000). Alternative herbicide programs for control of propanil-resistant *Echinochloa crus-galli* (L.) Beauv. in drill-seeded rice. In *Abstracts of the Third International Weed Science Congress, Foz do Iguassu, Brazil,* June 6-11 (p. 149). [CD-ROM.] Corvallis, OR: International Weed Science Society.

Thill, D.C. and D. Lemerle (2001). World wheat and herbicide resistance. In S.B. Powles and D.L. Shaner (Eds.), *Herbicide resistance and world grains* (pp. 165-194). New York: CRC Press.

Thill, D.C., J.T. O'Donovan, and C.A. Mallory-Smith (1994). Integrated weed management strategies for delaying herbicide resistance in wild oats. *Phytoprotection* 75(Supplement): 61-70.

Thomas, A.G. and J.Y. Leeson (1999). Persistence of volunteer wheat and canola using weed survey data. In W. Asbil and D. Cloutier (Eds.), *Expert Committee on Weeds: Proceedings of the 1999 National Meeting* (p. 88). Sainte-Anne-de-Bellevue, Quebec: Expert Committee on Weeds.

Valverde, B.E. (1996). Management of herbicide resistant weeds in Latin America: The case of propanil-resistant *Echinochloa colona* in rice. In H. Brown, G.W. Cussans, M.D. Devine, S.O. Duke, C. Fernandez-Quintanilla, A. Helweg, R.E. Labrada, M. Landes, P. Kudsk, and J.C. Streibig (Eds.), *Proceedings, Second International Weed Control Congress, Copenhagen, Denmark* (pp. 415-420). Flakkebjerg, Slagelse, Denmark: Dept. Weed Control and Pesticide Ecology.

Valverde, B.E., I. Garita, E. Vargas, L. Chaves, F. Ramirez, A.J. Fischer, and H. Pabon (1999). Anilofos as a synergist to propanil for controlling propanil-resistant junglerice, *Echinochloa colona*. *Weed Science Society of America Abstracts* 39: 159.

Valverde, B.E., C.R. Riches, and J.C. Caseley (2000). *Prevention and management of herbicide resistant weeds in rice: Experiences from Central America with* Echinochloa colona. San José, Costa Rica: Cámara de Insumos Agropecuarios.

VanGessel, M.J. (2001). Glyphosate-resistant horseweed from Delaware. *Weed Science* 49: 703-705.

Warwick, S.I., H.J. Beckie, and E. Small (1999). Transgenic crops: New weed problems for Canada? *Phytoprotection* 80: 71-84.

Werck-Reichhart, D., A. Hehn, and L. Didierjean (2000). Cytochromes P450 for engineering herbicide tolerance. *Trends in Plant Science* 5: 116-123.

Westra, P., S. Nissen, and D. Karam (2000). Risk factors that impact resistant weed selection. In *Abstracts of the Third International Weed Science Congress, Foz do Iguassu, Brazil,* June 6-11 (pp. 154-155). [CD-ROM.] Corvallis, OR: International Weed Science Society.

Willis, A.D., A.M. Mortimer, P.D. Putwain, and S.R. Moss (1997). Half-sib analysis of the genetic basis of graminicide resistance in blackgrass, *Alopecurus myosuroides* L. (Huds.) in the UK. In *Resistance '97: Integrated approach to combating resistance, abstracts*. Harpenden, Hertsfordshire, UK: IACR-Rothamsted.

Wrubel, R.P. and J. Gressel (1994). Are herbicide mixtures useful for delaying the rapid evolution of resistance? A case study. *Weed Technology* 8: 635-648.

Zand, E. and H.J. Beckie (2002). Competitive ability of hybrid and open-pollinated canola *(Brassica napus)* with wild oat *(Avena fatua)*. *Canadian Journal of Plant Science* 82: 473-480.

Chapter 21

Recent Advances in Parasitic Weed Research: An Overview

Jamal R. Qasem

PARASITIC SPECIES: AN INTRODUCTION

Parasitic flowering plants (parasitic weeds) are species unique and unusual in their growth habit and habitats. Because these are not in direct contact with soil and soil factors necessary for growth, development, and existence, these species, for almost the entire growth cycle, have evolved nutritional modes that are clearly different from normal green plants. They have been a peculiar subject since ancient times. Parasitic species are usually attached to other plant species of different genera and plant families (with some species exhibiting self-parasitism) through a haustorium—an absorptive organ developed upon attachment to host tissues. A haustorium (pl. haustoria) is a bridge of tissue connecting the host and parasite, usually a swollen mass consisting of both host and parasite tissue. This acts as a conduit for the flow of water and nutrients from host to parasite. It is a suckerlike structure varied in size, shape, and morphology for different parasitic species.

Parasitic habit is a widespread phenomenon among angiospermae, and at least five orders have parasitic species (Musselmann, 1982). It is reported only from dicotyledons. Kuijt (1969) reported 17 plant families having parasitic species, of which six were considered of great economic importance, including Scrophulariaceae, Orobanchaceae, Santalaceae, Cuscutaceae, Viscaceae, and Loranthaceae. At present, 20 plant families are known to contain parasitic plant species numbering between 3000 to 5000 (http://www.uni-hohenheim.de/~www380/parasite/general.htm). However, approximately 30 genera of parasitic angiosperms have been reported to negatively impact cultivated crop species, while only about 11 percent of all genera have members that could be considered pathogens. However, most of the damage to economic host crops is caused by *Cuscuta, Arceuthobium, Orobanche,* and *Striga.*

Parasitic species are greatly varied in their degree of dependence on host plants. Some are holoparasites, totally lack chlorophyll (or nearly so), are nonphotosynthetic, and obtain all their water and nutrients from host plants; most are root parasites, whereas others are hemiparasites with green pigments, engage in photosynthesis, and can produce their own food (at least during some portion of the life cycle), but in addition they parasitize other plant species from which they obtain partial or all of their food and/or water. Some species of *Cuscuta* (e.g., greater dodder, *C. europaea* L.) are stem holoparasites that have lost RUBISCO, thylakoids, chlorophyll, and light-dependent CO_2 fixation (Machado and Zetsche, 1990). Hemiparasites can be further divided into two groups, facultative and obligate, depending upon their degree of dependence on host species. Facultative hemiparasites do not require a host to complete their life cycle and are photosynthetic (e.g., *Osyris alba* L.); when found with host roots, they invariably form haustorial connections. Upon attachment to host roots, these parasites extract water and dissolved minerals. They can be found in several root-parasitic families such as Olacaceae, Opiliaceae, Santalaceae, Krameriaceae, and Scrophulariaceae (Nickrent, 2002). However, for almost the entire life cycle, all parasitic species appear to lack a normal root system which was replaced by an absorptive organ or structure that provides physical support to the germinated seedlings, or they form an absorptive haustorium attached to the host tissues. On the other hand, certain parasitic species subsist on the root systems of host plants, while others are stem or shoot parasites.

Parasitic species are different in their seed production, longevity, dispersability, and dispersing agents. Birds and animals are the main dispersing agents for *Loranthus, Osyris,* and *Viscum,* but wind, seed contamination, water, farm machinery, and probably other means are important for *Orobanche, Striga, Cuscuta* (except wind), *Cistanche,* and *Thesium* seed dissemination. Seed germination of certain parasites requires a preconditioning period and host stimulant(s) before start *(Orobanche, Striga, Cistanche).* Others do not require for such stimulant(s) (*Cuscuta, Thesium, Osyris, Viscum,* and *Loranthus*) but appear to have a limited and specific host range and are not able to attack other plant species, which may reflect host preference/specificity or certain host mechanical, physiological, or chemical resistance of the nonhost species. The mechanism behind failure of these parasites from attacking any plant species is unknown and may deserve attention. Differences in parasite requirements for host stimulants have created a wide range of host species. Certain parasitic species are mainly confined to specific host growth forms and appear to be highly specialized to woody shrubs, fruit trees/shrubs, or forestry species *(Loranthus, Viscum, Osyris, Cistanche),* others are restricted to herbaceous and field crops *(Thesium, Striga),* while a third group can attack both woody and herbaceous species (certain species of *Orobanche* and *Cuscuta*). Differences among species of the same genera in their host ranges have been already noted (Parker and Riches, 1993).

Differences in the attachment of different parasitic species and their abilities to resume growth after being mechanically removed from host plants have been also noted. Certain species such as *Orobanche, Cistanche, Cynomorium,* and *Thesium* are fragile, weakly connected to host tissues, and unable to revegetate after being detached from host plants; others such as *Cuscuta, Viscum, Osyris,* and *Loranthus* are difficult to withdraw from the host tissues, and some can revegetate and resume growth from suckers or haustoria imbeded inside tissues. This may imply differences in reproduction methods and life cycle of different parasitic species. However, all are generally referred to as annuals and reproduce by seeds, although certain species are able to reproduce vegetatively from the over-wintered haustoria inserted in host tissues (certain *Cuscuta* species and *Loranthus*) and therefore behave as perennials. However, all parasitic plant species have evolved under the constraint that they do not kill their hosts prior to successful reproduction.

Differences among parasitic species in their seed production and germination, attachment to host plants, morphology and growth habit, and physiological and anatomical relationships between these and their hosts make their control a real challenge. Selective chemical control using herbicides is far from successful, difficult to achieve in most cases, and depends on differences in tolerance between parasites and host plants, while effective rates of herbicides may be injurious to the parasite as well as to the host plant. In addition, certain parasitic species are not visible before emergence (root parasites), the stage at which most damage to host plant has already occurred. However, different methods of control are recommended and employed against many parasitic species, while some other promising results have been also reported.

OROBANCHE *SPECIES (BROOMRAPES)*

Orobanchaceae is primarily of the northern warm and temperate zones family. About 10 percent of the species occur in the tropics, while only one reaches the arctic (Eltayeb et al., 2000). It includes 15 genera and more than 200 species (Kabulov, 1978). Among the genera *Orobanche* is the most important and destructive parasite. The genus includes more than 160 angiosperm parasitic species (Oliveira Velloso, 1990), the majority of which (about 90 percent) are old world natives. However, five (branched broomrape, *O. ramosa* L., *O. aegyptiaca* Pers.; nodding broomrape, *O. cernua*

Loefl.; bean broomrape, *O. crenata* Forssk.; and small broomrape, *O. minor* Sm.) are considered the most problematic species to many of the economic crops of temperate and semiarid regions in the world.

Orobanche species are thermophilic and often highly demanding in their habitat preferences. These often require dry, open, nutrient-poor soils, where they are vulnerable to agricultural and scrub development or other types of habitat loss (http://www-aes.tamu.edu/mary/brmrape/brmrape.htm). Some also appear sensitive to minor climatic changes and may disappear from sites that still appear suitable. Many are extremely local throughout their entire geographical ranges. *Orobanche* spp. are flowering holoparasites devoid of chlorophyll and thus are not able to produce their own food, and their leaves are vestigial. Aboveground stems are produced only for the purpose of flowering and seed production. They attach themselves to the roots of broadleaf plants from which they draw nutrients and water. They complete their life cycle within a year and are considered to be annuals in duration. *Orobanche* species are prolific seed producers; seed production varied per plant and among species in number of seeds produced and seed longivity in the soil. Seeds are quite small, measuring 0.2 by 0.3 mm, with a minute amount of reserved food. Some species bear seeds in huge numbers and of near-microscopic size (e.g., *O. minor*). In addition, 1,000 seeds of knapweed broomrape (*O. elatior* Sut.) weigh 0.0049 g, and one plant produces 270,000 seeds. Picris broomrape (*O. picridis* F. Schultz.) produces from 94,000 to 116,000 seeds per plant, and 1,000 seeds weigh 0.0029 g. Seeds are widely disseminated by wind and, though minute, are very long-lived, having been known to survive in the soil for 20 years, and germinating in succession but under low moisture level, dormancy may be extended up to 30 years. *Orobanche* species are also different in branching habit and in presence or absence of bracteoles, characters that distiguish species and are regarded as useful tools in identification. While *O. ramosa* and *O. aegyptiaca* are branched and bracteoles, *O. cernua, O. minor,* and *O. crenata* are usually unbranched and lack bracteoles (see Photo 21.1). However, differences among species of each group in stature, flower size and color, and host range are evident (Parker and Riches, 1993). Different species also differ in distribution, parasite forms or variations, and losses caused to crop plants.

Germination and Development

Seeds of *Orobanche* are dormant and in general require a period of after ripening in warm, dry storage, followed by conditioning in a warm, moist environment, before they will respond to germination stimulants. Seeds of

PHOTO 21.1. Different *Orobanche* spp. (*O. cumana, O. crenata, O. cernua, O. aegyptiaca, O. ramosa, O. minor, O. lutea, O. gracilis,* and *O. foetida* (from left to right and top to bottom). (*Source:* Adapted from Linke, Sauerborn, and Saxena, 1996. Used by permission of International Center for Agricultural Research in the Dry Areas.) (See corresponding plate in central color section.)

certain species (e.g., *O. crenata*) in particular must undergo after ripening dormancy which may extend from 18 to 24 months of storage. However, the length and effectiveness of the after-ripening period may vary with temperature, humidity, and other environmental factors. The nature of the changes occurring in the seeds during after ripening are unknown, though various structural and metabolic changes in the seedcoat, embryo, and respiratory substrate are involved. Mature seeds require a conditioning period between one and three weeks, during which seedcoat permeability may increase, water absorption occurs, and/or changes in the levels or activities of endogenous germination promoters or inhibitors may occur. The duration of conditioning is dependent on certain factors including temperature, origin, and age of the seeds. The period of time necessary for conditioning ranges from several days to several weeks.

Conditioning temperature varies according to species but ranges between 18 and 25°C. Fluctuation between conditioning and germination temperature enhances germination. However, seeds germinate when in contact with host roots, triggered by chemical recognition; thus seeds should receive germination stimulants from root exudate of host species, while host roots should be within a few milimeters of germinated seeds. In general, germination occurs in the immediate vicinity of the host root. It appears that the chemical stimulant secreted by the host root system can cause germination up to a distance of 1 cm from the root, but only seeds within 2 to 3 mm of the root can actually infest the host. Germination in the field starts five to seven days after exposure to the host root exudate. Temperature between 20 and 25°C is optimal for germination, wheares soil alkalinity tends to inactivate germination stimulants, thus reducing infestation. Conflicting reports on the effect of light during conditioning or germination phases of *Orobanche* seeds are available. Light enhanced *O. ramosa* seed germination but inhibited germination of *O. crenata*. On the other hand, changes in pH or presence of calcium or phosphate ions that modify the effect of pH had a significant effect on germination stimulants, while inactivation of germination stimulants in the soil was due to microorganisms, and the effect varied with soil type. Exudate concentration is another factor affecting germination, with higher concentration reducing germination due to components rather than stimulants. It seems that plant exudates contain more than one stimulant, and some trigger particular species to germinate but not others. This explains the variation in host specificity of different *Orobanche* species and differences in the degree of specificity shown by seeds of the same species but of different origin. Different materials were found to enhance seed germination of *Orobanche* species which include nonhost exudates of flax, maize, and cotton. In addition, ethylene gas is reported to enhance seed germination of *O. ramosa,* but conflicting reports have already been released on the effect of this gas. Among other chemicals reported are gibberellic acid, strigols, and calcium hypochlorite. In contrast, phosphate fertilizer decreases *Orobanche* incidence.

Contact and Attachment

Germination of *Orobanche* seeds is assured by the extension of a radicle (germination tube) that is capable of growing for 1 to 2 mm before the seeds' stored energy is depleted. Parasite seedlings must find a suitable host plant directly after germination. Radicle movement toward the host root system is by chemotropic response, or by bending movement in response to inhibitory chemical substances in the root exudates of host plants. This germination feature is probably highly evolved and acts to enhance seedling survival. However, for successful contact between the parasite and host the radicle must find the host root, followed by development of the haustoria, and finally penetration into the host root through the epidermis and cortex or connection between the vascular systems of the parasite and the host must be developed. Upon contact, sticky papillae are formed which may have resulted from adherance to the root surface. The symbiotic relationship between microorganisms and host root system may enhace the penetration of *Orobanche* haustoria via rhizobium nodules on the roots of faba bean (*Vicia faba* L.). The primary haustoria and the tubercles are the major connecting organs (see Photo 21.2).

Once the connection is established between the vascular systems of the host and the parasite, the latter can start growth and development. Penetration of the developing haustoria into host root tissues mainly occurs through enzymatic action. When the attachment occurs, the germination tube withdraws the active meristem of the embryo from the seed to the root surface of host plant and the radicle does not act anymore. Development of the haustoria takes place and the haustoria swells, forming tubercles or nodules, establishing connections with the vascular system of host plants (Dorr and Kollmann, 1975, 1976). An underground tuber develops, from which, eventually, a shoot bud may differentiate or develop, elongate, and emerge to produce the flowering shoot. However, the *Orobanche* life cycle consists

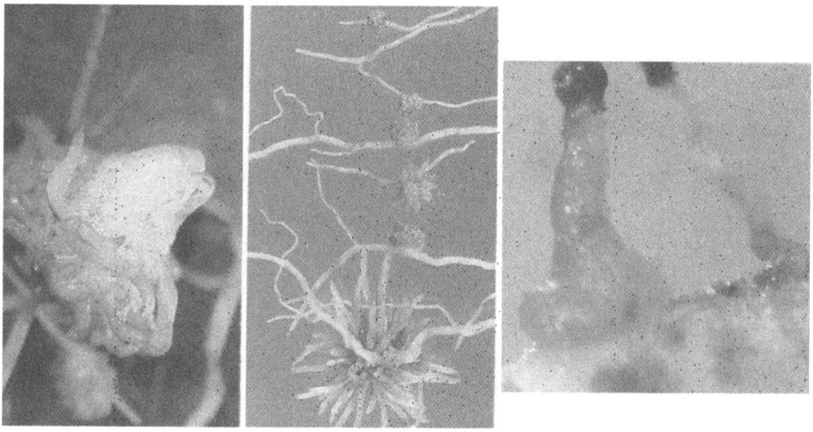

PHOTO 21.2. *Orobanche* haustoria at different developmental stages. (Used by permission of International Center for Agricultural Research in the Dry Areas, Aleppo, Syria.) (See corresponding plate in central color section.)

of two phases. The first is the preconditioning phase during which seed embryo conditioning occurs under suitable temperature and moisture. This phase may range from several days to a few weeks depending on the species and is considered important for leaching of germination inhibitors from *Orobanche* seeds (Edwards, 1972), while other researchers suggested that germination stimulants or related substances are accumulating during this period (Jain and Foy, 1992; Ibrahim et al., 2000). The second phase is the attachment of the seedlings to the host and the start of true parasitism. The latter phase can be further divided into two stages: the *hypogeal* stage lasting for 60 days (Mukumov, 1974) during which the vegetative subterranean organ develop and food materials, and the *epigeal* stage during which the vegetative organ grows rapidly and produces reproductive organs above the ground. Emergence, flowering, and seeding of the parasite may last for only 15 to 18 days.

Biology, Ecology, and Physiology

Orobanche species depend completely on their hosts. Trends in metabolic reduction and dependence of parasitic angiosperms on their hosts for acquisition of resources is reflected in their reduced ability to incorporate inorganic carbon and nitrogen. These parasites lack nitrate reductase, and therefore their ability to utilize ammonium-N depends on the availability of glutamine synthase. *Orobanche* species also depend on hosts for their water supply since they lack normal root systems, while water flow from host to parasite is enhanced by the high osmotic pressure that can be attributed to transformation of disaccharides to monosaccharides by invertase enzyme. The parasite tends to have a high level of mannitol, especially under moisture stress, which may provide a suitable solute that protects enzymes from denaturation (Parker and Riches, 1993). *Orobanche* spp. accumulate high amounts of mannitol, whereas hexitol is absent from the major host crops (Simier et al., 2001), and some inhibitors of the mannose-6-phosphate reductase (M6PR, EC 1.1.1.224), the key enzyme of mannitol synthase, have recently been identified (Robert et al., 1999). In addition, the parasite is rich in starch and can grow slowly for several weeks after the death of its host.

Orobanche affects its host without killing it. It can reduce P, K, RNA, protein, and total carbohydrate levels and glutamic acid but increases Ca, Na, Mg, and DNA concentration of host plants. Reduction in carbohydrate level may seriously affect the host's ability to absorb water under moisture stress conditions and can lead to stomata closure. Intake of water and minerals from the host causes visible damage, wilting, and reduction in plant size, and decreases yield and lowers quality. With legumes, the most intensive growth of *Orobanche* usually coincides with flower and pod development of the host. This results in flower dropping and low podding due to increased competition from the parasitic sink. The wilting symptoms occur due to reduced water uptake because of a deficiency of energy in the host roots, as a result of the flow of carbohydrates to the parasite.

With root parasites, damage is already done before the parasites emerge. More than 80 percent of *Orobanche* dry matter is already deposited before flowering. Damage is greatest where water and nutrients are limited, which occurs in most of the regions where parasites are present. The initiation of *Orobanche* parasitism and parasite growth and development are affected by different environmental factors and the development stage of host plants at the time of infestation. These factors are discussed next.

Temperature

This factor affects parasites at conditioning and germination stages as well as during growth and development.The effects of temperature on germination of different *Orobanche* species indicate differences in their optimum temperature for germination and development which may be due to differences in their geographical distribution. The lower limit of temperature is found to be 8ºC, optimum is about 20ºC or slightly higher, and the highest is around 30ºC. However, some intraspecific differences exist; e.g., *O. aegyptiaca* and *O. crenata* require higher temperature at all levels. For *O. crenata* optimum temperature for germination is lower (15/5°C) than optimum temperature (20/10°C) for attachment. However, germination of *Orobanche* can also occur at temperatures as low as 5°C but is very much delayed. For *O. crenata* and *O. cumana* attachment is observed at this low temperature, but this is not true for *O. aegyptiaca*. The lower the temperature, the slower the *Orobanche* development. At 30/20°C a cardinal temperature is reached where development is also delayed. *O. crenata* prefers to germinate at an alternating temperature of 15/5°C, while the other two species germinate in a broader range of temperature (10/5 to 25/15°C). *Orobanche crenata* mainly attacks winter crops, while *O. aepyptiaca* parasitize summer and winter species and sunflower broomrape (*O. cumana* Wallr.) prefers summer crops. Poor germination at low and high temperatures may be due to the lack of host root exudate, inactive stimulants, or failure of seeds to germinate under these conditions.

The lower and upper limits of germination temperature may explain some field observations. The time of germination, attachment, and further development of *O. crenata* were found to be related to soil temperature and not to broadbean growth stage (Manschadi, Sauerborn, and Stutzel, 2001). In the Mediterranean region, *Orobanche* species are weeds of winter crops. In areas where midwinter temperatures are below the limits of germination temperature, *Orobanche* germination may be reduced or completely delayed until the following spring. Therefore, low temperature decreases the number of germinating parasites and also the effect on the host because the position in the host root system is also affected. Reduction in germination at high temperatures may also play a role in this context. Seeds of *O. ramosa* can tolerate temperature up to 50°C for 35 days while dry, but wet seeds were killed at 40°C within a short time. Exposure to 30°C in 100 ppm GA solution appeared to be more damaging than exposure to 35°C in water (Drennan and Mohamed-Ahmed, 1992).

Light

Light intensity has a significant influence on infestation of a host by *Orobanche* spp. Reduction in intensity of light reduced *O. crenata* and *O. aegyptiaca* infestation to broadbean. However, day length or host flowering appeared to have no direct effect on the flowering of *Orobanche* plants. In vitro studies on the effect of light on germination of *Orobanche* species showed an indirect intraspecific variation. Certain species seem to be indifferent whereas others were completely inhibited. Factors other than light also differ and affect *Orobanche* infestation. Low light intensities often coincide with high air humidity. Under humid conditions transpiration rate of the host plant is lower, resulting in decreased uptake of water and nutrients to the shoots. Mohamed-Ahmed and Drennan (1994) reported that a 20 h photoperiod resulted in earlier emergence of *O. ramosa,* and plants of broadbean, lentil, or peas were larger compared with plants grown under a 10 h photoperiod.

Moisture

Moisture is another environmental factor affecting *Orobanche* infestation and prevalence. In well-drained and well-textured soils, most of the root parasites find optimal conditions for germination, while in wet soils, germination is poor because of a lack of oxygen for respiration. Germination stimulants may be diluted or washed off under these conditions. However, *Orobanche* is adapted to dry conditions. Germination can take place even at an osmotic potential of –1.29 MPa. Normal angiosperm species are able to germinate up to an osmotic potential of only about –0.4 to 0.5 Mpa. Although certain *Orobanche* species appeared under both rainfed and irrigation conditions, differences also exist between different species. *Orobanche cumana* and *O. minor* are mostly associated with rainfed conditions, *O. ramosa* and *O. aegyptiaca* are common in irrigated fields but not exclusively so, while *O. crenata* is found higher under sufficient than under limiting water supply conditions (Manschadi, Sauerborn, and Stutzel, 2001). In addition, *Orobanche* spp. may be favored by low relative humidity (Parker and Riches, 1993).

Flooding

Sometimes flooding reduces *Orobanche* infestation in certain crops such as rice. It may be due to either disappearance of black pigments of *Orobanche* seed coat or the germination stimulants are dissolved and diluted and then inactivated. Flooding may also have an indirect effect by reducing oxygen supply which may cause seed dormancy or reduce attachment and establishment of the seedlings. Light also affects growth phase since the parasite cannot complete photosynthesis and produce oxygen necessary for respiration. Respiration is high in the haustorial region of the host root system, and if oxygen supply via the host is limited, flooding will be fatal for *Orobanche,* especially during the hypogeal phase. The *Orobanche* problem is less in moist soils and indeed in most areas around the Mediterranean where *O. crenata* on broadbean is a real problem under rainfed conditions. Mohamed-Ahmed and Drennan (1994) reported that waterlogging is an effective cultural practice for *O. ramosa* control. They found that the number of emerged *Orobanche* shoots decreased with longer waterlogging period. Studies on the optimal use of water and its level indicate that for broadbean during pod filling and good yield, soil moisture is essential, but the number of attachments was higher at high moisture levels.

Nutrients

In general, *Orobanche* species prefer poor soils. Parasitism was found to occur more readily in potting media low in fertility (Jain and Foy, 1992). These conditions are considered to be an important factor associated with severe infestation in fields (Gharib, 1973). Animal manure and fertilizers reduce *Orobanche* problems and increase crop yield (Pesch and Pieterse, 1982; Agabawi and Younis, 1985). It was found that ammonium-N and urea-N in vitro had negative effects on germination and growth of *O. crenata* (Pieterse, 1991a). However, nitrate-N had no such effect and the effect was absent under soil conditions tested (Jain and Foy, 1992). The effect of superphosphate on *O. ramosa* germination in soil with *Carum ajowan* as phosphorus stimulated germination but hampered growth of procaulome and thus greatly reduced *Orobanche* attachment to host plants. The effect of fertilizer may be indirectly due to reduction in the host root system, while other mechanisms may be active as well. There might be a reduction in root exudates of host plants or better defense of the host roots under a suitable supply of minerals. It is sugested that urea-N influences two physiological processes (Pieterse, 1991b). Parasite penetration into broadbean roots was stimulated by the presence of rhizobium which may be the reason for the low number of *Orobanche* found at high nitrogen levels. The presence of bacteria is necessary for germination of different *Orobanche* species. Rhizobium and mycorrhiza are essential symbionts for Leguminosae and other species. Germination seems to be best in slightly acid soils, which may be due to the interaction with germination stimulants.

The effect of organic (manure and compost) and inorganic fertilizer on *Orobanche* infestation has long been reported (Popova, 1942) and the idea is not new. The primary fertilization effects are due to nitrogen, especially to ammonium sulfate and urea, while the effect of nitrate was found to be small. However, recent studies on the role of chemical fertilizers are in conflict and show that extremely high rates of fertilizers (e.g., ammonium sulfate) should be applied to achieve an even level of control, suggesting that fertilizers affect parasites by dehydrating the germination tube in a plasmolysis process. Westwood and Foy (1999) concluded that ammonium-N was more inhibitory than a nitrate form of nitrogen, while the elongation of the seedling radicle is primarily inhibited by ammonium rather than seed germination, and the effect was varied between species. One proposed mechanism of action of nitrogen may be due to direct interaction with parasite metabolism and the alteration in osmotic potential (Ernst, 1986). However, other researchers reported that using fertilizer as a control method revealed inconsistent results with respect to parasite growth, with nitrogen treatment often having no descernible effect (Smaling, Stein, and Sloot, 1991) or even increasing parasite growth (Pieterse and Verkleij, 1991). Osmotic stress has been suggested as another possible reason for inhibition of *Orobanche* germination by nitrogen (Ernst, 1986), but no studies were carried out to correlate this factor to nitrogen inhibition of *Orobanche* germination.

Soil Type

It has been indicated that *Orobanche* species favor alkaline soils in Mediterranean regions (e.g., *O. minor* and *O. ramosa*). However, other contradictory reports indicate that both *O. minor* and *O. ramosa* were in worse conditions on alkaline soils. In addition, no strict link was found between prevalence of *Orobanche* species and high pH, while poor soils are considered the most infested.

Salinity

Few reports showed the effect of salinity on prevalence of *Orobanche* in certain saline regions. In a recent study, Al-Khateeb (2002) reported that NaCl reduced *O. cernua* germination when seeds were exposed during the conditioning period. NaCl solution (75 mM conc.) added to pot soil inoculated with *Orobanche* seeds and grown with tomato completely prevented parasite infestation. Results of this work may support the idea that fertilizers act against *Orobanche* by affecting or killing the highly sensitive germination tube in a physical manner rather than any other mechanism.

Insects and Pathogens

Orobanche species are attacked by the insect *Phytomyza orobanchia* (Diptera, Agromyzidae) (Hassanein et al., 1998). The insect damages the stalk of the parasite and attacks seed capsules (Jafarzadeh and Pourmirza, 1999) causing great reduction in seed production of the parasite, ranging between 29 and 94 percent depending on *Orobanche* species and location (Tsybul'skaya and Skoklyuk, 1978). However, this monophagous insect is influenced by different factors including the presence of its natural enemies in the field and the use of insecticides which may limit its efficiency. In addition, other insects have also been reported to attack *Orobanche* plants including *Psila* species (at larval stage) which attack stem of *O. minor* and *Chyliza extenuata*. Other insects that parasitize *Orobanche* spp. are polyphagous and often damaging to the crop. Various insects that attack *Orobanche* spp. are listed in Table 21.1.

On the other hand, 62 fungi species are reported to attack different *Orobanche* species (including *O. aegyptiaca, O. cernua, O. crenata,* and *O. ramosa*) (Thomas, Heller, et al., 1999) (see Table 21.2). However, only four genera proved to be highly effective and selective to *Orobanche*. These are *Ulocladium* species, *Fusarium* species (*F. oxysporum* var. *orthoceras, F. orobanche*), and *Alternaria* sp., while *Rhizoctonia solani* was found to be effective against *O. ramosa* in tomato and reduced its infestation by 80 percent. *Ulocladium atrum* is effective as a biocontrol agent for *O. crenata* at high humidity.

TABLE 21.1. Phytophagous insects of *Orobanche* spp.

Insect species	*Orobanche* spp.	Damaged plant parts	Reference(s)
Acrotylus humbertianus	*O. cernua*	Capsules	Manjunath and Nagarkatti, 1977
Agriotes sp.	*O. cernua*	Shoots	Mihajlovic, 1986; Lekic, 1970a,b
Agrotis sp.	*O. crenata*		Martelli, 1933a,b
Agrotis ypsilon	*O. cernua*	Shoot	Manjunath and Nagarkatti, 1977
Agryrotaenia pulchella	*O. cernua*	Larvae in flowers and capsules	Mihajlovic, 1986
Aiolopus simulatrix	*O. cernua*	Capsules	Manjunath and Nagarkatti, 1977
Aphis fabae	*O. cernua*	Sucking	Mihajlovic, 1986; Lekic, 1970a,b
Bagous sumatrensis	*O. cernua*	Seed capsules	Manjunath and Nagarkatti, 1977
Carpophilus mutilatus	*O. cernua*	Seed capsules	Manjunath and Nagarkatti, 1977
Cetonia sp.	*O. crenata*	—	Martelli, 1933a,b
Chortophila sp.	*O. crenata*	—	Martelli, 1933a,b
Chyliza atriseta	*O. crenata*	—	Martelli, 1933a,b
Chyliza extenuata	*O. crenata*	Shoots	Grandi, 1951b
Dysmicoccus brevipes	*O. cernua*	Sucking	Manjunath and Nagarkatti, 1977
Epicometis sp.	*O. crenata*	—	ICARDA, 1986
Eumerus sp.	*O. aegyptiaca*	—	Al-Khazraji and Abdel Wahid, 1987
Frankliniella paucispinosa	*O. cernua*	Flowers	Manjunath and Nagarkatti, 1977
Gonocephalum dorsogranosum	*O. cernua*	—	Manjunath and Nagarkatti, 1977
Heliothis armigera	*O. cernua*	Shoot and capsules	Manjunath and Nagarkatti, 1977
Luperomorpha bombayensis	*O. cernua*	Seed capsules	Manjunath and Nagarkatti, 1977
Lygus pratensis	*O. cumana*	—	Mihajlovic, 1986; Lekic, 1970a,b
Meloe majalis	*O. crenata*	Shoots, flowers, seed capsules	Klein, 1995
Mesomorphus striolatus	*O. cernua*	Superficial	Manjunath and Nagarkatti, 1977
Microtermes sp.	*O. cernua*	Shoot	Manjunath and Nagarkatti, 1977

Morphacris fasciata	*O. cernua*	Capsules	Manjunath and Nagarkatti, 1977
Mylabris olea	*O. crenata*	Shoots, flowers, seed capsules	Klein, 1995
Myllocerus blandus	*O. cernua*	Seed capsules	Manjunath and Nagarkatti, 1977
Myzus persicae	*O. cernua*	Sucking	Manjunath and Nagarkatti, 1977
Neocleonus sannio	*O. cernua*	Seed capsules	Manjunath and Nagarkatti, 1977
Nesidiocoris tenuis	*O. cernua*	Sucking	Manjunath and Nagarkatti, 1977
No specification	*O. cernua*	Larvae in shoots	Manjunath and Nagarkatti, 1977
Opatroides punctulatus	*O. cernua*	Superficial	Manjunath and Nagarkatti, 1977
Oxythyrea funesta	*O. crenata*	—	Martelli, 1933a,b
Pachycera buprestoides	*O. cernua*	Superficial	Manjunath and Nagarkatti, 1977
Pachychila salzmanni	*O. crenata*	Young shoots	Klein, 1995
Pentodon punctatus	*O. crenata*	—	Martelli, 1933a,b
Phytomyza orobanchia	*Orobanche* spp.	Shoots and capsules	Klein, Kroschel, and Sauerborn, 1999
Plusia sp.	*O. crenata*	—	Martelli, 1933a,b
Potosia sp.	*O. crenata*	—	Martelli, 1933a,b
Psila sp.	*O. minor*	Larvae	Parker and Riches, 1993
Scotia segetum	*O. cernua*	Larvae	Mihajlovic, 1986
Scotia segetum	*O. cernua*	Larvae	Lekic, 1970a,b
Sipha maidis	*O. crenata*	—	Grandi, 1951a
Siphonella sulcicollis	*O. crenata*	Larvae in shoots	Martelli, 1933a,b; Klein, 1995
Smycronyx cyaneus	*O. crenata*	Seed capsules	Martelli, 1933 a,b; Zermane et al., 1999
Smycronyx cyaneus var. *opacus*	*O. crenata*	Seed capsules	Klein, own observation, 1996
Spodoptera litura	*O. cernua*	Shoot and capsules	Manjunath and Nagarkatti, 1977
Trilophidia annulata	*O. cernua*	Capsules	Manjunath and Nagarkatti, 1977
Tropinota hirta	*O. crenata*	—	Martelli, 1933a,b
Tropinota squalida	*O. crenata*	Flowers	Klein, 1995
Truxalis indica	*O. cernua*	Capsules	Manjunath and Nagarkatti, 1977

Source: Adapted from Klein and Kroschel, 2002.

TABLE 21.2. Major fungi reported as natural enemies to *Orobanche* species.

***Orobanche* species**	**Pathogen**	**References**
O. alba	*Alternaria infectoria*, *A. tenuis*, *Hyphoderma roseum*	Agricultural Research Service, 2000
O. aegyptiaca	*Acremonium fusidioides*, *Alternaria alternata*, *Cladosporium cladosporioides*	Thomas, Heller, et al., 1999
	Colletotriuchum lagenarium	Agricultural Research Service, 2000
	Epicoccum nigrum	Thomas, Sauerborn, et al., 1999
	Fusarium oxysporum	Agricultural Research Service, 2000
	Fusarium oxysporum f.sp. *orthoceras*	Bedi, 1994
	F. solani	Agricultural Research Service, 2000
	Mortierea alpina, *Papulaspora* sp., *Phoma* spp., *Rhizoctonia* sp., *Sordaria fimicola*, *Trichoderma* spp., *T. roseum*	Thomas, Sauerborn, et al., 1999
O. cernua	*Cladosporium herbarum*	Mazaheri et al., 1991
	Fusarium oxysporum, *F. oxysporum* f.sp. *orthoceras*, *F. sambucinum*	Agricultural Research Service, 2000
	Monilia humicola	Thomas et al., 1998
	Sclerotium rolfsii	Raju, 2001
O. crenata	*Alternaria* sp.	Agricultural Research Service, 2000
	Alternaria alternata	Abdel-Kader et al., 1998
	A. infectoria, *Aspergillus niger*, *Azotobacter chroococcum*, *Cladosporium herbarum*, *Cochliobolus spicifer*, *Cylindrocarpon* sp., *Escherichia coli*, *Fusarium compactum*, *F. equiseti*, *F. oxysporum*, *F. solani*	Agricultural Research Service, 2000
	Fusarium spp.	Abdel-Kader et al., 1998

	Mucor odoratus, *Penicillium cyclopodium*	Khalaf, Ali, et al., 1997; Khalaf, Youssef, et al., 1997
	Phomopsis sp., *Sclerotinia* sp.	Agricultural Research Service, 2000
	Trichoderma sp.	Abdel-Kader et al., 1998
	Ulocladium atrum	Agricultural Research Service, 2000
	U. botrytis	Linke et al., 1992
	U. consortial	Abdel-Kader et al., 1998
O. cumana	*Cladosporium herbarum*, *Fusarium oxysporum*	Agricultural Research Service, 2000
	F. oxysporum f.sp. *orthoceras*	Bedi, 1992,1994; Thomas et al., 1998; Bedi and Sauerborn, 1999; Shabana and Sauerborn, 2001
	F. sambucinum	Bedi and Donchev, 1991, 1995
O. minor	*Alternaria* sp.	Linke et al., 1992
	Fusarium sp., *Ulocladium atrum*	Agricultural Research Service, 2000
O. muteli	*Cladosporium fasciculatum*	Agricultural Research Service, 2000
	Fusarium gibbosum, *F. lateritium*, *Rhizopus mocrosporus*, *R. nigricans*, *Verticillium microsporum*	Agricultural Research Service, 2000
O. nana	*Cladosporium fasciculatum*	Agricultural Research Service, 2000
O. orientalis	*Aspergillus glaucus*, *A. versicolor*	Agricultural Research Service, 2000
O. ramosa	*Fusarium lateritium*	Bozoukov and Kouzmanova,1993
	F. oxysporum, *F. oxysporum* f.sp. *orthoceras*	Agricultural Research Service, 2000
	F. solani	Gold, Sagen, and Wilhelm,1979
	Gliocladium varians	Bedi, 1994
	Rhizoctonia solani	Agricultural Research Service, 2000
O. reticulata	*Conoisporium arundinis*, *Fusarium lateritium*, *F. moniliforme*	Agricultural Research Service, 2000

TABLE 21.2 *(continued)*

Orobanche species	Pathogen	References
O. ramosa, O. cernua, O. crenata, O. aegyptiaca	*Alternaria alternata, Chetomium* sp., *Dendrophora* sp., *Fusarium* sp.	Hameed and Al-Shyab, 2001
	Fusarium gibbosum, F. sambucinum, F. orobanche, F. oxysporum f.sp. *orthoceras*	Barloy and Pelhate, 1962; Talsakh'yan and Grigoryan, 1978; Murasheva, 1995; Raju et al., 1995
	Rhizoctonia sp.	Hameed and Al-Shyab, 2001
	Sclerotium rolfsii, Verticillium microsporum	Barloy and Palhate, 1962; Talsakh'yan and Grigoryan, 1978; Murasheva, 1995; Raju et al., 1995

Distribution and Host Range

Most of the Orobanchaceae belong to the northern hemisphere, and most species occur in the temperate zone. Several hundred thousand hectares are infested with *Orobanche* from Europe to the Middle East, Russia and China, as well as Cuba and California (Parker and Riches, 1993). Areas of origin can be found in Turkey, Italy, Spain, and Morocco, and *Orobanche* species are distributed from the Atlantic up to China *(O. cernua),* Cape Verde through Europe and North Africa up to Kaukasus and Mozambique *(O. minor),* south of the equator to east Africa and New Zealand *(O. minor),* west Australia *(O. cernua),* and to South Africa *(O. muteli).* Natural occurrence of *Orobanche* coincides with the vegetation zone which is characterized by sclerophyllous woodland with winter rains, natural savanna or grassland, and areas with 250 to 500 mm precipitation. *Orobanche* species are presently reported from more than 58 countries.

Orobanche spp. are essentially restricted to dicot hosts. Most species are highly host specific, sometimes restricted to a single species or genus. Others are capable of parasitizing a number of unrelated plants but usually show strong preferences in certain areas. However, there are several species of *Orobanche,* each of which has a wide host range. They are most damaging to the members of Asteraceae, Fabaceae, Lamiaceae, Brassicaceae, Umbellifereae, and Cucurbitaceae, thus they directly impact food destined for human consumption (Jain and Foy, 1992; Parker and Riches, 1993; Diaz and Normabuena, 2001). The host range of certain important *Orobanche* species is summarized in Table 21.3.

TABLE 21.3. Hosts of different *Orobanche* species.

***Orobanche* species**	**Host species**	**Reference**
Orobanche spp.	*Brassica juncea, Eruca sativa*	Bedi, Kapur, and Mohan, 1997
O. aegyptiaca	*Amoracia lapathifolia, Angelica archangelica, Anthriscus cerefolium*	Foy, Jain, and Jacobsohn, 1989
	Apium graveolens	Americanos, 1991
	Arachis hypogaea	Foy, Jain, and Jacobsohn, 1989; Parker and Riches, 1993
	Aster spp.	Foy, Jain, and Jacobsohn, 1989
	Brassica sp.	Parker and Riches, 1993
	B. napus	Nandula, Foy, and Orcutt, 1999
	B. oleracea var. *botrytis*	Musselman, 1993; Americanos and Vouzounis, 1995; Korashi et al., 1996
	B. oleracea var. *capitata*	Qasem (unpublished data)
	B. rapa	Foy, Jain, and Jacobsohn, 1989
	Cannabis sativa	Musselman, 1993
	Capsicum annuum	Hershenhorn et al., 1996
	C. frutescens	Foy, Jain, and Jacobsohn, 1989; Parker and Riches, 1993
	Carthamus tinctorius	Foy, Jain, and Jacobsohn, 1989
	Cicer arietinum	Foy, Jain, and Jacobsohn, 1989; Parker and Riches, 1993
	Citrullus vulgaris	Parker and Riches, 1993
	Cucumis melo	Musselman, 1993
	C. sativus	Foy, Jain, and Jacobsohn, 1989; Parker and Riches, 1993
	Cucurbita pepo	Kabulov and Kabulova, 1978; Parker and Riches, 1993
	Daucus carota	Jacobsohn and Kelman, 1980; Musselman, 1993
	Ephedra sp.	Nickrent, 2002; University of Hohenheim, 2002; Texas Cooperative Extension, Agricultural and Environmental Safety, n.d. <http://www.aes.tamu.edu/mary/brmrape/brmrape.htm>
	Foeniculum vulgare	Parker and Riches, 1993
	Gossypium hirsutum	Cited by Kasasian, 1971
	Helianthus annuus	Herzlinger and Kleifeld, 1985; Foy, Jain, and Jacobsohn, 1989
	Lactuca sativa	Foy, Jain, and Jacobsohn, 1989; Musselman, 1993; Korashi et al., 1996

TABLE 21.3 *(continued)*

Orobanche species	Host species	Reference
	Lens culinaris	Bayaa et al., 1998; Uludag, Demir, and Akosy, 2001
	Linum usitatissimum	Kleifeld et al., 1994
	Lycopersicon esculentum	Vouzounis and Americanos, 1998
	Matricaria chamomilla	Foy, Jain, and Jacobsohn, 1989
	Nicotiana tabacum	Faizieva, 1977; Musselman, 1982; Sahu and Sinha, 1983; Foy, Jain, and Jacobsohn, 1989; Parker, 1994
	Ocimum basilicum	Foy, Jain, and Jacobsohn, 1989
	Pastinaca sativa	Parker and Riches, 1993
	Petroselinum crispum, Phoeniculum vulgare	Foy, Jain, and Jacobsohn, 1989
	Pisum sativum	Foy, Jain, and Jacobsohn, 1989; Parker and Riches, 1993
	Salvia sclarea, Sinapis alba	Foy, Jain, and Jacobsohn, 1989
	Solanum melongena	Korashi et al., 1996; Vouzounis and Americanos, 1998
	S. tuberosum	Joel et al., 1999
	Spinacia oleracea	Goldwasser, 1995
	Trifolium alexandrinum	Foy, Jain, and Jacobsohn, 1989; Kleifeld et al., 1994
	T. pratense	Morozov, Foy, and Westwood, 2000
	Vicia faba	Geipert, 1997
	V. sativa	Nandula, Foy, and Orcutt, 1999
	Vigna radiata	Kleifeld et al., 1994
O. alba	*Thymus serpyllum*	Rosvall, 1979
O. cernua	*Amygdalus communis*	Qasem (unpublished data)
	Artemisia barrelieri, A. campestris, A. gallica, A. herba-alba	Pujadas and Thalouarn, 1998
	Helianthus annuus	Castejon et al., 1991; Munoz, Munoz, and Torres, 1993; Parker, 1994; Pujadas and Thalouarn, 1998; Sukno, Melero-Vara, and Fernandez-Martinez, 1999
	Launaea lanifera	Pujadas and Thalouarn, 1998
	Lupinus sp.	Korashi et al., 1996
	Lycopersicon esculentum	Musselman, 1982; Parker, 1994; Korashi et al., 1996

	Nicotiana tabacum	Musselman, 1982; Palled et al., 1985; Meti and Hosmani, 1994; Parker, 1994; Dhanapal, ter Borg, and Struik, 1998
	Ornamentals	Kleifeld, 1996
	Pisum sativum	Jacobsohn et al., 1982
	Solanum melongena	Rao, Joshi, and Sarma,1982; Kleifeld, 1996
	S. tuberosum	Qasem (unpublished data)
	Vicia faba	Korashi et al., 1996
	Xanthium spinosum	Qasem (unpublished data)
O. crenata	*Amaranthus* sp.	Korashi et al., 1996
	Ammi majus	Qasem (unpublished data)
	Anethum graveolens	Kleifeld, 1996; Korashi et al., 1996
	Angelica archangelica, Anthriscus cerfolium	Foy, Jain, and Jacobsohn, 1989
	Apium graveolens	Musselman, 1993; Kleifeld, 1996
	Arachis hypogaea	Jacobsohn et al., 1991
	Beta vulgaris	Korashi et al., 1996
	Carthamus tinctorius	Torres and Granados, 1991
	Carum carvi	Korashi et al., 1996
	Cicer arietinum	Linke, Schnell, and Saxena, 1991; Kleifeld, 1996; Korashi et al., 1996
	Cucumis melo	Parker and Riches, 1993
	Cuminum cyminum	Korashi et al., 1996
	Daucus carota	Foy, Jain, and Jacobsohn, 1989; Musselman, 1993; Korashi et al., 1996
	Foeniculum vulgare, Gazania longiscapa	Foy, Jain, and Jacobsohn, 1989
	Helianthus annuus	Herzlinger and Kleifeld, 1985; Foy, Jain, and Jacobsohn, 1989; Parker and Riches, 1993
	Lactuca sativa	Parker and Riches, 1993
	Lathyrus odoratus	Sillero, Cubero, and Rubiales, 2001
	Lens culinaris	Torres and Granados, 1991; Musselman, 1993; Jurado-Exposito, Garcia-Torres, and Castejon-Munoz, 1997; Uludag, Demir, and Akosy, 2001
	Lycopersicon esculentum	Parker and Riches, 1993
	Matricaria chamomile	Korashi et al., 1996
	Medicago sativa	Parker and Riches, 1993
	Nicotiana tabacum	Zazzerini, Torre, and Tosi, 1981

TABLE 21.3 *(continued)*

***Orobanche* species**	**Host species**	**Reference**
	Ornamentals	Hassan, 1998
	Pelargonium sp.	Kleifeld, 1996
	Pelargonium graveolens	Foy, Jain, and Jacobsohn, 1989
	Petroselinum crispum	Musselman, 1993; Kleifeld, 1996
	Pisum sativum	Torres and Granados, 1991; Jurado-Exposito, Castejon-Munoz, and Garcia-Torres, 1996
	Polygonum sp.	Korashi et al., 1996
	Salvia sclaria	Foy, Jain, and Jacobsohn, 1989
	Sonchus oleraceous, Trigonella foenum-graecum	Korashi et al., 1996
	Verbena officinalis	Foy, Jain, and Jacobsohn, 1989
	Vicia sativa	Torres and Granados, 1991; Kleifeld, 1996
	V. faba	Torres and Granados, 1991; Kharrat, Halila, and Beniwal, 1993
O. cumana	*Carthamus tinctorius*	Kleifeld, 1996
	Helianthus annuus	Petzoldt, Nemli, and Sneyd , 1993; Bedi, 1994; Thomas et al., 1998; Yakutkin and Tavolzhanskri, 1999
	Lycopersicon esculentum	Kleifeld, 1996
	Nicotiana tabacum	Musselman, 1993
O. foetida	*Astragalus lusitanicus*	Pujadas, 1999
	Cicer arietinum	Roman et al., 2001
	Lotus spp.	Pujadas, 1999
	Medicago sativa L.	University of Hohenheim, 2002
	Melilotus spp., *Scorpius muricatus, Trifolium angustifolium*	Pujadas, 1999
	Vicia faba L.	Kharrat et al., 1992; Zermane et al., 2001
O. hederae	*Hedera helix*	Conaghan, 1998
O. minor	*Apium graveolens*	Parker and Riches, 1993
	Burley tobacco	Frost and Musselman, 1979
	Carthanus tinctorius	Foy, Jain, and Jacobsohn, 1989; Korashi et al., 1996; Hassan, 1998
	Carum ajowan	Cited by Kasasian, 1971
	Daucus carota	Frost and Musselman, 1979; Parker and Riches, 1993

	Eryngium maritimum	Hipkin, 1992
	Guizotia abyssinica	Foy, Jain, and Jacobsohn, 1989; Parker and Riches, 1993
	Helianthus annuus	Foy, Jain, and Jacobsohn, 1989
	Hypochoeris spp.	Dhanapal, 1996
	Hypochoeris radiata	Hipkin, 1992
	Lactuca sativa	Foy, Jain, and Jacobsohn, 1989; Parker and Riches, 1993
	Lathyrus odoratus, Lens culinaris	Foy, Jain, and Jacobsohn, 1989
	Lotus corniculatus L.	Parker and Riches, 1993
	Lupinus sp.	Korashi et al., 1996
	Medicago sativa, Nicotiana tabacum	Foy, Jain, and Jacobsohn, 1989; Parker and Riches, 1993
	Pisum sativum	Hassan, 1998
	Shrubs and herbs	Frost and Musselman, 1979
	Sonchus oleraceus	Korashi et al., 1996; Hassan, 1998
	Trifolium spp.	ter Borg, 1994; Suzaki et al., 1994; Korashi et al., 1996; Yoneyama, Ogasowara, et al., 1998; Yoneyama, Takeuchi, et al., 1998
	T. pratense	Hipkin, 1992; ter Borg, 1994; Morozov, Foy, and Westwood, 2000; Colquhoun, Mallory-Smith, and Suverly, 2001
	T. hybridum	Hipkin, 1992
	T. repens	Suzaki et al., 1994; Dale and Press, 1998
	T. subterraneum	Foy, Jain, and Jacobsohn, 1989
	Vicia spp.	Foy, Jain, and Jacobsohn, 1989; Dhanapal, 1996
O. ramosa	*Allium sativum*	Parker and Riches, 1993
	Amaranthus retroflexus	Mitich, n.d.; INRA-Dijon, 2000
	Ammi majus	Qasem (unpublished data)
	Apium graveolens	Americanos, 1991
	Arachis hypogaea	Parker and Riches, 1993
	Artemisia campestris	Mitich, n.d.; INRA-Dijon, 2000
	Beta vulgaris	Mitich, n.d.
	Brassica spp.	Collin, 1999
	B. napus	Gibot-Leclerc et al., 2001
	B. oleracea var. *bortytis*	Musselman, 1993; Qasem (unpublished data)
	B. oleracea var. *capitata*	Musselman, 1993

TABLE 21.3 *(continued)*

***Orobanche* species**	**Host species**	**Reference**
	Cannabis sativa	Musselman, 1993; McPartland, 1996; Rubiales, Pujadas and Roman, 2001
	Capsella bursa-pastoris	Boulet et al., 2001
	Capsicum fruitecens	Eplee, 1984
	Carthamus tinctorius	Cited by Kasasian, 1971
	Cicer arietinum	Parker and Riches, 1993
	Citrullus vulgaris	Parker and Riches, 1993
	Cucumis sp.	Cited by Kasasian, 1971
	Cucumis melo	Musselman, 1993
	C. melo var. *flexuosus*	Qasem (unpublished data)
	C. sativus	Mitich, n.d.
	Cucurbita pepo	Qasem (Unpublished data)
	Daucus carota	Musselman, 1993
	Engelmannia pinnatifida	Musselman, 1982
	Epilobium tetragonum	Boulet et al., 2001
	Eryngium creticum	Qasem (unpublished data)
	Foeniculum vulgare	Parker and Riches, 1993
	Galium aparine	Boulet et al., 2001
	G. tricorne	Qasem (unpublished data)
	Geranium dissectum	Boulet et al., 2001
	Gossypium hirsutum	Cited by Kasasian, 1971
	Helianthus annuus	Eplee, 1984; Parker and Riches, 1993
	Lactuca sativa	Musselman, 1993
	Lens culinaris	Mohamed-Ahmed and Drennan, 1994
	Linum usitatissimum	Musselman, 1982
	Lycopersicon esculentum	Parker, 1994; Korashi et al., 1996; Vouzounis and Americanos, 1998
	Malva parviflora	Qasem (unpublished data)
	Melilotus indicus	Qasem (unpublished data)
	Nicotiana tabaccum	Zazzerini, Torre, and Tosi, 1981; Lolas, 1994; Parker, 1994; Sandri, Sandri, and Martini, 1998
	Oxalis corniculata	Qasem (unpublished data)
	Pastinaca sativa	Parker and Riches, 1993
	Pisum sativum	Mohamed-Ahmed and Drennan, 1994
	Prunus armeniaca	Al-Khazraji, Ishaak, and Hameed, 1989
	Raphanus raphanistrum	Boulet et al., 2001

	Rhagadiolus stellatus	Qasem (unpublished data)
	Ridolfia segetum	Boulet et al., 2001
	Senecio vulgaris	
	Solanum spp.	Mitich, n.d.; INRA-Dijon, 2000
	S. melongena	Braun, Burgstaller, and Walter, 1984; Braun, Koch, and Stiefvater, 1987
	S. nigrum	Qasem (unpublished data)
	S. tuberosum	Haidar, Bibi, and Abdel-Khalek, 1995
	Sonchus asper	Boulet et al., 2001
	Tecurium sp.	Mitich, n.d.; INRA-Dijon, 2000
	Trifolium spp.	Parker and Riches, 1993
	Veronica persica	Boulet et al., 2001
	Vicia faba	Eplee, 1984; Mohamed-Ahmed and Drennan, 1994; Korashi et al., 1996
	Xanthium spinosum	Mitich, n.d.; INRA-Dijon, 2000
	Zea mays	Quoted by Kasasian, 1971
O. uniflora	*Solidago canadensis*, *S. rugosa*, *S. uliginosa*	Nelson and Mohamed, 2001

Orobanche aegyptiaca and *O. ramosa* are more specialized in solanaceous crops, *O. crenata* and *O. minor* in grain and fodder legumes, while *O. cumana* is highly restricted to sunflower. *Orobanche crenata* is the most destructive to legumes and is mainly found in the Mediterranean region, but it is not important in cucurbits or composite species. As *O. aegyptiaca* requires higher temperatures for germination and development than *O. crenata,* it is therefore more often found parasitizing tobacco or tomato instead of the legumes.

Although *Orobanche* is spread all over the world, losses occur mainly in southern and eastern Europe, northern Africa and western Asia. In the Mediterranean area and western Asia, about 16 million ha of arable land, which is about 1.2 percent of the world's arable land, are threatened by *Orobanche.* Parasitism leads in dependency on the number of parasite attachments per host plant and the time of infestation to more or less intensive yield reduction. Damage by the parasite is dependent on the infestation level in the field. Yield loss is also dependent on the time of infestation and, therefore, on the duration of infection. Most of the attached parasites are not visible aboveground, since usually only about 1 to 30 percent of the attached plants emerge.

Economic Uses

A few species in Orobanchaceae are used in folk medicines (http://www.wssa.net/photo&info/larrymitich_info/orobanche.html). During the sixteenth and seventeenth centuries, great broomrape *(O. major)* of Europe and Russia was prescribed as a medicinal herb throughout most of Europe to remove stones in the bladder and kidneys and as a provoker of lustry urine. Applied externally, the juice was regarded as "a singular remedy" in the treatment of "old green wounds—the runnings of ulcers and sores-malignant and scabby ulcers, those that be hollow also" and "for fretting sores." The decocted flower spikes were used as a wash for "cleansing the skin" and "for freckles, black or blue spots or pushes thereof" (http://www.wssa.net/photo&info/larrymitich_info/orobanche.html).

Control

It seems that the autotrophic period of *Orobanche* and establishment on a host is the best stage for control of these parasites (Salle, Aber, and Le-Coz, 1984). However, control options for *Orobanche* include quarantine, crop rotation with trap and catch crops, planting date, mineral fertilizer, flooding, organic material, managed fallow, resistance/partial resistance, soil disinfection, herbicides, germination-inducing compounds, cleaning of crop seed, hand weeding, weeding after harvest, burning, deep plowing, soil solarization, insects, fungi, and integrated methods of control. Conventional methods for controlling *Orobanche,* such as the use of herbicides, have not generally proven successful. Alternatives involve varying agronomic methods (integrated control), the development of resistant crops, and biological control agents such as *Fusarium oxysporium* (Cubero et al., 1999; Lu et al., 1999; Thomas, Heller, et al., 1999; Thomas, Sauerborn, et al., 1999), and it has even been suggested that eating broomrape might be considered as part of the integrated control package (Rubiales, 1999). Ultimately, development of effective and low-cost control measures for *Orobanche* remains the main goal for weed control specialists, plant pathologists, agronomists, and biotechnologists.

More recently, it has been suggested that genetic engineering can be used to debilitate *Striga* by incorporating deleterious genes (Gressel, 1999; Joel et al., 1995), and similar approaches are also being explored for *Orobanche.* Management of *Orobanche* includes prevention, sanitation, and control (Kleifeld, 1998).

Prevention Methods

Orobanche seeds spread by irrigation water and can transfer with infested seeds, soil, organic materials, transplants, agriculture machinery, contaminated containers, wind, and workers (Kleifeld, 1998; Karnkowski, 1999).

The best control remedy is prevention, i.e., sowing clean seed. *Orobanche* seeds are among the tiniest in existence, so they are easily removed from large-seeded crops. Cleaning of crop seed to remove the minute seeds of *Orobanche* will not normally be feasible, but where absolutely necessary, risks could be greatly reduced by very thorough washing. Any bag or other container that is known or thought to have been contaminated with seeds of *Orobanche* should not be used for the transport of crop seeds from infested to uninfested areas. One of the most important control methods is to prevent the distribution of parasite seeds from one field to another and from infested touninfested areas (http://www.uni-hohenheim.de/~www380/parasite/oro_path.htm). Different mechanisms are responsible for the dispersal of the seeds. National and international trade of crop seeds contributes to seed dispersal over long distances.

Other methods of prevention of seed dispersal are to prevent animals from feeding on parasitic weeds and to clean shoes and tools when leaving parasitic-weed-infested fields. Moreover, it is important to recognize and to eradicate parasitic weeds in new infested fields to prevent a buildup of the parasitic weed population. Hand pulling and burning of weeded parasitic plants is important to prevent the reproduction of thousands of seeds (Miller et al., 1997).

Cultural and Mechanical Methods

Land preparation. Position of *Orobanche* seeds in the soil plays a major role in determining parasitism. Seeds of *O. ramosa* placed 10 cm deep resulted in heavy infestation compared to those placed at 15 cm. Thus, transplanting tomato seedlings at 15 cm reduced infestation in the field (Eltayeb et al., 2000). Likewise, Mohamed-Ahmed and Drennan (1994) found that *O. ramosa* failed to emerge in significant numbers when crop plants were grown from seeds planted at 11 cm or deeper compared with seeds planted at 3 to 4 cm depth. On the other hand, zero and minimum tillage increased broadbean infested by *O. crenata* (Kukula and Masri, 1984), while earlier reports mentioned that deep plowing (40 to 50 cm deep) reduced *Orobanche* by 80 to 90 percent in a three-year period (cited by Kasasian, 1971).

Planting or sowing date. Delaying the planting date affects *Orobanche* more than its hosts; the delay should be only two weeks from the date optimal for sowing in an uninfested field (http://www.uni-hohenheim.de/~www380/parasite/oro_path.htm). This will result in partial avoidance of infestation, as environmental factors affect germination and development of the parasite. It was found that for effective control of *O. crenata,* broadbean cv. Giza 402 should be sown around mid-November in Egypt without glyphosate application and with zero tillage (Nassib, Hussein, and Rayes, 1984). According to Munoz, Munoz, and Torres (1993), the duration of *O. cernua* emergence period increased as sowing date was delayed. Because sunflower yield was higher at earlier sowing dates, early sowing of this crop may be recommended in *Orobanche*-infested areas under mild Mediterranean climates. Early sowing of peas was found to be effective in reducing *Orobanche* incidence (Pirani et al., 1993), and early sowing of sunflower may be recommended in *O. cumana*–infested areas under Mediterranean climates (Castejon-Munoz, Romero-Munoz, and Garcia-Torres, 1993). The change of the sowing date seems not to be very promising under certain conditions but is still recommended for different regions. *Orobanche cumana* on sunflower appeared earlier at later planting dates as compared to early ones. *Orobanche* shoots per plant were fewer when sunflower planting was late (Aydin and Mutlu, 1996). In another study, parasitism of *O. crenata* was found to be decreased with later sowing dates, while better results were obtained with resistant varieties and late sowing (Sillero et al., 1999). Similar observations have been made by Perez-de-Luque, Sillero, and colleagues (2001).

Hand weeding. Clean cultivation followed by hand weeding or hoeing of *Orobanche* plants before they produced and disseminated seeds was commonly recommended for control. This is especially important with new infestations. The attachment with the host roots is not very firm, and the whole parasite breaks away readily. It is necessary to remove broomrape plants systematically for several years until the dormant seeds have all developed, taking into account that survival of only a very few plants is sufficient to reinfest a large area. Weeding of parasitic plants is not very promising in highly infested areas, but it is an alternative method available to all farmers and should be first used, though its exact function and limitations should be clearly understood. In the short term, economic analysis of weeding is likely to show no benefit because of little or no yield increase. Weeding is time-consuming, labor input is too high in certain regions, and the success is only in the long term and can cause injuries to the root system, allowing invasion of certain soilborne pathogens. But because no single method is reliable enough, hand pulling of remaining *Orobanche* shoots as an accompanying method is most important to avoid reinfestation. In combination with other methods it can reduce the seed bank effectively, and therefore weeding is an economical method on a longer-term basis (http://www.uni-hohenheim.de/~www380/parasite/oro_path.htm). Knowing more about the reproduction of *Orobanche* by innumerable seeds will lead to a better acceptance of hand weeding as an important tool of control, especially in fields with starting infestation. In tobacco, using a spear (iron blade stick) which is worked manually to cut vegetative shoots of *O. cernua* at ground level or 2 to 3 cm deep in the soil and repeated three to four times each season at four- to five-day intervals, is effective in eradicating this parasite within five to six years (Krishna-Murty, 1992).

Crop rotation with trap/catch crop. Quick dispersal of *Orobanche* into new fields is related to crop rotation. Monocroping or frequent cropping with host crops speeds up the process (Kleifeld, 1998). Cultural methods generally involve the use of trap or catch crops, which are alternated in rotation with crop species.

Trap crops are plants that stimulate the germination of parasitic weed seeds but are not infected by them. In this way trap crops reduce the seed bank in the soil and cause a decrease in infestation. Depending upon the agricultural system, different crops can be used as trap species (see Table 21.4). The efficacy of a trap crop might not exceed 10 to 40 percent in one year, i.e., up to 10 to 40 percent of the *Orobanche* seed bank in soil might become stimulated. These false hosts stimulate only germination, without becoming parasitized and do not require any additional labor when in-

TABLE 21.4. Trap crops for *Orobanche* spp.

Orobanche species	**Trap crop species**	**Common name**	**References**
O. aegyptiaca	*Allium sativum*	Garlic	University of Hohenheim, n.d.
	Arabidopsis thaliana	Mouseear cress	Goldwasser, Plakhine and Yoder, 2000
	Capsicum annuum	Sweet pepper	Hershenhorn et al., 1996
	Carum ajowan	Caraway	University of Hohenheim, n.d.
	Glycine max	Soybean	University of Hohenheim, n.d.
	Lablab purpureus	Hyacinth bean	University of Hohenheim, n.d.
	Lathyrus ochrus	Ochrus vetch	University of Hohenheim, n.d.
	Linum usitatissimum	Linseed	Kleifeld et al., 1994
	Medicago sativa	Alfalfa	University of Hohenheim, n.d.
	Phaseolus vulgaris	French bean	University of Hohenheim, n.d.
	Pisum sativum	Pea	University of Hohenheim, n.d.
	Ricinus communis	Castor bean	University of Hohenheim, n.d.
	Sesamum indicum	Sesame	University of Hohenheim, n.d.
	Vigna radiata	Green gram	Kleifeld et al., 1994

TABLE 21.4 *(continued)*

Orobanche species	Trap crop species	Common name	References
	Vigna unguiculata spp. *unguiculata*	Cowpea	University of Hohenheim, n.d.
O. cernua	*Bidens pilosa*	Spanish needle	University of Hohenheim, n.d.
	Capsicum annuum	Sweet chilies	Cited by Kasasian, 1971; University of Hohenheim, n.d.
	Cichorium intybus	Chicory	University of Hohenheim, n.d.
	Crotalaria juncea	Sunhemp	Dhanapal and Struik,1996; Dhanapal, Mallory-Smith and ter Borg, 2001
	Gossypium spp.	Cotton	University of Hohenheim, n.d.
	Guizotia abyssinica	Nugcoat buttons	University of Hohenheim, n.d.
	Illacrotvloma uniflorum	Horsegram	University of Hohenheim, n.d.
	Linum usitatissimum	Linseed	University of Hohenheim, n.d.
	Medicago sativa	Alfalfa	University of Hohenheim, n.d.
	Phaseolus aureus	Greengram	Dhanapal, Mallory-Smith and ter Borg, 2001
	Ricinus communis	Castor bean	University of Hohenheim, n.d.
	Sinapis alba	Wild mustard	Cited by Kasasian, 1971
	Tridax procumbens	Buttons	University of Hohenheim, n.d.
	V. mungo	Black gram	University of Hohenheim, n.d.
	V. radiatia	Green gram	Dhanapal and Struik,1996
	V. aconitifolia	Moth bean	University of Hohenheim, n.d.
O. crenata	*Allium sativum*	Garlic	Hassan, 1998
	Astragalus boeticus	Vetch	Schnell, Linke, and Sauerborn,1994
	Brassica rapa	Turnip	Al-Menoufi and Adam, 1996
	Capsicum annuum	Pepper	Al-Menoufi and Adam, 1996
	Coriandrum sativum	Coriander	Al-Menoufi and Adam, 1996; Zemrag and Bajja, 2001
	Glycine max	Soybean	Schnell, Linke, and Sauerborn,1994
	Hedysarum coronarium	Sulla	Schnell, Linke, and Sauerborn,1994
	Helianthus annuus	Sunflower	University of Hohenheim, n.d.

	Hordeum vulgare	Barley	University of Hohenheim, n.d.
	Lablab purpureus	Hyacinth bean	Schnell, Linke, and Sauerborn,1994
	Lathyrus ochrus	Ochrus vetch	Schnell, Linke, and Sauerborn,1994
	Linum usitatissimum	Linseed	Khalf, 1992; Abou-Salama, 1995
	Lupinus termis	Lupinus	Al-Menoufi and Adam, 1996
	Phaseolus vulgaris	French bean	Schnell, Linke and Sauerborn,1994
	Pisum sativum	Pea	Hassan, 1998
	Saccharum officinarum	Sugarcane	Abou-Salama, 1995
	Sesamum indicum	Sesame	Al-Menoufi, 1991
	Trifolium alexandrinum	Berseem	Schnell, Linke, and Sauerborn,1994; Al-Menoufi and Adam, 1996
	Trigonella foenum graecum	Fenugreek	Al-Menoufi and Adam, 1996; Zemrag and Bajja, 2001
	Vicia narbonensis	Narbonne vetch	Schnell, Linke, and Sauerborn,1994
	Vigna radiata	Green gram	Schnell, Linke, and Sauerborn,1994
	V. unguiculata	Cowpea	Schnell, Linke, and Sauerborn,1994
O. cumana	*Bidens pilosa*	Spanish needle	University of Hohenheim, n.d.
	Capsicum annuum	Sweet chillies	University of Hohenheim, n.d.
	Cichorium intybus	Chicory	University of Hohenheim, n.d.
	Crotalaria juncea	Sunhemp	University of Hohenheim, n.d.
	Gossypium spp.	Cotton	University of Hohenheim, n.d.
	Guizotia abyssinica	Nugcoat	University of Hohenheim, n.d.
	Illacrotvloma uniflorum	Horsegram	University of Hohenheim, n.d.
	Linum usitatissimum	Linseed	University of Hohenheim, n.d.
	Medicago sativa	Alfalfa	University of Hohenheim, n.d.
	Ricinus communis	Castor bean	University of Hohenheim, n.d.
	Tridax procumbens	Buttons	University of Hohenheim, n.d.

TABLE 21.4 *(continued)*

Orobanche species	Trap crop species	Common name	References
	Vigna aconitifolia	Moth bean	University of Hohenheim, n.d.
	V. mungo	Black gram	University of Hohenheim, n.d.
	V. radiatia	Green gram	University of Hohenheim, n.d.
O minor	*Allium sativum*	Garlic	Hassan, 1998
	Arabidopsis thaliana	Mouseear cress	Goldwasser, Plakhine, and Yoder, 2000
	Linum usitatissimum	Linseed	Brown et al., 1952
	Pisum sativum	Pea	Hassan, 1998
	Sorghum bicolor	Sorghum	Brown et al., 1952
	Zea mays	Corn	Brown et al., 1952
O. ramosa	*Allium sativum*	Garlic	University of Hohenheim, n.d.
	Arabidopsis thaliana	Mouseear cress	Goldwasser, Plakhine, and Yoder, 2000
	Brassica sp.	Mustard	Cited by Kasasian, 1971
	B. napus	Rape	Cited by Kasasian, 1971
	B. rapa	Turnip	Al-Menoufi and Adam, 1996
	Capsicum spp.	Capsicums	Cited by Kasasian, 1971; Sand, 1983
	Carum ajowan	Caraway	University of Hohenheim, n.d.
	Coriandrum sativum	Coriander	Al-Menoufi and Adam, 1996
	Glycine max	Soybean	University of Hohenheim, n.d.
	Lablab purpureus	Hyacinth bean	University of Hohenheim, n.d.
	Lathyrus ochrus	Ochrus vetch	University of Hohenheim, n.d.
	Linum usitatissimum	Linseed	Eplee, 1984; Khalaf, 1992
	Lupinus termis	Lupinus	Al-Menoufi and Adam, 1996
	Mediacago sativa	Lucerne	University of Hohenheim, n.d.
	Pennisetum sp.	Millet	Cited by Kasasian, 1971
	Phaseolus aureus	Mung bean	Kleifeld, 1996
	P. vulgaris	French bean	University of Hohenheim, n.d.
	Pisum sativum	Pea	University of Hohenheim, n.d.
	Ricinus communis	Castor bean	Cited by Kasasian, 1971; University of Hohenheim, n.d.
	Sesamum indicum	Sesame	Cited by Kasasian, 1971; University of Hohenheim, n.d.

Setaria sp.	Foxtail	Cited by Kasasian, 1971
Trifolium sp.	Clover	Cited by Kasasian, 1971
Trigonella foenum-graecum	Fenugreek	Al-Menoufi and Adam, 1996
Vigna radiata	Green gram	University of Hohenheim, n.d.
V. unguiculata ssp. *unguiculata*	Cowpea	University of Hohenheim, n.d.
Zea mays	Maize	Cited by Kasasian, 1971

cluded in crop rotation. Heavily infested fields should be planted for two or three years with crops that are not attacked by the parasite (http://www.wssa.net/photo&info/larrymitich_info/orobanche.html). Severe infestation by *Orobanche* builds up only where susceptible crops come frequently at short intervals in crop rotation. Therefore, rotation with nonhost crops, especially with trap crops, is of great advantage (Eltayeb et al., 2000). When broomrape seed occurs close to the roots of such species, 13 to 15 percent of these seeds germinate but do not develop haustorial connections. In this regard, peppers (*Capsicum annuum* L.) grown in rotation with tobacco tend to reduce broomrape seed infestations in the soil (http://www.wssa.net/photo&info/larrymitich_info/orobanche.html). Likewise, growing flax in two successive winter seasons or one summer cropping with mungbean reduced early infestation of *O. aegyptiaca* and significantly increased tomato growth and production (Kleifeld et al., 1994).

On the other hand, catch crops are host plants of parasitic weeds (possible fodder harvest or green manure) that induce germination and get attacked. These crops are real hosts, become parasitized, and have to be harvested, plowed, or destroyed after six to eight weeks, before the parasite emerges and reproduces, or, at the latest, before blooming of the parasite. It is safe and fast and can efficiently reduce the parasitic weed seed bank in the soil. Suicidal germination of the parasitic seeds triggered by growing trap crops reduced the weed population and the growth of the host plants was hastened due to green manuring effect of trap crops. Therefore, including a trap crop in the rotation may reduce the problem. Sunhemp (*Crotalaria juncea* L.) and greengram (*Vigna radiata* L.) are promising trap crops in a cropping system containing bidi tobacco in areas where tobacco is grown in a long growing season.

Use of trap crops effectively reduces the *Orobanche* seed bank, but to a certain extent these crops could be used as a part of integrated *Orobanche* control approach. Crop rotation alone will not solve the problem, but it can contribute to reduce the infestation level to a degree that renders control methods such as hand pulling feasible. Crop rotation is already practiced by farmers, and susceptible crops are not cultivated year by year on the same fields. However, different workers reported successful control of different *Orobanche* species following this method. Schnell, Linke, and Sauerborn (1994) reported legume crops to cause greater reduction in *O. crenata* seed banks. Resistance to *O. crenata* has been found both in cultivated and wild-grown legumes (Rubiales, Sillero, and Cubero, 1998, 1999; Sillero, Moreno, and Rubiales, 2001). More recently, resistance to *O. crenata* has been identified in different species of *Lathyrus,* and these could be used as trap crops under field conditions (Sillero, Cubero, and Rubiales, 2001). In vitro flax seed exudates induced germination of *O. crenata* and *O. ramosa,* and similar results were obtained on *O. aegyptiaca* with flax root exudates, GR-7 and GR-24 (Matthews, Joel, and Steffens, 1991). However, flax as trap crop for eliminating *Orobanche* infection in fields producing broadbean is reported as showing limited potential (Khalaf, 1992). In another study, flax and sugarcane in a rotation with broadbean were observed to significantly reduce *O. crenata* (Abou-Salama, 1995). Hershenhorn and colleagues (1996) observed that few attached and developed *O. aegyptiaca* shoots are found on sweet pepper. Pepper (*Capsicum annuum* L.) roots stimulated germination of 22 to 26 percent of nodding *Orobanche* seeds but without the formation of parasitic attachments. A fourfold increase in the number of nodding *Orobanche* was observed on tomato roots interplanted with pepper than without pepper. Likewise, hairy vetch (*Vicia villosa* ssp. *dasycarpa*) in rotation with lentil, chickpea, and broadbean reduces *O. crenata* infestation to a low level (Schnell et al., 1996).

Intercropping can effectively control *O. crenata* and *O. minor* infestation in pea. Garlic in mixture with peas resulted in the best root system of peas and least *Orobanche* emergence, while higher seed yield was obtained when pea was mixed with onion. Complete elimination of *Orobanche* emergence was achieved in mixture with black cumin [*Bunium persicum* (Boiss.) Fedtsch.] (Hassan, 1998). Maximum percentage of germination of *O. cernua* was obtained when the parasite seeds were exposed to greengram, sunhemp, and sesamum crops followed by blackgram and sunflower crops both in incubator and glasshouse conditions (Dhanapal, Struik, and ter Borg, 1998). *Arabidopsis thaliana* also inducses seed germination of *O. aegyptiaca, O. minor,* and *O. ramosa* at a rate of 87, 72, and 67 percent of maximum seed germination, respectively (Goldwasser, Plakhine, and Yoder, 2000). Rodriguez-Ojeda, Alonso, and Fernandez-Escobar (2001) reported root exudates of corn, sorghum, millet, cotton, rice, eggplant, and cauliflower to induce seed germination of *O. cumana* under laboratory conditions while none of the crops mentioned was infected with this parasite, suggesting that some of these crops could be used as trapping species.

Certain weeds including wild oat (*Avena sterilis* L.), Canadian horseweed [*Conyza canadensis* (L.) Cronq.], large bullwort (*Ammi majus* L.), jimsonweed (*Datura stramonium* L.), endive (*Cichorium endivia* L.), scarlet pimpernel (*Anagallis arvensis* L.), and black nightshade (*Solanum nigrum* L.) support *O. ramosa* attachment, but different resistance levels are observed and later lead to the death of parasite (Boulet et al., 2001). These weeds could be used as trap species for *O. ramosa* control (Boulet et al., 2001).

However, the success of trap crops in exhausting the soil seed bank of the parasite may be highly questionable under severe infestation or may require a fairly long period for effective control which may not be economically feasible under certain growing systems. In general, good results with an effective trap crop may not be efficient and possibly impractical for ecological/economical reasons. Even important trap crops to reduce the seed bank for different parasitic weeds are not available. Catch crops are likely to be used as a preceding crop, and they have to be closely planted. This method of control, however, is costly due to additional labor, not usable if the growing season is short, and needs good mechanization because of possible loss of a growing period. However, cultivation of all hosts should be easy and inexpensive, germination stimulation should be high, their elimination is not problematic, and they are high in yield as fodder or green manure.

Plant residues and soil cover. Surface- and preplant-incorporated mulches of wheat and barley straw residues used at different rates significantly reduced *O. ramosa* infestation and growth in potato (Haidar, Bibi, and Abdel-Khalek, 1995). Olive (*Olea europea* L.) jift (a solid by-product of olive) added to pea growing soil prevented *O. crenata* infestation to this crop and reduced infection on broadbean and tomato (Ghosheh et al., 1999). Qasem and Foy (2001) reported high allelopathic activity of shoot residues of certain weed species on seed germination of *O. ramosa* seeds in tomato grown under glasshouse conditions. Later, Qasem (2002) observed an inhibitory effect of dried shoot residues and root exudates of different weed and crop species to *O. ramosa* infestation on tomato plants. Plastic mulch was found to be effective in controlling *Orobanche* spp. (Tsybul'skaya and Skoklyuk, 1978), and black plastic mulch completely controlled *O. ramosa* and *O. aegyptiaca* on tomato and eggplant (Vouzounis and Americanos, 1998).

Soil solarization. Solarization is a method of disinfecting by heating moist soil using clear polyethylene (PE) sheets and natural sunlight. Soil solarization significantly reduced population and incidence of *Orobanche* (Sauerborn et al., 1989). Solar heating followed by mulching was promising for the control of *O. ramosa* and increased eggplant yield with glyphosate used at 40 g·ha^{-1} (Braun et al., 1985).

The relatively high cost of this method can be partly compensated by the control of difficult pests such as nematodes, most other weeds, fungal and bacterial pathogens, and an improved nutrient status of the soil. Futhermore, there is a residual effect over several years following the solarization treatment, resulting in increased crop yields. This method, however, is not applicable in general for economic reasons, but under certain conditions, where water is available for irrigation, soil solarization with plastic sheets could be an additional measure to reduce the *Orobanche* infestation in small plots. *Orobanche aegyptiaca* can be controlled by solar heating using polyethylene clear film 0.03 mm thick left in place for 30 to 50 days in the summer (Cartia, 1985). Soil solarization was also found to be effective in controlling *O. cernua* in tobacco (Meti and Hosmani, 1994) and *O. crenata* in broadbean (Abdalla, 1999). Soil solarization alone killed *Orobanche* seeds at 0 cm depth, but in combination with chicken manure, *Orobanche* seeds at all soil depths from 0 to 10 cm were killed. Solarization for two to six weeks with or without chicken manure significantly reduced weed growth and infestation in cabbage (Haidar and Sidahmed, 2000).

Mineral fertilization/organic matter. The essential element for reversing this effect is nitrogen. *Orobanche* tends to be present in association with less fertile soil conditions. High levels of fertilizer, especially nitrogen, have a suppressive effect. Nitrogen fertilizer reduced *O. crenata* (Kukula and Masri, 1984). Ammonium nitrate applied alone or in combination with potassium phosphate completely inhibited infection of tomato by *O. aegyptiaca* (Jain and Foy, 1987). Low rates of ammonium sulphate reduced *O. crenata* infestation without having an adverse effect on broadbean (Jain and Foy, 1992). High levels of nitrogen reduced infestation of *O. ramosa* in tobacco and tomato, but it also reduced tomato yield (Qasem, personal observations). When ammonium-N based liquid fertilizer was in direct contact with *O. crenata* it adversely affected parasite germination potential. The adverse effect was dependent on the rate of N used (Eplee, Norris, and Merritt, 1994). Number of emerged and dry weight of *O. ramosa* in tomato decreased with increase in N level, while an increase in the level of K had the opposite effect (Demirkan and Nemli, 1994).

In vitro *Orobanche* emergence could be reduced by applying nitrogen, but only a few field studies confirm the reduction of *Orobanche* infestation by nitrogen application. Ammonium sulfate (8 mM) applied during the conditioning period reduced germination of *O. crenata;* 4 mM ammonium sulfate strongly inhibited germination when applied after conditioning during the germination phase. Urea at 8 mM reduced germination to only a limited extent (van Hezewijk and Verkleu, 1996). Radicle lengths of *O. aegyptiaca* and *O. ramosa* were inhibited more by ammonium nitrate and ammonium chloride than by potassium nitrate, and both parasite species differ in their sensitivity to nutrients (Nandula, Foy, and Westwood, 1996). Ammonium form was more inhibitory to *Orobanche* spp. than nitrate (Westwood and Foy, 1999). The input of mineral fertilizer to the cropping system under consideration is rather low, probably just sufficient to replace the nutrients lost by harvest. For a considerable reduction in *Orobanche* infestation, the N supply by input of mineral fertilizers is by far not sufficient, and for economic reasons it is not very promising to increase the input of mineral fertilizers to reduce *Orobanche* infestation. Germination of *O. aegyptiaca* and *O. oxyloba* was also strongly inhibited by copper and vanadium, while boron inhibited radicle elongation of *O. oxyloba* 25 $\mu g \cdot L^{-1}$ (Zaitoun, Saleh, and ter Borg, 1996). Foliar spray of 0.5 percent boric acid 60 days after *Vicia faba* sowing completely controlled *O. crenata* (Hassan and Farrag, 1982). Elongation of *O. aegyptiaca* and *O. oxyloba* was inhibited by Zn and Mo at all concentrations used (Zaitoun, Saleh, and ter Borg, 1996). Metham sodium was also found to affect *O. aegyptiaca* germination (Goldwasser et al., 1995)

Flooding. If done for an extended period of time (weeks), flooding kills parasite seeds in the soil (http://www.uni-hohenheim.de/~www380/parasite/oro_path.htm). Under rainfed agriculture water shortage usually could not permit this method. Regular irrigation will support the host to compensate for the attack of the parasite to a certain extent. However, waterlogging inhibited *Orobanche* emergence when applied for more than six weeks (Mohamed-Ahmed and Drennan, 1994).

Resistance and Tolerance

Research on crop resistance to *Orobanche* has not received much attention (Cubero, 1991). However, good levels of resistance have been found in several host/parasite systems including sunflower to *O. cernua*, broadbean to *O. crenata,* common vetch to *O. crenata*, eggplant to *O. aegyptiaca,* and several cucurbits to *O. ramosa* and *O. aegyptiaca*. Observations on collection of peas and chickpeas also indicate strong levels of resistance (Alonso, 1998). A summary of resistant/tolerant varieties or cultivars of different crop species along with the specific *Orobanche* spp. follows.

Broadbean (O. crenata). An accession from the *paucijuga* group (VF 172) was reported to have certain level of resistance (Cubero, 1973). A study by Zaitoun and Sabra (2000) confirmed that the new broadbean resistant varieties ('Giza 667' and 'Giza 843') were less infected by *O. crenata.* The host resistance is related to the level of phenolic compounds, especially in its roots. The degree of variation in susceptibility within one species is less than that between species. Lines with partial resistance have also been identified (http://www.uni-hohenheim.de/~www380/parasite/oro_path.htm), including 'Line 561' (cross between 'Giza 402' and 'Seville giant'); lines 667 and 674 ('Giza 402' as one parent, composite crosses). Various lines (F8 generation) in Spain having 'Giza 402' as a parent among *paucijuga* types are usually more resistant. The line F402 was the first reported to show a high resistance and to be acceptable agronomical characters, and thus the 'Giza 402' variety was derived from this line. In turn, the line 'TVF 1071' was derived from 'Giza 402' and became more resistant than the mother variety (Cubero and Hernandez, 1991), and later the cultivar Baraca was obtained and regarded as a high-yielding and well-adapted cultivar (Cubero et al., 1992). 'Giza 402' from Egypt could be also classified as tolerant to *O. crenata* since parasite had no significant effect on its pod and seed production (Zaitoun, Al-Menoufi, and Weber, 1991). Among other reported resistant lines were cvs. 18009, 18035, and 18105 (ICARDA, 1992).

Lentil (O. crenata). In general, low intraspecific variation in susceptibility to *Orobanche* exists in lentil compared to broadbean. Screening wild lentil species therefore might be promising. Line 'ILL 5748' showed 50 percent fewer infections in laboratory screening. Although line 'ILL 8' is not less susceptible, it is suitable for late planting because it escapes the attack.

Chickpea (O. crenata). Indication of resistance was found in 'ILL 3279' and some other lines; however, chickpea, in general, is much less infested (by a factor of about eight) than broadbeans. Variety ILC-280 is mentioned as a resistant one with a substantial amount of phytoalexins (Wegmann et al., 1991).

Field pea (O. crenata). Low intraspecific variation has been found so far; a lower infestation in the field, however, was noticed in 'Facima' and the ICARDA accession '290'.

Fodder legumes (O. crenata/O. aegyptiaca). Resistance was found in Spain with common vetch (*Vicia sativa* L.) lines including lines 473A, 485B, 486A, and 417B (Gil, Martin, and Cubero, 1987), and at ICARDA, Syria, in *V. sativa* lines 715 and 2541, purple broad vetch (*V. narbonensis* L.) line 578, *V. villosa* ssp. *dasycarpa,* winged vetchling [*Lathyrus ochrus* (L.) DC.], and pea vetch (*L. sativus* L.) line 201. The cultivar Sadot, developed from purple vetch (*Vicia atropurpurea* Desf.), was reported to be highly resistant to both *Orobanche* species (Goldwasser et al., 1996, 1997).

Cucurbits (O. aegyptiaca). Summer varieties of muskmelon, watermelon, and cucumber were most resistant to *Orobanche* spp. due to the short growing season. 'Kuzibai' was the most resistant watermelon cultivar (Mukumov and Faizieva, 1977). In oilseed pumpkin (*Cucurbita pepo* L.), *O. aegyptiaca* formed tracheids in the haustora, but these did not penetrate the vessels of the host plant, indicating considerable resistance to the parasite. However, in resistant muskmelon varieties cvs. Kukkalapush and Umyrvaki, the haustoria did not form tracheids and derived nutrients from parenchyma cells. In resistant cucumber variety cv. Margelanskii, *Orobanche* formed single vessels in haustoria, but these failed to penetrate vessels of the host plant, and growth of the parasite was greatly stunted (Kabulov and Kabulova, 1977).

Sunflower (O. cernua/O. cumana). Races of broomrape have been identified and sources of resistance appear to have stabilized sunflower production in certain European countries, but racial evolution of broomrape has been very rapid during the past few years, and all resistance has been overcome. Three sets of cultivated plant material were tested in Spain, including 160 USDA breeding lines. Of these lines, 5 percent were completely resistant and 19 percent segregated for resistance. These lines trace back to crosses of two USDA restorer lines with Russian, Turkish, and Romanian hybrids. Also, PI accessions from the USDA Plant Introduction Station at Ames, Iowa, that segregated for resistance were primarily derived from the former USSR and from Romania (Dominguez et al., 1993).

Varieties with various degrees of resistance are now available, but mainly in oil sunflowers. Line no. 16693 cv. Armavirskii 3497 was completely resistant to different races of *O. cumana;* 'Start' cultivar was found to be 98 percent resistant to different races of *Orobanche* (Mamonov, Poplaukhin, and Nazarko, 1978), and 'Tubro' cultivar showed resistance to *O. cumana* (Vranceanu et al., 1990). One important source of resistance is Jerusalem antichoke (*Helianthus tuberosus* L.). All perennial *Helianthus* spp. were resistant to different populations of *O. cernua* (Ruso et al., 1994) and have been frequently used. In a more recent study on resistance among 23 wild species of perennial *Helianthus,* all were found resistant/immune to two widely spread *Orobanche* isolates. Slender sunflower (*H. gracilentus* A. Gray) and Nuttall's sunflower (*H. nuttallii* Torr. & Gray) showed resistance segregation (Ruso et al., 1994, 1996; Dominguez et al., 1996; Suknol et al., 1996). Serpentine sunflower (*H. exilis* Gray), western sunflower (*H. anomalus* Blake), and southeastern sunflower (*H. agrestis* Pollard) were the only resistant annual species (Sukno et al., 1996). Sunflower cultivars SW-501, RW-637, SW-501, Sunbred-254, and RW-637 were also reported as resistant to *O. cumana* (Ish-Shalom-Gordon, Jacobsohn, and Cohen, 1993). The 'R-41' inbred line was found to be resistant to *O. cumana* which is conferred by two independent dominant genes (Dominguez, 1996). Sunflower inbred line 'KI-374' showed high resistance to *O. cernua* and *O. cumana* (Rodriguez-Ojeda, Alonso, and Fernandez-Escobar, 2001). Lines 'MX199' and 'MX655' were reported to have complete resistance to the most virulent race of *O. cumana* (Batchvarova et al., 2001).

Eggplants (O. cernua). Recent reports on the resistance of this crop to *Orobanche* are lacking. However, results reported by Dalela and Mathur (1971) are still regarded as the only source of information on this crop. These authors mentioned that out of the 128 cultivars tested for resistance to *O. cernua*, the highly resistant cultivars were the hybrids of cv. Pusa Purple Long with cv. Manjri Gota or with cvs. Nurki, Running King, Verma's Giant, D-12-2-66, DC-4-1-67, and E-147.

Tomato (O. ramosa/O. aegyptiaca). Some resistance was reported from the Russian lines 'Ora', 'Gibrid-88', and 'Bolgarskii' of tomato which were the most tolerant to *Orobanche* infestation (Mukumov and Faizieva, 1977). Avdeev and Scherbinin (1983) reported some important resistant varieties to *O. cernua* including wild tomato (*L. hirsutum* L.), garden tomato (*L. pyriforme* Dunal), and plum tomato (*L. pruniforme* L.). Avdeev and Scherbinin (1977, 1978) found 'UZ-1' to be an almost immune line to *O. aegyptiaca,* and later study proved that 'PZU-11' is resistant to the same parasitic species (Avdeev, 1986). However, conflicting results have been obtained by Cubero (1991) on the same line. Kasrawi and Abu-Irmaileh (1989) reported the wild accession 'LA 372' of Peruvian tomato (*L. peruvianum* Mill.) to be a resistant line to *O. ramosa*. Tomato line CE-TO-32-2/82 was also reported to be the most resistant to *Orobanche* infestation, with only 3 percent infection (Sobrino Vesperinas, 1985). In more recent studies, high to moderate resistance to *O. ramosa* was obtained in the cultivars Tiny Tim, Acora, Cestor, Promodora, Orient, Red Alert, and cherry tomato [*L. pimpinellifolium* (Jusl.) Mill] LA 1478, with 'Tiny Tim' showing the highest level of resistance (Qasem and Kasrawi, 1995). Avdeev (1998) reported that tomato var. 'Astrakhan' and 'Bakhtemir' were not infested by *O. aegyptiaca,* and no effect on yield was noted. 'PZU-11' was also reported to be a resistant line to *O. aegyptiaca*.

Tobacco (O. cernua/O. crenata/O. ramosa). Wide variations are reported among tobacco varieties, cultivars, and accessions in their susceptibility/tolerance/resistance to *Orobanche* species infestation (Racovitza, 1973). Moderate resistance was detected in different tobacco lines and cultivars (Vinogradov et al., 1981; Sarychev et al., 1982; Vinogradov, Mironov, and Sarychev, 1984; Raju, 1996). Among tolerent and resistant lines already mentioned were cvs. Trebizond-285, Corojo, Florida-513, NS-2512, Hicks Resistant, Khatana, R X T, and Jamaica K-444 (Vinogradov et al., 1981).

In general, resistant varieties will be accepted only if they meet all the requirements requested by farmers and the market.

Biological Methods

Phytophagous insects on *Orobanche,* including *Phytomyza orobanchia* (Diptera), occurs in most *Orobanche*-infested regions. *Phytomyza* is a black monophagous fly 2.5 to 3 mm long, and it lays eggs (one to three) into the bud, flower, or stem. It was reported to attack a wide range of *Orobanche* populations either on crops or in the wild (Rubiales, Pujadas, and Roman, 2001). The female can lay 180 to 200 eggs in her lifetime. The duration of the life cycle is 30 to 35 days. Depending on the climatic conditions and the vegetative stage of the host, two to three generations occur in Eastern Europe and four to five in North Africa and West Asia. *Phytomyza* hibernates as pupae in the underground parts of the stem, and the fly emerges together with the *Orobanche* plant. The larvae feed on the shoots and especially on the capsules of *Orobanche* spp. (Klein, Kroschel, and Sauerborn, 1999). Depending on the insect population, they can damage *Orobanche* seriously and may reduce seed production by 30 percent under natural infestation in Syria and by 89 percent in Egypt. Natural enemies can reduce the fly population considerably: 12 species in 5 families have been reported to parasitize *Phytomyza* (Eltayeb et al., 2000).

At present, the only way to make use of *Phytomyza orobanchia* is to collect *Orobanche* infested with the fly's larvae and to distribute these shoots in the field at the time when *Orobanche* is expected to emerge. This technique was first developed in Eastern Europe for *O. cernua* control on sunflower and has been applied against *O. crenata* in Morroco. Preliminary field releases have shown an efficacy of more than 95 percent. *Phytomyza puparia* is also reported to feed on *Orobanche* seeds and damage the stalk of the parasite (Tsybul'skaya and Skoklyuk, 1978; Tsybul'skaya et al., 1978). It has also been reported on *O. crenata* (Hassanein et al., 1998).

The efficacy of *Phytomyza orobanchia* as a biological control agent of *Orobanche* is well established, and the fly is frequently occurring on *Orobanche* in different Eastern Europian countries. Under natural conditions the seed production is already reduced by about 41 percent on an average. In another study, Elzein and colleagues (1999) reported that *P. orobanchia* destroyed *O. ramosa* and *O.cernua* seed capsules by 81.4 and 71.7 percent, respectively, in tomato fields. Mass rearing of *Phytomyza* and the yearly release could be a possible approach in *Orobanche* control for the Mediterranean area, and adapting the regular form of aphid control by insecticides to include this approach should be considered.

On the other hand, several fungi species were also reported to be virulent on *Orobanche,* in addition to mycoplasma-like organisms detected in *O. cernua* sap stream (Singh and Joshi, 1992). Except for Russia, where good results were obtained by adding *Fusarium oxysporum* into the planting holes of watermelon ("Product F"), no practical use could be made so far of these fungi. However, current investigations concentrate mainly on *Fusarium oxysporum* f. sp. *orthoceras,* a pathogen with high potential to control *O. cumana* parasitizing sunflower. The fungus was reported to control *O. cumana* in sunflower by 90 percent, and *O. ramosa* and *O. aegyptiaca* were found to be suceptible as well (Bedi, 1994). In other studies, the fungus controlled *O. cumana* by 74 to 90 percent in naturally infested fields (Bedi, 1992), decreased *Orobanche* infestation in tobacco by 75.23 percent, and increased tobacco yield by 80.5 percent (Mazaheri et al., 1991). It has been recommended for *O. cumana* control in sunflower (Bedi and Donchev, 1995; Thomas et al., 1998; Bedi and Sauerborn, 1999). It was found to be effective at a concentration of 69 conidia·mL^{-1} water (Thomas et al., 1998), and under field conditions the fungus at 10 g·kg^{-1} soil resulted in more than 90 percent control of *O. cumana* (Bedi and Donchev, 1991). Research focuses on suitable mass-production techniques and appropriate formulations for field application of the pathogen. *Fusarium solani* is another fungus reported to supress the growth of *O. ramosa* in tomato (Gold, Sagen, and Wilhelm, 1979).

Ulocladium atrum is reported to infect and destroy underground tubercles and emerged shoots of *O. crenata* at 20°C and RH 50 to 80 percent (Linke et al., 1992). Other plant pathogens including *Alternaria, Fusarium,* and *Trichoderma* have also been found infesting *O. crenata* in Egypt (Abdel-Kader et al., 1998). Selective active *Azotobacter chroococcum* and *Escherichia coli* transformants added to soil infested with *O. crenata* seeds showed promising results for controlling *O. crenata* parasitism on broadbean (Khalaf, Ali, et al., 1997; Khalaf, Youssef, et al., 1997). Of the 62 fungi species isolated from different *Orobanche* species (Thomas, Sauerborn, et al., 1999), eight show pathogenicity to *Orobanche* and four of them belong to the genus *Fusarium,* among which *F. oxysporum* f.sp. *orthoceras* (Thomas, et al., 1998; Thomas, Sauerborn, et al., 1999) and *F. arthrosporioides* (Amsellem et al., 2000) are still to be developed as mycoherbicides.

Pathogens play a certain role in the regulation of *Orobanche* under natural conditions. Some research work has already focused on the use of these natural antagonists for the development of a mycoherbicide. This could be a future tool in *Orobanche* control.

Chemical Methods

Herbicides. Chemical control usually consists of treating the area with low-rate glyphosate (Roundup) and planting false host crops (http://www-aes.tamu.edu/mary/brmrape/brmrape.htm). Chemical control by systemic herbicides is also an option. The following are herbicide treatments recommended for *Orobanche* control in different crop species.

Broadbean: Trifluralin applied presowing and linuron postsowing gave the highest seed yield of broadbean infected with *O. crenata* (Zahran, 1982). Nearly complete control of *Orobanche* was achieved with chlorsulfuron, imazaquine, and glyphosate. A similar study by Ibrahim and colleagues (2000) revealed that glyphosate resulted in significant reduction in *Orobanche* spikes, ranging from 92 to 100 percent compared to untreated plots. This high efficiencey against *Orobanche* increased seed yield by 3,296 and 2,550 t·ha^{-1} in two locations, respectively, over mean of all sites (Dongola, Mohamed, and Ahmed, 2000). Similar results were obtaned by Salem, El-Shiny, and Ahmed (1989), with glyphosate used at 52.5 and 178.5 mL·ha^{-1} and at 60 g·ha^{-1} at visible shoot buds or shoots and when vestigial roots were well developed at 130 and 452.4g a.i.·ha^{-1} (El-Hattab et al., 1982; Mesa-Garcia and Garcia-Torres, 1985). Americanos (1983) confirmed the effectiveness of glyphosate acid at 60 to 120 g a.i.·ha^{-1} in controlling *O. crenata* in broadbean. Hamid, Shebl, and Ghalwash (2000) reported that the application of butralin at 2.85 kg a.i.·ha^{-1} followed by glyphosate at 58 g a.i.·ha^{-1}, imadazolinone at 214.2 g a.i.·ha^{-1}, and hand hoeing (twice) are very effective against *Orobanche*. These treatments reduced fresh weight of *Orobanche* spikes by 95, 79, and 84 percent, respectively. Both imidazolinone and sulfunylurea herbicides have been recommended for *O. crenata* control (Garcia-Torres and Lopez-Granados, 1991a). Glyphosate proved to be effective against *O. crenata* in broadbean in Morrocco (Zermane, 2000), and when applied at 0.12 kg·ha^{-1} the herbicide delayed parasite emergence and increased yield, but two applications were required in early-sown crop and high parasite infestation (Mesa-Garcia and Garcia-Torres, 1982). Complete control of *O. crenata* was acheived with glyphosate used at 60 mL a.i.·ha^{-1} once or twice (13 to 16 days between treatments), and bean yield increased from 0 to 1.6 t·ha^{-1} and from 0.3 to 2.3 t·ha^{-1} in plots treated with glyphosate (Schluter and Aber, 1980). When two sprays of glyphosate were applied at 64 g·ha^{-1} at the start of flowering and at the end of flowering three or six weeks, seed and straw yields of cv. 'Giza 2' were increased by 1 and 1.18 t·ha^{-1}, respectively (Nassib, Hussein, and Rayes, 1984), and very good results of *O. crenata* control were obtained when the herbicide was sprayed at 74 g·ha^{-1} at start of flowering (Miccolis and Bianco, 1984). However, the right time for glyphosate application has to be determined by inspections of the broadbean roots twice per week for *Orobanche* tubercules. The exact dosage of glyphosate is important for efficiency and to avoid phytotoxicity to the crop. Two treatments in an interval of 14 days have to be done. Even considering these points, the success of the application is not guaranteed. The effective dose to control *Orobanche* varies according to the numbers of tubercules attached and the weather conditions. For the late-planting/low-yield cropping systems, which normally face short and slight *O. crenata* infestations, crop preemergence treatments of imazethapyr at 75 to 100 g·ha^{-1} and imazypyr at 12.5 g·ha^{-1} resulted in efficient control without damaging broadbean. Imazethapyr, imazapyr, and chlorsulfuron were the most effective for *O. crenata* control in legumes (Garcia-Torres and Lopez-Granados, 1991b), and imazethapyr was also effective against *O. aegyptiaca* and *O. foetida* (Geipert, 1997). According to Garcia-Torres and Lopez-Granados (1991b), early-planting/high-yield cropping systems which are normally subjected to heavy and lasting parasite infestations would normally require a combination of preemergence treatments. Soaking broadbean and lentil seeds for five minutes in 0.01 percent herbicide solution or coating at 20 to 40 g·ha^{-1} (at sowing rate of 160 kg·ha^{-1}) with imazethapyr did not affect seed germination and crop growth but resulted in 60 to 80 percent *O. crenata* control. Treatment using imazethapyr followed by late postemergence application of imazapyr (5 g·ha^{-1}) resulted in more than 95 percent weed control. Effective seed treatments were with imazapyr for broadbean and imazethapyr in lentils (Jurado-Exposito, Garcia-Torres, and Castejon-Munoz, 1997).

Lentils: Effective seed treatment is with imazethapyr (Jurado-Exposito, Garcia-Torres and Castejon-Munoz, 1997). Glyphosate at 40 or 60 g·ha^{-1} controlled *Orobanche* and increased yield of this crop by 117 and 51 percent, respectively (Arjona-Berral and Garcia-Torres, 1983).

Peas: Orobanche crenata causes yield loss in this crop by 100 percent at high infestation, and when *O. crenata* and *O. aegyptiaca* are stabilized it was at about 50 percent for moderate infestation (Bernhard, Jensen, and Andreasen, 1998). Application of imazethapyr alone or imazethapyr with imazapyr at 5 g·ha^{-1} to pea seeds controlled *O. crenata,* and seeds treated at late postemergence resulted in excellent

Orobanche control (Jurado-Exposito, Castejon-Munoz, and Garcia-Torres, 1996). Ethylene dibromide alone or with chloropicrin controlled *O. crenata* and *O. cernua* (Jacobsohon et al., 1982). It has been previosly reported that two or three application of glyphosate at 150 $g{\cdot}ha^{-1}$ gave the highest yield under light infestation of *O. crenata* (Jacobsohn and Kelman, 1980), while preemergence tank mixture of propyzamide at 0.5 $kg{\cdot}ha^{-1}$ with methabenzthiazuron at 2.5 kg was also effective against *O. crenata* (Schlingloff and Alkamper, 1992).

Cabbage: Orobanche aegyptiaca is effectively controlled with glyphosate at 60 to 100 g a.i.$\cdot ha^{-1}$ or imazaquin at 5 to 10 g a.i.kg$\cdot ha^{-1}$. Trifluralin at 0.9 kg a.i.$\cdot ha^{-1}$ has a beneficial effect in reducing *Orobanche* infestation (Americanos and Vouzounis, 1995).

Carrots: Glyphosate at 1 to 1.5 $kg{\cdot}ha^{-1}$ resulted in moderate control of *Orobanche* spp. (Jacobshon and Kelman, 1982) and was also recommended for *O. crenata* and *O. aegyptiaca* control (Jacobsohn and Kelman, 1980).

Celery: Infestation of both *O. ramosa* and *O. aegyptiaca* was reduced by spraying 20 to 50 $g{\cdot}ha^{-1}$ glyphosate. The most effective treatment was with 50 g glyphosate$\cdot ha^{-1}$ (Americanos, 1991).

Sunflower: Soaking sunflower seeds in 0.1 percent boric acid or copper sulfate solutions increased the resistance of the plants to herbicides and to attack by *Orobanche* spp. (Aziz, 1989). Treatment of sunflower seeds with 40 ppm of benzothiadiazole Bion for 36 hours completely prevented infection of *O. cumana* (Buschmann, Giashi, and Sauerborn, 2001). Propyzamide applied to sunflower plants in the greenhouse followed by sprinkler irrigation reduced or delayed *O. crenata* and *O. aegyptiaca* infestation. Injection of the herbicide into irrigation water was the most effective method of control. Direct application of glyphosate and propazamide to the soil via irrigation system was effective in controlling *O. cernua* (Kleifeld and Herzlinger, 1984). Presowing incorporation of propyzamide plus thiram into the soil prevented herbicide decomposition and kept the sunflower free of *Orobanche* until the end of flowering (Herzlinger and Kleifeld, 1985). Preemergence application of oxyflourfen at 1 and 2 $L{\cdot}ha^{-1}$ reduced *O. cumana* density, and further reduction was obtained with presowing treatment of trifluralin at 3.5 $L{\cdot}ha^{-1}$ Oxyflourfen inhibited the formation of secondary haustorium of *O. cumana* (Horvath and Osztrogonac, 1991). Imazethapyr, imazapyr, and chlorsulfuron were the most effetive for *O. cernua* control (Garcia-Torres and Lopez-Granados, 1991b; Garcia-Torres et al., 1995, 1998). In a postemergence treatment, imazethapyr at 26.6 to 53.2 $g{\cdot}ha^{-1}$ killed *Orobanche* nodules (Alonso et al., 1998). Imazapic reduced *O. cumana* infestation in postemergence and in sequential treatments in sunflower in both irrigated and nonirrigated fields, while sunflower crops were not affected (Aly et al., 2001).

Tobacco: Higher tobacco yields were obtained with 0.25 kg a.i.$\cdot$ha^{-1} maleic hydrazide (MH), which was on par with hand weeding treatment both in infested and noninfested tobacco plants. However, higher concentrations of MH were toxic to tobacco crops. The foliage spray of maleic hydrazide triethanolamine salt at 6 and 9 kg$\cdot$ha^{-1} completely killed *Orobanche* spp., and the herbicide increased leaf yield and quality (Darbinyan, Saakyan, and Gabrielyan, 1977). MH reduced *Orobanche* spikes at 0.25 to 0.75 kg a.i.$\cdot$ha^{-1} applied at 30 or 40 days after transplanting tobacco and up to the flowering stage (Papyan and Arutyunyan, 1981; Dhanapal, 1996). Two applications of MH at 1 to 8 L$\cdot$ha^{-1} resulted in 90 percent control of *O. ramosa* in this crop (Danko, 1993a), while imazapyr and EPTC were less effective. Imazaquin at 0.07 + 0.1 kg, maleic hydrazide at 0.45 + 0.45 kg, and sulfosate at 0.2 + 0.3 kg all reduced number of *O. ramosa* plants per pot compared with untreated control (Lolas, 1994). Glyphosate and touchdown at 100 to 150 mL$\cdot$ha^{-1} were also used to control *O. ramosa* in tobacco (Musselman, 1993). Glyphosate applied at 0.035 percent to the lower parts of tobacco plants in spot treatment was effective against *O. ramosa* (Danko, 1993b), and under field conditions, glyphosate in two, three, or four treatments with total dose of 1000 mL$\cdot$ha^{-1} gave good control of *O. ramosa* (Sandri, Sandri, and Martini, 1998). At 0.5 kg a.i.$\cdot$ha^{-1} glyphosate applied 60 days after transplanting and imazaquin at 0.01 kg a.i.$\cdot$ha^{-1} applied at 30 days after transplanting reduced *Orobanche* population by almost 80 percent and increased tobacco leaf dry weight by more than 40 percent compared with the control (Dhanapal, ter Borg, and Struik, 1998). Tobacco yields were similar to weedfree treatment when 1.5 kg fluchloralin, 0.75 kg bentiocarb, or 0.1 kg oxyfluorfen and four-inter-row cultivation were done (Metha, Trivedi, and Patel, 1985). Preplant application of fluchloralin at 2 kg$\cdot$ha^{-1} gave high yield of tobacco equal to the weedfree treatment. Fluchloralin and diphenamid at 4 kg$\cdot$ha^{-1} were highly selective to tobacco, while pepulate and fluchloralin each at 3 kg$\cdot$ha^{-1} were highly effective against *O. cernua* (Palled et al., 1985). Allyl alcohol at 0.1 to 0.2 percent concentration sprayed four times at weekly intervals reduced infestation of *Orobanche* (India, Central Tobacco Research Insitute, 1979). Chlorsulfuron (2 g$\cdot$ha^{-1}), imazaquin (70 and 100 g$\cdot$ha^{-1}), and imazapyr (20 to 30 g$\cdot$ha^{-1}) applied pretransplanting, and glyphosate (200 to 300 g$\cdot$ha^{-1}), glyphosate-trimesium (300 and 400 g$\cdot$ha^{-1}), imazaquin (70 and 100 g$\cdot$ha^{-1}), and maleic hydrazide (450 g·ha^{-1}) applied over tobacco plants are effective in controlling *Orobanche* with no phytotoxicity in cv. Oriental (Lolas, 1997).

Tomato: In pot experiments, chlorsulfuron applied at 37 and 180 g a.i.·ha^{-1} was the most effective treatment for *O. ramosa* control and the least toxic to tomato (Syka and Eleftherohorinos, 1991). *Orobanche aegyptiaca* control in tomato was achieved with three-split application of chlorsulfuron at 2.5 g a.i.·ha^{-1} and triasulfuron at 7.5 g a.i.·ha^{-1} (Hershenhorn et al., 1998). Qasem (1998) reported excellent control of *O. ramosa* using chlorsulfuron at 2.44 g a.i.·ha^{-1} added with irrigation water or directly incorporated into the soil. Chlorsulfuron at 2.38 g, 7.14 g, and 11.9 g·ha^{-1} gave 63.45 percent, 100 percent, and 81.81 percent control, respectively (Dongola, Mohamed, and Ahmed, 2000). Imazaquin and glyphosate applied at 37 and 180 g a.i.·ha^{-1}, respectively, controlled *Orobanche,* but imazaquin reduced crop yield (Syka and Eleftherohorinos, 1991). Trials for chemical control of *O. ramosa* in tomato showed that imazethapyr and chlorsulfuron are effective treatments for both crop/parasite situations (Dongola, Mohamed, and Ahmed, 2000). Imazethapyr at the rate of 20, 40, and 80 g·ha^{-1} resulted in 77.72, 72.7, and 78 percent control, respectively. Oxadiazon controlled *Orobanche* spp. when applied in sprinkler irrigation to the root zone of tomato (Kleifeld et al., 1982). Deep incorporation of dazomet at three weeks before transplanting tomato prevented or delayed *Orobanche* emergence. Metham-sodium in drip-chemigation controlled *O. aegyptiaca* at 20 to 100 L·ha^{-1} (Kleifeld et al., 1991), and similar results were obtained by Goldwasser and colleagues (1993). Polyethylene mulching of the metam-treated soil drastically improved metam performance and *O. aegyptiaca* control. Three applications (at bud formation, near flowering, and beginning of ovary formation) of glyphosate at 150 g·ha^{-1} resulted in the highest tomato yield and lowest *O. ramosa* number (V'lchev et al., 1995). Glyphosate and sulfosate applied twice at 30 to 50 g a.i.·ha^{-1} were effective against *O. ramosa* and *O. aegyptiaca* on tomato and eggplant but reduced tomato yield. However, wide variations among tomato cultivars were detected for glyphosate treatment at 150 ppm glyphosate in 250 L water·ha^{-1} applied at two-true leaf stage (Foy and Jacobsohn, 1983; Foy, Jacobsohn, and Jain, 1988). Under in vitro conditions, effect of bensulfuron, chlorsulfuron, nicosulfuron, primsulfuron, trimsulfuron, thifensulfuron, and triasulfuron were tested for their effects on seed germination of *O. aegyptiaca* at preconditioning and germination stages. All herbicides reduced radicle elongation of the parasite. Chlorsulfuron and triasulfuron applied at 2.5 and 25 μM a.i. and rimsulfuron at 5 μM a.i. almost completely inhibited parasite development (Hershenhorn et al., 1998). Single application of rimsulfuron (25 g a.i.·ha^{-1}) to tomato at 10 days after planting and split applications at 10 and 20, and 10, 20, and 30 days after planting significantly reduced number of *O. aegyptiaca* inflorescences. Chlorsulfuron and triasulfuron applied directly to soil at rates of 3.75 g a.i.·ha^{-1} completely controlled *O. aegyptiaca*. Rimsulfuron at 10 to 20 g a.i.·ha^{-1} reduced *Orobanche* weight and number of shoots (Vouzounis and Americanos, 1998). Primisulfuron and chlorimuron each at 22.5 g a.i.·ha^{-1} reduced the number of parasite inflorescence (Hershenhorn et al., 1998).

Natural Products

Eizenberg and colleagues (2001) reported that root extracts of resistant sunflower cultivars inhibited development of *O. cumana,* while methalonic crude extract exhibited the strongest phytotoxic activity. These findings indicate possible involvement of phytoalexins in the resistant reaction (Eizenberg et al., 1999). Sunflower coumarins (ayapin and scopoletin) were found to act as toxic allelochemicals inhibiting *O. cumana* seed germination and seedling development (Perez-de-Luque, Rubiales, et al., 2001).

Application of plant oils (gigelly, groundnuts, palm, sunflower, safflower, niger, castor, linseed, neem, coconut, or tobacco seed oils), two or three drops applied to the top of vegetative *O. cernua* killed shoots in two to four days. Oil applications need to be repeated every four to five days and three to four times on emerging shoots (Krishna-Murty, 1992). Botanical pesticides from neem (*Azadirachta indica* A. Juss.), *Pongamia* spp., Inguilba tobacco (*Nicotiana gossei* L.), and nicotine sulfate for postemergence control of *O. cernua* with plant oils have also been recommended (Chari, Prabhu, and Ramaprasad, 1993).

Swabbing natural plant oils killed the bud and stem parts of the parasite by suffocation. Neem, coconut, and sunflower oils showed quick knockdown effects in killing the bud part, whereas neem oil did not kill the stem part of the parasite. Niger [*Guizotia abyssinica* (L.f.) Cass.], castorbean, and mustard oils appeared to be somewhat less effective (Dhanapal and Struik, 1998). In another study, neem, coconut, and sunflower oils affected the buds of *O. cernua* within three days; castor and niger oils killed buds in three to four days; and mustard oil took five days to kill buds. Coconut and sunflower oils killed *Orobanche* stems more rapidly than niger and castorbean oils, with no phytotoxicity on tobacco observed (Dhanapal et al., 1998). Volatile oils and crude acetone extracts of different plant species were found to be effective against *O. ramosa* in a greenhouse experiment (Solymosi, 1998). Orobanchol and alectrol produced by red clover are natural germination stimulants of *O. minor* (Yoneyama, Takeuchi, and Yokota, 2001), indicating that strigolactones are likely to be natural germination stimulants for *Orobanche* spp.

The role of microorganisms (e.g., *Streptomyces*) has been also implicated in *Orobanche* seed germination. Christeva and Naumova (1998) reported different strains of *Streptomyces* to induce germination of *O. ramosa* and concluded that microorganisms may take part in germination processes of the parasite. In another study, Yoneyama, Takeuchi, et al. (1998) reported stimulation of *O. minor* seed germination (up to > 50 percent) by certain fungal metabolites including cotylenins and fusicoccins at concentrations as low as 10^{-5} M.

Germination Stimulants

Calcium hypochlorite and GA in combination or used separately enhanced *O. crenata* seed germination (Pieterse, 1981). Immersion of 6- to 12-month-old seeds of *O. crenata* in 0.5 to 5 N H_2SO_4 solution for 1 to 20 minutes increased germination by 20 percent (Lopez-Granados and Garcia-Torres, 1996). Yoneyama, Ogasawara, et al. (1998) reported that jasmonates and related compounds elicited seed germination of *O. minor,* and methyl jasmonate was the most active stimulant. Synthetic germination stimulants, strigol analogs (GR compounds) were found to stimulate seed germination of different *Orobanche* species. Strigol has been reported for *Orobanche* stimulation (Vail et al., 1985), and its analogs GR7 and GR45 stimulated germination of *O. minor.* It was found effective at 10^{-10} g/seed and GR45 was effective at 10^{-6} g/seed (Spelce and Musselman, 1981). The highest activity occurred at concentrations of 0.1 to 1 ppm. GR7, GR 24, and GR 28 at 1 ppm were the most active of these compounds in inducing *O. ramosa* seed germination in vitro. Studies on the activity of various strigol analogs in different soils revealed that GR 7, GR 24, and GR 45 at 1 ppm were most active. In pot experiments, GR7 applied at 1 to 3 ppm 6 weeks before transplanting tomatoes resulted in significant reduction in the dry weight and number of shoots of *O. ramosa* plants and increased tomato yield (Saghir, Janudi, and Shafyuddin, 1983). Under field conditions, GR7 at 0.3 $kg \cdot ha^{-1}$ resulted in good control of *O. crenata* in broadbean grown in acidic soils. The application of 1.5 $kg \cdot ha^{-1}$ of the stimulant controlled *O. crenata* in alkaline soils (Saghir, 1994). Strigol and sorgolactone are also highly potent germination stimulants of *Orobanche* seeds (Willem et al., 1997). Perez-de-Luque and colleagues (2000) reported that parthenolide and 3,5-dihydroxydehydrocostus-lactone significantly increased *O. cumana* germination and presented higher activity than GR24.

Dhanapal (1996) reported that germination phase of *O. cernua* in tobacco is a critical period. The seed bank of the parasite can be reduced by stimulating the germination through chemicals, natural host stimulants, or both. GR 24 at 1.0 and 0.1 ppm was the standard to assess potential germination. Among other chemicals, gibberellic acid at 10 and 20 ppm was most effective. The stimulating effects of host plants were significant even when GR24 was applied. The role of this hormone was further confirmed by Joel (1998), who reported that application of uniconazole as an inhibitor of gibberellin synthesis in preplanting treatment to sunflower significantly reduced the number of infections by *O. cumana* and allowed the development of normal sunflower heads. Application of NAA and IBA induced increase in root growth of eggplant and enabled *O. cernua* to be maintained successfully. NAA at 60 ppm gave the best results (Rao, Joshi, and Sarma, 1982), and a recent study carried out by Harb (2002) confirmed the role that IAA plays in maintanace of *Orobanche* on tomato roots. The activity of germination stimulants for suicidal germination of *Orobanche* seeds under field conditions could be substantially enhanced by applying brassinolide (2α,3α,22R, 23R-tetrahydroxy-24S-methyl-B-homo-7-oxa-5α-cholestan-6-one) and related compounds to the infested soils (Takeuchi et al., 1995).

Genetic Engineering

Transgenics resistant to the herbicide Basta (glufosinate) could be used as a postemergence agent for *Orobanche* spp. control (Valkov et al., 1998). Transgenic potatoes resistant to assulam have been recommended as a method of weed eradication. *Orobanche* was completely suppressed on the best potato transformant when assulam was used at different concentrations (0 to 700 mM) at different times after shoot emergence, and potato viability was preserved (Surov et al., 1998). Tomato plants engineered with the *bar* gene conferring metabolic resistance to glufosinate, an inhibitor of glutamate synthase, tobacco plants containing the *SuRB-Hra* gene encoding a medified aetolactate synthase *(ALS)* gene conferring target-site resistance to chlorsulfuron, rape plants containing the *aro-A* gene encoding a modified enolphosphate-shikimate phosphate synthase conferring target-site resistant to glyphosate, and tobacco plants containing the *sul-1* gene conferring target-site resistance to assulam can be used for controlling *Orobanche* spp. (Joel et al., 1995). *Orobanche* control in transgenic tobacco cultivars ('Elenski 817', 'Melnik 812', 'Djebel', and 'Jaka') resistant to glufosinate ammonium was achieved at 2 to 20 $L{\cdot}ha^{-1}$ and with chlorsulfuron 60 to 180 g·ha^{-1} (Batchvarova et al., 1998). Promising results on tobacco resistance to *O. ramosa* were also reported after treating tobacco seeds with N-nitrosoethylurea or ethylmethanesulphonate solutions (Slavov, Batchvarova, and Valkov, 2001). Glyphosate proved to be excellent in suppressing *O. aegyptiaca* in both genetically engineered vetch and rapeseed (Nandula, Foy, and Orcutt, 1999). In transgenic potato plants, *O. aegyptiaca* was completely controlled by assulam without damaging potato plants (Joel et al., 1999).

Integrated Control Methods

Integrated control of parasitic weeds means to integrate different preventive measures and control instruments into the given farming system. These include cultural methods of *Orobanche* control which help to avoid germination and improve crop tolerance (sowing date, plowing, altering planting dates, crop rotation, trap and catch crops, resistant varieties, screening for field tolerance and identification of trap crops, and fertilization and varying ferilizer application); mechanical methods (hand weeding, tillage, cutting); physical methods (soil solarization, mulching and soil covers, flooding, irrigation, burning); biological methods (insects and pathogens); and future research (new herbicide treatments, pathogens, transgenic plants, germination inhibitors, and allelochemicals and herbicides applied with crop seeds); and chemical methods of control (glyphosate, preemergence herbicides, new postemergence herbicides, and soil fumigants/sterilents, germination stimulants) (Garcia-Torres, 1993; Krishna-Murty, 1994; Dhanapal et al., 1996; Fer and Thalouarn, 1997).

Not all satisfactory means of control have been found so far; therefore, a combination of suitable measures for a given situation in the sense of an integrated approach seems to be the most appropriate way to deal with the problem (Dominguez et al., 1993). The results obtained within single control methods showed insufficient control. Combining two or more of the single methods can tackle the problem better. However, in this regard analysis of the farming system is necessary (http://www.uni-hohenheim.de/~www380/parasite/oro_path.htm).

At present, control procedures consist of a combination of mechanical, chemical, and cultural procedures. Generally control consists of hand removal of the branched *Orobanche* plants, treating the area with low rates of glyphosate (Roundup), and planting false host crops (http://www-aes.tamu.edu/mary/brmrape/brmrape.htm). For an *O. cumana* control program in sunflower, resistant cultivars, deep plowing, and seed cleaning have been suggested (Petzoldt, Nemli, and Sneyd, 1993). Early sowing (mid-October) and annual application of 75 g·ha^{-1} imazethapyr and 40 g·ha^{-1} glyphosate is the best method for weed control in broadbean (Lopez-Granados, Diaz Sanchez, and Garcia-Torres, 1997). Glyphosate with Nuflim-P (14.4 g + 40 cm^3) resulted in the greatest seed yield of broadbean, glyphosate plus Greenzit-NPK (21.6 g + 60 cm^3) gave the greatest straw yield (Nassib, Hassanein, and Hussein, 1990), and glyphosate 34 g a.i.·ha^{-1} with NPK fertilizer effectively controlled *O. crenata* (Hussein, El-Deeb, and Saleib, 1998).

In tobacco, hand weeding, crop rotation, trap crops, catch crops with solarization, *Phytomyza orobanchia,* resistant varieties, fumigants, stimulants (ethylene), and herbicides (maleic hydrazide, glyphosate, sulfosate, imazaquin, and imazapyr) were all recommended for integrated control of *O. ramosa* and *O. cernua* (Lolas, 1998). In tomato, mulching with polyethylene sheets for 7 to 10 days after chemigation with metham sodium improved *O. aegyptiaca* control and gave effective control in the upper soil layer (Kleifeld et al., 1991).

In legume crops, the team work of ICARDA suggested an effective control program of different *Orobanche* species using integrated control methods. These are summarized in Table 21.5.

TABLE 21.5. Promising combinations for integrated *Orobanche* control in legume crops in Syria.

Crops	Management practices
Chickpea	Less infected variety (ILC 482); herbicide (imazethapyr 50g a.i./ha^{-1} postsowing), only if infestation of the site is heavy
Faba bean	Less infected or resistant variety (Spanish material) + imazethapyr 2 × 30g a.i./ha; imazethapyr 2 × 20g a.i./ha^{-1}, or delayed sowing + herbicide
Field pea	Delayed sowing + herbicide (imazaquin 2 × 20g a.i./ha^{-1} in spring)
Forage legumes	Less/not infected species *(Vicia dasycarpa, Lathyrus ochrus),* or delayed sowing + herbicide, or mowing for hay production followed by soil tillage
Lentil	Delayed sowing (December/January) + genotype, adapted for late sowing (+ herbicide, depending on severity of field infestation)
Narbonne vetch	Less infected genotype (e.g., ICARDA acc. 578) + sowing delayed for two weeks (under heavy infestation: imazaquin 2 × 20g a.i./ha^{-1}) in spring
Vetch	Less infected genotype (e.g., ICARDA acc. 715) + sowing delayed for two weeks

Source: Adapted from <www.uni-hohenheim.de/~www380/parasite/oro_path.htm>.

CUSCUTA *SPECIES (DODDERS)*

Cuscuta are usually regarded as annuals, but they may be perennial in warm climates, produce from seeds, and are the most problematic parasitic weeds in agricultural systems in the semihumid and semiarid areas of Africa and Asia (Sauerborn, 1994). The variation in seed dormancy enables these parasites to adapt to different environmental conditions. Dodders are classified as members of the morning glory family (Convolvulaceae) in old references and as members of dodder family (Cuscutaceae) in the more recent publications. Botanists recognize some 150 species of dodder, though some authors place a few of them into the genus *Grammica*. The number of species occuring in crops is at least 14, among which *C. campestris* Yunck. (common field dodder) is the most widespread and aggressive (Yang-han, 1994; Floridata, 2000; Swift, 2001). *Cuscuta* is a twining yellow or orange plant sometimes tinged with purple or red color, ocasionally white. The stems can be very thin and threadlike or relatively stout (species characteristics).

Germination and Development

Cuscuta seeds are easily confused with alfalfa and clover seeds. The fruit is about 1/8th inch in diameter, with thin papery walls, and contains one to four seeds. Seeds are yellow to brown or black, nearly round, and have a fine rough surface with one round and two flat sides. However, some species have a distinguishing rough seed coat and can be borne in tight balls or in loose clusters (Ristau, 2001; Swift, 2001). Seed dormancy in *Cuscuta* is mainly due to the seedcoat itself. *Cuscuta* seed therefore may stay dormant for several years until the testa is rotted. On the other hand, part of the seeds may germinate directly after maturity on the mother plant. Dormant seeds, which are called hard seeds, require 90 percent sulfuric acid treatment for 30 minutes for effective break of dormancy (Zaki, El-Metwaly, and Hassan, 1998). However, potassium nitrate and mechanical scarification can break dormancy in all species. In contrast, soft seeds are ready to germinate if favorable environmental conditions are available. Light also can induce dormancy breaking. Far-red and blue light are important for seed germination. Far-red and blue light induce mutual and self-parasitism of Japanese dodder (*Cuscuta japonica* Choisy) (Furuhashi, Kanno, and Morita, 1995; Tada, Suqai, and Furuhashi, 1996). The parasitism of *C. japonica* is controlled by the coordinated effects of two physical signals, far-red light and appropriate tactile pressure. Seed treatment of clover dodder (*C. trifolii* Bab.) and common field dodder (*C. campestris* Yunck.) was effective in increasing germination rate. The removal of the hard seedcoat by sulfuric acid and seed conditioning promoted *C. trifolii* germination (Lados, 1998).

The parasite, however, does not require any host stimulation or the presence of a host for seed germination to occur. Anatomical study of *Cuscuta* spp. grown separately in pots with the host crops mungbean, black gram, pigeonpea, sunflower, sesame, castorbean, rice, sorghum, and maize indicated that the parasite did not require any germination stimulus from the host. Germination occurs (permeability of seed testa provided) when temperature and moisture conditions are favorable, even immediately after harvest. However, seedling emergence of *Cuscuta* spp. (e.g., Chinese dodder, *Cuscuta chinensis* Lam.) depends on temperature, rainfall, weed population, cultivation, and cropping system. The optimal temperature ranges from 20 to 26°C. Sandy soil with sufficient moisture and acidic pH range is important for germination, but there is a variation among species in response to pH. A higher germination ratio for this spcies was found in the acidic pH range, around 5.5, while changes in pH had no effect on *C. campestris* (Lados, 1998). However, optimum conditions for seed germination were found in sandy rather than clay soils (Zaki, El-Metwaly, and Hassan, 1998).

Contact and Attachment

Dodder is a parasitic weed that grows only by penetrating tissues of host plants to obtain water and nutrients. The seedling usually emerges from shallow soil depths as a leafless shoot with a swollen, stubby terminal appendage. At the time of germination, the root has no apical meristem and the vascular tissue is exposed at the apex of the seedling. Nodes and small scalelike leaves are also present, and branching occurs at the nodes. Three days after emergence, many of the root cells, with the exception of the vascular tissue, have begun to lose their cellular integrity, with wall loosening and swelling of the cytoplasmic component crushing neighboring cells.

Immunocytochemical investigations indicate that both cellulose and xyloglucans in the cell wall are being degraded, leaving just the middle lamellae holding cells together. By five days of growth, even more dramatic loss of cellular integrity occurs, progressing from the radicle end. The shoot end of the plant, however, sustains normal growth. *Cuscuta* roots undergo rapid senescence/apoptosis, perhaps shunting the metabolites to the shoot, sustaining growth there until successful parasitism occurs. The parasite loses its root tissue within three to five days after emergence.

Seedlings grow upright for 2 to 4 cm before attaching to the host (Ristau, 2001). It emerges as an arch, and the tip extends and straightens upward after emergence. The entire seedling looks like a yellow thread with no stem-root differences. To survive, *Cuscuta* seedlings must attach to a suitable host within a few days of germination or they die. The epicotyl (3 to 6 cm long) searches for a host in a slow counterclockwise rotation that occurs after extention; it is even able to move for some distance to reach a host and establish contact with it.

The stem entwines around any elongated object. It is sensitive to touch, and its yellowish stem gropes in the air until it makes contact within one to three inches of the seedling. Stems grow and branch up to three inches per day of contact with a plant. The contact is made firm by one or more coils around the stem. The degree of twining, branching, and attachment varies among species. Some species have tendrils for twining, while shading can inhibit initial growth and attachment (Ristau, 2001). The yellow or orange vine strands entwine around stems and leaves of plants. The growing tips reach out and attack adjacent plants, forming a gradually enlarging circle of infestation. Patches of 3 m in diameter can be formed by a single plant of *Cuscuta* (see Photo 21.3). Upon flowering, clusters of white, pink, or yellowish flowers are produced, which soon form large numbers of seeds.

The parasite is more likely to coil on a host of high nutritional status and to grow away from hosts of poor quality than vice versa. They do this before taking up any food from the host to dissociate active choice from the negative effect of growth and mortality. If this host happens to contain food suitable to *Cuscuta* then a secondary stimulus is aroused which causes rootlike branches (haustoria) to form and penetrate the stem. The basal part of the

PHOTO 21.3. *Cuscuta campestris* spread on weed species along roadsides. (See corresponding plate in central color section.)

parasite soon shrivels away so that no soil connection exists (Swift, 2001), and then a dense mat will develop over the host as *Cuscuta* spreads (see Photo 21.4).

Cuscuta achieve greater growth volume when simultaneously infecting two hosts of different species rather than one species, which depends on the order in which the parasite encounters the hosts. This relationship is parasitic, not symbiotic. Xylem bridges between the host and *Cuscuta* extend throughout the length of the haustorium. Haustorial treachery elements in contact with host bundles form only after cells extended from the front of the haustorium become attached to the xylem tissue in all the dicots examined and also interfasicular tissues (Tsiovion, 1980).

Resistance to parasite penetration through host tissues is different for different plant species.The resistance mechanisms of Graminae and tomato to *C. campestris* were different. There was a mechanical barrier in Graminae, but in tomato infection is partially successful. On the other hand, there was a reaction between beans and tomato toward *C. campestris* infection. In tomato, if the haustorium manages to penetrate the outer layers of the undamaged cortex, a functional connection with the transport element is achieved, whereas in beans haustorial cells were damaged inside the host stem, thereby preventing their connection with host transport elements (Tsiovion, 1979). However, two bands of a vinyl group compound were

PHOTO 21.4. *Cuscuta monogyna* Vahl. (eastern dodder) on lemon [*Citrus limon* (L.) Burm.f.] in Jordan. (See corresponding plate in central color section.)

found in green bean extracts. The haustoria of *Cuscuta* did not succeed in entering the beans, resulting in the death of the parasite (Zaki, El-Metwaly, and Hassan, 1998).

In general, host resistance to the parasite or infection is considered at two levels, which were very well demonstrated in the resistance of sweet potato [*Ipomoea batatas* (L.) Lam.] and tomato in which establishment of haustorium connection was difficult. Host resistance to giant dodder (*C. reflexa* Roxb.) or infection was at the surface of the host and during the haustorial bursting and its connection with the vascular bundles of the host. The two hosts contained a strong inhibitor of cell-wall-degrading enzyme. This inhibited the cell-wall-degrading enzyme activity of the parasite and provided a defense mechansim to the host against haustorial penetration. Although haustorial penetration to host tissues becomes comparatively easy, the haustorial bursting inside the host tissue and subsquent connection with the host vascular bundles is prevented. Thus, the defense mechanism appears to spread throughout the resistant host tissues (Archana and Madhav, 1997).

In nonsusceptible species of sweet potato and potato the inhibition mechanism is related to different cations found in the host cell wall. From the divalent ions tested, Ca^{2+} proved to be the most potent inhibitor of *Cuscuta* cellulase, followd by Mg^{2+}, Mn^{2+}, and Fe^{2+}. Five- to tenfold Ca content was found in nonsusceptible plants (Nagar and Sanwal, 1984). Xylanase and cellulase activity of the host increased, while cellulase activity of the parasite decreased as the result of infection (Nagar, Singh, and Sanwal, 1984). In cotton (*Gossypium hirsutum* L.), resistance was based on the formation of suberized layers of cells from a secondary meristem resulting in isolation of the haustorium development of secondary parenchyma and expulsion of the parasite (Capderon, Fer, and Ozenda, 1985).

Tsiovion (1981) found that there was competition between the sink of the host and *C. campestris* for a common pool of photosynthate produced by the host, which reduced the shoot length of the host. In such competition conditions, the less efficient sink is weakened. He concluded that the parasite influences the host through the xylem. In addition to reduction in height, infection of the parasite reduced the chloroplast content by 24 percent while it increased carotenoid content by 44 percent. It was suggested that chlorophyll acts as a reactive site of the host-parasite interaction in cases of angiosperm parasites. There was high concentration of cytokinin in the haustorium-bearing region in *C. reflexa,* which might be due to high RNAase activity in the region or translocation of RNAase from the host to the parasite or both (Gupta and Singh, 1985). When coleus (*Coleus blumei* Benth.) is grown in quartz sand culture and fed with different rates of nitrate, parasitism by *Cuscuta* leads to a substantial increase in the chlorophyll concentration of host leaves and nitrate uptake of parasite plants. Increases in photosynthesis by the host plant and accelarated nitrogen uptake by the parasite are related to increased sink for assimilate and higher demand for nitrogen in the parasitized plant (Jeschke, Baig, and Hilpert, 1997).

The induction of haustorial formation in *C. reflexa* proved to be independent of the presence of a suitable host. During haustorial develoment, the growth rate of this parasite is retarded and an interaction of IAA and cytokinins is possibly involved in the regulation of these processes (Ihl and Wiese, 2000). *Cuscuta* species haustoria fused with the vascular bundle of purple foxglove (*Digitalis purpurea* L.) but avoided contact with the host of woolly foxglove (*Digitalis lanata* Ehrh.), although both species could be infected (Rothe et al., 1999). The cardenolide concentration decreased from the haustorium regions to the shoot tip (Rothe et al., 1999).

Cuscuta plants are yellow due to a limited amount of chlorophyll. Stems appear as leafless, threadlike branching vines. Older plants produce a mat of stems over the host and entwine and attach to the host with suckerlike organs called haustoria (Ristau, 2001). Roots are nonexistent or very reduced. They support the seedlings until atachment is made to the host. The root system is not suitable for nutrient uptake but can absorb water. However, *Cuscuta* is succulent and seldom needs water as a seedling (Ristau, 2001).

Many clusters of flowers develop after a short period of vegetative growth. Each flower is about 3 mm long (Salimi, 2000; Ristau, 2001; Swift, 2001; University of California, 2002a), white to pink, bell shaped, inconspicuous, in small numerous compact clusters. Flowering occurs from early summer to early autumn. Many capsules with two to four seeds are produced and fall to the ground or are gathered during harvest. Fruits are two to four globular capsules with a slight depression on top. *Cuscuta* is said to contain some chlorophyll in the buds, fruits, and stems, but the amount of food manfactured in this tissue is of little significance to the survival of the plant (Swift, 2001).

The *Cusuta* stems, nodes, tendrils, and even haustoria have regenerative cells that are capable of developing new shoots. Twenty or more shoots can develop from one tendril attached to a host. Even when all stems are removed from the host, the haustoria can reintroduce the infestation from within the host. Spread of *Cuscuta* is essentially by seeds. Dispersal is not typically by wind due to the seed weight. Most seeds spread in contaminated seeds, soil, and hay by humans, livestock, and machinery. Irrigation water can also spread *Cuscuta* seeds. Each plant produces thousands of hard seeds that drop to the ground and germinate during the next growing season if a suitable host is present; otherwise the seeds may stay dormant in the soil for years. The cause of dormancy exists in the seed coat, and it probably causes dormancy by production of mechanical barriers.

Biology, Ecology, and Physiology

Parasitic seed plants acquired host plants as food sources during the course of evolution, enabling them to survive in nutrient-poor sites. An adequate habitat for a host plant is also sufficient for *Cuscuta*.

Most *Cuscuta* spp. depend on the sucrose unloaded from the phloem of the host plant. In a study carried out on greater dodder (*C. europaea* L.), it has been shown that the parasite obtains sucrose from the apoplast of the host plant by releasing invertase enzyme that degrade carbohydrates into sucrose. When apoplast is tested it is found that intensive resortion of sucrose occurs within the stem after the release of the enzyme into the apoplast of the host (Wolswinkel and Ammerlaan, 1983). Phloem unloading in seed is similar to that of the stem, which is sucrose specific. In development, legume seed phloem transport into the strong sink was turgor sensetive (water turgor) (Wolswinkel, 1984). Some *Cuscuta* species (*C. campestris,* swamp dodder—*C. gronovii* Willd. ex Schultes, *C. reflexa,* and canyon dodder—*C. subinclusa* Dur. & Hilg.) possess thylakoids, which contain both chlorophyll a and b. Light-driven electron transport is present in all chlorophyll-containing species. The photosystem II efficiency was low and ranged from 0.511 to 0.687. Apparently the genus *Cuscuta* comprises species with different degrees of plastids with impaired protein production and gene expression-to-plastogene content (Kooij et al., 2000). The plastid *rbc* L could not be detected in *C. odorata* Ruiz. & Pav. but is present in all others. Neither *rbc* L nor the large subunit of ribulose-1,5-bisphosphate carboxylase-oxygenase (RUBISCO) could be decteted in *C. odorata* and *C. grandiflora* Kunth. RUBISCO and autofluoresence of chlorophyll are largely restricted to a band of cells adjacent to the vascular bundles. Both light and dark reactions in the chloroplast are functional. Carbon dioxide for carbon assimilation is derived from the internally transpired CO_2, and this layer of photosynthetic cells makes a positive contribution to the carbon budget of the parasite (Hibberd et al., 1998).

Thylakoid membranes of the preparasitic phase of five-angled dodder (*C. pentagona* Engelm.) are almost all organized into long, overlapping grana stacks of mainly two to five thylakoids with little space between adjacent stacks. Chloroplast proteins are associated with PSI and PSII as well as cytochrome and plastocyanin; stroma extracts contain rubisco and phosphoribulokinase. Rubisco is located in the stroma, whereas antibodies to the PSII protein light harvesting chlorophyll a/b complex and the oxygen evolving complex PSII are concentrated over the thylakoid. A limited capacity for CO_2 fixation is found in seedlings. The chloroplast from this species contains a number of proteins required for success of fixation of CO_2, and the proteins in the thylakoid are organized much like other higher plants with the exception of the large percentage of the thylakoid organized into grana (Sherman, Pettigrew, and Vaughn, 1999).

The fluxes and partitioning of nitrogen within the host *Lupinus* was found to be altered due to the presence of *C. reflexa* (Jeschke et al., 1994; Jeschke, Baumel, and Rath, 1995). When this species is grown with the host the quinolizidine alkaliod content and synthesis in the host decreased by about 50 percent, while the host-parasite system showed about 1.3-fold higher alkaloid. Sixty-three percent of the alkaloid was attracted by the parasite. This indicated that there is (1) a restriction of catabolic process within the infected lupins, (2) a massive shift of nitrogen metabolism in the direction of alkaloid, and (3) an enormous sink potential of *Cuscuta* for nitrogen compounds. They concluded that *Cuscuta* is likely to overcome the chemical barrier of alkaloid by accumulation and catabolic process.

Most of the nitrogen attracted by the parasite comes from the host phloem. *Cuscuta* derive only about 4.5 percent of their total alkaloids from xylem, while the rest are from the phloem (Baumel et al., 1995). The C:N ratio found in *C. reflexa* derived from phloem of the host was smaller in the parasite than in the host phloem, indicating the specific withdrawal of solute with a distinct reference to nitrogen. *C. reflexa* extracted C and N mainly from the host phloem, but only 0.5 percent of the carbon demand (Jescheke et al., 1994). Jeschke, Baumel, and Rath (1995) found that this parasite derived 99.5 percent of its carbon and 93.6 percent of its N demand from the host phloem. Fer, Painset, and Rey (1994) also suggested that *Cuscuta* obtain inorganic nutrients from the host phloem and exerted a very strong sink and competition efficiently with roots. It attracts about 81 percent of the current photosynthate and more than 23 percent of nitrogen from the host root. This causes stress on the plant root by changing resource allocation toward the parasite; water and nutrient acquisition of the host may then be affected.

Apart from carbon losses, it appeared to induce a sink-dependent stimulation of photosynthesis. More carbon compound mobilization toward the parasite stimulates more photosynthesis to overcome the demand. In contrast, the nitrogen fixation is inhibited in the roots of infected plants. As the parasite obtains most of its carbon and nitrogen demand from the phloem, the host requires massive transfer of nitrogen from xylem to phloem. Higher uptake of carbon compounds by the parasite affects the nodulating roots of white lupin (*Lupinus albus* L.).

Castor bean grown under limited N supply in quartz sand culture indicated that the dry matter production and growth of both the host and parasite were severely depressed with decreasing N supply to the host. Infection by the parasite resulted in increased tissue N levels in the host. However, nitrogen content of the system (castor bean + *C. reflexa*) was the same or more than uninfected plants because sink-dependent simulation increases nitrogen uptake by the host to satisfy the demand of the parasite. Increased nitrate uptake was reflected particularly at low N supply.

In the xylem transport, nitrate concentration was substantially elevated when the host is infected. In contrast, amino acids are decreased. This shows an inhibition of nitrate assimilation in the roots of parasitized plants under limited nitrogen supply. Because of high demand from the parasite sink, the nitrates taken up by the host are mobilized to the shoot where haustoria are. In addition, *C. reflexa* induced a substantial sink-dependent stimulation of the photosynthsis in the host leaves and a concomitant increase in stomatal opening and transpiration. This stimulated dependence on the relative sink size induced by the *C. reflexa* on the nitrogen nutrition uptake leads to leaf age, indicating that delayed senescence of leaves contributes to the oveall effect of parasite on its host by accumulating more nitrogen in the leaves. The *Cuscuta* induced inhibition on nitrate assimilation in the roots, and the increase in the nitrate uptake suggests that nitrate reduction was shifted toward the leaves in the presence of *C. reflexa* (Jeschke and Hilpert, 1997). Fer, Painset, and Rey (1994) found that the mineral compostion of vetch *(Vicia sativa)* was not modified despite the transfer of potassium toward *C. lupuliformis,* suggesting certain regulatory mechanism in legume. *C. campestris* infection decreased on plots fertilized with nitrogen. Lower rates of an inadequate ratio of mineral, particularily in dry years, accounted for high infestation of the parasite (Stojsin, Maric, and Jocic, 1991).

In *C. europaea,* stolon diameter is an indicator of resource availability. The greater probability of ramet initiation for resource-poor stolons (inter ramet) distance was noted in the poor patches. In constrast, the only formal model previously proposed assumed that resource availability is negatively associated with ramet length, generally, even in the face of contradiction. There is interspecific variation in stolon responses to the resource. It is postulated that likely causes are variation among species in sun/shade adaptation, commonly encountered resource patch size, and/or pressure of herbivory or ramet aggregation (Kelly, 1994).

Soil moisture is a prerequisite for germination and development of parasites; it acts directly and indirectly by affecting host physiology and growth. Germination trials indicated toward adaptation of parasite germination under dry conditions and took place even at an osmotic potential of –1.29 MPa.

Natural Enemies

Seven species of the weevil *Smicronyx* spp. are recorded as feeding on 12 *Cuscuta* species (Khan and Zafar, 1981). *Smicronyx jungermanniae,* a gall-forming weevil, and the fungus *Colletotricum gloeosporoides* are the most used natural enemies (Stroud, 1989). *Smicronyx tartaricus* was also reported from fields of *C. campestris* and *C. europaea* (Shinkarenko, 1982). Other natural enemies attacking different *Cuscuta* species are *Melanagromyza* sp., *Eupoecilia ambiguella,* and *Celastrina argiolus* (Protsenko, 1981). Biological control of *C. australis, C. chinensis,* and *C. maritima* was achieved using Lu Bao No. 1 which contains the fungus *Colletotrichum gloeosporioides* f. sp. *cuscutae* (Gao and Gan, 1992). Phytophages of *Celastrina argiolus* destroyed the fruits of *C. campestris, C. lupuliformis, C. monogyna,* and *C. lehmanniana. Smicronyx coecus, S. albapiches,* and *Smicronyx* spp. cause stem galls, and *Melanagromyza cuscutae* showed a similar effect. The larvae of the first generation mine the stems, while those of the second destroy the seeds.

The most interesting pathogens of *Cuscuta* are *Alternaria* spp. and *Cladosporium* spp., *Alternaria cuscutacidae* having been already used in biological control (Volkov, 1989). Wang (1990) reported *Prococidochares utilis, Agasicles hygrophila,* and *Zygogramma suturalis* and the fungus *Colletorichum gloeosporioides* to be good potential agents for dodder control. *C. gloeosporioides* f. sp. *cuscutae* can control *Cuscuta* in soybean (Templeton, 1992). In a more recent study, Hopen, Caruso, and Bewick (1997) reported control of *C. gronovii* in cranberry by *Fusarium tricinctum* and *Alternaria conjuncta/A. infectoria* applied preemergence or postemergence to the parasite in established cranberry beds.

Distribution and Host Range

Cuscuta spp. are distributed worldwide and attack many different host plants. International trade, mainly with contaminated crop seed, has led to the wide distribution of this parasite. *Cuscuta* species are native to almost every part of the temperate and tropical world. Habitat varies with the host plants that grow in disturbed areas. Some species are salt tolerant or pesticide resistant. *Cuscuta* seem to grow in full sun. Moisture requirements vary according to host plants. Some species grow in marsh lands and others in arid marsh lands (Floridata, 2000).

Greenhouse experiments conducted by Kelly (1992) revealed that when individual stems of *C. europaea* were transplanted on the host plant hawthorn (*Crataegus monogyna* Jacq.) of varying nutritional status, the pararasite was more likely to coil on (accept) a host with high nutritional status and to grow away from (reject) hosts of poor quality than vice versa. *Cuscuta europaea* exhibited this accepetance-rejection response before taking up any food from the host, so it was possible to dissociate active chioce from the negative effect of growth and mortality. The probabilty of the acceptance was guaged to expected reward (Kelly, 1992).

In its native Texas grassland, *C. attenuata* grows more vigorously in patches of mixed host species than the nonspecific host patches. Field experiments with naturally occurring host individuals demonstrated that the parasite achieved greater volume when it simultaneously infected two hosts of different species than the same species, and this effect depends on the order in which the parasite encounter those hosts (Kelly and Horning, 1999). Sodium-acting osmoticum was found to be the factor responsible for this response (Kelly and Horning, 1999).

Cuscuta parasitizes various kinds of wild and cultivated plants. It is especially destructive to alfalfa, lespedeza [*Lespedeza cuneata* (Dumont) G. Don.], flax, clover, and potato. Ornamentals attacked include *Chrysanthemum, Dahlia, Helinium,* Virginia creeper [*Parthenocissus quinquefolia* (L.) Planch.], trumpet vine [*Campsis radicans* (L.) Seem.], English ivy (*Hedera helix* L.), and *Petunia. Cuscuta* is particularly troublesome where alfalfa, clover, and onion are grown for seed because *Cuscuta* seeds are difficult to remove from the desired seed crop and can be spread with infested seeds (Ristau, 2001; Swift, 2001).

Cuscuta is a well-known parasitic plant that is sometimes seen in deserts, when its yellow, tangled stems climb over desert shrubs such as creosote bush [*Larrea tridentata* (DC) Coville], mesquite [*Prosopis juliflora* (Sw.) DC], and burrobush [*Ambrosia dumosa* (Gray) Payne], etc. Many species of *Cuscuta* are found in several parts of the world. They are difficult to tell apart, but all have the same characteristic feature of being parasites on higher plants. Some species of *Cuscuta* can parasitize a wide range of host plants, while others may be more host restricted (see Table 21.6). In any case, a heavy infestation by *Cuscuta* can be expected to cause significant damage to the host plants, especially in marginal conditions when water is in a short supply. *Cuscuta* spp. are not host specific. *C. europaea* has been reported from 237 host species, *C. epithymum* from 147, and *C. campestris* from 69. Host susceptibility can vary with age. *Cuscuta campestris* can attack young tomato, but not plants over 21 days old (Yang-han, 1994; University of Calfornia, 2002b).

TABLE 21.6. Economically important *Cuscuta* species and some of their main host plants.

Cuscuta species	Hosts	References
Cuscuta spp.	*Acacia arabica, A. auriculiformis*	Banerjee, Khatua, and Mukherjee, 1993
	Allium cepa	Qadri et al., 1985; Kamlesh, Gurjar, and Chitle, 1995
	A. sativum	Kamlesh, Gurjar, and Chitle, 1995
	Capsicum annum	Bewick et al., 1987
	Capsicum sp.	Mikha'lova and Genov, 1978
	Casuarina equistifolia	Banerjee, Khatua, and Mukherjee, 1993
	Daucus carota	Shlevin and Golan, 1982
	Digitalis lanata, D. purpurea	Rothe et al., 1999
	Glycine max	Templeton, 1992; Smith, 1993
	Glycosmis pentaphylla	Banerjee, Khatua, and Mukherjee, 1993
	Hevea brasiliensis	Thankamma, 1990
	Inga dulcis	Banerjee, Khatua, and Mukherjee, 1993
	Lotus corniculatus	Saric, 1981
	Lycopersicon esculentum	Mullen et al., 1996
	Medicago sativa	Saric, 1981; Young, 1984; Dawson, 1992; Canevari and Sieckert, 1993; Sarpe et al., 1996; Dimitrova, 1998; Faghih and Narimani, 1998
	Millingtonia hortensis	Banerjee, Khatua, and Mukherjee, 1993
	Nicotiana tabbacum	Diana and Castelli, 1994
	Parthenium hysterophorus	Dhopte, 1998
	Pelargonium zonale	Berg, van der Kooij, and Krupinska, 2001
	Solanum tuberosum	Bewick et al., 1987
	Spathodes campanulata, Tecoma stans	Banerjee, Khatua, and Mukherjee, 1993
	Vicia faba	Khallida et al., 1993
	Vitis vinifera	Bahat, 1985
C. approximata	*Medicago sativa*	Tepe, Deveci, and Keskiin, 1997
C. arabica	*Vitex agnus-castus, Zilla spinosa*	Karim, 1978
C. arvensis	*Beta vulgaris*	Minakova, 1983

C. australis	*Allium cepa, A. porrum, A. sativum, Amaranthus blitoides, A. gracilis*	Al-Haddar, 1989
	Amaranthus hybridus, Aster sp.	Zermane and Saghir, 1995
	Beta maritima	Al-Haddar, 1989
	B. vulgaris	Giannopolitis and Paspatis, 1980; Al-Haddar, 1989
	Brassica amplexicaulis, Brassica rapa, Calendula arvensis, Capsicum sp.	Al-Haddar, 1989
	Chenopodium album	Al-Haddar, 1989; Zermane and Saghir, 1995
	Chenopodium ambrosioides, C. murale, C. opulifolium, Chrysanthemum coronarium, Citrullus vulgaris, Convolvulus althaeoides, C. arvensis, Cucumis melo, Cucumis sativus, Cucurbita pepo, Cynara scolymus	Al-Haddar, 1989
	Cynodon dactylon	Al-Haddar, 1989; Zermane and Saghir, 1995
	Daucus carota	Al-Haddar, 1989
	Delphinium sp.	Zermane and Saghir, 1995
	Diplotaxis erucoides, Ecballium elaterium, Emex spinosa, Erodium malacoides, Euphoprbia helioscopia, E. peplus, Foeniculum vulgare, Hordeum murinum, Lactuca sativa, Lolium rigidum, Lycopersicon esculentum, Malva sylvestris, Mercurialis annua, Oxalis cernua, Picris echioides	Al-Haddar, 1989
	Polygonum aviculare	Al-Haddar, 1989; Zermane and Saghir, 1995

TABLE 21.6 *(continued)*

***Cuscuta* species**	**Hosts**	**References**
	Portulaca oleracea, Pteroselinium vulgaris, Raphanus raphnestrum, Senecio leucanthemifolius, Sinapis arvensis, Solanum melongena, S. nigrum, Sonchus oleraceus, S. tenerrimus, Spinacia oleracea, Vigna sinensis	Al-Haddar, 1989
	Vitis vinifera	Zermane and Saghir, 1995
C. babylonica	*Anagallis foetida Zilla spinosa*	Karim, 1978
C. campestris	*Alhagi maurorum*	Fayad, Hameed, and Al-Ani, 1990
	Allium cepa, A. sativum, Amaranthus angustifolius, A. hybridus, Ammi majus, Anacyclus clavatus, Anagallis arvensis, Andryala integrifolia	Zermane and Saghir, 1995
	Aster tripolium	Fayad, Hameed, and Al-Ani, 1990
	Beta vulgaris	Stojsin, Maric, and Jocic, 1991; Proctor, 1993; Zermane and Saghir, 1995
	Calendula arvensis	Zermane and Saghir, 1995
	Cannabis sativa	Bhellum and Mangotra, 1996
	Capsicum annuum, Chenopodium album, Chenopodium vulvaria, Chrysamthemum coronarium, Cicer arietinum, Cichorium intybus, Citrulus vulgaris, Citrus sp., *Convolvulus arvensis*	Zermane and Saghir, 1995
	Corchorus olitorius	Qasem (unpublished data)
	Coriandrum sativum	Zermane and Saghir, 1995
	Cucumis melo	Joel and Goldman-Guez, 2001
	Cucurbita maxima	Graph et al., 1985

Cyperus rotundus, Datura stramonium, Daucus carota, Erigeron bonariensis, Foeniculum vulgare, Fumaria officinalis, F. parviflora	Zermane and Saghir, 1995
Glycine max	Graph, Herzlinger, and Fridmann, 1985
Hevea brasiliensis	Thankamma and Marattukalam, 1995
Ipomoea palmata	Bhellum and Mangotra, 1996
Koeleria phleoides, Lamium amplexicaule	Zermane and Saghir, 1995
Lantana camara	Bhellum and Mangotra, 1996
Lavatera cretica, Lolium multiflorum	Zermane and Saghir, 1995
Lycopersicon spp., *L. esculentum*	Al-Menoufi and Ashton, 1991
Malus pumila, Medicago hispida	Zermane and Saghir, 1995
Medicago sativa	Dawson and Saghir, 1982; Heap, 1992; Sarpe, Dinu, and Lungulescu, 1992; Al-Juboory and Shati, 1996; Lados, 1998
Meliotus indica, Mentha rotundifolia	Zermane and Saghir, 1995
Mucuna bracteata	Thankamma and Marattukalam, 1995
Nicotiana tabacum	Zermane and Saghir, 1995
Ocimum basilicum	Joel and Goldman-Guez, 2001
Papaver hyridum	Zermane and Saghir, 1995
Parthenium hysterophorus	Bhellum and Mangotra, 1996
Petroselinum sativum	Zermane and Saghir, 1995
Phaseolus sp.	Tsiovion, 1979
Picris echioides	Zermane and Saghir, 1995
Pisum sativum	Tsiovion, 1981
Polygonum aviculare, Portulaca oleracea	Zermane and Saghir, 1995
Prosopis fracta	Qasem (unpublished data)
Prunus cerasus	Zermane and Saghir, 1995
Pueraria phaseoloides	Thankamma and Marattukalam, 1995

TABLE 21.6 *(continued)*

***Cuscuta* species**	**Hosts**	**References**
	Pyrus communis, *Ranunculus sardous,* *Rosa* sp., *Rumex crispus,* *Sinapis nigra,* *Solanum lycopersicum*	Zermane and Saghir, 1995
	Solanum melongena	Al-Juboory and Shati, 1996
	Solanum tuberosum, *Sonchus oleraceus*	Zermane and Saghir, 1995
	Trifolium sp.	Qasem (unpublished data)
	Trigonella foenum-graecum, *Vicia faba*	Zermane and Saghir, 1995
	Vigna sinensis	Fayad, Hameed, and Al-Ani, 1990
	Xanthium cavanillesii, *X. spinosum*	Zermane and Saghir, 1995
C. chinensis	*Ammi visnaga*	Karim, 1978
	Guizotia abyssinica	Tosh, Patro, and Misra, 1978
	Mentha longifolia	Karim, 1978
	Nicotiana tabbacum	Yong-han et al., 1999
	Solanum nigrum	Karim, 1978
	Vigna mungo, *V. radiata*	Rao and Rao, 1991
C. epilinum	*Linum usitatissium*	Zermane and Saghir, 1995; Mirek, 1997
C. epithymum	*Achillea millefolium*	Doyle, 1993
	Allium cepa	Zermane and Saghir, 1995
	Artemisia herba-alba	Qasem (unpublished data)
	Asperula cyanchia	Doyle, 1993
	Asphodelus sp., *Bupleurum spinosum*	Zermane and Saghir, 1995
	Capparis spinosa	Qasem (unpublished data)
	Capsicum annuum, *Chenopodium album,* *Cloeme arabica,* *Convolvulus arvensis,* *Coronopus squamatus,* *Dekerra racemosa*	Zermane and Saghir, 1995
	Heliotropium europaeum	Kutbay, Karaer, and Klinc, 1996
	Hormathophylla spinosa	Gomez, 1994
	Lotus corniculatus	Doyle, 1993
	Matthiola longipetala, *Medicago sativa*	Zermane and Saghir, 1995

	Medicago sativa	Garcia and Zaragoza, 1983
	Nicotiana tabacum	Qasem (unpublished data)
	Plantago coronopus, Scabiosa crenata, Solanum lycopersicum, Sonchus oleraceus	Zermane and Saghir, 1995
	Thymus praecox, Trifolium pratense, Viola sp.	Doyle, 1993
C. europaea	*Aegopodium podograria*	Gal'vidis, 1993
	Beta vulgaris	Rapparini and Campagna, 1998
	Crataegus monogyna	Kelly, 1992
	Fraxinus excelsior, Lamium album	Gal'vidis, 1993
	Lamium sp.	Zermane and Saghir, 1995
	Rubus sp., *Solidago canadensis*	Gal'vidis, 1993
	Urtica dioica	Gal'vidis, 1993; Zermane and Saghir, 1995
C. gronovii	*Red potatoes, Allium* spp., *Brassica oleracea*-Italica group	Gaertner, 1950
	Citrus sp.	–
	Daucus carota	Bewick et al., 1987
	Lactuca sativa	Gaertner, 1950
	Vaccinium sp.	Hopen, Caruso, and Bewick, 1997
	Vitis vinifera	–
C. indecora	*Medicago sativa*	Dawson and Saghir, 1982; Cudney, Orloff, and Reints, 1992; Cudney, Orloff, and Demason, 1993
C. japonica	*Nicotiana tabacum*	Liu, 1992
	Salix purpure	Huang and Li, 1991
	Vigna radiata	Tada, Suqai, and Furuhashi, 1996
C. kotschyana	*Marrubium* sp., *Peganum harmala, Phlomis* sp.	Karim, 1978
C. kurdica	*Campanula syspirensis, Cousinia* sp., *Eryngium billardieri, Galium kurdicum, Thymus* sp.	Karim, 1978
C. lehmanniana	*Populus euphratica, Quercus aegilops, Rosa canina*	Karim, 1978
C. lupuliformis	*Pelargonium zonale*	Fer, 1983

TABLE 21.6 *(continued)*

Cuscuta species	Hosts	References
C. monogyna	*Citrus* sp.	Qasem (unpublished data)
	Echinochloa colonum	Karim, 1978
	Pistacia lentiscus	Cited by Zermane and Saghir, 1995
	Rubus coriaria	Karim, 1978
	Vitis vinifera L.	Qasem (unpublished data)
C. palaestina	*Alhaji maurorum*	Karim, 1978
	Bupleurum spinosum	Zermane and Saghir, 1995
	Tragopogon major	Karim, 1978
C. planiflora	*Alhaji maurorum*, *Allium cepa*	Mamluk and Weltzien, 1978
	Artemisia scoparia, *Astragalus spinosa*	Karim, 1978
	Beta vulgaris	Mamluk and Weltzien, 1978
	Bubleurum spinosum	Zermane and Saghir, 1995
	Capsicum annuum	Mamluk and Weltzien, 1978
	Cistus heterophyllus	Zermane and Saghir, 1995
	Cucumis sativus	Mamluk and Weltzien, 1978
	Cynoglossum cheirifolium	Zermane and Saghir, 1995
	Elaeoselinum fontanesii	Zermane and Saghir, 1995
	Erucaria sp.	Karim, 1978
	Globularia alypum	Zermane and Saghir, 1995
	Helianthamum lipii	Karim, 1978
	Lagonychium fractum	Mamluk and Weltzien, 1978
	Launaea sp., *Lotus creticus*, *L. palustris*	Zermane and Saghir, 1995
	Lycopersicon esculentum, *Medicago sativa*	Mamluk and Weltzien, 1978
	Mentha pulegium, *Moricandia alypifolia*	Zermane and Saghir, 1995
	Phagnalon rupestre	Karim, 1978
	Salvia balansae	Zermane and Saghir, 1995
	Solanum melongena, *S. nigrum*	Mamluk and Weltzien, 1978
	Thymus sp.	Zermane and Saghir, 1995
	Trifolium alexandrinum	Qasem (unpublished data)
C. reflexa	*Aegle marmulos*	Rajak, Rai, and Panday, 1985
	Ancistrocladus heyneanus	Bringmann et al., 1999
	Bougainvillea spectabilis	Debabkata et al., 1999
	Brassica campestris	Mishra and Sanwal, 1994

	Brassica juncea	Mishra and Sanwal, 1992
	Clerodendron inermis	Rajak, Rai, and Panday, 1985
	Coleus blumei	Jeschke, Baig, and Hilpert, 1997
	Crotalaria juncea	Rao, Rao, and Reddy, 1985
	Dauranta repens	Rawat, Sachan, and Sachan, 1994
	Digitaria ciliaris	Dey and Pati, 1998
	Foeniculum vulgare	Bhati, 1994
	Hamelia patens	Rajak, Rai, and Panday, 1985
	Helianthus annuus	Sharma, Khanna, and Sanwal, 1985
	Hibiscus cannabinus	Rao and Gupta, 1981
	Ipomoea fistalosa	Rawat, Sachan and Sachan, 1994
	Lagerstroemia parviflora	Rajak, Rai, and Panday, 1985
	Lantana camara	Sharma, Khanna, and Sanwal, 1985; Rawat, Sachan, and Sachan, 1994
	Lupinus angustifolius	Baumel et al., 1993
	Medicago sativa	Sharma, Khanna, and Sanwal, 1985
	Pelargonium zonale	De Bock, Fer, and Marigo, 1995
	Phaseolus trilobus	Rao and Gupta, 1981
	Pisum sativum	Sharma, Khanna, and Sanwal, 1985
	Plumeria rubra	Rajak, Rai, and Panday, 1985
	Ricinus communis	Jeschke and Hilpert, 1997
	Thevetia neriifolia	Debabkata et al., 1999
	Vigna mungo	Rao, Rao, and Reddy, 1985; Rao and Rao, 1991
	V. radiata, *V. triloba*	Rao, Rao, and Reddy, 1985
	Vitex negundo	Reddy, Nalini, and Prabhakar, 1990
C. santapaui	*Scurrula cordifloia*	Pundir, Purohit, and Promod, 1994
C. trifolii	*Medicago sativa*	Sarpe, Dinu, and Lungulescu, 1992; Lados, 1998
	Solanum melongena	

Economic Impact

Cuscuta infestation can recur for many years once seeds enter the soil. A portion of the *Cuscuta* seeds from the soil seed bank is typically always available for germination under environmental conditions suitable for the host plant. A mature *Cuscuta* plant develops a mat up to 30 feet in diameter over and around the host plant. Yield and stand reduction can occur from severe infestations which can stunt, smother, and kill host plants. The moisture and density of *Cuscuta* can cause harvesting problem. *Cuscuta* is not considered to be either poisonous or nutritious to livestock. The similarity in appearance and size of *Cuscuta* seed to that of the crop seed causes difficulty in cleaning seeds. Roughness of the *Cuscuta* seed is used to remove the parasite from contaminated seeds, yet some contamination commonly remains. *Cuscuta* is the only weed legally restricted as a contaminant to crops in each state of the United States (Ristau, 2001). For obvious reasons, the U.S. Department of Agriculture classifies dozens of *Cuscuta* species as noxious weeds. American dodder (*C. americana* L.) and *C. campestris* are particularly damaging agricultural weeds.

Cuscuta is a major economic problem for such crops as alfalfa, clover, and flax. Other plants commonly parasitized by *Cuscuta* include many daisy relatives (especially *Chrysanthemum*), virginia creepers, trumpet vines, English ivies, petunias, camellias, citrus, beets, redbuds, hollies, sumaes, button bushes, cucumbers, raspberries, potatoes, tomatoes, roses, coleus, impatiens, and various legumes. *Cuscuta* can also carry plant viruses, including phytoplasma, which is responsible for many of the yellow diseases. Although *Cuscuta* is not thought of as a poisonous plant, cows and horses have shown colicky symptoms after eating it (Floridata, 2002).

In Algeria *Cuscuta* causes 5 to 100 percent loss in such crops as potato, carrot, tomato, and sugar beets. In fruit tree nurseries, *Cuscuta* considerably delays seedling growth. In Morocco, it causes 20 to 50 percent loss in vegetables, and it can reach up to 100 percent in pepper mint (Zermane, 2000). In Egypt, *Cuscuta* is a serious problem in some fields of forage and vegetable crops, fruit trees, and ornamental plants (Hassanein and Ibrahim, 2000).

Economic Uses

Various *Cuscuta* species have been used to treat a great variety of different ailments in numerous cultures. Some are thought to take on and enhance the medicinal properties of the plants they parasitize. European/clover dodder [*C. epithymum* (L.) Merr.] that grows on thyme (*Thymus* spp.) has been a favorite in European folk medicine. The following species are also frequently mentioned as medicinal herbs: *C. americana, C. chinensis, C. europaea, C. japonica,* bigfruit dodder (*C. megalocarpa* Rydb.), and *C. reflexa* (Floridata, 2002).

Control

Different methods are effective in controlling *Cuscuta* either before or after emergence. The autotrophic period of the *Cuscuta* life cycle, i.e., the germination and establishment on the host, seems to the best time to apply any methods of control. Xylem bridges between the host and the parasite extends throughout the length of the haustorium at which the host and parasite make contact to exchange food and nitrogen. There are different mechanisms of resistance in hosts. For example, when a solution of propyzamide, the fumigants metham-Na, and methylbromide as well as treament stress were applied to soft and hard seeds, hard seeds germinated after they were treated with acid scarification. Chemicals were not effective in controlling hard-seeded *Cuscuta* because these seeds are nongerminating. Therefore, seed hardness is an effective form of resistance to chemical control (Horowitz, Gevelberg, and Bucsbaum, 1983). Control methods of *Cuscuta* are discussed next.

Prevention Methods

As *Cuscuta* seeds stay dormant for several years, it is important to control broadleaf weeds; otherwise, *Cuscuta* can reproduce on them. Distribution and infestation by *Cuscuta* species throughout the world is mainly due to planting contaminated crop seed. Therefore, the best method for *Cuscuta* control is by preventing it to infest any field, starting with the use of clean crop seeds. Seed cleaning today is easily done by mechanized cleaning with felt rollers, which trap the rough-surfaced *Cuscuta* seed.

Cultural and Physical Methods

Hand pulling and burning, plowing, crop rotation with cereals, resistant varieties, and trap and catch crops have been used in some North African countries (Zermane, 2000) and in the United States (Ristau, 2001) to control *Cuscuta*. Flail mowing and burning along with preemergence herbicides are found to control *Cuscuta* in established alfalfa (Cudney, Orloff, and Reints, 1992). These treatments are effective in controlling attached *C. indecora* are economical, and caused less reduction to lucerne yield and stand density. Such cereals as sorghum, rice, and maize are never infected when grown in mixture of pulse and oil crops than when grown separately. It is suggested that infestation may be minimized by the cultivation of these crops in highly infested regions where possible (Kumar et al., 1993).

Mineral fertilizers. Spraying the infected plants with cellulase inhibitors Ca^{2+} selectively controlled *C. reflexa* in greenhouse experiment (Nagar and Sanwal, 1984). *Cuscuta campestris* infestation on sugar beet decreased on plots fertilized with N and P alone or with NPK combination (Stojsin, Maric, and Jocic, 1991). Low nitrogen supply severely depressed *C. reflexa* on castorbean (Jeschke and Hilpert, 1997).

Hand weeding. Weeding of parasitic weeds is one option open to all farmers, and its use should be the first priority of extension work, though its exact function and limitations should be clearly understood. Even in fields with dense infestation weeding should be applied. Although time-consuming, in combination with other methods, it can reduce the seed bank very efficiently and is economical on a longer-term basis. However, it is efficient only if different instructions are followed such as time of weeding with the beginning of the flowering stage (several repetitions of weeding are necessary) and burning of weeded plants.

Flaming/burning. For *Cuscuta* control, flaming is practical with a propane- or butane-fueled burner (instead of contact herbicides); however, the crop will also be damaged by this treatment. When *Cuscuta* has already set seeds, it has to be burned at the site where it grows to avoid any spread of viable seeds and to kill seeds that already dropped on the soil surface. Large seed dodder (*C. indecora* Choisy) in lucerne fields was controlled effectively by flail mowing and burning using a hand-held propane gas burner. Burning *Cuscuta* patches at the end of the season reduced seed viability by an average of 99 percent. It is proposed that the use of a three-tiered integrated approach consisting of preemergence herbicides followed by flail mowing in midseason to control escaped *Cuscuta,* and burning at the end of the season to reduce *Cuscuta* seeds would provide an effective control of the parasite (Cudney, Orloff, and Reints, 1992). Similar results were reported with *C. campestris* in lucerne using glyphosate as post-emergence, flail mowing, and gas burning (Heap, 1992).

Deep plowing. This method may reduce the infestation if the top soil layer is properly turned to the bottom. Where equipment is available and soil permits, this technique can form a useful part of an integrated control management. Seedlings of *Cuscuta* which are just germinated will be killed when buried in the soil. Sanding may be used as a part of a program consisting of a combinations of multiple management methods directed toward the control of *Cuscuta* (Sandler, Else, and Sutherland, 1997). This method was effective in reducing seedling emergence of *C. gronovii,* and at least 2.5 cm of sand was needed to reduce parasite emergence (Sandler, Else, and Sutherland, 1997).

Soil solarization. During hot weather, moist soil is covered by transparent PE sheets to increase soil temperature. The mode of action on seed during solarization is a combined effect of increased temperature, high moisture and modified gas composition in soil, and microroganisms attacking the parasite seeds. The measure might be effective in killing seeds of *Cuscuta* species. Soil solarization with chicken manure were effective against *C. campestris*. Solarization reduced seed germination of this species by 95 percent (Haidar et al., 1999). The relatively high cost of the method can be partly compensated by the control of difficult-to-control pests including nematodes, most other weeds, fungal and bacterial pathogens, and an improved nutrient status of the soil. Futhermore, there is a residual effect over several years following the solarization treatment, resulting in increased crop yields. However, it is applicable only in certain geographical regions with high temperatures.

Resistant Cultivars

Bean cultivars, namely 'Bulgarian' and 'Tenderette', show a very high degree of resistance to *C. campestris* (Tsiovion, 1979). Black gram cv. T9 and mungbean cv. M2 were tolerent to *Cuscuta* infestation due to their early vigor and hard, woody stem (Kumar and Kondap, 1993). In another study, two cultivars of *V. mungo* (cv. LBG-628 and AKU-3) showed some tolerance, while mungbean cv. PDM-85-182, PDM-85-186, RMG-131, RMG-146, and Lam-M2 showed considerable resistance to *C. chinensis* (Rao and Rao, 1991). The mechanism of resistance to attack by *Cuscuta* spp. exhibited by *Gossypium hirsutum* var. *latifolium* cv. Rex and as stated before is based on the formation of a suberized layer of cells from a secondary meristem resulting in isolation of the haustoria, development of secondary parenchyma, and explosion of the parasite (Capderon, Fer, and Ozenda, 1985). In more recent glasshouse and field work carried out to evaluate the tolerance of commercial hybrid tomato varieties to *C. campestris,* Goldwasser, Lanini and Worbel (2001) found that Heinz varieties 9492, 9553, and 9992 exhibited tolerance to the parasite and the parasite failed to penetrate into the stems of these varieties. Different crop species including okra [*Abelmoschus esculentus* (L.) Moench.], sugar beet (*Beta vulgaris* L. ssp. *cicla*), Asian pigeonwings (*Clitoria ternatea* L.), muskmelon (*Cucumis melo* L.), sea island cotton (*Gossypium barbadense* L.), pea vetch, tomato, mint, French bean, and potato are reported as being resistant to *C. campestris* (Farah, 2001).

Biological Methods

Several insects restrict *Cuscuta* growth, as already mentioned, the most promising being a small weevil, *Smicronyx* sp., which feeds on growing points and forms galls on the young *Cuscuta* shoots. *Smicronyx jungermanniae* Reich., a gall-forming weevil, has been used successfully for controlling *C. campestris* in Hungary and the former USSR (Stroud, 1989; Parsons and Cuthbertson, 1992). Different *Cuscuta* species are attacked by *Fusarium tricinctum* and *Alternaria* sp., and crops tested including carrots, cranberry, celery, mint, potato, and alfalfa showed no susceptibilty to either of the fungi (Bewick et al., 1987). *Cuscuta campestris* is also attacked by the fungal pathogens *Alternaria cuscutacidae* in southern portions of the former Soviet Union, *Colletotrichum destructivum* in the United States, and a form of *Colletotrichum gloeosporioides* in China.

The fungi species *Fusarium tricinctum* and *Alternaria conjuncta* showed excellent efficacy when used singly or in combination at several concentrations for the control of *C. gronovii* in cranberry beds (Hopen, Caruso, and Bewick, 1997). These mycoherbicides controlled well when they are applied either pre- or postemergence. The autotrophic period of the parasite seems to be the best period to apply any methods of control (Salle, Aber, and Le-Coz, 1984). Recently, Fayad, Hameed, and Al-Ani (1990) found that *Alternaria alternata* and *Geotrichum candidum* are both effective in controlling *C. campestris* on alfalfa, and treatment with these fungi resulted in 95 percent control of the parasite. Fenggen, Yanghan, and Fuzhen (1998) isolated *Fusarium semitectum, Fusarium solani, Pestalotiopsis guepinii,* and *Alternaria tenuis* from infested stems of *C. reflexa* and were able to reinfest and kill both *C. japonica* and *C. reflexa*.

Although several insects occurring on *Cuscuta* were investigated for their potential use as biological control agents, no promising method has been developed so far. On the other hand, based on the fungus *Colletotrichum gloeosporioides* f. sp. *cuscutae,* the product Luboa II has been developed against *C. chinensis* in China.

Chemical Methods

Herbicides. Herbicides are mainly applied either before or after parasite emergence. Effective control of *Cuscuta* sp. was obtained with 18 to 22.5 kg·ha^{-1} maleic hydrazide applied at the flowering stage (Kolomiets et al., 1982). Soft seeds of *C. campestris* were found to be affected by propyzamide, metham-Na, methyl bromide, and heat stress at 45°C (Horowitz, Gevelberg, and Bucsbaum, 1983). Complete control of *Cuscuta* sp. in newly established alfalfa was achieved using 3 kg·ha^{-1} propyzamide in preemergence treatment (Kontsiotou, 1981). This herbicide reduced *Cuscuta* spp. on mixtures of established lucerne, redclover, or birds foot trefoil in mixture (Kontsiotou, 1979; Saric, 1981). Propyzamide, perflaidone, and chlorpropham were the most effective against *C. campestris* on lucerne (Paspates and Damanakes, 1979). Propyzamide at 2.38g·ha^{-1} applied at 30 days from sowing and glyphosate at 75 ppm gave good control of *Cuscuta* on alfalfa without any adverse effect on the crop (Hassanein and Ibrahim, 2000). Pronamide 50 percent at a rate of 1.1 kg a.i.·ha^{-1}, glyphosate 48 percent at rate of 86 or 56 g a.i.·ha^{-1} and imadazolinone at 171.5 g a.i.·ha^{-1} were also effective in controlling *Cuscuta* in Egyptian clover (Abd El-Hamid and Shebl, 2000). At 1.25 and 2.5 kg·ha^{-1} propyzamide completely controlled *Cuscuta* in mungbean, blackgram, mothbean, kenaf (*Hibiscus cannabinus* L.), and sunhemp (Rao and Gupta, 1981).

Propyzamide was highly effective against *C. australis* in sugar beet at an intermediate stage of *Cuscuta* (Giannopolitis and Paspatis, 1980), and at 3.5 kg·ha^{-1} decreased *C. arvensis* var. *calycina* infestation in this crop by 89 percent (Minakova, 1983). When used at 1.5 kg·ha^{-1} at the stage between attachment to the host and before flowering, the herbicide showed effective control of *C. australis* (Giannopolitis and Paspatis, 1980). At 3 to 4 kg·ha^{-1} it gave the best control and reduced *Cuscuta* spp. and their contamination (Popovic and Stjepanovic, 1995). Satsifactory control of *C. europaea* in beet was achieved by split use of reduced doses of propyzamide from the first stages of develoment of the crop, together with the normal postemergence treatment during products containing ethofumerate. Optimal control of *C. europaea* led to an increase in the productivity of the beet crop (Faghih and Narimani, 1998; Rapparini and Campagna, 1998). Preemergence application of propyzamide prevented emergence of *C. campestris* in chickpeas, tomato, and pumpkins. The herbicide was also active by postemergence treatment; when applied at 0.5 kg·ha^{-1} it provided early control of *C. campestris* in chickpeas (Graph, Herzlinger, and Fridmann, 1985). Kontsiotou (1979) found that propyzamide at 300 mL·ha^{-1} and chlorathal-dimethyl at 1.1 kg·ha^{-1} applied preemergence were effective in field experiment. Control of *Cuscuta* in the chickpea was also selectively obtained by propyzamide or chlorathal-dimethyl in preemergence treatment or propyzamide postemergence.

Chlorathal-dimethyl at 10.5 kg·ha^{-1} provided the most effective control of *Cuscuta* species in carrots (Shlevin and Golan, 1982) and when used at 11.25 kg·ha^{-1} controlled *C. campestris* in carrots and onions (Beuret and Neury, 1979). Ethofumisate had strong inhibitory effect on germination of the same parasite in carrots, onions, and redbeet (Beuret, 1981). Preemergence application of this herbicide at 8 or 9 kg·ha^{-1} resulted in complete control of the parasite in newly established alfalfa (Kontsiotou, 1981).

DCPA or H26905 killed *C. campestris* and *C. indecora* on lucerne in a selective application (Dawson and Saghir, 1983) and gave good control of *Cuscuta* in alfalfa (Faghih and Narimani, 1998). Trifluralin selectively controlled *C. indecora* in alfalfa (Cudney, Orloff, and Demason, 1993), and thiazoy applied preemergence selectively controlled *Cuscuta* in this crop (Cudney, Orloff, and Demason, 1993). Preemergence application of linuron at 0.75 or 1.5 kg·ha^{-1} gave better control than the same rate of prometryn and was effective against *C. chinensis* at 0.75 or 1.5 kg·ha^{-1}. Preemergence application of dibutalen at 1.43 to 2.15 kg a.i.·ha^{-1} in early season gave complete or high level of control of the same species when applied at an approporiate time (Qian, 1996).

Successful use of herbicides has been limited, and repeated application is common. Herbicide recommendations for *Cuscuta* in forage crops can be more complicated than for other row crops. Soil-applied herbicides offer the most effective means of control, because control occurs before parasite attachment to the host. Unfortunately, these herbicides have a limited longevity, making repeated treatment necessary. In summary, potential soil-acting herbicides include DCPA, chlorpropham, protamide, dichlobenil, and trifluralin (Ristau, 2001). Dawson (1990) suggested pendimethalin and prodiamine at 2.2 to 3.4 kg·ha^{-1} and trifluralin at 4.5 kg·ha^{-1} as suitable for practical *Cuscuta* control. Pendimethalin at 1.25 kg or higher rates was effective in reducing the infestation of *Cuscuta* in *Vigna mungo*. Postemergence application of this herbicide at 1.5 kg·ha^{-1} alone or with fluchloralin followed by water spray also gave the lowest levels of *Cuscuta* infestation (Rao and Rao, 1991). Preplant application of pendimethalin at rates of 0.5, 1, or 1.5 kg a.i.·ha^{-1} in 200 L spraying liquid completely prevented *Cuscuta* spp. infestation in legume crops and up to 42 days, the infestation was between 2 to 6 percent (Zaki, El-Metwaly, and Hassan, 1998). Fluchloralin at 1.87 kg·ha^{-1} incorporated presowing, nitrofen and pronamide in *Vigna* spp. and sunhemp completely inhibited seed germination of *C. reflexa* at 15 days after sowing and markedly decreased populations at 30 days after sowing (Rao, Rao, and Reddy, 1985). Cuscutex (pendimethalin, linuron, and diuron mixture) was recommended for *Cuscuta* spp. control in forage crops (Fer, 1981). Thiazopyr applied preemergence was effective for selective control of *Cuscuta* in alfalfa (Cudney, Orloff, and Demason, 1993). Isopropalin at 2.5 L·ha^{-1} gave the best control of *Cuscuta* sp. when applied immediately after *Capsicum* sowing with irrigation (Mikha'lova and Genov, 1978). Chloropropham at 6.7 kg·ha^{-1} controlled *C. campestris* and *C. indecora* under glasshouse conditions, and the addition of insecticide (0.8 kg carbaryl·ha^{-1}) and of p-chlorophenyl N-methylcarbamate

(PCMC) at 0.8 kg·ha^{-1} enhanced parasite control in the field (Dawson, 1981, 1984). Chlorthiamid and dichlobenil prevented *Cuscuta* germination completely (Hacquet et al., 1983). However, Eplee and Norris (1995) have developed a list of soil-applied herbicides for various crops under *Cuscuta* infestations; chlorpropham at 4 to 6.7 kg a.i.·ha^{-1} (in most crops), DCPA (chlorthal dimethyl) (in most crops especially for tomato and ornamentals), dichlobenil (due to its cost only in high-value plantings, mainly ornamentals and blueberries) and pronamide for alfalfa, chlorthaldimethyl, trifluralin for lucerne, ethofumesate for sugar beets, CDEC for tomatoes, chlordimethyl for carrot and chickpea, pendimethalin for onion and carrot, butralin for soybean, and pronamide for chickpea, sugar beet, and cranberry.

Other workers mentioned trifluralin, pendimethalin, and prodiamine to be effective soil-applied herbicides for *Cuscuta* control. However, *Cuscuta* control diminished throughout the growing season. A high, early-season application rate or a split application improved season long control (Ristau, 2001).

Herbicides applied after attachment have limited success. General contact herbicides can be effective but also will injure the forage plants. Among herbicides used, glyphosate at a rate of 0.075 kg·ha^{-1} reduced the growth of new stems and tendrils and caused progressive necrosis in both *C. campestris* and *C. indecora* in lucerne without affecting the host (Dawson and Sagrir, 1982). *Cuscuta campestris* was susceptible to glyphosate applied twice at 0.05 kg·ha^{-1}, and regrowth of treated *Cuscuta* was abnormal (Lange and Saghir, 1982). Glyphosate and dicamaba were strongly accumulated by the parasite, and glyphosate was promising for selective control of *Cuscuta* (Fer, 1984). In alfalfa, *Cuscuta* was controlled with glyphosate sprayed in 400 L·ha^{-1} immediately after cutting. Glyphosate at 50 cc per 10 L of water and clorthal-dimethyl at 20 L·ha^{-1} provided good control of *Cuscuta* in vines (Cyprus Ministry of Agricultural and Natural Resources, 1984). Bewick, Binning, and Dana (1988) reported that isopropylamine salt of glyphosate at a rate of 1.7 kg a.i.·ha^{-1} and sulphosate at rate of 1.1 or 1.7 kg a.i.·ha^{-1} resulted in satisfactory control of *C. gronovii* in cranberry. These scientists have also reported that glyphosate at 0.4 kg·ha^{-1} applied late in carrot when swamp *Cuscuta* was in full flowering controlled the parasite and increased yield of carrot roots. Glyphosate was used at 75 to 150 g·ha^{-1} for *Cuscuta* control in lucerne but before bud stage (Dawson, 1992). Two applications of glyphosate at 150 to 300 g a.i.·ha^{-1} at a one-week interval resulted in *Cuscuta* control in alfalfa by 86 percent and increased crop yield by an average of 272 percent without any phytotoxic effect on the crop (Salimi, 1998). The same herbicide applied at 0.125, 0.25, and 0.5 kg a.i.·ha^{-1} two weeks after emergence of different legume crops effectivelly controlled *Cuscuta,* and at the highest rate used parasite infestation was reduced by 92 percent (Zaki, El-Metwaly, and Hassan, 1998). At 0.8 L·ha^{-1} Round-up and bioforce (glyphosate) were most effective (98 percent reduction) against *Cuscuta* spp. in various crops, at 0.6 L·ha^{-1} 97 percent reduction in *Cuscuta* growth was obtained (AlJuboory and Shati, 1996; Molnar, Gyulai, and Czepo, 1998). Treatment of *C. lupuliformis* seedlings with glyphosate decreased tryptophan content, anthocyanin levels, and the inhibition of protein synthesis. It is concluded that the biosynthesis of three aromatic amino acids is the sole site of action of glyphosate in *C. lupuliformis* (Zhiqian, 1994).

Paraquat is a herbicide in this classification. Herbicide selectivity is difficult to accomplish at this stage. Selective herbicides can control the stems, but often the haustoria escapes (Ristau, 2001). Among herbicides used, trimsulfuron plus surfactant was effective in controlling *Cuscuta* spp. in processing tomato in postemergence application by 95 to 100 percent. A split application of 0.021 kg·ha^{-1} + 0.035 kg·ha^{-1} three days later resulted in 100 percent control (Mullen et al., 1996). Dinoseb, dinoseb acetate, and diquat were also effective against *Cuscuta* sp. in lucerne (Sarpe et al., 1996). 2,4-DB, dicamba, and glyphosate controlled *Cuscuta* in lucerne (Fer, 1983), while application of orthofumesate at 4 kg·ha^{-1} decreased the infestation by 88 percent (Minakova, 1983). In vineyards *Cuscuta* spp. control was successfully achieved using glyphosinate ammonium at 1 kg·ha^{-1} (Bahat, 1985). Postemergence application of carbetamide reduced infestation of *Cuscuta* in chickpeas (Graph, Herzlinger, and Fridmann, 1985) and chlorathal-dimethyl sprayed at 8.75 kg·ha^{-1} on chickpea foliage parts controlled *Cuscuta* development (Graph, Herzlinger, and Fridmann, 1985).

Imazethapyr at rates of 0.5, 0.75, and 1 lb per acre resulted in excellent control of parasites in alfalfa (Canevari and Sieckert, 1993). It was effcetive against *C. trifolii* and *C. campestris* at 75 to 200 g·ha^{-1} in lucerne (Sarpe, Dinu, and Lungulescu, 1992). When applied at 100 g·L^{-1}, at 100 to 1500 mL·ha^{-1}, and at 0.125 to 0.75 kg·ha^{-1} the herbicide was highly selective for *Cuscuta* control in alfalfa (Dimitrova, 1998; Faghih and Narimani, 1998). At 75 g a.i.·ha^{-1} the same herbicide was highly effective in controlling *Cuscuta* species in broadbean, and its postemergence application at 20 g a.i.·ha^{-1} at the preflowering stage provided good control of *Cuscuta* without phytotoxic effect on crop plants (Khallida et al., 1993).

Eplee and Norris (1995) listed paraquat, diquat, dinoseb, DNOC, and glyphosate (50 to 150 g a.i.·ha^{-1} depending on the respective crop) for postattachment application in alfalfa. For host denial they have listed diuron, metribuzin and terbacil, and glyphosate for postattachment application in lucerne. Nonselective foliage-applied herbicides including, paraquat, diquat, dinoseb, 2,4-D, etc., kill parasite and host, thereby preventing any futher seed production of the parasite. Gasoline is also reported to give good control of *C. campestris* in eggplant fields. The same parasite was selectively controlled in lucerne and eggplant with detergents used at 10 $L{\cdot}ha^{-1}$ and growth was reduced by 50 percent (Al-Juboory and Shati, 1996).

Other chemicals (plant hormones and plant materials). Paliyath, Maheshwari, and Mahadevan (1978) reported that 6-benzylaminopurine (BA) or kinetin 0.5 percent in linolin in ring application around the stem of *C. chinensis* and 1 cm below the apex decreased elongation and induction of coiling preceded radial expansion of the stem and the appearance of haustoria 72 h after cytokinin application. Fresh weight of *C. reflexa* was inhibited and seedlings looked thinner after gibberellic acid treatment compared with the control. Ethylene, cytokinin, and high concentration of auxin (IAA) exhibited very marked growth inhibition accompained by swelling of the stem (Rajput, Bharti, and Laloraya, 1990). Zeatin (phytohormones) stimulated coiling whereas IAA inhibited zeatin-induced coiling. It is suggested that Zeatin-induced coiling and prehaustoria development may be mediated by phytochrome (Haidar, Orr, and Westra, 1998).

Plant extracts are also reported to control *Cuscuta*. Green bean extract showed strong seed germination inhibition of different *Cuscuta* spp. (Zaki, El-Metwaly, and Hassan, 1998). In contrast, Egyptian clover extract induced germination by 98 percent followed by chickpea extract (95 percent) and lupin extract (91 percent) and seed germination was better in sandy soils than in clay soils.

VISCUM *SPECIES (MISTLETOES)*

Mistletoe has a misleading tendency in the taxonomy of *Loranthus* (family Loranthaceae) and *Viscum* (family Viscaceae) (Steuber, n.d.). Sometimes the latter are referred to as the dwarf mistletoes to distinguish them from the genus *Loranthus*. However, both genera are perennial leafy mistletoes of six to seven years life span, morphologically and taxonomically related, and similar in parasitic habit and host plants they attack. They mostly occur in the tropics, and their families are quite primitive, containing tree- or bushlike plants. Both are hemiparasites and can complete photosynthesis. Mistletoes represent some, but far from all, of the flowering plant of order Santalales. The order is named for the renowned fragrance source, sandalwood, which is a parasite but not a mistletoe. Different mistletoe species, their number, and distribution are listed in Table 21.7.

The family Viscaceae, which is widespread in the tropics and somehow to the north temperate zone, contains seven genera and about 400 species (Barlow, 1983). *Viscum* refers generally to those shrubby, parasitic

TABLE 21.7. Pathogenic parasitic angiosperms (the mistletoes, order Santalales).[a]

Family	Genera	Species	Parasitism type	Example Genera	Distribution
Loranthaceae	74	910	stem and root, hemi	*Loranthus, Amyema, Phthirusa, Psittacanthus, Tapinanthus*	Pantropical
Misodendraceae	1	8	stem, hemi	*Misodendrum*	South America
Olacaceae	29	193	root, hemi	*Schoepfia, Ximenia*	Pantropical
Opiliaceae	10	32	root, hemi	*Agonandra, Opilia*	Pantropical
Viscaceae	7	350	stem, hemi	*Viscum, Arceuthobium, Phoradendron*	Worldwide
Santalaceae	38	490	stem and root, hemi	*Osyris, Comandra, Santalum, Thesium*	Worldwide

[a]*Pathogenic* is defined as negatively impacting a host plant that is cultivated or harvested by humans. Adapted from Nickrent (2002). Data compiled from Musselman (1980); Gill and Hawksworth (1961); Hawksworth and Wiens (1996); Riches and Parker (1995).

species that attack stems instead of roots. According to Kuijt (1987), certain mistletoes such as European mistletoe (*Viscum album* L.), rank among the world's most beautiful plants that give flowers no bigger than a pea and sprout white berries. Leafy mistletoes have green stems with thick leaves that are nearly oval in shape (see Photo 21.5). Plants often develop a roundish form up to 2 feet or more in diameter. The small, sticky, whitish or red berries (depending on the species) are produced from October to December. Evergreen clumps of mistletoe are readily observed on deciduous trees in winter when leaves are off.

Biology, Ecology, and Physiology

Mistletoe plants are either female (produce berries) or male (produce only pollen). The berries of the female plant are small, sticky, and whitish for *V. album* but turn red for *V. cruciatum.* They are very attractive to birds

PHOTO 21.5. Oriental mistletoe (*Viscum cruciatum* Sieb.) on wild almond (*Amygdalus communis* L.). (See corresponding plate in central color section.)

that feed on and digest their pulps, excreting the living seeds that stick tightly to any branch on which they land. In most cases, the initial infestation occurs on larger or older trees because birds prefer to perch in the tops of taller trees. A heavy buildup of mistletoe often occurs within an infested tree because birds are attracted to the berries which they feed on. In addition, seeds may fall from mistletoe plants in the upper part of the tree, creating new infestations on the lower branches. The rapidity with which mistletoe spreads is directly related to the proximity and severity of established infestations, and newly planted trees can be quickly infested if they are growing near old, heavily infested trees. Hawksworth, Scharpf, and Marosy (1991) reported that *V. album* spreads approximately 0.35 km per year.

Contrary to other parasitic genera, *Viscum* or *Loranthus* seeds do not require any special germination stimulant from the host. After seed germinates, the parasite radicle grows through the bark to the cambial layer and into the tree's water-conducting tissues, where haustoria develop. The haustoria gradually extend up and down within the branch as the mistletoe grows. Initially, the parasitic plant grows slowly; it may take years before the plant blooms and produces seeds. Broadleaf mistletoes have succulent stems that become woody at the base. Old, mature mistletoe plants may be several feet in diameter, and, on some host species, large swollen areas develop at the point of parasite attachment on the infected branches where the mistletoe penetrates. Swollen points vary from moderate swelling to gall formation. If the visible portion of the mistletoe is removed, new plants often resprout from the haustoria (Geils, n.d.).

Mistletoes absorb water, mineral nutrients, and possibly some photosynthate from host trees. Under moisture stress, severe damage occurs to host plants due to competition between the host and the parasite for water. However, healthy host trees can tolerate a few mistletoe branch infections, but individual branches may be weakened or sometimes killed. Heavily infested trees may be reduced in vigor, stunted, or even killed, especially if they are stressed by other problems such as drought or disease. *Viscum album* plants are able to photosynthesize all year round, but at slower rates than their hosts (Garcia-Plazaola, Hernandez, and Becerril, 2001). In addition, mistletoes are able to change their photochemical effeciency daily depending on the climatic conditions as other temperate evergreen plants do (Adams and Demming-Adams, 1995).

Distribution and Host Range

Viscum is considered to be an Old World genus of about 60 species (Kuijt, 1987). It is reported to have up to 185 hosts, including fruit and forest trees and ornamentals (Kuijt, 1987). Details on some of its host range are shown in Table 21.8. They are reported from Italy, Turkey, England, China, the Middle East, Africa, Australia, and Asia.

Economic Uses

European mistletoe *(V. album)* has been documented as a traditional treatment for diabetes. Gray and Flat (1999) reported that aqueous extract of this species evokes insulin secretion in clonal pancreatic B-cells. The result demonstrated the presence of insulin-releasing natural products in *V. album,* which may contribute to the reported antidiabetic properties of the plant. In addition, the same species received much attention in the past and at present for its value as a source of lectins and arginine for anticancer research and therapy.

Control

In newly developed areas or in older established areas where trees are being replaced, the ideal method of controlling or preventing mistletoe is to plant trees believed to be resistant or moderately resistant to this parasite. Where many new trees are being planted, control of mistletoe in any surrounding infected trees is always recommended to reduce the infection of

TABLE 21.8. Host range of some *Viscum* species.

***Viscum* species**	**Host species**	**Reference**
Viscum album	*Abies alba*	Nierhaus-Wunderwald and Lawrenz, 1998
	Abies spp.	Barney, Hawksworth, and Geils, 1998; Milius, 2000
	Acer rubrum	Tylkowski, 1982; Barney, Hawksworth, and Geils, 1998; Scharpf, Weber and Dufft, 2001
	Acer spp.	Barney, Hawksworth, and Geils, 1998; Regents of the University of California, 2001; Scharpf, Weber, and Dufft, 2001
	Alnus rhombifolia	Scharpf, Weber, and Dufft, 2001
	Alnus spp.	Regents of the University of California, 2001
	Aucuparia sp.	Kubus, 1998
	Betula ermanii	Scharpf, Weber, and Dufft, 2001
	B. pendula	Tylkowski, 1982; Kubus, 1998
	Carya ovata	Scharpf, Weber, and Dufft, 2001
	Citrus orange	Weber, 1993
	Corylus cornuta var. *californica*,	
	Crataegus mollis	Scharpf, Weber, and Dufft, 2001
	Crataegus monogyna	Garcia-Plazaola, Hernandez, and Becerril, 2001
	Crataegus spp.	Barney, Hawksworth, and Geils, 1998; Kubus, 1998
	Diopyros kaki	Milius, 2000
	Fraxinus latifolia, *F. pennsylvanica*, *Gleditsia triacanthos* f. *inermis*	Scharpf, Weber, and Dufft, 2001
	Larix sp.	Milius, 2000
	Malus spp.	Langbehn and Weber, 1997; Barney, Hawksworth, and Geils, 1998; Kubus, 1998; Milius, 2000; Dorka and Hellrung, 2001; Regents of the University of California, 2001; Scharpf, Weber, and Dufft, 2001
	Pinus nigra	Lopez-Saez, 1992; Vallauri, 1998
	P. pinaster	Lopez-Saez, 1992; Barney, Hawksworth, and Geils, 1998
	P. sylvestris	Schulze, Turner, and Glatzel, 1984; Lopez-Saez, 1992; Mertzig and Prien, 1996; Milius, 2000

TABLE 21.8 *(continued)*

Viscum species	Host species	Reference
	Populus alba, P. euramericana	Kubus, 1998
	P. nigra	Ronewicz, 1997
	Populus spp.	Salle, Armillota, and Frochot, 1984; Barney, Hawksworth, and Geils, 1998; Milius, 2000; Regents of the University of California, 2001
	Prunus cerasifera	Scharpf, Weber, and Dufft, 2001
	Prunus sp.	Barney, Hawksworth, and Geils, 1998
	Pyrus sp.	Kubus, 1998; Milius, 2000
	Quercus acutissima	WonYung, 1996
	Q,. petraea	Hariri et al., 1992
	Q. robur	Schulze, Turner, and Galtzel, 1984; Hariri et al., 1992
	Q. rubra	Hariri et al., 1992; Milius, 2000
	Robinia pseudoacacia	Kubus, 1998; Regents of the University of California, 2001; Scharpf, Weber, and Dufft, 2001
	Salix alba	Kubus, 1998
	Salix spp.	Barney, Hawksworth, and Geils, 1998; Scharpf, Weber, and Dufft, 2001
	Sorbus sp.	Barney, Hawksworth, and Geils, 1998; Kubus, 1998
	Spireax vanhaouttei	Tylkowski, 1982
	Tilia platyphyllos	Kubus, 1998
	Ulmus spp.	Milius, 2000
V. cruciatum	*Amygdalus communis*	Qasem (unpublished data)
	Crataegus monogyna	Gomez et al., 1997; Saenz et al., 1997
	Olea europea	Milius, 2000
	Punica granatum, Quercus sp., *Retama raetam*	Qasem (unpublished data)
	Retama sphaerocarpa	Cordero, Serrano, and Gonzalez, 1993
	Rhamnus palaestina	Qasem (unpublished data)
	Robinia pseudoacacia	Milius, 2000
V. laxum	*Pinus sylvestris*	Schulze, Turner, and Glatzel, 1984; Hartmann, 1997

V. rotundifolium	*Acacia karroo, A. nebrownii, Catophractes alexandri, Euphorbia virosa, Grewia flava, Salvadora persica, Tamarix usneoides, Tapinanthus oleifolius, Zizphus mucronata*	Popp et al., 1995

new trees. For treatment of existing trees, it is important to remove mistletoe before it produces seeds and spreads to other limbs or trees. Mechanical control through pruning is the most effective method for removal. Growth regulators provide a degree of temporary control, but repeated applications are required. Severely infested trees should be removed and replaced with less-susceptible species to protect surrounding trees.

Mechanical Methods

The most effective way to control mistletoe and prevent its spread is to prune out infected branches, if possible, as soon as the parasite appears. Using thinning-type pruning cuts, remove infected branches at their point of origin or back to large lateral branches. Infected branches need to be cut at least one foot below the point of mistletoe attachment in order to completely remove embedded haustoria and to avoid regrowth. If done properly, limb removal for mistletoe control can maintain or even improve tree structure. Severe heading (topping) is often used to remove heavy tree infestations; however, such pruning weakens a tree's structure and destroys its natural form. In some cases it is best to remove severely infested trees entirely because they are usually a source of mistletoe seed.

Mistletoes infecting a major branch or the trunk where it cannot be pruned may be controlled by cutting off the mistletoe flush with the limb or trunk. Then wrap the area with a few layers of wide, black polyethylene to exclude light. Use twine or tape to secure the plastic to the limb, but do not wrap it too tightly or the branch may be damaged. In some tree species, callus tissue will form under the plastic, further weakening the limb. Broadleaf mistletoe requires light and will die within a couple of years without it. It may be necessary to repeat this treatment, especially if the wrapping becomes detached or if the mistletoe does not die. The use of isolation strips has been recommended for protection of regenerating or new plantations from mistletoe attack in the United States. Simply cutting the mistletoe out of an infested tree each winter, even without wrapping, is better than doing nothing at all. Even though the parasite will grow, spread is reduced because broadleaf mistletoe must be several years old before it can bloom and produce seed.

Shading. Shading from a vigorous host plant may restrict growth of the parasite and may be the main factor responsible for variations in growth of the parasite in different host species. It is noteworthy that growth of the parasite is much better and more developed on decideous hosts as compared to evergreen species. Because the parasite depends mainly on its own photosynthesis, any limitation in light may seriously affect its growth. Therefore, parasite covering especially at early stage of parasitizm by any obage materials may be a useful agricultural method of parasite control. In this regard, any growth promoter of the host plant may lead to a well-developed host plant with enough vegetative growth to supress the parasite; therefore, any management practice that enhances host growth (fertilizer, irrigation, weed control, plant growth regulators, and others) would supress the parasite growth.

Chemical Methods

The plant growth regulator ethephon (Monterey Florel brand) may be used as directed by the label to control mistletoe in dormant host trees. To be effective, the mistletoe must also be dormant and the spray must thoroughly wet the mistletoe foliage. Mistletoe comes out of dormancy before most trees, so the ideal time to treat is when trees are dormant; the tree foliage will not get in the way of the treatment and the mistletoe clumps are more visible. Spraying provides only temporary control, especially on well-established infestations, by causing some of the mistletoe plant to fall off. The mistletoe will soon regrow at the same point, requiring retreatment. MCPB 8 percent a.i. and 2,4-5T were used as foliage spray to control *V. album* var. *abietis* (Frochot and Delabraze, 1979). Control of the same species in poplar trees was also acheived with aqueous solution of 80 g asulam per L, 21.6 or 43.2 g glyphosate per L (Frochot, Pitsch, and Wehrlen, 1983). 2,4-D and ethephon have been reported to be effective against *V. album* on *Abies* spp. and *Salix* spp. (Delabraza and Lanier, 1972), while injection method using toxic salts such as copper sulfate solutions or herbicides such as 2,4-D or metribuzin are also effective at times but have not been followed on any significant scale (Parker and Riches, 1993).

Biological Control

Carulaspis sp. was found on *V. album* in Spain (Soria, Cadahia, and Munoz, 1993). Control of the same parasite was also reported by a pathogen *Sphaeropsis visci* and *Plectophomella visci* (Fischl, 1996). In addition, two fungi have also been reported that kill mistletoes but do not hurt trees. These are *Colletotrichum gloeosporiodes,* which strikes the leaves and stems, and *Nectria neomacrospora,* which attacks the parasite's tissues inside the host (Milius, 2000).

Resistant Species

Some tree species appear resistant to broadleaf mistletoe. Bradford flowering pear (*Pyrus calleryana* Decne), Chinese pistachio (*Pistacia chinensis* Bunge), crape myrtle, eucalypts (*Eucalyptus* spp.), *Ginkgo,* golden rain tree, *Liquidambar* spp., sycamore (*Acer pseudoplatanus* L.), and conifers such as redwood *(Sequoia sempervirens)* and cedar (*Cedrus* sp.) are rarely infested. These or other resistant species should be considered when planting in infested areas or when replacing infested trees (Perry and Elmore, n.d.). Moreover, varietal resistance has been noted for certain black poplar (*Populus nigra* L.) and *Quercus* spp. (Salle, Hariri, and Andry, 1991).

Integrated Pest Management

An effective mistletoe control program in a community requires a combination of methods and the cooperation of developers, homeowners, businesses, and public agencies. Property owners can substantially reduce mistletoe infestations in their own trees, but without community cooperation, infestations will recur. Public wooded areas, such as parks and stream banks adjacent to urban areas, can be a continual source of seed and, therefore, mistletoe infestation. For this reason, the planting of tree species not susceptible to mistletoe infestation should be a part of every city and park plan (Perry and Elmore, n.d.).

The most drastic and possibly the best control measure is to remove severely infested trees and replace them with less-susceptible species as mentioned before. Economically, tree removal could be a practical approach for both public agencies and landowners, in addition to providing a source of firewood. To assist citizens in removing mistletoe from less severely infested trees on their property, some cities loan removal tools. In other cases, neighborhood residents may pool resources to hire a tree service to remove all mistletoe in their neighborhood (Perry and Elmore, n.d.).

LORANTHUS *SPECIES (MISTLETOES)*

Loranthaceae is mainly a tropical family in distribution, having about 1,000 species belonging to 74 genera (Barlow et al., 1989), of which at least 30 have been recorded as parasites of economically valuable tree crops. The genus *Loranthus,* also called mistletoe, encompasses parasitic evergreen shrubs, typically producing fruit berries or drupe (Karim, 1978) and includes 600 species (Sengbusch, n.d.). It is a hemiparasite (see Photo 21.6) with a branch system, living on ligneous plants, gabling branched out, which is embedded with haustorium at the plant host.

Biology, Ecology, and Physiology

Loranthus spp. produces tough green leaves on woody, forked stems and seems to rely upon its host plant for the water and salts that normal plants derive from the soil through their roots. The plants are dispersed by birds. The seeds, surrounded by a coating of viscous slime, are left on the bark of trees.

On germination a disklike structure (sucker) is formed first. From this a fine peg grows out that penetrates by way of a crack in the bark and grows

PHOTO 21.6. *Loranthus acaciea* Zucc. parasitizing Christ's thorn [*Zizyphus spinachristi* (L.) Desf.]. (See corresponding plate in central color section.)

through the tissues of the host until it finally reaches the surface of the wood. From the disk, a leafy shoot is also produced, but at first it shows very little growth. The peg formed by the sucker develops a core of lignified wood and in the following year, as the host-stem undergoes radial growth, this peg becomes buried in the host wood; each year, from the neck of the original sucker, other branches are formed which grow along the surface of the wood for some distance and become buried by radial growth of the host. The neck of the sucker itself does not become buried, and it has a growing region which is active at about the same level as the cambium of the host. From the original sucker the mistletoe continually makes new unions with the wood of the host (see Photo 21.7). A sort of "woody crown" is built up in the position of the original sucker, from which the woody shoot arises externally and internally (Karim, 1978).

Hyperparasitism

Two types of coevolutionary relationships are known to exist between parasitic species, facultative (hyperparasites) and obligate (epiparasites). Hyperparasitism mainly describes association between different parasitic species (*Cuscuta* and *Cassytha*) that parasitize a variety of plants, including

PHOTO 21.7. *Loranthus acaciea* parasitizing Chinese date (*Zizyphus jujuba* Mill.). (See corresponding plate in central color section.)

woody root hemiparasites (e.g., *Ximenia, Santalum,* and others), and also as hyperparasites of mistletoes such as *Phoradendron* and *Struthanthus* (Kuijt, 1964; Wiens and Calvin, 1987). In contrast, epiparasitism is known for mistletoes of Loranthaceae, Viscaceae, and Santalaceae, and also Santalaceae represented by *Phacellaria*—the obligate parasite of Loranthaceae of southeast Asia (Danser, 1939). However, all possible host-parasite relationships have been reported between loranth and viscoid, although certain combinations are more common in particular regions. At least ten species of epiparasitic *Viscum* are reported from Australia and Asia (examples include *V. articulatum* Burm.f. and *V. loranthi* Elmer) and Africa (e.g., *V. loranthicola*), with Loranthaceae being the most frequent hosts.

In Africa, Loranthaceae are frequently found parasitizing other species of Loranthaceae and Viscaceae (Polhill and Wiens, 1998). More remarkable are tripartite epi- or hyperparasitic associations, such as *Scurrula ferrugineus* on *Viscum articulatum* on *Elytranthe barnesii* which is itself parasitic on *Durio* (Sands, 1924, cited by Nickrent, 2002). Likewise, *Viscum verrucosum* on *Tapinanthus quequensis* (= *T. leendertziae*) upon *Agelanthus natalitius* (= *T. natalitius*) upon *Combretum apiculatum* has been reported in South Africa (Visser, 1982).

Distribution and Host Range

Mistletoes are widespread in Central and South America, Africa, Asia, and Australia. In Indonesia alone, three species are recorded, namely *Loranthus elasticus* Desr., *L. longiflorus* Desr., and *L. pulverulentus* Wall. Grassland and savannah ecosystems that receive less precipitation harbor diverse species of the family Loranthaceae. Their distribution is said to be related to their physiological adaptation (McKinney and Hemmerly, 1977). As parasitic plants, they must have lower water potentials than their hosts to ensure the flow of water and nutrients moving through the haustorial connection. It is accomplished by maintaining a higher transpiration rate than that of host plants. This may be the reason that one does not encounter large numbers of stem parasitic Loranthaceae and Viscaceae in dense, dark tropical rainforest conditions. Savannas, with their high solar incidence and numerous host trees, provide ideal situations for these mistletoes, and significant adaptive radiations in these families have taken place in such dry ecosystems in Central and South America, Africa, and Australia. The mistletoes tend to have adaptations to human disturbance and thereby increase in population size. According to McKinney and Hemmerly (1977), the heaviest concentration of mistletoe was in residential areas of central Tennessee, and this was likely the case throughout much of the distribution of this parasite. A patchy distribution of host trees appeared to limit and concentrate the movements of seed-dispersing birds such as waxwing, thus allowing the development of large parasite populations (Skeate, 1987). Some Loranthaceae have especially wide host ranges, as exemplified by *Dendrophthoe falcata* which has been recorded from over 350 species, and additional reports continue to increase this number (Narayanasamy and Sampathkumar, 1981). Sidahmed (1984) reported that a survey of some fruit orchards in the central region of the Sudan revealed severe infestation of citrus and guava trees by mistletoe. According to his observations, these obligate parasites cause severe damage to fruit trees, which include growth loss, reduced yield, and mortality. Some relatively recently reported hosts of *Loranthus* spp. are listed in Table 21.9.

TABLE 21.9. Hosts of certain *Loranthus* spp.

Loranthus spp.	Host species	Reference
L. acaciae	*Acacia* spp., *Acacia erioloba*	Sengbusch, n.d.
	Zizyphus jujuba, Z. spina-christi	Qasem (Unpublished data)
L. dregei	*Carya illinoensis, Citrus spp.*	De Villiers, 1997
L. elasticus	*Mangifera indica*	David and Regu, 1991
L. europaeos	*Aesculus hippocastanum*	Harvala, Exner, and Becher, 1984
	Quercus cerris, Q. petraea	Mayer, 1983
	Q. robur	Mayer, 1983; Schulze, Turner, and Glatzel, 1984
L. ferrugineus / L. pentandrus	*Adenanthera pavonina, Artocarpus* sp., *Ceiba pentandra, Enterolobium saman, Grewia tementosa, Lagerstroemia flos-reginae, Pongamia pinnata, Pterocarpus indicus*	Lim, 1984
L. yadoriki	*Neolitsea sericea*	WonYung, 1996
	Loranthus sp., *Citrus aurantium, C. grandis, Mangifera indica, Punica granatum, Achras sapota, Syzygium jambos*	Mishra, Tosh, and Mohanty, 1982

Control

Loranthus spp. on different fruit trees are controlled by a mixture of sodium salt of 2,4-D and $CuSO_4$ (1:4), which prevented regeneration of the parasite on all trees when applied at cut-end treatment (Mishra, Tosh, and Mohanty, 1982). Harten (1996) reported an aphid, *Sitobion loranthi,* to live on several species of the genus *Loranthus.*

OSYRIS *SPECIES*

Biology, Ecology, and Physiology

Osyris generally reaches a height of 20 to 60 cm with alternate dark green long leaves of 3 to 8 cm size. The flower is yellow and introduces along a thin spur. Plants are similar to linen plants; the seeds are enclosed in a berry of red color that on maturity becomes dark brown. In Europe, it blooms in May and autumn.

Osyris alba *L.*

Osyris alba (see Photo 21.8) is a Medeterranean basin species that belongs to the Santalaceae family. It is a shrub of 0.4 to 1 m height and of average growth, with foliage persisting, coriaceous, and alternate. Sessile leaves are green, clearly narrow, and lengthened. Flowers are small, dioecous, and scented (honey odor) in spring, with three yellowish petals; the females are final and solitary; males form inside bunches. Fruits are drupes, globulous, red, and nonedible. They are found in sandy soils and common in scrublands regions. It is a hemiparasite containing chlorophyll, but even if it can complete photosynthesis, it takes advantage of the roots of other plants.

Osyris lanceolata *Hochst. & Steud.*

This is a subshrub, long-lived, radicular, hermaphrodite, ranging from 40 to 150 cm with a green and angular stem. Leaves are alternate, simple, linear-lanceolate-acute, coriaceous, uninervate, and dark green (Watson and Dallwitz, 1992). Flowers are yellowish, scented (with honey odor), dioiceous, males in small side bunches, and solitary females with the extremity unbroken into leaf branch. Fruit is a drupe that is red with maturity. This species is mainly found in arid places as undergrowth. It flowers from April to June.

PHOTO 21.8. *Osyris alba* parasitizing almond *(Amygdalus communis).* (See corresponding plate in central color section.)

Osyris quadripartita *Salzm. ex Decne*

This species is an evergreen root hemiparasitic dioecious shrub up to 3 m in height occurring in a variety of Mediterranean-type scrublands, ranging from the Canary Islands, northwest Africa, and the Iberian Peninsula (Herrera, 1988a). *Osyris quadripartita* parasitizes evergreen and summer drought-decidous shrubs of many species. The flowering period lasts for nearly six months for females and nearly the entire year for males, with a peak in May-June for both sexes. The interval between anthesis and fruit ripening is quite variable (13 to 54 weeks), due to within-plant variance in developmental rates. As a consequence, ripe fruits are produced throughout the year, with a major peak in winter and a minor one in spring (Herrera, 1984, 1985).

Males and females are similar in their proportional use of host-plant species at the two study sites, and differ significantly at the other locality. This result provides no evidence for niche differentiation of the sexes of *O. quadripartita* along an environmental gradient defined by abiotic factors (moisture, soil nutrients), but favor the view of small-scale segregation based on differential utilization of host plants (Herrera, 1988b).

Distribution and Host Range

Osyris spp. are temperate to tropical parasites, cosmopolitan in distribution, except in cold regions (Watson and Dallwitz, 1992). Distribution of different *Osyris* species is shown in Table 21.10.

Economic Uses

Some allied species in Santalaceae produce edible fruit, and sandalwood (*Santalum album* L.) is the source of timber and perfume (sandalwood/sandalwood oil) (Watson and Dallwitz, 1992).

Control

Reported control methods for *Osyris* are using natural enemies as indicated in Table 21.11.

CISTANCHE *SPECIES*

Biology, Ecology, and Physiology

Cistanche (family Orobanchaceae) is a root parasite of arid regions distributed in warmer regions of the world, desert tropics, and subtropics,

TABLE 21.10. Distribution of some *Osyris* species.

Osyris species	Distribution	Reference
Osyris abyssinica Hochst. ex A. Rich.	Africa	Nickrent, 2003
O. alba L.	Greece, Spain	New York Botanical Garden, n.d.; Nickrent, 2003
O. arborea Wall.	Nepal	New York Botanical Garden, n.d.
O. compressa Berg. A. DC	S. Africa	Nickrent, 2003
O. lanceolata Hochst. & Steud.	Guzal la Vigie; S. Africa	New York Botanical Garden, n.d.; Nickrent, 2003
O. pendula Balf. F.	Yemen	New York Botanical Garden, n.d.
O. quadripartita Salzm. Ex Decne	Spain	Nickrent, 2003
O. wightiana Wall.	India, Peninsula	New York Botanical Garden, n.d.

TABLE 21.11. Natural enemies of some *Osyris* species.

Osyris species	Enemy species	Reference
Osyris abyssinica	*Mylothris agathina*	<http://www.funet.fi/pub/sci/bio/life/warp/food-plants-o.htm>
O. alba	*Orgyia trigotephras*	<http://www.funet.fi/pub/sci/bio/life/warp/food-plants-o.htm>
O. compressa	*Charaxes pelias*	<http://www.funet.fi/pub/sci//bio/life/warp/food-plants-o.htm>
O. lanceolata	*Charaxes hansahli* *C. pelias*	<http://www.funet.fi/pub/sci/bio/life/insecta/lepidoptra/ditrysia/papilionoidea/nymphalidae/charaxes/index/index.html#hansali>
O. quadripartita	*Charaxes jasius jasius*	<http://www.funet.fi/pub/sci/bio/life/warp/food-plants-o.htm>

southern Spain, southern Portugal, Crete, North Africa, and Arabia (see Table 21.12) and is represented by 16 species. It parasitizes higher life forms, mostly including trees and shrubs. *Cistanche tubulosa* is the most common species. It is a root parasite, annual or perennial, 30 to 65 cm long, fleshy herb, nongreen (commonly lacking chlorophyll), and brownish (see Photo 21.9). *Cistanche deserticola* is a desert-inhabiting root parasitic plant parasitizing the desert shrub saxual [*Haloxylon ammodendron (*CA Mey) Bunge.].

Distribution and Host Range

The distribution and host range of different *Cistanche* species are shown in Table 21.12.

Economic Uses

Cistanche deserticola, better known in its pharmaceutical name as Herba Cistanche, is used to treat several ailments. This herbal medicine consists of the pulpy stem with scales of *C. deserticola,* produced mainly in inner Mongolia, Gansu, Xinjiang, and Qinghai (China). It is collected in spring, dried in the sun, and then sliced for use. It is characterized as sweet and salty in flavor, warm in property, and acting on the kidney and large intestine channels. Its effect is reported to be tonifying the kidney, reinforcing yang, replenishing the vital essence and blood, and moistening the bowels to relieve constipation (http://asiachi.com/cistanche.html). Due to its high potential as herbal medicine *C. deserticola* is at present included in the list of highly endangered plant species requiring urgent conservation.

TABLE 21.12. Distribution and host range of different *Cistanche* species.

Cistanche spp.	Distribution	Host species	Reference
C. deserticola Ma	China, Mongolia	*Haloxylon ammodendron*	
C. lanzhouensis ZY Zhang	China	–	
C. phelypaea [L.] P. Cout.	Spain, Iraq, Pakistan, Arabian Gulf	*Hammada elegans* (Bunge) Botsch.	Fahmy, El-Tantawy, and El-Ghani, 1996; Azmeh and Musselman, 1999
C. salsa (CA Mey) Beck.	Jordan, Kazakhstan, Turkmenistan, Kyrgystan, Tajikstan, China, Southwest Asia	*Atriplex* spp., *Salsola* spp., Chenopodiaceae	Qasem (unpublished data);
C. tubulosa, (Schenk) Wight	India, Pakistan, Iraq, Arabian Gulf, Jordan, Syria	*Aerva persica* (Burm. f.) Merr.; *Anabasis articulata* (Forssk.) Moq.; *throcnemum mactrostachyum* (Moric) Moris; *Atriplex* spp., *Calligonum polygonoides* L.; *Calotropis procera* L.; *Capparis spinosa* L.; *Haloxylon salicornicum* Bunge ex Boiss.; *Hammada elegans*; *Salsola* spp.; *Zygophyllum coccineum* L.	Karim, 1978; Holm et al., 1979; Agrawal, 1984; Al-Khazarji and Abdel Wahid, 1987; Fahmy, El-Tantawy, and El-Ghani, 1996

Control

As such *Cistanche* is not a plant destined for destruction. Instead, it is labeled as one of the most valuable endangered species (Hua and Zaifu, n.d.). However, for certain less desirable species research has been under way to develop means to control.

Biological Control

In Safawa (southern desert of Iraq), Al-Khazraji and Abdel Wahid (1987) reported insect pests attacking parasitic flowering plants. The insect was

PHOTO 21.9. *Cistanche tubulosa.* (See corresponding plate in central color section.)

identified to be *Eumerus* bulb flies (Syrphidae). The larvae were found feeding on *Cistanche* and then transfer to *Orobanche,* as well as to *Cynomorium,* in a cultivated area dominated by tomato plants located 30 to 50 km northwest of El Basra. There were two species identified as large type and small type. According to the authors, both species are symmetrical, have the same development time, and the nature of their feeding is the same. There was high infestation rate and sharing by both larvae on both plants. In addition, an unidentified thrip species has also been found attacking *C. tubulosa* seeds and appeared widely spread in Al-Azraq area in Jordan (Qasem, unpublished data); the larvae was found feeding on seeds in the capsules.

CYNOMORIUM *SPECIES*

Biology, Ecology, and Physiology

According to Watson and Dallwitz (1992), the Cynomoriaceae (Balanophoraceae) are described as red herbs with thick, simple flowering stems, and attack shrubs and trees in wetlands (humid tropics and subtropics) including *Hevea, Atriplex,* and *Coffea*. These are plants of very unusual vegetative form (without chlorophyll, the unbranched stems with leaves alternate, membranous and reduced to deltoid scales, and terminating in inflorescences); more or less fungoid; with roots; succulent; totally parasitic on roots of the host; rhizomatous (the rhizomes bearing haustoria).

Primary vascular tissue centrifugal. Secondary thickening absent. Vessel end-walls are simple. Unisexual flowers present. Plants monoecious, or polygamomonoecious. Gynoecium of male flowers pistillodial (small). Flowers are minute and aggregated in inflorescences in spikes, or in heads. Inflorescences are terminal, crimson, clavate (phalloid), composed of numerous suppressed false capitula (Terekhin, 1975). Fruit is a nonfleshy, small, indehiscent nut. Seeds are endospermic with a testa and embryo is rudimentary at the time of seed release.

Watson and Dallwitz (1992) described *Cynomorium* as root parasite, fleshy herb, annual or perennial, 25 to 40 cm long without chlorophyll and stomata. Peduncle, cylindrical, brownish-red, 18 to 30 cm long. Flowering part spadixlike, 7 to 12 cm long; flowers male and female, very rarely hermaphroditic, densely crowded into unisexual or androgynous inflorescences; flowers zygomorphic, with valvate perianth 5 to 6 lobed, inner perianth fleshy, thick, larger than the others 4 to 7 mm long. Female flower with one pistil. *Cynomorium* ovary is unilocular; fruit is indehiscent, very small nutlike, frequently multiple fruit or fleshy aggregate (http://www.unex.es/botanica/rosidae/rosid-tl.html).

Distribution and Host Range

The genus *Cynomorium* is generally distributed in temperate to sub-tropical regions, mainly in the Mediterranean region and West Asia (Watson and Dallwitz, 1992). It has two species and is generally spread in southern Europe, northern Africa, central Asia, and probably other places in the world. The representative species is *Cynomorium coccineum* L. (see Photo 21.10).

The main host species include *Atriplex dimorphostegia* Kar. & Kir., tree purlane (*A. halimus* L.), tumbling saltweed (*A. rosea* L.); Tatarian orach (*A. tatarica* L.) (Gill and Hawksworth, 1961; Musselman, 1980; Riches and Parker, 1995; Hawksworth and Wiens, 1996). *Cynomorium coccineum* is also reporteed to parasitize on salt cedar (*Tamarix tetragyna* C. Ehrenb.) (Pazy, Plitmann, and Cohen, 1996) and Russian thistle (*Salsola vermiculata* L.) in Jordan (Qasem, unpublished data). *Cynomorium songaricum* Rupr. is reported to attack *Nitraria* sp. (Luo et al., 1993).

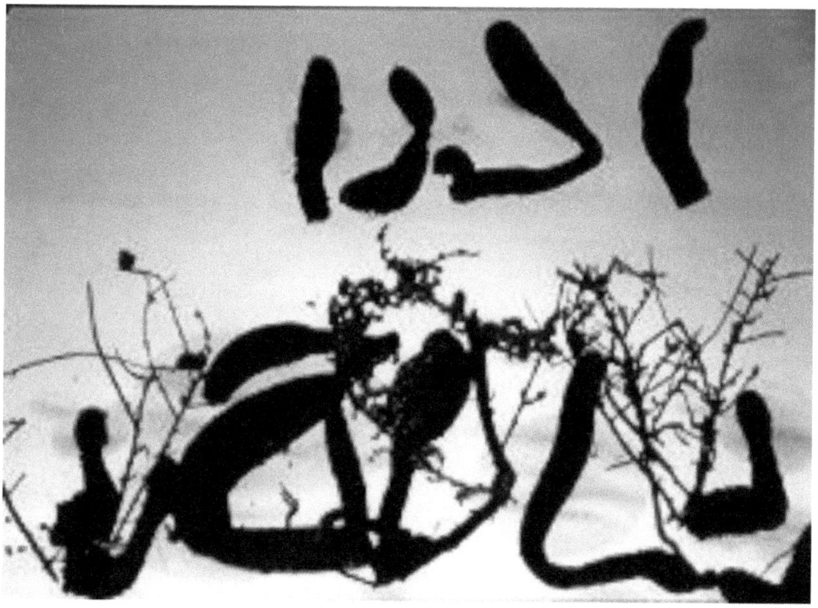

PHOTO 21.10. *Cynomorium coccineum* collected from *Salsola vermiculata* plants. (See corresponding plate in central color section.)

Economic Uses

Watson and Dallwitz (1992) compiled the following information indicating that *Cynomorium coccineum* is regarded as a famine food along with the host species it parasitizes. It is cited in the Bible (Job 30:4) as an emergency food. The parasitic scarlet plant attaches itself to the roots of saltwort (*Atriplex* sp.). The host plants—probably *A. dimorphostegia, A. halimus, A. rosea,* and *A. tatarica* were also gathered and eaten for their soft, mucilaginous leaves (http://www.hort.purdue.edu/newcrop/faminefoods/ff_families/CYNOMORIACEAE.html).

Control

Reports on control methods are very limited may be due to that the parasite is not attacking more species of economic importance. However, Luo and colleagues (1993) reported some information on germination of *C. songaricum* and found that a mixture extracted with organic solvents and water from the bast of *Nitraria* sp. roots, auxins (IAA and NAA) and cytokinins (Benzyladenine) affects germination. Treatment of *Cynomorium* seeds with the acetone and ether extracts resulted in 54 percent and 16 percent seed germination, respectively, whereas treatment with NAA (1 ppm) with benzyladenine (0.1 ppm) resulted in 68 percent seed germination of *C. songaricum*. Optimum conditions for germination of this species are 25 to 30°C, darkness and pH 6 to 9. A low temperature treatment resulted in a high germination rate and short germination time.

CONCLUSIONS

Parasitic flowering plants are of widespread significance in different parts of the world and represent a major problem to agrosystems and other ecosystems. Certain species of different plant families have been considered to be serious threats to food production and food security, since they parasitize a wide range of host plant species. Many host species are of economic importance, including herbs, shrubs, and trees from which parasites obtain food and/or water. Parasitic weeds are different in their growth habit and habitats, requirements for germination, attachment to host plants and host parts they attack, degree of dependence upon their hosts, and presence or absence of chlorophyll for photosynthesis. Reviewing their ecological, biological, and physiological requirements and behavior revealed great differences among species in their responses. Ecological and biological factors affecting germination, growth, and development of these parasites were reviewed. All possible control measures for each species were carefully considered and analyzed. This review indicated some promising herbicides and synthetic and natural chemicals, and emphasized the importance of some physical, mechanical, agricultural, and biological measures of control. However, an integrated control approach has been strongly recommended for parasitic weed control by different workers.

REFERENCES

Abdalla, M.M.F. (1999). No tillage sweet corn production following faba bean and effect on *Orobanche* seeding depth. *Bulletin of Faculty of Agriculture, University of Cairo* 50: 416-435.

Abd El-Hamid, M.M. and S.M. Shebl (2000). Dodder (*Cuscuta* spp.) control in Egyptian clover *(Trifolium alexandrium)*. In P.G. Americanos, B.E. Abu-Irmaileh, and A.R. Saghir (Eds.), *Proceedings of the First Near East Conference on Improved Weed Management* (pp. 271-274). Khartum, Sudan.

Abdel-Kader, M.M., R. Badiaa, R. Ismail, and M.M. Diab (1998). Preliminary evaluation of some soilborne fungi parasitizing *Orobanche crenata* in greenhouse. In J. Maillet (Ed.), *6th EWRS Mediterranean Symposium* (pp. 127-132). Montpellier, France: INRA.

Abou-Salama, A.M. (1995). Utilization of crop rotation for the control of *Orobanche crenata* Forsk. *Assiut Journal of Agricultural Sciences* 26: 245-252.

Adams, W.W. and B. Demming-Adams (1995). The xanthophyll cycle and sustained thermal energy dissipation activity in *Vinca minor* and *Euonymus kiauschovicus* in winter. *Plant, Cell and Environment* 18: 117-127.

Agabawi, K.M. and A.E. Younis (1985). Effect of nitrogen application on growth and nitrogen content of *Striga hermonthica* (Del.) Benth. and *Sorghum vulgare* Lur. grown for forage. *Plant and Soil* 23: 295-304.

Agrawal, K.C. (1984). Ecological studies of *Cistanche tubulosa* Wight. In C.L. Parker, L.J. Musselman, R.M. Polhill, and A.K. Wilson (Eds.), *Proceedings of the Third International Symposium on Parasitic Weeds* (pp. 31-39). Aleppo, Syria: International Center for Agricultural Research in the Dry Areas (ICARDA).

Agricultural Research Service (2000). Sunflowers resistant to parasitic plants. Available at <http://www.ars.usda.gov/is/pr/2000-000606.htm>.

Al-Haddar, T. (1989). *Cuscuta australis* R.Br., a new record in Tunisia. *Arab Journal of Plant Protection* 7: 179-181.

Al-Juboory, B.A. and R.K Shati (1996). Control of dodder *(Cuscuta campestris)* growing on alfalfa (*Medicago sativa* L.) and eggplant (*Solanum melongena* L.) plants. *Arab Journal of Plant Protection* 14: 36-40.

Al-Khateeb, W.M. (2002). *Orobanche* and tomato interaction as influenced by salinity and water deficit under in vivo and in vitro conditions. Master's of science thesis, Jordan University of Science and Technology, Irbid.

Al-Khazraji, T.O. and A.U. Abdel Wahid (1987). Insect pests attacking parasitic flowering plants in the southern desert of Iraq. *Iraqi Journal of Agricultural Science "Zanco"* 5: 197-216.

Al-Khazraji, T.O., E.Y. Ishaak, and K.M. Hameed (1989). Broomrape *(Orobanche ramosa)* on apricot *(Prunus armeniaca)* in Iraq. *Arab Journal of Plant Protection* 7: 147-152.

Al-Menoufi, O.A. (1991). Crop rotation as a control measure of *Orobanche crenata* in *Vicia faba* fields. In K. Wegmann and L.J. Musselman (Eds.), *Orobanche research: Proceedings, International Workshop on Orobanche Research* (pp. 241-247). Tuebingen, Germany: Eberhard-Karls-Universitat.

Al-Menoufi, O.A. and M.A. Adam (1996). Biological and chemical inhibition of *Orobanche* seed germination. In M.T. Moreno, J.I. Cubero, D. Berner, D. Joel, L.J. Musselman, and C. Parker (Eds.), *Advances in parasitic plant research: Proceedings of Sixth International Parasitic Weed Symposium* (pp. 418-423). Cordoba, Spain: Junta de Andalucia, Direccion General de Investigacio Agraria.

Al-Menoufi, O.A. and F.M. Ashton (1991). Studies on the parasitism of *Cuscuta* spp. series 8: Susceptibility and resistance of some *Lycopersicon* species to *Cuscuta campestris* infection. In J.K. Ransom, L.J. Musselman, A.D. Worsham, and C. Parker (Eds.), *Proceedings of the Fifth International Symposium of Parasitic Weeds* (pp. 293-297). June 24-30. Nairobi, Kenya: CIMMYT.

Alonso, L.C. (1998). Resistance to *Orobanche* and resistance breeding: A review. In K. Wegmann, L.J. Musselman, and D.M. Joel (Eds.), *Current problems of* Orobanche *researches: Proceedings of the Fourth International Workshop on* Orobanche (pp. 233-257). Albena, September 23-26. General Toshevo, Bulgaria: Institute for Wheat and Sunflower.

Alonso, L.C., M.I. Rodriguez-Ojeda, J. Fernandez-Escobar, and G. Lopez-Ruiz-Calero (1998). Chemical control of broomrape (*Orobanche cernua* Loefl.) in sunflower (*Helianthus annuus* L.) resistant to imazethapyr herbicide. *Helia* 21: 45-53.

Aly, R., Y. Goldwasser, H. Eizenberg, J. Hershenhorn, S. Golan, and Y. Kleifeld (2001). Broomrape *(Orobanche cumana)* control in sunflower *(Helianthus annuus)* with imazapic. *Weed Technology* 15: 306-309.

Americanos, P.G. (1983). Control of *Orobanche* in broadbeans. Technical Bulletin of Cyprus Agricultural Research Institute No. 50. Nicosia, Cyprus: Agricultural Research Institute.

Americanos, P.G. (1991). Control of *Orobanche* in celery. Technical Bulletin of Cyprus Agricultural Research Institute No. 137. Nicosia, Cyprus: Agricultural Research Institute and Ministry of Agriculture and Natural Resources.

Americanos, P.G. and N.A. Vouzounis (1995). Control of *Orobanche* in cabbage. Technical Bulletin of Cyprus Agricultural Research Institute No. 170. Nicosia, Cyprus: Agricultural Research Institute.

Amsellem, Z., N.K. Zidack, P.C. Quimby Jr., and J. Gressel (2000). Long-term dry preservation of viable mycelia of two mycoherbicidal organisms. *Crop Protection* 18: 643-649.

Archana, S. and S. Madhav (1997). Incompatibility of *Cuscuta* haustoria with the resistant hosts—*Ipomoea batatas* L. and *Lycopersicon esculentum* Mill. *Journal of Plant Physiology* 150: 592-596.

Arjona-Berral, A. and L. Garcia-Torres (1983). Broomrape control in lentils with glyphosate. *LENS* 10: 20-22.

Avdeev, Y. (1986). Identification of a dominant gene for resistance to broomrape in tomato and its uses in breeding. *Plant Breeding Abstracts* 50: 3944.

Avdeev, Y.I. (1998). The variety 'Astrakhan' is resistant to broomrape. *Kartofel' i ovoshchi* 2: 40.

Avdeev, Y.I. and B.M. Scherbinin (1977). Tomato resistance to the broomrape *Orobanche aegyptiaca* Pers. *Tsitologiya i Genetika* 11: 491-496.

Avdeev, Y.I. and B.M. Scherbinin (1978). Investigating tomato resistance to broomrape and the inheritance of this trait. *Doklady Vsesoyuznoi Akademii Sel'skokhozyaistevennykh Nauk imeni V.I. Lenina* 1: 14-16.

Avdeev, Y.I. and B.M. Scherbinin (1983). Resistance of tomato to sunflower broomrape *Orobanche cumana. Plant Breeding Abstracts* 56: 2308.

Aydin, A. and H. Mutlu (1996). Broomrape development on sunflower planted at different dates. *Helia* 19: 105-110.

Aziz, K.M. (1989). The use of microelements and herbicides. *Zashchita Rastenii (Moskva)* 10: 30.

Azmeh, M.F. and L.J. Musselman (1999). *Cistanche phelypaea,* a native root parasite attacking introduced shrubs. In K. Wegmann, L.J. Musselman, and D.M. Joel (Eds.), *Current problems of* Orobanche *researches: Proceedings of the Fourth International Workshop on* Orobanche. General Toshevo, Bulgaria: Institute for Wheat and Sunflower.

Bahat, A. (1985). Glufosinate ammonium for general weed control in vineyards, citrus and other fruit orchards, and uncultivated areas. *Phytoparasitica* 13: 239 (Abstract).

Banerjee, K., D.C. Khatua, and N. Mukherjee (1993). Some new hosts of *Cuscuta* sp. *Indian Forester* 119: 760-761.

Barlow, B.A. (1983). Biogeography of Loranthaceae and Viscaceae. In M. Calder and P. Bernhardt (Eds.), *The biology of mistletoes* (pp. 19-46). Sydney, Australia: Academic Press.

Barlow, B.A., F.G. Hawksworth, J. Kuijt, R.M. Polhill, and D. Wiens (1989). Genera of mistletoes. In R.M. Polhill (Ed.), *The golden bough,* Volume 1 (pp. 1-4). Kew, England: Royal Botanic Gardens.

Barloy, J. and J. Pelhate (1962). Preliminary observations on the parasites and diseases of hemp in Anjou. *Annals of Epiphyties* 13: 117-149. (*Horticultural Abstracts* 32: 7160, 1963).

Barney, C.W., F.G. Hawksworth, and B.W. Geils (1998). Hosts of *Viscum album. European Journal of Forest Pathology* 28: 187-208.

Batchvarova, R., M. Hristov, S. Slavov, and V. Valkov (2001). Development of sunflower lines resistant to *Orobanche cumana* Wall. through mutagenesis and interspecific hybridization. In A. Fer, P. Thalouarn, D.M. Joel, L.J. Musselman, C. Parker, and J.A.C. Verkleij (Eds.), *Proceedings of the Seventh International Parasitic Weed Symposium* (pp. 233-234). Nantes, France: Faculté des Sciences, Université de Nantes.

Batchvarova, R., S. Slavov, V. Valkov, S. Atanassova, and A. Atanassova (1998). Control of *Orobanche* spp. by herbicides resistant crops: An example with transgenic tobacco. In J. Maillet (Ed.), *Comptes-rendus 6eme Symposium Mediterraneen EWRS* (pp. 153-154). Montpellier, France, May 13-15. Montpellier, France: ENSA.

Baumel, P., W.D. Jeschke, L. Witte, F.C. Czygan, and P. Proksch (1993). Uptake and transport of quinolizidine alkaloids in *Cuscuta reflexa* parasitizing on *Lupinus angustifolius. Zeitschrift für Naturforschung* 48c: 436-443.

Baumel, P., W.D. Jeschke, N. Rath, F.C. Czygan, and P. Proksch (1995). Modelling of quinolizidine alkaloid net flows in *Lupinus albus* and between *L. albus* and the parasite *Cuscuta reflexa:* New insights into the site of quinolizidine alkaloid synthesis. *Journal of Experimental Botany* 45: 1721-1730.

Bayaa, B., S.G. Kumari, A. Akkaya, W. Erskine, K.M. Makkouk, Z.O. Turk, and I. Zberk (1998). Survey of major biological stresses of lentil in South-East Anatolia, Turkey. *Phytopathologia Mediterranea* 37: 88-95.

Bedi, J. (1992). Studies on *Fusarium oxysporum* for control of sunflower broomrape. *Haustorium* 27: 1-2.

Bedi, J.S. (1994). Further studies on control of sunflower broomrape with *Fusarium oxysporum* f.sp. *orthoceras*—A potential mycoherbicides. In A.H. Pieterse,

J.A.C. Verkleij, and S.J. ter Borg (Eds.), *Biology and management of* Orobanche: *Proceedings of the Third International Workshop on* Orobanche *and Related* Striga *Research* (pp. 539-544). Amsterdam, Netherlands, November 8-12, 1993. Amsterdam, Netherlands: Royal Tropical Institute.

Bedi, J.S. and N. Donchev (1991). Results on mycoherbicide control of sunflower broomrape (*Orobanche cumana* Wall.) under glasshouse and field conditions. In J.K. Ransom, L.J. Musselman, A.D. Worsham, and C. Parker (Eds.), *Proceedings of the Fifth International Symposium of Parasitic Weeds* (pp. 76-82). Nairobi, Kenya: CIMMYT.

Bedi, J.S. and N. Donchev (1995). A technique for macroconidia production by *Fusarium oxysporum* f.sp. *orthoceras,* a biocontrol agent for *Orobanche. Plant Disease Research* 10: 62-63.

Bedi, J.S., S.P. Kapur, and C. Mohan (1997). *Orobanche*—Threat to raya and taramira in Punjab. *Journal of Research, Punjab Agricultural University* 34: 149-152.

Bedi, J.S. and J. Sauerborn (1999). A new technique for chlamydospore production by *Fusarium oxysporum orthoceras,* a mycoherbicidal agent for *Orobanche cumana. Plant Disease Research* 14: 207-209.

Berg, S., T.A.W. van der Kooij, and K. Krupinska (2001). Analysis of a transcriptional apparatus in the holoparasitic flowering plant genus *Cuscuta.* In A. Fer, P. Thalouarn, D.M. Joel, L.J. Musselman, C. Parker, and J.A.C. Verkleij (Eds.), *Proceedings of the Seventh International Parasitic Weed Symposium* (p. 78). Nantes, France: Faculté des Sciences, Université de Nantes.

Bernhard, R.H., J.E. Jensen, and C. Andreasen (1998). Prediction of yield loss caused by *Orobanche* spp. in carrots and pea crops based on the soil seedbank. *Weed Research* 38: 191-197.

Beuret, E. (1981). Field dodder (*Cuscuta campestris* Yunker) a parasite of vegetable crops in the Valais. *Revue Suisse de Viticulture, Arboriculture, Horticulture* 13: 141-148.

Beuret, E. and G. Neury (1979). Dodder control in carrots and onions in the Valais. In *Compte Rendu de la 10e Conference du Columa* (pp. 1035-1043). Nyon, Switzerland: Station Fédérale de Recherche Agronomiques de Changins.

Bewick, T.A., L.K. Binning, and M.N. Dana (1988). Post attachment control of swamp dodder *(Cuscuta gronovii)* in cranberry *(Vacinium macrocarpa)* and carrot *(Daucas carota). Weed Technology* 2: 166-169.

Bewick, T.A., L.K. Binning, W.R. Stevenson, and J. Stewart (1987). A mycoherbicide for control of swamp dodder (*Cuscuta gronovii* Willd.). In H.C. Weber and W. Forstreuter (Eds.), *Parasitic flowering plants: Proceedings of the Fourth International Symposium on Parasitic Flowering Plants* (pp. 93-104). Marburg, Germany: Philipps-Universität.

Bhati, D.S. (1994). Dodder (*Cuscuta reflexa* Roxb.), a severe parasitic weed on fennel (*Foeniculum vulgare* Mill.). *Journal of Spices and Aromatic Crops* 3: 152-154.

Bhellum, B.L. and R. Mangotra (1996). *Cuscuta campestris* Yuncker—A new record for the flora of Jammu and Kashmir State. *Indian Journal of Forestry* 19: 103-104.

Boulet, C., P. Labrousse, M.C. Arnaud, N. Zehhar, and A. Fer (2001). Weed species present various responses to *Orobanche ramosa* L. attack. In A. Fer, P. Thalouarn, D.M. Joel, L.J. Musselman, C. Parker, and J.A.C. Verkleij (Eds.), *Proceedings of the Seventh International Parasitic Weed Symposium* (pp. 228-231). Nantes, France: Faculté des Sciences, Université de Nantes.

Bozoukov, H. and I. Kouzmanova (1993). Biological control of tobacco broomrape (*Orobanche* spp.) by means of some fungi of the genus *Fusarium*. In A.H. Pieterse, J.A.C. Verkleij, and S.J. ter Borg (Eds.), *Biology and management of* Orobanche: *Proceedings of the Third International Workshop on Orobanche and Related Striga Research.* November 8-12. Amsterdam, Netherlands: Royal Tropical Institute.

Braun, M., H. Burgstaller, A.M. Eldin, and G. Fadl (1985). Approaches for the control of the parasitic weed *Orobanche ramosa* L. with special references to the use of glyphosate spraying and solar heating techniques on eggplant under the conditions of Sudan. *Acta Horticulturae* 158: 335-345.

Braun, M., H. Burgstaller, and H. Walter (1984). Critical evaluation of control methods for *Orobanche ramosa* L. occuring in smallholder vegetable farms of the Khartoum Province, Sudan. In C.L. Parker, J. Musselman, R.M. Polhill, and A.K. Wilson (Eds.), *Proceedings of the Third International Symposium on Parasitic Weeds* (pp. 245-249). Aleppo, Syria: International Center for Agricultural Research in the Dry Areas (ICARDA).

Braun, M., W. Koch, and M. Stiefvater (1987). Solarization possibilities and limitations demonstrated in trials in southern Germany and the Sudan. *Gesunde Pflanzen* 39: 301-309.

Bringmann, G., J. Schlauer, M. Ruckert, B. Wiesen, K. Ehrenfeld, P. Proksch, and F.C. Czygan (1999). Host-derived acetogenins involved in incompatible parasitic relationship between *Cuscuta reflexa* (Convolvulaceae) and *Ancistrocladus heyneanus* (Ancistroladaceae). *Plant Biology* 1: 581-584.

Brown, R., A.D. Greenwood, A.W. Johnson, A.G. Long, A.R. Lansdowne, and N. Sunderland (1952). The *Orobanche* germination factor: III. Concentration of the factor by counter-current distribution. *Biochemical Journal* 52: 557-574.

Buschmann, H., K. G. Giashi, and S. Sauerborn (2001). Induced resistance in sunflower against *Orobanche cumana*. In A. Fer, P. Thalouarn, D.M. Joel, L.J. Musselman, C. Parker, and J.A.C. Verkleij (Eds.), *Proceedings of the Seventh International Parasitic Weed Symposium* (p. 232). Nantes, France: Faculté des Sciences, Université de Nantes.

Canevari, M. and E. Sieckert (1993). Timing and rate of Mon 13200 for dodder control in alfalfa. *Proceedings of the Western Society of Weed Science* 46: 20-21.

Capderon, M., A. Fer, and P. Ozenda (1985). An unreported system of expulsion of a parasite: Dodder on cotton (*Cuscuta lupuliformis* on *Gossypium hirsutum*). *Comptes Rendus de l'Academie des Sciences* 6: 227-232.

Cartia, G. (1985). Solar heating of the soil for the control of soil pests and perennial weeds. *Colture Protette (Protected Cultivations)* 14: 37-42.

Castejon-Munoz, M., F. Romero-Munoz, and L. Garcia-Torres (1993). Effect of planting date on broomrape (*Orobanche cernua* Loefl.) infections in sunflower (*Helianthus annuus* L.). *Weed Research* 33: 171-176.

Castejon-Munoz, M., M.J. Suso, F. Romero-Munoz, and L. Garcia-Torres (1991). Isoenzymatic study of broomrape *(Orobanche cernua)* populations infesting sunflower *(Helianthus annuus)*. In J.K. Ransom, L.J. Musselman, A.D. Worsham, and C. Parker (Eds.), *Proceedings of the Fifth International Symposium of Parasitic Weeds* (pp. 313-319). Nairobi, Kenya: CIMMYT.

Chari, M.S., S.R. Prabhu, and G. Ramaprasad (1993). Role of botanicals in pest management of tobacco in India. In M.S. Chari and G. Ramprasad (Eds.), *Botanical pesticides in integrated pest management* (pp. 128-133). Rajahmundry, India: Indian Society of Tobacco Science.

Christeva, T. and S. Naumova (1998). Stimulation of broomrape seed germination by soil microorganisms. In K. Wegmann, L.J. Musselman, and D.M. Joel (Eds.), *Current problems of* Orobanche *researches: Proceedings of the Fourth International Workshop on* Orobanche (pp. 73-77). Albena, September 23-26. General Toshevo, Bulgaria: Institute for Wheat and Sunflower.

Collin, J.M. (1999). *Orobanche ramosa*—Its development and continued cause for concern on rapeseed crops in the Poitou-Charentes region. *Phytoma* 515: 19-20.

Colquhoun, J., C. Mallory-Smith, and L. Suverly (2001). Distribution and importance of *O. minor* in Oregon. In A. Fer, P. Thalouarn, D.M. Joel, L.J. Musselman, C. Parker, and J.A.C. Verkleij (Eds.), *Proceedings of the Seventh International Parasitic Weed Symposium* (p. 43). Nantes, France: Faculté des Sciences, Université de Nantes.

Conaghan, J. (1998). *Orobanche hederae* Duby. in East Donegal (H34). *Irish Naturalists Journal* 26: 136.

Cordero, C.M., A.M. Serrano, and M.J.A. Gonzalez (1993). Transfer of bipiperidyl and quinolizidine alkaloids to *Viscum cruciatum* Sieber (Loranthaceae) hemiparasitic on *Retama sphaerocarpa* Boissier (Leguminosae). *Journal of Chemical Ecology* 19: 2389-2393.

Cubero, J.I. (1973). Resistance to *Orobanche crenata* Forsk. in *Vicia faba* L. In *Proceedings of the Symposium on Parasitic Weeds* (pp. 205-217). Wageningen, the Netherlands: The European Weed Research Council, Parasitic Weeds Research Group.

Cubero, J.I. (1991). Breeding for resistance to *Orobanche* species: A review. In K. Wegmann and L.J. Musselman (Eds.), *Progress in* Orobanche *research: Proceedings of International Workshop on* Orobanche *Research* (pp. 257-277). Obermarchtal, 1989. Tübingen, Alemania: Eberhard-Karls-Universität.

Cubero, J.I. and L. Hernandez (1991). Breeding faba bean (*Vicia faba* L.) for resistance to *Orobanche crenata* Forsk. In J.I. Cubero and M.C. Saxena (Eds.), *Present status and future prospects of faba bean production and improvement in the Mediterranean countries* (pp. 51-57). Options Méditérranéens, Série A. Séminaires Méditerranéens n. 10. Zaragoza, Spain: CIHEAM/ECC/ICARDA.

Cubero, J.I., M.T. Moreno, D. Rubiales, and J. Sillero (Eds.) (1999). *Resistance to* Orobanche: *The state of art*. Sevilla, Spain: Dirección General de Investigación y Formación Agraria. Servicios de Publicaciones y Divulgación; Junta de Andalucía; Consejería de Agricultura y Pesca.

Cubero, J.I., A.H. Pieterse, S.A. Khalil, and J. Sauerborn (1992). Screening techniques and source of resistance to parasitic angiosperms. In F.J. Muehlbauer and

W.J. Kaiser (Eds.), *Expanding the production and use of cool season food legumes: Proceedings of the Second International Food Legume Research Conference on Pea, Lentil, Faba Bean, Chickpea and Grasspea* (pp. 333-345). Dordrecht, Netherlands: Kluwer Acadmic Publishers.

Cudney, D.W., S.B. Orloff and D.A. Demason (1993). Effects of thiazopyr and trifluralin on dodder *(Cuscuta indecora)* in alfalfa *(Medicago sativa). Weed Technology* 7: 860-864.

Cudney, D.W., S.B.Orloff, and J.S. Reints (1992). An integrated weed management procedure for the control of dodder *(Cuscuta indecora)* in alfalfa *(Medicago sativa). Weed Technology* 6: 603-606.

Cyprus Ministry of Agricultural and Natural Resources (1984). *Annual report of the Department of Agriculture for the year 1983.* Nicosia, Cyprus: Ministry of Agriculture and Natural Resources.

Dale, H. and M.C. Press (1998). Elevated atmospheric CO_2 influence the interaction between the parasitic angiosperm *Orobanche minor* and its host *Trifolium repens. New Phytologist* 140: 65-73.

Dalela, G.G. and R.L. Mathur (1971). Resistance of varieties of eggplant, tomato and tobacco to broomrape (*Orobanche cernua* Loefl.). *PANS* 17: 482-483.

Danko, J. (1993a). Effectiveness of preparations based on the hydrazide of maleic acid against *Orobanche ramosa* L. in tobacco. *Polnohospodarstvo* 37: 923-931.

Danko, J. (1993b). Use of herbicides for control of broomrape (*Orobanche ramosa* L.) on tobacco. *Ochrana Rostlin* 29: 303-307.

Danser, B.H. (1939). A revision of the genus *Phacellaria* (Santalaceae). *Blumea* 3: 212-235.

Darbinyan, G.A., R.I. Saakyan, and Z.P. Gabrielyan (1977). Testing of triethanolamine salt of maleic hydrazide (MH-T) against *Orobanche* in tobacco plantations. In *Tezisy 8-1 sessi Zakavkaziskogo soveta po koordinatsii nauchno-issledovatel'skikh rabot po zashchite rastenii* (pp. 473-474). Erevan, Armenian SSR.

David, B.V. and K. Regu (1991). A new record and description of *Rachisphora rutherfordi* (Quaintance and Barker) (Homoptera: Aleyrodidae) from India. *Journal of Insect Science* 4: 60-70.

Dawson, J.H. (1981). Microbial inhibitor extends activity of chloropropham in soil. *Proceedings of the Western Society of Weed Science* 34: 57-58.

Dawson, J.H. (1984). Effect of carbaryl and PCMC on dodder (*Cuscuta* spp.) control with chlorpropham. *Weed Science* 32: 290-292.

Dawson, J.H. (1990). Dodder (*Cuscuta spp.*) control with dinitroaniline herbicide in alfalfa *(Medicago sativa). Weed Technology* 4: 341-348.

Dawson, J.H. (1992). Response of alfalfa *(Medicago sativa)* grown for seed production to glyphosate and SC-0224. *Weed Technology* 6: 378-381.

Dawson, J.H. and A.R. Sagrir (1982). Low rates of glyphosate controlled dodder selectively in alfalfa. *Haustorium* 8: 2-3.

Dawson, J.H. and A.R. Saghir (1983). Herbicides applied to dodder (*Cuscuta* spp.) after attachment to alfalfa *(Medicago sativa). Weed Science* 31: 465-471.

Debabkata, D., R.B. Ghosh, D. Avik, and U.K. Maji (1999). Acensus on the host of *Cuscuta reflexa* Roxb. in the district of lateritic Bankura, West Bengal. *Environment and Ecology* 17: 763-764.

De Bock, F., A. Fer, and G. Marigo (1995). Effect of fusicoccin on sucrose exchange between a host, *Pelargonium zonale,* and a higher parasitic plant, *Cuscuta reflexa. Australian Journal of Plant Physiology* 22: 553-559.

Delabraza, P. and L. Lanier (1972). Chemical control of *Viscum album. European Journal of Forestry* 2: 95-103.

Demirkan, H. and Y. Nemli (1994). Effects of some fertilizers on *Orobanche ramosa* L. on tomato. In A.H. Pieterse, J.A.C. Verkleij, and S.J. ter Borg (Eds.), *Biology and management of* Orobanche: *Proceedings of the Third International Workshop on Orobanche and Related* Striga *Research* (pp. 499-501). Amsterdam, Netherlands: Royal Tropical Institute.

De Villiers, E.A. (1997). Remove mistletoe from trees. *Inligtings bulletin-Instituut vir Tropiese en Subtropiese Gewasse* 295: 6-7.

Dey, D.K. and B.R. Pati (1998). Record of parasitism of *Cuscuta reflexa* Roxb. on *Digitaria ciliaris* Pers. *Journal of Economic and Taxonomic Botany* 22: 235-236.

Dhanapal, G.N. (1996). Management of broomrape *(Orobanche cernua)* in tobacco *(Nicotiana tabacum).* Dissertation, Department of Agronomy, Wageningen Agricultural University, Wageningen, the Netherlands.

Dhanapal, G.N., P.C. Mallory-Smith, and S.J. ter Borg (2001). Interactions between nodding broomrape and bidi tobacco in India. In A. Fer, P. Thalouarn, D.M. Joel, L.J. Musselman, C. Parker, and J.A.C. Verkleij (Eds.), *Proceedings of the Sevententh International Parasitic Weed Symposium* (p. 42). Nantes, France: Faculté des Sciences, Université de Nantes.

Dhanapal, G.N. and P.C. Struik (1996). Broomrape *(Orobanche cernua)* control before attachment to host through chemically or biologically manipulating and germination. *Netherlands Journal of Agricultural Science* 44: 279-291.

Dhanapal, G.N. and P.C. Struik (1998). Natural plant oils: Do they kill broomrape spikes?. In K. Wegmann, L.J. Musselman, and D.M. Joel (Eds.), *Current problems of* Orobanche r*esearches: Proceedings of the Fourth International Workshop on Orobanche* (pp. 177-180). General Toshevo, Bulgaria: Institute for Wheat and Sunflower.

Dhanapal, G.N., P.C. Struik, and S.J. ter Borg (1998). Effect of natural stimulants with and without GR24 on broomrape germination. In K. Wegmann, L.J. Musselman, and D.M. Joel (Eds.), *Current problems of* Orobanche *researches: Proceedings of the Fourth International Workshop on Orobanche* (pp. 51-57). General Toshevo, Bulgaria: Institute for Wheat and Sunflower.

Dhanapal, G.N., P.C. Struik, P.C.J.M. Timmermans, and S.J. ter Borg (1998). Postemergence control of broomrape with natural plant oils. *Journal of Sustainable Agriculture* 11: 5-12.

Dhanapal, G.N., P.C. Struik, M. Udayakumar, and P.C.J.M. Timmermans (1996). Management of broomrape (*Orobanche* spp.)—A review. *Journal of Agronomy and Crop Science* 176: 335-359.

Dhanapal, G.N., S.J. ter Borg, and P.C. Struik (1998). Post-emergence chemical control of nodding broomrape *(Orobanche cernua)* in bidi tobacco *(Nicotiana tabacum)* in India. *Weed Technology* 12: 652-659.

Dhopte, A.M. (1998). Inhibition of *Cuscuta* growing on *Parthenium. Annals of Plant Physiology* 12: 80-81.

Diana, G. and F. Castelli (1994). Adversities to the tobacco crop: Plants. *Informatore Fitopatologico* 44: 11-17.

Diaz, J. and H. Norambuena (2001). Parasitism and phenology of *Orobanche ramosa* L. on tomato. In A. Fer, P. Thalouarn, D.M. Joel, L.J. Musselman, C. Parker, and J.A.C. Verkleij (Eds.), *Proceedings of the Seventh International Parasitic Weed Symposium* (p. 47). Nantes, France: Faculté des Sciences, Université de Nantes.

Dimitrova, T. (1998). Study of the herbicide Pivot 100 EK for *Cuscuta* spp. control in alfalfa stand establishment (in Bulgarian). *Plant Science* 35: 651-655.

Dominguez, J. (1996). R-41, a sunflower restorer inbred line, carrying two genes for resistance against a highly virulent Spanish population of *Orobanche cernua. Plant Breeding* 115: 203-204.

Dominguez, J., J.M. Melero-Vara, J. Ruso, J.F. Miller, and J.M. Fernandez-Martinez (1993). *Screning for resistance to broomrape* (Orobanche cerenua) *in cultivated sunflower.* USDA Animal and Plant Health Inspection Service, USDA Plant Protection and Quarantine, Georgia. Department of Agriculture, Georgia Cooperative Extension Service, and Georgia Department of Transportation.

Dominguez, J., J.M. Melero-Vara, J. Ruso, J.F. Miller, and J.M. Fernandez-Martinez (1996). Screening for resistance to broomrape *(Orobanche cernua)* in cultivated sunflower. *Plant Breeding* 115: 201-202.

Dongola G.M., A.K. Mohamed, and M.K. Ahmed (2000). Effect of some control methods for Broomrape Haalouk *(Orobanche ramosa)* in the Sudan. In P.G. Americanos, B.E. Abu-Irmaileh, and A.R. Saghir (Eds.), *Proceedings of the First Near East Conference on Improved Weed Management* (pp. 176-180). Khartoum, Sudan.

Dorka, R.R. and W. Hellrung (2001). The rhythms of nutritional-movements in *Viscum album* L. under constant conditions. In A. Fer, P. Thalouarn, D.M. Joel, L.J. Musselman, C. Parker, and J.A.C. Verkleij (Eds.), *Proceedings of the Seventh International Parasitic Weed Symposium* (pp. 139-142). Nantes, France: Faculté des Sciences, Université de Nantes.

Dorr, I. and R. Kollmann (1975). Structural features of parasitism of *Orobanche:* II. The differentiation of assimilate conducting elements within the haustorium. *Protoplasma* 83: 185-199.

Dorr, I. and R. Kollmann (1976). Structural features of parasitism of *Orobanche:* III. The differentiation of xylem connection of *O. crenata. Protoplasma* 89: 235-239.

Doyle, G.J. (1993). *Cuscuta epithymum* (L.) L. (Convolvulaceae), its hosts and associated vegetation in limestone pavement habitat in the Burren Lowlands in County Clare (H9), western Ireland: Biology and environment. *Proceedings of the Royal Irish Academy, Section B* 93B: 61-67.

Drennan, D.S.H. and A.G. Mohamed-Ahmed (1992). Some effects of temperature on germination of *Orobanche ramosa* L. In *IXeme Colloque International sur la biologie des mauvaises herbes, 16-18 September 1992* (pp. 117-120). Dijon, France, Paris: ANPP.

Edwards, W.G.H. (1972). *Orobanche* research at the Royal University of Malta. *PANS* 18: 19-22.

Eizenberg, H., D. Plakhine, E. Dor, J. Hershenhorn, Y. Kleifeld, and B. Rubin (2001). Phytotoxic root extract from resistant sunflower (*Helianthus annuus* L. cv. Ambar) inhibits *Orobanche cumana* development. In A. Fer, P. Thalouarn, D.M. Joel, L.J. Musselman, C. Parker, and J.A.C. Verkleij (Eds.), *Proceedings of the Seventh International Parasitic Weed Symposium* (pp. 190-191). Nantes, France: Faculté des Sciences, Université de Nantes.

Eizenberg, H., D. Plakhine, Y. Hershenhorn, Y. Kleifeld, and B. Rubin (1999). Resistance mechanism of sunflower *(Helianthus annuus)* cv. Amber to *Orobanche cumana.* In Proceedings of XIVth International Plant Protection Congress (IPPC) (p. 77), Jerusalem, Israel, July 25-30.

El-Hattab, A.H., M.M.F. Abdalla, A.A. Metwally, and D.S. Darwish (1982). *Effect of glyphosate on* Orobanche *parasitism in a collection of faba beans.* Research Bulletin no. 2101. Cairo, Egypt: Faculty of Agriculture, Ain Shams University.

Eltayeb, D.M., P. Thalouran, A. Fer, A.G.T. Babiker, Y.F. Mohamed, A.M. Hamdoun, and G.A. Dafalla (2000). Effect of broomrape *(O. ramosa)* seed placement on parasitism of tomato (*L. esculentum* Mill) plants. In P.G. Americanos, B.E. Abu-Irmaileh, and A.R. Saghir (Eds.), *Proceedings of the First Near East Conference on Improved Weed Management* (pp. 194-198). Khartoum, Sudan.

Elzein, A.E.M., J. Kroschel, A. Assefa, and F. Masresha (1999). Preliminary evaluation of *Phytomyza orobanchia* (Diptera: Agromyzidae) as a controller of *Orobanche* spp. in Ethiopia. *Sinet, an Ethiopian Journal of Science* 22: 271-282.

Eplee, R.E. (1984). *Orobanche ramosa* in the United States. In C.L. Parker, L.J. Musselman, R.M. Polhill, and A.K. Wilson (Eds.), *Proceedings of the Third International Symposium on Parasitic Weeds* (pp. 40-42). Aleppo, Syria: International Center for Agricultural Research in the Dry Areas (ICARDA).

Eplee, R.E. and R. Norris (1995). Control of parasitic weeds. In M.C. Press and J.D. Groves (Eds.), *Parasitic plants* (pp. 256-278). London: Chapman and Hall.

Eplee, R.E., R. Norris, and J. Merritt (1994). Effect of nitrogen on parasitic weeds. In A.H. Pieterse, J.A.C. Verkleij, and S.J. ter Borg (Eds.), *Biology and management of* Orobanche: *Proceedings of the Third International Workshop on Orobanche and Related Striga Research.* Amsterdam, Netherlands, November 8-12, 1993. Amsterdam, Netherlands: Royal Tropical Institute.

Ernst, W.H.O. (1986). Mineral nutrition of *Nicotiana tabacum* cv. Bursana during infection by *Orobanche ramosa.* In S.J. ter Borg (Ed.), *Proceedings of the Workshop on Biology and Control of* Orobanche (pp. 80-85). Wageningen, the Netherlands: LH/VPO.

Faghih, A. and V. Narimani (1998). Investigation efficacy of propyzamide and imazethapyr for the control of dodder (*Cuscuta* spp.) and other weeds in East Azerbaijan—Iran. In H. Darmency, C. Fernandez-Quintanilla, J. Maillet, M.L.

Navas, E. Tei, G. Zanin, and C. Zaragoza (Eds.), *Proceedings of 6th EWRS Mediterranean Symposium* (pp. 156-157). INRA: Montpellier, France.

Fahmy, G.M., H. El-Tantawy, and M.M.A. El-Ghani (1996). Distribution, host range and biomass of two species of *Cistanche* and *Orobanche cernua* parasitising the roots of some Egyptian xerophytes. *Journal of Arid Environment* 34: 263-276.

Faizieva, S.Z. (1977). The influence of Egyptian broomrape on the chemical composition of tobacco varieties. *Trudy Samarkandskogo Gosudarstvennogo Universiteta imeni A. Navoi* 336: 78-81.

Farah, A.F. (2001). Variation in response of some legume crops to field dodder (*Cuscuta campestris*). In A. Fer, P. Thalouarn, D.M. Joel, L.J. Musselman, C. Parker, and J.A.C. Verkleij (Eds.), *Proceedings of the Seventh International Parasitic Weed Symposium* (pp. 186-189). Nantes, France: Faculté des Sciences, Université de Nantes.

Fayad, A.H., K.M. Hameed, and H.A. Al-Ani (1990). Blight of dodder (*Cuscuta campestris* Yunck.) caused by *Alternaria alternata* (Fr.) Keissler and Geotrichum candidum Link. ex Pers. *Arab Journal of Plant Protection* 8: 55-59.

Fenggen, G., L. Yanghan, and D. Fuzhen (1998). Screening biological control fungi for dodders on woody hosts. *Chinese Journal of Biological Control* 14: 159-162.

Fer, A. (1981). Knowledge of the biology and physiology of *Cuscuta* spp. as a basis for research into a rational method of controlling these parasites. In *Compte Rendu de la 11e Conference du COLUMA*. Paris, France: Comite Francias de Lutte contre les Mauvaises Herbes.

Fer, A. (1983). Investigation of a new control strategy for dodder using systemic herbicides. In *Compte Rendu de la 12e Conference du COLUMA,* I (pp. 179-186). Paris, France: Comite Francias de Lutte contre les Mauvaises Herbes.

Fer, A. (1984). Physiological approach to the chemical control of *Cuscuta:* Experiments with ^{14}C-labelled herbicides. In C.L. Parker, L.J. Musselman, R.M. Polhill, and A.K. Wilson (Eds.), *Proceedings of the Third International Symposium on Parasitic Weeds* (pp. 164-174). Aleppo, Syria: International Center for Agricultural Research in the Dry Areas (ICARDA).

Fer, A., J.C. Painset, and L. Rey (1994). Characterstics of mineral nutrition in some types of parasitic plants *(Cuscuta, Orobanche, Thesium)* and their development on crops. *Bulletin de la Societe Botanique de France, Actualites Botaniques* 139: 111-121.

Fer, A. and P. Thalouarn (1997). *Orobanche:* A threat to our crops. *Phytoma* 50: 34-36, 38-40.

Fischl, G. (1996). A leaf-spot disease of mistletoe (*Viscum album* L.) *Levelfoltossaga Novenyvedelem* 32: 181-183.

Floridata (2000). *Cuscuta* spp. Available online at <http://www.floridata.com/ref/c/cuscuta-spp.cfm>.

Foy, C.L. and R. Jacobsohn (1983). Screening tomato lines for glyphosate tolerance. *Proceedings of the Southern Weed Science Society* 36: 165.

Foy, C.L., R. Jain, and R. Jacobsohn (1989). Recent approaches for chemical control of broomrape (*Orobanche* spp.). *Reviews of Weed Science* 4: 123-152.

Foy, C.L., R. Jacobsohn, and R. Jain (1988). Screening of *Lycopersicon* spp. for glyphosate and/or *Orobanche aegyptiaca* Pers. resistance. *Weed Research* 28: 383-291.

Frochot, H. and P. Delabraze (1979). Effects of aryloxyacid herbicides on mistletoe in firs. In *Compte Rendu de la 10th Conference du COLUMA* (pp. 798-804). Nyon, Switzerland: Station Fédérale de Recherche Agronomiques de Changins.

Frochot, H., M. Pitsch, and L. Wehrlen (1983). Herbicide efficiency in mistletoe (*Viscum album* Mali) growing on poplar. In *Compte Rendu de la 12e Conference du COLUMA,* Tome 1 (pp. 157-165). Paris, France: Comite Francais de Lutte contre les Mauvaises Herbes.

Frost, C.C. and L.J. Musselman (1979). Studies of the distribution and floral biology of *Orobanche minor.* In L.J. Musselman, A.D. Worsham, and R.E. Eplee (Eds.), *Proceedings of the 2nd International Symposium on Parasitic Weeds.* Raleigh: North Carolina State University.

Furuhashi, K., M. Kanno, and J. Morita (1995). Photocontrol of parasitism in a parasitic flowering plant, *Cuscuta japonica* Chois, cultured in vitro. *Plant and Cell Physiology* 36: 533-536.

Gaertner, E.E. (1950). Studies of seed germination, seed identification and host relationships in dodders, *Cuscuta* spp. *Cornell University Agricultural Experiment Station Memoirs* 294: 56.

Gal'vidis, I. (1993). Biological features of great dodder (*Cuscuta europea* L.) in eastern Lithuania. *Ekologija* 3: 65-69.

Gao, Z.Y. and J. Gan (1992). Biological control of dodder- a review on research progress of the herbicides "Lu Bao No.1." *Chinese Journal of Biological Control* 8: 173-175.

Garcia, C. and L.C. Zaragoza (1983). Weeds in lucerne and their control in the north of Spain. *Informacion Tecnica Economica Agraria* 14: 27-34.

Garcia-Plazaola, J.I., A. Hernandez, and J.M. Becerril (2001). Seasonal differences in photosynthwtic performance in european mistletoe (*Viscum album* L.) plants. In A. Fer, P. Thalouarn, D.M. Joel, L.J. Musselman, C. Parker and J.A.C. Verkleij (Eds.), *Proceedings of the Seventh International Parasitic Weed Symposium* (pp. 135-138). Nantes, France: Faculté des Sciences, Université de Nantes.

Garcia-Torres, L. (1993). Progress in *Orobanche* control, an overview. In A.H. Pieterse, J.A.C. Verkleij, and S.J. ter Borg (Eds.), *Biology and management of* Orobanche: *Proceedings of the Third International Workshop on* Orobanche *and Related* Striga *Research* (pp. 390-399). Amsterdam, Netherlands, November 8-12. Amsterdam, Netherlands: Royal Tropical Institute.

Garcia-Torres, L., M. Castejon-Munoz, F. Lopez-Granados, and M. Jurado-Exposito (1995). Imazapyr applied postemergence in sunflower *(Helianthus annuus)* for broomrape *(Orobanche cernua)* control. *Weed Technology* 9: 819-824.

Garcia-Torres, L. and F. Lopez-Granados (1991a). Control of broomrape (*Orobanche crenata* Forsk.) in broad bean (*Vicia faba* L.) with imidazolinones and other herbicides. *Weed Research* 31: 227-235.

Garcia-Torres, L. and F. Lopez-Granados (1991b). Progress of herbicide control of broomrape (*Orobanche* spp.) in legumes and sunflower (*Helianthus annuus* L.).

In J.K. Ransom, L.J. Musselman A.D. Worsham, and C. Parker (Eds.), *Proceedings of the Fifth International Symposium of Parasitic Weeds* (pp. 306-309). June 24-30. Nairobi, Kenya: CIMMYT.

Garcia-Torres, L., F. Lopez-Granados, M. Jurado-Exposito, and J. Diaz-Sanchez (1998). The present state of *Orobanche* spp. infestations in Andalusia and the prospects for its management. In J. Maillet (Ed.), *Proceedings of the Sixth EWRS Mediterranean Symposium* (pp. 141-145). Montpellier, France: INRA.

Geils, B. (n.d.). Forest Service's Rocky Mountain Research Station at Flagstaff, Arizona. Agriculture Handbook 709. Available online at <http://www.rms.nau.edu/mistletoe>.

Geipert, S. (1997). *Potentials and constraints for the control of* Orobanche crenata *Forsk. in faba beans (*Vicia faba *L.) in Morocco.* Stuttgart, Germany: Institute fur Pflanzenproduktion in den Tropen and Subtropen, Universitat Hohenheim.

Gharib, M.S. (1973). Biological and economical aspects of the broomrapes (*Orobanche* spp.) in Northern Iraq. In *Proceedings of European Weed Research Council Symposium on Parasitic Weeds* (pp. 44-47). Wageningen, The Netherlands: Malta University Press.

Ghosheh, H.Z., K.M. Hameed, M.A. Turk, and A.F. Al-Jamali (1999). Olive *(Olea europea)* jift suppresses broomrape (*Orobanche* spp.) infections in faba bean *(Vicia faba),* pea *(Pisum sativum)* and tomato *(Lycopersicon esculentum). Weed Technology* 13: 457-460.

Giannopolitis, C.N. and E.A. Paspatis (1980). Seed production of propyzamide- or ethofumesate-treated dodder. *Annales de l'Institut Phytopathologique Benaki* 12: 236-243.

Gibot-Leclerc, S., C. Tuquet, F. Corbineau, G. Arjaure, and G. Salle (2001). New insights on *Orobanche ramosa* L. parasitizing oilseed rape in western part of France. In A. Fer, P. Thalouarn, D.M. Joel, L.J. Musselman, C. Parker, and J.A.C. Verkleij (Eds.), *Proceedings of the Seventh International Parasitic Weed Symposium* (p. 45). Nantes, France: Faculté des Sciences, Université de Nantes.

Gil, J., L.M. Martin, and J.I. Cubero (1987). Genetics of resistance in *Vicia sativa* L. to *Orobanche crenata* Forsk. *Plant Breeding* 99: 134-143.

Gill, L.S. and. F.G. Hawksworth (1961). The mistletoes: A literature review. *U.S.D.A. Forest Service Technical Bulletin* 1242: 1-87.

Gold, A.H., J.E. Sagen, and S. Wilhelm (1979). California soils suppressive to branched broomrape (*Orobanche ramosa* L.). In L.J. Musselman, A.D. Worsham, and R.E. Eplee (Eds.), *Proceedings of the Second International Symposium on Parasitic Weeds.* Raleigh: North Carolina State University.

Goldwasser, Y. (1995). Spinach (*Spinacia oleracea* var. Attica) as a host of Egyptian broomrape *(Orobanche aegyptiaca). Phytoparasitica* 23: 357-358.

Goldwasser, Y., Y. Kleifeld, S. Golan, A. Bargutti, and B. Rubin (1993). Metham sodium's dissipation from the soil and its effect on the control of Egyptian broomrape (*Orobanche aegyptiaca* Pers.). In A.H. Pieterse, J.A.C. Verkleij, and S.J. ter Borg (Eds.), *Biology and management of* Orobanche: *Proceedings of the Third International Workshop on* Orobanche *and Related* Striga *Research.* Amsterdam, Netherlands, November 8-12. Amsterdam, Netherlands: Royal Tropical Institute.

Goldwasser, Y., Y. Kleifeld, S. Golan, A. Bargutti, and B. Rubin (1995). Dissipation of metham-sodium from soil and its effect on the control of *Orobanche aegyptiaca*. *Weed Research* 35: 445-452.

Goldwasser,Y., Y. Kleifeld, D.M. Joel, and D. Plakhine (1996). Variations in vetch (*Vicia* spp.) response to *Orobanche aegyptiaca* Pers. In M.T. Moreno, J.I. Cubero, D. Berner, D. Joel, L.J. Musselman, and C. Parker (Eds.), *Advances in parasitic plant research: Proceedings of the Sixth International Parasitic Weed Symposium* (pp. 615-624). Cordoba, Spain: Junta de Andalucia, Direccion General de Investigacio Agraria.

Goldwasser, Y., Y. Kleifeld, D. Plakhine, and B. Robin (1997). Variation in vetch (*Vicia* spp.) response to *Orobanche aegyptiaca*. *Weed Science* 45: 756-762.

Goldwasser, Y., W.H. Lanini, and R.T. Worbel (2001). Tolerance of tomato varieties to lespedeza dodder. *Weed Science* 49: 520-523.

Goldwasser, Y., D. Plakhine, and J.I. Yoder (2000). *Arabidopsis thaliana* susceptibility to *Orobanche* spp. *Weed Science* 48: 342-346.

Gomez, J.M. (1994). Importance of direct and indirect effects in the interaction between a parasitic angiosperm *(Cuscuta epithymum)* and its host plant *(Hormathophylla spinosa)*. *Oikos* 71: 97-106.

Gomez, M.A., M.T. Saenz, M.D. Garcia, M.C. Ahumada, and R. De La Puerta (1997). Cytostatic activity against Hep-2 cells of methanol extracts from *Viscum cruciatum* Sieber parasitic on *Crataegus monogyna* Jacq. and two isolated principles. *Phytotherapy Research* 11: 240-242.

Granados, F.L. and L.G. Torres (1993). Seed bank and other demographic parameters of broomrape (*Orobanche crenata* Forsk.) populations in fababean (*Vicia faba* L.). *Weed Research* 33: 319-327.

Grandi, G. (1951a). Vol. I, Ed. Agricole. Bologna, Italia.

Grandi, G. (1951b). Vol. II, Ed. Agricole. Bologna, Italia.

Graph, S., G. Herzlinger, and Y. Fridmann (1985). Absorption and activity of the herbicide propyzamide and its role in the interrelation between dodder and chickpeas. *Phytoparasitica* 13: 243 (Abstract).

Graph, S., G. Herzlinger, Y. Kleifeld, A. Bargutti, B. Retig, and W. Lehrer (1985). Control of dodder in chickpeas, tomatoes and pumkins. *Phytoparasitica* 13: 243-244.

Gray, A.M. and P.R. Flat (1999). Insulin secreting activity of the traditional antidiabetic plant *Viscum album* (mistletoe). *Journal of Endocrinology* 160: 409-414.

Gressel, J. (1999). Tandem constructs: preventing the rise of superweeds. *Trends in Biotechnology* 17: 361-366.

Gupta, A. and M. Singh (1985). Mechanism of parasitism by *Cuscuta reflexa:* Distribution of cytokinins in different regions of the parasite vine. *Physiologia Plantarum* 63: 76-78.

Hacquet, J., R. Dumont, A. Fer, B. Cadot, and J. Rousselot (1983). Account of trials in 1982 and 1983 for dodder control in lucerne. In *Compte Rendu de la 12e Conference du COLUMA,* Tome 1 (pp. 167-178). Paris, France: Comite Francais de Lutte contre less Mauvaises Herbes.

Haidar, M.A., W. Bibi, and N. Abdel-Khalek (1995). Effect of wheat and barley residues on branched broomrape *(Orobanche ramosa)* growth and development in potatoes. In *Proceedings of 1995 Brighton Crop Protection Conference—Weeds,* Volume 3 (pp. 871-876). Farnham, UK: British Crop Protection Council.

Haidar, M.A., N. Iskandarani, M. Sidahmed, and R. Baal-Baki (1999). Response of of field dodder *(Cuscuta campestris)* seeds to soil solarization and chicken manure. *Crop Protection* 4: 253-258.

Haidar, M.A., G.L. Orr, and P. Westra (1998). The response of dodder (*Cuscuta* spp.) seedlings to phytohormones under various light regimes. *Annals of Applied Biology* 132: 331-338.

Haidar, M.A. and M.M. Sidahmed (2000). Soil solarization and chicken manure for the control of *Orobanche crenata* and other weeds in Lebanon. *Crop Protection* 19: 169-173.

Hameed, K.M. and Z. Al-Shyab (2001). Potential biological control of *Orobanche* by fungi isolated from diseased specimens in Jordan. In A. Fer, P. Thalouarn, D.M. Joel, L.J. Musselman, C. Parker, and J.A.C. Verkleij (Eds.), *Proceedings of the Seventh International Parasitic Weed Symposium* (p. 298). Nantes, France: Faculté des Sciences, Université de Nantes.

Hamid, A.E., M.M. Shebl, and A.M. Ghalwash (2000). New herbecide for control broomrape and other annual weeds in peas *(Pisum sativum)* at Kafr El-Sheikh. In P.G. Americanos, B.E. Abu-Irmaileh, and A.R. Saghir (Eds.), *Proceedings of the First Near East Conference on Improved Weed Management* (pp. 245-250). Khartum, Sudan.

Harb, A. (2002). The role of indole-3-acetic acid in host/*Orobanche* interaction. Master's of science thesis, Jordan University of Science and Technology, Irbid, Jordan.

Hariri, E.B., B. Jeune, S. Baudino, K. Urech, and G. Salle (1992). Development of a coefficient of resistance to mistletoe for oak. *Canadian Journal of Botany* 70: 1239-1246.

Harten, A.V. (1996). *Sitobion loranthi*, a new aphid from Angola living on mistletoes (Homoptera: Aphididae). *Entomologische Berichten* 56: 76-79.

Hartmann, T. (1997). Pine mistletoe contrary to ecological silviculture? *Allgemeine Forst Zeitschrift for Waldwirtschaft und Umweltvorsorge* 52: 52-53.

Harvala, E., J. Exner, and H. Becher (1984). Flavonoids of *Loranthus europaeus. Journal of Natural Products* 47: 1054-1055.

Hassan, E.A. (1998). Broomrape species in Egypt, a recent survey in relatioa to geographical distribution. In J. Maillet (Ed.), *Proceedings of the Sixth EWRS Mediterranean Symposium* (p. 155). Montpellier, France: INRA.

Hassan, E.A. and A.A. Farrag (1982). Host-parasite relationship. The influence of boron application at earlier stages of infection on growth behaviour of broad bean *(Vicia faba)* and *Orobanche* plants. *Research Bulletin, Faculty of Agriculture, Ain Shams University* 2017: 1-19.

Hassanein, E.E., Y.H. Fayad, F.F. Shalaby, and A.S. Kkolosy (1998). Natural role of *Phytomyza orobanchia* Kalt., a beneficial fly against the parasitic weeds *Orobanche* spp. infesting legumes and carrots in Egypt. *Annals of Agricultural Science (Cairo)* 43: 201-206.

Hassanein, E.E. and H.M. Ibrahim (2000). Preliminary study on dodder control in alfalfa. In P.G. Americanos, B.E. Abu-Irmaileh, and A.R. Saghir (Eds.), *Proceedings of the First Near East Conference on Improved Weed Management* (pp. 275-277). Khartum, Sudan.

Hawksworth, F.G., R.F. Scharpf, and M. Marosy (1991). European mistletoe continues to spread in Sonoma County. *California Agriculture* 45: 39-40.

Hawksworth, F.G., and D. Wiens (1996). *Dwarf mistletoes: Biology, pathology and systematics.* Washington, DC: USDA Forest Service Handbook 709.

Heap, J.W. (1992). *Golden dodder and silverleaf nightshade.* Study tour to USA. Technical Report. Department of Agriculture, South Australia No. 190: 34.

Herrera, C.M. (1984). The annual cycle of *Osyris quadripartita,* a hemiparasitic dioecious shrub of Mediterranean scrublands. *Journal of Ecology* 72: 1065-1078.

Herrera, C.M. (1985). Predispersal reproductive biology of female *Osyris quadripartita* (Santalaceae), a hemiparasitic dioecious shrub of Mediterranean scrublands. *Botanical Journal of the Linnean Society* 90: 113-127.

Herrera, C.M. (1988a). The fruiting ecology of *Osyris quadripartita:* Individual variation and evolutionary potential. *Ecology* 69: 233-249.

Herrera, C.M. (1988b). Plant size, spacing patterns and host plant selection in *Osyris quadripartita,* a hemiparasitic dioecious shrub. *Journal of Ecology* 76: 995-1006.

Hershenhorn, J., Y. Goldwasser, D. Plankhine, R. Ali, T. Blumenfeld, H. Bucsbaum, G. Herzlinger, S. Golan, T. Chelf, H. Eizenberg, et al. (1998). *Orobanche aegyptiaca* control in tomato fields with sulfonylurea herbicides. *Weed Research* 38: 343-349.

Hershenhorn, J., Y. Goldwasser, D. Plakhine, G. Herzlinger, S. Golan, R. Russo, and Y. Kleifeld (1996). Role of pepper *(Capsicum annuum)* as a trap and catch crop for control of *Orobanche aegyptiaca* and *O. cernua. Weed Science* 44: 948-951.

Herzlinger, G. and Y. Kleifeld (1985). Effect of the herbicide propyzamide on broomrape damage to specific host plants. *Phytoparasitica* 13: 242 (Abstract).

Hibberd, J.M, R.A. Bungard, M.C. Press, W.D. Jeschke, J.D. Scholes, and W.P. Quick (1998). Localization of photosynthetic metabolism in the parasitic angiosperm *Cuscuta reflexa. Planta* 205: 506-513.

Hipkin, C.R. (1992). Host range and specificity of *Orobanche minor* Sm. on Crymlyn Burrows. *Watsonia* 9: 113-120.

Holm, L., J. Pancho, J. Herberger, and D. Plucknett (1979). *A geographical atlas of world weeds.* New York: Wiley-Interscience.

Hopen, H.J.I., F.L. Caruso, and T.A. Bewick (1997). Control of dodder in cranberry *Vaccinium macrocarpon* with a pathogen based bioherbicide. *Acta Horticulturae* 446: 427-428.

Horowitz, M., A. Gevelberg, and H. Bucsbaum (1983). Dodder seeds: Characteristics and implications for the dissemination and control of the parasite. *Hassadah* 63: 1776-1779.

Horvath, Z. and J. Osztrogonac (1991). Regulating effect of goal 2E herbicide (Oxyfluorfen) on the populations of sunflower broomrape (*Orobanche cumana* Wallr.). *Novenyvedelem* 27: 128-132.

Hua, P. and Zaifu, Xu. (n.d.). *The threatened wild plants used for medicine as Chinese medicinal herbs*. Xishuangbanna, Yunnan: Chinese Academy of Sciences. Available online at <http://monkey.ioz.ac.cn/bwg-cciced/english/bwg-cciced/tech-33.htm>.

Huang, J.Z. and Y.H. Li (1991). Calmodulin and mechanism of haustorial formation induced by exogenous cytokinin application. In J.K. Ransom, L.J. Musselman, A.D. Worsham, and C. Parker (Eds.), *Proceedings of the Fifth International Symposium of Parasitic Weeds* (pp. 14-16). June 24-30. Nairobi, Kenya: CIMMYT.

Hussein, A.H.A., M.A. El-Deeb, and S.R. Saleib (1998). Effect of number and timing application of glyphosate on *Orobanche* infestation and faba bean yields in middle Egypt. *Arab Universities Journal of Agricultural Sciences* 6: 437-444.

Ibrahim, T.S.I., A.A. Salim, I. El-Sherif, M.M. Abd El-Hamid, and A.A. Shalaby (2000). Demonstration plots for broomrape control in fababean. In P.G. Americanos, B.E. Abu-Irmaileh, and A.R. Saghir (Eds.), *Proceedings of the First Near East Conference on Improved Weed Management* (pp. 186-190). Khartum, Sudan.

ICARDA (1986). Food legume improvement. In *Annual progress report for 1986* (pp. 204-206). Aleppo, Syria: International Center for Agricultural Research in the Dry Areas (ICARDA).

ICARDA (1992). *ICARDA annual report 1991*. Aleppo, Syria: International Centre for Agricultural Research in the Dry Areas (ICARDA).

Ihl, B. and K. Wiese (2000). Studies on *Cuscuta reflexa* Roxb: VIII. Mechanical induction of haustoria formatioin in non-twinning stems of the parasite. *Flora (Jena)* 195: 1-8.

India, Central Tobacco Research Insitute (1979). *Control of* Orobanche [leaflet]. Central Tobacco Research Institute No.1, Revised.

INRA-Dijon, Unité de Malherbologie et Agronomie (2000). *Orobanche crenata*. Available at <http://www.dijon.inra.fr/malherbo/hyppa/hyppa-a/oracr_ah.htm>.

Ish-Shalom-Gordon, N., R. Jacobsohn, and Y. Cohen (1993). Inheritance of resistance to *Orobanche cumana* in sunflower. *Phytopathology* 83: 1250-1252.

Jacobsohn, R., B. Bolinger, E. Elder, and V.P. Agrawal (1991). Crop host range of *Orobanche* species in an experimental field. In J.K. Ransom, L.J. Musselman, A.D. Worsham, and C. Parker (Eds.), *Proceedings of the Fifth International Symposium of Parasitic Weeds* (pp. 176-179). June 24-30. Nairobi, Kenya: CIMMYT.

Jacobsohn, R. and Y. Kelman (1980). Effectiveness of glyphosate in broomrape (*Orobanche* spp.) control in four crops. *Weed Science* 28: 692-698.

Jacobsohn, R. and Y. Kelman (1982). Proper timing of glyphosate application for broomrape control in carrot and celery. *Phytoparasitica* 10: 268 (Abstract).

Jacobsohn, R., Y. Kelman, R. Shaked, and L. Klein (1982). Control of broomrape in peas with ethylene dibromide and chlorpicrin. *Phytoparasitica* 10: 267 (Abstract).

Jafarzadeh, N. and A.A. Pourmirza (1999). A study on the biology of *Phytomyza orobanchia* Kalt. under laboratory and field conditions in Urmia (Iran). *Iranian Journal of Agricultural Sciences* 30: 791-798.

Jain, R. and C.L. Foy (1987). *Parasitic flowering plants.* In H.C. Weber and W. Forstreuter (Eds.), *Proceedings of the Fourth International Symposium on Parasitic Flowering Plants* (pp. 427-436). Marburg, Germany: Philipps-Universität.

Jain, R. and C.L. Foy (1992). Nutrient effects on parasitism and germination of Egyptian broomrape (*Orobanche aegyptiaca*). *Weed Technology* 6: 269-275.

Jeschke, W.D., A. Baig, and A. Hilpert (1997). Sink-stimulated photosynthesis, increased transpiration and increased demand-dependent stimulation of nitrate uptake, nitrogen and carbon relation in the prasitic association *Cuscuta reflexa-Coleus blumei. Journal of Experimental Botany* 48: 915-925.

Jeschke, W.D., P. Baumel, and N. Rath (1995). Partitioninig of nutrients in *Cuscuta reflexa-Lupinus albus* association. *Aspects of Applied Biology* 42: 71-79.

Jeschke, W.D. and A. Hilpert (1997). Sink simulated photosynthesis and sink dependent increase nitrate uptake, nitrogen and carbon relations of the parasitic association *Cuscuta reflexa-Ricinus communis. Plant, Cell and Environment* 20: 47-56.

Jeschke, W.D., N. Rath, P. Baumel, F.C. Gzyga, and P. Proksch (1994). Modeling the flow and partitioning of carbon and nitrogen in the holoparasite *Cuscuta reflexa* Roxb. and its host *Lupinus albus* L.: I. Methods for estimating net flows. *Journal of Experimental Botany* 45: 791-800.

Joel, D.M. (1998). Key developmental processes in parasitic weeds as potential targets for novel control methods. In J. Maillet (Ed.), *Comptes-rendus 6eme Symposium Mediterraneen EWRS* (pp. 135-140). Montpllier, France: INRA.

Joel, D.M., D. Aviv, T. Surov, V. Portnoy, T. Goldman-Guez, and J. Gressel (1999). Transformation of crops to herbicide-resistance and their use against parasitic weeds. In A. Altman, S. Izhar, and M. Ziv (Eds.), *Plant biotechnology and in vitro biology in the 21st century: Proceedings of the Ninth International Congress of the International Association of Plant Tissue Culture and Biotechnology* (pp. 499-502). Dordrecht, the Netherlands: Kluwer Academic Publishers.

Joel, D.M. and T. Goldman-Guez (2001) The mode of *Cuscuta* haustorial penetration. In A. Fer, P. Thalouarn, D.M. Joel, L.J. Musselman, C. Parker, and J.A.C. Verkleij (Eds.), *Proceedings of the Seventh International Parasitic Weed Symposium* (pp. 32-35). Nantes, France: Faculté des Sciences, Université de Nantes.

Joel, D.M., Y. Kleifeld, D. Losner-Goshen, G. Herzlinger, and J. Gressel (1995). Transgenic crops against parasites. *Nature* 374: 220-221.

Jurado-Exposito, M., L. Garcia-Torres, and M. Castejon-Munoz (1997). Broad bean and lentil seed treatments with imidazolinones for the control of broomrape *(Orobanche crenata). Journal of Agricultural Science* 129: 307-314.

Jurado-Exposito, M., M. Castejon-Munoz, and L. Garcia-Torres (1996). Broomrape *(Orobanche crenata)* control with imazethapyr applied to pea *(Pisum sativum)* seed. *Weed Technology* 10: 774-780.

Kabulov, D.T. (1978). *Broomrape of central Asia.* Zarazikhovya srednei Azii. Tashkent, Uzbek SSR, Fan.

Kabulov, D.T. and A.D. Kabulova (1977). The anatomical structure of broomrape and the vegetative organs of parasitized plants. *Trudy Samarkandskogo Gosudarstvennogo Universiteta imeni A. Navoi* 336: 92-108.

Kamlesh, M., R.B. Gurjar, and K. Chitle (1995). Onion and garlic-two new hosts of dodder. *Indian Journal of Mycology and Plant Pathology* 25: 335.

Karim, F.M. (1978). *Flowering parasitic plants of Iraq*. Ministry of Agriculture and Agrarian Reforms, Republic of Iraq.

Karnkowski, W. (1999). Notifiable weeds and parasitic-plants in the plant material imported into poland during 1996-99. *Ochrona Roslin* 43: 15-33.

Kasasian, L. (1971). *Orobanche* spp. *Plant Analysis and News Summary (PANS)* 17: 35-37.

Kasrawi, M.A. and B.E. Abu-Irmaileh (1989). Resistance to branched broomrape *(Orobanche ramosa)* in tomato germplasm. *HortScience* 24: 822-824.

Kelly, C.K. (1992). Resource chioce in *Cuscuta europaea. Proceedings of the National Academy of Science of the United States of America* 89: 12194-12197.

Kelly, C.K. (1994). On the economics of plant gowth: Stolon length and ramet initiation in the parasitic clonal plant *Cuscuta europaea. Evolutionary Ecology* 8: 459-470.

Kelly, C.K. and K. Horning (1999). Acquisition order and resource value in *Cuscuta attenuata. Proceedings of the National Academy of Sciences of the United States of America* 96: 13219-13222.

Khalaf, K.A. (1992). Evaluation of the biological activity of flax as a trap crop against *Orobanche* parasitizm of *Vicia faba. Tropical Agriculture* 69: 35-38.

Khalaf, K.A., A.M. Ali, K.A. Youssef, and S.A. Abd-Alaziz (1997). Studies on the control of *Orobanche crenata:* I. Use of *Azotobacter* spp. and *Escherichia coli* transformants to break dormancy of *Orobanche crenata. FABIS Newsletter* 40: 25-30.

Khalaf, K.A., K.A. Youssef, A.M. Ali, and S.A. Abd-Alaziz (1997). Studies on the control of *Orobanche crenata:* II. effectiveness of *Azotobacter* spp. and *Escherichia coli* transformants in biological control of *Orobanche crenata* on *Vicia faba* under soil conditions. *FABIS Newsletter* 40: 31-35.

Khallida, R., S.P.S. Beniwal, Z. Fatemi, and M.C. Saxena (1993). Imazethapyr can control dodder (*Cuscuta* spp.) infestation in faba bean. *FABIS Newsletter* 33: 30-32.

Khan, A.G. and T. Zafar (1981). *Smicronyx* spp. (Col.: Curculionidae) attacking *Cuscuta* spp. in Pakistan. *Biologia* 27: 265-273.

Kharrat, M., M.H. Halila, and S.P.S. Beniwal (1993). Parasitism of two faba bean varieties as affected by different seed inoculum levels of *Orobanche crenata* and *O. foetida.* In A.H. Pieterse, J.A.C. Verkleij, and S.J. ter Borg (Eds.), *Biology and management of* Orobanche: *Proceedings of the Third International Workshop on* Orobanche *and Related* Striga *Research.* Amsterdam, the Netherlands, November 8-12. Amsterdam, the Netherlands: Royal Tropical Institute.

Kharrat, M., M.M. Halila, K.H. Linke, and T. Haddar (1992). First report of *Orobanche foetida* Poiret on faba bean in Tunisia. *FABIS Newsletter* 30: 46-47.

Kleifeld, Y. (1996). Role of pepper *(Capsicum annum)* as a trap and catch crop for control of *Orobanche aegyptiaca* and *O. ceruna. Weed Science* 44: 948-951.

Kleifeld, Y. (1998). Progress in *Orobanche* control. In K. Wegmann, L.J. Musselman, and D.M. Joel (Eds.), *Current problems of* Orobanche *researches: Proceedings of the Fourth International Workshop on* Orobanche (pp. 327-330). Albena, September 23-26. General Toshevo, Bulgaria: Institute for Wheat and Sunflower.

Kleifeld, Y., Y. Goldwasser, G. Herzlinger, S. Golan, and A. Baraghutti (1991). Broomrape control with metham sodium. In J.K. Ransom, L.J. Musselman, A.D. Worsham, and C. Parker (Eds.), *Proceedings of the Fifth International Symposium of Parasitic Weeds* (pp. 382-390). June 24-30. Nairobi, Kenya: CIMMYT.

Kleifeld, Y., Y. Goldwasser, G. Herzlinger, D.M. Joel, S. Golan, and D. Kahana (1994). The effects of flax *Linum usitatissimum* (L.) and other crops as trap and catch crops for control of Egyptian broomrape (*Orobanche aegyptiaca* Pers.). *Weed Research* 34: 37-44.

Kleifeld, Y. and G. Herzlinger (1984). Chemical control of broomrape in sunflowers. *Hassadeh* 64: 2011.

Kleifeld, Y., Y. Regev, G. Herzlinger, and A. Bargutti (1982). Experiments in chemical control of broomrape. *Phytoparasitica* 10: 268 (Abstract).

Klein, O. (1995). Untersuchungen zur Populationsdynamik und zur Verwendung von *Phytomyza orobanchia* in der biologischen Bekämpfung von *Orobanche* spp. in Marokko. *M.Sc. thesis, Department of Plant Production and Agroecology in the Tropics and Subtropics.* Hohenheim, Germany: University of Hohenheim, p. 99.

Klein, O. and J. Kroschel. (2002). Biological control of *Orobanche* spp. with *Phytomyza orobanchia,* a review. *BioControl* 47: 245-277.

Klein, O., J. Kroschel, and J. Sauerborn (1999). Potential of *Phytomyza orobanchia* for the biological control of *Orobanche* spp. and its possible application. In J.I. Cubero, M.T. Moreno, D. Rubiales, and J.C. Sillero (Eds.), *Resistance to broomrape: The state of art* (pp. 179-185). Sevilla, Spain: Junta de Andalucia.

Kolomiets, A.P., S.I. Matushkin, A.B. Aleibrnova, A.I. Nedashkovskiibr, and M.F. Gordov (1982). Application of herbicides to sugarbeet on irrigated lands. *Sakharnaya Svekla* 4: 36-38.

Kontsiotou, E.K. (1979). Dodder control in alfalfa stand by herbicides. *Georgike Ereuna* 3: 250-258.

Kontsiotou, E.K. (1981). Dodder control with herbicides in newly sown lucerne. In *Panhellenic Congress of Geotechnical Research* (p. 32 [abstract]). Larissa, Greece: Institouton Ktenotrofikon Futon kai Boskon.

Kooij, T.A., W. Vander, K. Krause, I. Dorr, and K. Krupinska (2000). Molecular, functional and ultrastructural characterisation of plastids from six species of the parasitic flowering plants genus *Cuscuta. Planta* 210: 701-707.

Korashi, A.A., M.M. El-Borollosy, E.A. Hassan, M.R. El-Suoud, M.R. Zein-El-Deen, and A. Koraim (1996). Hosts of *Orobanche* spp. and yield losses in delta and upper Egypt. In M.T. Moreno, J.I. Cubero, D. Berner, D. Joel, L.J. Musselman, and C. Parker (Eds.), *Advances in parasitic plant research: Proceedings of Sixth International Parasitic Weed Symposium* (pp. 488-493). Cordoba, Spain: Junta de Andalucia, Direccion General de Investigacio Agraria.

Krishna-Murty, G.V.G. (1992). Two new methods to control *Orobanche* on tobacco. *Indian Farming* 42: 7-8.

Krishna-Murty, G.V.G. (1994). *Orobanche* problem and management in India. In A.H. Pieterse, J.A.C. Verkleij, and S.J. ter Borg (Eds.), *Biology and management of* Orobanche: *Proceedings of the Third International Workshop on* Orobanche *and Related* Striga *Research* (pp. 591-598). Amsterdam, the Netherlands, November 8-12. Amsterdam, the Netherlands: Royal Tropical Institute.

Kubus, M. (1998). The mistletoe *Viscum album* L. on the area of the right-site part of Szczecin. *Folia Universitatis Agriculturae Stetinensis, Agriculturae* 71: 51-62.

Kuijt, J. (1964). Critical observations on the parasitism of New World mistletoes. *Canadian Journal of Botany* 42: 1243-1278.

Kuijt, J. (1969). *The biology of parasitic flowering plants.* Berkeley: University of California Press.

Kuijt, J. (1987). Novelities in mesoamerican mistletoes (Loranthaceae and Viscaceae). *Annals of Missouri Botanical Garden* 74: 511-532.

Kuijt, J. (1990). New species and combinations in neotropical mistletoes (Loranthaceae and Viscaceae). *Proceedings of the Koninklijke Nederlandse Akademie van Wetenschappen, Amsterdam* 93: 113-162.

Kukula, S.T. and H. Masri (1984). Integrated cultural practices and chemical control of *Orobanche crenata* in faba bean. In C.L. Parker, L.J. Musselman, R.M. Polhill, and A.K. Wilson (Eds.), *Proceedings of the Third International Symposium on Parasitic Weeds* (pp. 256-261). Aleppo, Syria: International Center for Agricultural Research in the Dry Areas (ICARDA).

Kumar, R.M. and S.M. Kondap (1993). Response of greengram and blackgram cultivars in *Cuscuta* infestation. *Indian Journal of Plant Protection* 21: 167-171.

Kumar, R.M., S.M. Kondap, N.M. Rao, and A. Ramakrishna (1993). An anatomical study on host specificity of a parasitic weed—*Cuscuta. Indian Journal of Weed Science* 2: 61-63.

Kutbay, H.G., F. Karaer, and M. Klinc (1996). Nutrient relationships of *Cuscuta epithymum* var. *epithymum.* In H. Brown, G.W. Cussans, M.D. Devine, S.O. Duke, C. Fernandez-Quintanilla, A. Helweg, R.E. Labrada, M. Landes, P. Kudsk, and J.C. Streibig (Eds.), *Proceedings of the Second International Weed Control Congress* (pp. 91-94). Copenhagen, Denmark: Slagelse, Denmark, Department of Weed Control and Pesticide Ecology.

Lados M. (1998). Effect of temperature, pH and host plant extracts on the germination of *Cuscuta trifolii* and *Cuscuta campestris. Acta Agronomica Hungarica* 46: 317-325.

Langbehn, A. and H.C. Weber (1997). Further observations of growth rates and the development of *Viscum album* L. (Viscaceae) growing on apple trees (*Malus* sp.). *Beitrage zur Biologie der Pflanzen* 69: 141-145.

Lange, A.H. and A.R. Saghir (1982). Post attachment control of dodder in tomato. *Proceedings of the Western Society of Weed Science* 35: 51-54.

Lekic, M.B. (1970a). Phytophagous insects observed on parasitic phanerogams of the genera *Orobanche* and *Cuscuta* in 1968. In F.J. Simmonds (Ed.), *Proceedings of the First International Symposium on Biological Control of Weeds* (pp.

21-24). Ascot, UK: Commonwealth Institute of Biological Control, Miscellaneous Publication No. 1.

Lekic, M.B. (1970b). Role of *Phytomyza orobanchia* Kalt. (Agromyzidae) in reducing the populations of parasitic angiosperms of the *Orobanche* genus on territory of Vojvodina. *Savremena Poljoprivreda* 18: 627-637.

Lim, W.C. (1984). Incidence of parasitic higher plants and strangling figs on ornamental trees in Penang, Malaysia. In C.L. Parker, L.J. Musselman, R.M. Polhill, and A.K. Wilson (Eds.), *Proceedings of the Third International Symposium on Parasitic Weeds* (pp. 1-5). Aleppo, Syria: International Center for Agricultural Research in the Dry Areas (ICARDA).

Linke, K.H., J. Sauerborn, and M. Saxena (1996). *Field guide for broomrapes identification.* Aleppo, Syria: ICARDA (in Arabic).

Linke, K.H., C. Scheibel, M.C. Saxena, and J. Sauerborn (1992). Fungi occurring on *Orobanche* spp. and their preliminary evaluation for *Orobanche* control. *Tropical Pest Management* 38: 127-130.

Linke, K.H., H. Schnell, and M.C. Saxena (1991). Factors affecting the seed bank of *Orobanche crenata* in fields under lentil based cropping systems in northern Syria. In J.K. Ransom, L.J. Musselman, A.D. Worsham, and C. Parker (Eds.), *Proceedings of the Fifth International Symposium of Parasitic Weeds,* June 24-30 (pp. 321-327). Nairobi, Kenya: CIMMYT.

Liu, X.M. (1992). Discovery of *Cuscuta japonica* Choisy on tobacco in Jilin Province. *Journal of Jilin University* 14: 88-99.

Lolas, P. (1994). Herbicides for control of broomrape (*Orobanche ramosa* L.) in tobacco (*Nicotiana tabacum* L.). *Weed Research* 34: 205-209.

Lolas, P. (1997). Sub-group collaborative study on broomrape. 1996-1997 Report. *Bulletin d'Information CORESTA* 3: 79-85.

Lolas, P.C. (1998). Methods and strategies for control of broomrape in tobacco. In *Proceedings of 1998 Tobacco Symposium, Indian Tobacco—Problems and Prospects* (pp. 33-42). Rajahmundry, India: Indian Society of Tobacco Science.

Lopez-Granados, F., J. Diaz Sanchez, and L. Garcia-Torres (1997). A bioeconomic model for broomrape control *(Orobanche crenata)* in faba beans *(Vicia faba)* under different management strategies. In *Proceedings of the 1997 Congress of the Spanish Weed Science Society: La Malherbología en la producción integrada,* Valencia, Spain, November 24-26 (pp. 29-32). Cordoba, Spain: Instituto de Agricultura/CSIC and Sociedad Espanola de Malherbologia.

Lopez-Granados, F. and L. Garcia-Torres (1996). Effect of environmental factors on dormancy and germination of crenata broomrape (*Orobanche crenata*). *Weed Science* 44: 284-289.

Lopez-Granados, F. and L. Garcia-Torres (1997). Modelling the demography of crenate broomrape (*Orobanche crenata* Forsk.) as affected by broad bean *(Vicia faba)* cropping frequency and planting date. *Weed Science* 47: 261-268.

Lopez-Saez, J.A. (1992). Contribution to the study of *Viscum album* L. subsp. Austriacum (Wlesb.) Vollman populations in forest areas of Avila Province (Spain). *Anales de Biologia* 18: 77-80.

Lu, Y., G. Gagne, B. Grezes-Besset, and P. Bianchard (1999). Integration of a molecular linkage group containing the broomrape resistance gene *Or5* into an RFLP map in sunflower. *Genome* 42: 453- 456.

Luo, M., G.E. Su, J.H. Liu, and T.G. Ao (1993). Study on the seed germination of *Cynomorium songaricum* without host. *Grasslands of China* 2: 62-64.

Machado, M.A. and K. Zetsche (1990). A structural, functional and molecular analysis of plastids of the holoparsites *Cuscuta reflexa* and *Cuscuta europaea. Planta* 181: 91-96.

Mamluk, O.F. and H.C. Weltzien (1978). Verbreitung and Wirtsspektrum einiger *Cuscuta*-Herkunfte aus dem Vorderen and Mittleren Oriten [Geographic distribution and host range of some *Cuscuta* strains (widely spread on sugarbeets) in the Near and Middle East. *Zeitschrift fur Pflanzenkrankheiten und Pflanzenschutz* 85: 102-107.

Mamonov, I.F., V.P. Poplaukhin, and T. Nazarko (1978). The results of sunflower selection for resistance to broomrape. *Selektsiya I Semenovodstvo, USSR* 1: 23-25.

Manjunath, T.M. and S. Nagarkatti (1977). Natural enemies of *Orobanche* in India and possibilities of its biological control. *Technical Bulletin of the Commonwealth Institute of Biological Control* 18: 75-83.

Manschadi, A.M., J. Sauerborn, and H. Stutzel (2001). Quantitative aspects of *Orobanche crenata* infestation in faba beans as affected by abiotic factors and parasite soil seedbank. *Weed Research* 41: 311-321.

Martelli, G.M. (1933a). Nota preliminare sui parassiti animali dell *Orobanche* della Fava *Orobanche* speciosa. D.C. *Rivista di Patologia Vegetale* 23: 233-240.

Martelli, G.M. (1933b). Primo contributo alla conoscenza di alcuni parassiti dell' *Orobanche* della Fava (*Orobanche speciosa* D.C.). *Bolletino del Laboratorio de Zoologia Generale e Agraria in Portici* 28: 11-36.

Matthews, D.E., D.M. Joel, and J.C. Steffens (1991). Species-specific response of *Orobanche* to host-derived and synthetic germination stimulants. In J.K. Ransom, L.J. Musselman, A.D. Worsham, and C. Parker (Eds.), *Proceedings of the Fifth International Symposium of Parasitic Weeds.* June 24-30. Nairobi, Kenya: CIMMYT.

Mayer, H. (1983). Attack by *Loranthus europaeus* in the Weinviertel. *Schweizerischen Zeitschrift fur Forstwesen* 134: 847-858.

Mazaheri, A., N. Moazami, M. Vaziri, and N. Moayed-Zadeh (1991). Investigation of *Fusarium oxysporum* a possible biological control of broomrape (*Orobanche* spp.). In J.K. Ransom, L.J. Musselman, A.D. Worsham, and C. Parker (Eds.), *Proceedings of the Fifth International Symposium of Parasitic Weeds*. Nairobi, Kenya: CIMMYT.

McKinney, L.E. and T.E. Hemmerly (1977). Host specificity of mistletoe in middle Tennessee: III. Davidson County. *Journal of the Tennessee Academy of Science* 52: 40.

McPartland, J.M. (1996). A review of *Cannabis* diseases. *Journal of the International Hemp Association* 3: 19-23.

Mertzig, C. and S. Prien (1996). Occurrence of pine mistletoe in forest stands in Niederlausitz. *Allgemeine Forst Zeitschrift fur Waldwirtschaft und Umweltvorsorge* 51: 160-162.

Mesa-Garcia, J. and L. Garcia-Torres (1982). Broomrape (*Orobanche crenata* Forsk.) control in bean (*Vicia faba* L.) with glyphosate affected by infection intensity. In *Proceedings of 1982 British Crop Protection Conference—Weeds* (pp. 765-770). Farnham, UK: British Crop Protection Council.

Mesa-Garcia, J. and L. Garcia-Torres (1985). *Orobanche crenata* (Forsk) control in *Vicia faba* (L.) with glyphosate as affected by herbicide rates and parasite growth stages. *Weed Research* 25: 129-134.

Metha, H.M., G.C. Trivedi, and H.R. Patel (1985). Study to find out effective method of weed control in bidi tobacco. In *Abstracts of papers, Annual Conference of Indian Society of Weed Science* (p. 80). Anand, Gujarat, India: Department of Agronomy, Gujurat Agricultural University and Indian Society of Weed Sciences.

Meti, S.S. and M.M. Hosmani (1994). Broomrape control in bidi tobacco by soil solarization. *Tobacco Research* 20: 60-70.

Miccolis, V. and V.V. Bianco (1984). The influence of glyphosate, propyzamide and some cultural techniques on the infestation of *Orobanche crenata* Forsk. in broad beans for fresh consumption. *Difesa delle Piante* 7: 161-175.

Mihajlovic, L. (1986). Results of investigation on *Orobanche* spp. entomofauna in Yogoslavia and the possibility of using insects for biological control. In S.J. ter Borg (Ed.), *Proceedings of the Workshop on Biology and Control of* Orobanche (pp. 118-126). Wageningen, the Netherlands: LH/VPO.

Mikha'lova, A. and G. Genov (1978). Preliminary results of spraying 2,4-D on young carnation plants. *Bulletin of the Colorado Flower Growers' Association* 341: 1-2.

Milius, S. (2000). Botany under the mistletoe: Twisters, spitters, and other flowery thoughts for romantic moments. *Science News Online*. Available online at <http://www.sciencenews.org/20001223/bob2.asp>.

Miller, A.E., G.K. Douce, T.R. Murphy, and J.T. Waston (1997). *Cooperative extension.* Deparment of Entomology, Cooperative Extension Service, College of Agricultural and Environmental Sciences, The University of Georgia, April. GACAPS 0497.

Minakova. T.B. (1983). Control measures against *Cuscuta arvensis*. *Sakharnaya Svekla* 6: 35.

Mirek, Z. (1997). Extinction of flax-weeds in Sweden. *Acta Societatis Botanicorum Poloniae* 66: 221-222.

Mishra, A., G.C. Tosh, and D.C. Mohanty (1982). Preliminary study on the control of *Loranthus* from fruit trees. In *Abstracts of Papers, Annual Conference of Indian Society of Weed Science.* Bhubneshwar, Orissa, India: Department of Botany, Orissa University of Agriculture and Technology and Indian Society of Weed Science, p. 45.

Mishra, S. and G.G. Sanwal (1992). Alterations in lipid composition of seed oil from *Brassica juncae* upon infection with *Cuscuta reflexa*. *Journal of Agricultual and Food Chemistry* 40: 52-55.

Mishra, S. and G.G. Sanwal (1994). Effect of *Cuscuta* infection on chloroplast lipid composition of *Brassica* leaves. *European Journal of Plant Pathology* 100: 61-70.

Mitich, Larry (n.d.). *Orobanche:* The broomrapes. Available at <http://www.wssa.net/photo/info/larrymitich_info/orobanche.html>.

Mohamed-Ahmed, A.G. and D.S.H. Drennan (1994). Factors affecting establishment of *Orobanche* spp. in legumes. In A.H. Pieterse, J.A.C. Verkleij, and S.J. ter Borg (Eds.), *Biology and management of* Orobanche: *Proceedings of the Third International Workshop on* Orobanche *and Related* Striga *Research* (pp. 312-319). Amsterdam, the Netherlands: Royal Tropical Institute.

Molnar, F., B. Gyulai, and M. Czepo (1998). A new possibility of controlling dodder (*Cuscuta* spp.) in the field. *Novenyvedelem* 34: 379-383.

Morozov, I.V., C.L. Foy, and J.H. Westwood (2000). Small broomrape *(Orobanche minor)* and Egyptian broomrape *(Orobanche aegyptiaca)* parasitization of red clover *(Trifolium pratense). Weed Technology* 14: 312-320.

Mukumov, Kh. (1974). The degree of germination of broomrape seed in relation to the growth stage of host plants. *Nauchnye Trudy Biologicheskogo Fakulteta, Samarkandskii Gosudarstennye University, Imeni A. Navoi Botanika* 207: 134-140.

Mukumov, Kh.M. and S.Z. Faizieva (1977). Results of container trials investigating the broomrape resistance of vegetable and melon crops. *Trudy Samarkandskogo Gosudarstvennogo Universiteta imeni A. Navoi* 336: 73-78.

Mullen, R.J., J.P. Orr, J. Caprile, T.C. Viss, and R.W. Whiteley (1996). Preemergence and postemergence studies with rimsulfuron for the control of *Solanum* and other weed species in processing tomatoes. In *Proceedings of the First International Conference on the Processing Tomato and the First International Symposium on Tropical Tomato Diseases* (pp. 63-66). Alexandria, VA: American Society for Horticultural Science.

Munoz, M.C., F.R. Munoz, and L.G. Torres (1993). Effect of planting date on broomrape (*Orobanche cernua* Loefl.) infection in sunflower (*Helianhus annuus* L.). *Weed Research* 33: 171-176.

Murasheva, V.N. (1995). Influence of *Fusarium oxysporum* var. *orthoceras* (Appel et Wr.) Bilai toxic properties on its vitality in soil and pathogenicity. *Mikologiya i Fitopatologiya* 29: 53-85.

Musselman, L.J. (1980). The biology of *Striga, Orobanche*, and other root-parasitic weeds. *Annual Review of Phytopathology* 18: 463-489.

Musselman, L.J. (1982). Parasitic weeds of the arable land. In W. Holzner and M. Numata (Eds.), *Biology and ecology of weeds* (pp. 175-185). The Hague, Netherlands: Dr. W. Junk.

Musselman, L.J. (1993). Alien broomrape (*Orobanche* species) of potential danger to American crops. *Proceedings of the Southern Weed Science Society* 46: 244-247.

Nagar, N. and G.C. Sanwal (1984). Biochemical aspects of parasitism in *Cuscuta reflexa:* Inhibition of cell wall degrading enzymes of *Cuscuta* by nonsusceptible plants. In C.L. Parker, L.J. Musselman, R.M. Polhill, and A.K. Wilson (Eds.), *Proceedings of the Third International Symposium on Parasitic Weeds* (pp. 175-183). Aleppo, Syria: International Center for Agricultural Research in the Dry Areas (ICARDA).

Nagar, N., M. Singh, and G.G. Sanwal (1984). Cell wall degrading enzymes in *Cuscuta reflexa* and its hosts. *Journal of Experimental Botany* 35: 1104-1112.

Nandula, V.K., C.L. Foy, and D.M. Orcutt (1999). Glyphosate for *Orobanche aegyptiaca* control in *Vicia sativa* and *Brassica napus. Weed Science* 47: 486-491.

Nandula, V.K., C.L. Foy, and J.H. Westwood (1996). Environmental influences on germination of *Orobanche*. In M.T. Moreno, J.I. Cubero, D. Berner, D. Joel, L.J. Musselman, and C. Parker (Eds.), *Advances in parasitic plant research: Proceedings of Sixth International Parasitic Weed Symposium* (pp. 409-416). Cordoba, Spain: Junta de Andalucia, Direccion General de Investigacio Agraria.

Narayanasamy, C. and R. Sampathkumar (1981). Host parasite relationships of *Dendrophthoe falcata* (Linn. f.) Bettingh. (*Loranthus longiflorus* Desr.). *Journal of the Bombay Natural History Society* 78: 192-193.

Nassib, A.M., E.E. Hassanein, and A.H.A. Hussein (1990). Broomrape control in faba (*Vicia faba* L.) with reduced rates of glyphosate mixed with three adjuvants. *Annals of Agricultural Science, Moshtohor* 28: 17-25.

Nassib, A.M., A.H. Hussein, and F.M.El. Rayes (1984). Effect of variety, chemical control, sowing date and tillage on *Orobanche* spp. infestation and faba bean yield. *FABIS Newsletter* 10: 11-15.

Nelson, A.P. and K.I. Mohamed (2001). Growth and reproductive biology of *Orobanche uniflora* L. In A. Fer, P. Thalouarn, D.M. Joel, L.J. Musselman, C. Parker, and J.A.C. Verkleij (Eds.), *Proceedings of the Seventh International Parasitic Weed Symposium* (p. 40). Nantes, France: Faculté des Sciences, Université de Nantes.

New York Botanical Garden (n.d.). Herbarium specimen catalog. Available at <http://scisun.nybg.org:8890/searchdb/owa/wwwcatalog.searchform>.

Nickrent, D.L. (2002). Parasitic plants of the world. In J.A. López-Sáez, P. Catalán, and L. Sáez (Eds.), *Parasitic plants of the Iberian Peninsula and Balearic Islands* (pp. 7-27). Madrid: Mundi-Prensa.

Nickrent, D. (2003). Santlaceae description. In "Parasitic Plant Connection." Available at <http://www.science.siu.edu/parasitic-plants/index.html>.

Nierhaus-Wunderwald, P.D. and P. Lawrenz (1998). A note on the biology of mistletoe (in French). *ForÃt* 51: 5-9.

Oliveira Velloso, J. (1990). El genero *Orobanche* en los cultivos en el sur de Espaha, taxonomie, corologia, ecologia. Doctoral thesis, Univesidada de Cordoba.

Paliyath, G., R. Maheshwari, and S. Mahadevan (1978). Initiation of haustoria in *Cuscuta* (*chinensis*) by cytokinin application. *Current Science* 47: 427-429.

Palled, Y.B., Y.S. Path, Y.C. Panchal, and M.M. Hosmani (1985). Studies on chemical weed control in irrigated bidi tobacco (*Nicotiana tabacum* L.) with special reference to *Orobanche* (*Orobanche cernua* Loefl.). *Indian Journal of Weed Science* 17: 6-11.

Papyan, A.G. and Kh.M. Arutyunyan (1981). Harmful effect of broomrape on tobacco plants. *Izvestiya Sel'skokhozyaistvennoi Nauki Armyanskoi SSR* 5: 33-37.

Parker, C. (1994). *Orobanche* control: The state of art. In A.H. Pieterse, J.A.C. Verkleij, and S.J. ter Borg (Eds.), *Biology and management of* Orobanche: *Proceedings of the Third International Workshop on* Orobanche *and Related* Striga

Research. Amsterdam, the Netherlands, November 8-12, 1993. Amsterdam, the Netherlands: Royal Tropical Institute.

Parker, C. and C.R. Riches (1993). *Parasitic weeds of the world: Biology and control.* Wallingford, UK: CAB International.

Parsons, W.T. and E.G. Cuthbertson (1992). *Noxious weeds of Australia.* Melbourne, Australia: Indata Press. Available online at <http://www.hear.org>.

Paspates, E.A. and M.E. Damanakes (1979). Control of dodder (*Cuscuta campestris* Yuncker) with pre-emergence herbicides. *Geoponika* 25: 147-152.

Pazy, B., U. Plitmann, and O. Cohen (1996). Bimodal karyotype in *Cynomorium coccinum* L. and its systematic implications. *Botanical Journal of Linnean Society* 120: 279-281.

Perez-de-Luque, A., J.C.G. Galindo, F.A. Macias, and J. Jorrin (2000). Sunflower sesquiterpene lactone models induce *Orobanche cumana* seed germination. *Phytochemistry* 53: 45-50.

Perez-de-Luque, A., D. Rubiales, J.C.G. Galindo, F.A. Macias, and J. Jorrin (2001). Allelopathy and allelochemicals within the plant-parasitic weed interaction. Studies with the sunflower-*Orobanche cumana* system. In A. Fer, P. Thalouarn, D.M. Joel, L.J. Musselman, C. Parker, and J.A.C. Verkleij (Eds.), *Proceedings of the Seventh International Parasitic Weed Symposium* (pp. 196-199). Nantes, France: Faculté des Sciences, Université de Nantes.

Perez-de-Luque, A., J.C. Sillero, D. Rubiales, and J.I. Cubero (2001). Effect of sowing date on the establishemnt of *O. crenata* in susceptible and resistant cultivar of four legume species. In A. Fer, P. Thalouarn, D.M. Joel, L.J. Musselman, C. Parker, and J.A.C. Verkleij (Eds.), *Proceedings of the Seventh International Parasitic Weed Symposium* (p. 244). Nantes, France: Faculté des Sciences, Université de Nantes.

Perry, E.J. and C.L. Elmore (n.d.). Mistletoe. In B. Ohlendorf (Ed.), *Pest notes.* Department of Agricultural and Natural Resources Publication no. 7437. University of California, Davis: IPM Education and Publications, University of California Statewide IPM Project.

Pesch, C. and A.H. Pieterse (1982). Inhibition of germination in *Striga* by means of urea. *Experientia* 38: 559-560.

Petzoldt, K., Y. Nemli, and J. Sneyd (1993). Integrated control of *Orobanche cumana* in sunflower. In A.H. Pieterse, J.A.C. Verkleij, and S.J. ter Borg (Eds.), *Biology and management of* Orobanche: *Proceedings of the Third International Workshop on* Orobanche *and Related* Striga *Research.* Amsterdam, the Netherlands, November 8-12. Amsterdam, the Netherlands: Royal Tropical Institute.

Pieterse, A.H. (1981). Germination of *Orobanche crenata* Forsk. seeds in vitro. *Weed Research* 21: 279-287.

Pieterse, A.H. (1991a). The effect of nitrogen fertilizers on the germination of seeds of *Striga* hermonthica and *Orobanche crenata.* In K. Wegmann and L.J. Musselman (Eds.), *Progress in* Orobanche *research: Proceedings, International Workshop on* Orobanche *Research* (pp. 115-124). Tuebingen, Germany: Eberhard-Karls-Universitat.

Pieterse, A.H. (1991b). Research on broomrape and *Striga* at the Royal Tropical Institute and cooperating institute. In S.K. Kim (Ed.), *Combating* Striga *in Africa,*

Proceedings of the International Workshop held in Ibadan, Nigeria (pp. 69-72). Ibadan, Nigeria: International Institute of Tropical Agriculture.

Pieterse, A.H. and J.A.C. Verkleij (1991). Genetic variability in *Orobanche* (broomrape) and *Striga* (witchweed) and its implications for host resistance breeding. In J. Rozema and J.A.C. Verkleij (Eds.), *Ecological responses to environmental stresses* (pp. 209-302). Dordrecht, the Netherlands: Kluwer Academic Publishers.

Pirani, V., G. Capriotti, A. Del Gatto, and M. Tiberi (1993). Sowing date, anticryptogam treatments, *Orobanche* damage and losses at harvesting in dry pea crops. *Informatore Agrario* 49: 30-37.

Polhill, R. and D. Wiens (1998). *Mistletoes of Africa*. Kew, Richmond-Surrey, UK: The Royal Botanic Gardens.

Popova, A.A. (1942). The effect of fertilizers on the infection of *Nicotiana rustica* with wildfire and *Orobanche ramosa* L. *Vses. Inst. Tabachn. I Makhov. Promyshl. Krasnodar* 144: 158-175.

Popovic, S. and M. Stjepanovic (1995). Production of alfalfa seed is limited by presence of *Cuscuta* spp. *Sjemenarstvo* 12: 135-149.

Popp, M., R. Mensen, A. Richter, H. Buschmann, and D.J. von Willert (1995). Solutes and succulence in southern African mistletoes. *Trees, Structure and Function* 9: 303-310.

Proctor, G. (1993). IIRB weed control study group-Italy. *British Sugar Beet Review* 61: 12-14.

Protsenko, N.G. (1981). Prospects of using biological method for the control of dodder in Chuisk valley of Kirgizia. *Entomologicheskie Issledovaniya v Kirgizii* 14: 104-109.

Pujadas, A. (1999). Species of the family *Orobanchaceae* parasitic of cultivated plants and its relatives growing on wild plants, in the south of the Iberian Peninsula. In J.I. Cubero, M.T. Moreno, D. Rubiales, and J. Sillero (Eds.), *Resistance to* Orobanche: *The state of the art* (pp. 187-193). Sevilla, Spain: Junta de Andalucía, Consejería de Agricultura y Pesca.

Pujadas, A. and P. Thalouarn (1998). *Orobanche cernua* Loefl and *O. cumana* Waller. in the Iberian Peninsula. In J. Maillet (Ed.), *Sixth EWRS Mediterranean Symposium* (pp. 159-160). Montpellier, France: INRA.

Pundir, Y.P. S., M.C. Purohit, and K. Promod (1994). Hyper and triple-parasitism of *Cuscuta* on mistletoes. In K.C. Sahni (Ed.), *Advances in plant science research,* Volume 2 (pp. 69-72). Dehra Dun, India: International Book Distributors.

Qadri, S.M.H., K.J. Srivastava, B.K. Tiwari, and S.R. Bhonde (1985). Dodder injury of onion. *Indian Horticulture* 30: 9.

Qasem, J.R. (1998). Chemical control of branched broomrape *(Orobanche ramosa)* in glasshouse grown tomato. *Crop Protection* 17: 625-630.

Qasem, J.R. (2002). Plants as sources of natural herbicides against branched broomrape (*Orobanche ramosa* L.). In M.J. Reigosa and N. Pedrol (Eds.), *Alleopathy: From molecule to ecosystem* (pp. 153-182). Enfield, NH: Science Publisher, Inc.

Qasem, J.R. and C.L. Foy (2001). Weed allelopathy, its ecological impact and future prospects: A review. In R.K. Kohli, H.P. Singh, and D.R. Batish (Eds.),

Allelopathy in agroecosystems (pp. 43-119). Binghamton, NY: The Haworth Press.

Qasem, J.R. and M.A. Kasrawi (1995). Variation of resistance to broomrape *(Orobanche ramosa)* in tomatoes. *Euphytica* 81: 109-114.

Qian, Xi. (1996). Studies of the biological characteristics and control of *Cuscuta chinensis* in soyabean fields in the reclamation area of north Jiangsu. *Soybean Science* 15: 62-68.

Racovitza, A. (1973). Research on the susceptibility of various species of *Nicotiana* and various cultivars of tobacco to *Orobanche ramosa*. In *Proceedings of European Weed Research Council Symposium on Parasitic Weeds* (pp. 194-204). Wageningen, the Netherlands: Malta University Press.

Rajak, R.C., M.K. Rai, and A.K. Panday (1985). New host records of *Cuscuta reflexa*. *Journal of Tropical Forestry* 1: 166-167.

Rajput, B.S., S. Bharti, and M.M. Laloraya (1990). Effect of hormones on the growth of autotrophic seedlings of *Cuscuta*. *Indian Journal of Plant Physiology* 23: 300-305.

Raju, C.A. (1996). Variability in tobacco germplasm toward *Orobanche* infections. In M.T. Moreno, J.I. Cubero, D. Berner, D. Joel, L.J. Musselman, and C. Parker (Eds.), *Advances in parasitic plant research: Proceedings of Sixth International Parasitic Weed Symposium* (pp. 609-614). Cordoba, Spain: Junta de Andalucia, Direccion General de Investigacio Agraria.

Raju, C.A. (2001). Biological control of parasitic weeds: An appraisal. In A. Fer, P. Thalouarn, D.M. Joel, L.J. Musselman, C. Parker, and J.A.C. Verkleij (Eds.), *Proceedings of the Seventh International Parasitic Weed Symposium* (p. 297). Nantes, France: Faculté des Sciences, Université de Nantes.

Raju, C.A., G.V.G. Kishna-Murty, K. Nagarajan, and M.S. Chari (1995). A new disease of *Orobanche cernua* parasitizing tobacco, caused by *Sclerotium rolfsii*. *Phytoparasitica* 23: 235-237.

Rao, K.N. and K.M. Gupta (1981). Chemical control of *Cuscuta* in pulses and other crops. In *Proceedings of the Eighth Asian-Pacific Weed Science Society Conference,* Bangalore, India, November 22-29 (pp. 215-217). Bapatla, Andhara Pradesh, India: Weed Science Division, Andhra Pradesh Agricultural University.

Rao, K.N. and R.S.N. Rao (1991). Studies on tolerance of black gram and green gram varieties to dodder (*Cuscuta chinensis* Lam.). In J.K. Ransom, L.J. Musselman, A.D. Worsham, and C. Parker (Eds.), *Proceedings of the Fifth International Symposium of Parasitic Weeds* (pp. 170-175). June 24-30. Nairobi, Kenya: CIMMYT.

Rao, K.N., A.S. Rao, and R. Reddy (1985). Effect of different herbicides on control of *Cuscuta* in legumes. *Pesticides* 19: 61-62.

Rao, R., B.G. Joshi, and C.B. Sarma (1982). Maintenance of *Orobanche cernua* on brinjal plants. *Indian Journal of Agricultural Sciences* 52: 764-767.

Rapparini, G. and G. Campagna (1998). Update on the control of *Cuscuta*. *Informatore Agrario* 54: 61-67.

Rawat, A., S. Sachan, and S.N. Sachan (1994). Host range of *Cuscuta reflexa* Roxb. from Doon Valley. *Indian Journal of Forestry* 17: 73-77.

Reddy, P.P., A.S. Nalini, and A.S. Prabhakar (1990). *Vitex negundo* Linn., a new host for *Cuscuta reflexa* Roxb. *Current Research* 19: 55-56.

Regents of the University of California (2001). State integrated pest management project. PMG/PESTNOTES/pn7437.

Riches, C.R. and C. Parker (1995). Parasitic plants as weeds. In M.C. Press and J.D. Graves (Eds.), *Parasitic plants* (pp. 226-255). London: Chapman and Hall.

Ristau, R.J. (2001). *Dodder.* Crop series production no. 3.112. Fort Collins, CO: Colorado State University Cooperative Extension.

Robert, S., D. Dubreuil, P. Simier, J.P. Praderre, and A. Fer (1999). Inhibition studies on mannose 6-phosphate reductase purified from *Orobanche ramosa. Carbohydrate Letters* 3: 231-238.

Rodriguez-Ojeda, M.I., L.C. Alonso, and J. Fernandez-Escobar (2001). Effect of different crops on the germination of *Orobanche cernua* Loelf. (*O. cumana* Wallr). seeds. In A. Fer, P. Thalouarn, D.M. Joel, L.J. Musselman, C. Parker, and J.A.C. Verkleij (Eds.), *Proceedings of the Seventh International Parasitic Weed Symposium* (p. 124). Nantes, France: Faculté des Sciences, Université de Nantes.

Roman, B., C. Alfaro, A. Torres, Z. Satovic, M. Kharrat, A. Pujadas, and D. Rubiales (2001). An analysis of genetic variation in natural populations of *Orobanche foetida* from Spain and Tunisia. In A. Fer, P. Thalouarn, D.M. Joel, L.J. Musselman, C. Parker, and J.A.C. Verkleij (Eds.), *Proceedings of the Seventh International Parasitic Weed Symposium* (pp. 57-60). Nantes, France: Faculté des Sciences, Université de Nantes.

Ronewicz, P. (1997). The mistletoe *Viscum album* L. on the area of Miedzyodrze in Szczecin. *Zeszyty Naukowe Akademii Rolniczej w Szczecinie, Rolnictwo* 66: 63-68.

Rosvall, T. (1979). *Orobanche alba* on Gotland, Sweden. *Svensk Botanisk Tidskrift* 73: 1-6.

Rothe, K., B. Diettrich, B. Rahfeld, and M. Luckner (1999). Uptake of floem specific cardenolides by *Cuscuta* sp. growing on *Digitalis lanata* and *Digitalis purpurea. Phytochemistry* 51: 357-361.

Rubiales, D. (1999). Eating broomrape? In J.I. Cubero, M.T. Moreno, D. Rubiales, and J. Sillero (Eds.), *Resistance to* Orobanche: *The state of the art* (pp. 195-199). Sevilla, Spain: Junta de Andalucía, Consejería de Agricultura y Pesca.

Rubiales, D., A. Pujadas, and B. Roman (2001). Occurrence and distribution of *Phytomyza orobanchia* feeding on broomrapes (*Orobanche* spp.) in southern spain. In A. Fer, P. Thalouarn, D.M. Joel, L.J. Musselman, C. Parker, and J.A.C. Verkleij (Eds.), *Proceedings of the Seventh International Parasitic Weed Symposium* (pp. 254-257). Nantes, France: Faculté des Sciences, Université de Nantes.

Rubiales, D., J.C. Sillero, and J.I. Cubero (1998). Broomrape (*Orobanche crenata* Forsk.) resistance in peas (*Pisum sativum* L.). In *Third European Conference on Grain Legumes: Opportunity for high quality, healthy and added-value crops to meet European demands* (p. 238). November 14-19. Valladolid, Spain.

Rubiales, D., J.C. Sillero, and M.T. Moreno (1999). *Orobanche crenata* in winter chickpea. *Grain Legumes* 26: 7.

Ruso, J., J.M. Melero-Vara, J. Dominguez, J.Miller, and J.M. Fernandez-Martinez (1994). Survey of broomrape (*Orobanche cernua* Loefl.) resistance to collections of cultivated sunflower inbred lines and wild species of *Helianthus*. In A.H. Pieterse, J.A.C. Verkleij, and S.J. ter Borg (Eds.), *Biology and management of* Orobanche: *Proceedings of the Third International Workshop on* Orobanche *and Related* Striga *Research.* Amsterdam, the Netherlands, November 8-12, 1993. Amsterdam, the Netherlands: Royal Tropical Institute.

Ruso, J., S. Sukno, J. Dominguez-Gimenez, J.M. Melero-Vara, and J. Fernandez-Matinez (1996). Screening of wild *Helianthus* species and derived lines for resistance to several populations of *Orobanche cernua*. *Plant Disease* 80: 1165-1169.

Saenz, M.T., M.D. Garcia, M.C. Ahumada, and M.A. Gomez (1997). Antimitotic effects of hexane extracts from *Viscum cruciatum* Sieber parasitic on *Crataegus monogyna* Jacq. and *Crataegus monogyna* Jacq.: Parasitized and non-parasitized and triterpenic fractions isolated from them, by the *Allium* test. *Phytopathology Research* 11: 158-160.

Saghir, A.R. (1994). An overview of new techniques in *Orobanche* control. *Plant Protection Quarterly* 9: 77-80.

Saghir, A.R., A. Janudi, and M. Shafyuddin (1983). The use of germination stimulants for the control of *Orobanche* in tomato. In *Proceedings of Tenth International Congress of Plant Protection* (p. 282). Croyden, UK: British Crop Protection Council.

Sahu, R. and R. Sinha (1983). Occurrence of *Orobanche* in tobacco areas of Nepal and central north-eastern region of India. *Tobacco Research* 9: 49-51.

Salem, S., M., O.M.A. El-Shiny, and A.H.H. Ahmed (1989). Comparative studies of the effectiveness of fusilade and glyphosate in controlling *Orobanche crenata* parasitizing *Vicia faba* plants. *Bulletin of Faculty of Agriculture, University of Cairo* 40: 633-650.

Salimi, H. (1998). Possibility of dodder control *(Cuscuta planiflora)* on alfalfa with glyphosate in Iran. In J. Maillet (Ed.), *Sixth EWRS Mediterranean Symposium* (p. 161). Montpellier, France: INRA.

Salimi, H. (2000). A study on comparison of seed dormancy and germination in three species of dodder *(Cuscuta campestris, C. monogyna, C. planiflora)*. In P.G. Americanos, B.E. Abu-Irmaileh, and A.R. Saghir (Eds.), *Proceedings of the First Near East Conference on Improved Weed Management* (pp. 278-293). Khartum, Sudan.

Salle, G., M. Aber, and S. Le-Coz (1984). Comparative histo-physiological study of the germination and establishment of *Cuscuta*. In C.L. Parker, L.J. Musselman, R.M. Polhill, and A.K. Wilson (Eds.), *Proceedings of the Third International Symposium on Parasitic Weeds* (pp. 10-21). Aleppo, Syria: International Center for Agricultural Research in the Dry Areas (ICARDA).

Salle, G., A. Armillota, and H. Frochot (1984). Mechanism of reaction of four cultivars of poplar against *Visvum album* L. In C.L. Parker, L.J. Musselman, R.M. Polhill, and A.K. Wilson (Eds.), *Proceedings of the Third International Symposium on Parasitic Weeds* (pp. 22-30). Aleppo, Syria: ICARDA.

Salle, G., E.B. Hariri, and C. Andry (1991). Some mechanisms involved in resistance of poplar (*Populus* spp.) to mistletoe (*Viscum album* L.). In J.K. Ransom, L.J. Musselman, A.D. Worsham, and C. Parker (Eds.), *Proceedings of the Fifth International Symposium of Parasitic Weeds.* June 24-30. Nairobi, Kenya: CIMMYT.

Sand, P.F. (1983). *Orobanche ramosa* in Texas. In Meeting of Weed Science Society of America, 1983. Animal and Plant Health Inspection Servey, Plant Protection and Quarantine, Hyattsville, Maryland. (Abstract no. 47).

Sandler, H.A., M.J. Else, and M. Sutherland (1997). Application of sand for inhibition of swamp dodder *(Cuscuta gronovii)* seedling emergence and survival on cranberry *(Vaccinium macrocarpon)* bogs. *Weed Technology* 11: 318-323.

Sandri, G., A. Sandri, and G. Martini (1998). Protection of tobacco against *Orobanche*. *Informatore Agrario* 54: 74-75.

Sands, W.N. (1924). Mistletoes attacking cultivated trees in Malaya. *Malayan Agricultural Journal* 12: 64-76.

Saric, T. (1981). The study of soil-acting foliar herbicides in perennial forage legumes and their mixtures with grasses. *Radovi Poljoprivrednog Fakulteta Univerziteta u Sarajevo* 29: 21-31.

Sarpe, N., C. Dinu, and I. Lungulescu (1992). Efficacy of the butylate, bentazone and imazethapyr herbicides in the control of dodder and other weeds in seedling alfalfa, in first year of vegetation. *Mededelingen van de Faculteit Landbouwwetenschappen, Rijksuniversiteit Gent* 57: 1117-1126.

Sarpe, N., I. Moga, M. Pocacean, I. Lungulescu, C. Roibu, and G.H. Buroi (1996). Control of annual weeds, *Elymus repens* and *Cuscuta* species in alfalfa crops. In H. Brown, G.W. Cussans, M.D. Devine, S.O. Duke, C. Fernandez-Quintanilla, A. Helweg, R.E. Labrada, M. Landes, P. Kudsk, and J.C. Streibig (Eds.), *Proceedings of the Second International Weed Control Congress*. Copenhagen, Denmark: Slagelse, Denmark, Department of Weed Control and Pesticide Ecology.

Sarychev, Yu., L.I. Trofimova, N.I. Strelyaeva, and V.A. Vinogradov (1982). Phenotypic expression of genes for resistance to broomrape in tobacco. *Genetika (Moscow)* 18: 114-148.

Sauerborn, J. (1994). Natural antagonists for parasitic weed control in agroecosystem. In R.K. Malik (Ed.), *Integrated weed management for sustainable agriculture: Proceedings of International Symposium of Indian Society of Weed Science,* Volume 1 (pp. 29-33). Hisar, Haryana, India: Indian Society of Weed Science.

Sauerborn, K.H., M.C. Linke, M.C. Saxena, and W. Koch (1989). Solarization, a physical control method for weed and parasitic plants (*Orobanche* spp.) in Mediterranean agriculture. *Weed Research* 29: 391-397.

Scharpf, R.F., H.C. Weber, and K. Dufft (2001). *Viscum album* in California, a century after its introduction. In A. Fer, P. Thalouarn, D.M. Joel, L.J. Musselman, C. Parker, and J.A.C. Verkleij (Eds.), *Proceedings of the Seventh International Parasitic Weed Symposium* (pp. 11-14). Nantes, France: Faculté des Sciences, Université de Nantes.

Schlingloff, S. and J. Alkamper (1992). Competition for mosture between dry peas and weeds at a semi-arid site in Syria. In Proceedings of 16th German Conference on Weed Biology and Control. *Journal of Plant Disease and Plant Protection, Sonderheft* 13: 221-228.

Schluter, K and M. Aber (1980). Chemical control of *Orobanche crenata* in field beans in Morocco. *Zeitschrift fur Pflanzenkrankheiten und Pflanzenschutz* 87: 433-438.

Schnell, H., M. Kunisch, M.C. Saxena, and J. Sauerborn (1996). Simulation of the seed bank dynamics of *Orobanche crenata* Forsk. in some crop rotations common in northern Syria. *Experimental Agriculture* 32: 395-403.

Schnell, H., K.H. Linke, and J. Sauerborn (1994). Trap cropping and its effect on yield and *Orobanche crenata* Forsk. infestation on following pea (*Pisum sativum* L.) crops. *Tropical Science* 34: 306-314.

Schulze, E.D., N.C. Turner, and G. Glatzel (1984). Carbon, water and nutrient relations of two mistletoes and their hosts: A hypothesis. *Plant, Cell and Environment* 7: 293-299.

Sengbusch, P.V. (n.d.). Santalas. In "Botanik online–Botany online–The Internet Hypertextbook." Available at <http://www.biologie.uni-hamburg.de/b-online/d51/51h.htm>.

Shabana, Y. and J. Sauerborn (2001). Evaluation of pesta-pelletized *Fusarium oxysporum* f.sp. *orthoceras* as a potential mycoherbicide for *Orobanche cumana*. In A. Fer, P. Thalouarn, D.M. Joel, L.J. Musselman, C. Parker, and J.A.C. Verkleij (Eds.), *Proceedings of the Seventh International Parasitic Weed Symposium* (pp. 296-297). Nantes, France: Faculté des Sciences, Université de Nantes.

Sharma, S., R. Khanna, and G.G. Sanwal (1985). Lipids of *Cuscuta reflexa* and changes in lipids of its host plants after infection. *Physiologia Plantarum* 63: 315-321.

Sherman, T.D., W.T. Pettigrew, and K.C. Vaughn (1999). Structural and immunological characterization of *Cuscuta pentagona* L. chloroplast. *Plant and Cell Physiology* 40: 592-603.

Shinkarenko, V.A. (1982). A weevil and the control of dodder. *Zashchita Rastenii* 7: 30-31.

Shlevin, E and D. Golan (1982). Selective control of dodder in carrot. *Phytoparasitica* 10: 267.

Sidahmed, O.A. (1984). Incidence of mistletoe (*Loranthus* spp.) on citrus and guava trees in the central region of the Sudan. *Acta Horticulturae* 143: 417-419.

Sillero, J.C., J.I. Cubero, and D. Rubiales (2001). Resistance to broomrape (*Orobanche crenata*) in *Lathyrus*. In A. Fer, P. Thalouarn, D.M. Joel, L.J. Musselman, C. Parker, and J.A.C. Verkleij (Eds.), *Proceedings of the Seventh International Parasitic Weed Symposium* (pp. 224-227). Nantes, France: Faculté des Sciences, Université de Nantes.

Sillero, J.C., M.T. Moreno, and D. Rubiales (2001). Low induction of *Orobanche crenata* seed germination in wild legume species. In A. Fer, P. Thalouarn, D.M. Joel, L.J. Musselman, C. Parker, and J.A.C. Verkleij (Eds.), *Proceedings of the Seventh International Parasitic Weed Symposium* (p. 234). Nantes, France: Faculté des Sciences, Université de Nantes.

Sillero, J.C., A. Perez-de-Luque, D. Rubiales, and M.T. Moreno (1999). Effect of the sowing date on the *Orobanche crenata* infection on susceptible and resistant faba bean and vetch varieties. In *Proceedings of SEMh Congreso 1999:*

Sociedad Espanola de Malherbologia (pp. 401-405). Lerida, Spain: Departamento Hortofruticultura, Botanica i Jardineria.

Simier, P., P. Delavault, S. Thoiron, C. Veronesi, A. Fer, and P. Thalouarn (2001). Changes in M6PR expression and enzyme activity in *Orobanche ramosa* during its development on tomato: Consequences in mannitol level. In A. Fer, P. Thalouarn, D.M. Joel, L.J. Musselman, C. Parker, and J.A.C. Verkleij (Eds.), *Proceedings of the Seventh International Parasitic Weed Symposium* (pp. 147-150). Nantes, France: Faculté des Sciences, Université de Nantes.

Singh, S.J. and H.K. Joshi (1992). A new record of witches' broom on *Orobanche cernua* parasitising little-leaf diseased eggplant. *International Journal of Tropical Plant Disease* 10: 73-77.

Skeate, S.T. (1987). Interactions between birds and fruits in a northern Florida hammlock community. *Ecology* 68: 297-309.

Slavov, S.B., R.B. Batchvarova, and V.T. Valkov (2001). Possibilities for obtaining tobacco to *Orobanche* spp. by chemical mutagenesis. In A. Fer, P. Thalouarn, D.M. Joel, L.J. Musselman, C. Parker, and J.A.C. Verkleij (Eds.), *Proceedings of the Seventh International Parasitic Weed Symposium* (pp. 88-91). Nantes, France: Faculté des Sciences, Université de Nantes.

Smaling, E.M.A., A. Stein, and P.H.M. Sloot (1991). A statistical analysis of the influence of *Striga hermonthica* on maize yields in fertilizer trails in southwestern Kenya. *Plant and Soil* 138: 1-8.

Smith, R.J. Jr. (1993). Biological controls as components of integrated weed management for rice in the United States. *Extension Bulletin-ASPAC, Food and Fertilizer Technology Center* 367: 1-9.

Sobrino Vesperinas, E. (1985). Search for resistance to *Orobanche ramosa* L. in rapeseed. *Cruciferae Newsletter* 10: 120-121.

Solymosi, P. (1998). Biology of broomrape (*Orobanche ramosa* L.) and possible control methods. *Novenyvedelem* 34: 469-475.

Soria, S., D. Cadahia, and A. Munoz (1993). The genus *Carulaspis* Mac. Gillivray, 1921 (Homoptera, Diaspididae) in the gardens of Partrimonio. *Nacional. Boletin de Sanidad Vegetal, Plagas* 19: 273-284.

Spelce, D.L. and L.J. Musselman (1981). *Orobanche minor* germination with strigol and GR compounds. *Zeitschrift fur Planzenzenphysiologie* 104: 281-283.

Steuber, K. (n.d.). *MPI for breeding research.* Kurt Stuebers online library of historic biological books. Available online at <http://www.mpiz-koeln.mpg.de/~Stueber/stueber_library.html>.

Stojsin, V., A. Maric, and B. Jocic (1991). Harmfulness of *Cuscuta campestris* Yunck on sugarbeet under varying mineral nutrition. *Zastita Bilja* 42: 357-363.

Stroud, A. (1989). *Weed management in Ethiopia: An extension and training manual.* Rome, Italy: Food and Agriculture Organization of the United Nations.

Sukno, S., J.M. Melero-Vara, and J.M. Fernandez-Martinez (1999). Inheritance of resistance to *Orobanche cernua* Loefl. in six sunflower lines. *Crop Science* 39: 674-678.

Sukno, S., J. Ruso, J. Melero-Vara, and J. M. Fernandez-Martinez (1996). Evaluation of wild *Helianthus* germplasm for resistance to highly virulent races of *Orobanche cernua* Loefl. and transfer to cultivated sunflower. In M.T. Moreno,

J.I. Cubero, D. Berner, D. Joel, L.J. Musselman, and C. Parker (Eds.), *Advances in parasitic plant research: Proceedings of Sixth International Parasitic Weed Symposium* (pp. 626-632). Cordoba, Spain: Junta de Andalucia, Direccion General de Investigacio Agraria.

Surov, T., D. Aviv, R. Aly, D.M. Joel, T. Goldman-Guez, and J. Gressel (1998). Germination of transgenic asulam-resistant potatoes to facilitate eradication of parasitic broomrapes (*Orobanche* spp.), with the *sul* gene as the selectable marker. *Theoretical and Applied Genetics* 96: 132-137.

Suzaki, Y., Y.H. Zhang, N. Murofushi, and Y. Takeuchi (1994). Endogenous gibberellins in clover broomrape, a parasitic plant, and its host, clover: Dependency of the parasite on the host for gibberellin production. *Journal of Plant Growth Regulation* 13: 63-67.

Swift, C.E. (2001). Cuscuta *and* Gramicca *species: Dodder—A plant parasite.* Fort Collins: Colorado State University Cooperative Extension, TRI River Area.

Syka, E.K. and I.G. Eleftherohorinos (1991). *Orobanche ramosa* L. (broomrape) control in tomato (*Lycopersicon esculentum* Mill.) with chlorsulfuron, glyphosate and imazaquin. *Weed Research* 31: 19-27.

Tada, Y., M. Suqai, and K. Furuhashi (1996). Haustoria of *Cuscuta japonica* a holoparasitic flowering plant, are induced by the cooperative effects of far-red light and tactile stimuli. *Plant and Cell Physiology* 37: 1049-1053.

Takeuchi, Y., Y. Omigawa, M. Ogasawara, K. Yoneyama, M. Konnai, and A.D. Wosham (1995). Effects of brassinosteroids on conditioning and germination of clover broomrape *(Orobanche minor)* seeds. *Plant Growth Regulation* 16: 153-160.

Talsakh'yan, M.G. and S.V. Grigoryan (1978). Fungi found on broorape in the Armenian-SSR, USSR. *Mikologiya i Fitopatologiya* 12: 112-114.

Templeton, G.E. (1992). Use of *Colletotrichum* strains as mycoherbicides. In J.A. Bailey, and M.J. Jeger (Eds.), Colletotrichum: *Biology, pathology and control* (pp. 358-380). Wallingford, UK: CAB International.

Tepe, I., M. Deveci, and B. Keskiin (1997). Studies on damage and parasite levels of dodder (*Cuscuta approximata* Bab.) on some alfalfa *(Lucerne)* cultivars. In Y. Nemli and H.J. Demirkan (Eds.), *Proceedings of Second Turkish Weed Science Congress,* September 1-4 (pp. 353-260). Bornova, Izmir, Turkey: Turkish Weed Science Society.

ter Borg, S.J. (1994). General aspects of taxonomy, distribution and ecology: State of the art after the Third International Workshop on *Orobanche.* In A.H. Pieterse, J.A.C. Verkleij, and S.J. ter Borg (Eds.), *Biology and management of* Orobanche: *Proceedings of the Third International Workshop on Orobanche and Related Striga Research.* November 8-12, 1993. Amsterdam, The Netherlands: Royal Tropical Institute.

Terekhin, E.S. (1975). Development of microsporangium, pollen grains, ovule, embryo sac in *Cymomorium* (Cymomoriaceae). *Botanical Zhurnal (Leningrad)* 60: 153-162.

Texas Cooperative Extension, Agricultural and Environmental Safety (n.d.). Keep watch for branched broomrape *(Orobanche ramosa),* a major threat to U.S. crops. Available at <http://www-aes.tamu.edu/mary/brmrape/brmrape.htm>.

Thankamma, L. (1990). Dodder menace: A threat to rubber plantations. *Rubber Board Bulletin* 26: 27.

Thankamma, L. and J.D. Marattukalam (1995). Dodder (*Cuscuta campestris* Yuncker) infestation in rubber plantations. *Rubber Board Bulletin* 27: 15-18.

Thomas, H., A. Heller, J. Sauerborn, and D. Muller-Stover (1999). *Fusarium oxysporum* f.sp. *orthoceras*, a potential mycoherbicide, parasitize seeds of *Orobanche cumana* (sunflower broomrape): A cytological study. *Annals of Botany* 83: 453-458.

Thomas, H., J. Sauerborn, D. Muller-Stover, and J. Kroschel (1999). Fungi of *Orobanche aegyptiaca* in Nepal with potential as biological control agents. *Biocontrol Science and Technology* 9: 379-381.

Thomas, H., J. Sauerborn, D. Muller-Stover, A. Ziegler, J.S. Bedi, and J. Kroschel (1998). The potential of *Fusarium oxysporum* f.sp. *orthoceras* as a biological control agent of *Orobanche cumana* in sunflower. *Biological Control* 13: 41-48.

Torres, L.G. and F.L Granados (1991). Control of broomrape (*Orobanche crenata* Forsk.) in broad bean (*Vicea faba* L.) with imidazolinones and other herbicides. *Weed Research* 31: 227-235.

Tosh, G.C., G.K. Patro, and A. Misra (1978). Control of *Cuscuta* in niger seed. *PANS* 24: 167-171.

Tsiovion, Y. (1979). Resistance reaction of tomato and bean plants to infection by *Cuscuta*. *Phytoparasitica* 7: 141.

Tsiovion, Y. (1980). Host tissue determination of xylem formation in the haustoria of *Cuscuta*. *Weed Abstracts*. 2578: 285.

Tsiovion, Y. (1981). Suppression of axillary buds of its host by parasitizing *Cuscuta:* 1. Competition among sinks and indirect inhibition. *New Phytologist* 87: 91-99.

Tsybul'skaya, G.N., B.G. Degtyarev, N.A. Fedoryak, and A.N. Skoklyuk (1978). Determination of viability of *Phytomyza puparia*. *Zashchita Rastenii* 5: 29-30.

Tsybul'skaya, G.N. and A.N. Skoklyuk (1978). Calculating release of *Phytomyza*. *Zashchita Rastenii* 11: 49.

Tylkowski, T. (1982). New hosts for mistletoe *(Viscum album)* in the Kornik Arboretum. *Arboretum Kornickie* 26: 31-34.

Uludag, A., A. Demir, and E. Akosy (2001). *Orobanche* species in red lentil areas of Turkey. In A. Fer, P. Thalouarn, D.M. Joel, L.J. Musselman, C. Parker, and J.A.C. Verkleij (Eds.), *Proceedings of the Seventh International Parasitic Weed Symposium* (p. 41). Nantes, France: Faculté des Sciences, Université de Nantes.

University of California (2002a). Dodder (*Cuscuta* spp.). Available online at <http://helios.bto.ed.ac.uk/>.

University of California (2002b). *Pest management and identification*. Statewide Integrated Pest Management Project. Available online at <http://www.ipm.ucdavis.edu/pmg/crops-agriculture.html>.

University of Hohenheim, Ecology and Management of Parasitic Weeds (n.d.). Pathogens on *Orobanche*. Available at <http://www.uni-hohenheim.de/~www380/parasite/oro_path.htm>.

University of Hohenheim (380) (2002). *Ecology and management of parasitic weeds.* University of Hohenheim, Stuttgart, Hohenheim, Germany. Available online at <http://www.uni-hohenheim.de/~www380/parasite/CourseHH.htm>.

Vail, S.L., O.D. Dailey Jr., W.J. Connik, and A.B. Pepperman (1985). Strigol synthesis and related structure-bioactivity studies. In A. Thompson (Ed.), *The chemistry of allelopathy: Biochemical interactions among plants* (pp. 445-456). American Chemical Society Symposium Series 268. Washington, DC: American Chemical Society.

Valkov, V.T., R.B. Bachvarova, S.B. Slavov, S.I. Atanassova, and A.I. Atanassov (1998). Genetic transformation of tobacco for resistance to Basta. *Bulgarian Journal of Agricultural Science* 4: 1-7.

Vallauri, D. (1998). Dynamics of *Viscum album* L. on Austrian black pine in Saignon watershed (Southern French Alps). *Annales des Sciences Forestieres* 55: 823-835.

van Hezewijk, M.J. and J.A.C. Verkleu (1996). The effect of nitrogenous compounds on *in vitro* germination of *Orobanche crenata* Forsk. Weed Research 36: 395-404.

Vinogradov, V.A., E.K. Mironov, and Yu.F. Sarychev (1984). Degree of field resistance to *Orobanche* in tobacco mutants, hybrid and varieties. *Plant Breeding Abstracts* 54: 3464.

Vinogradov, V.A., E.K. Mironov, N.I. Strelyaeva, and Y.F. Sarychev (1981). Breeding tobacco for resistance to *Orobanche. Tabak* 4: 55-57.

Visser, J.H. (1982). Epiparasitism—How many mistletoes on one another? *The Golden Bough (Royal Botanic Gardens, Kew)* 1: 4.

V'lchev, N., D. Bakhariev, S. Masheva, and K. Toskov (1995). Possibilities for the use of chemicals in the control of *Orobanche (Orobanche ramosa)* in tomato stands. *Rasteniev'dni Nauki* 32: 62-64.

Volkov, O.G. (1989). Prospects for biological regulation of *Cuscuta. Zashchita Rastenii (Moskava)* 11: 18-19.

Vouzounis, N.A. and P.G. Americanos (1998). Control of *Orobanche* (broomrape) in tomato and eggplant. *Technical Bulletin of Cyprus Agricultural Research Institute* 196: 1-7.

Vranceanu, A.V., G. Soare, D. Craiciu, and G. Voinescu (1990). Adaptability of some Romanian sunflower hybrids in Europe. *Analele Institutului de Cercetari Pentru Cereale si Plante Tehnice, Fundulea* 58: 61-80.

Wang, R. (1990). Biological control of weeds in China: A status report. In E.S. Delfosse (Ed.), *Proceedings of the Seventh International Symposium on Biological Control of Weeds* (pp. 689-693). Rome, Italy: Istituto Sperimentale per la Patologia Vegetale, Ministero dell'Agricoltura e delle Foreste.

Watson, L. and M.J. Dallwitz (1992). *The families of flowering plants: Descriptions, illustrations, identification, and information.* Available online at <http://biodiversity.uno.edu/delta/>.

Weber, H.C. (1993). Observation on the establishment and growth rates in the first stage of development of *Viscum album* L. (Viscaceae). *Beitrage zur Biologie der Pflanzen* 67: 319-331.

Wegmann, K., E. von Elert, H.J. Harloff, and M. Stadler (1991). Tolerance and resistance to *Orobanche*. In K. Wegmann and L.J. Musselman (Eds.), *Progress in* Orobanche *research: Proceedings of the International Workshop on* Orobanche *Research* (pp. 318-321). Tuebingen, Germany: Eberhard-Karls-Universitat.

Westwood, J.H. and C.L. Foy (1999). Influence of nitrogen on germination and early development of broomrape (*Orobanche* spp.). *Weed Science* 47: 2-7.

Wiens, D. and C.L. Calvin (1987). Epiparasitism in mistletoes. *The Golden Bough (Royal Botanic Gardens, Kew)* 9: 3-5.

Willem, J., J.F. Thuring, N.W.J.T. Heinsman, R.W.A.W.M. Jacobs, G.H.L. Nefkens, and B. Zwanenburg (1997). Asymmetric synthesis of all stereoisomers of demethylsorgolactone: Dependence of the stimulatory activity of *Striga hermonthica* and *Orobanche crenata* seed germination on the absolute configuration. *Journal of Agricultural and Food Chemistry* 45: 507-513.

Wolswinkel, P. (1984). Recent developments in phloem transport research: The parallel between the site of attachment of *Cuscuta* and developing seeds of legumes. In C.L. Parker, L.J. Musselman, R.M. Polhill, and A.K. Wilson (Eds.), *Proceedings of the Third International Symposium on Parasitic Weeds* (pp. 156-163). Aleppo, Syria: International Center for Agricultural Research in the Dry Areas (ICARDA).

Wolswinkel, P. and A. Ammerlaan (1983). Sucrose and hexose release by excised stem segments of *Vicia faba* L. The sucrose-specific stimulating influence of *Cuscuta* on sugar release and the activity of acid invertase. *Journal of Experimental Botany* 34: 1516-1527.

WonYung, A. (1996). Analysis of chemical constituents of saccharides and triterpenoids in the Korean native mistletoes. *Journal of the Korean Weed Science and Technology* 24: 27-33.

Yakutkin, V.I. and N.P. Tavolzhanskri (1999). Disease control in sunflower crops in Russia. In China Association of Agricultural Societies (Ed.), *Research progress in plant protection and plant nutrition* (pp. 214-218). Beijing, China: China Agriculture Press.

Yang-han, L. (1994). *Cuscuta* species. In R. Labrada and C. Parker (Eds.), *Weed management for developing countries* (pp. 143-147). FAO Plant Production and Protection Paper 120. Rome, Italy: Food and Agriculture Organization of the United Nations.

Yoneyama, K., M. Ogasawara, Y. Takeuchi, M. Konnal, Y. Sugimoto, T. Sassa, and S. Yoshida (1998). Effect of jasmonates and related compounds on seed germination of *Orobanche minor* Smith and *Striga hermonthica* (Del.) Benth. *Bioscience, Biotechnology and Biochemistry* 62: 1448-1450.

Yoneyama, K., Y. Takeuchi, M. Ogasawara, M. Konnal, Y. Sugimoto, and T. Sassa (1998). Cotylenins and fusicoccins stimulate seed germination of *Striga hermonthica* (Del.) Benth and *Orobanche minor* Smith. *Journal of Agricultural and Food Chemistry* 46: 1583-1586.

Yoneyama, K., Y. Takeuchi, and T. Yokota (2001). Natural germination stimulants for *Orobanche minor*. In A. Fer, P. Thalouarn, D.M. Joel, L.J. Musselman, C. Parker, and J.A.C. Verkleij (Eds.), *Proceedings of the Seventh International*

Parasitic Weed Symposium (p. 123). Nantes, France: Faculté des Sciences, Université de Nantes.

Yong-han, C., Li DongGuang, H. Zhifu, and Y. Meil (1999). Catalogue of infective tobacco disease in Jilin province. *Journal of Jilin Agricultural University* 21: 30-33.

Young, S.A. (1984). Seed certification in Utah. *Utah Science* 45: 79-83.

Zahran, M.K. (1982). Weed and *Orobanche* control in Egypt. In G. Hawtin and C. Webb (Eds.), *Faba bean improvement* (pp. 191-198). The Hague, Netherlands: Martinus Nijhoff.

Zaitoun, F.M.F., O.A. Al-Menoufi, and H.C. Weber (1991). Susceptibility of five varieties of *Vicia faba* to infection with *Orobanche crenata*. In J.K. Ransom, L.J. Musselman, A.D. Worsham, and C. Parker (Eds.), *Proceedings of the Fifth International Symposium of Parasitic Weeds* (pp. 209-216). June 24-30. Nairobi, Kenya: CIMMYT.

Zaitoun, F.M.F. and F.S. Sabra (2000). Resistance of the new Egyptian faba bean cultivars against *Orobanche crenata*. In P.G. Americanos, B.E. Abu-Irmaileh, and A.R. Saghir (Eds.), *Proceedings of the First Near East Conference on Improved Weed Management* (pp. 225-230). Khartum, Sudan.

Zaitoun, F.M.F., M.E. Saleh, and S.J. ter. Borg (1996). Effect of some trace elements on germination and seedling development of *Orobanche*. In H. Brown, G.W. Cussans, M.D. Devine, S.O. Duke, C. Fernandez-Quintanilla, A. Helweg, R.E. Labrada, M. Landes, P. Kudsk, and J.C. Streibig (Eds.), *Proceedings of the Second International Weed Control Congress*. Copenhagen, Denmark: Department of Weed Control and Pesticide Ecology.

Zaki, M.A., H.S. El-Metwaly, and R.A. Hassan (1998). Studies on dodder (*Cuscuta* spp.) control. In J. Maillet (Ed.), *Sixth EWRS Mediterranean Symposium* (pp. 147-152). Montpellier, France: INRA.

Zazzerini, A., G.D. Torre, and L. Tosi (1981). Broomrape of tobacco: Epidemiology and control. *Informatore Fitopatologico* 31: 15-23.

Zemrag, A. and M. Bajja (2001). Characterization of *Orobanche* spp. in Morocco and the effect of some trap crops on *Orobanche crenata* Forsk in faba bean (*Vicia faba* L.). In A. Fer, P. Thalouarn, D.M. Joel, L.J. Musselman, C. Parker, and J.A.C. Verkleij (Eds.), *Proceedings of the Seventh International Parasitic Weed Symposium* (p. 300). Nantes, France: Faculté des Sciences, Université de Nantes.

Zermane, N. (2000). Parasitic weeds of economic importance in North Africa: Problems and management. In P.G. Americanos, B.E. Abu-Irmaileh, and A.R. Saghir (Eds.), *Proceedings of the First Near East Conference on Improved Weed Management* (pp. 257-260). Khartum, Sudan.

Zermane, N., J. Kroschel, G. Sallé and Z. Bouznad. (1999). Prospects for biological control of the parasitic weed *Orobanche* in Algeria. In *Advances in Parasitic Weed Control at on-farm level,* eds. J. Kroschel, H. Betz, and M. Abderahibi, Vol. II. Joint Action to control *Orobanche* in the Wana region. Weikersheim, Germany: Margraf Verlag, pp. 173-184.

Zermane, N., J. Kroschel, T. Souissi, and M. Kharrat (2001). Investigations on the *Orobanche* problem in Tunisia. In A. Fer, P. Thalouarn, D.M. Joel, L.J. Musselman, C. Parker, and J.A.C. Verkleij (Eds.), *Proceedings of the Seventh Interna-*

tional Parasitic Weed Symposium (p. 46). Nantes, France: Faculté des Sciences, Université de Nantes.

Zermane, N. and A.R. Saghir (1995). The genus *Cuscuta* in Algeria. *Arab Journal of Plant Protection* 13: 69-75.

Zhiqian, L. (1994). Mechanism of action of glyphosate on dodder control. *Acta Phytophylacica Sinica* 21: 369-372.

Web Sites

Agricultural Research Service, News and Events: <http://www.ars.usda.gov/is/pr/2000/000606.htm>

AsiaChi.com, Chinese Traditional Medicine, Raw Herbs, "Cisthanche": <http://asiachi.com/cisthanche.html>

Colorado State University Cooperative Extension: <http://www.coopext.colostate.edu/TRA/dodder.html>

Colorado State University Cooperative Extension, Crop Series–Production: <http://www.ext.colostate.edu/pubs/crops/03112.pdf>

E-floras.org: <http://hua.huh.harvard.edu/china/mss/volume18/OROBANCHACEAE.published.pdf>

Larval Food Plant Information: <ftp://www.funet.fi/pub/sci/bio/life/warp/food-plants-o.htm>

New York Botanical Garden Herbarium Specimen Catalog: <http://scisun.nybg.org:8890/searchdb/owa/wwwcatalog.searchform>

Northern Prairie Wildlife Research Center, "Native Wildflowers of the North Dakota Grasslands": <http://www.npwrc.usgs.gov/resource/literatr/wildflwr/species/orobfasc.htm>

Old Dominion University, Haustorium Parasitic Plants Newsletter: <http://www.odu.edu/webroot/instr/sci/haustorium.nsf>

Purdue University, Department of Horticulture and Landscape Architecture, "Famine Foods": <http://www.hort.purdue.edu/newcrop/ faminefoods/ff_families/Cynomoriaceae.html>

Southern Illinois University–Carbondale, College of Science, Parasitic Plant Connection: <http://www.science.siu.edu/parasitic-plants/santalaceae/description.html>

Texas A&M University, Texas Cooperative Extension, Agricultural and Environmental Safety, "Weed Alert: Branched Broomrape": <http://www-aes.tamu.edu/mary/brmrape/brmrape.htm>

Unité de Malherbologie & Agronomie, Weed Science & Agronomy, INRA-Dijon: <http://www.dijon.inra.fr/malherbo/hyppa/hyppa-a/oracr_ah.htm>

Universidad de Extremadura, Lecciones hipertextuales de Botánica: <http://www.unex.es/botanica/rosidae/rosid-tl.htm>

University of California–Davis, Statewide Integrated Pest Management Program: <http://www.ipm.ucdavis.edu/PMG/P/W-LR-PSPP-IF.001.HTML>

University of Hohenheim, Ecology and Management of Parasitic Weeds: <http://www.uni-hohenheim.de/~www380/parasite/>

University of Paisley, Biological Sciences, "Biodiversity Reference": <http://www-biol.paisley.ac.uk/bioref/Plantae/Orobanche.html>

USDA, Natural Resources Conservation Service, Plants Database: <http://plants.usda.gov>

Weed Science Society of America, "Orobanche–The Broomrapes": <http://www.wssa.net/photo&info/larrymitich_info/orobanche.html>

Chapter 22

Management of Weeds in Pasture Systems

Brian M. Sindel

INTRODUCTION

More than 40 percent of the world's agricultural land is covered by grassland communities used for grazing livestock (Zimdahl, 1999). These complex plant communities are typically denoted as either rangelands or pastures, depending on the level of management input that is employed to obtain some productive output from them. Rangelands have been defined as seminatural ecosystems that are grazed with relatively little other management input, while at the other end of the management spectrum, intensively managed pastures may be planted down to improved species, irrigated, and fertilized. Grazing may also be highly regulated in pastures. For the purposes of this discussion, all grazed plant communities will be referred to as pastures, and an ability to manage these pastures in some way and the animals that graze them will be assumed.

What then is a pasture weed? This is not an easy question to answer precisely. In an annual crop a weed will probably be any species other than the cultivated crop plant, while in a natural ecosystem a weed may be any plant not native to that particular region and vegetation type; however, weeds of these systems may be highly desirable species within a pasture. A good example of this is *Lolium rigidum* Gaudin (annual ryegrass), which is one of the worst weeds of cereal crops in southern Australia and has also developed widespread resistance to herbicides (Powles and Holtum, 1990). In a survey covering most agricultural areas of Western Australia, *L. rigidum* was ranked by farmers as both the worst weed on their farms and the most beneficial plant, reflecting the plant's good qualities for grazing as well as its competitiveness in crops. Likewise, in that state, crop weeds such as *Arctotheca calendula* (L.) Levyns (capeweed), *Hordeum* spp., and even the

P. M. Dowling, K. C. Harrington, R. S. Jessop, D. R. Kemp, J. M. Scott, and R. D. B. Whalley provided helpful comments on earlier versions of this manuscript.

potentially poisonous plant *Echium plantagineum* L. (Paterson's curse) are considered to be desirable pasture plants in most situations (Maddin and Moore, 1993).

Although some weedy species can contribute significantly to livestock production in many farming systems, they are unlikely to be nutritionally optimal. For example, ample evidence suggests that pastures based on legumes provide better overall nutrition for livestock than do nonlegume species (Maddin and Moore, 1993). However, in the higher-rainfall regions of southern Australia, pastures based upon a mixture of legumes such as *Trifolium repens* L. (white clover), *T. subterraneum* L. (subterranean clover), and *Medicago* spp. as well as palatable perennial grasses are regarded as the best option. In general, pasture ecosystems are likely to be more productive and stable in the long term (i.e., sustainable) where they are composed of multiple desirable species and filling different ecological niches (Smith and Martin, 1995). This concept is developed later in this chapter.

The aim of this chapter is to identify some of the major problems caused by weeds in pastures, explore the reasons for weed invasions in pastures, and present some of the principles of weed management solutions for pastures. Weed control is inherently more complex in pastures than in crops because pasture weeds often keep germinating and competing throughout the year and because most herbicides are likely to cause collateral damage to useful species within the pasture community. An integrated approach, one in which all appropriate weed control techniques are brought together to place the greatest stress on weeds, will, of necessity, be developed on a case-by-case basis. Such an approach will depend on the weeds present, the condition of the pasture, topography, and other local conditions. Nevertheless, where different techniques can be easily combined, these will be emphasised in this discussion. The chapter has been written largely from a temperate Australian perspective, and because of the diversity of pasture and grazing systems worldwide, readers will need to determine how the outlined principles can be applied to their particular situations.

IMPACT OF PASTURE WEEDS

The first thing to be said about weed management in pastures is that weed control (or weed management) per se is not the desired outcome of pasture management but rather the maintenance of a productive pasture that adequately feeds the livestock that graze it. The importance of any pasture weed, therefore, should be gauged by the realized or potential impact that it might have on livestock production. Weeds may affect livestock *directly,* for example, through poisoning, or they may have an *indirect* impact by altering the quality, quantity, and availability of the pasture on which livestock graze. Michael (1970) classified weeds of Australian pastures and grazing lands according to the type of impact that they have and the situations in which they have their impact. His categories are by no means mutually exclusive, but they serve to demonstrate the types of impacts that weeds can have in pastures.

Weeds of Established Pastures

Plants Harmful to Livestock or Which Lower the Value of Livestock Products

This group of pasture weeds includes a wide range of poisonous plants, for example, *Heliotropium europaeum* L. (common heliotrope), *Senecio jacobaea* L. (tansy ragwort), *Oxalis pes-caprae* L. (Bermuda buttercup soursob), *Salvia reflexa* Hornem. (mintweed), and *Euphorbia drummondii* Boiss. (caustic weed). They contain various toxic substances such as alkaloids, fluoroacetate, oxalate, nitrate, hydrocyanic acid, and selenium, but in many cases the toxic principles of poisonous plants are unknown. Although potentially toxic, the risk of poisoning from these plants can be dramatically affected by factors such as the palatability of the weed, availability of alternative feed, herbicide damage, age of the plant, variation in animal tolerance, and prevailing weather conditions.

For example, conflicting opinions as to the risk to livestock grazing pastures in southeastern Australia infested with *Senecio madagascariensis* Poir. (fireweed), which contains the liver-damaging pyrrolizidine alkaloid senecionine, arise primarily from the fact that *S. madagascariensis* is unpalatable to cattle and horses (Sindel et al., 1998). Where its ingestion cannot be avoided, such as in pasture heavily infested with young plants or poor-quality pasture where there is a shortage of alternative feed, or in hay or silage, then poisoning is more likely to occur. In these situations, substantial field mortalities of cattle have been recorded. Sheep and goats, on the other hand, are much less susceptible than cattle to poisoning from this weed (Sindel et al., 1998).

In the case of *Hypericum perforatum* L. (St. John's wort), which occurs widely in temperate areas of the world, the toxic component hypericin causes photosensitization in livestock with light-colored skin. However, because of lower hypericin levels during the winter and spring months in Australia (compared with autumn and summer) (Southwell and Bourke, 2001), sheep can be grazed on infested pastures over this period with less deleterious effects on animal health. Likewise, hypericin tolerance in Merino sheep appears to be strongly related to wool coverage and protection from light, such that shorn sheep were far more intolerant of hypericin in *H. perforatum* than wool-protected sheep. Differences in tolerance were also detected between fine and coarse wool sheep types, while cattle tolerance varied with breed and was as much as four to six times higher than wool-protected Merino sheep (Bourke, Southwell, and Mayo, 2002).

Many plants cause mechanical injuries to livestock, both external and internal; for example, the pointed fruit of the low-growing *Tribulus terrestris* L. (caltrop/puncturevine) may damage the feet of livestock, particularly horses, sheep, cattle, and dogs (Parsons and Cuthbertson, 1992). The spikelets of many grasses, such as the *Austrostipa* spp. and *Aristida* spp., can penetrate the skins of livestock and enter their eyes, causing blindness. *Chondrilla juncea* L. (skeleton weed), abundant in some parts of Europe, North and South America, and southern Australia, may cause choking in cattle because of its fibrous stems and low proportion of digestible leaf material at maturity (Currie, 1936).

Many plants have burrs that adhere to wool, for example, *Xanthium occidentale* Bertol. (Noogoora burr) and *X. spinosum* L. (Bathurst burr/ spiny cocklebur), which make shearing more difficult and which downgrade the value of the wool. Similarly, meat and milk products may be rejected or downgraded due to tainting where livestock have consumed pungent weeds such as *Melilotus indicus* (L.) All. (Hexham scent/annual yellow sweetclover) or *Parthenium hysterophorus* L. (parthenium weed).

Plants Occupying Ground That Could Be Utilized by More Useful Pasture Species

Although not causing much direct harm to livestock, this group of plants may be less palatable, nutritious, or productive than other more desirable species or may not be eaten by livestock at all. Valuable herbage production may be lost through plant interference; in addition, there may be lower efficiency of conversion of herbage to livestock weight gain. For sheep and cattle, this group includes the annual and biennial thistles, for example, *Carduus pycnocephalus* L. (slender thistle), *C. tenuiflorus* Curtis (winged slender thistle), *C. nutans* L. (nodding thistle), *Cirsium vulgare* (Savi) Ten. (spear thistle), *Onopordum acanthium* L. (Scotch thistle), *Silybum marianum* (L.) Gaertn. (variegated thistle), and *Cynara cardunculus* L. (artichoke thistle). It also includes unpalatable perennial grasses such as *Nassella neesiana* (Trin. & Rupr.) Barkworth (Chilean needle grass) and *N. trichotoma* (Nees) Hack. ex Arechav. (serrated tussock), annual grasses that set seed quickly such as *Hordeum* and *Vulpia* spp., and woody perennial shrubs such as *Rubus fruticosus* L. agg. (blackberry), *Lycium ferocissimum* Miers (African boxthorn), *Rosa rubiginosa* L. (sweet briar), and *Lantana camara* L. (lantana), all widespread in different parts of the world. Weeds in this group can sometimes be utilized for part of their life cycle (so-called transitional weeds), for example, early in the vegetative stage of annual grasses, or be managed in such a way as to keep weeds in a palatable state for a longer period, which usually involves removing old growth and then regular defoliation to keep regrowth young and nutritious.

Weeds Interfering with the Establishment of Pastures

A range of weeds cause problems for clearing of land (where this is allowed) and the subsequent establishment of sown pastures. For example, many of the native woody shrubs in Australia have hindered the establishment of pastures on low-fertility soils. Where land is naturally fertile or where the fertility has been increased through the application of fertilizer and introduction of pasture legumes, weeds of high fertility become important when pastures are being replanted. This group includes the annual grasses (e.g., *Hordeum* spp., *Bromus* spp., *Vulpia* spp.), nitrophilous thistles [e.g., *Silybum marianum* (L.) Gaertn; milk thistle], and many other pioneer weeds that accompany the establishment of sown pastures.

WHY ARE THERE WEEDS IN PASTURES?

"It is a fact of biological life that if one organism occupies a particular space in an ecosystem, then another organism cannot do so" (Hill, 1977, p. 10). Pastures are dynamic plant communities often comprising a suite of 10 to 20 species per square meter (Friend and Kemp, 2000), but often only a few of these species contribute to the bulk of pasture growth and livestock production. Furthermore, the species composition in the soil seed bank below the surface is usually markedly different from that above ground (Lodge, 2001). The seed bank of weeds can be of the order of tens of thousands of seeds per square meter (Andrews, Whalley, and Jones, 1996; Gardener, Whalley, and Sindel, 2003) whereas that of the desirable perennial grasses is usually of the order of tens or hundreds of seeds per square meter (Whalley, Gardener, and Earl, 1999; Lodge, 2001). Where there is a gap in the pasture canopy, weeds are the most likely to germinate and become established. Likewise, weeds such as *Vulpia bromoides* (L.) Gray (squirrel-tailed fescue) can recover more rapidly than other more desirable pasture species following pasture disturbance and suppression, for example, from drought, because of the higher finite rates of population increase of the weeds (Freckleton et al., 2000).

All other factors being the same, the greater the proportion of bare ground within a pasture, the more susceptible it is to invasion by weeds. Because annual species leave ground bare for part of the year, pastures based on annual plants are generally more prone to invasion by weeds than those with a more stable background matrix of perennial species. Moreover, once annual weeds have become established in a pasture, they are likely to be self-promoting because when they die they leave bare ground that can again result in recruitment of weedy species (Panetta et al., 1993).

In temperate Australian pastures, some regenerative spaces are often needed for the establishment of annual legumes such as *Trifolium subterraneum,* which play a crucial role in fixing nitrogen in the soil. However, these regenerative spaces can be filled by a variety of other species from the pasture community and, if perennial grasses decline over time, as they do under certain management systems and environmental conditions, other more weedy annual plants such as *Vulpia* spp. and *Bromus* spp. and various thistles are able to establish and dominate (Dowling et al., 1998). It is also evident that not all plants occupy the same ecological niche. For example, in an annual pasture dominated by the two grasses *Lolium rigidum* and *V. bromoides* and the legume *T. subterraneum,* competitive interactions between the grasses were generally strong so that one or the other tended to dominate, while interactions between the grasses and legumes were weaker. As a result, the annual pasture community was predicted to tend toward a mixture of one grass and *T. subterraneum*. The grass to dominate was dependent on the founding composition of the pasture, but also overwhelmingly influenced by disturbance effects such as drought (Freckleton et al., 2000).

Where there is sufficient year-round soil moisture and temperatures to support perennial pasture growth (e.g., see Kemp and Dowling, 1991), weed invasion may indicate a problem affecting the growth of the desirable perennial pasture plants that reduces their competitiveness. This problem may be a pest such as red-legged earth mite or aphids, disease such as clover scorch, fertilizer imbalance, inappropriate grazing strategies, or a combination of these factors. Identification and correction of the problem will often lead to reduced levels of weeds on a long-term basis. However, such pasture renovation may need to be supplemented with management methods aimed directly at weeds, for example, herbicides, particularly while population levels are being brought down to a manageable level.

In temperate Australian pastures, the common cause of perennial plant decline and weed invasion by annual grasses and many nongrass weeds is overgrazing (Dowling et al., 1998), and this is often exacerbated by drought. Overgrazing leads to severe defoliation of the pasture which in turn results in the weakening and even death of desirable plants which can create gaps and allow weeds to invade (Friend and Kemp, 2000). For this reason, it is essential that the pasture manager should understand the grazing responses of the predominant pasture species and that grazing is managed accordingly. Nevertheless, the fact that pastures seldom enjoy ideal climatic and seasonal conditions for perennial plants to dominate means that weeds will always have the potential to invade and that management inputs will always be needed. Competition plays a key role in determining the composition of plant communities, but the outcome of competition will be modulated by disturbances such as drought and overgrazing (Freckleton et al., 2000). The implications of overgrazing are discussed in the next section.

STRATEGIES FOR CONTROL

Pasture Competition

Pasture weeds are often best controlled by the pasture plants themselves, which act as effective competitors, reducing light (because of a high proportion of ground cover), moisture, and nutrient availability to germinating weed seedlings (Michael, 1970). Therefore, the essential principle of any weed management program must be the provision of a dense, vigorous, and competitive pasture, particularly in those periods that coincide with the bulk of weed emergence and seedling establishment. This time will vary with species, climate, and seasonal conditions.

Pasture species for sowing should be selected on the basis of their adaptation to the climatic and soil conditions, their persistence and productivity, and the niche that they may fill within the pasture sward. Where the rainfall is sufficiently distributed throughout the year to support perennial grasses, for example, *Phalaris aquatica* L. (phalaris/tuberous canary grass) or *Festuca arundinacea* Schreb. (tall fescue), then these should dominate the pasture mix. Likewise, *Medicago sativa* L. (lucerne or alfalfa) is well-known for its ability to suppress the growth of some weeds (Michael, 1968). Often, only two or three species are sown in pastures and yet, in time, these pastures are found to be comprised of 20 or more species. This fact suggests that the sown species are not always able to exploit the available resources or niches in the pasture and that a wider mix of species, particularly desirable broadleaf plants, may be warranted at sowing as a means of reducing the rate of invasion by weeds (Kemp et al., 2003). The timing of reseeding of pastures (if that is required) needs to be carefully planned to avoid germination of and early competition from annual weeds; for example, late autumn or winter sowing can often avoid the problem of germinating annual grasses and thistles in autumn in some areas of Australia (J. Scott, personal communication, January 2004).

Weed competition will also be influenced by plant density. Sown pastures can have weed problems because sowing rates are too low for the conditions. Plus, insufficient attention may be paid to reducing the soil weed seed bank before reseeding. Short cropping phases have potential advantages in offsetting the costs of reestablishing a perennial pasture, but the effectiveness of the rotation in reducing the weed seed bank will depend on good weed control and prevention of weed seed set in each phase of the rotation (Dowling et al., 2003).

Established pastures must be carefully managed to prevent their degradation and reinfestation by weeds. Any strategy that favors the growth of the desirable pasture component over the weedy component is likely to be successful. For instance, spelling in the spring can favor the survival of perennial grasses such as *P. aquatica* and *F. arundinacea* (Fitzgerald and Lodge, 1997). Also, appropriate fertilizer applications can promote pasture growth and lead to reduced weed numbers (e.g., Cook, Blair, and Lazenby, 1978), although applications need to be timed to coincide with appropriate phenological stages of the desirable pasture components and seasonal potential for growth. For example, *Vulpia myuros* (L.) C.Gmel. (rat's tail fescue) grows well under low phosphate conditions (Loneragan and Asher, 1967), while the weed is less competitive in well-nourished pastures. Some weeds will also become more palatable under higher nutrient conditions and therefore be more likely to be removed by grazing animals. Lodge and Whalley (1985) found that a strategy of heavy grazing of the weedy grass *Aristida ramosa* R.Br. (purple wiregrass) during its flowering period, combined sometimes with burning to remove dead low-quality material, and resting during the period when the native grass *Danthonia linkii* Kunth (wallaby grass) was flowering, over a number of years eliminated the weed and promoted the desirable grass.

Grazing

Livestock, with the exception of goats (Beskow, 2001), will rarely select weedy species over the more desirable pasture component, and so the effect of selective grazing is usually to the disadvantage of desirable species with respect to the weeds. Controlling selective grazing by manipulating the time of grazing, the grazing pressure, the method of grazing, and the type of livestock is a key aspect of managing weeds with grazing, and each approach can influence botanical composition of the pasture (Friend and Kemp, 2000).

The two goals of grazing management for weed control are either to encourage livestock to consume the weeds or to encourage the growth of other desirable plants in the pasture community that in turn suppress the weeds through competition. One strategy is to manage grazing to provide lower stocking rates prior to and during peak germination flushes of weeds in areas known to have high weed populations. The reduced stocking rates can promote pasture growth and significantly increase the level of weed seedling mortality (Forcella and Wood, 1986). For example, deferred autumn grazing (until winter or spring) has been used successfully to control *Arctotheca calendula, Echium plantagineum,* and some thistles in southern Australia (e.g., Bendall, 1973). In general, weeds are weakest, or most susceptible to control, at this early seedling stage. In Mediterranean climates with mild wet winters and hot dry summers, grazing may need to be deferred over the previous spring period when there is active growth of perennial grasses and an excess of feed on offer to livestock to ensure that a minimum level of ground cover is present over the dry summer period and at the start of the autumn break to reduce the recruitment of annual weeds (Dowling et al., 1998). Likewise, deferment from grazing over winter has increased the legume content (mostly *Trifolium subterraneum*) of a pasture at the expense of *Hordeum leporinum* Link (barley grass), while resting another pasture from grazing over summer when conditions for growth favored *Dactylis glomerata* L. (cock'sfoot), and led to a reduction in the percentage of *Vulpia* spp. in the pasture (Dowling et al., 1993). The timing of deferment will vary with the species concerned.

Apart from resting the pasture from grazing to control weeds through increased pasture competition, extra grazing pressure for short periods may be used to force stock to eat relatively unpalatable plants that may otherwise be left. Such high stocking rates (mob stocking) may be part of a grazing system, or it may be used independently in a tactical manner. Resting the pasture prior to mob stocking may make plants more susceptible to grazing damage as with thistles that then grow more upright (e.g., *Carthamus lanatus* L., saffron thistle) (Grace et al., 2002). However, care needs to be taken when forcing stock to eat poisonous plants (Friend and Kemp, 2000). Although heavy grazing at the vegetative stage of plant growth may reduce the vigor and seed production of a weed, intense grazing at the reproductive stage is likely to cause more damage, particularly in annual weeds, which often rely on high seed production for survival.

Running different types of grazing animals can also be exploited to apply grazing pressure to certain weeds. In particular, goats graze a variety of weeds such as the woody rangeland shrubs *Dodonaea attenuata* A.Cunn.

(narrowleaf hop bush), *Cytisus scoparius* (L.) Link (Scotch broom), *Eucalyptus* spp., and *Acacia* spp. (Allan, Holst, and Campbell, 1997), often reducing plant numbers and preventing seed production. Although alternative more palatable pasture may be plentiful, their preferences for browse and other species, often those taller plants in the canopy (Beskow, 2001), means that less-palatable species, such as thistles, will still be eaten (Allan and Holst, 1995). Moreover, integration of goats into cattle grazing systems may result in not only a more desirable botanical composition of the pasture but also a more uniform and competitive pasture sward (Beskow, 2001).

Chemical Control

Chemical control of weeds in pastures should always be seen as only a short-term solution to a weed problem, because herbicide effects are usually only temporary if the pasture remains in a susceptible condition to weed invasion. Nevertheless, weeds can become established in newly sown pastures and to a limited extent even in the presence of properly grazed and highly competitive pastures, and so may need to be controlled through chemical means. However, care should be taken to avoid repeated annual applications of herbicides that may weaken pastures, create gaps in the canopy, and make the pasture more susceptible to invasion, which could lead to the development of herbicide resistance, as has occurred with *Carduus nutans* in New Zealand (Harrington, 1990) and with serrated tussock grass (*Nassella trichotoma* Hackel ex Arech.) in southern Australia (D. A. McLaren, personal communication, January 2004).

The aim of chemical control, except where pastures are about to be resown, is to selectively control the undesirable weeds but to leave the desirable pasture plants relatively unharmed and sufficiently competitive to permanently replace the weeds. Selective broadleaf weed control can be achieved using herbicides such as 2,4-DB [4-(2,4-dichlorophenoxy) butanoic acid] followed by resting of the pasture and fertilizing. Paraquat (1:1-dimethyl-4,4'-bipyridinium dichloride) can be used to selectively remove annual grass weeds, but there may be some temporary injury to the perennial grass and annual legume pasture component. Removal of perennial grass weeds from perennial grass-based pastures is more difficult but flupropanate (sodium-2,2,3,3-tetra fluoropropionate), for example, has been used to help control *Eragrostis curvula* (Schrad.) Nees (African lovegrass) and *N. trichotoma*.

Pasture weeds are usually best sprayed when growth is rapid and annual weeds are small and thus more susceptible to chemical control. Once plants enter their reproductive stage of development or begin to elongate, they become much more difficult to control by chemical means. The choice of herbicide and application rate depends largely on the age and type of pasture and the stage, density, and species of weeds present (Sindel, 1991).

Selectivity can be achieved using nonselective herbicides if the desirable pasture species (often perennial grasses) are in a dormant state and the weeds are still actively growing, or when there is a height differential that can be exploited using some type of blanket or wick wiper (Kent and Preston, 2000); Rayner and Peirce (2002). Woody shrub weeds can be selectively controlled by spot spraying or stem application of the chemical. For example, scattered infestations of *Cryptostegia grandiflora* Roxb. ex R.Br. (rubber vine) in Australian tropical rangelands can be controlled with 2,4-D by either spraying the bark around the base of the plant, on the foliage of plants less than 2 m high, or by cutting the stems of larger plants close to the ground and immediately saturating the cut surfaces with herbicide (Tomley, 1995).

Spray topping (or pasture topping), the application of low rates of herbicide to weeds prior to or in early flowering, can substantially reduce the number of viable seeds set by plants and hence reduce the soil seed bank for future years. If spraying is correctly timed (e.g., from green bud to early flowering stage in *Carthamus lanatus*), viable seed production can be reduced by 90 to 100 percent. As the majority of seeds of most annual grass weeds are viable for only one or two years, spray topping with glyphosate [N-(phosphonomethyl) glycine] or paraquat during flowering can be very effective in reducing seed carryover to the following year (Campbell, 1997). However, the sensitivity of spray topping to time of application reduces its versatility, especially when more than one weed species is present. Hence, spray topping may need to be repeated where weeds are maturing at different times. Even after spray topping, for example in dense thistle stands, grazing capacity can be reduced for some time because of the remaining standing plants. In addition, in complex pastures, perennial grasses that are effective competitors of the annual weed may be very susceptible to the herbicide, e.g., *Austrodanthonia* spp. to glyphosate (Keys and Simpson, 1993), so that the spray topping is counterproductive.

Spray grazing, on the other hand, is the application of sublethal doses of herbicides such as 2,4-D, MCPA [(4-chloro-2 methylphenoxy) acetic acid] or glyphosate to modify the metabolism of weeds and increase their palatability, thereby encouraging livestock to graze them in preference to other pasture plants (Campbell, 1997). This technique has proved to be effective in controlling the poisonous plant *Echium plantagineum* when sheep are heavily grazed on the pasture ten days after spraying with MCPA. However, there is a penalty incurred in stock health (Dowling, Michalk, and Sindel, 2000) and the practice has not been widely adopted.

Physical Control

The effectiveness of mowing (or slashing) on pasture weeds is variable and will depend on the weed species concerned, their ability to reshoot, and the timing and number of passes over the pasture. One approach is to treat weeds that are poorly adapted to regrowing after cutting compared with the resident pasture species. This cutting then slows their regrowth rate relative to the pasture species, shifting the competitive balance in favor of the pasture. Cutting may also reduce flowering and seeding as well as weaken some weeds by reducing their underground storage reserves. Weed growth is removed altogether where hay and silage is cut. Although there is a risk of dispersal of viable weed seeds with the feeding out of hay, most weed seeds are killed in the ensilage process (Tildesley, 1937, cited by Zimdahl, 1999; Blackshaw and Rode, 1991).

Mowing is usually effective in preventing seeding of thistles, for example, only if done either immediately prior to or when plants are just starting to flower. Otherwise, flowering may only be delayed (Sindel, 1991). If left too late in the flowering process, viable seeds may still develop in capitula following cutting as in *Silybum marianum* (Dodd, 1989). Mowing may be made more difficult because of the sequential flowering of different species. One benefit of slashing in dense weed stands is that grazing stock can gain better access to pastures and watering points. Slashing can, however, induce annual plants to survive through to the next year, and these can become even more resistant to chemical control. Slashing of weeds such as *Pteridium esculentum* (G.Forst.) Cockayne (common bracken), which has large underground reserves in its roots and rhizomes, is largely ineffective.

Plowing would be used only in conjunction with the sowing of a new pasture, and in that case, disc plowing has been found to better control perennial weeds than tined implements by cutting below the plant crown (Campbell, 1997). However, disc plowing of soils with poor structure can lead to severe soil degradation and so should be avoided.

Hand weeding and chipping or roguing with a hoe are practiced for some weeds of pastures, particularly where weed numbers are relatively small and on small-sized holdings, for example, *Senecio madagascariensis* on beef and dairy farms in coastal eastern Australia (Sindel and Michael, 1988).

Burning

Burning is potentially very useful for controlling woody weeds (shrubs and trees) and some herbaceous species, particularly in regions where techniques other than grazing are unlikely to be economical. The degree of reduction of woody weeds can be related to the proportion of resprouting species present and the level of recruitment of hard-seeded species (Hodgkinson et al., 1984). However, because of the associated environmental risks, widespread burning is not readily adopted in many countries.

Biological Control

Classical biological control has a vital role to play in the management of pasture weeds. Most classical biocontrol agents (insects or diseases introduced from the weed's regions of origin) have been aimed at broadleaf weeds of rangelands and pastures. Grasses have not received much attention until now because most biocontrol agents that attack weedy grasses would not be sufficiently selective to avoid damaging the bulk of desirable perennial species. However, biological control agents are now being sought in parts of South America for control of the persistent and troublesome weeds *Nassella trichotoma* and *N. neesiana* (Briese and Evans, 1998).

Biological control has not been used commonly in annual crops because no host plants are available for much of the year and insect life cycles are likely to be disrupted by cultivation and other cropping operations. Such constraints are not so important in perennial-based pastures, but untimely pasture management operations such as grazing and slashing can have a negative impact on the populations of biological control agents if the land manager is not familiar with the life cycles of the organisms concerned. The potential for integration with other techniques is discussed briefly later in this chapter. Most land managers will have little input into the evaluation, importation, or release of biological control agents, as this is largely done at a government level. However, they can play an important function in the redistribution of organisms once released, and it is to their benefit that they become familiar with what agents are likely to be present in their area and how best to encourage their control of local weeds. The speeding up of delivery of biological control to pasture managers in southern Australia is meeting with success for weeds such as *Echium plantagineum, Onopordum* spp., and *Carduus nutans* (Swirepik and Smyth, 2002). However, despite dramatic successes of some biological control agents such as *Cactoblastis cactorum* (Berg) (cactoblastis moth) against *Opuntia stricta* (Haw.) Haw. (common prickly pear), biological control is rarely effective alone, and more and more it is being regarded as just one part of an overall weed management strategy. For a good overview of the rationale and operational framework for classical biological control, the reader is directed to Briese (2000).

Despite the apparent effectiveness of inundative biocontrol agents (bioherbicides or mycoherbicides), they have not been commercialized widely, often because of small market potential and difficulties in mass production, meeting moisture requirements, and ensuring sufficient stability and shelf life. These factors remain a stumbling block to their adoption. Where bioherbicides have been commercialized, for example, DeVine (Kenny, 1986) and Collego (Bowers, 1986), they have principally been applied to high-value annual cropping and horticultural systems to which they are more suited rather than to broad-scale pasture systems (McRae and Auld, 2000), although this situation may change in the future.

Farm Hygiene

The prevention of the spread of weeds through hygienic farm management practices is of vital importance in the fight against weeds, particularly for those species that may not yet be widespread. Weeds are often easily spread with the movement of fodder and hay in times of drought and by livestock introduced from weed-infested areas. Ideally, drought feeding should occur in fenced, flat areas (to avoid water runoff and seed spread) where the pasture is due for reseeding or cropping so that new emerging weeds can be monitored, contained, and controlled relatively easily. The prevention of spread of *Sporobolus fertilis* (Steud.) Clayton (giant Parramatta grass) illustrates the subtleties of farm hygiene. Seeds of this species become very sticky when wet and easily adhere to the coats of animals that brush past seeding plants. If seed spread is to be avoided when the plant is in its reproductive stage, then livestock must not be moved from infested to uninfested areas on a farm during the morning when plants are wet with dew or after rain.

INTEGRATED MANAGEMENT OF PASTURE WEEDS

Various techniques have been developed to manage weeds in pastures, but in most cases they are ineffective in the long term if used alone. Integrated weed management (IWM) recognizes the complex nature of pasture systems and the need to be able to flexibly substitute management inputs where appropriate for each specific weed situation. One of the major barriers to the development and adoption of IWM strategies has been the vast array of potential combinations of techniques and the lack of tools to predict the likely effects of a proposed set of management options on the pasture resource base and the economics of the enterprise. Although large, multisite field trials (e.g., Huwer et al., 2002) linked to the development of dynamic pasture models (that include weed control practices) (e.g., Kriticos et al., 2002) are likely to help identify optimal management technologies, it is evident from the foregoing discussion that certain techniques combine well to improve the efficiency of weed management, for example, competitive pastures, herbicides, and grazing. Of foremost importance is the need to establish and/or promote and maintain a competitive perennial pasture, where climate allows, as the basis for weed management. Other techniques will fit well with this strategy but will be effective in the long term only where such a pasture is present. Likewise, farm hygiene, to keep unwanted plants out, is an essential component of any IWM strategy. Most other weed management techniques, perhaps with the exception of burning and plowing, can be added to these two with little conflict of interest, but some problems do arise when attempting to combine biological control with practices such as grazing, herbicide application, and physical techniques where these practices interfere with the survival and life cycles of the biological control agents. For example, in a study that looked to combine spray grazing with the crown weevil *Mogulones larvatus* Schultze on the pasture weed *Echium plantagineum,* intense simulated sheep grazing reduced the biological control agent's numbers by 85 percent (Smyth and Sheppard, 1996). However, some damage can be achieved in this situation by the biological control agent and its population may be able to persist. Preliminary results with other agents indicate that beneficial biological control impacts on target weeds are not necessarily incompatible with other management techniques, although each case needs to be addressed separately. For example, plants of *Onopordum* spp. that missed being sprayed with herbicide were damaged by the weevils *Lixus cardui* Olivier (stem-borer weevil) and *Larinus latus* Herbst (capitulum weevil) even though sprayed plants died before these biological control agents could utilize them (Huwer et al., 2002). The progressive increase in populations of insect biological control agents for weeds such as *Carduus, Onopordum,* and *Echium* spp. (Swirepik and Smyth, 2002) means that another management input needs to be integrated in farmers' weed control strategies.

An integrated weed management strategy will often involve removal of existing mature weeds using physical or chemical techniques, and then maintaining competitive pressure with desirable pasture plants on those remaining individuals, in addition to emerging seedlings (Dowling, Michalk, and Sindel, 2000). Biological control agents that weaken the weeds will then heighten the competitive pressure to which those weeds are subjected (McLaren, 1993); Smyth and Sheppard (2002).

CONCLUSIONS

Weeds in pastures can be both a *symptom* and *cause* of pasture decline in that weeds will invade gaps in the pasture canopy and also promote and sustain gaps in the sward where the invading weeds are annual species (Dowling, Michalk, and Sindel, 2000). To be effective, weed management strategies that utilize a range of *weed removal* techniques must be integrated or combined with *pasture management* methods that enable the sward to resist the ingress of weeds and exert maximum ground cover and competitive pressure, particularly at times when weeds are emerging. The precise strategy that is designed will depend on the life cycle and ecology of the dominant weeds and pasture species and on environmental conditions and other local factors.

REFERENCES

Allan, C. and P. Holst (1995). Goats in fight against difficult weeds. In M. Casey (Ed.), *Pasture plus: The complete guide to pastures* (pp. 326-328). Belmont, Western Australia: Kondinin Group.

Allan, C., P. Holst, and M. Campbell (1997). *Weed control using goats: A guide to using goats for weed control in pastures.* Orange, New South Wales: NSW Agriculture.

Andrews, T.S., R.D.B. Whalley, and C.E. Jones (1996). Seed production and seedling emergence of giant Parramatta grass on the North Coast of New South Wales. *Australian Journal of Experimental Agriculture* 36: 299-308.

Bendall, G.M. (1973). The control of slender thistle, *Carduus pycnocephalus* L. and *C. tenuiflorus* Curt (Compositae), in pasture by grazing management. *Australian Journal of Agricultural Research* 24: 831-837.

Beskow, W.B. (2001). Integration of goats into sheep and cattle grazing systems as a permanent weed control tool. Doctoral thesis, Palmerston North, New Zealand, Massey University.

Blackshaw, R.E. and L.M. Rode (1991). Effect of ensiling and rumen digestion by cattle on seed viability. *Weed Science* 39: 104-108.

Bourke, C.A., I.A. Southwell, and G.M. Mayo (2002). Sheep as biological control agents against St. John's wort (*Hypericum perforatum* L.): Factors affecting hypericin variation and hypericin tolerance. In H. Spafford Jacob, J. Dodd, and J.H. Moore (Eds.), *Proceedings of the Thirteenth Australian Weeds Conference* (pp. 398-401). Perth: Plant Protection Society of Western Australia.

Bowers, R.C. (1986). Commercialization of Collego—An industrialist's view. *Weed Science* 34(Supplement): 24-25.

Briese, D.T. (2000). Classical biological control. In B.M. Sindel (Ed.), *Australian weed management systems* (pp. 161-192). Melbourne: R.G. and F.J. Richardson.

Briese, D.T. and H.C. Evans (1998). Biological control of serrated tussock *(Nassella trichotoma):* Is it worth pursuing? *Plant Protection Quarterly* 13: 94-97.

Campbell, M. (1997). Pasture weeds. In J.V. Lovett and J.M. Scott (Eds.), *Pasture production and management* (pp. 254-268). Melbourne: Inkata Press.

Cook, S.J., G.J. Blair, and A. Lazenby (1978). Pasture degeneration: II. The importance of superphosphate, nitrogen and grazing management. *Australian Journal of Agricultural Research* 29: 19-29.

Currie, G.A. (1936). *A report on a survey of weed problems in Australia.* CSIR Research Pamphlet No. 60. Melbourne, Victoria: Council of Scientific and Industrial Research (Australia).

Dodd, J. (1989). Phenology and seed production of variegated thistle, *Silybum marianum* (L.) Gaertn., in Australia in relation to mechanical and biological control. *Weed Research* 29: 255-263.

Dowling, P.M., D.R. Kemp, D.L. Michalk, and G.D. Millar (1998). Changes in pasture management needed for improved control of weeds in pastures. In D.L. Michalk and J.E. Pratley (Eds.), *Proceedings of the Ninth Australian Agronomy Conference* (pp. 559-562). Wagga Wagga: Australian Society of Agronomy.

Dowling, P.M., D.R. Kemp, D.L. Michalk, G.D. Millar, and M. Goodacre (1993). Manipulating pasture composition—A matter of time, dollar$ and sense. In G.K. McDonald and W.D. Bellotti (Eds.), *Proceedings of the Seventh Australian Agronomy Conference* (pp. 248-251). Adelaide: Australian Society of Agronomy.

Dowling, P., D. Michalk, A. Mead, G. Carney, and S. Betts (2003). Effect of cropping rotation on weed incidence and botanical composition of ensuing pastures at Oolong in SE NSW. In Australian Society of Agronomy (Eds.), *Solutions for a better environment: Proceedings of the Eleventh Australian Agronomy Conference.* Geelong, Victoria: Australian Society of Agronomy. Available online at <http://www.regional.org/au/au/asa/2003/p/7/dowling.htm>.

Dowling, P.M., D.L. Michalk, and B.M. Sindel (2000). Weed management in pasture systems. In B.M. Sindel (Ed.), *Australian weed management systems* (pp. 307-328). Melbourne: R.G. and F.J. Richardson.

Fitzgerald, R.D. and G.M. Lodge (1997). *Grazing management of temperate pastures: Literature reviews and grazing guidelines for major species.* Technical Bulletin No. 47. Orange, New South Wales: NSW Agriculture.

Forcella, F. and H. Wood (1986). Demography and control of *Cirsium vulgare* (Savi) Ten. in relation to grazing. *Weed Research* 26: 199-206.

Freckleton, R.P., A.R. Watkinson, P.M. Dowling, and A.R. Leys (2000). Determinants of the abundance of invasive annual weeds: Community structure and non-equilibrium dynamics. *Proceedings of the Royal Society of London B* 267: 1153-1161.

Friend, D.A. and D.R. Kemp (2000). Grazing management methods. In B.M. Sindel (Ed.), *Australian weed management systems* (pp. 139-159). Melbourne: R.G. and F.J. Richardson.

Gardener, M.R., R.D.B. Whalley, and B.M. Sindel (2003). Ecology of *Nassella neesiana,* Chilean needle grass, in pastures on the Northern Tablelands of New

South Wales: II. Seedbank dynamics, seed germination, and seedling recruitment. *Australian Journal of Agricultural Research* 54: 621-626.

Grace, B.S., R.D.B. Whalley, A.W. Sheppard, and B.M. Sindel (2002). Managing saffron thistle in pastures with strategic grazing. *Rangeland Journal* 24: 313-325.

Harrington, K.C. (1990). Spraying history and fitness of nodding thistle, *Carduus nutans,* populations resistant to MCPA and 2,4-D. In J.W. Heap (Ed.), *Proceedings of the Ninth Australian Weeds Conference* (pp. 201-204). Adelaide: Crop Science Society of South Australia.

Hill, T.A. (1977). *The biology of weeds.* London: Edward Arnold (Publishers) Ltd.

Hodgkinson, K.C., G.N. Harrington, G.F. Griffith, J.C. Noble, and M.D. Young (1984). Management of vegetation with fire. In G.N. Harrington, A.D. Wilson, and M.D. Young (Eds.), *Management of Australia's rangelands* (pp. 141-156). Melbourne: CSIRO.

Huwer, R.K., M.J. Neave, P.M. Dowling, W.M. Lonsdale, A.W. Sheppard, D.T. Briese, and D.L. Michalk (2002). Integrated weed management (IWM) in perennial pasture using pasture and grazing management, herbicide strategies and biological control. In H. Spafford Jacob, J. Dodd, and J.H. Moore (Eds.), *Proceedings of the Thirteenth Australian Weeds Conference* (pp. 727-730). Perth: Plant Protection Society of Western Australia.

Kemp, D.R. and P.M. Dowling (1991). Species distribution within improved pastures over central N.S.W. in relation to rainfall and altitude. *Australian Journal of Agricultural Research* 42: 647-659.

Kemp, D.R., W.McG. King, G.M. Lodge, S.R. Murphy, P.E. Quigley, and P. Sanford (2003). Plant species diversity and productivity in grazed permanent grasslands. In Australian Society of Agronomy (Eds.), *Solutions for a better environment: Proceedings of the Eleventh Australian Agronomy Conference.* Geelong, Victoria: Australian Society of Agronomy. Available online at <http://www.regional.org.au/au/asa/2003/p/17/kemp.htm>.

Kenny, D.S. (1986). DeVine—The way it was developed: An industrialist's view. *Weed Science* 34(Supplement): 15-16.

Kent, J.H. and C. Preston (2000). Application and fate of herbicides in the environment. In B.M. Sindel (Ed.), *Australian weed management systems* (pp. 227-250). Melbourne: R.G. and F.J. Richardson.

Keys, M. and P. Simpson (1993). Herbicide tolerance of two native grasses. In D. Michalk (Ed.), *Proceedings of the Eighth Annual Conference of the Grassland Society of New South Wales* (pp. 103-105). Orange, NSW: The Grassland Society of New South Wales.

Kriticos, D.J., P.M. Dowling, W.McG. King, A.W. Sheppard, and R.E. Jones (2002). 4SPaM: A process-based and weather-driven population model for managing weeds in temperate pastures. In H. Spafford Jacob, J. Dodd, and J.H. Moore (Eds.), *Proceedings of the Thirteenth Australian Weeds Conference* (pp. 244-247). Perth: Plant Protection Society of Western Australia.

Lodge, G.M. (2001). Studies of soil seedbanks in native and sown pastures in northern New South Wales. *Rangeland Journal* 23: 204-223.

Lodge, G.M. and R.D.B. Whalley (1985). The manipulation of species composition of natural pastures by grazing management on the northern slopes of New South Wales. *Australian Rangeland Journal* 7: 6-16.

Loneragan, J.F. and C.J. Asher (1967). Response of plants to phosphate concentration in solution culture: II. Rate of phosphate absorption and its relation to growth. *Soil Science* 103: 311-318.

Maddin, R.W. and J.H. Moore (1993). Weeds in pastures. In J. Dodd, R.J. Martin, and K.M. Howes (Eds.), *Management of agricultural weeds in Western Australia* (pp. 121-122). Western Australia: Department of Agriculture.

McLaren, D.A. (1993). Overview and use of biological control in pasture situations. *Plant Protection Quarterly* 8: 159-162.

McRae, C.F. and B.A. Auld (2000). Inundative biological control of weeds—The bioherbicide tactic. In B.M. Sindel (Ed.), *Australian weed management systems* (pp. 193-208). Melbourne: R.G. and F.J. Richardson.

Michael, P.W. (1968). Perennial and annual pasture species in the control of *Silybum marianum. Australian Journal of Experimental Agriculture and Animal Husbandry* 8: 331-339.

Michael, P.W. (1970). Weeds of grasslands. In R.M. Moore (Ed.), *Australian grasslands* (pp. 349-360). Canberra: Australian National University Press.

Panetta, F.D., T.J. Ridsdill-Smith, M.J. Barbetti, and R.A.C. Jones (1993). The ecology of weeds, invertebrate pests and diseases of Australian sheep pastures. In E.S. Delfosse (Ed.), *Pests of pastures: Weed, invertebrate and disease pests of Australian sheep pastures* (pp. 87-114). Melbourne: CSIRO Information Services.

Parsons, W.T. and E.G. Cuthbertson (1992). *Noxious weeds of Australia.* Melbourne: Inkata Press.

Powles, S.B. and J.A.M. Holtum (1990). Herbicide resistant weeds in Australia. In J.W. Heap. (Ed.), *Proceedings of the Ninth Australian Weeds Conference* (pp. 185-193). Adelaide: Crop Science Society of South Australia.

Rayner, B.J. and J.R. Peirce (2002). Control of weeds using an experimental blanket wiper. In H. Spafford Jacob, J. Dodd, and J.H. Moore (Eds.), *Proceedings of the Thirteenth Australian Weeds Conference* (pp. 158-160). Perth: Plant Protection Society of Western Australia.

Sindel, B.M. (1991). A review of the ecology and control of thistles in Australia. *Weed Research* 31: 189-201.

Sindel, B.M. and P.W. Michael (1988). Survey of the impact and control of fireweed (*Senecio madagascariensis* Poir.) in NSW. *Plant Protection Quarterly* 3: 22-28.

Sindel, B.M., I.J. Radford, R.H. Holtkamp, and P.W. Michael (1998). The biology of Australian weeds. 33. *Senecio madagascariensis* Poir. *Plant Protection Quarterly* 13: 2-15.

Smith, A.E. and L.D. Martin (1995). Weed management systems for pastures and hay crops. In A.E. Smith (Ed.), *Handbook of weed management systems* (pp. 477-517). New York: Marcel Dekker, Inc.

Smyth, M.J. and A.W. Sheppard (1996). The effect of simulated spray-grazing on the Paterson's curse crown weevil, *Mogulones larvatus* Schultze. In R.C.H.

Shepherd (Ed.), *Proceedings of the Eleventh Australian Weeds Conference* (pp. 291-293). Melbourne: Weed Society of Victoria.

Smyth, M.J. and A.W. Sheppard (2002). *Longitarsus echii* and its impact on *Echium plantagineum* (Paterson's curse): The insect for the Mediterranean rainfall range of the weed? In H. Spafford Jacob, J. Dodd, and J.H. Moore (Eds.), *Proceedings of the Thirteenth Australian Weeds Conference* (pp. 422-425). Perth: Plant Protection Society of Western Australia.

Southwell, I.A. and C.A. Bourke (2001). Seasonal variation in hypericin content of *Hypericum perforatum* L. (St. John's wort). *Phytochemistry* 56(5): 437-441.

Swirepik, A. and M. Smyth (2002). Biological control of broad-leafed pasture weeds (Paterson's curse, *Onopordum* and nodding thistles). What have we achieved and where to from here? In H. Spafford Jacob, J. Dodd, and J.H. Moore (Eds.), *Proceedings of the Thirteenth Australian Weeds Conference* (pp. 373-376). Perth: Plant Protection Society of Western Australia.

Tildesley, W.T. (1937). A study of some ingredients found in ensilage juice and its effect on the vitality of certain weed seeds. *Scientific Agriculture* 17: 492-501.

Tomley, A.J. (1995). The biology of Australian weeds 26. *Cryptostegia grandiflora* R.Br. *Plant Protection Quarterly* 10: 122-130.

Whalley, R.D.B., M.R. Gardener, and J.E. Earl (1999). Pasture management of reproductively efficient grassy weeds. In A.C. Bishop, M. Boersma, and C.D. Barnes (Eds.), *Proceedings of the Twelfth Australian Weeds Conference* (pp. 174-175). Hobart, Tasmania: Tasmanian Weed Society.

Zimdahl, R.L. (1999). *Fundamentals of weed science*. San Diego: Academic Press.

Chapter 23

Integrated Turfgrass Weed Management

Rakesh S. Chandran

INTRODUCTION

The concept of integrated pest management (IPM) evolved in the early 1970s. Although IPM has lacked a single commonly accepted definition since it first appeared in the literature in 1967, two key elements involve the use of multiple control tactics and the integration of pest biology knowledge into the management system (Smith and van den Bosch, 1967; Kogan and Bajwa, 1999; Buhler, 2002). Integrated weed management (IWM), an offshoot of IPM or a component of whole-crop management systems, is gaining renewed interest in all areas of crop production. For simplicity, and for the subject of reference in this chapter, namely turfgrasses, IWM could be considered as a proactive effort to manage weeds by providing good growth conditions to turf, thereby reducing the opportunities for weed infestation, and, when necessary, implement a broad array of control strategies to resolve a weed problem. Elmore (1996) considers IWM to be a reintroduced practice that has become widespread across all crop production areas. With a goal of equipping cropping systems unfavorable for weeds and to minimize the effect of those that survive, IWM may be best achieved by integrating cropping-system design and weed management strategies to an economically and environmentally viable practice (Buhler, 2002).

The status of weeds in turf as a pest has not changed considerably over the past four decades despite modern techniques to establish a turf and to manage weeds. Schery's (1961) portrayal of turf weeds, "lurking in every soil, with compatriots for every season, weeds await any turn of events that gains for them a chink of bare ground; a foothold achieved, rampant weeds slither among grass shoots, thinning the turf" holds true even today (p. 115). As Bingham, Chism, and Bhowmik (1995) stated, weeds not only affect the aesthetics of a turf but may also reduce such other attributes as erosion control, cooling effects of turf, and playability of recreational turf. Turf weeds are opportunists and are well adapted to meet the rigors of a closely mowed lawn. If adequate growing conditions are not provided for optimum turf growth, weeds emerge and may eventually dominate. Maintaining a dense turf with a competitive ability to reduce the emergence and establishment of weeds is perhaps the best strategy to minimize weed infestation in turf. The success of integrated weed management in turf, therefore, depends largely upon the maintenance of a vigorous turfgrass.

IDENTIFICATION AND THRESHOLD LEVELS

Identification of a weed is the first step toward executing a management program. A good understanding of weed biology, life cycle, and weed ecology is needed to make sound management decisions. Based on morphology, common turf weeds are classified as broadleaves, grasses, and sedges. Based on life cycle, they can be categorized into annuals, biennials, or perennials. Depending on the season of active growth, annuals can be classified as winter or summer annuals. Plants that regenerate from a single growing point year after year, and hence are incapable of spreading vegetatively, are referred to as simple perennials, whereas those that regenerate and spread through perennating organs such as stolons, rhizomes, or tubers are called creeping perennials. Creeping perennials are usually more troublesome and difficult to control. Apart from numerous digital resources on the Internet, several excellent publications are available for keying and identifying turf weeds (see Table 23.1).

A weed-free turf is not always feasible and in many cases is not practical. The threshold level of weeds in turf is subjective because of the unique characteristics of turf and the purpose it serves. Presence of excessive weeds is not tolerated in certain recreational turfs, such as golf courses, athletic fields, and high-quality turfgrasses, where it is imperative to keep weeds at a low level. On the other hand, homeowners have a wide range of tolerance toward weeds in their turfgrasses. Some like to maintain their turf as an immaculate monoculture, whereas the presence of a few weeds—especially such flowering weeds as dandelion—interspersed with turf is tolerated by others. Cook (1992) suggests planting such flowering broadleaf species along with the turfgrass and argues that it would be similar to a natural ecosystem that is more stable than turfgrass planted alone. However, for some, such turf systems require a change in one's visual concept of what constitutes an ideal lawn.

Regardless of management practices, it is prudent to adopt a predetermined threshold level for a well-planned integrated weed management program. The goal should be to keep the weeds from building beyond a certain level and to prevent them from dominating the turfgrass (Uhler, 1992). Es-

TABLE 23.1. Selected references for weed identification in turfgrasses.

Publication	Author(s)	Year of publication and source
Weeds of the Northeast	Uva, R.H., J.C. Neal, and J.M. DiTomaso	1997, Ithaca, NY: Cornell University Press
Weeds of the West	Whitson, T.D., L.C. Burril, S.A.Dewey, D.W. Cudney, B.E. Nelson, R.D. Lee, and R. Parker	2000, Laramie, WY: Western Society of Weed Science–Univ. of Wyoming
Weeds of the North Central States	Wax, L.M., R.S. Fawcett, and D. Isley	1981, Urbana, IL: Univ. Illinois Ag. Experiment Station, Bulletin 772
Weeds of Southern Turfgrasses	Murphy, T.R., D.L. Colvin, R. Dickens, J.W. Everest, D. Hall, and L.B. McCarty	1996, Gainesville, FL: IFAS Publication Center, University of Florida
SWSS Weed Identification Guide	Various authors	1998, Champaign, IL: Southern Weed Science Society
Color Atlas of Turfgrass Weeds	McCarty, L.B., J.W. Everest, D.W. Hall, T.R. Murphy, and F. Yelverton	2002, Hoboken, NJ: John Wiley & Sons
Virginia Tech Weed Identification Guide	Bradley, K.W and E.S. Hagood	2003, <http://www.ppws.vt. edu/weedindex.htm>
Midwestern Turfgrass Weed Identification and Control	T. Voigt, T. Fermanian, B. Branham, B. Spangenberg, L. Cella, and D. Weisenberger	2003, <http://www.turf.uiuc. edu/weed_web/index.htm>
Weeds in Turf	T.R. Murphy and B.J. Johnson	1998, <http://floridaturf.com/weeds/weherbic. htm>
WSSA Photo Herbarium	Various authors	2003, <http://www.wssa.net/photo&info/weed frame. htm>

tablishing threshold levels can be achieved by monitoring total weeds in a given unit area or by noting levels of weeds in representative areas of the lawn. A simple method found to be particularly useful is called the transect-line method which utilizes transects (e.g., tape measure) laid across the turf at randomly chosen points. At predetermined unit lengths along each transect, the presence or absence of a weed is recorded within a 2.5 cm radius. The presence of a void should also be considered as a weed due to its potential to become a weedy spot. The percent cover of weeds could then be determined based on the ratio of weedy spots to total spots sampled. Usually a cover of weeds less than 5 to 10 percent is considered to be acceptable in home lawns. Golf courses and other high-quality turf communities may warrant much lower threshold levels. Lower tolerance thresholds require higher levels of maintenance. Proper scouting and keeping records of weeds in a site map with estimates on population density are useful to keep track of population shifts (Neal, 1994).

CULTURAL METHODS

Establishing the Turfgrass

Species and cultivar selection is one of the basic factors that determines the vigor of a turfgrass community. Selection should be based on the performance of a turf under the prevailing environmental conditions. Important considerations should include dominant seasons of the location (e.g., cool-season turf species for higher latitudes and warm-season turf species for lower latitudes), presence of trees (shade tolerance), soil salinity (salt tolerance), and traffic over the turf (tolerance to wear and tear, and compaction). Several choices of species and cultivars are available today that are adapted to most environmental conditions. Using a blend of three or more cultivars that are adapted to the local conditions is recommended as opposed to planting a turf from a single cultivar (Harivandi, 2000). This variety allows the turf to better tolerate a broader range of conditions. The success of this strategy would depend upon uniform spread and germination of the various cultivars used to establish the turf. It should be noted that certain species, such as perennial ryegrass, tall fescue, bermudagrass, and St. Augustinegrass, establish faster than other species, such as bentgrass, Kentucky bluegrass, carpetgrass, and zoysiagrass (Turgeon, 1980) (see Table 23.2 for Latin binomials). In a blend, the species that establishes faster and relatively earlier eventually dominates the turf.

The need for establishing turf in a proper medium is paramount to weed management. In modern residential and urban environments, turf is com-

TABLE 23.2. Common names and Latin binomials of crops and weeds discussed in the text.

Common name	Scientific name
alligator weed	*Alternanthera philoxeroides* (Mart.) Griseb.
artichoke, Jerusalem	*Helianthus tuberosus* L.
bahia grass	*Paspalum notatum* Fluegge
barnyard grass	*Echinochloa crus-galli* (L.) Beauv.
bentgrass, colonial	*Agrostis tenuis* Sibth.
bentgrass, creeping	*Agrostis stolonifera* L.
bermudagrass	*Cynodon dactylon* (L.) Pers.
bluegrass, annual	*Poa annua* L.
bluegrass, Kentucky	*Poa pratensis* L.
broomsedge	*Andropogon virginicus* L.
buttonweed, Virginia	*Diodia virginiana* L.
canola	*Brassica napus* L. ssp. *napus*
carpetgrass	*Axonopus affinis* Chase
centipede grass	*Eremochloa ophiuroides* (Munro) Hack.
corn	*Zea mays* L.
clover, red	*Trifolium pratense* L.
clover, white	*Trifolium repens* L.
crabgrass, large	*Digitaria sanguinalis* (L.) Scop.
crabgrass, smooth	*Digitaria ischaemum* (Schreb. ex Schweig) Schreb. ex Muhl.
creeper, Virginia	*Parthenocissus quinquefolia* (L.) Planch.
dallisgrass	*Paspalum dilatatum* Poir.
dandelion	*Taraxacum officinale* Weber in Wiggers
fescue, tall	*Festuca arundinacea* Schreb.
foxtail, yellow	*Setaria glauca* (L.) Beauv.
fumitory	*Fumaria officinalis* L.
goosegrass	*Eleusine indica* (L.) Gaertn.
groundsel, common	*Senecio vulgaris* L.
hawkweed	*Hieracium* spp.
ivy, ground	*Glechoma hederacea* L.
ivy, poison	*Toxicodendron radicans* (L.) Ktze.
johnsongrass	*Sorghum halepense* (L.) Pers.
knapweed, spotted	*Centaurea maculosa* Lam.
knotweed, prostrate	*Polygonum aviculare* L.
kyllinga, false green	*Kyllinga gracillima* Miq.
kyllinga, green	*Kyllinga brevifolia* Rottb.
lespedeza, common	*Lespedeza striata* (Murray) H. & A.
marigold, corn	*Chrysanthemum segetum* L.

TABLE 23.2 *(continued)*

Common name	Scientific name
medic, black	*Medicago lupilina* L.
nettle, small	*Urtica urens* L.
nimblewill	*Muhlenbergia schreberi* J. F. Gmel.
nutsedge, purple	*Cyperus rotundus* L.
nutsedge, yellow	*Cyperus esculentus* L.
parsley, fool's	*Aethusa cynapium* L.
pea	*Pisum sativum* L.
pigweed, prostrate	*Amaranthus blitoides* S. Wats.
plantain, broadleaf	*Plantago major* L.
plantain, buckhorn	*Plantago lanceolata* L.
radish, wild	*Raphanus raphanistrum* L.
ragweed, common	*Ambrosia artemisiifolia* L.
ryegrass, perennial	*Lolium perenne* L.
sesbane, hemp	*Sesbania exaltata* (Raf.) Cory
shepherd's-purse	*Capsella bursa-pastoris* (L.) Medikus
sneezeweed, bitter	*Helenium amarum* (Raf.) Rock.
sorrel, red	*Rumex acetosella* L.
speedwell, corn	*Veronica arvensis* L.
spurge, prostrate	*Euphorbia humistrata* (Engelm. ex Gray) Small
spurge, spotted	*Euphorbia maculata* L.
St. Augustinegrass	*Stenotaphrum secundatum* (Walt.) Kuntze.
strawberry, mock	*Duchesnea indica* (Andr.) Focke
stiltgrass, Japanese	*Microstegium viminieum* (Trin.) Camus
thistle, Canada	*Cirsium arvense* (L.) Scop.
trefoil, bird'sfoot	*Lotus corniculatus* L.
woodsorrel, creeping	*Oxalis corniculata* L.
woodsorrel, yellow	*Oxalis stricta* L.
zoysiagrass	*Zoysia japonica* Steud.

monly established on subsoils or poor topsoils that are not well suited for plant growth. Such turfgrasses, often established from sods, are predisposed to adverse growing conditions, and they eventually result in turf that serves as a host for not only weeds but also other pests. Topsoils used in new lawns, often two to four inches deep, have an inherent weed seed bank in them (Ashton and Monaco, 1991). Two ways to reduce the size of a seed bank are to (1) obtain the topsoil from fields relatively free of perennial turf weeds and (2) use a fumigant to kill the weed propagules present. Another alternative would be to establish the lawn in subsoil with a minimal seed bank, which could then be amended with materials high in organic matter such as well-decomposed sawdust or manure followed with a proper fertilization program. Norrie and Gosselin (1996) analyzed the effect of paper sludge amendments on soil and found that they were effective to improve soil quality and ground cover, and to decrease competition from weeds. Turf grown on amended sand had higher quality and tolerated drought better compared to unamended turf, which had an indirect effect on weed pressure (Miller, 2000). Turgeon (1999) suggested summer fallowing followed by application of a nonselective systemic herbicide after weed emergence to reduce weed competition. If amendments are used to establish a turf, checking them for weed propagules is a good sanitary practice. This could be carried out by assaying representative samples for germination of weed propagules. Once a vast deposit of seeds is stored in the growing medium, it is very difficult to get rid of it even after growing grass for several years. Chancellor (1986), in a long-term study to analyze soil under a grass cover, concluded that planting a weedy arable field to grass for 20 years did not eradicate the seeds of most weed species. In his study, corn marigold and wild radish declined at a rate of about 30 percent per year, whereas fumitory and fool's parsley had a half-life of more than 20 years.

When establishing a turf from seed, sanitary practices such as the use of pure seed and cleaning equipment that has been used in other sites will help reduce future weed infestations. Hanson (1990) recommended mixing turf seed uniformly, especially if a blend is used, with straw mulch and other ingredients that enhance proper germination and establishment to reduce weed infestation. Use of pregerminated seeds especially for turf species that are slow to germinate provides an early start for turf, thereby minimizing competition from weeds (Lush and Birkenhead, 1987). In their study, partially germinated seeds of colonial bentgrass, perennial ryegrass, and Kentucky bluegrass emerged up to ten days before ungerminated seeds of the same species. The authors observed fewer weeds in turf established thus, and generalized that this technique is beneficial to slow germinating turf species where competition from weeds is expected. Separately, using higher seeding rates also enhanced the establishment of turf (Elmore and Boquist, 1988). Seeding perennial ryegrass at 4.8 kg·100 m^{-2} resulted in rapid turf establishment and fewer weeds than seeding it at 1.6 kg·100 m^{-2}. Optimum seeding rates recommended for other common turf species are 3.2 to 4.8 kg·100 m^{-2} for bluegrasses and 12.8 to 16.0 kg·100 m^{-2} for fescues (Hanson, 1990). Excessive seeding rates may occasionally trigger the onset of seedling diseases, which could delay turfgrass establishment and have an indirect effect on weed competition (Turgeon, 1999). It is also important not to seed, reseed, or overseed a turfgrass within the prescribed period after applying a preemergence (PRE) herbicide. However, it is a common practice to apply siduron, a PRE herbicide, at the time of seeding cool- season grasses, especially if annual monocot weeds such as crabgrass and bluegrass are prevalent.

Hensler, Baldwin, and Goatley (2001) compared the use of a bioorganic fiber-seeding mat to traditional seeding into a prepared soil to determine any differences in weed levels and turfgrass establishment between the two planting methods. Bahiagrass, bermudagrass, carpetgrass, centipedegrass, St. Augustinegrass, and zoysiagrass were seeded at recommended rates using bioorganic mats under both irrigated and nonirrigated conditions. They found that the weed counts in mat-seeded plots were lower than those in direct-seeded plots and concluded that a bioorganic fiber seeding mat is a viable method of establishing the turfgrasses tested, especially to reduce weed populations.

Soil Conditions

Soil pH and the inorganic amendments used to alter the pH may have a role in the presence of certain weeds and their management. Red sorrel and broomsedge are indicative of acidic soils (Derr, 2000), whereas plantains (*Plantago* spp.) are favored by high soil pH (McCarty, 1995). Similarly, soils low in nitrogen are indicated often by the presence of such weeds as black medic, birdsfoot trefoil, white clover, hawkweed, and broomsedge (Derr, 2000). Increasing soil pH and a corresponding reduction of exchangeable aluminum (Al) and increase of extractable phosphorus (P) by adding P fertilizer to limed soils also benefited annual bluegrass more than creeping bentgrass (Shiou, 1993). Magnesium carbonate ($MgCO_3$) used to raise the pH reduced the germination of large crabgrass, whereas a similar pH increase using calcium carbonate ($CaCO_3$) did not affect germination rates of this weed (Pierce et al., 1999). The researchers also found that increasing soil pH reduced shoot and root dry weights of large crabgrass seedlings regardless of material used to raise the soil pH and concluded that raising soil pH may reduce the ability of large crabgrass to compete with turfgrass. Obviously, managing soil pH may be of value if the dominant weed species and the cultivated turf respond differently to changes in pH.

Soil compaction may result in a shift toward plant species that can better resist poor aeration or structure. Prostrate knotweed thrives in turf subjected to high levels of traffic. Derr (2000) mentioned other weeds including goosegrass, annual bluegrass, spotted spurge, corn speedwell, broadleaf plantain, and common lespedeza as possible indicators of compacted soils. Aerating compacted soils up to four times a year is recommended to alleviate such problems. Harivandi (1994) recommends aerating soils during the time of the year when the growth rate of cultivated turf is high. This would correspond to late spring and early fall for cool-season grasses and summer months for warm-season grasses. Although these time periods may coincide with the germination of certain weeds, the turf may outcompete the germinating weed seedlings due to its vigorous growth. Bigelow and colleagues (2001) determined that a forced-air system in conjunction with soil amendments had significant effects on turfgrass establishment and quality and to decrease the competitive ability of weeds. High earthworm populations in the soil can help improve aeration naturally.

MECHANICAL METHODS

Mowing and Weeds

Mowing frequency and mowing height closely govern the presence of weeds in turf. A general model cannot be devised based on the available information on weed control by manipulating turfgrass mowing height alone because of differential species responses. In a three-year study on weed management in tall fescue, Dernoeden, Carroll, and Krouse (1993) concluded that the turfgrass mowed at 8.8 cm resisted smooth crabgrass invasion and exhibited best visual summer turf quality compared to the turf mowed at 3.2 or 5.5 cm. However, the researchers noted, the incidence of white clover was high in turf mowed at 8.8 cm compared to lower mowing heights.

In most cases, mowing the turf too close to the ground increases weed competition. Mock strawberry was least invasive in a tall fescue turf left unmowed and survived best when the turf was mowed at 6 cm (Gray and Call, 1993). Several sedges can also respond differently to mowing heights. When the mowing height of a Kentucky bluegrass turf was reduced from 7.6 cm to 2.2 cm, the density of yellow nutsedge increased (Black, 1975). However, when grown independently, sedges respond to mowing heights similar to turfgrasses. The growth of purple and yellow nutsedge, and green and false green kyllinga was adversely affected by low mowing heights in the absence of turf (Summerlin, Coble, and Yelverton, 2000). In their study, purple nutsedge shoot number was reduced six to nine weeks after initiating a mowing height of 1.3 cm and 14 to 18 weeks after mowing them at height of 3.8 cm. Both mowing heights reduced rhizome length, tuber number, and tuber size in purple nutsedge and inhibited tuber production by yellow nutsedge. The two mowing heights did not cause such dramatic effects on the kyllingas, although their growth attributes were generally suppressed by mowing at 1.3 cm compared to nonmowed control. Interestingly, Lowe and colleagues (2000) documented a two- to fivefold increase in green kyllinga when grown and mowed alongside bermudagrass turf at a mowing height of 2.5 cm, compared to the two plant species mowed at a height of 5.0 cm. In their study, higher fertilization rates during the establishment phase of bermudagrass helped reduce invasion by green kyllinga. Turf species also respond similarly to mowing heights. Kentucky bluegrass mowed at 2.5 cm produced 14 percent fewer roots and 41 percent fewer rhizomes compared to the same grass mowed at 5 cm (Juska and Hanson, 1961). Lush and Rogers (1992) documented that the total biomass of perennial ryegrass turf species increased with increasing mowing height whereas the tiller density was inversely proportional to mowing height.

These findings clearly indicate the relationship between mowing height and the development of the root system and underground storage organs, regardless of the plant species. Frequency of mowing and fate of grass clippings may also affect weed buildup in turf. A turf mowed twice a week received higher quality ratings than one mowed at weekly or biweekly intervals (Haley et al., 1985). In their study, when the clippings were returned to the turf, the quality was enhanced compared to a turf where the clippings were discarded. As expected, higher rates of fertilizer application (300 kg $N \cdot ha^{-1}$) improved turf quality compared to lower fertilization rates. As the turf vigor increased with higher nitrogen application rates, encroachment by weeds was also reduced. A similar interaction between mowing height and fertilizer application was also reported by Hanson (1990). Two times more fertilizer was required to maintain a bluegrass free of weeds when the mowing height was reduced from two inches to one inch. This could be attributed to the additional nitrogen requirements to keep up the shoot density while compensating for the reduced mowing heights. The recommended mowing heights of common turfgrass species were also discussed by Hanson (1990).

Nutrition, Irrigation, and Other Factors Affecting Weed Control

From the previous discussion, it is clear that the nutritional requirements of turf are quite demanding. High levels of shoot regrowth associated with mowing frequency of turfgrass increases its demand for fertilizers for a given unit area. The nutrient required in largest quantity by turfgrasses is nitrogen, which should therefore be applied in sufficient quantities at appropriate times of the year (Harivandi, 1994). Cool-season grasses exhibit periods of growth spurts during spring and fall months. Fertilizer application in spring and fall is therefore recommended for such grasses. A higher dose of fall fertilizer also helps the grasses overwinter better. Split application of fertilizer at monthly intervals throughout the growing period is recommended for warm-season grasses. Applying small quantities of essential nutrients every week or two during the growing period and adjusting the rates according to turfgrass response is the ideal practice (Turgeon, 1999). The application of fertilizer during dormant periods or periods of low growth (summer for cool-season grasses, and winter for warm-season grasses) may encourage weed growth. Fertilizer applications also enhance quick recovery of turfgrasses treated with herbicide for weed control (Johnson, 1984). Such a recovery can reduce competition from weeds that germinate and establish while the turf is under stress. Fertilizer or lime application during establishment of turfgrass is usually based on soil tests to determine deficiencies in phosphorus and potassium.

Irrigation can have direct or indirect effects on the prevalence of turfgrass weeds. A study to assess the effect of irrigation on turf pest incidence indicated that irrigation did not influence weed pressure as much as it did affect turf diseases (Jiang, Fry, and Tisserat, 1998). In their field study, irrigating a perennial ryegrass turf either daily with 7.6 mm of water to replace approximately 80 percent of evapotranspiration had no effect on the incidence of smooth crabgrass or dandelion. Nevertheless, weeds compete with turf for nutrients and water. Fry and Butler (1989) documented that white clover used more water than a Kentucky bluegrass turfgrass. Interestingly, barnyardgrass used less water than the turf, whereas the monocot weeds yellow foxtail and smooth crabgrass exhibited evapotranspiration rates similar to that of the turf.

Elmore and Boquist (1988) stated that excessive irrigation can favor weeds such as crabgrass, nutsedge, and annual bluegrass, whereas light irrigation may subject the turf to be prone to weeds such as prostrate knotweed, goosegrass, and clovers (*Trifolium* spp.) (prostrate knotweed and goosegrass are also prevalent in compacted turf). They also noted some interactions between irrigation and turf species on the prevalence of certain weeds that indicated the ability of ryegrass and tall fescue to suppress weeds. When bluegrass was irrigated to replace 50 percent of evapotranspiration, a fivefold increase in prostrate pigweed was seen compared to that in ryegrass and tall fescue. All three grass species resisted this weed completely when irrigated to replace all the water lost to the atmosphere. Gibeault, Meyer, and Younger (1985) reported increased incidence of smooth crabgrass, with 100 percent of evapotranspiration and a decreased incidence of spotted spurge when three different turf types were irrigated using subterranean irrigation versus sprinkler irrigation. In contrast, yellow woodsorrel, broomsedge, spotted knapweed, bitter sneezeweed, and black medic thrive well under drought conditions (Derr, 2000). Derr (2000) observed that the presence of rushes (*Juncus* spp.), green kyllinga, barnyardgrass, alligatorweed, pennywort (*Hydrocotyle* spp.), Virginia buttonweed, and bittercress (*Cardamine* spp.) in turf might indicate excessive soil moisture because these weeds prefer moist environments.

The presence of an undesirable turf could pose a pest problem at times. After renovation of turf, plants of the previous turf may invade the new turf. A study indicated that cultural practices suited to the new turf could encourage the invading turf (Brede, 1992). Competition from volunteer bermudagrass from prior turf in a tall fescue turf could be reduced by seeding tall fescue at a higher rate, mowing at a higher height, and applying lower amounts of fertilizer.

Apart from hand weeding and maintaining proper mowing heights, mechanical methods have limited implications in turfgrass weed management. A belief that cultivation would reduce PRE herbicide efficacy by disrupting the herbicide barrier was disproved (Branham and Rieke, 1986). In a different study, it was noted that thatch buildup can have indirect adverse effects due to increased irrigation needs as a result of turfgrass root orientation toward the thatchy layer (Hanson, 1990). This would not only attract weed species that are favored by excessive moisture levels but also lead to higher levels of injury in turf treated with the dinitroanaline family of herbicides. Weeds such as ground ivy and crabgrass may become dominant as a result of attempting to eliminate thatch, by scalping the turf using a conventional mower or due to improper use of a verti-cut mower, exposing the soil and providing an opportunity for weeds to germinate.

The concept of allelopathy, or biochemical exudates of one plant suppressing the emergence of another, has gained renewed interest in turfgrass weed control. Elmore (1990) suggests a novel system where a turfgrass possessing allelopathic properties could be effectively utilized to manage weeds due to the unique growth habits of turf. He determined that 'Olympic' and 'Falcon' cultivars of tall fescue and 'Gator' and 'Repell' cultivars of perennial ryegrass reduced large crabgrass and creeping woodsorrel populations. 'America' cultivar of Kentucky bluegrass had significantly more crabgrass than other cultivars tested. Bertin, Rossi, and Weston (2002) noted differences in the weed-suppressive abilities of fine fescues (*Festuca rubra* L.) and observed the release of exudates by seedling fescues that caused growth reductions and chlorosis of weed seedlings.

A few weeds may be managed effectively through cultural and mechanical measures while certain others cannot be controlled by these means. In certain instances, the turf species play a role in the prevalence of a given weed. The percent cover of prostrate spurge was 22.3 percent in perennial ryegrass, 15.3 percent in Kentucky bluegrass, and 8.5 percent in turf-type tall fescue (Elmore and Boquist, 1988). A review of the genus *Oxalis* revealed that woodsorrels, perennial weeds prevalent in turfgrasses, could not be controlled using cultural methods (Marshall, 1987).

The amount of sunlight reaching the ground also influences species composition (Derr, 2000). Weeds such as ground ivy, Japanese stiltgrass, poison ivy, violets (*Viola* spp.), and Virginia creeper flourish well under shaded conditions. Letting in more sunlight by pruning excessive tree branches may help suppress such weed problems. A few novel mechanical methods are being tested by researchers as alternate options to control weeds. Andreasen, Hansen, and Streibig (1999) determined that annual bluegrass, common groundsel, shepherd's purse, small nettle, canola, and pea differed in their sensitivity to ultraviolet (UV) radiation. The authors indicate the potential for using UV radiation to control weeds after further research to determine its selectivity and potential health hazards. Chandran and Singh (2000), determined that heating an artificial seed bank under simulated winter conditions (–12°C) resulted in the germination of johnsongrass, barnyardgrass, and hemp sesbane, and the seedlings were killed when the heat source was removed to revert the ambient temperatures to winter conditions. The applicability of these techniques to manage weeds in turf is yet to be explored.

BIOLOGICAL METHODS

Using biological control agents to manage weeds in turf is less popular than using them to control other pests such as nematodes and insects. The current economics and short shelf life of these agents make it difficult for the user to successfully implement these as practical tools. Different isolates of *Xanthomonas campestris* have been successfully tested as biological control agents against annual bluegrass (Johnson, 1994; Zhou and Neal, 1995; Imaizumi et al., 1997, 1999; Nishino, Morita, and Fujimori, 1997). Imaizumi and colleagues (1997) found one isolate of *Xanthomonas campestris* pv. *poae.,* JT-P482, to be an effective pathogen of annual bluegrass. Temperature following inoculation was seen to have an effect on its pathogenicity. They noted severe wilting of bluegrass at day/night temperatures of 25-20°C in seven to ten days after inoculating, with a loss of efficacy at lower temperatures. Other researchers (Nishino, Morita, and Fujimori, 1997) have determined that the population of this bacterium rapidly declined after application and that the number of other soil bacteria and soil fungi did not change drastically after its application. This strain of bacteria was, therefore, considered not to be able to persist in a natural ecosystem and to have little effect on other soil microbial organisms.

The disease pathogen *Phoma herbarum* was determined to be a potential biological control agent for dandelion in turf in a controlled environment study (Neumann and Boland, 1999). The severity of disease was enhanced in the field by formulating and applying the fungal pathogen with different adjuvants. In a separate study, Johnson, Wyse, and Jones (1996) documented that the phytopathogenic bacteria *Pseudomonas syringae* pv. *tagetis* was more effective on annual bluegrass if there were injuries in the host to gain entry. In their experiments, spray application with a surfactant produced severe disease in Canada thistle, common ragweed, and Jerusalem artichoke, and application of the bacterium without surfactant was ineffective on all reported hosts. They also controlled annual bluegrass in bermudagrass golf greens by applying *Xanthomonas campestris* pv. *poannua* as spray while mowing the grass. Application of the bacterium to annual bluegrass in the absence of fresh mowing injuries failed to produce symptoms.

Apart from traditional biological control agents, arbuscular mycorrhizal fungi were also considered to be useful to reduce annual bluegrass levels in turf (Gange, Lindsay, and Ellis, 1999). The researchers observed a negative relation between mycorrhizae levels and annual bluegrass abundance in 18 golf greens sampled. In an experiment, the mycorrhizal inoculum, when added to a golf green, enhanced the colonization levels of the creeping bentgrass turf roots, thereby improving their competitive ability. Limited reports are available on the use of insects as biological control agents for selective weed management in turf, and their effectiveness as selective feeders are yet to be documented. Choo, Woo, and Bum (1992) collected 36 insect species that fed on monocots. *Lema concinnipennis* was identified as a potential biological agent for control of dayflower (*Commelina* spp.). The insect chafer *Adoretus tenuimaculatus* also fed this weed; however, it was noted to be a pest of turfgrass and trees as well. A study in Kentucky (Crutchfield and Potter, 1995) reported that root-feeding white grubs (Coleoptera: Scarabaeidae) fed upon and reduced root biomass of white clover, red clover, large crabgrass, annual bluegrass, buckhorn plantain, and dandelion. However, it also feeds on roots of Kentucky bluegrass. The host specificity of insects must be favorable before they could be considered as successful biological control agents.

Biological control agents are known to synergize the activity of chemical herbicides. In a study to determine the effect of adding a biological control agent to the commonly used herbicide 2,4-D, Schnick, Stewart-Wade, and Boland (2002) determined that the fungal pathogen *Sclerotinia minor* when added to 25 or 50 percent of the normal use rates of 2,4-D synergized the activity of the herbicide to control dandelion. Oka and Pimentel (1976) had noted that 2,4-D increased the susceptibility of plants to disease pathogens. Reports on the effects of adding herbicides to biological control agents have also been documented in other weed-crop systems (Charudattan, 1986; Grant et al., 1990). The role of plant pathogens in integrated weed management in modern agroecosystems is also discussed elsewhere (Charudattan, 2001).

CHEMICAL METHODS

Herbicides are effective tools for weed management, especially when high-quality turf is desired. However, they must be used in conjunction with proper turfgrass management for best results. Herbicides are applied either as a spot treatment or over the entire turf as a broadcast application. To minimize costs and conserve chemical use, spot application is recommended if weed infestation is localized. This would be applicable primarily for postemergence (POST) herbicides, since PRE herbicides are often applied as a broadcast treatment. However, with experience and good record keeping, PRE herbicides may also be applied in selected areas prone to weeds. This may include edges of the lawn, compacted regions, or any other spot of a turf with a history of specific weed problems. The label has to be consulted for determining tolerance level of a given turf species to a herbicide or if the herbicide is registered for use in a given situation.

Most PRE herbicides need rainfall within a week after treatment to activate the active ingredient. If rainfall does not occur, adequate irrigation may be provided to activate the chemical. Application timing of PRE herbicides is critical to obtain successful weed management. They have to be applied before the weed seeds germinate, namely spring, if summer annuals are targeted, and fall for winter annuals. Some annuals, however, germinate earlier during the season. A PRE herbicide with extended residual activity when applied during fall may provide adequate levels of residues to control early-germinating summer annuals. This strategy would be especially useful if the given turf has a history of winter annual weeds and summer annuals that germinate earlier during the season. In field studies, Chandran, Derr, and Bingham (1998) determined that fall application of isoxaben, a herbicide with a three- to five-month half-life, offered control of both winter annuals and certain summer annuals that germinated in early spring.

Postemergence herbicides are most effective if applied when weeds are young and actively growing. Early summer application is recommended for grasses (two- to three-leaf stage) and broadleaf annuals and fall application for winter annuals. For effective control of perennials, a systemic herbicide applied in late summer or early fall would be more effective than one applied in spring/early summer, or a contact herbicide. Postemergence herbicides should not be applied when rainfall is expected since the chemical has to be absorbed by the foliage to be effective. Other environmental factors such as temperature and drought also influence the efficacy of POST herbicides. Application of herbicides, especially systemic chemicals that have to be absorbed and translocated to the target site, when plants are under stress, usually leads to decreased weed control efficacy. Some difficult-to-control weeds such as yellow nutsedge or ground ivy may require sequential applications with three- to four-week intervals between applications.

Herbicides used in turf come either as a sprayable formulation or as granules. Commonly used sprayable formulations include dry flowables (DF) or water dispersible granules (WDG), liquids (L), emulsifiable concentrates (EC), and wettable powders (WP). If tank mixing more than one sprayable formulation, fill the tank about halfway through with the carrier and follow the proper sequence, i.e., WP (or any fertilizers if any) followed by WDG or DF, L, and finally the EC formulation. If granular (G) formulations are used, splitting the dose into halves and making two passes perpendicular to each other would ensure proper coverage.

Continuous use of the same herbicides may cause tolerant weeds to multiply, eventually resulting in buildup of weed species tolerant to the herbicide program. Therefore, rotation of chemicals is recommended for weed management. This also reduces chances for evolution of resistant biotypes of weeds. Based on the intensity of management efforts and acceptable level of weeds, chemical weed control could be simplified as three different programs, namely low input, medium input, and high input.

In a *low-input* program, POST herbicides alone are used on a need basis. Herbicide is applied as a spot application only to areas that have weeds growing. Commonly used POST herbicides for grass control include dithiopyr, diclofop, DSMA (disodium salt of methylarsonic acid), fenoxaprop, imazapic, metribuzin, MSMA (monosodium salt of methylarosonic acid), napropamide, quinclorac, and sethoxydim; those for broadleaf weeds include carfentrazone-ethyl, clopyralid, 2,4-D, dicamba, dichlorprop, mecoprop (MCPP), triclopyr, and two-, or three-way combinations of these herbicides (Bingham, Chism, and Bhowmik, 1995; Monaco, Weller, and Ashton, 2002). Most commercial formulations of broadleaf herbicides have two or three active ingredients in them. Sedges are more susceptible to such POST herbicides as bentazon, imazapic, imazaquin, halosulfuron, and MSMA. While applying certain POST herbicides, for example, dicamba, for weed control in turf, precautions should be taken to avoid herbicide spray drift to susceptible ornamental plantings and trees. Also, do not spray such herbicides into the drip lines of susceptible ornamentals.

A *medium-input* program utilizes broadcast application of a PRE herbicide before the dominant weeds germinate during the season (Bhowmik and Bingham, 1990). This would prevent the establishment of summer annuals, most of which germinate during late spring or early summer. Some of the PRE grass herbicides include dinitroanalines such as benefin, oryzalin, pendimethalin, prodiamine, trifluralin, and others including bensulide, dithiopyr, DCPA (dimethyl 2,3,5,6-tetrachloro-1,4-benzenedicarboxylate), ethofumesate, metolachlor, napropamide, oxadiazon, and siduron. Isoxaben is specifically formulated to control a wide range of broadleaf weeds PRE, while atrazine, DCPA, oxadiazon, simazine, and certain dinitroanalines such as benefin, oryzalin, and pendimethalin may offer PRE control of certain broadleaf weeds.

A *high-input* program uses both PRE and POST herbicide applications for weed management. This program is suitable for golf courses, athletic fields, and professional grounds maintenance. Preemergence herbicides may be applied during spring or fall months either alone or as a tank mix with POST herbicides. Sequential treatments of PRE or POST herbicides or both are carried out in certain instances. Some of the POST grass herbicides may cause temporary injury to turf, but turf usually greens up within three to four weeks.

Spot treatments with systemic nonselective herbicides such as glyphosate or glufosinate may be needed for controlling certain difficult-to-control weed species including bermudagrass, dallisgrass, tall fescue, and nimblewill. In certain instances, wick application using such herbicides to tall growing clumps may be effective to achieve selectivity. Avoid wetting turf with drip or equipment leak between spot applications. Bare spots emerging due to weed control may have to be disturbed sufficiently to alleviate soil compaction before subsequent reseeding. While using chemicals for weed control, good record keeping is warranted. Knowledge of different modes of action and families of herbicides allows the user to make informed decisions about rotating and/or tank mixing chemicals to minimize buildup of resistance to herbicides among weeds. The use of reduced-risk crop protection chemicals is gaining popularity among IPM practioners, and most future pesticides will probably belong to this category (Baron, 2002).

A few organic and inorganic compounds conventionally used for domestic purposes have been shown to possess herbicidal properties on turf weeds. Hatterman, Owens, and Christians (1996) demonstrated that ground ivy in a Kentucky bluegrass turf could be selectively controlled by applying liquid borax at 61 $kg{\cdot}ha^{-1}$. They noticed temporary injury to the turfgrass. Herbicidal soaps can be used as a nonselective herbicide or for spot treatments (Christians, 1999). Corn gluten, a natural product based on corn flour, is known to have inhibitory effects on weed seed germination (Christians, 1999; McDade and Christians, 2000). Gardner, Christians, and Bingaman (1997) also documented that adding corn gluten at 470 $kg{\cdot}ha^{-1}$ with 0.29 $kg{\cdot}ha^{-1}$ of pendimethalin at 15 weeks after herbicide application provided crabgrass control similar to that of the recommended rate of pendimethalin applied alone at a six times higher rate of application. Other naturally occurring chemicals that may potentially possess herbicidal properties are discussed extensively elsewhere (Duke et al., 2002).

CONCLUSIONS

A weed-free turf is difficult to achieve, maintain, and manage. The turf ecosystem is dynamic by nature. The complexity of turf systems and interactions between flora and fauna within a turfgrass community makes it challenging to elucidate strategies to manage weeds in totality. Compared to other crops, turfgrasses are unique since weed interference primarily causes aesthetic losses to most turfgrass communities. This also allows us to have an extended window to manage these pests, unlike in annual food and fiber crops. The unique attributes of turf, therefore, permit integrated weed management to be practiced to greater extents than in other crops. Proper management methods and providing an environment that encourages the growth of a healthy and competitive turf is the cornerstone for weed management in turf. Choosing a suitable strategy follows monitoring and diagnosing the weed problem. Selecting the management strategy depends upon the urgency to manage weeds and the control method most suited for the situation. Apart from proper management (cultural techniques), other tools such as mechanical and biological control methods are slowly becoming available to manage weeds in turf. Herbicides are effective tools to manage weeds in major crops and will continue to play an important role in future weed management (Buhler, Liebman, and Obrycki, 2000).

Buhler, Liebman, and Obrycki (2000) also indicated that the basic principles of IPM have been successfully applied to many agricultural pests; however, its application to manage weeds lags behind that of other pest management disciplines. Many of the concepts and approaches of IPM are relevant to weed management, but these were not developed specifically for weed management and are not sufficient to address it adequately. Integrated weed management in turfgrass is, therefore, a discipline still in its infancy. The complexity of a turf ecosystem and the various interactions between biotic and abiotic factors influencing turf-weed dynamics makes weed management decisions in turf more challenging than in other crops. Genetically engineered turf, which is already in the pipeline, capable of resisting herbicides may offer use of reduced-risk herbicides for weed control. Using molecular approaches to breed turfgrasses expressing allelopathic characteristics may also be used to manage weeds using IPM tactics. Research to elucidate interactions among turfgrasses, weed species, and the environment will also be helpful to gain a better understanding of turf-weed dynamics and to delineate an integrated approach to manage weeds in turf.

REFERENCES

Andreasen, C., L. Hansen, and J.C. Streibig (1999). The effect of ultraviolet radiation on fresh weight of some weeds and crops. *Weed Technology* 13: 554-560.

Ashton, F.M. and T.J. Monaco (1991). *Weed science: Principles and practices.* New York: John Wiley and Sons, Inc.

Baron, J.J. (2002). The role of reduced risk crop protection chemicals on the success of IPM in the US. In M. Appleby (Ed.), *Proceedings of International IPM Conference: Exploring New Frontiers in IPM* (pp. 60-62). Ontario, Canada: Ontario Ministry of Agriculture, Food and Rural Affairs.

Bertin, C., F. Rossi, and L.A. Weston (2002). Further evaluation of the allelopathic potential of fine leaf fescue. *Proceedings of Northeastern Weed Science Society* 56: 116.

Bhowmik, P.C. and S.W. Bingham (1990). Preemergence activity of dinitroaniline herbicides used for weed control in cool-season turfgrasses. *Weed Technology* 4: 387-393.

Bigelow, C.A., D.C. Bowman, D.K. Cassel, and T.W. Rufty Jr. (2001). Creeping bentgrass response to inorganic soil amendments and mechanically induces subsurface drainage and aeration. *Crop Science* 41: 797-805.

Bingham, S.W., W.J. Chism, and P.C. Bhowmik (1995). Weed management systems for turfgrasses. In A.E. Smith (Ed.), *Handbook of weed management systems* (pp. 603-665). New York: Marcel Dekker, Inc.

Black, D.W. (1975). Chemical and cultural control of yellow nutsedge in turf. Master's thesis, Urbana Champaign, University of Illinois.

Branham, B.E. and P.E. Rieke (1986). Effect of turf cultivation practices on the efficacy of preemergence grass herbicides. *Agronomy Journal* 78: 1089-1091.

Brede, A.D. (1992). Cultural factors for minimizing bermudagrass invasion into tall fescue turf. *Agronomy Journal* 84: 910-922.

Buhler, D.D. (2002). Challenges and opportunities for integrated weed management. *Weed Science* 50: 273-280.

Buhler, D.D., M. Liebman, and J.J. Obrycki (2000). Theoretical and practical challenges to an IPM approach to weed management. *Weed Science* 48: 274-280.

Chancellor, R.J. (1986). Decline of arable weed seeds during 20 years in soil under grass and the periodicity of seedling emergence after cultivation. *Journal of Applied Ecology* 23: 631-638.

Chandran, R.S., J.F. Derr, and S.W. Bingham (1998). Effect of isoxaben application rate and timing on residual broadleaf weed control in turf. *Weed Technology* 12: 569-574.

Chandran, R.S. and M. Singh (2000). Depletion of an artificial weed seed bank during the dormant period via heating and subsequent chilling of soil. *HortScience* 35: 1092-1094.

Charudattan, R. (1986). Integrated control of water hyacinth *(Eichhornia crassipes)* with a pathogen, insects and herbicides. *Weed Science* 34: 26-30.

Charudattan, R. (2001). Biological control of weeds by means of plant pathogens: Significance for integrated weed management in modern agro-ecology. *BioControl* 46: 229-260.

Choo, H.Y., K.S. Woo, and K.J. Bum (1992). Phytophagous insect fauna of Monocotyledoneae (Tracheophyta: Angiospermae) weeds. *Korean Journal of Applied Entomology* 31: 170-173.

Christians, N.E. (1999). Using biological control strategies for turf, Part 3: Weeds. *Grounds Maintenance* 34: 28-32.

Cook, T. (1992). Can we maintain turf without synthetic chemicals? *Journal of Pesticide Reform* 12: 26-30.

Crutchfield, B.A. and D.A. Potter (1995). Feeding by Japanese beetle and southern masked chafer grubs on lawn weeds. *Crop Science* 35: 1681-1684.

Dernoeden, P.H., M.J. Carroll, and J.M. Krouse (1993). Weed management and tall fescue quality as influenced by mowing, nitrogen and herbicides. *Crop Science* 33: 1055-1061.

Derr, J.F. (2000). Weeds as indicators of environmental conditions. *Turf Grass Trends* 8: 9-13.

Duke, S.O., F.E. Dayan, A.M. Rimando, K.K. Schrader, G. Aliotta, A. Olivia, and J.G. Romagni (2002). Chemicals from nature for weed management. *Weed Science* 50: 138-151.

Elmore, C.L. (1990). Potential for weed control with allelopathy in turfgrass. *Proceedings of Californian Weed Conference* 42: 214-216.

Elmore, C.L. (1996). A reintroduction to integrated weed management. *Weed Science* 44: 409-412.

Elmore, C.L. and D.J. Boquist (1988). New techniques for non-chemical weed control in the landscape. *Proceedings of Californian Weed Conference* 40: 34-38.

Fry, J.D. and J.D. Butler (1989). Evapotranspiration rates of turf weeds and groundcovers. *HortScience* 24: 73-75.

Gange, A.C., D.E. Lindsay, and L.S. Ellis (1999). Can arbuscular mycorrhizal fungi be used to control the undesirable grass *Poa annua* on golf courses? *Journal of Applied Ecology* 36: 909-919.

Gardner, D.S., N.E. Christians, and B.R. Bingaman (1997). Pendimethalin and corn gluten meal combinations to control turf weeds. *Crop Science* 37: 1875-1877.

Gibeault, V.A., J.L. Meyer, and V.B. Younger (1985). Irrigation of turfgrass below replacement *ET* as a means to conserve water: Weed invasion in three cool season turfgrasses. In F. Lemaire (Ed.), *Proceedings of the Fifth International Turfgrass Research Conference* (pp. 365-372). Avignon, France: INRA.

Grant, N.T., E. Prusinkiewicz, K. Mortensen, and R. Makowski (1990). Herbicide interactions with *Colletotrichum gloeosporioides* f. sp. *malvae* a bioherbicide for round-leaved mallow *(Malva pusilla)* control. *Weed Technology* 4: 716-723.

Gray, E. and N.M. Call (1993). Fertilization and mowing persistence of Indian mockstrawberry *(Duchesnea indica)* and common blue violet *(Viola papilionacea)* in a tall fescue *(Festuca arundinacea)* lawn. *Weed Science* 41: 548-550.

Haley, J.E., D.J. Wehner, T.W. Fermanian, and A.J. Turgeon (1985). Comparison of conventional and mulching mowers for Kentucky bluegrass *Poa pratensis* cultivar 'Baron' maintenance. *HortScience* 20: 105-107.

Hanson, D.L. (1990). Systems approach to weed management in turfgrass. *Proceedings of Californian Weed Conference* 42: 94-104.

Harivandi, M.A. (1994). Preventing weed invasion in lawns through cultural management. *Proceedings of Californian Weed Science Society* 46: 15-17.

Harivandi, M.A. (2000). ITM (Intelligent turf management) to prevent weed invasion. *Proceedings of Californian Weed Science Society* 52: 9-10.

Hatterman, V.H., M.D.K. Owen, and N.E. Christians (1996). Ground ivy (*Glechoma hederacea* L.) control in a Kentucky bluegrass turfgrass with borax. *Journal of Environmental Horticulture* 14: 101-104.

Hensler, K.L., B.S. Baldwin, and J.M. Goatley Jr. (2001). Comparing seeded organic-fiber mat with direct soil seeding for warm-season turfgrass establishment. *HortTechnology* 11: 243-248.

Imaizumi, S., T. Nishino, K. Miyabe, T. Fujimori, and M. Yamada (1997). Biological control of annual bluegrass (*Poa annua* L.) with a Japanese isolate of *Xanthomonas campestris* pv. *poae* (JT-P482). *Biological Control* 8: 7-14.

Imaizumi, S., A. Tateno, D. Miyamoto, K. Morita, and T. Fujimori (1999). Seasonal factors affecting the control of annual bluegrass (*Poa annua* L.) with *Xanthomonas campestris* pv. *poae* (JT-P482). *Biological Control* 16: 18-26.

Jiang, H., J. Fry, and N. Tisserat (1998). Assessing irrigation management for its effect on disease and weed levels in perennial ryegrass. *Crop Science* 38: 440-445.

Johnson, B.J. (1984). Influence of nitrogen on recovery of bermudagrass *Cynodon dactylon* cultivar 'Tifway' treated with herbicides. *Weed Science* 32: 819-823.

Johnson, B.J. (1994). Biological control of annual bluegrass with *Xanthomonas campestris* pv. *poannua* in bermudagrass. *HortScience* 29: 659-662.

Johnson, D.R., D.L. Wyse, and K.J. Jones (1996). Controlling weeds with phytopathogenic bacteria. *Weed Technology* 10: 621-624.

Juska, F.V. and A.A. Hanson (1961). Effects of interval and height of mowing on growth of 'Marion' and common Kentucky bluegrass (*Poa pratensis* L.). *Agronomy Journal* 53: 385-388.

Kogan, M. and W.I. Bajwa (1999). Integrated pest management: A global reality? *Annals of the Society of Entomology, Brasil* 28: 1-25.

Lowe, D.B., T. Whitwell, L.B. McCarty, and W.C. Bridges (2000). Mowing and nitrogen influence green kyllinga *(Kyllinga brevifolia)* infestation in Tifway bermudagrass (*Cynodon dactylon* × C. *transvaalensis*) turf. *Weed Technology* 14: 471-475.

Lush, W.M. and J.A. Birkenhead (1987). Establishment of turf using advanced pregerminated seeds. *Australian Journal of Experimental Agriculture* 27: 323-327.

Lush, W.M. and M.E. Rogers (1992). Cutting height and the biomass and tiller density of a *Lolium perenne* amenity turfs. *Journal of Applied Ecology* 29: 611-618.

Marshall, G. (1987). A review of the biology and control of selected weed species in the genus *Oxalis: O. stricta* L., *O. latifolia* H.B.K. and *O. pes-caprae* L. *Crop Protection* 6: 355-364.

McCarty, L.B. (1995). Winter weed control in southern turf: Early detection, recognition and action are key. *Turf Grass Trends* 4: 9-14.

McDade, M.C. and N.E. Christians (2000). Corn gluten meal—A natural preemergence herbicide: Effect on vegetable seedling survival and weed cover. *American Journal of Alternative Agriculture* 15: 181-191.

Miller, G.L. (2000). Physiological response of bermudagrass grown in soil amendments during drought season. *HortScience* 35: 213-216.

Monaco, T.J., S.C. Weller, and F.M. Ashton (2002). *Weed science: Principles and practices,* Fourth edition (pp. 500-521). New York: John Wiley and Sons, Inc.

Neal, J.C. (1994). Using IPM to manage weeds in turfgrass. *Turf Grass Trends* 3: 1-3, 6-7.

Neumann, S. and G.J. Boland (1999). Influence of selected adjuvants on disease severity by *Phoma herbarum* on dandelion *(Taraxacum officinale). Weed Technology* 13: 675-679.

Nishino, T., K. Morita, and T. Fujimori (1997). Fate of *Xanthomonas campestris* pv. *poae*, a biological control agent for annual bluegrass, in soil. *Journal of Pesticide Science* 22: 326-330.

Norrie, J. and A. Gosselin (1996). Paper sludge amendments for turfgrass. *HortScience* 31: 957-960.

Oka, I.N. and D. Pimentel (1976). Herbicide (2,4-D) increases insects and pathogen pests on corn. *Science* 193: 239-240.

Pierce, G.L., S.L. Warren, R.L. Mikkelsen, and H.M. Linker (1999). Effects of soil calcium and pH on seed germination and subsequent growth of large crabgrass *(Digitaria sanguinalis). Weed Technology* 13: 421-424.

Schery, R.W. (1961). *The lawn handbook.* New York: The MacMillan Co.

Schnick, P.J, S.M. Stewart-Wade, and G.J. Boland (2002). 2,4-D and *Sclerotinia minor* to control common dandelion. *Weed Science* 50: 173-178.

Shiou, K. (1993). Effect of lime and phosphate on the growth of annual bluegrass and creeping bentgrass in two acid soils. *Soil Science* 156: 94-100.

Smith, R.F. and R. van den Bosch (1967). Integrated control. In W.W. Kilgore and R.L. Doutt (Eds.), *Pest control: Biological, physical, and selected chemical methods* (pp. 295-340). NewYork: Academic Press.

Summerlin, J.R. Jr., H.D. Coble, and F.H. Yelverton (2000). Effect of mowing on perennial sedges. *Weed Science* 48: 501-507.

Turgeon, A.J. (1980). *Turfgrass science.* Englewood Cliffs, NJ: Prentice Hall.

Turgeon, A.J. (1999). *Turfgrass management.* Upper Saddle River, NJ: Prentice Hall Publishing Co.

Uhler, B. (1992). Lawn care without pesticides. *Journal of Pesticide Reforms* 12: 38-39.

Uva, R.H., J.C. Neal and J.M. DiTomaso (1997). *Weeds of the Northeast.* Ithaca, NY: Cornell University Press.

Zhou, T. and J.C. Neal (1995). Annual bluegrass *(Poa annua)* control with *Xanthomonas campestris* pv. *poannua* in New York State. *Weed Technology* 9: 173-177.

Chapter 24

Approaches to Integrated Weed Management

Douglas D. Buhler

INTRODUCTION

Weeds have a significant economic impact on agricultural production. In the United States alone, it was estimated that weeds and weed control have an annual economic impact of more than $15 billion (Bridges, 1994), with even greater relative costs in developing countries (Akobunudu, 1991). Concerns over the economic costs and environmental impacts of current practices (National Research Council, 1989; Radosevich and Ghersa, 1992; Flora, 1995) have led many weed scientists and crop producers to seek alternative strategies for weed control (Gressel, 1992; Wyse, 1992). Herbicides are important tools for modern agriculture but have become the dominant weed control practice in many production systems. New control options and management knowledge will give producers more flexibility as well as help preserve the effectiveness of herbicides. Using herbicides in a more integrative manner will help forestall the development of weed resistance and reduce the potential for environmental contamination.

Herbicides are valuable management tools and have provided benefits to the agricultural and urban communities; however, weed management should be viewed as an integrated science (Burnside, 1993). Increased attention must be given to biological, cultural, mechanical, and preventive tools and techniques to manage weeds. Refocusing from dependence on a single technology (herbicides) in simplified cropping systems will require a greater understanding of biological systems than we currently possess (Navas, 1991; Holt, 1994). The progression from a focus on tools targeted at the prevailing weed population in a single year to the adoption of a holistic approach to crop and pest management will require analysis, theory, and information to support implementation at the ecosystem level. Such sys-

---

The author thanks those who have written previously about the future of weed science. Their views have been invaluable in the development of this paper.

tems must also consider the interactions of long-term management practices, populations of crops, and populations of pest, beneficial, and innocuous organisms.

WEED MANAGEMENT

Weed management implies a shift away from reliance on control of existing weed problems and places greater emphasis on prevention of propagule production, reduction of weed emergence in a crop, and minimization of weed competition with the crop (Zimdahl, 1991; Buhler, Liebman, and Obrycki, 2000). Weed management emphasizes integration of techniques to anticipate and manage problems rather than reacting to them after they are present. The goal is to maximize crop production where appropriate, and optimize grower profit by integrating preventive techniques, scientific knowledge, management skills, and the best control techniques. Although additional knowledge is needed in all areas of weed management, the most important task of weed science is to increase knowledge of weed biology and ecology, creating a better understanding of weediness. This knowledge will lead to the use of appropriate management techniques that will not only produce short-term results but also develop long-term solutions to persistent weed problems.

In modern times, much weed control technology has been directed at weed eradication (Liebman, Mohler, and Staver, 2001). This may be a worthy goal, but it is not a realistic possibility because of the nature of cropping systems and the capacity of weed communities to persist and rapidly adapt to selection pressures imposed upon them. Diversifying weed management systems to take better advantage of ecological processes such as resource competition, allelopathy, seed predation, natural enemies, and weed response to soil disturbance may present a better option for the long-term management of weeds. It should be noted that this concept does not exclude the use of herbicides but considers herbicides as one of many tools that can be used to manage the crop-weed system.

WEED CONTROL SCIENCE AND PRINCIPLES OF WEED POPULATION DYNAMICS

Weed science research has been separated into two major categories: weed control science and technology, and principles of weed population dynamics (Wyse, 1992; Buhler, 1996). Research on weed control science and technology involves herbicides, tillage, biological control, and other methods to remove unwanted vegetation from a disturbed system (see Exhibit 24.1). Research on principles of weed population dynamics focuses on weed biology and ecology and examining why weeds are present and their impacts on crop yields and the agroecosystem. If we are to view weed management as an integrative science, we must also include a category for integrating the science and technology generated into new management systems.

Principles of weed population dynamics provide an understanding for the basis of weed problems. This research is the basis for new, ecologically based weed management systems. The knowledge gained will provide the foundation for development of new strategies and more efficient techniques, resulting in more reliable weed management systems that are cost-effective and pose less of a threat to the environment. This approach is also consistent with the principles of integrated pest management, which center on knowledge of pest biology and ecology (Smith, 1978). Weed science has lagged behind other pest management disciplines in developing integrated management systems, largely due to our limited understanding of the principles of weed population dynamics (Burnside, 1993).

The term *integrated pest management* (IPM) first appeared in the literature in 1967 (Smith and van den Bosch, 1967) and has roots in the concept of integrated control (Stern et al., 1959). Although many definitions of IPM

EXHIBIT 24.1. Examples of weed science research categorized as (1) weed control science and technology and (2) principles of weed population dynamics.

Weed control science and technology	**Principles of weed population dynamics**
• Herbicide development	• Seed and bud dormancy mechanisms
• Biological control	• Seed development and production
• Allelopathic cover crops	• Population genetics
• Tillage	• Seed banks and emergence dynamics
• Herbicide selection decision aids	• Weed/crop yield responses
• Managed competition	• Population shifts
• Competitive cultivars	• Modeling weed/crop systems
• Risk management	• Spatial distributions

Source: Modified from Wyse, 1992.

have been proposed, they all contain two key elements: (1) the use of multiple control tactics and (2) the integration of knowledge of pest biology into the management system (Bottrell, 1979). As such, developing integrated management strategies for weeds calls for broader approaches that move beyond control of existing weed populations (Liebman and Gallandt, 1997). For example, integrating cropping system design and weed science could lead to systems that best utilize resources, diversify the selection pressure on weed communities, and provide producers a broader range of management options.

Changing the focus of weed science from tools targeted at the weed population present at a given time to the adoption of a holistic approach to crop and weed management will require analysis, theory, and information to support implementation at the cropping system and ecosystem levels (Cardina et al., 1999). Integrated weed management must be developed within the context of the entire cropping system, with the farm and the surrounding area being considered as part of a larger ecosystem. Weeds should not be considered in isolation, and a better understanding of the interactions of ecosystem health, population dynamics of weeds, and weed and ecosystem responses to management practices are needed (Liebman and Gallandt, 1997). We need to integrate cropping systems and IPM concepts to include comprehensive theories that include all management variables, building on the foundations provided by the theories and practices of plant ecology, population management, plant protection, and cropping systems.

APPROACHES TO DEVELOPING INTEGRATED WEED MANAGEMENT SYSTEMS

Significant new challenges are developing for individuals involved in managing weeds in agronomic crops. Environmental regulations, reduced-tillage systems, increasing farm size, economic pressure, and industry consolidation and product cancellations are restricting the weed control options available to crop producers. In addition, weed populations continue to adapt to production practices through herbicide resistance and population shifts. Limited crop choice is reducing crop rotation and intensifying the selection pressure on weed communities.

Fundamental changes in weed management are complex and will need to take the unique nature of individual sites and systems into account. We need both short- and long-term approaches to developing integrated weed management systems. Short-term approaches apply existing knowledge, offer immediate benefits, and provide transition to new systems (see Exhibit

24.2). Long-term approaches require basic changes in crop management systems and must provide unique methods to manage weeds.

Short-Term Approaches

Short-term approaches to integrated weed management center around increasing the efficiency of herbicide use and maintaining weed control in current cropping systems. Many methods may be used to diversify weed control systems and reduce herbicide use (see Exhibit 24.2). However, the ability to implement these practices varies among growers. Growers with good management skills and fields with low weed pressure can reduce herbicide use and incur only a slight risk of weed control failure. For others, the savings achieved by reducing herbicide use may be small compared with the risk of control failures, crop yield losses, the additional time needed for weed control, and increased weed densities in future years.

EXHIBIT 24.2. Potential approaches for short- and long-term development of integrated weed management strategies.

Short-term

Application of existing knowledge
- Scouting
- Increased management

Refinements of current systems
- Combining control tactics
- Treatment thresholds
- Appropriate use of tillage
- Decision aids
- Reduced rates of herbicides
- Application technology

Long-term

Weed biology and ecology
- Population biology
- Weed/crop interactions
- Emergence and growth predictors

Site-specific management
- Herbicide application
- Soil and crop management

Redesign cropping systems
- Increase rotation and diversity
- Improve resource utilization
- New tillage systems

New control methods
- Biological control
- Allelopathy
- Managed competition
- Competitive cultivars

Mechanical weed control to supplement or replace herbicides remains an option in many cropping systems (Gunsolus, 1990). Properly designed and executed mechanical control systems are effective on most weed species. The increased use of reduced-tillage systems has lessened the applicability and effectiveness of some mechanical control operations (Springman et al., 1989). However, interrow cultivation is an integral component of ridge tillage (Forcella and Lindstrom, 1988) and has been effective in combination with reduced rates of herbicides in *Zea mays* planted into untilled seed beds (Buhler et al., 1995).

Banding of herbicide over the crop row and using cultivation to control weeds between the rows is a proven method to reduce herbicide use (Hartzler, 1993). Banding is an excellent tool for reducing herbicide use for growers who already practice interrow cultivation. Applying herbicides at rates below those listed on product labels is another method to reduce total use. Although legal, growers using reduced rates assume responsibility for herbicide efficacy.

Scouting fields and maintaining accurate field records are critical components of efficient weed management programs (Hartzler, 1993). Maintaining records for several years allows for assessment of weed populations and distribution patterns within fields. Fields where excellent control has been maintained for several years should have low weed densities and provide opportunities for reducing herbicide use. Emerging technologies such as global positioning systems (GPS) and geographic information systems (GIS) may improve weed mapping, record keeping, and scouting (Mortensen, Johnson, and Young, 1993).

Weed thresholds (O'Donovan, 1996) and associated models (Swinton and King, 1994; Wiles et al., 1996) may provide a mechanism to improve weed control decisions. Threshold models are based on an assessment of weed pressure and the relationship of this weed pressure to immediate and future yield losses and costs of control. Economic thresholds are being used in insect and disease management and should find greater application in weed management as existing knowledge is applied and new knowledge is developed (O'Donovan, 1996).

The widespread reliance on chemical weed control has been attributed to the ability of herbicides to reduce variability in economic returns (Olson and Eidman, 1992). Gunsolus and Buhler (1999) stated that a producer's perception of risk was the key limiting factor to the adoption of integrated weed management systems. Weed management programs that relied on a combination of mechanical and herbicidal tactics were found to be less risky than a program relying solely on herbicides or mechanical tactics (Gunsolus and Buhler, 1999). This is not surprising because it is well documented that diverse management strategies can provide as effective, or more effective, weed control as herbicide-based systems (Gunsolus, 1990; Hartzler et al., 1993).

The likelihood of being able to implement a control tactic during its peak effectiveness is a critical factor in a producer's decision-making process. Mechanical control tactics generally have a narrower window of opportunity and take more time than herbicides (Gunsolus, 1990). The time constraints of specific control tactics become more critical with increasing farm size. In contrast, farmers less reliant on herbicides generally reduce risks through use of more diverse management systems (Liebman and Gallandt, 1997).

Long-Term Approaches

Long-term approaches to integrated weed management (see Exhibit 24.2) are more difficult to define than short-term herbicide use reduction. Currently, weed control practices are justified by the knowledge that weed populations are very difficult to eliminate once established and cause direct economic losses when left uncontrolled; uncontrolled plants also contribute to future infestations. Nonetheless, we seldom examine the causes of the perpetual presence of weeds. Understanding the fundamental ecological and management processes that provide weeds the opportunity to exist are critical to integrated management systems.

Herbicides are often used as a solution to weed problems created through herbicide use. Although it is often convenient to use a different herbicide to control weeds not killed by the herbicide previously used, attempts to solve a problem with the same technology that generated the problem usually result in a similar, but more difficult, problem (Ferre, 1988). For example, addressing herbicide resistance by using different herbicides may create multiple and cross-resistance, eventually destroying the control efficacy of each chemical on that pest (LeBaron and Gressel, 1982).

A more integrative approach to weed management considers that weed populations have specific environmental niches that mimic the crops grown or take advantage of the conditions created in establishing the crop. For example, summer annual weed species predominate in a *Z. mays*/*Glycine max* rotation because both crops are summer annuals. In addition, growing crops in rows provides space and light for weeds. Cropping systems that discourage the buildup of adapted weed species and biotypes must be as diverse and resource efficient as possible. Diversity may be attained by including crops with differing life cycles and differing tillage intensity and timing, as well as by using all available control methods. Resource use may be improved by changing planting dates and patterns, using crop cultivars with rapid early growth, and planting cover crops. A new perspective might also consider that a diverse, low-density weed population may have a positive role in the agroecosystem (Odum and Biever, 1984).

New methods of weed control may also play an important role in the development of integrated weed management systems. Private industry may continue to provide new chemical technology, but biologically based control tactics are needed to facilitate integrated weed management systems. Currently, weed science has few, if any, economically and environmentally desirable alternatives to tillage and herbicides.

Research has developed technologies that may provide new methods for weed control (see Exhibit 24.2). Biological control is an established practice in some systems but has had minimal impact on weed management in agronomic crops (Gressel, 1992). Most instances where biological control of weeds has been successful are where there is a single target weed in a stable environment, such as pasture or rangeland (Strobel, 1991). Biological control of weeds in annual cropping systems is complicated by the instability of the environment, multispecies weed complexes, and the desire for near-complete control. Innovative approaches, such as microorganisms that attack the seeds or seedlings in the soil (Kremer, 1993; Harris and Stahlman, 1996), may provide new biological control options. Other biologically based control tactics, such as allelopathic crops or crop residues (Duke, 1990), smother plants (DeHaan et al., 1994), and crop cultivars with enhanced competitiveness with weeds (Pester, Burnside, and Orf, 1999) also offer potential as components of integrated weed management systems.

Weeds are a persistent problem in crop production, and it is the dynamic nature of weed populations that results in the perpetual encounters with this pest complex. The nature of a weed infestation in a field is a product of the production practices (rotation, tillage, fertilization, row spacing, and herbicides) used in recent history. Although new tactics and approaches for weed management are important, we need to understand the fundamental elements of agroecosystems and the relationship of weeds with edaphic factors. We also need to gather feedback on the effectiveness of control tactics through better monitoring and assessment methods because of the importance of the reproductive output of surviving weeds on future weed populations.

The adoption of herbicide-based weed management systems has allowed the development of relatively simple cropping systems and has facilitated earlier crop planting and increased farm size. With the availability of herbicides, weed management is no longer an integral component of the cropping system. Although herbicide use has benefited the farm community in many ways, the heavy reliance on herbicides creates an environment favorable for weed population shifts and resistence development and creates the potential for environmental contamination. The current challenge for producers is to manage herbicides and other weed control tools in a manner that prevents adapted species from reaching troublesome population levels and to minimize off-site movement. The challenge for weed scientists is to develop innovative, alternative management techniques that can be integrated into current and future production systems that will bring a more diverse and integrative approach to weed management.

CONCLUSION

The basis of weed science has been control technology rather than understanding weedy species and their roles in the agroecosystem. Weeds have been present since the beginnings of agriculture and are not likely to disappear in the near future. All forms of disturbance result in survival and selection of the best-adapted weed species (Holt, 1994). Any cropping system that exerts a continuous, strong selection pressure will cause a buildup of the best-adapted weed species and biotypes. Development of integrated weed management systems will require an approach that considers the processes and patterns that link fundamental scientific, economic, and sociological disciplines to agricultural systems.

The scientific disciplines of weed science and production agriculture must change by taking a broader view of weeds as part of the crop ecosystem and by addressing the long-term questions regarding weed perpetuation, crop production practices, and weed control technology. Agricultural systems are composed of interacting production, environmental, biological, economic, and social components. These interactions require the study of not only the parts but also the entire system (Holling, 1978). Long-term improvements in weed management and agricultural production systems will require a convergence of traditional agriculture, ecological theory, economics, and sociology (Levins, 1986; Radosevich and Ghersa, 1992). Linkages among these disciplines will form the basis for successful, stable, and profitable cropping systems.

REFERENCES

Akobunudu, I.O. (1991). Weeds in human affairs in sub-Saharan Africa: Implications for sustainable food production. *Weed Technology* 5: 680-690.

Bottrell, D.R. (1979). Integrated pest management: Definition, features, and scope. In *Integrated pest management* (pp. 19-26). Washington, DC: Council on Environmental Quality, U.S. Government Printing Office.

Bridges, D.C. (1994). Impacts of weeds on human endeavors. *Weed Technology* 8: 392-395.

Buhler, D.D. (1996). Development of alternative weed management strategies. *Journal of Production Agriculture* 9: 501-505.

Buhler, D.D., J.D. Doll, R.T. Proost, and M.R. Visocky (1995). Integrating mechanical weeding with reduced herbicide use in conservation tillage corn production systems. *Agronomy Journal* 87: 507-512.

Buhler, D.D., M. Liebman, and J.J. Obrycki (2000). Theoretical and practical challenges to an IPM approach to weed management. *Weed Science* 48: 274-280.

Burnside, O.C. (1993). Weed science: The step child. *Weed Technology* 7: 515-518.

Cardina, J., T.M. Webster, C.P. Herms, and E.E. Regnier (1999). Development of weed IPM: Levels of integration for weed management. *Journal of Crop Production* 2: 239-267.

DeHaan, R.L., D.L. Wyse, N.J. Ehlke, B.D. Maxwell, and D.H. Putnam (1994). Simulation of spring-seeded smother plants for weed control in corn *(Zea mays)*. *Weed Science* 42: 35-43.

Duke, S.O. (1990). Natural pesticides from plants. In J. Janick and J.E. Simon (Eds.), *Advances in new crops* (pp. 511-518). Portland, OR: Timber Press.

Ferre, F. (1988). *Philosophy of technology*. Englewood Cliffs, NJ: Prentice Hall, Inc.

Flora, C.B. (1995). Social capital and sustainability: Agriculture and communities in the Great Plains and Corn Belt. *Research in Rural Sociological Development* 6: 227-246.

Forcella, F. and M.J. Lindstrom (1988). Movement and germination of weeds in ridge-till crop production systems. *Weed Science* 36: 56-59.

Gressel, J. (1992). Addressing real weed science needs with innovations. *Weed Technology* 6: 509-525.

Gunsolus, J.L. (1990). Mechanical and cultural weed control in corn and soybeans. *American Journal of Alternative Agriculture* 5: 114-119.

Gunsolus, J.L. and D.D. Buhler (1999). A risk management perspective on integrated weed management. *Journal of Crop Production* 2: 167-187.

Harris, P.A. and P.W. Stahlman (1996). Soil bacteria as selective biological control agents of winter annual grass weeds in winter wheat. *Applied Soil Ecology* 3: 275-281.

Hartzler, R.G. (1993). *Reducing herbicide use in field crops*. University Extension Bulletin IPM-36. Ames: Iowa State University.

Hartzler, R.G., B.D. Van Kooten, D.E. Stoltenberg, E.M. Hall, and R.S. Fawcett (1993). On-farm evaluation of mechanical and chemical weed management practices in corn *(Zea mays)*. *Weed Technology* 7: 1001-1004.

Holling, C.S. (1978). The nature and behavior of ecological systems. In *Adaptive environmental assessment and management* (pp. 25-37). New York: John Wiley and Sons.

Holt, J.S. (1994). Impact of weed control on weeds: New problems and research needs. *Weed Technology* 8: 400-402.

Kremer, R.J. (1993). Management of weed seed banks with microorganisms. *Ecological Applications* 3: 42-53.

LeBaron, H.M. and J. Gressel (1982). *Herbicide resistance in plants.* New York: John Wiley and Sons.

Levins, R. (1986). Perspectives in integrated pest management: From an industrial to ecological model of pest management. In M. Kogan (Ed.), *Ecological theory and integrated pest management practices* (pp. 1-18). New York: John Wiley and Sons.

Liebman, M. and E.R. Gallandt (1997). Many little hammers: Ecological management of crop-weed interactions. In L. Jackson (Ed.), *Ecology in agriculture* (pp. 291-341). New York: Academic Press.

Liebman, M., C.L. Mohler, and C.P. Staver (2001). Weed management: A need for ecological approaches. In *Ecological management of agricultural weeds* (pp. 1-39). New York: The Cambridge University Press.

Mortensen, D.A., G.A. Johnson, and L.J. Young (1993). Weed distribution in agricultural fields. In P.C. Robert, R.H. Rust, and W.E. Larson (Eds.), *Soil specific crop management* (pp. 113-123). Madison, WI: American Society of Agronomy.

National Research Council Committee on the Role of Alternative Farming Methods in Modern Production Agriculture (1989). *Alternative agriculture.* Washington, DC: National Academy Press.

Navas, M.L. (1991). Using plant population biology in weed research: A strategy to improve weed management. *Weed Research* 31: 171-179.

O'Donovan, J.T. (1996). Weed economic thresholds: Useful agronomic tool or pipe dream? *Phytoprotection* 77: 13-28.

Odum, E.P. and L.J. Biever (1984). Resource quality, mutualism, and energy partitioning in food chains. *American Naturalist* 124: 360-376.

Olson, K.D. and V.R. Eidman (1992). A farmer's choice of weed control method and the impacts of policy and risk. *Reviews of Agricultural Economics* 14: 125-137.

Pester, T.A., O.C. Burnside, and J.H. Orf (1999). Increasing crop competitiveness to weeds through crop breeding. *Journal of Crop Production* 2: 31-58.

Radosevich, S.R. and C.M. Ghersa (1992). Weeds, crops, and herbicides: A modern-day "neckriddle." *Weed Technology* 6: 788-795.

Smith, R.F. (1978). History and complexity of integrated pest management. In E.H. Smith and D. Pimentel (Eds.), *Pest control strategies* (pp. 41-53). New York: Academic Press.

Smith, R.F. and R. van den Bosch (1967). Integrated control. In W.W. Kilgore and R.L. Doutt (Eds.), *Pest control: Biological, physical, and selected chemical methods* (pp. 295-340). New York: Academic Press.

Springman, R., D. Buhler, R. Schuler, D. Mueller, and J. Doll (1989). *Row crop cultivators for conservation tillage systems.* Madison: University of Wisconsin Extension Publication A3483.

Stern, V.M., R.F. Smith, R. van den Bosch, and K.S. Hagen (1959). The integrated control concept. *Hilgardia* 29: 81-101.

Strobel, G. (1991). Biological control of weeds. *Scientific American* 265: 72-78.

Swinton, S.M. and R.P. King (1994). A bioeconomic model for weed management in corn and soybean. *Agricultural Systems* 44: 313-315.

Wiles, L.J., R.P. King, E.E. Schweizer, D.W. Lybecker, and S.M. Swinton (1996). GWM: General weed management model. *Agricultural Systems* 50: 355-376.
Wyse, D.L. (1992). Future of weed science research. *Weed Technology* 6: 162-165.
Zimdahl, R.L. (1991). *Weed science: A plea for thought.* Washington, DC: United States Department of Agriculture, Cooperative State Research Service.

Chapter 25

Learning Groups for Implementation of Integrated Weed Management: Principles and Practical Guidelines

N. Jordan
H. Niemi
S. Simmons
R. Becker
J. Gunsolus
S. White

INTRODUCTION

The chapters of this handbook present a rich and promising range of options for ecologically based weed management. These chapters show that many or most farming operations can be planned and practiced—i.e., fine-tuned—in ways that take account of weed ecology. Of course, this fine-tuning of farming operations in recognition of weed ecology is subject to many constraints, as noted in this chapter, yet for many farming operations such fine-tuning is feasible and highly effective. The operations will then have a heightened negative effect, or at least a reduced positive effect, on populations of undesirable weed species. Many chapters of this book give at least broad guidelines for fine-tuning a number of aspects of farming operations. A sampling of these aspects includes spatial and temporal diversification on a variety of scales, primary and secondary tillage operations, applications of nutrients and irrigation, selection of crop varieties, fallow management, and development of decision trees to guide in-season management. Overall, we can conclude that a significant knowledge base exists that can support implementation of holistic, integrated weed management schemes (IWM).

These schemes are based on a conception of the farming operation itself as the basic instrument of weed management. That is, weed management is seen as an *emergent property of the farming operation,* rather than the result of a small number of specific control operations such as herbicide applications or mechanical weed control. Farmers' experiences show that this holistic approach can be highly cost-effective (Fernholz, 1992; Bender, 1993; Jordan, 1996) and can help achieve basic requirements of profitable production without degrading the on-farm resource base or off-farm environmental quality. Moreover, a similar approach can be taken in other managed ecosystems where weed management is needed, e.g., forests, lakes, and nature reserves. However, we must recognize that land managers—farmers, ranchers, foresters, etc.—are primary actors in implementing such IWM. Land managers often have detailed knowledge of the interacting and ever-shifting components of the ecosystems that they manage, and of the fixed or fluctuating economic, social, and environmental factors that constrain their operations. Given time and support, they can design and operate these ecosystems to achieve as much fine-tuning as possible in their circumstances. However, implementation of holistic and integrated weed management remains rare in mainstream farming of developed countries and in most settings in developing countries.

Unfortunately, many major barriers to implementation exist (Liebman, Mohler, and Staver, 2001). First, social and economic factors operate at very large scales. These include economic and other policies that favor commodity production over other farming outcomes such as increases in environmental quality or resource-use efficiency, so that there is little incentive for undertaking a transition to IWM. Also, social, economic, and regulatory trends are reducing the farm labor supply while sharply increasing the overall complexity of farm management. Second, many farm-scale factors create barriers. The process of fine-tuning is necessarily highly farm specific. Each farm will have a particular set of problematic weed species, and the range of farm operations that can practically be modified for weed management purposes will depend on the particular circumstances of a given farm (Jordan, 1996). The farmer is best positioned to identify these site-specific problems and potential solutions. Unfortunately, time and labor constraints are often severe, making it difficult for farmers to find time for stages in the process of fine-tuning that appear necessary to reduce risks and increase rewards. These stages include determining opportunities for cost-effective modifications of current farming operations, identifying key unknowns, searching for information to reduce uncertainty, planning modifications to current practices, and evaluating outcomes of these modifications.

A third major barrier arises from the general paucity of well-organized supporting "social infrastructures" that can support farmers as they carry out necessary stages of transition to IWM (Shaver, 2001). That is to say, it is crucial to recognize that implementation of IWM will require new or improved practices in a multiplicity of relevant sectors, including extension workers, farm advisors, input suppliers, researchers of many sorts, equipment makers, and plant breeders (Svejcar, 1996). Each sector has something to contribute; a clear example is use of extension approaches that emphasize increasing farmers' capacity to play the "designer/operator" role described previously (Röling and Jiggins, 1998). An instructive model is the multisector innovation that enabled the rapid adoption of synthetic fertilizers (Engel, 1997), which depended on innovation not only in farmers' practices, but equally so by equipment makers, input suppliers, grain marketers, agronomists, and extension workers. Therefore, a broad sweep of innovation to enable IWM implementation is needed among diverse professionals organized into a supportive social infrastructure. Such an infrastructure will be comprised of mutually supportive and cooperative members of interacting professions, such as extension workers, farming advisors, research scientists, nongovernmental organizations (NGOs), and state agencies that play regulatory, service, and educational roles.

In what follows, we suggest means of breaching IWM implementation barriers that arise from two of these challenges: fine-tuning farm operations and developing supportive social infrastructures. The existing knowledge base relevant to these challenges is inadequate to allow widespread implementation of IWM. That is to say, there is not an adequate base of farmer knowledge to guide implementation of IWM via fine-tuning approaches. In addition, we claim that an adequate knowledge base is lacking among the farmers' "collegial professions," i.e., those professions that are necessary to a social infrastructure that can effectively support IWM implementation. All elements of these knowledge bases need expansion, and the integration among these elements needs to be intensified. How can this knowledge expansion occur? We argue that existing *knowledge production processes themselves* are decidedly inadequate for the purpose. In particular, the central source or "expert-model" approaches (Biggs, 1990) are not suitable. These approaches view research scientists in centralized research facilities as the preeminent source of relevant knowledge. This is problematic because such approaches do not produce a broad enough range of knowledge forms. For example, the need for innovation and knowledge production by extension workers is not recognized in central source approaches, which see agricultural science knowledge as more important than other knowledge forms.

In recognition of the primary role of farmers in meeting the challenge of developing site-specific IWM, a set of strategies that can help farmers make progress on that challenge is presented first. These strategies are based on a conception of the farming operation itself as the basic instrument of weed management.

Next, we introduce a knowledge production process—termed *coordinated innovation*—that can create the enlarged knowledge base and social infrastructure needed to move IWM across implementation barriers. Such a process is decentralized and cooperative, and based on the collaborative learning of interacting professionals. We focus on cross-sector learning groups, which provide effective settings for coordinated innovation. Guidelines, principles, and practices are proposed for learning-group participation by professionals crucial to IWM implementation—farmers, extension workers, and research scientists—based on the experiences of ourselves and others.

APPROACHES FOR DEVELOPING SITE-SPECIFIC IWM

As noted, farmers have a primary role in meeting the challenge of developing site-specific IWM. Recently, a set of concepts and strategies has emerged that can guide planning, implementation, and refinement of site-specific IWM systems. The most important idea is that many or most farming operations—e.g., tillage, fertilization, irrigation, and spatial and temporal crop rotations—can be planned and practiced in ways that take account of weed ecology. As noted, we can speak of this as a fine-tuning of farming operations. The desired outcome of this fine-tuning process is that these farming operations will then become an instrument of IWM.

One key consideration is how to fine-tune farming practices so as to achieve *agroecological selectivity,* i.e., promotion of crop growth and yield while disadvantaging undesirable weed species (Fernholz, 1992; Bender, 1993; Liebman, Mohler, and Staver, 2001). For example, early-emerging weeds and ridge-tillage cultivation operations were used to select against large-seed dicotyledonous weeds that were problematic in a low-input conservation-tillage cropping system (Thompson, 1991). In general, the sequencing of cash and cover crops and the nature and timing of fertilizer, irrigation, and tillage application all have strong effects on weed community dynamics (Liebman, Mohler, and Staver, 2001). Therefore, each of these management factors will likely afford some opportunity to fine-tune the farm operation for IWM purposes. For instance, manipulations of nutrient cycling by nutrient application and tillage, say by using organic N sources or high C residues to reduce early-season nitrogen availability, may have strong effects in reducing the growth and reproduction of certain problematic weed species (Dyck and Liebman, 1995). It is likely that the spatial and temporal placement of these nutrient-cycling and tillage interventions will be critical to their contribution to IWM (Nordell, 1993).

A second key consideration is gaining potential ecological benefits from certain weed species or communities while reducing costs from weeds to tolerable levels. Under certain circumstances, weeds can provide a range of ecological benefits (Jordan and Vatovec, 2004) including soil protection, nutrient conservation, and provision of habitats and resources for a variety of beneficial biota. In order to increase benefits and reduce costs of weeds, IWM should aim to create a weed community that consists of a series of weed "patches" that appear at opportune locations in space and time. Of course, this is an ideal which can never be fully achieved and which will require ongoing management. However, recognition of this goal suggests that fine-tuning of farming operations should aim to "confine" the weed community largely to spatial/temporal patches where the weeds produce ecological benefits.

The value of manipulating spatial and temporal placement of weed patches for ecological benefit is well illustrated by the use of weed patches to guide the population processes (Andow and Hidaka, 1989) of arthropod pests and their natural enemies (Bugg, 1992). By considering factors such as relative effects of weed patches on size, dispersal, and resilience to disturbance of populations of both herbivorous arthropods and their natural enemies, it is possible to reduce pest abundance in crops during vulnerable periods of crop growth and development (Bugg, 1992; Barbosa and Benrey, 1998; Barbosa and Wratten, 1998). Manipulations of temporal distributions include modifying weed phenology (by stimulating weed emergence or by destroying weed patches) to attract pests and beneficials only at opportune times. For example, weed patches that have attracted pests might be destroyed at the most vulnerable pest life-cycle stages, or permitted to flower at times that will keep pests away from crops (Bugg, 1992). Spatial manipulations include planting weed strips as barriers to pest dispersal or as refuges for natural enemies (Bugg, 1992).

At this point, we propose the following as important considerations in designing and implementing IWM that aims to confine weeds to patches where they produce net benefits.

1. *The spatial and temporal location of weeds is determined by the set of ecological benefits that is desired from weeds in a particular cropping system.* The particulars of a farming operation will determine where, when, and how weeds might feasibly provide ecological benefits. For example, certain weeds may be desired to improve the nutritional quality of corn harvested for silage production, to suppress other less-desirable weeds, to support specific beneficial organisms, or as com-

ponents of permanent vegetated walkways in horticultural crop production.

2. *Presence of weeds along ecological boundaries may often provide benefits.* Ecological boundaries (so-called ecotones at various scales) are locations where movement of organisms, energy, and materials are controlled and concentrated (Risser, 1995). For example, situating weed patches at field margins may increase landscape connectivity, allowing a variety of beneficial organisms, e.g., natural enemies of pest insects, to disperse more readily, thus helping to sustain and increase populations of these species (Fry, 1994). These patches may have minimal cost. To date, studies of weed populations in field margins have generally found that these populations do not increase within-field weed control problems (Blumenthal and Jordan, 2001).
3. *Refuges from various disturbances may help maintain desirable weed species.* Studies of a variety of plant communities are demonstrating the importance of so-called biological legacies to the maintenance of ecosystem functioning in response to disturbance (Perry, 1995). Agroecosystems are subject to many forms of disturbance on an ongoing basis, e.g., tillage, harvesting, and nutrient inputs. Biological legacies are components of biological diversity that survive a disturbance event and subsequently provide some benefit to an agroecosystem (Perry, 1995). Legacies persist through disturbance only if some form of protective refuge exists in the agroecosystem, within which necessary biodiversity is sufficiently sheltered from disturbance to be able to exert desired effects on ecosystem functioning that can continue after disturbance. Accordingly, an important consideration in IWM is creation of refugia from disturbance. In agroecosystems, refugia and biological legacies have received relatively little research attention, but clear examples exist, such as conservation-tillage methods that provide semipermanent protection from disturbances to certain soil sectors (e.g., ridge tillage) (Jordan, 1993) and deliberately sown weed strips (Nentwig, Frank, and Lethmayer, 1998).
4. *Data-poor and active adaptive management are necessary to develop IWM.* In principle, successful implementation of IWM is indicated by a trend of decreased harm from weeds, decreased damage to the agroecosystem from weed management (e.g., damage to soil quality from overuse of mechanical weed control methods), and increased levels of ecological benefits from weeds. Unfortunately, accurate direct measurement of agronomic costs and benefits of the weed flora of a particular agroecosystem is impossible in practice. Therefore, we believe that the only viable approach is to use practical on-farm moni-

toring techniques along with data-poor and adaptive management methods (Johannes, 1998). Data-poor management recognizes that detailed prediction of ecosystem dynamics (e.g., weed population and community dynamics) is almost never cost-effective and may be impossible in principle (Johannes, 1998). Instead, it aims to develop broad rule of thumb guidelines for precautionary use of particular management practices. Precautionary management seeks adequate—but not necessarily maximum—production levels while emphasizing protection of an agroecosystem's resource base, e.g., soil quality and beneficial biodiversity such as natural enemies of pests (Altieri, 1999). By definition, data-poor guidelines require only moderate amounts of quantitative information about the agroecosystem to guide their application. Such guidelines typically indicate how management actions should be taken in time and over space (Johannes, 1998). For example, current evidence suggests that certain weeds can have strongly valuable effects on the symbiosis between crops and arbuscular mycorrhizal fungi (AMF) (Jordan and Vatovec, 2004). This evidence suggests a rule of thumb that weed control practices should aim to leave modest residual populations of weeds that host these fungi and that this practice will maintain AMF communities that are strongly beneficial to crop growth.

Active adaptive management techniques (Shea et al., 2002) involve linking management to an active process of inquiry into the outcomes of management. Active inquiry entails intentional appraisal of important uncertainties regarding management and its outcomes, analysis of how these uncertainties could be lessened, and then use of these considerations in planning of management. Management processes and outcomes are then monitored to gather information needed to reduce uncertainty, and this information is then used to refine management practice and plan further active inquiries. Practical, cost-effective on-farm monitoring techniques are essential for necessary data-poor and active adaptive management. Such techniques have been developed for soil quality monitoring (Roming, Garlynd, and Harris, 1996). Weed monitoring could proceed by observing and mapping patterns of weeds and noting their relative abundance on an ongoing basis. There must also be some means of assessing benefits associated with weed patches, probably through various easily observed indirect measures, e.g., noting trends in soil quality attributes, or practical monitoring that farmers can use to apply the described principles to develop and refine site-specific IWM. One valuable approach is careful observation and mapping of spatial and temporal patterns of relevant factors. "Pattern analysis" (Conway and Barbier, 1990) is a systematic approach to such mapping. Useful focal points for the observation, mapping, and experimental fine-tuning of management practices include patterns of spatial/temporal diversity in the cropping system (Liebman and Dyck, 1993), applications of factors that appear to affect soil quality and biotic integrity, including large synthetic fertilizer inputs and intense soil disturbance (Bethlenfalvay and Linderman, 1992), and the coincidence of time/labor shortages with critical periods in weed life cycles. Mapping is also helpful in considering when and how weeds provide habitats or other resources to valued biodiversity in an agroecosystem (Bugg, 1992; Zavaleta, Hobbs, and Mooney, 2001). Observation, mapping, and fine-tuning of management practices must encompass both the within-farm scale (e.g., addressing weeds in field margins or in permanent walkways) and the scale of the landscape that embeds that farm (e.g., addressing hedgerows or other nonfield landscape elements) (Fry, 1994).

Of course, this agroecosystem tuning approach is constrained by lack of site-specific knowledge, time and labor demands, and other farm management challenges. However, it is important to recognize that the disciplines and habits of mind of farmers who practice agroecosystem tuning are likely to be valuable to farmers for other reasons. For example, mapping and other pattern analysis techniques are likely to be essential to developing site-specific practices for integrated soil management to maintain soil quality and fertility and increase the cost-effectiveness of nutrient applications. Farmers are also likely to require these same disciplines and habits of mind to respond to mounting regulatory pressure on them to reduce environmental impacts of their farming operations (e.g., N emissions to surface waters).

A CONCEPTUAL MODEL OF COORDINATED INNOVATION IN SUPPORT OF IWM IMPLEMENTATION

We will introduce our discussion of coordinated innovation for IWM by considering principles for organizing and supporting such innovation processes. Coordinated innovation and learning as means for developing ecological approaches to agriculture and natural resources management have received much practical and scholarly attention in recent years. A central premise is the idea that coordinated innovation arises from situations that enable social or collaborative learning, i.e., collective cognitive and learning processes that occur at the scale of a group (Jiggins and Röling, 2000). Social learning requires that all group members contribute their perceptions and experiences to development of their individual and/or joint understanding of complex situations. Social learning processes are now seen as crucial to significant and effective work on complex problem situations such as IWM implementation (Lappe and DuBois, 1994; Woodhill and Röling, 1998; Steins and Edwards, 1999; Daniels and Walker, 2001).

In its most ideal form, social learning leads to this outcome via the interplay of many different kinds and sources of knowledge, cultural perspectives, and ethical principles (Steins and Edwards, 1999; Daniels and Walker, 2001). In principle, such social learning is more effective in addressing complex problems than less-integrative knowledge-creation approaches, e.g., scientists working in isolation from other knowledge forms (Funtowicz and Ravetz, 1993; Holling, 1995; Röling and Maarleveld, 1999; Blumenthal and Jannink, 2000; Hesterman and Fisk, 2000; Ison et al., 2000; Ison and Russell, 2000; King, 2000). This superiority is seen to result from taking effective account of the interwoven resources and relationships needed to create and sustain a significant transformation in agricultural practice in a given local setting (Lappe and DuBois, 1994; Engel, 1997; Campbell, 1998; Scott, 1998; Röling and Jiggins, 1998). Such a setting will have a unique profile of historical, cultural, and geographical dimensions that must be considered in order to carry out a significant change, such as widespread implementation of IWM. Guided by these precepts, situations and settings that enable effective social learning and coordinated innovation are being developed on many fronts. In each, efforts are underway to develop specific means of social learning by a group of people who have a common interest in some agricultural or natural-resource issue, often through participatory or collaborative projects (Rocheleau, 1994; Röling and Wagemakers, 1998; Blumenthal and Jannink, 2000; Cerf et al., 2000). For example, one notable area of active practical and theoretical work is the development of forums and other settings for creation of improved management approaches for multiple-use common pool resources, such as watersheds, nature preserves, forests, or fisheries. These settings are intended to enable social learning about the resource by members of various different, sometimes conflicting, user groups (Campbell, 1998; Steins and Edwards, 1999; Hesterman and Fisk, 2000; William, 2000). Pest management examples include Indonesian farmer field schools (Röling and van der Fliert, 1998), areawide cotton pest management schemes (Smith, 1998), and Central American participatory IPM and IWM programs (Nelson, 1994; Shaver, 2001). Clearly, promising cases abound, and the following principles for organizing and supporting social learning and coordinated innovation processes have emerged (Engel, 1997).

1. *Certain developmental stages should be recognized in organizing these processes.* To begin, individuals or organizations must perceive

critical gaps in their experience or knowledge, and be willing to enter interactive relationships to form a broader and deeper understanding of a common problem and develop possible solutions (Kline and Rosenberg, 1986; Moss Kanter, 1989; Engel, 1997). A stage of planned activism ensues, in which individuals and/or organizations intentionally make contact with others to assess their interest in forming relationships. Then, probing occurs, in which individuals exchange views of the situation and how it might be improved. Next, narrowing is needed, in which ideas about the ends and means of the network converge enough to enable effective collaboration, At this point, collaborative structures such as resource coalitions (agreements to pool or exchange resources) and communication networks (situations in which social learning can occur) will typically be formed. Finally, to sustain learning and innovation processes, ongoing mutual adjustments among members must be negotiated and implemented.

2. *Organizing and facilitating work are indispensable to collaborative learning and innovation processes.* Potential participants are likely to have more or less divergent ideas about what is problematic about a situation and how it could be improved (Checkland and Scholes, 1999). To benefit from collective cognition, it is important that dialogue between divergent viewpoints occurs *before* convergence to a shared idea of how to improve the situation and organize useful learning/innovation processes (Engel, 1997; Checkland and Scholes, 1999). Skillful organizing and facilitation are essential to support the creative tensions inherent in such situations (Daniels and Walker, 2001). The tensions are exemplified by conflicts between land managers and scientists with regard to standards of evidence, timetables for arriving at conclusions, and methods for quality assurance about knowledge (Raedeke and Rikoon, 1997). As reported next, we found our learning groups effective in managing these tensions, although participating extension workers noted that skillful facilitation was required to do so. It is crucial to create and support an effective organizing "agency," i.e., a formal or informal organization that takes responsibility for bringing a diverse range of participants together for learning/innovation processes. Fortunately, a variety of tested tools are available for these purposes. One is the soft-systems methodology (Checkland and Scholes, 1999; developed by Engel, 1997, as "rapid appraisal of agricultural knowledge systems"). A second is civic organizing (Michels and Massengale, 2003), which applies community-based organizing methods to the broader-scale challenge of organizing collaboration among multiple professional sectors. Finally, mediation methods

have shown much promise in addressing conflicts between different sectors, which can be a major barrier to network formation (Daniels and Walker, 2001). Of course, extension workers represent a major resource for the necessary organizing and facilitation processes. For many extension workers, assuming these roles will require significantly different ways of working (Ison and Russell, 2000). Many difficulties will arise in organizing and facilitating networking and collaborative learning among a range of professionals who are at best not accustomed to such activities, and at worst have a history of conflict (Daniels and Walker, 2001). The work is time-consuming and may seem unproductive because of long lead times. Methods for evaluation and quality assurance of this sort of work are poorly developed. Despite these and other difficulties, extension workers in many different situations around the world are experimenting with organizing/facilitating practices (Campbell, 1998; King, 2000; Ison and Russell, 2000). Extension educators who are participating in our experimental knowledge network have found their experience to be very gratifying and fulfilling (Jordan et al., 2000). In particular, they report that extension work in this mode can alleviate some major professional frustrations that they have experienced with technology-transfer approaches.

3. *Social learning processes are most effective within the context of intentional, ongoing learning groups.* Collay and colleagues (1998) offer a conceptual model of such groups. Groups initially begin with a stage of construction, including agreements about goals, relationships, and governing principles. Second, new knowledge is constructed from individual and/or collective experiences. For this purpose, practical learning settings for so called discovery learning are very valuable, wherein learning can take place through shared experiences, observations, and experiments (Röling and Jiggins, 1998). Social learning is promoted by sincere and open deliberation within the group, by documented and shared reflection on experience, and by evaluation of progress toward shared goals.

We now describe our experiences with a set of experimental learning groups that were designed to support IWM implementation by enabling collaborative learning and innovation among farmers, extension workers, and research scientists.

RESULTS FROM EXPERIMENTAL IWM LEARNING GROUPS

First, we briefly describe the source of these tentative recommendations: our experimental learning group project. Our work began in 1997 and continues now in expanded form (Jordan et al., 2000, 2003). During summer 1998, two experimental groups were organized and began to meet in November 1998. One group addressed soybean production, the other vegetables, and each had 18 members: 3 research scientists, 2 extension workers and 13 farmers or farm management advisors. Group membership was stable; a few farmers left the groups or joined them in progress. Each group met five times at monthly intervals between November and April of 1998-1999 and 1999-2000, once at a member's farm during summers of 1999 and 2000, and twice in winter 2000-2001. Activities in each group were planned through discussion during the initial meetings of each year.

After this initial planning, the groups turned to discussions of weed problems facing the farmers and their current approaches to weed management. The groups then agreed to devote several meetings to the biology and ecology of these weed species and to experiences with their management, successful and otherwise. In these meetings, weed science discussed research findings relevant to the weed ecology issues identified by the group. These research discussions generated extensive dialogue among group members. Dialogue generally took the form of probing for insights from the day's theme into how the farm as a whole presented both *opportunities for* and *challenges to* integrated weed management. For example, in one meeting a weed researcher described research insights into sources of risk in weed management, such as that arising from reliance on mechanical weed control. This presentation spurred a long discussion about other sources of risk and ways of assessing and reducing risk by fine-tuning farm operations. Another meeting was devoted to the subject of systematic record keeping and monitoring in farming, with farmers taking the lead in presenting their approaches to the group. Still another focused on beneficial soil biota, interactions between weeds and these biota, and IWM implications of these interactions. Several meetings were focused on particularly difficult weed species, e.g., certain perennial weeds. At the final meeting in each year, each farmer identified an on-farm project investigating some aspect of weed management on their farm and presented their plans to the group. These projects involved monitoring, record keeping, and deliberate evaluation of some aspect of weed control.

The following sections discuss experiences of each of a set of professions in an IWM learning group.

Farmer Participants

Perceived Changes in Ability to Practice IWM

Almost all farmers reported that group participation had empowered them to become better practitioners of IWM, both by changing their basic understanding of weed management and by enabling new insights into IWM practice. An important theme in interviews was expression of an increased sense of self-reliance and self-confidence to manage weeds using on-farm resources. Many participants reported an increase in the value they placed on their own role and importance as skilled managers; therefore, they felt less dependent on outside inputs. A member of the vegetable learning group told us that she feels "more like a manager than I used to. I used to view myself as almost only a person that did the handwork on the hoe and then helped the kids pick." Moreover, many farmer participants reported perceiving a link between group participation and their sense of increased empowerment to practice IWM. Specifically, participants noted the influence of the groups on thinking and problem-solving approaches to IWM, and on specific management practices. An organic farmer reported that he "thought differently" after being a part of the learning group. He reported a new ability to "think like a weed." He described this as a way that he could manage weeds more proactively by making better predictions about weed behavior. He perceived that he "did a much better job handling the situations that the weeds presented" by being better able to cope with variable circumstances such as weather events. A vegetable farmer explained that she "thinks differently because I remember the talks on the biology of the weeds. I think differently, and more than that, I feel differently." Another reported that he "just totally thinks more about weeds. I even look differently at articles and notice things I never did before"; consequently, he perceived that his "entire thought process is different."

Perceptions About Collaborative Learning Processes and IWM

Many farmers related their perceptions of *how* collaborative learning processes had been instrumental to changes in their thought and actions regarding IWM. First, several participants noted that, prior to their group experiences, they had felt *disempowered* by competitiveness and lack of communication among farmers. As one vegetable farmer explained,

> Farmers are isolated from each other primarily around competition in the marketplace. You don't want to tilt your hand too much. We hold our cards in tight and we close ourselves in and it's in that isolation that we cut ourselves off from learning.

This feeling of isolation was reported as being common among group members. They spoke of "loneliness" and how other people have "walls up."

Participants reported feeling empowered because the group effectively alleviated these feelings of isolation and competitiveness. One stated: "I prioritized those meetings; that's how important they got to me. . . . I think we built camaraderie." He noted that he, an organic farmer, had formed close relationships with others who used very different weed management strategies. Group members reported gaining additional empowerment because they felt that being recognized as integral to a group learning process validated their own knowledge. One member spoke very emotionally about the group's effect on the way he viewed himself and his fellow farmer participants: "We are starting to honor the collective knowledge that farmers have and to treat that knowledge in a professional way. We are creative people by nature and that was very, very clear to me in this group." Another member explained how such validation created feelings of empowerment for him: "The group was all about acceptance and nonjudgment, support, validation, vulnerability, and humility. It gave me that positive push that I needed to keep being innovative." Still another reported, "[The members] establish a trust level, confidence in ourselves to bare our souls, you might say—not only on things that were successful, but failures as well."

Many participants commented on the importance of group diversity to the empowering outcomes of the learning groups. Included in each group were farmers with greatly varying farm sizes and who farmed with different practices including both "organic" and "conventional" approaches. As the groups went on, learning in a group of diverse management systems became important to the members because it allowed them all the opportunity to think together and come to "some type of common ground." One farmer reported that he had anticipated a "defensive posture on both sides" but found that largely absent. An organic farmer reflected: "Too many like things, you have a dull group. Too many opposite things and all you have is a big argument. So it takes a mix to make the group work."

In addition to a diverse range of farmers, group members included scientists and extension educator members. Farmer participants perceived that this additional diversity was essential to their learning process. One observed how the learning groups contrasted favorably with past experiences in which "PhDs were the experts and the farmers were the passive recipients of their knowledge." A farming couple noted their appreciation for the "mix of information and science to back it up and the opportunity for everybody to have dialogue around a particular scientific aspect of the problem." In their view, it was very important to this process that the discussion was not solely based on practitioner knowledge, or solely on scientific information.

University Extension and Research Workers

Learning-group experience appears to have strongly increased these participants' sense of empowerment to help implement IWM. In interviews, these participants described new understandings of the broad goals of their work and new professional capabilities that increased their perceived ability to attain these goals. They ascribed these new understandings and capabilities to their participation in the learning groups and in many cases related these outcomes to particular experiences in the groups.

Community-Based Extension Educators

All three educators found that the learning groups empowered them to make progress on several values that they held highly. First, each reported that the learning groups were a very effective vehicle for bringing diverse viewpoints together. One educator emphasized his commitment to creating programming that helped farmers gain skills for managing complex farm challenges; this goal required that he bring together people who viewed these challenges quite differently, because "they need each other" to make progress in building their skills. For another educator, it was of great importance to develop practical knowledge while also asking how does practice guided by that knowledge "play out in the rest of life, in community and in the broader world?" In his view, such practical knowledge was best created when principles and practices are "shaken through more screens"; these "screens" are contrasting viewpoints on an issue.

A second broad outcome reported by the extension educators is new understanding of the role of "expertise," be it their own or others'. In their previous extension work, each educator had viewed the knowledge of scientific or technical experts as largely sufficient to guide practice. In their experience with the learning groups, the educators found that scientific or technical expertise could function quite differently: as one ingredient in an interactive, interdependent process of knowledge construction. As one educator put it, "I think the best adult learning is where there's enough of a common base of knowledge and a recognition that people can learn a little bit in a half an hour to get them learning more through a conversation."

A second educator described his dissatisfaction with the use of expertise to prescribe practice thusly:

> If I think that I had five hundred people in a room and I taught them some particular concept about dairy nutrition or weed control and that they all went home and made great strides with it, then I'm kidding myself in most cases.

This educator was also strongly conscious of the role of farmers' expertise, and how to promote and elicit this expertise. He was particularly concerned with protecting and sponsoring the contributions of farmers who held a minority viewpoint in his group: two soybean farmers who had large-scale farms, making heavy use of pesticide and synthetic fertilizers. He learned, over the course of the learning groups' meetings, techniques for including the knowledge of these farmers in the group's discussions.

Campus-Based Extension/Research Faculty

Both of these faculty members reported that their learning groups' experience empowered them to better stimulate and support transformational change in how farmers thought and acted about weed management. These two faculty members had mixed responsibilities: to create and teach educational programming regarding weed management, and to do research in support of these programs. Over their careers, both had experienced major problems with technology-transfer approaches, and each reported that the learning groups provided a setting where they could develop and refine alternatives to these approaches. One had shifted from updating farmers on herbicide developments to a much broader focus on integration of multiple methods of weed control. In recent years, he had concentrated on helping his audience gain key insights that they could apply to the ongoing design and management of their farm operations. For example, one emphasis has been on the exponential growth process of weed seedlings, with its phase of extremely rapid growth when weeds can very quickly grow into and beyond controllable size ranges. His intention is to motivate and enable farmers to begin to plan and manage their labor schedules with an eye to the exponential growth phases of weeds. He has found that his pedagogical program of "opening eyes" to characteristic features of weed growth and other aspects of weed ecology to be powerful and well received. He reported that he valued the learning groups as a safe setting to develop and refine this pedagogy, where he could "hone skills" and test his "ideas and suspicions" about what audiences might need, and discover "new possibilities" for programming that creates awareness about the interplay of weed management and weed biology.

The other faculty member had a different unresolved problem in his extension practice. Over his career, he found himself faced with expectations to provide detailed weed management recommendations for an increasingly broad range of managed ecosystems and ecosystem managers, which presently range from horticultural crops to wetlands to pastures to roadside areas. He found these expectations impossible to satisfy: little relevant research had been done, few herbicides were available, and relatively few effective weed control options of any sort were known. Given this deficiency of knowledge, this faculty member had grown increasingly uncomfortable with acting in the role of "expert." Moreover, he perceived a deeper problem with functioning as an "answer provider." When he played that role, he feared that his extension clientele does not develop—or even see a need for—a capacity to "think differently about problems" and therefore to "pull more resources in" to contribute to improving problem situations. Even when relevant weed management information is available, he has concluded that "handing information to them" is often undesirable, because such information is frequently "not used effectively." In contrast, he reported that working in his learning group led to "giving farmers tools so that they can approach problems in different ways and are able to deal with the next problems on their own."

Campus-Based Researchers

Both of these faculty members reported that the learning groups had empowered them to make significant shifts in the objectives and methodologies of their research programs. Over time, each had become convinced that their research programs must expand to address the problem of creating a knowledge base for sustainable agriculture practices (e.g., IWM). For both, this represented a significant transformation in their research; previously, they had emphasized crop physiology and plant ecology, respectively.

One researcher has recently redirected his research effort to test the value of learning theory for understanding and facilitating farmer learning. He had been particularly motivated by a perception that much agronomy research was not accepted or noticed by farmers because of a failure to account for the thinking and learning of farmers. He now focuses on collaborating with colleagues in other disciplines to better understand farmers' learning. In this work, he has noticed that he, as an agronomist, can "hear things" in work with farmers that colleagues in other disciplines cannot.

The second researcher reported that he had experienced a major shift in his scholarship because of his learning group experience. He reports that his work with the groups has lead him to work with a new set of colleagues in several disciplines to develop a working model of agricultural science that is more directly engaged with complex problems—such as IWM implementation—that require integrating the work of multiple professions. At the heart of this model is scientists' participation in multiprofessional groups that engage in ongoing collaborative learning and action to address a complex problem. In this model, this group's activities directly organize and structure scientific work. Thus, this researcher reported a major outcome from group participation: expansion of his research program into an entirely new domain.

Finally, we will present a set of suggested practices and guidelines for organizing, facilitating, and participating in IWM learning groups. These recommendations are the professional opinions of our participants; they are necessarily subjective and rooted in the context of our work.

RECOMMENDATIONS FOR PROFESSIONAL ROLES AND PRACTICES IN IWM LEARNING GROUPS

Farmers in Learning Groups

Certain farmers are recognized as especially skillful and successful in creating ecologically sound practices for the land that they manage. These individuals often approach the ideal of using the farm, ranch, or forestry operation as a whole as an instrument of weed management (Thompson, 1991; Fernholz, 1992; Bender, 1993). These operations and the ideas and methods of these farmers can fuel social learning by providing insightful models and case studies (Crookston, 1992) of management and decision making in complex situations. These farmers—and farmers collectively—possess a rich store of ideas and experiences for a learning group to examine, test, and develop. For example, in the weed management learning group, several farmers had well-developed, experience-based theories about weeds. We found great value in giving these members opportunities to explain their knowledge and perceptions, because these expressions were particularly generative of dialogue that identified opportunities and obstacles to achieving IWM by fine-tuning of farm operations.

Roles of Extension Workers in a Learning Group

We offer a number of guidelines, recommended from the experience of participating extension workers in our experimental groups.

1. *Emphasis must be placed on organizing learning groups.* As noted, dedicated organizing is crucial to identify participants in developing a learning group and to assure them that they will have willing and like-minded partners in a group. Our experimental learning groups have had the benefit of a dedicated organizer, and the following four principles for practice emerge from her work.

First, organizing work is often seen as the province of persons with certain skills for "working with people"; knowledge and credentials in a scientific or agricultural discipline have not been seen as necessary. However, our participants believe that our organizer's status as an experienced and recognized weed-science researcher, who had carried out many research projects on the farms of cooperating farmers, and her previous experience working in horticultural extension projects gave her abilities that were crucial to her effectiveness as an organizer. She believes that she was able, through her familiarity with the cultures of farmers, extension workers, and researchers, to effectively dialogue and communicate regarding the insights, interests, and constraints expressed by potential participants. In particular, she feels that she could listen with a well-tuned ear for issues and concerns of importance to potential participants that were not overtly expressed in initial conversations. In our experience, her qualities of multicultural awareness and sensitivity were very important to her efficacy as a bridging person who could organize professionals across major lines of difference. We believe that organizing work is typically underappreciated and undersupported in most extension and research organizations, and that it is crucial to respect, appreciate, and invest in building an organizational capacity for organizing.

Second, this multicultural awareness was seen as invaluable in assembling a diverse group that was able to make good use of that diversity. As noted, effective social learning is strongly dependent on exchanges across lines of difference. However, many participants reported experiences in diverse groups that were highly polarized and antagonistic. In our experience, taking time for careful one-on-one conversations with potential participants is crucial for identifying persons who are able and willing to work well in a highly diverse group.

Third, one crucial part of effective recruiting to learning groups is willingness to be honest about the open and indeterminate nature of such groups, and an ability to tolerate uncomfortable interpersonal interactions that may sometimes result from such frankness. In our experience, if potential participants are not fully informed about the unpredictable nature of social-learning processes in a highly diverse group, then they cannot make an informed decision to participate. Most important, they are unlikely to tolerate the early, relatively disorganized phases of the social-learning process, in which the group's ground rules and agenda are emerging, and participants are overcoming discomfort with speaking candidly about their experience, concerns, and insights. Therefore, an important issue in organizing is to be candid about these uncertain courses and outcomes of learning groups. In general, we have found that organizers have to be willing to tolerate substantial levels of ambiguity and interpersonal tensions. Such tensions arise from a variety of sources. Examples include situations in which individuals that represent "minority" perspectives in a group feel suspicious and doubtful that their concerns and perspectives will be respected, or when participants expect a role in determining the work of the group but in practice delegate these roles to project organizers and facilitators.

Finally, in our experience, ongoing organizing work is very valuable to the social-learning process itself. A number of the extension workers reported that they were very appreciative of how, through regular one-on-one conversations as part of ongoing organizing, we had enabled them to practice critical reflection on the learning groups and their own roles in them. We strongly recommend such conversations with all participants, conducted by the project organizer. In these conversations, each participant was asked to evaluate the work of the learning groups and articulate their insights regarding how the groups should work. Participants expressed a strong belief that this approach accelerated their personal learning processes as well as the social learning of the group.

2. *Facilitation is key to effective operation of learning groups.* Our participating extension workers have found that learning groups addressing implementation of IWM can be very successfully facilitated despite their initial feelings of inadequate knowledge of weed science. Relatively small amounts of technical knowledge and experiential learning (e.g., gained by farmers by conducting on-farm experiments) can enable extensive collaborative learning. They have found that "you do not need to know how to answer every question." Rather, what is important is their ability to introduce fundamental and provocative aspects of weed ecology into the social learning process, e.g., the exponential growth and seed-bank phenology issues mentioned previously. It is also necessary for extension workers to know how to facilitate use of "co-learning tools" (Hamilton, 1998) based on weed ecology. For example, we used computer software that demonstrates variability in weed seed-bank behavior as a function of weather (WEEDCAST, Archer et al., 2001). After introducing these elements into the group's work,

our participants found that it was very important to restrict and limit one's own comments during group work, even when this led to uncomfortable moments of silence and seemingly disorganized group process. Participants also report that the availability of consulting and training support from weed-science researchers in using these ecological fundamentals and co-learning tools has been essential to building their confidence, although in practice they have not needed to draw heavily on this support. Therefore, extension workers can facilitate learning groups with only moderate time commitments. Technical experts (e.g., researchers) can support groups by creating technical "modules" that assist in collaborative learning by effectively conveying simple but eye-opening concepts (e.g., the exponential growth of weed seedlings).

Participants have also found that the ability to create and facilitate shared experiential learning is an important skill. Such learning was seen as key to a valuable group process, in which participants report and reflect on their experiences to the group, followed by open dialogue leading to practical conclusions, leading in turn to a new round of experiences informed by these conclusions. Situations that were found to be conducive to such learning included group visits to participants' farms and formal "report backs" of outcomes from farmer management "inquiries," in which farmers carried out a planned examination or observation of some aspect of their farming operations that was relevant to IWM. Participants noted that group composition and ground rules were important to such learning; candid and informative reporting and reflection cannot occur in an atmosphere where participants fear hostile or judgmental responses from fellow group members. Especially, learning-group participants must feel a sense of safety and trust in order to share knowledge (e.g., of experiences with failure) that is critically important to collaborative learning. Therefore, facilitation must emphasize creation of such "safe" learning environments. For these purposes, participating extension workers were strongly impressed with the importance of ground rules, practices, and rituals (e.g., informal conversation while sharing food) that support and evoke values of tolerance, respect, outgoingness, and open-mindedness among group participants. Our participants acknowledge that to create and practice these values effectively required that they actively cultivate in themselves an attitude that every participant has some "credible" contribution to offer to the social learning of a group.

One important justification for the value of tolerance is that our participants found that "minority" members of diverse groups were suspicious and fearful that the group as a whole was not open to hearing their views. Extension participants therefore found it necessary to be highly proactive in pointing out how the contributions of minority group members connected to the groups' discussions. Moreover, our participants found that feelings of discomfort and inhibition were not confined to minority members of groups. Rather, they observed that most farmer members were reticent about revealing the ways in which their farming is "not pretty, is not a showcase," fearful of being asked questions they could not answer and uncomfortable about seeming to claim undue expertise and authority while reporting their accumulated knowledge and experience.

A very important and challenging aspect of group facilitation is the "staging of time-outs when you reflect and put into words" concepts and conclusions that are emerging from the work of the group. This must be done inclusively, and it requires considerable skill and practice to integrate seemingly diverging comments and perspectives that are offered by different group members. One strategy that may be effective for this purpose is to give participants an opportunity to reflect on interrelations that they perceive among their basic premises and values, the farming practices that they use, and the outcomes of those practices. These "time-outs" for making sense and identifying practical conclusions from group work are particularly important for retaining group members who may feel uncomfortable with the ambiguity and lack of definitive focus that can easily evolve when members of a group are candidly sharing diverse perspectives on a common theme.

Roles of Academic Weed Scientists in Learning Groups

For academic researchers, the challenge is to find ways to do weed science (in the most inclusive sense) in the context of knotty, ill-defined problems such as implementation of IWM, while still upholding roles, norms, and values that are crucial to the integrity of scientific research. In ecological science generally, sustainability concerns have spurred proposals for new forms of scholarly practice (Ison et al., 2000). Examples include "agroecosystem analysis" (Conway and Barbier, 1990), "postnormal science" (Funtowicz and Ravetz, 1993), "second-order R&D" (Ison and Russell, 2000), "ecological knowledge systems" (Röling and Jiggins, 1998; Röling and Wagemakers, 1998; Cerf et al., 2000), and certain aspects of "complexity theory" (Bradshaw and Bekoff, 2001). These proposals share a crucial premise: science is most helpful in meeting complex practical challenges when it is practiced in cooperation with other ways of knowing and reasoning (e.g., ethical, artistic, or spiritual rationalities) and supports the development of a coherent, shared, and effective worldview (Bradshaw and Bekoff, 2001). We offer a working model that integrates these sources with our own proposals, in the hope of assisting practical and conceptual development of guidelines for the practice of weed science in the broader context of a learning group (Peters, Jordan, and Lemme, 1999).

The heart of our model is the notion that the researcher is a full-fledged and ongoing participant in one or more IWM learning groups. We distinguish such participation from a more consultative role in which the scholar has some limited interaction with a group but then withdraws. To practice weed science in this manner, the researcher forms a research agenda in response to questions that emerge in the social learning of the group(s), and then contributes research findings to fuel further social learning by the group(s). We term such research "engaged science" or "public scholarship" (Jordan et al., 2002). Ison and Russell (2000) argue that the ultimate purpose of this form of science is to motivate collaborative innovation and action by a learning group. Thus, scientific research gains value by supporting social learning needed to create shared understanding and enable innovation and action by a group. In our view, ongoing participation in social-learning processes is necessary to allow the researcher to develop a research agenda that will fulfill this purpose. Equally, ongoing participation in social-learning processes is needed to draw forth the full meaning and implications of research findings (Chamala, 1995).

Our experience with our experimental learning groups is illustrative of the role and function of research and researchers. At the outset, we believed that weed scientists would impart carefully selected expert knowledge, e.g., concerning weed biology or management tools, which would be assimilated and adopted by the farmers. Although comments from the farmers suggest that this occurred to some extent, we found that our knowledge functioned quite differently. We found that relatively modest offerings of research-based knowledge served to elicit lively group discussions and co-learning. This experience suggests the importance of a different professional role: scientists making use of their knowledge to help the group pose—for itself—rich, evocative questions that "create curiosity, wondering and anticipation" (Brown and Bennett, 1995). Such questions are the essence of dialogue across lines of difference, social learning, and collective cognition (Brown and Bennett, 1995). Therefore, we recommend that weed scientists should consider their full participation in learning groups as a legitimate and highly important professional practice.

We note that, in our model, scientists are free to pursue *any* researchable question, as long as their work is an agreed-upon part of the social learning of a group. Thus, the entire scope of our discipline, from molecular studies to landscape ecology to studies of farmer learning and decision-making processes, may be appropriate, depending on the questions that arise from social-learning processes. For example, one sort of research effort that may meet these criteria is to search for design principles—rules of thumb—that can be used by farmers who are trying to design IWM systems based on fine-tuning of farm operations. For example, it is clear that using plant diversity to develop and maintain soil quality can facilitate IWM in a number of ways (e.g., Liebman, Mohler, and Staver, 2001). However, the necessary level of plant diversity and the spatial and temporal pattern of its deployment in a cropping system are unclear. Such questions call for tests of whether broad generalizations, i.e., rules of thumb, exist or not.

Finally, we note that such engaged science should facilitate the social learning of a particular learning group, but it should also transcend the particular interests and concerns of that group, so as to have a broader range of application. For example, the researcher's work may be informative to an entire class of groups working on similar projects.

CONCLUSIONS

We believe that coordinated innovation driven by collaborative learning among scientists, extension workers, farmers, and other agricultural professionals is essential for extensive implementation of IWM. Moreover, we share a broader challenge with all agriculturalists: the development an agricultural knowledge system (Engel, 1997) that can meet the enormous demands that will confront agriculture over the coming decades. These demands will come from an expanding population that increasingly insists that agriculture not damage environmental quality and public health. Agriculture will need to respond and adapt against a backdrop of climate change and other aspects of global change. In many areas, resource shortages (water, energy, fertility inputs) are or will be severe. It is very important to critically assess knowledge production processes in agriculture and to consider how new knowledge can be created that will enable effective and efficient responses to the intense pressures that agriculture will face. We believe that learning groups offer a mode of knowledge production that can help make progress in meeting these great challenges.

REFERENCES

Altieri, M.A. (1999). The ecological role of biodiversity in agroecosystems. *Agriculture, Ecosystems and Environment* 74: 19-31.

Andow, D.A. and K. Hidaka (1989). Experimental natural history of sustainable agriculture: Syndromes of production. *Agriculture, Ecosystems and Environment* 27: 447-462.

Archer, D.W., F. Forcella, J.J. Eklund, and J. Gunsolus (2001). WeedCast, Version 2.0. Available online at <http://www.morris.ars.usda.gov/morris/products/products.htm>.

Barbosa, P. and B. Benrey (1998). The influence of plants on insect parasitoids: Implications for conservation biological control. In P. Barbosa (Ed.), *Conservation biological control* (pp. 55-82). New York: Academic Press.

Barbosa, P. and S.D. Wratten (1998). The influence of plants on invertebrate predators: Implications to conservation biological control. In P. Barbosa (Ed.), *Conservation biological control* (pp. 83-100). New York: Academic Press.

Bender, J. (1993). *Future harvest*. Lincoln: University of Nebraska Press.

Bethlenfalvay, G.J. and R. Linderman (1992). Preface. In G.J. Bethlenfalvay and R. Linderman (Eds.), *Mycorrhizzae in sustainable agriculture* (pp. viii-xiii). ASA Special Publication No. 54. Madison, WI: American Society of Agronomy.

Biggs, S. (1990). A multiple source of innovation model of agricultural research and technology promotion. *World Development* 18: 1481-1499.

Blumenthal, D. and J.L. Jannink (2000). A classification of collaborative management methods. *Conservation Ecology* 4(2): article 13. Available online at <http://www.consecol.org/vol4/iss2/art13>.

Blumenthal, D. and N. Jordan (2001). Weeds in field margins—A simulation analysis. *Weed Science* 49: 509-519.

Bradshaw, G.A. and M. Bekoff (2001). Ecology and social responsibility: The re-embodiment of science. *Trends in Ecology and Evolution* 16: 460-465.

Brown, J. and S. Bennett (1995). Mindshift: Strategic dialogue for breakthrough thinking. In S. Chawla and J. Renesch (Eds.), *Learning organizations: Developing cultures for tomorrow's workplace* (pp. 167-184).

Bugg, R.L. (1992). Using cover crops to manage arthropods on truck farms. *HortScience* 27: 741-745.

Campbell, A. (1998). Fomenting synergy: Experiences with facilitating landcare in Australia. In N. Röling and M. Wagemakers (Eds.), *Facilitating sustainable agriculture* (pp. 232-249). Cambridge, UK: Cambridge University Press.

Cerf, M., D. Gibbon, B. Hubert, R. Ison, J. Jiggins, M. Paine, J. Proost, and N. Röling (Eds.) (2000). *Cow up a tree: Knowing and learning for change in agriculture*. Paris: INRA.

Chamala, S. (1995). Overview of participative action approaches in Australian land and water management. In S. Chamala and K. Keith (Eds.), *Participative approaches for landcare* (pp. 1-42). Brisbane: Australian Academic Press.

Checkland, P. and J. Scholes (1999). *Soft systems methodology in action*. New York: John Wiley and Sons.

Collay, M., D. Dunlap, W. Enloe, and G.W. Gagnon Jr. (1998). *Learning circles: Creating conditions for professional development*. Thousand Oaks, CA: Corwin Press.

Conway, G.R. and E.B. Barbier (1990). *After the Green Revolution: Sustainable agriculture for development*. London: Earthscan Publications.

Crookston, R.K. (1992). Decision cases as a research tool in agriculture. In M.J. Stanford, D.H. Davis, and S. Simmons (Eds.), *Decision cases for agriculture* (pp. 24-28). University of Minnesota, St. Paul: Decision Cases for Agriculture, College of Agriculture, Program for Decision Cases.

Daniels, S.E. and G.B. Walker (2001). *Working through environmental conflict: The collaborative learning approach.* Westport, CT: Praeger Press.

Dyck, E. and M. Liebman (1995). Crop-weed interference as influenced by a leguminous or synthetic fertilizer nitrogen source: II. Rotation experiments with crimson clover, field corn and lambsquarters. *Agriculture, Ecosystems and Environment* 56: 109-120.

Engel, P.G.H. (1997). *The social organization of innovation.* Amsterdam, the Netherlands: KIT Press.

Fernholz, C. (1992). *Sustainable management practices for the nineties.* Madison, MN: A Frame Press.

Fry, G.L.A. (1994). The role of field margins in the landscape. In N. Boatman (Ed.), *Field margins: Integrating agriculture and conservation* (pp. 31-40). BCPC Monograph No. 58. Surrey, UK: British Crop Protection Council.

Funtowicz, S.O. and J.R. Ravetz (1993). Science for the post-normal age. *Futures* 25: 739-755.

Hamilton, G. (1998). Co-learning tools: Powerful instruments of change in Southern Queensland, Australia. In N.G. Röling and M.A.E. Wagemakers (Eds.), *Facilitating sustainable agriculture* (pp. 172-190). Cambridge, UK: Cambridge University Press.

Hesterman, O.B. and J.W. Fisk (2000). Catalyzing systems change through community-based learning. In M. Cerf, D. Gibbon, B. Hubert, R. Ison, J. Jiggins, M. Paine, J. Proost, and N. Röling (Eds.), *Cow up a tree: Knowing and learning for change in agriculture* (pp. 369-380). Paris: INRA.

Holling, C.S. (1995). What barriers? What bridges? In L.H. Gunderson, C.S. Holling, and S.S. Light (Eds.), *Barriers and bridges to the renewal of ecosystems and institutions* (pp. 3-34). New York: Columbia University Press.

Ison, R.L., C. High, C.P. Blackmore, and M. Cerf (2000). Theoretical frameworks for learning-based approaches to change in industrialized-country agricultures. In M. Cerf, D. Gibbon, B. Hubert, R. Ison, J. Jiggins, M. Paine, J. Proost, and N. Röling (Eds.), *Cow up a tree: Knowing and learning for change in agriculture* (pp. 32-53). Paris: INRA.

Ison, R. and D. Russell (2000). *Agricultural extension and rural development: Breaking out of traditions.* Cambridge, UK: Cambridge University Press.

Jiggins, J. and N. Röling (2000). Toward capacity building for complex systems management: Imagining three dimensions. In M. Cerf, D. Gibbon, B. Hubert, R. Ison, J. Jiggins, M. Paine, J. Proost, and N. Röling (Eds.), *Cow up a tree: Knowing and learning for change in agriculture* (pp. 429-440). Paris: INRA.

Johannes, R.J. (1998). The case for data-less marine resource management: Examples from tropical nearshore finfisheries. *Trends in Ecology and Evolution* 13: 243-246.

Jordan, N. (1993). Simulation analysis of weed population dynamics in ridge tillage. *Weed Science* 41: 468-474.

Jordan, N. (1996). Weed prevention: Priority research for alternative weed management. *Journal of Production Agriculture* 9: 485-490.

Jordan, N., R. Becker, J. Gunsolus, S. White, and S. Damme (2003). Knowledge networks: An avenue to ecological management of invasive weeds. *Weed Science* 51: 271-277.

Jordan, N., J. Gunsolus, R. Becker, and S. White (2002). Public scholarship: Linking weed science with public work. *Weed Science* 50: 547-554.

Jordan, N. and C. Vatovec (2004). Agroecological benefits from weeds. In Inderjit (Ed.), *Weed biology and management* (pp. 137-158). Boston: Kluwer Academic Publishers.

Jordan, N., S. White, J. Gunsolus, R. Becker, and S. Damme (2000). Learning groups developing collaborative learning methods for diversified, site-specific weed management: A case study from Minnesota, USA. In M. Cerf, D. Gibbon, B. Hubert, R. Ison, J. Jiggins, M. Paine, J. Proost, and N. Röling (Eds.), *Cow up a tree: Knowing and learning for change in agriculture* (pp. 85-95). Paris: INRA.

King, C. (2000). Moving from natural to systemic social learning through systematic reflection and dialogue. In M. Cerf, D. Gibbon, B. Hubert, R. Ison, J. Jiggins, M. Paine, J. Proost, and N. Röling (Eds.), *Cow up a tree: Knowing and learning for change in agriculture* (pp. 205-225). Paris: INRA.

Kline, S. and N. Rosenberg (1986). An overview of innovation. In R. Landau and N. Rosenberg (Eds.), *The positive sum strategy: Harnessing technology for economic growth* (pp. 275-306). Washington, DC: National Academy Press.

Lappe, F.M. and P.M. DuBois (1994). *The quickening of America: Rebuilding our nation, remaking our lives.* San Francisco, CA: Jossey-Bass, Inc.

Liebman, M. and E. Dyck (1993). Crop rotation and intercropping strategies for weed management. *Ecological Applications* 3: 92-122.

Liebman, M., C.L. Mohler, and C.S. Staver (Ed.) (2001). *Ecological management of agricultural weeds.* Cambridge, UK: Cambridge University Press.

Michels, P. and A. Massengale (2003). *Civic organizing framework.* Available online at <http://www.activecitizen.org/CivOrg.html>.

Moss Kanter, R. (1989). *When giants learn to dance: Mastering the challenges of strategies, management and careers in the 1990s.* New York: Simon and Schuster, Inc.

Nelson, K.C. (1994). Participation, empowerment and farmer evaluations: A comparative analysis of IPM technology generation in Nicaragua. *Agriculture and Human Values* 11: 109-125.

Nentwig, W., T. Frank, and C. Lethmayer (1998). Sown weed strips: Artificial ecological compensation areas as an important tool in conservation biological control. In P. Barbosa (Ed.), *Conservation biological control* (pp. 133-154). New York: Academic Press.

Nordell, E. (1993). A few long furrows on horsedrawn tillage. *Small Farmers' Journal* 17: 17-23.

Perry, D.A. (1995). Self-organizing systems across scales. *Trends in Ecology and Evolution* 10: 241-244.

Peters, S., N. Jordan, and G. Lemme (1999). Toward a public science: Building a new social contract between science and society. *Higher Education Exchange:* 34-47.

Raedeke, A.H. and J.S. Rikoon (1997). Temporal and spatial dimensions of knowledge: Implications for sustainable agriculture. *Agriculture and Human Values* 14: 145-158.

Risser, P. (1995). The status of the science examining ecotones. *Bioscience* 45: 318-325.

Rocheleau, D.E. (1994). Participatory research and the race to save the planet: Questions, critique, and lessons from the field. *Agriculture and Human Values* 11: 4-25.

Röling, N.G. and J. Jiggins (1998). The ecological knowledge system. In N.G. Röling and M.A.E. Wagemakers (Eds.), *Facilitating sustainable agriculture* (pp. 283-311). Cambridge, UK: Cambridge University Press.

Röling, N.G. and M. Maarleveld (1999). Facing strategic narratives: An argument for interactive effectiveness. *Agriculture and Human Values* 16: 295-308.

Röling, N.G. and E. van der Fliert (1998). Introducing integrated pest management in rice in Indonesia: A pioneering attempt to facilitate large-scale change. In N.G. Röling and M.A.E. Wagemakers (Eds.), *Facilitating sustainable agriculture* (pp. 153-171). Cambridge: Cambridge University Press.

Röling, N.G. and M.A.E. Wagemakers (Eds.) (1998). *Facilitating sustainable agriculture*. Cambridge, UK: Cambridge University Press.

Roming, D.E., M.J. Garlynd, and R.F Harris (1996). *Farmer-based assessment of soil quality: A soil health scorecard.* SSSA Special Publication 49. Madison, WI: Soil Science Society of America.

Scott, J.C. (1998). *Seeing like a state: How certain schemes to improve the human condition have failed.* New Haven, CT: Yale University Press.

Shaver, C.P. (2001). Knowledge, science and practice in ecological weed management: Farmer-extensionist-scientist interactions. In M. Liebman, C.L. Mohler, and C.S. Staver (Eds.), *Ecological management of agricultural weeds* (pp. 99-138). Cambridge, UK: Cambridge University Press.

Shea, K., H.P. Possingham, W.W. Murdoch, and R. Roush (2002). Active adaptive management in insect pest and weed control: Intervention with a plan for learning. *Ecological Applications* 12: 927-936.

Smith, J.W. (1998). Boll weevil eradication: Area-wide pest management. *Annual Entomological Society of America* 91: 239-247.

Steins, N. and V. Edwards (1999). Synthesis: Platforms for collective action in multiple-use common-pool resources. *Agriculture and Human Values* 16: 309-315.

Svejcar, T.J. (1996). What are working groups and why should scientists be involved? *Weed Technology* 10: 451-454.

Thompson, D. (1991). *Thompson on-farm research.* Greenbelt, MD: Wallace Institute.

William, R.D. (2000). Whole systems inquiry: Watersheds as integral systems for systemic learning and action. In M. Cerf, D. Gibbon, B. Hubert, R. Ison, J. Jiggins, M. Paine, J. Proost, and N. Röling (Eds.), *Cow up a tree: Knowing and learning for change in agriculture* (pp. 227-240). Paris: INRA.

Woodhill, J. and N. Röling (1998). The second wing of the eagle: The human dimension in learning our way to more sustainable futures. In N. Röling and

M. Wagemakers (Eds.), *Facilitating sustainable agriculture* (pp. 283-311). Cambridge, UK: Cambridge University Press.

Zavaleta, E.S., R.J. Hobbs, and H.A. Mooney (2001). Viewing invasive species removal in a whole-ecosystem context. *Trends in Ecology and Evolution* 16: 454-459.

Index

Page numbers followed by the letter "f" indicate figures; those followed by the letter "t" indicate tables.